DESIGN FOR EMBEDDED IMAGE PROCESSING ON FPGAs

DESIGN FOR EMBEDDED IMAGE PROCESSING ON FPGAs

Second Edition

Donald G. Bailey
Massey University, Palmerston North, New Zealand

WILEY

This second edition first published 2024

Edition History
John Wiley & Sons (Asia) Pte Ltd (1e, 2011)

Registered Office(s)
John Wiley & Sons, Inc., 111 River Street, Hoboken, NJ 07030, USA
John Wiley & Sons Ltd, The Atrium, Southern Gate, Chichester, West Sussex, PO19 8SQ, UK

For details of our global editorial offices, customer services, and more information about Wiley products visit us at www.wiley.com.

Wiley also publishes its books in a variety of electronic formats and by print-on-demand. Some content that appears in standard print versions of this book may not be available in other formats.

Library of Congress Cataloging-in-Publication Data

Names: Bailey, Donald G. (Donald Graeme), 1962- author.
Title: Design for embedded image processing on FPGAs / Donald G. Bailey.
Description: Second edition. | Hoboken, NJ : Wiley, 2024. | Includes index.
Identifiers: LCCN 2023011469 (print) | LCCN 2023011470 (ebook) | ISBN 9781119819790 (cloth) | ISBN 9781119819806 (adobe pdf) | ISBN 9781119819813 (epub)
Subjects: LCSH: Embedded computer systems. | Field programmable gate arrays.
Classification: LCC TK7895.E42 B3264 2024 (print) | LCC TK7895.E42 (ebook) | DDC 621.39/5–dc23/eng/20230321
LC record available at https://lccn.loc.gov/2023011469
LC ebook record available at https://lccn.loc.gov/2023011470

Cover Design: Wiley
Cover Image: © bestfoto77/Shutterstock

Set in 9/11pt TimesLTStd by Straive, Chennai, India

Contents

Preface

Image processing, and in particular embedded image processing, faces many challenges, from increasing resolution, increasing frame rates, and the need to operate at low power. These offer significant challenges for implementation on conventional software-based platforms. This leads naturally to considering field-programmable gate arrays (FPGAs) as an implementation platform for embedded imaging applications. Many image processing operations are inherently parallel, and FPGAs provide programmable hardware, also inherently parallel. Therefore, it should be as simple as mapping one onto the other, right? Well, yes … and no.

Image processing is traditionally thought of as a software domain task, whereas FPGA-based design is firmly in the hardware domain. There are a lot of tricks and techniques required to create an efficient design. Perhaps the biggest hurdle to an efficient implementation is the need for a hardware mindset. To bridge the gap between software and hardware, it is necessary to think of algorithms not on their own but more in terms of their underlying computational architecture. Implementing an image processing algorithm (or indeed any algorithm) on an FPGA therefore consists of determining the underlying architecture of an algorithm, mapping that architecture onto the resources available within an FPGA, and finally mapping the algorithm onto the hardware architecture. While the mechanics of this process is mostly automated by high-level synthesis tools, the underlying design is not. A low-quality design can only go so far; it is still important to keep in mind the hardware that is being implied by the code and design the algorithm for the underlying hardware.

Unfortunately, there is limited material available to help those new to the area to get started. While there are many research papers published in conference proceedings and journals, there are only a few that focus specifically on how to map image processing algorithms onto FPGAs. The research papers found in the literature can be classified into several broad groups.

The first focuses on the FPGA architecture itself. Most of these provide an analysis of a range of techniques relating to the structure and granularity of logic blocks, the routing networks, and embedded memories. As well as the FPGA structure, a wide range of topics are covered, including underlying technology, power issues, the effects of process variability, and dynamic reconfigurability. Many of these papers are purely proposals, or relate to prototype FPGAs rather than commercially available chips. While they provide insights as to some of the features which might be available in the next generation of devices, most of the topics within this group are at too low a level.

A second group of papers investigates the topic of reconfigurable computing. Here, the focus is on how an FPGA can be used to accelerate some computationally intensive task or range of tasks. While image processing is one such task considered, most of the research relates more to high-performance computing rather than low-power embedded systems. Topics within this group include hardware and software partitioning, hardware and software co-design, dynamic reconfigurability, communications between an FPGA and central processing unit (CPU), comparisons between the performance of FPGAs, graphics processing units (GPUs) and CPUs, and the design of operating systems and specific platforms for both reconfigurable computing applications and research. Important principles and techniques can be gleaned from many of these papers even though this may not be their primary focus.

The next group of papers considers tools for programming FPGAs and applications, with a focus on improving the productivity of the development process. A wide range of hardware description languages have been proposed, with many modelled after software languages such as C, Java, and even Prolog. Many of these are developed as research tools, with very few making it out of the laboratory to commercial availability. There has also been considerable research on compilation techniques for mapping standard software languages to hardware (high-level synthesis). Techniques such as loop unrolling, strip mining, and pipelining to produce parallel hardware are important principles that can result in more efficient hardware designs.

The final group of papers focuses on a range of applications, including image processing and the implementation of both image processing operations and systems. Unfortunately, as a result of page limits and space constraints, many of these papers give the results of the implementation of various systems but present relatively few design details. This is especially so in the case of many papers that describe deep learning systems. Often the final product is described, without describing many of the reasons or decisions that led to that design. Many of these designs cannot be recreated without acquiring the specific platform and tools that were used or inferring a lot of the missing details. While some of these details may appear obvious in hindsight, without this knowledge, many are far from obvious just from reading the papers. The better papers in this group tended to have a tighter focus, considering the implementation of a single image processing operation.

So, while there may be a reasonable amount of material available, it is quite diffuse. In many cases, it is necessary to know exactly what you are looking for, or just be lucky to find it. The intention of this book, therefore, is to bring together much of this diverse research (on both FPGA design and image processing) and present it in a systematic way as a reference or guide.

Intended Audience

This book is written primarily for those who are familiar with the basics of image processing and want to consider implementing image processing using FPGAs. Perhaps the biggest hurdle is switching from a software mindset to a hardware way of thinking. When we program in software, a good compiler can map the algorithm in the programming language onto the underlying computer architecture relatively efficiently. When programming hardware though, it is not simply a matter of porting the software onto hardware. The underlying hardware architecture needs to be designed as well. In particular, programming hardware usually requires transforming the algorithm into an appropriate parallel architecture, often with significant changes to the algorithm itself. This requires significant design, rather than just decomposition and mapping of the dataflow (as is accomplished by a good high-level synthesis tool). This book addresses this issue by not only providing algorithms for image processing operations but also discusses both the design process and the underlying architectures that can be used to implement the algorithms efficiently.

This book would also be useful to those with a hardware background, who are familiar with programming and applying FPGAs to other problems, and are considering image processing applications. While many of the techniques are relevant and applicable to a wide range of application areas, most of the focus and examples are taken from image processing. Sufficient detail is given to make many of the algorithms and their implementation clear. However, learning image processing is more than just collecting a set of algorithms, and there are any number of excellent image processing textbooks that provide these.

It is the domain of embedded image processing where FPGAs come into their own. An efficient, low-power design requires that the techniques of both the hardware engineer and the software engineer be integrated tightly within the final solution.

Changes Since the First Edition

Although many of the underlying design principles have not changed, the environment has changed quite significantly since the first edition. In general, there has been an increase in the requirements within applications. Resolutions are increasing, with high-definition television (HDTV) becoming ubiquitous, and growing demand within 4K and 8K resolutions (the 8K format has 33.2 Mpixels). Sensor resolutions are also increasing steadily, with sensors having more than 100 Mpixels becoming available. Frame rates have also been increasing, with up to 120 frames per second being part of the standard for ultra-high-definition television (UHDTV). The dynamic range is also increasing from commonly used 8 bits per pixel to 12–16 bits. All of these factors lead to more data to be processed, at a faster rate.

Against this, there has been an increasing awareness of power and sustainability issues. As a low-power computing platform, FPGAs are well placed to address the power concerns in many applications.

The capabilities of FPGAs have improved significantly as technology improvements enable more to be packed onto them. Not only has there been an increase in the amount of programmable logic and on-chip memory blocks, but FPGAs are becoming more heterogeneous. Many FPGAs now incorporate significant hardened logic blocks, including moderately powerful reduced instruction set computing (RISC) processors, external memory interfacing, and a wide range of communication interfaces. Digital signal processing (DSP) blocks are also improving, with the move towards supporting floating-point in high-end devices. Technology improvements have seen significant reductions in the power required.

Even the FPGA market has changed, with the takeover of both Altera and Xilinx by Intel and AMD, respectively. This is an indication that FPGAs are seen as a serious contender for high-performance computing and acceleration. The competition has not stood still, with both CPUs and GPUs increasing in capability. In particular, a new generation of low-power GPUs has become available that are more viable for embedded image processing.

High-level synthesis tools are becoming more mature and address many of the development time issues associated with conventional register transfer level design. The ability to compile to both software and hardware enables more complex algorithms to be explored, with faster debugging. They also allow faster exploration of the design space, enabling efficient designs to be developed more readily. However, the use of high-level synthesis does not eliminate the need for careful algorithm design.

While the use of FPGAs for image processing has not become mainstream, there has been a lot of activity in this space as the capabilities of FPGAs have improved. The research literature on programming and applying FPGAs in the context of image processing has grown significantly. However, it is still quite diffuse, with most papers focusing on one specific aspect. As researchers have looked at more complex image processing operations, the descriptions of the implementation have become higher level, requiring a lot of reading between the lines, and additional design work to be able to replicate a design.

One significant area that has become mainstream in image processing is the use of deep learning models. Deep learning was not around when the previous edition was written and only started becoming successful in image processing tasks in the early 2010s. Their success has made them a driving application, not only for FPGAs and FPGA architecture but also within computing in general. However, deep learning models pose a huge computational demand on processing, especially for training, but also for deployment. In an embedded vision context, this has made FPGAs a target platform for their deployment. Deep learning is a big topic on its own, so this book is unable to do much more than scratch the surface and concentrate on some of the issues associated with FPGA-based implementation.

Outline of the Contents

This book aims to provide a comprehensive overview of algorithms and techniques for implementing image processing algorithms on FPGAs, particularly for low- and intermediate-level vision. However, as with design in any field, there is more than one way of achieving a particular task. Much of the emphasis has been placed on stream-based approaches to implementing image processing, as these can efficiently exploit parallelism when they can be used. This emphasis reflects my background and experience in the area and is not intended to be the last word on the topic.

A broad overview of image processing is presented in Chapter 1, with a brief historical context. Many of the basic image processing terms are defined, and the different stages of an image processing algorithm are identified. The problem of real-time embedded image processing is introduced, and the limitations of conventional serial processors for tackling this problem are identified. High-speed image processing must exploit the parallelism inherent in the processing of images; the different types of parallelism are identified and explained.

FPGAs combine the advantages of both hardware and software systems, by providing reprogrammable (hence flexible) hardware. Chapter 2 provides an introduction to FPGA technology. While some of this will be more detailed than is necessary to implement algorithms, a basic knowledge of the building blocks and underlying architecture is important to developing resource efficient solutions. The synthesis process for building hardware on FPGAs is defined, with particular emphasis on the design flow for implementing algorithms.

Traditional hardware description languages are compared with high-level synthesis, with the benefits and limitations of each outlined in the context of image processing.

The process of designing and implementing an image processing application on an FPGA is described in detail in Chapter 3. Particular emphasis is given to the differences between designing for an FPGA-based implementation and a standard software implementation. The critical initial step is to clearly define the image processing problem that is being tackled. This must be in sufficient detail to provide a specification that may be used to evaluate the solution. The procedure for developing the image processing algorithm is described in detail, outlining the common stages within many image processing algorithms. The resulting algorithm must then be used to define the system and computational architectures. The mapping from an algorithm is more than simply porting the algorithm to a hardware description language. It is necessary to transform the algorithm to make efficient use of the resources available on the FPGA. The final stage is to implement the algorithm by mapping it onto the computational architecture. Several checklists provide a guide and hints for testing and debugging an algorithm on an FPGA.

Four types of constraints on the mapping process are limited processing time, limited access to data, limited system resources, and limited system power. Chapter 4 describes several techniques for overcoming or alleviating these constraints. Timing explores low-level pipelining, process synchronisation, and working with multiple clock domains. A range of memory and caching architectures are presented for alleviating memory bandwidth. Resource sharing and associated arbitration issues are discussed, along with reconfigurability. The chapter finishes with a section introducing commonly used performance metrics in terms of both system and application performance.

Chapter 5 focuses on the computational aspects of image processing designs. These help to bridge the gap between a software and hardware implementation. Different number representation and number systems are described. Techniques for the computation of elementary functions are discussed, with a particular focus on those that are hardware friendly. Many of these could be considered the hardware equivalent of software libraries for efficiently implementing common functions. Possible FPGA implementations of a range of data structures commonly found in computer vision algorithms are presented.

Any embedded application must interface with the real world. A range of common peripherals is described in Chapter 6, with suggestions on how they may be interfaced to an FPGA. Particular attention is given to interfacing cameras and video output devices. Interfacing with other devices is discussed, including serial communications, off-chip memory, and serial processors.

The next section of this book describes the implementation of many common image processing operations. Some of the design decisions and alternative ways of mapping the operations onto FPGAs are considered. While reasonably comprehensive, particularly for low-level image-to-image transformations, it is impossible to cover every possible design. The examples discussed are intended to provide the foundation for many other related operations.

Chapter 7 considers point operations, where the output depends only on the corresponding input pixel in the input image(s). Both direct computation and lookup table approaches are described. With multiple input images, techniques such as image averaging and background modelling are discussed in detail. The final sections in this chapter consider the processing of colour and hyperspectral images. Colour processing includes colour space conversion, colour balancing, and colour segmentation.

The implementation of histograms and histogram-based processing are discussed in Chapter 8. The techniques of accumulating a histogram, and then extracting data from the histogram, are described in some detail. Particular tasks are histogram equalisation, threshold selection, and using histograms for image matching. The concepts of standard 1-D histograms are extended to multi-dimensional histograms. The use of clustering for colour segmentation and classification is discussed in some detail. The chapter concludes with the use of features extracted from multi-dimensional histograms for texture analysis.

Chapter 9 considers a wide range of local filters, both linear and nonlinear. Particular emphasis is given to caching techniques for a stream-based implementation and methods for efficiently handling the processing around the image borders. Rank filters are described, and a selection of associated sorting network architectures reviewed. Morphological filters are another important class of filters. State machine implementations of morphological filtering provide an alternative to the classic filter implementation. Separability and both serial and parallel decomposition techniques are described that enable more efficient implementations.

Image warping and related techniques are covered in Chapter 10. The forward and reverse mapping approaches to geometric transformation are compared in some detail, with particular emphasis on techniques for stream processing implementations. Interpolation is frequently associated with geometric transformation. Hardware-based algorithms for bilinear, bicubic, and spline-based interpolation are described. Related techniques of image registration are also described at the end of this chapter, including a discussion of feature point detection, description, and matching.

Chapter 11 introduces linear transforms, with a particular focus on the fast Fourier transform (FFT), the discrete cosine transform (DCT), and the wavelet transform. Both parallel and pipelined implementations of the FFT and DCT are described. Filtering and inverse filtering in the frequency domain are discussed in some detail. Lifting-based filtering is developed for the wavelet transform. This can reduce the logic requirements by up to a factor of 4 over a direct finite impulse response implementation.

Image coding is important for image storage or transmission. Chapter 12 discusses the stages within image and video coding and outlines some of the techniques that can be used at each stage. Several of the standards for both still image and video coding are outlined, with an overview of the compression techniques used.

A selection of intermediate-level operations relating to region detection and labelling is presented in Chapter 13. Standard software algorithms for chain coding and connected component labelling are adapted to give efficient streamed implementation. These can significantly reduce both the latency and memory requirements of an application. Hardware implementations of the distance transform, the watershed transform, and the Hough transform are also presented, discussing some of the key design decisions for an efficient implementation.

Machine learning techniques are commonly used within computer vision. Chapter 14 introduces the key techniques for regression and classification, with a particular focus on FPGA implementation. Deep learning techniques are increasingly being used in many computer vision applications. A range of deep network architectures is introduced, and some of the issues for realising these on FPGAs are discussed.

Finally, Chapter 15 presents a selection of case studies, showing how the material and techniques described in the previous chapters can be integrated within a complete application. These applications briefly show the design steps and illustrate the mapping process at the whole algorithm level rather than purely at the operation level. Many gains can be made by combining operations together within a compatible overall architecture. The applications described are coloured region tracking for a gesture-based user interface, calibrating and correcting barrel distortion in lenses, development of a foveal image sensor inspired by some of the attributes of the human visual system, a machine vision system for real-time produce grading, stereo imaging for depth estimation, and face detection.

Conventions Used

The contents of this book are independent of any particular FPGA or FPGA vendor, or any particular hardware description language. The topic is already sufficiently specialised without narrowing the audience further! As a result, many of the functions and operations are represented in block schematic form. This enables a language-independent representation and places emphasis on a particular hardware implementation of the algorithm in a way that is portable. The basic elements of these schematics are illustrated in Figure P.1. I is generally used as the input of an image processing operation, with the output image represented by Q.

With some mathematical operations, such as subtraction and comparison, the order of the operands is important. In such cases, the first operand is indicated with a blob rather than an arrow, as shown on the bottom in Figure P.1.

Consider a recursive filter operating on streamed data:

$$Q_n = \begin{cases} I_n, & |I_n - Q_{n-1}| < T, \\ Q_{n-1} + k(I_n - Q_{n-1}), & \text{otherwise}, \end{cases} \tag{P.1}$$

where the subscript in this instance refers to the nth pixel in the streamed image. At a high level, this can be considered as an image processing operation, and represented by a single block, as shown in the top-left of Figure P.1. The low-level implementation is given in the middle-left panel. The input and output, I and Q, are

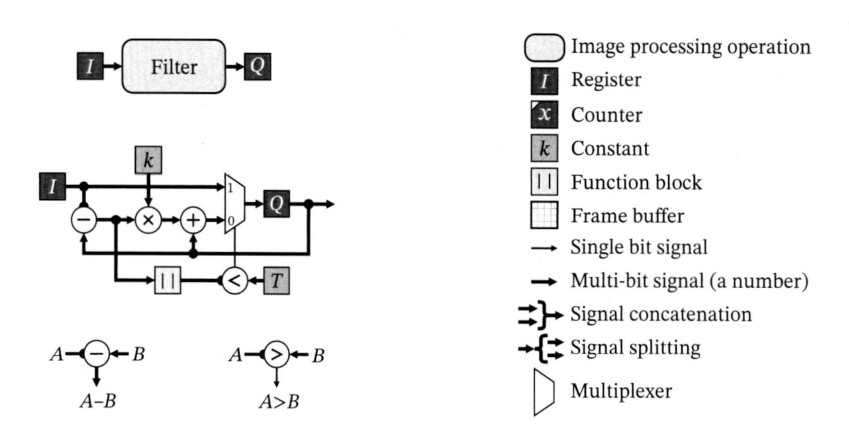

Figure P.1 Conventions used in this book. Top-left: representation of an image processing operation; middle-left: a block schematic representation of the function given by Eq. (P.1); bottom-left: representation of operators where the order of operands is important; right: symbols used for various blocks within block schematics.

represented by registers (dark blocks, with optional register names in white); the subscripts have been dropped because they are implicit with streamed operation. In some instances additional control inputs may be shown, for example CE for clock enable, RST for reset. Constants are represented as mid-grey blocks, and other function blocks with light-grey background.

When representing logic functions in equations, $\vee$ is used for logical OR, and $\wedge$ for logical AND. This is to avoid confusion with addition and multiplication.

Donald G. Bailey
Massey University
Palmerston North, New Zealand

Acknowledgments

I would like to acknowledge all those who have helped me to get me where I currently am in my understanding of field-programmable gate array (FPGA)-based design. In particular, I would like to thank my research students (David Johnson, Kim Gribbon, Chris Johnston, Aaron Bishell, Andreas Buhler, Ni Ma, Anoop Ambikumar, Tariq Khan, and Michael Klaiber) who helped to shape my thinking and approach to FPGA development as we struggled together to work out efficient ways of implementing image processing algorithms. This book is as much a reflection of their work as it is of mine.

Most of our early work used Handel-C and was tested on boards provided by Celoxica. I would like to acknowledge the support provided by Roger Gook and his team, first with Celoxica and later with Agility Design Solutions. Later work was on boards supplied by Terasic. I would like to acknowledge Sean Peng and his team for their on-going support and encouragement.

Massey University has provided a supportive environment and the freedom for me to explore this field. In particular, Serge Demidenko gave me the encouragement and the push to begin playing with FPGAs; he has been a source of both inspiration and challenging questions. Other colleagues who have been of particular encouragement are Gourab Sen Gupta, Richard Harris, Amal Punchihewa, and Steven Le Moan. I would also like to acknowledge Paul Lyons, who co-supervised several of my students.

Early versions of some of the material in this book were presented as half day tutorials at the IEEE Region 10 Conference (TENCON) in 2005 in Melbourne, the IEEE International Conference on Image Processing (ICIP) in 2007 in San Antonio Texas, and the 2010 Asian Conference on Computer Vision (ACCV) in Queenstown, New Zealand. It was then adapted to three-day workshops, which were held in Australia, New Zealand, Japan, and as a Masters Course in Germany. I would like to thank attendees at these workshops and courses for providing valuable feedback and stimulating discussion.

During 2008, I spent a sabbatical with the Circuits and Systems Group at Imperial College London, where I began writing the first edition. I would like to thank Peter Cheung, Christos Bouganis, Peter Sedcole, and George Constantinides for discussions and opportunities to bounce ideas off.

My wife, Robyn, has had to put up with my absence many evenings and weekends while working on the manuscripts for both the first and second editions. I am grateful for both her patience and her support. This book is dedicated to her.

Donald G. Bailey

About the Companion Website

This book is accompanied by a companion website.

www.wiley.com/go/bailey/designforembeddedimageprocessc2e

This website includes PowerPoint slides.

1

Image Processing

Vision is arguably the most important human sense. The processing and recording of visual data therefore has significant importance. The earliest images are from prehistoric drawings on cave walls or carved on stone monuments commonly associated with burial tombs. (It is not so much the medium that is important here; anything else would not have survived to today.) Such images consist of a mixture of both pictorial and abstract representations. Improvements in technology enabled images to be recorded with more realism, such as paintings by the masters. Images recorded in this manner are indirect in the sense that the light intensity pattern is not used directly to produce the image. The development of chemical photography in the early 1800s enabled direct image recording. This trend has continued with electronic recording, first with analogue sensors, and subsequently with digital sensors, which include analogue to digital (A/D) conversion on the sensor chip, to directly produce digital images.

Imaging sensors have not been restricted to the portion of the electromagnetic spectrum visible to the human eye. Sensors have been developed to cover much of the electromagnetic spectrum from radio waves through to gamma rays. A wide variety of other imaging modalities have also been developed, based on ultrasound, electrons, atomic force (Binnig et al., 1986), magnetic resonance, and so on. In principle, any quantity that can be sensed can be used for imaging, even dust rays (Auer, 1982).

Since vision is such an important sense, the processing of images has become important too, to augment or enhance human vision. Images can be processed to enhance their subjective content or to extract useful information. While it is possible to process the optical signals associated with visual images using lenses and optical filters, this book focuses on digital image processing, the numerical processing of images by digital hardware.

One of the earliest applications of digital image processing was for transmitting digitised newspaper pictures across the Atlantic Ocean in the early 1920s (McFarlane, 1972). However, it was only with the advent of computers with sufficient memory and processing power that digital image processing became widespread. The earliest recorded computer-based image processing was from 1957 when a scanner was added to a computer at the USA National Bureau of Standards (Kirsch, 1998). It was used for the early research on edge enhancement and pattern recognition. In the 1960s, the need to process large numbers of large images obtained from satellites and space exploration stimulated image processing research at NASA's Jet Propulsion Laboratory (Castleman, 1979). At the same time, research in high-energy particle physics required detecting interesting events from large numbers of cloud chamber photographs (Duff, 2000). As computers grew in power and reduced in cost, the range of applications for digital image processing exploded, from industrial inspection to medical imaging. Image sensors are now ubiquitous, in mobile phones, laptops, and in video-based security and surveillance systems.

1.1 Basic Definitions

More formally, an ***image*** is a spatial representation of an object, scene, or other phenomenon (Haralick and Shapiro, 1991). Examples of images include a photograph, which is a pictorial record formed from the light

Design for Embedded Image Processing on FPGAs, Second Edition. Donald G. Bailey.

Companion Website: www.wiley.com/go/bailey/designforembeddedimageprocessc2e

intensity pattern on an optical sensor; a radiograph, which is a representation of density formed through exposure to X-rays transmitted through an object; a map, which is a spatial representation of physical or cultural features; and a video, which is a sequence of two-dimensional images through time. More rigorously, an image is any continuous function of two or more variables defined on some bounded region of space.

A ***digital image*** is an image in digital format, so that it is suitable for processing by computer. There are two important characteristics of digital images. The first is spatial quantisation. Computers are unable to easily represent arbitrary continuous functions, so the continuous function is sampled. The result is a series of discrete picture elements, or ***pixels*** (for 2-D images) or volume elements, ***voxels*** (for 3-D images). Sampling does not necessarily have to be spatially uniform, for example ***point clouds*** from a LiDAR scanner. Sampling can represent a continuous image exactly (in the sense that the underlying continuous function may be recovered exactly), given a band-limited image and a sufficiently high sample rate (Shannon, 1949). The second characteristic of digital images is sample quantisation. This results in discrete values for each pixel, enabling an integer representation. Common bit widths per pixel are 1 (binary images), 8 (greyscale images), and 24 (3×8 bits for colour images). Modern high dynamic range sensors can provide 12–16 bits per pixel. Unlike sampling, value quantisation will always result in an error between the representation and true value. In many circumstances, however, this ***quantisation error*** or ***quantisation noise*** may be made smaller than the uncertainty in the true value resulting from inevitable measurement noise.

In its basic form, a digital image is simply a two (or higher)-dimensional array of numbers (usually integers), which represents an object or scene. Once in this form, an image may be readily manipulated by a digital computer. It does not matter what the numbers represent, whether light intensity, reflectance, attenuation, distance to a point (range), temperature, population density, elevation, rainfall, or any other numerical quantity.

Digital image processing can therefore be defined as subjecting such an image to a series of mathematical operations in order to obtain a desired result. This may be an enhanced image; the detection of some critical feature or event; a measurement of an object or key feature within the image; a classification or grading of objects within the image into one of two or more categories; or a description of the scene.

Image processing techniques are used in a number of related fields. While the principal focus of the fields often differs, many of the techniques remain the same at the fundamental level. Some of the distinctive characteristics are briefly outlined here.

Image enhancement involves improving the subjective quality of an image or the detectability of objects within the image (Haralick and Shapiro, 1991). The information that is enhanced is usually apparent in the original image but may not be clear. Examples of image enhancement include noise reduction, contrast enhancement, edge sharpening, and colour correction.

Image restoration goes one step further than image enhancement. It uses knowledge of the causes of how an image is degraded to create a model of the degradation process. This model is then used to derive an inverse process that is used to restore the image. In many cases, the information in the image has been degraded to the extent of being unrecognisable, for example severe blurring.

Image reconstruction involves restructuring the data that is available into a more useful form. Examples are image super-resolution (reconstructing a high-resolution image from a series of low-resolution images), image fusion (combining images from multiple sources), and tomography (reconstructing a cross section of an object from a series of projections).

Image analysis refers specifically to using computers to extract data from images. The result is usually some form of measurement. In the past, this was almost exclusively 2-D imaging, although with the advent of confocal microscopy and other advanced imaging techniques, this has extended to three dimensions.

Pattern recognition is concerned with identifying objects based on patterns in the measurements (Haralick and Shapiro, 1991). There is a strong focus on statistical approaches, although syntactic and structural methods are also used.

Computer vision tends to use a model-based approach to image processing. Mathematical models of both the scene and the imaging process are used to derive a 3-D representation based on one or more 2-D images of a scene. The use of models implicitly provides an interpretation of the contents of the images obtained.

The fields are sometimes distinguished based on application:

Machine vision is using image processing as part of the control system for a machine (Schaffer, 1984). Images are captured and analysed, and the results are used directly for controlling the machine while performing a specific task. Real-time processing is often critical.

Remote sensing usually refers to the use of image analysis for obtaining geographical information, either using satellite images or aerial photography (including from drones).

Medical imaging encompasses a wide range of imaging modalities (X-ray, ultrasound, magnetic resonance, positron emission, and others) concerned primarily with medical diagnosis and other medical applications. It involves both image reconstruction to create meaningful images from the raw data gathered from the sensors and image analysis to extract useful information from the images.

Image and video coding focuses on the compression of an image or video, so that it occupies less storage space or takes less time to transmit from one location to another. Compression is possible because many images contain significant redundant information. In the reverse step, image decoding, the full image or video is reconstructed from the compressed data.

1.2 Image Formation

While there are many possible sensors that can be used for imaging, the focus in this section will be on optical images captured from visible part of the electromagnetic spectrum. While the sensing technology may differ significantly for other types of imaging, many of the imaging principles will be similar.

The first requirement to obtaining an image is some form of sensor to detect and quantify the incoming light. This could be a single sensing element, with light from different directions reflected towards the sensor using a scanning mirror. This approach is commonly used with time-of-flight laser range scanners or LiDAR. The output of a LiDAR is typically a three-dimensional point cloud.

More commonly, a 2-D sensor array is used, which measures the light intensity at each sensing element and produces a 2-D image. In capturing an image, it is important to sample with at least two pixels across the highest spatial frequency (the Nyquist sampling criterion) to avoid aliasing. The two most common solid-state sensor technologies are charge coupled devices (CCDs) and complementary metal oxide semiconductor (CMOS) active pixel sensors (Fossum, 1993). The basic light sensing principle is the same: an incoming photon liberates an electron within the silicon semiconductor through the photoelectric effect. These photoelectrons are then accumulated during the exposure time before being converted into a voltage for reading out.

Although the active pixel sensor technology was developed before CCDs, the need for local transistors made early CMOS sensors impractical because the transistors consumed most of the area of the device (Fossum, 1993). Therefore, CCD sensors gained early dominance in the market. However, the continual reduction of feature sizes has meant that CMOS sensors became practical from the early 1990s. Early CMOS sensors had lower sensitivity and higher noise than a similar format CCD sensor, although with recent technological improvements there is now very little difference between the two families (Litwiller, 2005). Since CMOS sensors use the same process technology as standard CMOS devices, they also enable other functions such as A/D conversion to be directly integrated on the same chip. Most modern sensors are CMOS devices, with integrated digital conversion, providing a digital output stream.

A relative newcomer is an imaging sensor based on single-photon avalanche diodes (SPAD) (Morimoto et al., 2020). Each sensor element can detect single photons, which are amplified to give a pulse. These pulses can be counted, giving a direct digital measure of the intensity. The individual pulses can also be accurately timed, giving direct time-of-flight measurement for 3-D applications.

1.2.1 *Optics*

A camera consists of a sensor combined with a lens. The lens focuses a parallel beam of light (effectively from infinite distance) from a particular direction to a point on the sensor. For an object at a finite distance, as seen

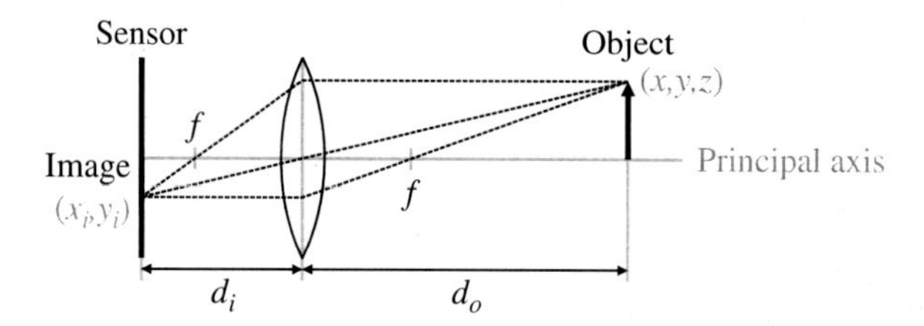

Figure 1.1 Optical image formation using a lens using geometric optics.

in Figure 1.1, geometric optics gives the relationship between the lens focal length, f, the object distance, d_o, and image distance, d_i as

$$\frac{1}{d_i} + \frac{1}{d_o} = \frac{1}{f}. \tag{1.1}$$

Since the ray through the centre of the lens is a straight line, this ray may be used to determine where an object will be imaged to on the sensor. This ***pinhole model*** of the camera will map a 3-D point (x, y, z) in camera coordinates (the origin is the pinhole location; the z-axis corresponds to the principal axis of the camera; and the x- and y-axes are aligned with the sensor axes) to the point

$$x_i = \frac{xd_i}{z}, \qquad y_i = \frac{yd_i}{z} \tag{1.2}$$

on the sensor plane. d_i is also called the effective focal distance of the lens. Note that the origin on the sensor plane is in the centre of the image (the intersection of the principal axis). This projective mapping effectively scales the object in inverse proportion to its distance from the camera ($z = d_o$). From similar triangles, the optical magnification is

$$m = \frac{d_i}{z} = \frac{d_i}{d_o}. \tag{1.3}$$

The image is generally focused by adjusting the image distance, d_i, by moving the lens away from the sensor (although in microscopy, d_i is fixed, and the image is focused by adjusting d_o). This can either be manual or, for an electronically controlled lens, an autofocus algorithm automatically adjusts d_i to maximise the image sharpness (see Section 6.1.7.2).

Objects not at the optimal object distance, d_o, will become blurred, depending on both the offset in distance from best focus, Δd_o, and the diameter of the lens aperture, D. The size of the blurred spot is called the ***circle of confusion*** and is given by

$$c = \frac{|\Delta d_o|}{d_o + \Delta d_o} \frac{Df}{d_o - f} = \frac{|\Delta d_o|}{d_o + \Delta d_o} \frac{f^2}{N(d_o - f)}, \tag{1.4}$$

with N being the numerical aperture of the lens ($N = \frac{f}{D}$).

The depth of field of a camera is the range of distances over which the degree of blurring is acceptable. This is often defined as the range where the size of the circle of confusion is less than one pixel. If the images are for human use, it is also necessary to take into account how the images are displayed (especially the degree of magnification), the viewing distance, and visual acuity. This is usually taken as 0.2 mm at 250 mm distance (Conrad, 2006).

A second factor which affects image sharpness (of an ideal lens) is diffraction. Rather than focusing a parallel beam to an ideal point, the focus pattern in the image plane with a circular aperture is an Airy disk. The Rayleigh criterion (Rayleigh, 1879) indicates that two points can be resolved if the peak of the diffraction pattern from one point is outside the first null of the diffraction pattern of the other point. The size of the blur spot from diffraction of an object at finite distance is therefore

$$d = 1.22\lambda \frac{d_i}{D} = 1.22\lambda N \frac{d_i}{f} = 1.22\lambda N(1 + m), \tag{1.5}$$

where λ is the wavelength of the light.

Equations (1.4) and (1.5) may be combined to give the effective size of the blur as (Hansma, 1996)

$$d_e = \sqrt{c^2 + d^2}. \tag{1.6}$$

Of course, any other lens aberrations or imperfections will further increase the blur.

1.2.2 Colour

Humans are able to see in colour. There are three types of colour receptors (cones) in the human eye that respond differently to different wavelengths of light. If the wavelength dependence of a receptor is $S_k(\lambda)$, and the light falling on the receptor contains a mix of light of different wavelengths, $C(\lambda)$, then the response of that receptor will be given by the combination of the responses of all different wavelengths:

$$R_k = \int C(\lambda)S_k(\lambda)d\lambda. \tag{1.7}$$

What is perceived as colour is the particular combination of responses from the three different types of receptors. Many different wavelength distributions produce the same responses from the receptors and therefore are perceived as the same colour (this is known as ***metamerism***). In practice, it is a little more complicated than this because of chromatic adaptation.

The incoming light depends on the wavelength distribution of the source of the illumination, $I(\lambda)$, and the wavelength-dependent reflectivity of the object we are looking at, $\rho(\lambda)$. So, Eq. (1.7) becomes

$$R_k = \int I(\lambda)\rho(\lambda)S_k(\lambda)d\lambda. \tag{1.8}$$

To reproduce a colour, it is only necessary to reproduce the associated combination of receptor responses. This is accomplished, for example in a television monitor, by producing the corresponding mixture of red, green, and blue light.

To capture a colour image for human viewing, it is necessary to have three different colour receptors in the sensor. Ideally, these should correspond to the spectral responses of the cones. However, since most of what is seen has broad spectral characteristics, a precise match is not critical except when the illuminant has a narrow spectral content (for example sodium vapour lamps and LED-based illumination sources). Silicon has a broad spectral response, with a peak in the near infrared. Therefore, to obtain a colour image, this response must be modified through appropriate colour filters.

Two approaches are commonly used to obtain a full colour image. The first is to separate the red, green, and blue components of the image using a prism and with dichroic filters. These components are then captured using three separate sensors, one for each component. The need for three chips with precise alignment makes such cameras relatively expensive. An alternative approach is to use a single chip sensor, with a mosaic of small filters, with one colour filter integrated with each pixel. Since each pixel senses only one colour channel, the effective spatial resolution of the sensor is decreased. A full colour image is created by interpolating the missing pixels of each component. The most commonly used colour filter array is the Bayer pattern (Bayer, 1976), with filters to select the red, green, and blue primary colours. Other patterns are also possible, for example filtering the cyan, magenta, and yellow secondary colours (Parulski, 1985). Cameras using the secondary colours have better low light sensitivity because each filter absorbs less of the incoming light. However, the processing required to produce the output gives a better signal-to-noise ratio from the primary filters than from secondary filters (Parulski, 1985; Baer et al., 1999).

One further possibility that has been considered is to stack the red, green, and blue sensors one on top of the other (Lyon and Hubel, 2002; Gilblom et al., 2003). The Foveon sensor relies on the fact that longer wavelengths of light penetrate further into the silicon before being absorbed. Thus, using photodiodes at different depths, a full colour pixel may be obtained without explicit filtering.

The camera output is usually streamed following a raster scan. For an analogue camera, this is a voltage waveform, often with combined synchronisation signals to indicate the end of each row and frame. A colour camera can output either component video (each colour channel has a separate signal) or composite video (with chrominance components modulating a colour sub-carrier). A ***frame grabber*** preprocesses the signal from the camera, amplifies it, separates the synchronisation signals, and if necessary decodes the composite colour signal into its components. The analogue video signal is digitised by an A/D converter (or three A/D converters for colour) and stored in memory for later processing. A digital camera has integrated A/D conversion and directly provides a digital pixel stream.

1.3 Image Processing Operations

After capturing the image, the next step is to process the image to achieve the desired result. The sequence of image processing operations used to process an image from one state to another is called an ***image processing algorithm***. The term algorithm can sometimes cause confusion, since each operation is also implemented through an algorithm in some programming language. This distinction between application- and operation-level algorithms is illustrated in Figure 1.2. Usually, the context will indicate the sense in which the term is used.

As the input image is processed by the application-level algorithm, it undergoes a series of transformations. These are illustrated for a simple example in Figure 1.3. The input image consists of an array of pixels. Each pixel, on its own, carries very little information, but there are a large number of pixels. The individual pixel values represent high volume but low value data. As the image is processed, collections of pixels are grouped together. At this intermediate level, the data may still be represented in terms of pixels, but the pixel values usually have more meaning; for example, they may represent a label associated with a region. Intermediate-level representations may depart from an explicit image representation, with regions represented by their boundaries (for example as a chain code or a set of line segments) or their structure (for example as a quad-tree). From each region, a set of features (a feature vector) may be extracted that characterise the region. Generally, the number of features is small (relative to the number of pixels), with each feature containing significant information that can be used to distinguish it from other regions or other objects that may be encountered. Intermediate-level data becomes significantly lower in volume but higher in quality. Finally, at the high level, the feature data is used to classify the object or scene into one of several categories or to derive a description of the object or scene.

Rather than focusing on the data, it is also possible to focus on the image processing operations. The operations can be grouped according to the type of data that they process (Weems, 1991). This grouping is sometimes referred to as an image processing pyramid (Downton and Crookes, 1998) as represented

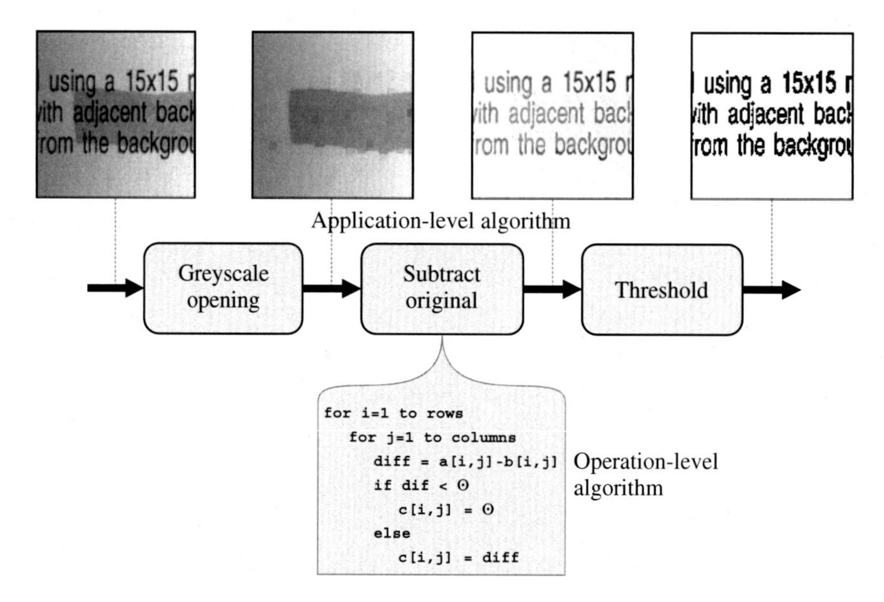

Figure 1.2 Algorithms at two levels: application level and operation level.

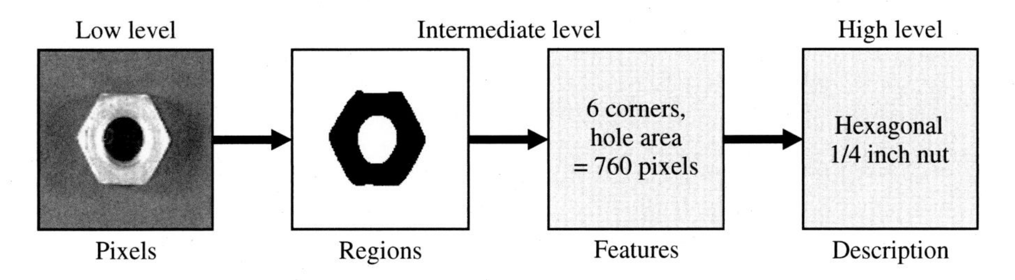

Figure 1.3 Image transformations as the image is processed.

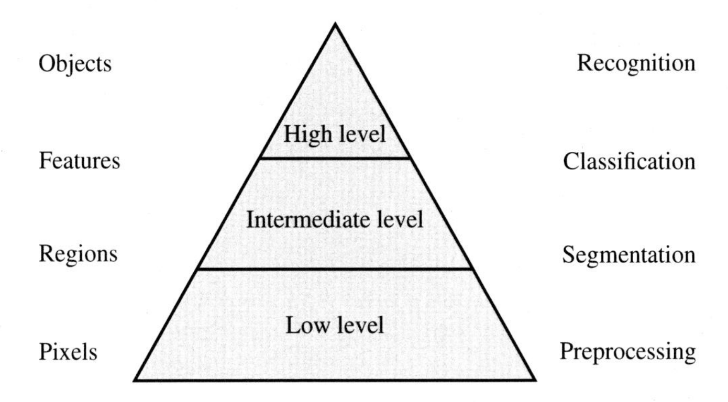

Figure 1.4 Image processing pyramid showing the form of the data and the type of operations at each level.

in Figure 1.4. At the lowest level of the pyramid are preprocessing operations. These are image-to-image transformations, with a purpose of enhancing the relevant information within the image, while suppressing any irrelevant information. Examples of preprocessing operations are distortion correction, contrast enhancement, and filtering for noise reduction or edge detection. Segmentation operations such as thresholding, colour detection, region growing, and connected component labelling occur at the boundary between the low and intermediate levels. The purpose of segmentation is to detect objects or regions in an image, which have some common property. Segmentation is therefore an image-to-region transformation. Numeric labels can be assigned to each segmented region to distinguish one region from another. After segmentation comes classification. Features of each region are used to identify objects or parts of objects or to classify an object into one of several predefined categories. Classification transforms the data from regions to a set of features extracted from each region (a feature vector). The data is no longer image based, but position information may be contained within the feature vectors. At the highest level is recognition, where descriptive labels may be assigned to the regions based on the features extracted. This leads to a description or some other interpretation of the scene.

The input bandwidth and processing complexity significantly depend on the level within the image processing pyramid (Johnston et al., 2004; Pohl et al., 2014). The operations applied to image data (preprocessing, segmentation, and feature detection) are generally at the low to intermediate level and can be divided into the following classes:

Point operations are the simplest class of operations, where each output pixel depends only on the corresponding input pixel value. Examples are brightness and contrast enhancement, colour space conversion, thresholding, and colour detection.

Temporal operations apply point operations to two or more images captured at different times. Consequently, they require a frame buffer to hold one or more images or models extracted from the images for later combination. Examples are background modelling, temporal noise averaging, change detection, and masking.

***Local filter or window filter* operations**, which require pixels from a small local region (window) to calculate the output pixel value. Example window operations are noise smoothing, edge enhancement, edge detection, texture detection, and feature detection.

Semi-global or patch operations require data from a patch within the input image to calculate the output value. Accessing pixels from the patches may require irregular memory access patterns. Examples are geometric transformation, texture classification, and feature descriptor calculation.

Global operations are those where the output potentially depends on any or all of the pixels within the input image. Examples are histogram accumulation, distance transform, connected component labelling, feature vector extraction, Hough transform, and Fourier transform.

Higher level operations (classification and recognition) often have application-specific data structures with irregular memory access patterns.

1.4 Real-time Image Processing

A real-time imaging system is one that regularly captures images, analyses those images to obtain some data, and then uses that data to control some activity. All of the processing must occur within a predefined time (often, but not always, the image capture frame rate). A ***real-time system*** is one in which the response to an event must occur within a specific time, otherwise the system is considered to have failed (Dougherty and Laplante, 1985).

Examples of real-time image processing systems abound. In machine vision systems, the image processing algorithm is used for either inspection or process control. Robot vision systems utilise vision for path planning or to control the robot in some way where the timing is critical. Autonomous vehicle control requires vision (or some other form of sensing) for vehicle navigation and collision avoidance in a dynamic environment. In video transmission systems, successive frames must be transmitted and displayed in the right sequence and with minimum jitter to avoid a loss of quality of the resultant video.

Real-time systems are categorised into either hard or soft real time. A ***hard real-time system*** is one in which the complete system is considered to have failed if the output is not produced within the required time. An example is using vision for grading items on a conveyor belt. The grading decision must be made before the item reaches the actuation point, where the result is used to direct the item one way or another. If the result is not available by this time, the system will fail to act correctly. On the other hand, a ***soft real-time system*** is one in which the complete system does not fail if the deadline is not met, but the performance deteriorates in some way. An example is video transmission via the internet. If the next frame is delayed or cannot be decoded in time, the quality of the resultant video deteriorates as a result of dropped frames. Such a system is soft real time, because although the deadline was not met, an output (although lower quality) could still be produced, and the complete system did not fail.

From a signal processing perspective, real time can mean that the processing of a sample must be completed before the next sample arrives (Kehtarnavaz and Gamadia, 2006). For video processing, this means that the total processing per pixel must be completed within a pixel sample time. Of course, not all of the processing for a single pixel can be completed before the next pixel arrives because many image processing operations require data from many pixels for each output pixel. However, this provides a limit on the minimum throughput, including any overhead associated with temporarily storing pixel values that will be used later (Kehtarnavaz and Gamadia, 2006).

A system that is not real time may have components that are real time. For instance, in interfacing with a camera, an imaging system must do something with each pixel as it is produced by the camera (either process it in some way or store it into a frame buffer) before the next pixel arrives. If not, then the data for that pixel is lost. Whether this is a hard or soft real-time process depends on the context. While missing pixels would cause the quality of an image to deteriorate (implying soft real time), the loss of quality may have a significant negative impact on the performance of the imaging application (implying that image capture is a hard real-time task). Similarly, when providing pixels for display, if the required pixel data is not provided in time, that region of the display would appear blank. In the case of image capture and display, the deadlines are in the order of tens of nanoseconds, requiring such components to be implemented in hardware.

The requirement for the whole image processing algorithm to have a bounded execution time implies that each operation must also have a bounded execution time. This characteristic rules out certain classes of operation-level algorithms from real-time processing. In particular, operations that are based on iterative or recursive algorithms can only be used if they can be guaranteed to converge satisfactorily within a predefined number of iterations, for all possible inputs that may be encountered.

One approach to guarantee a fixed response time is to make the imaging system synchronous. Such a system schedules each operation or step to execute at a particular time. This is suitable if the inputs occur at regular (or fixed) intervals, for example the successive frames from a video camera. However, such scheduled synchronous systems cannot be used reliably when events occur randomly, especially when the minimum time between events is less than the processing time for each event.

The time between events may be significantly less than the required response time. For example, on a conveyor-based inspection system, the time between items may be significantly less than the time it takes for the inspected object to move from the inspection point to the actuator. There are two ways of handling this situation. The first is to constrain all of the processing to take place during the time between successive items arriving at the inspection point, effectively providing a much tighter real-time constraint. If this new time constraint cannot be achieved, then the alternative is to use distributed or parallel processing to maintain the original time constraint but spread the execution over several processors. This can enable the time constraint to be met by increasing the throughput to meet the desired event rate.

A common misconception is that real-time systems, especially real-time imaging systems, require high speed or high performance (Dougherty and Laplante, 1985). The required response time depends on the application and is primarily dependent on the underlying process to which image processing is being applied. For example, a real-time coastal monitoring system looking at the movement of sand bars may have sample times in the order of days or even weeks. Such an application would probably not require high speed! Whether or not real-time imaging requires high-performance computing depends on the complexity of the algorithm. Conversely, an imaging application that requires high-performance computing may not necessarily be real time (for example complex iterative reconstruction algorithms).

1.5 Embedded Image Processing

An ***embedded system*** is a computer system that is embedded within a product or component. Consequently, an embedded system is usually designed to perform one specific task, or a few related specific tasks (Catsoulis, 2005), often with real-time constraints. An obvious example of an embedded image processing system is a digital camera, where the imaging functions include automatic exposure and focus control, displaying a preview and managing image compression and decompression.

Embedded vision is also useful for smart cameras, where the camera not only captures the image but also processes it to extract information as required by the application. Examples of where this would be useful are 'intelligent' surveillance systems, industrial inspection or control, and robot vision.

A requirement of many embedded systems is that they need to be of small size and light weight. Many run off batteries and are therefore required to operate with low power. Even those that are not battery operated usually have limited power available. All systems have a constrained operating temperature range. Cooling can also be an issue in systems with a high computational load. Most embedded systems try to avoid active cooling even if there is sufficient power available.

There is a growing interest within industry in embedded vision and applications. This is evidenced by the Edge AI and Vision Alliance (EAIVA, n.d.), which has over 100 participating companies, with a goal of 'inspiring and empowering innovators to add vision and AI to products'.

With the Internet of things, many sensors are distributed for surveillance or monitoring. The availability of low-cost, low-power cameras enables vision-based sensors to be integrated within an Internet-of-things framework. Such an embedded system could send video or, even better, perform processing locally on the video. If video is being transmitted, such processing could reduce the data rate by only transmitting the parts of the frame which have changed, significantly reducing the data rate and power required. Alternatively, it could process the images locally and only transmit the data extracted from the video. Such processing on network edge devices reduces both the communication overheads and processing required at the central processing side. This also has advantages both for privacy and security perspectives, by reducing access to the underlying images. A specific example of this is the use of cameras within smart home monitoring. Privacy concerns mean that the video should not be transmitted or recorded. Embedded image processing can be used to extract the required data features directly from the images, and these can be provided as output from the smart camera. The rich input of a vision sensor can provide robust high-level data, such as the position of occupants and detection of falls. When processing the images directly within the camera, there is inevitably a trade-off here between the power required to perform the processing (and data reduction) and the power required for communication.

While machine vision has been used for inspection and quality control for many years, the continuing advances in processing power and camera technology allow this to be embedded deeper within the manufacturing chain. In advanced manufacturing, faster image processing enables machine vision to be used for inspection during manufacturing rather than of the final product only at the end of the process. This allows tighter process control, the earlier detection of machine faults (such as tool wear and other breakage), and the earlier rejection of faulty parts. The end result is improved process yield and better quality assurance.

1.6 Computer Architecture

Traditional image processing platforms are based on a serial computer architecture. In its basic form, such an architecture performs all of the computation serially by breaking the operation-level algorithm down to a sequence of arithmetic or logic operations that are performed by the ALU (arithmetic logic unit). The rest of the CPU (central processing unit) is then designed to feed the ALU with the required data as quickly as possible. The algorithm is compiled into a sequence of instructions, which are used to control the specific operation performed by the CPU and ALU during each clock cycle. The basic operation of the CPU is therefore to fetch an instruction from memory, decode the instruction to determine the operation to perform, and execute the instruction. All of the advances in mainstream computer architecture over the past 50 years have been developed to improve performance by maximising the utilisation of the ALU and by squeezing more data through the narrow bottleneck between the memory and the ALU (the so-called von Neumann bottleneck (Backus, 1978)).

The obvious approach is to increase the clock speed and hence the rate at which instructions are executed. Tremendous gains have been made in this area, with top clock speeds of several GHz being the norm for high-end computing platforms. Such increases have come primarily as the result of reductions in propagation delay brought about through advances in semiconductor technology, reducing both transistor feature sizes and the supply voltage. This is not without its problems, however, and one consequence of a higher clock speed is significantly increased power consumption by the CPU.

While the speeds of bulk memory have also increased, they have not kept pace with increasing CPU speeds. This has caused problems in reading both instructions and data from the system memory at the required rate. Caching techniques have been developed to buffer both instructions and data in a smaller high-speed memory that can keep pace with the CPU. The problem then is how to maximise the likelihood that the required data will be in the cache memory rather than the slower main memory. Cache misses (where the data is not in cache memory) can significantly degrade processor performance because the CPU is sitting idle waiting for the data to be retrieved. This has also influenced algorithm design, to make the algorithms cache aware or cache friendly by maximising spatial and temporal locality of memory accesses.

Both instructions and data need to be accessed from the main memory. One obvious improvement is to use a Harvard architecture, with separate memories for instructions and data, which doubles the effective memory bandwidth. Note that a similar effect can be achieved by having separate caches for instructions and data.

Another approach is to increase the width of the ALU, so that more data is processed in each clock cycle. This gives obvious improvements for wider data word lengths (for example double-precision floating-point numbers) because each number may be loaded and processed in a fewer clock cycles. However, when the natural word length is narrower, such as typically encountered in image processing, the performance improvement is not as great. In such cases, it is also possible to design the ALU to operate as a vector processor. Multiple data words are packed into a single processor word, allowing the ALU to perform the same operation simultaneously on several data items (for example using the Intel multi-media extension (MMX) instructions (Peleg et al., 1997)). This can be particularly effective for low-level image processing operations where the same operation is performed on each pixel.

The speed of the fetch–decode–execute cycle may be improved by pipelining the separate phases. Thus, while one instruction is being executed, the next instruction is being decoded, and the instruction after that is being fetched. Such pipelining is effective when executing long sequences of instructions without branches. However, instruction pipelining becomes more complex with loops or branches that break the regular sequence. Branch prediction techniques try to anticipate which path will be taken at a branch and increase the likelihood that the correct instructions will be loaded. Speculative execution executes the instructions that have been loaded in the pipeline already, so that the clock cycles are not lost if the branch is predicted correctly. If the branch is incorrectly predicted, the speculatively executed results are discarded.

With serial processors, time is usually the critical resource. This is because serial processors can only do one thing at a time. More recently, multiple core architectures have become mainstream. These enable multiple threads within an application to execute in parallel on separate processor cores (Geer, 2005). These can give some improvement for image processing applications, provided that the application has been developed carefully to support multiple threads. If care is not taken, the memory bandwidth can still be a bottleneck. Even so, multi-core processors are still serially bound.

For low-power or embedded vision applications, the size and power requirements of a standard serial processor are impractical in many cases. Lowering the clock speed can reduce the power significantly but on a serial processor will also limit the algorithms that can be implemented in real time. To retain the computing power while lowering the clock speed requires multiple parallel processors.

In this regard, an important development is the GPU (graphics processing unit). Originally developed to accelerate graphics rendering, its primary function was to take vertex data representing triangular patches within the scene and produce the corresponding output pixels stored in a frame buffer. Early GPUs had dedicated pipelines for texture mapping, pixel shading, z-buffering and blending, and anti-aliasing, which restricted their use for wider application. Modern devices are programmable (general purpose GPU, or GPGPU), enabling them to be used for general purpose computing tasks. They are designed primarily for exploiting data parallelism, with a processing kernel replicated to run on many data items in parallel. Speed is also gained through data pipelining and lightweight multi-threading. Data pipelining reduces the need to write temporary results to memory only to read them in again to perform further processing. Instead, results are written to sets of local registers where they may be accessed in parallel. Multi-threading splits the task into several small steps that can be implemented in parallel, which is easy to achieve for low-level image processing operations where the independent processing of pixels maps well to the GPU architecture. Lightweight threads reduce the overhead of switching the context from one thread to the next. Therefore, when one thread stalls waiting for data, the context can be rapidly switched to another thread to keep the processing units fully utilised. GPUs can give significant improvements over CPUs for algorithms that are easily parallelised, especially where data access is not the bottleneck (Cope et al., 2005). While power requirements of GPUs have decreased over the years, power considerations still rule them out for many embedded vision applications.

1.7 Parallelism

In principle, every step within any algorithm may be implemented on a separate processor, resulting in a fully parallel implementation. However, if the algorithm is predominantly sequential, with every step within the algorithm dependent on the data from the previous step, then little can be gained in terms of reducing the response time. To be practical for parallel implementation, an algorithm has to have a significant number of steps that may be implemented in parallel. This is referred to as Amdahl's law (1967). Let s be the proportion of the algorithm that is constrained to run serially (the housekeeping and other sequential components), and p be the proportion of the algorithm that may be implemented in parallel over N processors. The best possible speedup that can be obtained is then

$$Speedup \leq \frac{s+p}{s+\frac{p}{N}} = \frac{N}{1+(N-1)s}. \tag{1.9}$$

Equality will only be achieved if there is no additional overhead (for example communication or other house-keeping) introduced as a result of parallelisation. This speedup is always less than the number of parallel processors and will be limited by the proportion of the algorithm that must be serially executed:

$$\operatorname*{Limit}_{N\to\infty} Speedup = \frac{1}{s}. \tag{1.10}$$

Therefore, to achieve any significant speedup, the proportion of the algorithm that can be implemented in parallel must also be significant. Fortunately, image processing is inherently parallel, especially at the low and intermediate levels of the processing pyramid. This parallelism shows in a number of ways.

1.7.1 Temporal or Task Parallelism

Virtually all image processing algorithms consist of a sequence of image processing operations. This is a form of ***temporal parallelism*** or ***task parallelism***. Such an algorithm structure suggests using a separate processor for each operation, as shown in Figure 1.5. This leads naturally to a pipelined architecture, with the data passing through each of the stages as it is processed, like a production line. Each processor applies its operation and passes the result to the next stage. If each successive processor has to wait until the previous processor completes its processing, this arrangement will not reduce the total processing time, or the response time. However, the throughput can increase, because all of the processors run in parallel. While the second processor is working on the output of operation 1, processor 1 can begin processing the next image. When processing images, data output from an operation can usually begin long before the complete image has been processed by that operation.

The time between data being input to an operation and the corresponding output is available is the ***latency*** of that operation. The latency is generally lowest when each operation only uses input pixel values from a small, local neighbourhood because it is necessary to load all of the input pixel values to calculate the output value. Operations that require the whole image to calculate an output pixel value will have a higher latency. Operation pipelining can give significant performance improvements when all of the operations have low latency because a downstream processor may begin performing its operation before the upstream processors have completely processed the whole image. This can give benefits not only for multi-processor systems but also for software-based systems using a separate thread for each process (McLaughlin, 2000) because the user can begin to see the results before the whole image has been processed. Of course, in a software system using a single core, the total response time will not normally be decreased, although there may be some improvement resulting from switching between threads while waiting for data from slower external memory when it is not available in the cache. In a hardware system, however, the total response time is given by the sum of the latencies of each of the stages plus the time to input in the whole image. If the individual latencies are small compared to the time required to load the image, then the speedup factor can be significant, approaching the number of processors in the pipeline.

Two difficulties can be encountered with using pipelining for more complex image processing algorithms (Duff, 2000). The first is algorithms with multiple parallel paths. If, for example, processor 4 in Figure 1.5 also takes as input the results produced by processor 1, then the data must be properly synchronised to allow for the latencies of the other processors in the parallel path. For synchronous processing, this is simply a delay line, but synchronisation becomes considerably more complex when the latencies of the parallel operations are variable (data dependent). A greater difficulty is handling feedback, either with explicitly iterative algorithms or implicitly where the parameters of the earlier processors adapt to the results produced in later stages.

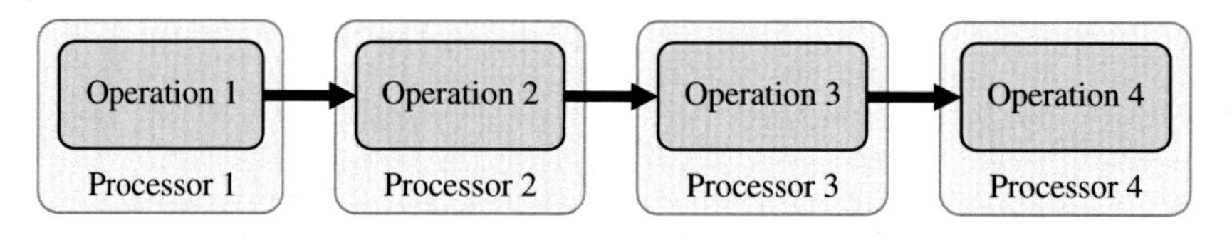

Figure 1.5 Temporal parallelism exploited using a processor pipeline.

1.7.2 Spatial or Data Parallelism

In software, many operation-level algorithms consist of nested loops. The outermost loop usually iterates over the pixels within the image, because many operations perform the same processing independently on many pixels. Unrolling (or partially unrolling) the outermost loop(s) enables parallelism to be applied to such operations. This is ***spatial parallelism***, or ***data parallelism***, which may be exploited by partitioning the image into several tiles and using a separate processor to perform the operation on each tile in parallel. Common partitioning schemes split the image into blocks of rows, blocks of columns, or rectangular blocks, as illustrated in Figure 1.6. For video processing, the image sequence may also be partitioned in time, by assigning successive frames to separate processors (Downton and Crookes, 1998).

In the extreme case, a separate processor may be allocated to each pixel within the image (for example the 'massively parallel processor' (Batcher, 1980)). Dedicating a processor to each pixel can facilitate some very efficient algorithms. For example, image filtering may be performed in only a few clock cycles. However, a significant problem with such parallelism is the overhead required to distribute the image data to and from each of the processors. Many algorithms for massively parallel processors assume that the data is already available at the start. Massively parallel processors also do not scale well with image size, so have not gained much interest outside of research circles.

The important consideration when partitioning the image is to minimise the communication between processors, which corresponds to minimising any data used from the other tiles. For low-level image processing operations, the performance improvement can approach the number of processors. However, the performance will degrade as a result of any communication overheads or contention when accessing shared resources. Consequently, each processor must have some local memory to reduce any delays associated with contention for global memory. Partitioning is therefore most beneficial when the operations only require local data, where local is defined by the tile boundaries.

If the operations performed within each partition are identical, this leads to a SIMD (single instruction, multiple data) parallel processing architecture according to Flynn's taxonomy (Flynn, 1972). In hardware, this corresponds to creating an instance of each processor for each tile. A scatter/gather data pattern is also required to distribute the incoming data to each tile processor and to gather the results.

With some intermediate-level operations, the processing time for each partition may vary significantly, depending on the image content within that partition. A simple static, regular, partitioning strategy will be less efficient in this case because the worst-case performance must be allowed for when allocating tiles to processors. As a result, many of the processors may be sitting idle for much of the time. In such cases, better performance may be achieved by having more tiles than processors and using a processor farm approach (Downton and Crookes, 1998). Each image partition, or tile, is then dynamically allocated to the next available processor. Again, it is important to minimise the communications between processors.

For high-level image processing operations, the data is no longer image based. However, data partitioning methods can still be exploited by assigning a separate data structure, region, or object to a separate processor. Such assignment generally requires the dynamic partitioning approach of a processor farm.

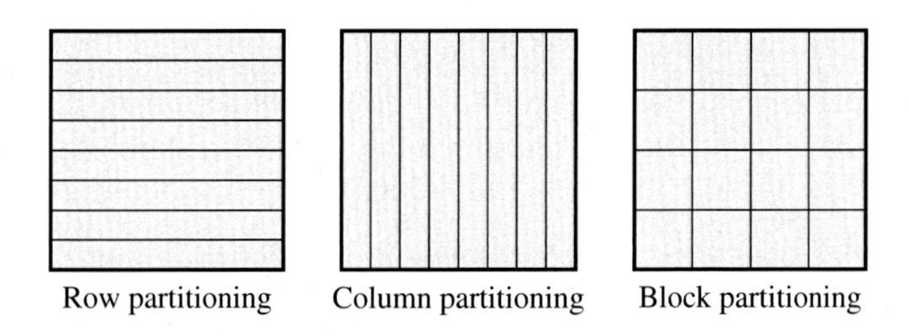

Figure 1.6 Spatial parallelism exploited by partitioning the image.

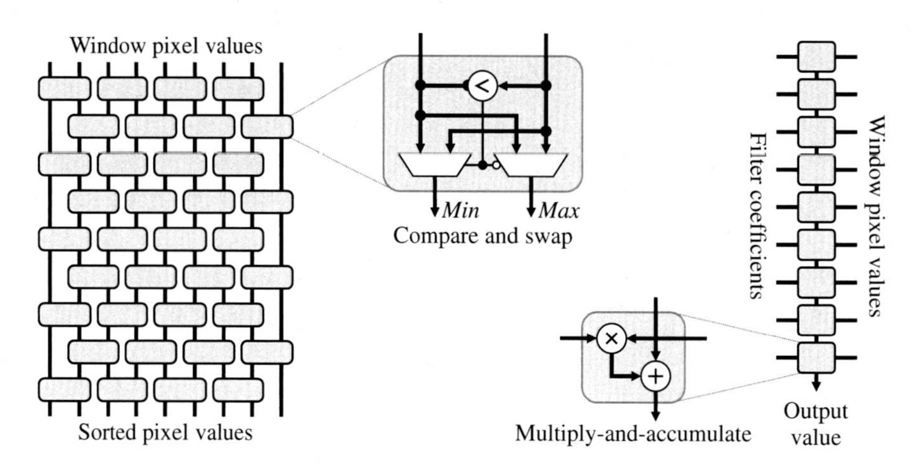

Figure 1.7 Examples of logical parallelism. Left: compare and swap within a bubble sorting odd–even transposition network; right: multiply-and-accumulate within a linear filter.

1.7.3 Logical Parallelism

Logical parallelism occurs when the same functional block is used many times within an operation. This often corresponds to inner loops within the algorithm implementing the operation. Logical parallelism is exploited by unrolling the inner loops and executing each instance of the function block in parallel. Often, the functions can be arranged in a regular structure, as illustrated by the examples in Figure 1.7. The odd–even transposition network within a bubble sort (for example to sort the pixel values within a rank window filter (Heygster, 1982; Hodgson et al., 1985)) consists of a series of compare and swap function blocks. Implementing the blocks in parallel gives a considerable speed improvement over iterating with a single compare and swap block. A linear filter or convolution multiplies the pixel values within a window by a set of weights or filter coefficients. The multiply-and-accumulate block is repeated many times and is a classic example of logic parallelism.

Often, the structure of the group of function blocks is quite regular, which when combined with pipelining results in a synchronous systolic architecture (Kung, 1985). When properly implemented, such architectures have a very low communication overhead, enabling significant speed improvements resulting from multiple copies of the function block.

1.7.4 Stream Processing

A bottleneck within image processing is the time, and bandwidth, required to read the image from memory and write the resultant image back to memory. Images are commonly streamed sequentially from cameras or to displays using a raster scan, often at a rate of one pixel per clock cycle (see Figure 1.8), although high-speed video interfaces may stream multiple pixels in each clock cycle. Many low- and intermediate-level image processing operations can be implemented with such a regular pattern of data accesses. ***Stream processing*** exploits this to convert spatial parallelism into temporal parallelism. It then performs all of its processing on the pixels on-the-fly as they pass through the processor. If processing multiple pixels per clock cycle, multiple copies of the hardware are required, although the control structures can usually be shared between the parallel hardware instances (Reiche et al., 2018). The individual operations within the operation-level algorithm are pipelined as necessary to maintain the required throughput. Data is cached locally to avoid the need for parallel external data accesses. This is most effective if each operation uses input from a small local neighbourhood, otherwise the caching requirements can become excessive. Stream processing also works best when the time taken to process each pixel is constant. If the processing time is data dependent, it may be necessary to insert FIFO (first-in, first-out) buffers on the input or output (or both) to maintain a constant data rate and manage less regular data

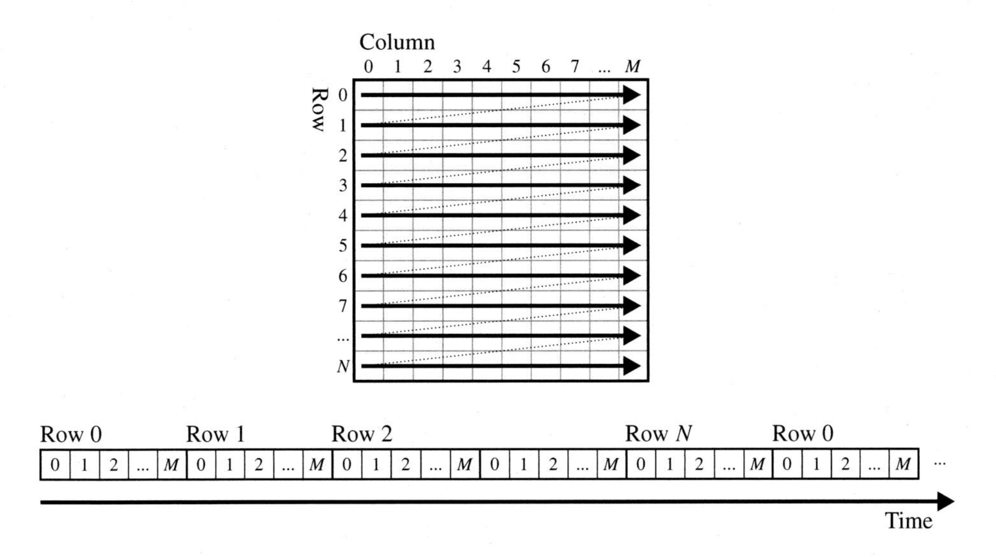

Figure 1.8 Stream processing converts spatial parallelism into temporal parallelism.

access patterns. In terms of Flynn's taxonomy (Flynn, 1972), stream processing is MISD (multiple instruction, single data) as the parallel processors perform separate instructions on the data as it is streamed through.

With stream processing, the response time is given by the sum of the time required to read or write the image and the processing latency (the time between reading a pixel and producing its corresponding output). For most operations, the latency is small compared with loading the whole image, so if the whole application-level algorithm can be implemented using stream processing, the response time is dominated by frame rate.

In this section, it was not made explicit whether the processors that made up the parallel systems were a set of standard serial processors or dedicated hardware implementing the parallel functions. In theory, it makes little difference, as long as the parallelism is implemented efficiently, and the communication overheads are kept low.

Hardware-based image processing systems are very fast, but their biggest problem is their relative inflexibility. However, field-programmable gate arrays (FPGAs) combine the inherent parallel nature of hardware with the flexibility of software in that their functionality can be reprogrammed or reconfigured. While early FPGAs were quite small in terms of the equivalent number of gates, modern FPGAs now have sufficient resources to enable whole applications to be implemented on a single FPGA, making them an ideal choice for embedded real-time vision systems.

1.8 Summary

Image processing covers a wide variety of applications, many of which require embedded or real-time operation, for example video processing, industrial inspection, and autonomous control. However, embedded real-time image processing is challenging for conventional serial processors. Fortunately, many image processing algorithms provide significant opportunities for exploiting parallelism, both at the operation and the application level. Sequences of operations may be pipelined using temporal or task-level parallelism. Images may be partitioned into tiles and processed using spatial or data-level parallelism. Inner loops may be unrolled to yield logical parallelism. Stream processing exploits the fact that images are naturally streamed from a camera, or to a display, to process images on-the-fly. Hardware-based processing (using digital logic) is inherently parallel, enabling FPGAs to exploit the parallelism within embedded image processing applications. Chapter 2 will examine FPGAs and explore how they can be applied to image processing.

References

Amdahl, G.M. (1967). Validity of the single processor approach to achieving large scale computing capabilities. *AFIPS Spring Joint Computer Conference*, Atlantic City, NJ, USA (18–20 April 1967), Volume **30**, 483–485. https://doi.org/10.1145/1465482.1465560.

Auer, S. (1982). Imaging by dust rays: a dust ray camera. *Optica Acta* **29** (10): 1421–1426. https://doi.org/10.1080/713820766.

Backus, J. (1978). Can programming be liberated from the von Neumann style? A functional style and its algebra of programs. *Communications of the ACM* **21** (8): 613–641. https://doi.org/10.1145/359576.359579.

Baer, R.L., Holland, W.D., Holm, J.M., and Vora, P.L. (1999). Comparison of primary and complementary color filters for CCD-based digital photography. *Sensors, Cameras, and Applications for Digital Photography*, San Jose, CA, USA (27–28 January, 1999). SPIE, Volume **3650**, 16–25. https://doi.org/10.1117/12.342859.

Batcher, K.E. (1980). Design of a massively parallel processor. *IEEE Transactions on Computers* **29** (9): 836–849. https://doi.org/10.1109/TC.1980.1675684.

Bayer, B.E. (1976). Color imaging array. United States of America Patent 3971065.

Binnig, G., Quate, C.F., and Gerber, C. (1986). Atomic force microscope. *Physical Review Letters* **56** (9): 930–933. https://doi.org/10.1103/PhysRevLett.56.930.

Castleman, K.R. (1979). *Digital Image Processing*. Englewood Cliffs, NJ: Prentice Hall.

Catsoulis, J. (2005). *Designing Embedded Hardware*, 2e. Sebastopol, CA: O'Reilly Media.

Conrad, J. (2006). *Depth of Field in Depth*. Large Format Photography. https://www.largeformatphotography.info/articles/DoFinDepth.pdf (accessed 18 April 2023).

Cope, B., Cheung, P.Y.K., Luk, W., and Witt, S. (2005). Have GPUs made FPGAs redundant in the field of video processing? *IEEE International Conference on Field-Programmable Technology*, Singapore (11–14 December 2005), 111–118. https://doi.org/10.1109/FPT.2005.1568533.

Dougherty, E.R. and Laplante, P.A. (1985). *Introduction to Real-time Imaging*. Bellingham, WA, USA: SPIE Optical Engineering Press.

Downton, A. and Crookes, D. (1998). Parallel architectures for image processing. *IEE Electronics and Communication Engineering Journal* **10** (3): 139–151. https://doi.org/10.1049/ecej:19980307.

Duff, M.J.B. (2000). Thirty years of parallel image processing. *Vector and Parallel Processing (VECPAR 2000)*, Porto, Portugal (21–23 June 2000), *Lecture Notes in Computer Science*, Volume **1981**, 419–438. https://doi.org/10.1007/3-540-44942-6_35.

EAIVA (n.d.). Edge AI and vision alliance. https://www.edge-ai-vision.com/ (accessed 18 April 2023).

Flynn, M. (1972). Some computer organizations and their effectiveness. *IEEE Transactions on Computers* **21** (9): 948–960. https://doi.org/10.1109/TC.1972.5009071.

Fossum, E.R. (1993). Active pixel sensors: are CCDs dinosaurs? *Charge-Coupled Devices and Solid State Optical Sensors III*, San Jose, CA, USA (2–3 February 1993), SPIE, Volume **1900**, 2–14. https://doi.org/10.1117/12.148585.

Geer, D. (2005). Chip makers turn to multicore processors. *Computer* **38** (5): 11–13. https://doi.org/10.1109/MC.2005.160.

Gilblom, D.L., Yoo, S.K., and Ventura, P. (2003). Real-time color imaging with a CMOS sensor having stacked photodiodes. *Ultrahigh- and High-Speed Photography, Photonics, and Videography*, San Diego, CA, USA (7 August 2003), SPIE, Volume **5210**, 105–115. https://doi.org/10.1117/12.506206.

Hansma, P.K. (1996). View camera focusing in practice. *Photo Techniques* (March/April 1996), 54–57.

Haralick, R.M. and Shapiro, L.G. (1991). Glossary of computer vision terms. *Pattern Recognition* **24** (1): 69–93. https://doi.org/10.1016/0031-3203(91)90117-N.

Heygster, G. (1982). Rank filters in digital image processing. *Computer Graphics and Image Processing* **19** (2): 148–164. https://doi.org/10.1016/0146-664X(82)90105-8.

Hodgson, R.M., Bailey, D.G., Naylor, M.J., Ng, A.L.M., and McNeill, S.J. (1985). Properties, implementations and applications of rank filters. *Image and Vision Computing* **3** (1): 3–14. https://doi.org/10.1016/0262-8856(85)90037-X.

Johnston, C.T., Gribbon, K.T., and Bailey, D.G. (2004). Implementing image processing algorithms on FPGAs. *11th Electronics New Zealand Conference (ENZCon'04)*, Palmerston North, NZ (15–16 November 2004), 118–123.

Kehtarnavaz, N. and Gamadia, M. (2006). *Real-Time Image and Video Processing: From Research to Reality, Synthesis Lectures on Image, Video and Multimedia Processing*. USA: Morgan and Claypool. https://doi.org/10.2200/S00021ED1V01Y200604IVM005.

Kirsch, R.A. (1998). SEAC and the start of image processing at the National Bureau of Standards. *IEEE Annals of the History of Computing* **20** (2): 7–13. https://doi.org/10.1109/85.667290.

Kung, S. (1985). VLSI array processors. *IEEE ASSP Magazine* **2** (3): 4–22. https://doi.org/10.1109/MASSP.1985.1163741.

Litwiller, D. (2005). CMOS vs. CCD: maturing technologies, maturing markets. *Photonics Spectra* **2005** (8): 54–59.

Lyon, R.F. and Hubel, P.M. (2002). Eyeing the camera: into the next century. *10th Color Imaging Conference: Color Science and Engineering Systems, Technologies, Applications*, Scottsdale, AZ, USA (12–15 November 2002), 349–355.

McFarlane, M.D. (1972). Digital pictures fifty years ago. *Proceedings of the IEEE* **60** (7): 768–770. https://doi.org/10.1109/PROC.1972.8775.

McLaughlin, J. (2000). The development of a Java image processing framework. Master of Technology thesis. Palmerston North, NZ: Massey University. https://doi.org/10179/12046.

Morimoto, K., Ardelean, A., Wu, M.L., Ulku, A.C., Antolovic, I.M., Bruschini, C., and Charbon, E. (2020). Megapixel time-gated SPAD image sensor for 2D and 3D imaging applications. *Optica* **7** (4): 346–354. https://doi.org/10.1364/OPTICA.386574.

Parulski, K.A. (1985). Color filters and processing alternatives for one-chip cameras. *IEEE Transactions on Electron Devices* **32** (8): 1381–1389. https://doi.org/10.1109/T-ED.1985.22133.

Peleg, A., Wilkie, S., and Weiser, U. (1997). Intel MMX for multimedia PCs. *Communications of the ACM* **40** (1): 24–38. https://doi.org/10.1145/242857.242865.

Pohl, M., Schaeferling, M., and Kiefer, G. (2014). An efficient FPGA-based hardware framework for natural feature extraction and related computer vision tasks. *24th International Conference on Field Programmable Logic and Applications (FPL)*, Munich, Germany (2–4 September 2014), 8 pages. https://doi.org/10.1109/FPL.2014.6927463.

Rayleigh, L. (1879). XXXI. Investigations in optics, with special reference to the spectroscope. *The London, Edinburgh, and Dublin Philosophical Magazine and Journal of Science* **8** (49): 261–274. https://doi.org/10.1080/14786447908639684.

Reiche, O., Ozkan, M.A., Hannig, F., Teich, J., and Schmid, M. (2018). Loop parallelization techniques for FPGA accelerator synthesis. *Journal of Signal Processing Systems* **90** (1): 3–27. https://doi.org/10.1007/s11265-017-1229-7.

Schaffer, G. (1984). Machine vision: a sense for computer integrated manufacturing. *American Machinist* **128** (6): 101–129.

Shannon, C.E. (1949). Communication in the presence of noise. *Proceedings of the IRE* **37** (1): 10–21. https://doi.org/10.1109/JRPROC.1949.232969.

Weems, C.C. (1991). Architectural requirements of image understanding with respect to parallel processing. *Proceedings of the IEEE* **79** (4): 537–547. https://doi.org/10.1109/5.92046.

2

Field-programmable Gate Arrays

In this chapter, the basic architectural features of field-programmable gate arrays (FPGAs) are examined. While a detailed understanding of the internal architecture of an FPGA is not essential to programme them (the vendor-specific place-and-route tools manage most of these details), a basic knowledge of the internal structure and how a design maps onto that architecture can be used to develop a more efficient implementation.

The approaches commonly used to programme FPGAs are then outlined. The concept of register transfer level (RTL) design is introduced, and the key characteristics of hardware description languages (HDLs) described. This culminates in a discussion of high-level synthesis (HLS) for algorithm implementation.

Finally, the opportunities for realising image processing algorithms on FPGAs are outlined, with advice on choosing an appropriate development board for image processing.

2.1 Hardware Architecture of FPGAs

The basic idea behind programmable hardware is to have a generic circuit where the functionality can be programmed for a particular application. Computers are based on this idea, where the arithmetic logic unit (ALU) can perform one of several operations based on a set of control signals. The limitation of an ALU is that it can only perform one operation at a time. Therefore, a particular application must be decomposed into the sequence of control signals for controlling the function of the ALU, along with the logic required to provide the appropriate data to the inputs of the ALU. The sequence of controls is provided by a sequence of programme instructions stored in memory.

Programmable logic, on the other hand, represents the functionality as a digital logic circuit, where the particular circuit can be programmed to meet the requirements of an application. The key distinction is that the logic functionality is implemented as a parallel system, rather than sequential. Any combinatorial logic function may be implemented using a two-level minterm (OR of AND) representation. Therefore, it made sense that early programmable logic devices consisted of a two-level programmable array architecture.

Early devices were either mask or fuse programmable. Mask-programmable devices were programmed at manufacture, effectively hardwiring the programming using a metal mask. Fuse-programmable devices could be programmed by the user, making such circuits ***field programmable***. The programmable sections can consist either of fuses where a higher-than-normal current is used to break a connection (blow the fuse in the conventional sense) or more usually anti-fuses, where a thin insulating layer is broken down by applying a high voltage across it, and it becomes conducting. Since blowing a fuse (or anti-fuse) is destructive, once programmed, such devices cannot be reprogrammed. Another advance, in the mid-1980s, was to control each programmable connection with an EEPROM (electrically erasable programmable read only memory) or flash memory cell rather than a fuse. This enables the devices to be cleared and reprogrammed. A key benefit is that the logic may be altered simply by reprogramming the device. This opens possibilities for remote upgrading, without necessarily having the product returned to the factory or service centre.

Design for Embedded Image Processing on FPGAs, Second Edition. Donald G. Bailey.

Companion Website: www.wiley.com/go/bailey/designforembeddedimageprocessc2e

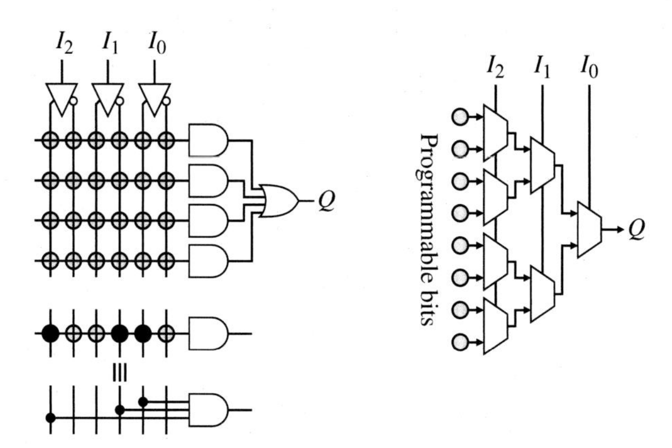

Figure 2.1 Programmable logic cell architectures. Left: two-level OR of AND; right: lookup table based on a multiplexer tree.

The next step in the development of programmable logic was to incorporate several blocks of logic on a single chip, with programmable interconnections to enable more complex logic functions to be created. Two classes of device resulted (see Figure 2.1): complex programmable logic devices (CPLDs) based on a two-level OR of AND architecture and FPGAs based on a lookup table architecture, usually implemented as a multiplexer tree. This architectural difference gives CPLDs a higher density of logic compared to interconnect, whereas the logic blocks within FPGAs are smaller, and the architecture is more dominated by interconnects. The other main difference between CPLDs and FPGAs is that CPLDs tend to be flash programmable, whereas the programme or configuration for FPGAs is usually held in static memory cells. This makes the FPGA configuration volatile, requiring it to be reloaded each time the device is powered on.

Both CPLDs and FPGAs were introduced in the mid-1980s and were initially used primarily as glue logic on printed circuit boards to flexibly interconnect several other components. However, as device densities increased, FPGAs dominated because of the increased flexibility that came primarily through the more flexible interconnect structure. Current FPGAs have sufficient logic resources (and associated interconnect routing) to implement even complex applications on a single chip. More recent trends have integrated hardened functional blocks within the FPGA, including multipliers, memories, high-speed input–output interfaces, and even central processing units (CPUs) (Leong, 2008; Trimberger, 2015). In this context, ***hardened*** means implemented in dedicated hardware, rather than out of the generic programmable logic (soft logic). Implementing such commonly used features as dedicated blocks frees up the programmable logic resources that would otherwise be required. It also increases the speed and reduces the power consumption compared with implementing those functions out of general-purpose programmable logic.

So, what exactly is inside an FPGA? Figure 2.2 shows the basic structure and essential components of a generic FPGA. The programmable logic consists of a set of fine-grained blocks that are used to implement the logic of the application. This is sometimes called the ***fabric*** of the FPGA. The logic blocks are usually based on a lookup table architecture, enabling them to implement any arbitrary function of the inputs. The logic blocks are typically tiled in a grid structure and interconnected via a programmable routing matrix that enables the blocks to be connected in arbitrary configurations. The input and output (I/O) blocks interface between the core of the FPGA and external devices. The routing means that virtually any signal can be routed to any I/O pin of the device.

In addition to these basic features, most FPGAs provide some form of clock synchronisation to control the timing of a clock signal relative to an external source. A clock distribution network provides clock signals to all parts of the FPGA while limiting the clock skew between different sections of a design. There is also some dedicated logic for loading the user's design or configuration into the FPGA. This logic does not directly form part of the user's design but is part of the overhead required to make FPGAs programmable.

Each of these components will be examined in a little more detail in Sections 2.1.1–2.1.8.

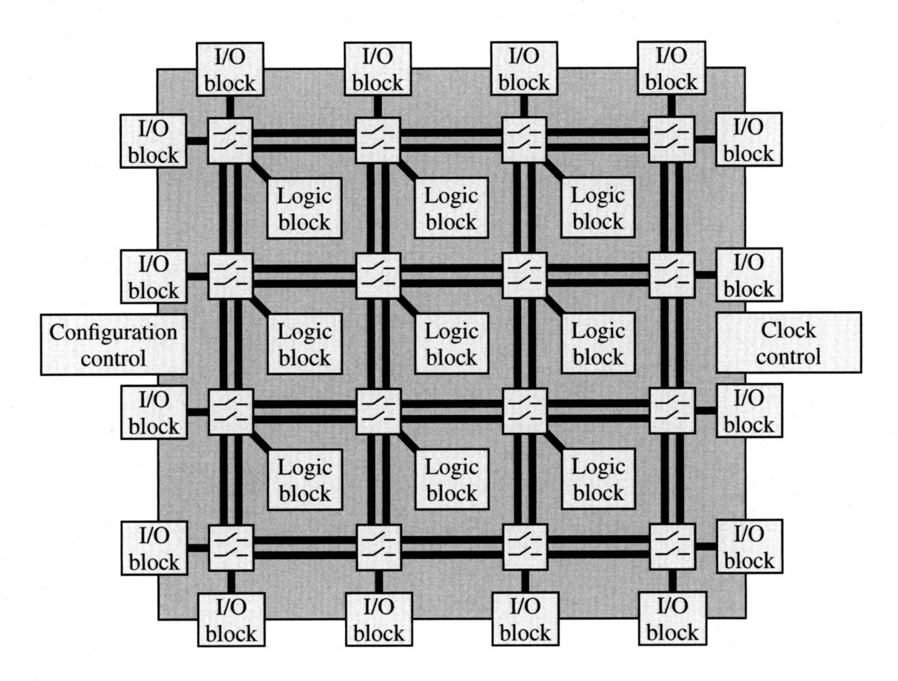

Figure 2.2 The basic architecture of a generic FPGA.

2.1.1 Logic

The smallest unit of logic within the FPGA producing a single bit output is a ***logic cell***. It typically consists of a small lookup table (LUT) and a 1-bit register (an edge-triggered flip-flop) on the output, as represented in Figure 2.3. A common terminology is to refer to a three-input LUT as a 3-LUT, and similarly for other LUT sizes. The flip-flop is used to register the output. On some FPGAs, both outputs may be provided, as in Figure 2.3; on others, a multiplexer may be used to select whether the registered or unregistered signal is output. The registers can be used for building finite state machines, counters, and to hold data between the stages of pipelined systems.

Each LUT can implement any arbitrary function of its inputs. More complex functions must be implemented by cascading multiple levels of LUTs by connecting the output, X, to an input of the next layer. FPGA manufacturers face a trade-off in choosing the size of the LUTs. Consider implementing a simple 2-input AND gate. With a large LUT, a significant fraction of the logic would be unused. Smaller LUTs would achieve a higher utilisation of resources. However, when implementing complex functions with small LUTs, multiple logic cells would have to be cascaded, with the increased routing significantly increasing the propagation delay. For complex functions, using a larger LUT would be preferable. To balance this, the silicon area of the table

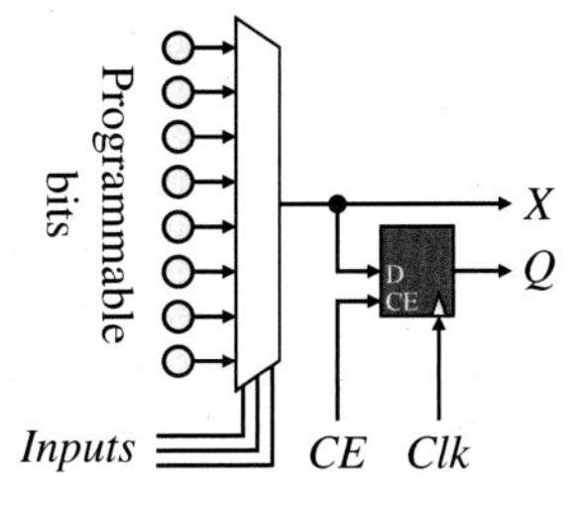

Figure 2.3 A logic cell is the basic building block of an FPGA; here, a 3-LUT is shown.

grows exponentially with the number of inputs, so at some stage it is more efficient to split a complex function over multiple LUTs. A larger LUT also has a longer propagation delay than a smaller table because of the extra levels in the multiplexer tree. Early studies (Kouloheris and El Gamal, 1991; Singh et al., 1992) showed that the optimum LUT size is four to six inputs. Early FPGAs used 3-LUTs and 4-LUTs. However, as device scales have shrunk, the proportion of the total propagation delay used by the routing matrix has increased. This has led to larger LUTs, with many more recent devices using 6-LUTs. Another approach taken is to combine two basic LUTs, sharing some of the inputs, but allow the logic to be partitioned between two (or more) outputs according to the application (Hutton et al., 2004).

It is also usual to combine multiple logic cells into a logic block or tile in Figure 2.2. The logic cells within a block share common control signals, for example clock and clock enable signals. In a logic block, the outputs are directly available as inputs to other logic cells within the same block. The resulting direct connections reduce the number of signals that need to be routed through the interconnect matrix, reducing the propagation delay for cascaded logic cells. Additional dedicated multiplexers allow more complex functions to be synthesised by combining adjacent LUTs to effectively create a larger LUT. Ahmed and Rose (2000) have shown that the optimal logic block size is 4–10 logic cells.

Many of the early FPGAs were homogeneous in that there was only one type of logic resource. However, as FPGAs developed and moved from being used primarily as glue logic to compute resources in their own right, implementing many functions using fine-grained logic was expensive. This has led to a more heterogeneous architecture, with several more complex coarser-grained building blocks (Trimberger, 2015).

Addition is a commonly performed operation. On an FPGA, a full adder would require two 3-LUTs, one to produce the sum and one to produce the carry. With a ripple adder, the propagation of the carry signal through the routing matrix is likely to form the critical path. Since addition is such a common operation, many manufacturers provide additional circuitry within logic cells (or logic blocks) to reduce the resources required and significantly reduce the critical path for carry propagation. This usually consists of dedicated logic to detect the carry (so only one LUT is needed per bit of addition) and a dedicated direct connection for the carry signal between adjacent logic cells.

2.1.2 DSP Blocks

Multiplication is also a common operation in digital signal processing (DSP) applications. A multiplier can be implemented using logic blocks to perform a series of additions, either sequentially or in parallel. Sequential multiplication is time consuming, requiring one clock cycle for each bit of the multiplier to perform the associated addition. Implementing the multiplication in parallel improves the timing but consumes significant resources. Therefore, it is common in FPGAs targeting DSP and other compute-intensive applications to have hardened multiplier or multiply-and-accumulate blocks.

A typical DSP block is shown in Figure 2.4. The pre-adder allows one of the multiplicands to be the sum of two terms. The final accumulator is helpful for multiply–add operations, as required for convolution, dot-product, and matrix multiplication. The cascade input to the output adder allows such operations to be pipelined, using the internal register as a pipeline register.

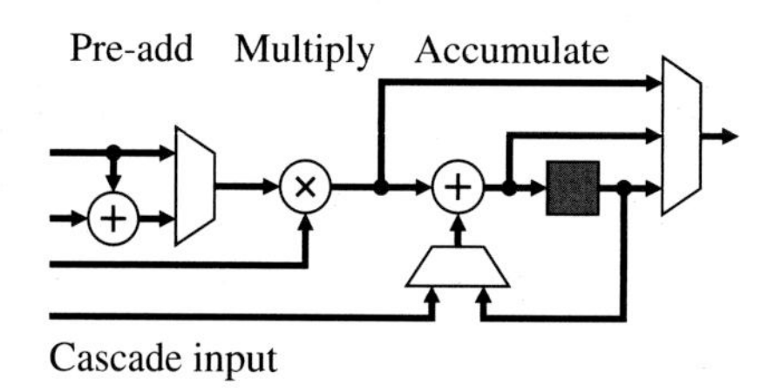

Figure 2.4 A typical DSP block, with pre-adder, multiplier, and accumulator.

In most FPGAs, the DSP block uses integer arithmetic. However, designs with a similar structure that can be used either as integer or floating-point numbers have been proposed (Langhammer and Pasca, 2015a,b). Such designs are starting to appear in high-end FPGAs (Sinha, 2017), significantly reducing the logic requirements for floating-point operations.

The Xilinx DSP48E block generalises the adder within the accumulator to give the ability to perform logical operations and pattern matching (Xilinx, 2018). As a result, some researchers have used such a DSP block as the ALU of a full function soft-core processor (Cheah et al., 2014; Cheah, 2016). Such a light-weight processor design allows many cores to be instantiated within an FPGA.

Recently, Intel has introduced a tensor block containing 30 8-bit integer multipliers designed for artificial intelligence and deep learning applications (Langhammer et al., 2021). Each tensor block has three columns, each capable of performing a 10-element dot product. The growing importance of deep learning is likely to see more dedicated resources such as this become available on FPGAs.

2.1.3 Memory

The flip-flops in the logic cells provide very limited storage space. While this is suitable for data that is accessed frequently, it is unsuited for storing larger volumes of data. Rather than require the use of off-chip memories, most FPGAs have added small blocks of memory onto the FPGA for intermediate storage and buffers. One approach is to adapt the structure of the logic cells to enable them to be used as small random access memories rather than as LUTs. In this way, a 4-LUT can hold 2^4 bits of data, rather than just the one bit in the output flip-flop. Such memory is called ***fabric RAM*** because it is made from the 'fabric' of the FPGA.

Larger blocks of RAM are implemented by adding dedicated blocks of static RAM, also referred to as ***block RAM***. A high bandwidth is maintained by keeping the memory blocks small and enabling independent access to each block. The memory blocks are usually dual-port, simplifying the construction of first-in first-out (FIFO) buffers and custom caches. ***True dual-port*** allows independent reading and writing on each of the ports. Some memories are only ***simple dual-port***, where one port is write-only, and one port is read-only. Most FPGAs allow the aspect ratio of block RAMs to be configured, for example an 18 Kbit block can be configured as 16,384 × 1, 8192 × 2, 4096 × 4, 2048 × 9, 1024 × 18, or 512 × 36.

Where more memory is required, this is usually off-chip static or dynamic RAM. Two disadvantages of external memory are the limited memory bandwidth and the significantly increased latency (which is often variable). Some recent high-end FPGAs (for example the Xilinx Ultrascale+ HBM (Wissolik et al., 2019)) integrate high-bandwidth memory consisting of stacks of dynamic random access memory (DRAM) chips within the same package as the FPGA.

2.1.4 Embedded CPU

Often, only parts of an application benefit from the parallelism offered by an FPGA, and other parts are more efficiently implemented using a serial processor. In these cases, a serial ***soft-core processor*** may be constructed from the FPGA fabric. However, the delays associated with programmable logic and interconnect will limit the clock speed of such a processor. Therefore, many modern FPGA families implement a hardened high-performance serial processor on chip, a ***hard processor system***. In early systems, a hardened PowerPC processor was used; in current systems, it is more usually a multicore Arm CPU. This combination of FPGA and CPU is often called ***system-on-chip*** (SoC) and can enable the complete design to be realised on a single chip or two chips if the CPU requires external memory. Using an SoC within an embedded system can reduce the overall system size, cost, and power (Trimberger, 2015).

SoC processors have high-speed connections between the CPU and FPGA fabric, which facilitates partitioning the design between hardware and software. The tightly coupled interface gives not only the CPU access to the FPGA logic but also access from the FPGA logic to the processor's address space. This can, for example, enable the FPGA to use the CPU memory as a frame buffer. Within a software-based system, the FPGA can be used to provide custom peripherals to the CPU or to accelerate compute-intensive tasks. If required, the accelerator can access data and return results directly into the processor cache (via the accelerator coherency port (ACP)), reducing the overheads of transferring data via external memory.

To summarise, fine-grained homogeneous FPGAs consist of a sea of identical logic blocks. However, current FPGAs have moved away from this homogeneous structure, with a heterogeneous mixture of both fine-grained logic blocks and larger hardened coarse-grained blocks targeted for accelerating commonly used operations. Heterogeneous systems will generally be faster because the coarse-grained blocks will not have the same level of overhead that comes with fine-grained flexibility. However, if the mix of specialised blocks is not right, much of the functionality that they provide may remain underutilised.

2.1.5 Interconnect

The programmable interconnect flexibly connects the different logic blocks to implement the desired functionality. Obviously, it is impractical to provide a dedicated direct connection from every possible output to every possible input. In any given application, only a very small fraction of these connections would be required; however, any of the connections potentially may be required. The solution is to have a smaller set of connection resources (routing lines) that can be shared flexibly to create the connections required for a given application, a little like the cables in a telephone network.

Since the interconnect takes up room on the silicon, there is a compromise. If only a few routing lines are provided, more space is available for logic, but complex circuits cannot be constructed because of insufficient interconnection resources. Alternatively, if too many routing lines are provided, many will not be used, and this reduces the area available for logic. There is an optimum somewhere in between these extremes, where sufficient routing lines are available for most applications; even so, interconnect requires significantly more space on the FPGA than the logic blocks (DeHon, 1999). This balance has shifted as device scales have reduced, and the number of logic blocks on an FPGA has increased (Trimberger, 2015).

The connection network for most FPGAs is based on a grid structure, similar to that shown in Figure 2.2. At each grid intersection, a form of crossbar switch enables programmable connection between the horizontal and vertical routing lines. While bus structures have been used on FPGAs in the past, the danger is connecting two outputs together. If one drives high, and the other low, this creates a low resistance path across the power supply, potentially damaging the chip.

Another factor that needs to be considered is the propagation delay through the interconnection network. Each switch that the signal passes through will add to the delay. This has led to a segmented structure, where not every routing line is switched at every junction. Also, dedicated direct connections can be provided between adjacent logic blocks to reduce the number of signals that need to go via the interconnect matrix. While having multiple different interconnection resources improves the flexibility and speed, it can make the process of finding good routes significantly more complex, especially for designs which use most of the logic available on the FPGA.

With large modern FPGAs, the interconnection network accounts for an increasing proportion of the system propagation delay. This has led Intel to introduce registers within the routing network (Hutton, 2017) to enable even the routing to be pipelined where required, allowing higher clock frequencies.

An alternative to pipelining the interconnections is to connect relatively independent blocks using ***network-on-chip*** (NoC) techniques. Rather than connect the individual computation blocks directly using point-to-point connections, a packet-switched network is used (Francis, 2013; Kundu and Chattopadhyay, 2015). In this way, a single shared network can connect multiple logic blocks, reducing the routing resources required. The network latency requires the blocks to operate relatively asynchronously. While the NoC can be implemented using soft logic, FPGAs with hardened NoCs are starting to appear (Swarbrick et al., 2019).

2.1.6 Input and Output

Input and output blocks connect between the FPGA logic and external devices. Most I/O pins can be programmed as inputs, outputs, or both and be connected directly to the internal logic of the FPGA or via registers. This range of functionality is provided by a generic I/O block such as that shown in Figure 2.5, with each

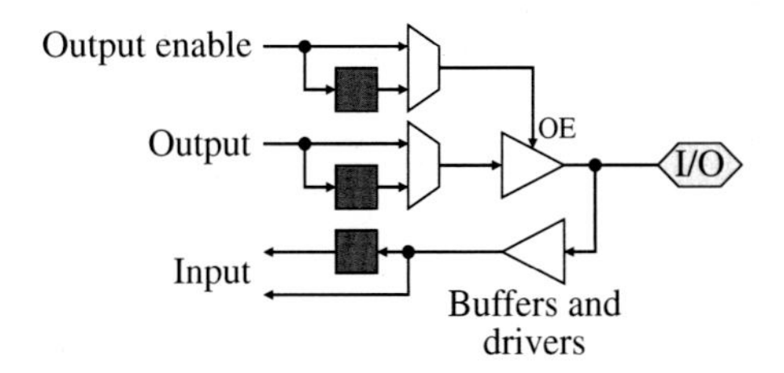

Figure 2.5 Generic input/output block.

multiplexer controlled by a programmable configuration bit. The main complexity of I/O arises from the large number of different signalling standards used by different devices and ensuring appropriate synchronisation between the FPGA and the off-chip components.

Perhaps, the most common signal standards are low-voltage transistor–transistor logic (LVTTL) and low-voltage complementary metal oxide semiconductor (LVCMOS), which are suitable for general-purpose 3.3 or 2.5 V signals. FPGAs require separate I/O power supplies because the I/O voltages are often larger than the core supply voltage. Some standards also require an external voltage reference to determine the threshold level between a logic '0' and '1'. The voltage supply and reference levels usually apply to a group or bank of I/O pins.

It is important when interfacing high-speed signals to consider transmission line effects. A general rule of thumb is that any connection longer than about 1/8 of the rise time should be appropriately terminated at each end to minimise signal reflections. As FPGAs have become larger and faster, it has become common practice to provide programmable on-chip termination. If not provided, it is necessary to ensure that appropriate termination is built off-chip for fast switching signals.

Just as dedicated logic blocks are provided in some FPGAs for commonly used functions, many newer FPGAs may also include hardened I/O blocks to manage some commonly used connections and protocols. One example is the logic required to interface to ***double data rate*** (DDR) memories. Two data bits are transferred per clock cycle on each I/O pin, one on clock high and one on clock low. Multiplexing and de-multiplexing are built into the I/O block, enabling the internal circuitry to operate at the clock rate rather than twice the clock rate.

Differential signalling is commonly used for high-speed communication because it is less sensitive to noise and cross-talk. This improved noise immunity enables a lower signalling voltage to be used, reducing the power requirements. Most FPGAs include the logic for managing ***low-voltage differential signalling*** (LVDS), which requires two I/O pins for each bit.

High-speed serial standards have signalling rates higher than the maximum clock speed supported by the core of the FPGA. Support for such standards requires incorporating parallel-to-serial converters on outputs and serial-to-parallel converters on inputs. The ***serialisation and deserialisation*** (SERDES) logic enables the FPGA to run at a lower speed than required by the communication data rate.

Many high-level communication protocols would require significant FPGA resources simply to manage the protocol. While this can be implemented in the fabric of the FPGA, the design is often quite complex and is better implemented using a dedicated interface chip. Such chips typically manage the media access control (MAC) and physical signalling (PHY) requiring only standard logic signals in their interface with the FPGA. Some FPGAs incorporate dedicated MAC and PHY circuitry for commonly used communication protocols such as PCI Express and gigabit Ethernet.

Another trend in recent years has also been to include hardened memory controllers for interfacing with high-speed external memory chips. These manage both the PHY level, providing accurate timing for the high-speed signals, and the high-level interface to the user's design. The memory-specific details such as activating rows, transferring data, and closing rows are managed by the controller (see Section 6.4.2 for more detail on DRAM operation). The user interface has multiple ports operating at a transaction level, where a memory request is made and is later acknowledged once completed. Such a transactional approach is required because the latency can vary depending on the current state of the memory and the controller.

2.1.7 Clocking

FPGAs are designed primarily as synchronous devices and so require a clock signal. Although multiple clocks may be used in a design, registers, synchronous memories, and I/O blocks can only be controlled by one clock. A ***clock domain*** consists of all the circuitry that is controlled by a particular clock.

Since a clock signal drives a lot of inputs, potentially over the whole of the FPGA, there are two special characteristics of clock lines that distinguish them from other interconnections on an FPGA. First, it is important that all registers in a clock domain are clocked simultaneously. ***Clock skew*** is the difference in arrival time of the clock signal at successive registers in a design (Friedman, 2001). Any skew will reduce the timing margin and impact the maximum clock speed of a design. Although skew can be minimised by distributing the clock using H- or X-tree networks, the overheads are reduced by dividing the chip into multiple regions (Lamoureux and Wilton, 2006). This leads to a two-level spine and ribs structure (see Figure 2.6), where the clock is distributed to regions using the top level, and then within each region, a spine distributes the clock to rows with ribs to provide the clock to each element on the row (Lamoureux and Wilton, 2006). A series of buffers is used within the distribution network to manage the high fan-out. In practice, it is a little more complex than this, as the network must cater for multiple clock domains, and many FPGAs also have a mixture of local and global clocks.

Delay-locked loops (DLLs) are used to minimise the skew and synchronise clocks with external sources. These dynamically adjust a programmable delay to align the clock edge with an external reference. A side effect is making the output duty cycle 50% and providing quadrature phase clock signals. While the DLL will match the input frequency, ***phase-locked loops*** (PLLs) can also be used to synthesise different clock frequencies by multiplying a reference clock by an integer ratio.

2.1.8 Configuration

All of the switches, structures, and options within the FPGA are controlled by configuration bits. Most commonly, the configuration data is contained within the FPGA in static random access memory (SRAM) cells, although flash-programmed devices are available from some manufacturers. Flash-programmed FPGAs are non-volatile, retaining their configuration after power-off, so they can immediately begin working after powering on. Static memory-based FPGAs lose their configuration on power-off, requiring the configuration file to be reloaded every time the system is powered on. Static memory has the advantage of being infinitely reprogrammable; flash memory typically has a limited number of write cycles.

The configuration data is contained within a ***configuration file*** (sometimes called a bit-file) which is used to programme the system. An FPGA is configured by loading the configuration file into the internal memory. This is accomplished by streaming the configuration file onto the FPGA either serially or in parallel 8, 16, or in some cases 32 bits at a time. In a stand-alone system, the configuration file is usually contained within an external flash memory, such as that shown in Figure 2.7, with the board set up to automatically configure the FPGA on power-on. FPGAs can also be programmed serially via a JTAG (Joint Test Action Group) interface, and this interface is commonly used when programming an FPGA from a host computer system.

The configuration file can be quite large for large FPGAs. Some FPGAs allow the configuration file to be compressed to reduce the time it takes to load. The need to transfer the configuration into the FPGA each time

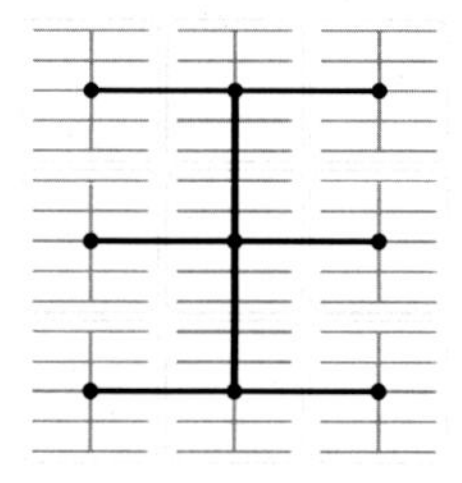

Figure 2.6 Clock distribution networks. Left: H-tree structure; right: two-level spine and ribs structure.

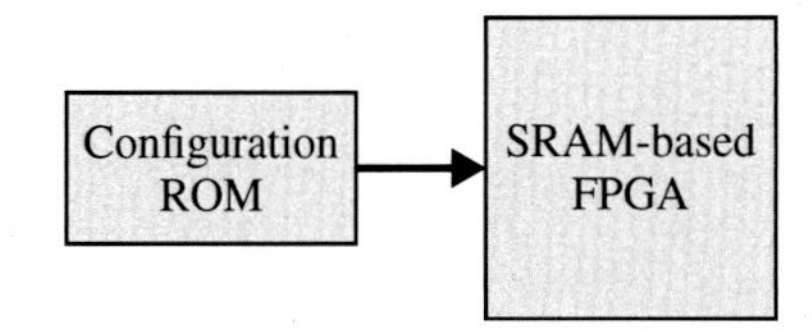

Figure 2.7 Configuration of static RAM-based FPGAs using an external ROM.

the system is powered on exposes the configuration file, making the intellectual property (IP) contained within vulnerable to piracy or reverse engineering. Higher end FPGAs overcome this limitation by encrypting the configuration file and decrypting the configuration once it is within the FPGA.

FPGAs can also be reconfigured at any time after power-on. Since the configuration controls all aspects of the operation of the FPGA, it is necessary to suspend operation and restart it after the new configuration is loaded. No state information can be retained from one configuration to the next; all of the flip-flops are reset, and all memories are usually initialised according to the new configuration. Therefore, any state information must be saved off-chip and reloaded after reconfiguration.

Some FPGAs can be partially reconfigured by directly addressing the internal configuration memory via internal registers. As before, that part of the FPGA being changed cannot operate during reconfiguration; however, the rest of the FPGA can continue executing normally. Partial reconfiguration requires the layout of the design to be structured in such a way that the blocks being reconfigured are within a small area on the FPGA to facilitate loading at runtime.

Some FPGAs also support read-back of the configuration from the FPGA. This feature can be useful for debugging a design, given appropriate tools, but is generally not used in normal operation.

2.1.9 FPGAs vs. ASICs

Programmability comes at a cost, however. If an FPGA-based implementation is compared with the same implementation in dedicated hardware as an application-specific integrated circuit (ASIC), the FPGA requires more silicon, is slower, and requires more power. This should not come as a surprise; a dedicated circuit will always be smaller than a generic circuit for several reasons. First, the dedicated circuit only needs the gates that are required for the application, whereas the programmable circuit will inevitably have components that are not used or not used in the best possible way. Second, a programmable interconnect is not required in a dedicated circuit; the logic is just wired where it is needed. The interconnect logic in FPGAs must also be sufficiently flexible to allow a wide range of designs, so again it will only be partially used in any particular configuration. Third, programmable hardware requires extra configuration logic, which also consumes a significant proportion of the chip real estate. Considering all these factors, an FPGA can require 20–40 times the silicon area of an equivalent standard-cell ASIC (Kuon and Rose, 2007). It is at the higher end of this range for designs dominated by logic, and at the lower end of the range for those applications that make significant use of multiplier blocks and memories, where the programmable blocks are bigger.

In terms of speed, the flexibility of programmable logic makes it slower than a dedicated circuit. Also, since programmable circuits are larger, the increased capacitance will reduce the maximum clock speed. On dedicated hardware, the wiring delays can be minimised by closely packing connected components. The need for programmable interconnect on FPGAs spaces the logic blocks further apart, so it takes longer for the signals to travel from one block to another. This is exacerbated by the fact that automated place-and-route tools tend to scatter the design over the available resources. Finally, every interconnection switch that a signal passes through also introduces a delay. Therefore, an ASIC will typically be three to four times faster than an FPGA (Kuon and Rose, 2007) for the same technology node.

An FPGA will also consume about 10–15 times more dynamic power than a comparable ASIC (Kuon and Rose, 2007). The main causes of the large disparity in power dissipation are the increased capacitance and the larger number of transistors that must be switched.

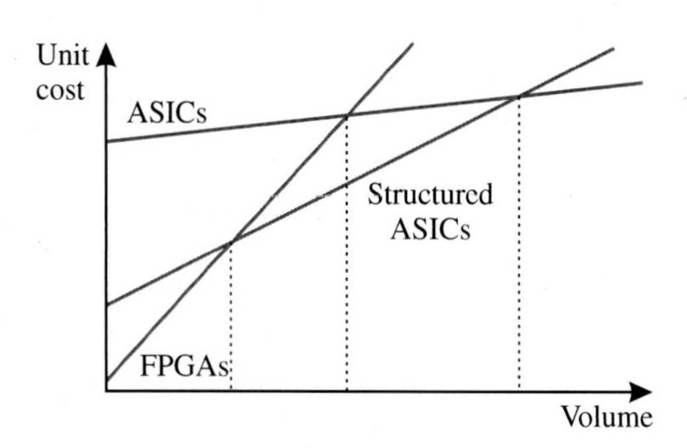

Figure 2.8 The relative costs of FPGAs, ASICs, and structured ASICs for the same technology node. The scales on the axes depend on the particular technology node.

In spite of these disadvantages, FPGAs have significant advantages over ASICs, particularly for digital systems. The mask and design costs of ASICs are significantly higher, although the cost per chip is significantly lower (see Figure 2.8). This makes ASICs economical only where high volumes are required, where the speed cannot be matched by an FPGA, or for applications where it is essential to integrate analogue components on the chip as part of the design. With each new technology generation, the capabilities of FPGAs increase, moving the crossover point to higher volumes (Trimberger, 2015). FPGAs also enable a shorter time to market, both because of the shorter design period, and the reconfigurability allows the design to be modified relatively late into the design cycle. The ability to reconfigure the design in the field can also extend the useful lifetime of a product based on FPGAs. The future is likely to see increased reconfigurability within ASICs to address these issues (Rutenbar et al., 2001).

Structured ASIC approaches, which effectively take an FPGA design and replace it with the same logic (hardwired) and fixed routing, can also overcome some of these problems. A structured ASIC consists of a predefined sea of gates, with the interconnections added later. This effectively makes them mask-programmed FPGAs. Fabricating the user's device consists of adding from two to six metal layers, forming the interconnection wiring between the gates. The initial cost is lower because the design and mask costs for the underlying silicon can be shared over a large number of devices, and only the metal interconnection layers need to be added to make a functioning device.

Both Intel (with eASIC) and Xilinx (with EasyPath (Krishnan, 2005)) provide services based on their high-end FPGA families for converting an FPGA implementation into a structured ASIC. The underlying sea of gates for these reflects the underlying architecture of the FPGA on which they are based. Such devices significantly reduce the cost for volume production, while achieving a 50% improvement in performance at less than half the power of the FPGA design they are based on (Leong, 2008). The FPGA design files can be directly used to create the structured ASIC (Altera, 2009).

For prototyping, and designs where a relatively small number of units is required, the programmability and low cost of FPGAs make them the device of choice.

2.2 Programming FPGAs

So far, a low-level view of the architecture of FPGAs has been presented. To programme an FPGA at this level, it is necessary to break down the design into the fine-grained logic blocks available on the FPGA and build up the logic required by the application from these blocks. There are two main problems with representing designs at this level. First, designing at such a low level is both tedious and error-prone. Second, the portability would be limited to FPGAs that had the same basic architecture and logic block granularity. While it is possible to programme FPGAs at this level, it is akin to programming a microprocessor in assembly language. It may be appropriate for the parts of the design where the timing and resource utilisation are critical. However, for most applications, this is too low level, and this is also the case for most image-processing designs based on FPGAs.

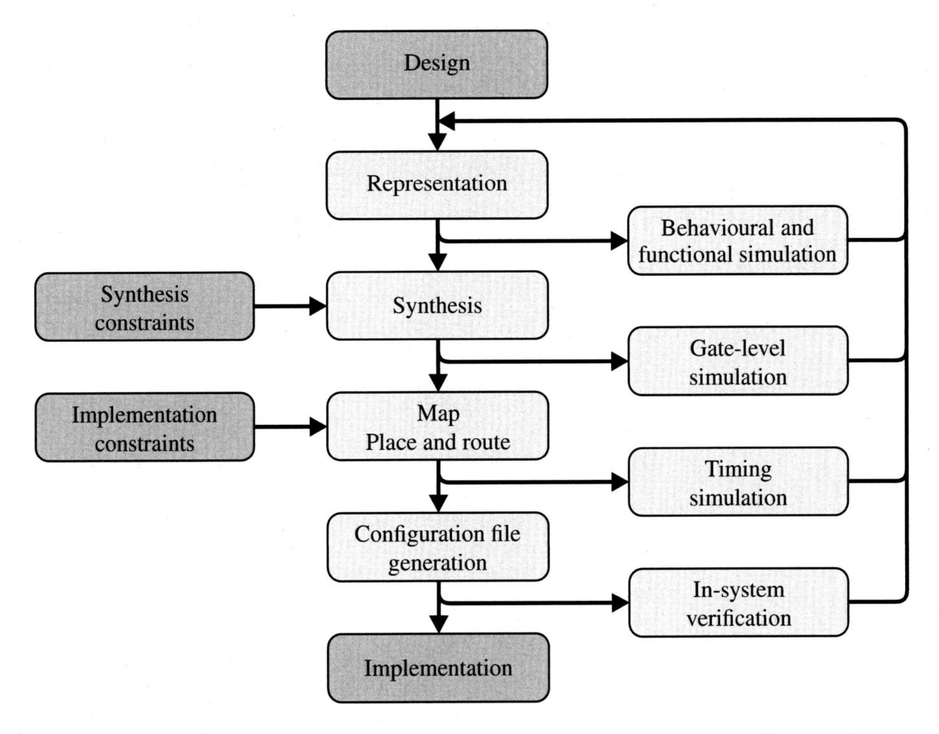

Figure 2.9 Steps to implementing a design on an FPGA.

Fortunately, the tools for implementing algorithms and systems on FPGAs enable us to work at a higher level. Implementing a working design requires several key steps, as illustrated by the generic design flow in Figure 2.9. First, the design must be coded in some form that is both human and machine readable. This represents the design using a ***hardware description language*** (HDL) and captures the essential aspects of the design that enables it to be both simulated and synthesised onto an FPGA. Although many HDLs are based on software languages, they are not software but instead describe the hardware required to implement the design. For an FPGA implementation of an algorithm, it is necessary to describe not only the algorithm but also the hardware used to implement it. Most development environments enable the hardware description to be compiled and simulated at the logical level to verify that it behaves in the intended manner.

Synthesis takes the logical representation describing the hardware and converts it to a net-list representing the actual hardware to be constructed on the FPGA. This form defines the circuits in terms of basic building blocks, both at the logic-gate level and in terms of the primitives available on the FPGA if they have been used within the description. It also defines the interconnectivity between the building blocks. Synthesis constraints control aspects of the synthesis process, such as whether to optimise for speed or area; automatically extract and optimise finite state machines; and force set and reset to operate synchronously. Gate-level functional simulation verifies that the design has been synthesised correctly and provides more information on the specific low-level characteristics of the circuit than the higher level simulation.

The next stage maps the net-list onto the target FPGA. There are two phases to this process. ***Mapping*** determines how the logic maps onto the resources available on the FPGA. This phase partitions the logic of the design into the logic cells, splitting complex logic over several logic cells or merging logic to fit within a single logic cell where appropriate. The ***place-and-route*** phase then associates these mapped components with particular logic blocks on the FPGA and determines the routing required to connect the logic, memories, and I/Os. Implementation constraints direct the placement of logic to specific blocks and, in particular, associate inputs and outputs with specified pins. They also specify the required timing, which is used to guide the placement and routing of critical nets. Once the design has been mapped onto the FPGA, more accurate estimates of timing are available based on the lengths of the connections, fan-outs, and particular resources that are used to implement the logic function.

The final stage is to generate the configuration file for programming the FPGA. During development, the configuration file can be loaded onto the target system to verify that the design works as intended, and that it correctly interacts with the other components within the system. In the deployed system, the FPGA is programmed by automatically loading the configuration file from flash memory on power-on.

As can be seen from Figure 2.9, implementing a design on an FPGA is quite different from implementing a design in software, even on a processor within an embedded system. It requires a hardware mindset, even during the initial design. However, image processing is usually considered to be a software development task. Successful implementation of algorithms on an FPGA therefore requires a mix of both software (algorithmic) and hardware (logic circuit design) skills.

2.2.1 Register Transfer Level

The traditional approach to FPGA-based design is carried out at an intermediate level that is a mixture of both algorithmic and circuit-level design. Synchronous systems are usually implemented at the ***register transfer level*** (RTL). As illustrated in Figure 2.10, this structures the application as a series of alternating blocks of logic and registers. The structure is not necessarily linear, as implied by Figure 2.10; from a computational perspective, the primary purpose of each block of logic is to determine the next value that will be stored into the registers on its outputs. It can take its inputs from anywhere: other registers, I/O pins, or internal memory blocks. As demonstrated in Figure 2.10, inputs, outputs, and memory accesses may be either direct (asynchronous) or via registers. The blocks of logic are not directly clocked; only the registers are clocked (this includes any registers as part of synchronous memories). Therefore, the signals must propagate through the logic before they can be clocked into the registers. This limits the clock speed of the design to the propagation delay through the most complex block.

The art of implementing an algorithm on an FPGA therefore falls to decomposing the algorithm into a sequence of register transfers and determining the appropriate logic between each register. Note that unlike implementing algorithms in software, each logic block operates in parallel. This usually requires adapting the algorithm from a conventional sequential form into one that can exploit the parallelism available from hardware.

Implicit within Figure 2.10 is the control logic required to sequence the steps within the algorithm. Depending on the algorithm and its implementation, not all of the registers may necessarily be updated at every clock cycle. The clock enable inputs for each register are used to control when they are updated. Also, the computation to produce a value for a register may not necessarily be the same every cycle. Multiplexers are used on the register inputs to select the appropriate computation, depending on the current stage within the algorithm. These are made more explicit in Figure 2.11, with control logic providing the clock enable and multiplexer selection signals. Implicit within the compute logic are any memory elements that may be used to perform part of the computation or provide cached input or output data.

The separation of control logic and compute logic can help with the transition from an algorithm to hardware. The control logic determines what is computed (or more precisely which computation is saved in which register), while the compute logic actually performs the computation. That is, the compute logic is the set of operations

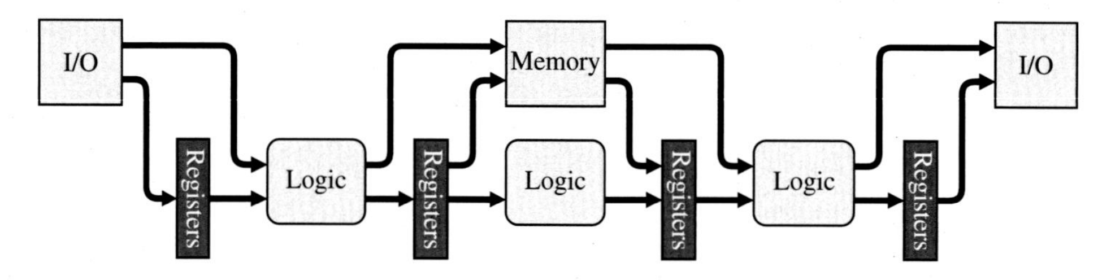

Figure 2.10 Basic structure of any application as logic interspersed with registers.

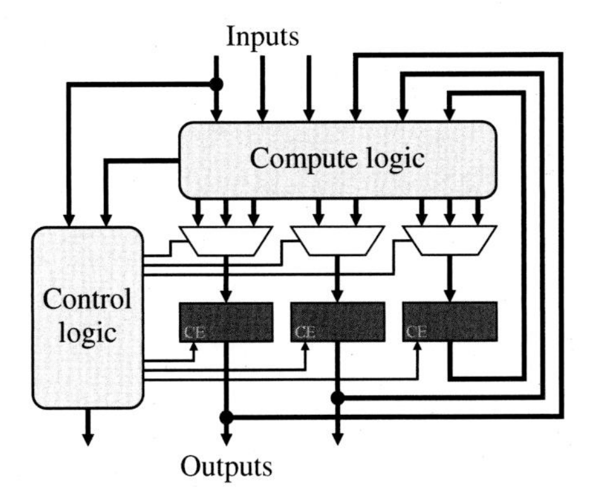

Figure 2.11 A rearrangement of Figure 2.10 to explicitly separate the control and compute logic.

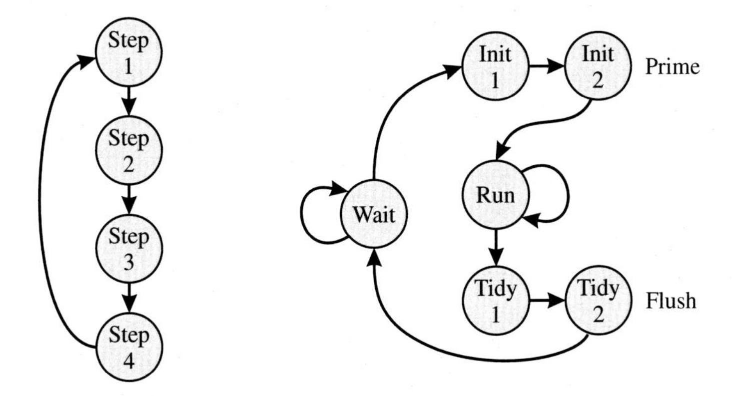

Figure 2.12 Examples of finite state machines. Left: a repeating sequential process; right: a pipelined stream process.

applied to the incoming data to provide the output data, whereas the control logic sequences these computations depending on the algorithm. Note that in hardware, the logic is always adjusting its output to reflect any changes on its inputs.

Since the computation is often context dependent, the control logic is commonly implemented as a finite state machine, with the context encoded in the states. Figure 2.12 gives two simple examples of finite state machines. The one on the left illustrates how a sequential algorithm may be implemented on parallel hardware. The algorithm consists of four steps, each of which take exactly one clock cycle, and when the algorithm has completed the fourth step, it returns to the start (perhaps operating on the next data item). The finite state machine on the right represents the typical structure for the control for a three-stage pipelined stream process with one clock cycle per pixel. The algorithm begins in the *Wait* state, waiting for the first pixel. When the first pixel arrives, the next two states are initialisation, which prime the processing pipeline. Once primed, the pipeline stays within the *Run* state while pixel data is arriving. At the end of each block of data, the state machine transitions to the *Tidy* states where the pipeline is flushed, completing the calculations for the last pixels to arrive. It then returns to the *Wait* state, waiting for the first pixel on the next block.

The transitions from one state to the next may be determined by input (control) signals, or for more complex algorithms by the data being processed, as reflected in Figure 2.11. In practice, a complex algorithm is often decomposed into a series of modules. Rather than control the complete design with a single large finite state machine, the control logic is also modularised, with a separate state machine for each module.

Having looked at the basic structure of an algorithm when implemented on an FPGA, the next step is to look at how to represent the algorithm being programmed onto the FPGA.

2.2.2 *Hardware Description Languages*

Electronic circuits have traditionally been represented through schematic diagrams. These represent the components and their connectivity graphically. Schematic diagrams are best used for representing smaller designs at the level of transistors and logic gates. For higher level designs, they represent modules and packages as block diagrams. Modern schematic capture tools integrate circuit simulation and printed circuit board layout, making them invaluable for the production of physical systems. High-end tools also integrate HDLs, allowing the complete development and modelling of systems containing FPGAs and other programmable hardware. However, the level of abstraction of schematic diagrams is typically too low for representing complex algorithms. At the RTL, a schematic representation is structural, as a collection of components or block diagram, rather than clearly representing the behaviour or algorithm. This makes them difficult to navigate for algorithmic designs, because the behaviour is not particularly transparent from a structural representation.

From these roots, HDLs evolved in the 1980s to address some of the limitations and shortcomings of schematic representations and to provide a standard representation of the behaviour of a circuit. Since hardware is inherently parallel, it is necessary for an HDL to be able to specify and model concurrent behaviour down to the logic-gate level. This makes HDLs quite different from conventional software languages, which are primarily sequential.

The two main HDLs are VHDL (very high-speed integrated circuit HDL) and Verilog, both of which have since been made into IEEE standards (the current standards are IEEE 1076 (2019) and IEEE 1364 (2006), respectively). Verilog has since been integrated into System-Verilog (IEEE standard 1800) to expand the tools for design verification (IEEE, 2018). A full description of either VHDL or Verilog is beyond the scope of this book. The interested reader should consult the standards (IEEE, 2006, 2018, 2019), the documentation for their synthesis tool, or other references or textbooks on the language (for example (Bhasker, 1999, 2005; Ashenden, 2008; Mealy and Tappero, 2012)).

The original purpose of both VHDL and Verilog was for documenting and modelling electronic systems. Therefore, entities or modules can readily be written in either language for testing the functionality of the circuits described. Through the design of appropriate testbenches, this allows the design to be thoroughly tested for correctness before it is synthesised. Many of the constructs of the languages are intended for simulation or verification and do not correspond to components that can be synthesised.

In terms of synthesis of logic circuits, there are three distinct coding styles: structural, concurrent dataflow, and behavioural. Both VHDL and Verilog are hierarchical structural languages in that they can describe the hardware structurally in terms of blocks of logic and their interconnections. Structural design enables complex designs to be modularised and built up from simpler components. A purely structural implementation would describe everything in this manner, using a hierarchy of connected components in much the same way that the schematic diagram would represent the circuit with boxes representing each component. For a low-level design, it is possible to directly instantiate primitives corresponding to the logic cells and other basic building blocks available on the FPGA, although designs at this level lack portability because of differences between the logic cells of different FPGA manufacturers. Structural descriptions provide the most control over the final synthesised circuit. Therefore, they will generally give the best performance, using the minimum of resources (if designed appropriately). This makes them good for developing libraries of intellectual property (IP) blocks that may be instantiated many times or in many designs. However, design at the structural level is quite low level and can be very tedious.

A concurrent dataflow coding style represents the design in terms of logic equations for signals. A signal is defined by an expression in terms of other signals; whenever one of the signals on the right-hand side changes, this will propagate through to give a new value for the signal on the left-hand side of the assignment. All such assignments are concurrent, with any changes to inputs propagating through to give a new value on the corresponding outputs. Expressions can include not only logic but also arithmetic operations, multiplexers, and even asynchronous memory accesses. It is the role of the synthesis tools to map these expressions onto the resources available on the target FPGA.

The third coding style describes the operation of a circuit in terms of its behaviour. Rather than directly specifying the functionality of the circuit in terms of logic, a behavioural description, as its name implies, describes how the circuit is to behave or respond to changes on its inputs. For example, an RTL design would specify how the circuit responds to the clock signal by specifying what gets written into which register. Behavioural descriptions also include sequential statements, making them more like conventional programming languages, even though they are describing hardware. A behavioural specification describes what the circuit does, without necessarily specifying how it is to be implemented. It is therefore the role of the synthesis tools to determine the actual logic required to implement the specified behaviour. The complexity of synthesising a circuit from its behavioural description limits what can be synthesised efficiently; generally, only the subset of behavioural descriptions that conform to an RTL style is synthesisable. Behavioural descriptions tend to be of higher level (although of course this depends on how they are written).

HDLs, by necessity, are general-purpose languages and allow considerable control over how the circuit is implemented. This is both an advantage and a disadvantage. The advantage is that the developer can target specific structures available on the FPGA or optimise the design for speed or resources required by programming in particular ways. By giving the designer control over every aspect of the design, fast and efficient designs are possible. The disadvantage is that the designer must control everything in quite fine detail, including both the data and control flows. The flexibility comes from the relatively low-level constructs, but programming at this level requires a lot of basic bookkeeping. This makes complex algorithmic programming more difficult.

The low level of traditional HDLs makes exploring the ***design space*** (the trade-offs between area, speed, latency, and throughput) more difficult. It is time consuming enough to programme one design, let alone several different designs to investigate the effects of different levels of loop unrolling, pipelining, and resource sharing. The low level of programming is also not always particularly transparent. Very similar behavioural constructs map to latches or multiplexers. While a structural design can specify registers explicitly, they tend to be implicit in the behavioural programming style. These characteristics require that close attention be paid at the low level and has a strong focus on the hardware rather than the algorithm. Compared to software-based languages, HDLs are more like programming in assembly language.

2.2.3 High-level Synthesis

In an attempt to overcome this problem, over the past couple of decades, considerable research has gone into the synthesis of hardware from high-level languages, in particular from C (and also more recently from Python, see, for example (Decaluwe, 2004; Clow et al., 2017), and other languages). The rationale behind HLS is twofold. First, it enables existing software to be easily ported or recompiled to a hardware implementation. Many algorithmic designs are first developed and tested using C or MATLAB®. Once working, the design is then manually converted into an HDL after which it is necessary to verify and refine the design manually. This conversion process is both time consuming and error prone. Having a single source that can be compiled to both software and hardware would speed the development process considerably. It would also make it easier to explore the partitioning of the final implementation between hardware and software (Cong et al., 2011). The time-critical and easily parallelisable components can be compiled to hardware, with the more sequential components compiled as software. This is increasingly important with SoC designs, where the FPGA is used to accelerate the key aspects of a design. A second motivation is to raise the level of abstraction to make the design process more efficient. Using a software-based language has the advantage of making hardware design more like software design, making it more accessible to software engineers (Alston and Madahar, 2002).

However, there are significant differences between hardware and software that pose significant challenges to describing a hardware implementation using a software language. Five areas of difference identified by Edwards (2006) are concurrency, having a model of time, data types, the memory model, and communication patterns.

Hardware is inherently concurrent, with each step or computation performed by separate hardware. In contrast, software performs its computation by reusing a central processor, making software inherently serial. As indicated in Section 1.7, software introduces parallelism at several levels: at the instruction level through wide instruction words (effectively issuing several instructions in parallel), through instruction pipelining (overlapping successive instructions in different phases of the fetch–decode–execute cycle), through coarse-grained parallelism introduced by splitting the application into several relatively independent processes, and using a separate thread for each. The fine-grained parallelism at the instruction level is closer to that of hardware but is controlled by the particular architecture of the processor. The compiler just needs to map the high-level statements to the given processing architecture. At this level, the algorithm is primarily sequential, although compilers can (and do) exploit dependencies between instructions to maximise parallelism. At the thread or process level, the coarse-grained model does not match the fine-grained concurrency of hardware but maps closer to having separate blocks of hardware for each thread or process.

Time and timing is absent from most software programming models. The sequential processing model of software ensures causality within an instruction sequence (Edwards, 2006), although this becomes more complex and less deterministic with multi-threaded execution and task switching. However, the time taken for each task is not considered by the language. In contrast, hardware, and especially synchronous hardware, is often governed by strict timing constraints. Many hardware interactions must be controlled at the clock level, and representing these with a high-level software language can be difficult. However, removing the explicit dependence on a clock enables HLS tools to both increase the abstraction level and optimise more aspects of the design (Takach, 2016).

The word width of data in a hardware system is usually optimised to the specific task being performed (Constantinides et al., 2002; Boland and Constantinides, 2012). In contrast, software languages have fixed word widths (8, 16, 32, and 64 bits) related to the underlying architecture rather than the computation being performed. Two approaches to this are to modify the language to allow the word width to be explicitly specified or to make use of the object-oriented features (Java or C++) to define new types and associated operators. Virtually all software languages support floating-point computation, and floating-point numbers have been the mainstay of scientific computing, including image and signal processing, for some time. However, until recently, floating-point number representations have been avoided on FPGAs because of the cost of implementing the more complex hardware. It is only with higher capacity FPGAs and higher level languages that floating-point calculations have become practical in hardware for other than very specialised tasks. Recently, high-end FPGAs have started to directly support floating-point operations in hardware.

The software memory model is very different to that used within hardware systems. Software treats memory as a single large block. The compiler may map some local variables to registers, but in general, everything is stored in memory. Memory can be allocated to a process dynamically, with the associated memory block referenced by its address. C even allows efficient memory traversal algorithms through pointer arithmetic. However, in hardware, many local variables are not implemented in memory but directly as registers. Memory is often fragmented into many small blocks, each with their own independent address space. In such an environment, pointers have limited meaning, and dynamic allocation of memory resources is difficult. This has a direct impact on the algorithms that can readily be used or compiled to hardware. Some attempts have been made to provide a form of dynamic memory allocation, both as part of hardware design (Dessouky et al., 2014; Ozer, 2014; Xue, 2019) and also in the context of HLS (Winterstein et al., 2013; Liang et al., 2018; Giamblanco and Anderson, 2019).

The differences in memory structure also directly impact communication mechanisms between parallel processes. Software generally uses shared memory (including mailboxes) to communicate between processes. Such a model makes sense where there is only a single CPU (or small number of CPUs), and the processes cannot communicate directly. More loosely coupled systems may communicate via a network. In hardware systems, parallel processes are truly concurrent and can therefore communicate directly. In data-synchronous systems, such as streamed pipelining, communication is either explicit through handshaking (for example advanced extensible interface (AXI)-type interfaces) or implicit through simple token passing. More complex systems can use FIFO buffers or other synchronising and communication mechanisms (especially when communicating between clock domains). In many hardware systems, the communication relies on dedicated hardware. Larger, more complex systems are beginning to see network concepts from software being applied to hardware communication through an NoC.

Another major difference between hardware and software relates to the representation and execution of an algorithm. In software, an algorithm is represented as a sequence of instructions, stored in memory. A program counter points to the current instruction in memory, and a memory-based stack can be used to maintain state information, including any local variables. This allows algorithms to be implemented using recursion where the state within a procedure can be saved, and the procedure called again. In hardware, the algorithm is represented by physical hardware connections. Saving the current algorithm state is difficult, making recursive algorithms difficult to implement in hardware (Skliarova and Sklyarov, 2009). Any such recursion within an algorithm must be remapped to iteration, or the data structures must be mapped to memory (Ninos and Dollas, 2008) to allow a practical implementation.

In spite of these significant differences, three approaches have been used to develop synthesis tools from high-level languages. One is to use the host language to create and manipulate a structural representation of the circuit constructed. This is effectively using the high-level language to simply model the hardware, rather than directly specifying it. The second is to modify or extend the syntax of a high-level language to provide constructs that enable it to be used to describe hardware (Todman et al., 2005; Edwards, 2006). The third approach is to use a (relatively) standard high-level language to provide a behavioural representation of the design and require the compiler to automatically extract parallelism from the design and appropriately map it to hardware. It is this third approach that is commonly called high-level synthesis. HLS compilers generally perform the following steps (see Figure 2.13) on the source code (Fingeroff and Bollaert, 2010; Coussy et al., 2009):

- Dataflow analysis, to determine what operations need to be performed on the data. The data dependencies are used to infer a processing sequence, and loops are analysed to determine the extent to which they can be pipelined or unrolled. A pure dataflow approach is limited by dynamic control structures (for example data-dependent loops), requiring this analysis to also consider control flow (Coussy et al., 2009).
- Resource allocation, to determine the number and type of hardware operators that need to be instantiated to perform all of the source code operations. Storage also needs to be allocated for intermediate data that may not be used immediately (either in memory or registers). Many HLS tools enable efficient implementation by having available an optimised resource library that is often specific to the target FPGA (Nane et al., 2016). Within a pipeline, the extent to which hardware operation blocks can be reused is explored. This depends on how often new data is applied to the pipeline (also called the initiation interval). Unrolling loops will result in multiple parallel copies of computational resources being allocated.
- Resource binding, to map the particular operations implied by the source code onto the particular instances of computational resources. This also includes allocating variables to storage locations where required.
- Scheduling, to determine when each source code operation is executed on its allocated hardware. This will often involve using state machines to control the operation of the hardware.
- RTL code generation, to represent the resulting design as a model that can be synthesised onto the target FPGA.

Note that the resource allocation, binding, and scheduling steps are not necessarily performed sequentially (as implied by Figure 2.13) but may be performed simultaneously, or in a different order, depending on the compiler (Coussy et al., 2009).

With HLS, the design flow of Figure 2.9 can be updated to that of Figure 2.14. Typically, a C-style software representation is used for the design. Vendors provide their own HLS language and tools, although the open standard OpenCL is increasingly being used. In OpenCL, kernel functions are programmed in a C-based language which can be executed either as software or on a graphics processing unit (GPU) or FPGA (Khronos, 2021). A host programme then executes the kernels on the acceleration platform. (OpenCL has also been extended to the SYCL specification, which integrates OpenCL with modern C++ (Khronos, 2020), with the associated programming language DPC++ (data parallel C++) (Reinders et al., 2021).)

The higher level representation of HLS generally makes the design process much easier. The resulting code is usually more compact and succinct than RTL, making it faster to enter and debug (Lahti et al., 2019). This is partly because the control path is often implicit in the high-level representation, whereas it must be designed

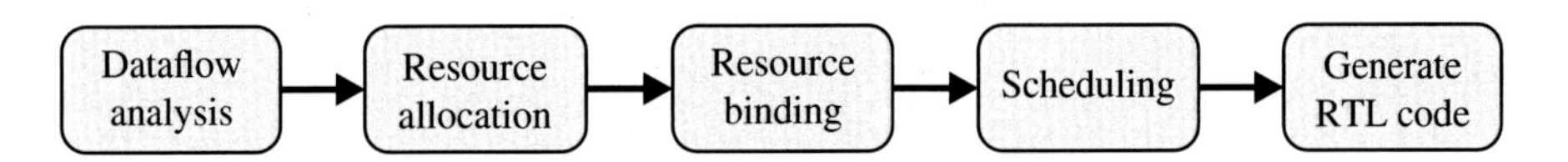

Figure 2.13 Steps performed by a high-level synthesis compiler.

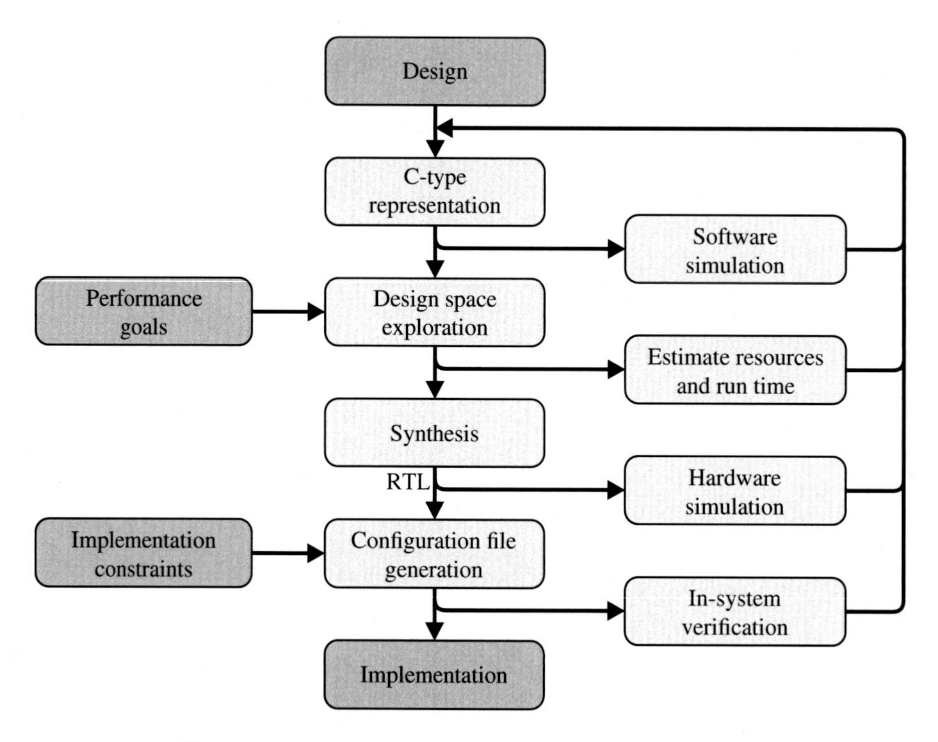

Figure 2.14 Simplified design flow for high-level synthesis.

and entered explicitly for an RTL design. The control path of a complex algorithm can take as much effort to design as the data path.

The high-level representation (whether OpenCL or other HLS language) enables the functionality of the design to be verified first as software. Simulation or verification is much faster because it is performed at a higher level (Sanguinetti, 2010). This is particularly important for image processing, where an image or image sequence provides a large dataset that is particularly time consuming to process using hardware simulators. The software-based source code can be used for debugging the algorithm and provide gold-standard processing results for the input images.

The next stage is to explore the design space (parallelisation, loop unrolling, bit-width optimisation, and so on) to try to satisfy the target performance goals. Design-space exploration is much more efficient in an HLS than at the RTL level. With HLS, this is either achieved through source code optimisation by the compiler or designer-inserted synthesis directives (Meeus et al., 2012). While some aspects of the data bit-width optimisation can be automated, it is mostly performed manually as it requires domain-specific knowledge (Nane et al., 2016). Different degrees of loop unrolling and pipelining can be controlled by adjusting compiler directives rather than manually recoding the control and data paths, a time-consuming and error-prone process. Researchers are even looking at automating design-space exploration within HLS tools (Schafer and Wang, 2020). For designs that push the envelope, they can optimise the design to maximise computational resources, given the available memory bandwidth (Siracusa et al., 2022). Most HLS tools can even provide a good estimate of resources required without having to actually synthesise the design (BDTI, 2010). This allows the effects of design-space exploration to be estimated without actually synthesising the hardware. Some HLS tools can also optimise the power requirements of a design by careful analysis of register enabling and clock gating (Takach, 2016).

HLS and partitioning between hardware and software is becoming increasingly important with SoC-based devices. This ability of HLS tools to compile the source code to either software or hardware enables the complete design to be partitioned more easily between hardware and software resources (see, for example (Ab Rahman et al., 2010; Vallina and Alvarez, 2013)). It is much easier to swap modules between hardware and software to find the best mix (although it is important to keep in mind any potential communication overhead between the FPGA and CPU).

The next step is to synthesise the design. When producing hardware, most HLS tools generate intermediate RTL code. The RTL code should be simulated against the software-based testbench to verify that the design is correctly represented in hardware. Verification is particularly important for non-behavioural aspects of the design, such as interfacing with other components or modules (Lahti et al., 2019). Debugging is made difficult by the generated RTL architecture often being quite different from the structure of the original C source code (Cong et al., 2011). The RTL code is finally compiled by the FPGA vendor's tools to give the hardware configuration file.

Given appropriately designed source code, modern HLS tools can produce hardware that is as efficient as hand-coded RTL in both resources and processing speed (BDTI, 2010; Meeus et al., 2012; Winterstein et al., 2013; Lahti et al., 2019). In some cases it is more efficient because the compiler may recognise different ways of arranging the operations to reduce the dependencies that may not necessarily be obvious to the designer. A more complete exploration of the design space (because the parameters are relatively easy to change) may enable more efficient designs to be found than would be achieved through the time-consuming RTL redesigns. Many of these transformations are increasingly being automated and implemented as compiler optimisations (Li et al., 2018). However, it has been shown that the order in which optimisations are applied can significantly influence the quality of the resulting hardware (Huang et al., 2015). The significant improvement is that development time from using HLS makes it well suited where time-to-market is critical (Takach, 2016; Lahti et al., 2019).

In spite of these advantages, HLS does not (yet) live up to the claim of allowing software engineers to design hardware (Bailey, 2015). While HLS can provide a hardware realisation of an algorithm, if the algorithm is largely sequential, then there may be limited opportunities for acceleration. In particular, the algorithm must be coded in a particular style to enable the compiler to identify parallelism (Fingeroff and Bollaert, 2010). If this style is not followed, the resulting hardware is inefficient, the resource requirements become bloated, and the resulting system performs poorly relative to the original software realisation (Herbordt et al., 2007). In particular, pointers and pointer arithmetic can hide data dependencies (Winterstein et al., 2013), leading to an inefficient dataflow analysis. Rather than use pointer arithmetic, the algorithm should be restructured to directly use arrays and array indexing. Another issue is recursion; recursion requires saving the context (local variables and caller location) on a stack. In hardware, local variables are usually stored in registers, rather than in a memory-based stack. Calling a procedure in software translates to providing inputs to a block of hardware; if the hardware is already in use, applying new inputs recursively will lead to corruption of register values. The only exception to this is tail recursion, where the values within all of the hardware registers are no longer required. Otherwise, recursion within an algorithm needs to be restructured as iteration.

Software has no explicit concept of timing. This flows into HLS where the scheduling manages local dependencies within the algorithm. Long-range synchronisation is more complex and difficult to manage. Consequently, where an algorithm has long-range dependencies, and in particular the synchronisation between parallel operations, HLS is forced to rely heavily on external memory for storage. Current HLS tools have an overly simplified model of external memory and associated accesses (Cong et al., 2011), which requires explicit data transfer and does not include automatic caching and prefetching.

Stream processing often has strict processing requirements. Cases where the processing for each pixel is data dependent makes synchronisation within stream processing more challenging. While this problem is not unique to HLS, the lack of explicit timing makes this situation more difficult for HLS. However, as long as the throughput requirement is met, inconsistencies in the dataflow on the input can usually be managed through the use of FIFO buffers.

HLS is only implementing the algorithm provided (and can apply transformations to exploit parallelism). However, the best algorithms for hardware are not necessarily the same as those for software (Lim, 2012; Lim et al., 2013). It needs to be remembered that the code is describing hardware. Simply porting or restructuring a software algorithm is often insufficient for an efficient hardware algorithm. Better results can generally be achieved by designing appropriate hardware and then encoding that design in whatever HDL is used. HLS has the advantage of being more efficient and easier to write than RTL code (Lahti et al., 2019), but this does not avoid the need for appropriate hardware design. Some examples Bailey (2015):

- Software algorithms can rely on random access of memory (subject to performance limitations resulting from cache misses). When working with stream processing data, such algorithms may need to be restructured. For example, chain coding a streamed image (Bailey, 2010) (see Section 13.3).
- HLS can identify the regular access patterns associated with filters and build the necessary row buffers for efficient streaming (Cong et al., 2012; Guo et al., 2004; Meeus and Stroobandt, 2014). However, software manages image borders using conditional code to manage the exceptions. This leads to naive HLS building a lot of extra hardware to handle these exceptions. With design, these irregular access patterns can be readily integrated into the design of the filter windows (Bailey, 2011; Rafi and Din, 2016; Al-Dujaili and Fahmy, 2017; Bailey and Ambikumar, 2018) (see also Section 9.1.1).
- For an FFT (fast Fourier transform, see Section 11.1), given a standard software algorithm performing the processing in place on data in memory, design-space exploration using HLS should be able to identify a good mix of parallel and pipelined stages. However, if a radix-2 algorithm is coded, it will not infer a more efficient higher radix algorithm. This requires the code to be written to imply such a structure (Bailey, 2015).
- A common approach to implementing a 2-D FFT is to perform a 1-D FFT on the rows and then on the columns of the result. The associated transpose buffer, due to its size, would need to be stored in external memory. If stored in dynamic RAM, then accessing by column can be inefficient, incurring significant overhead in activating and pre-charging each memory row (see Section 6.4). In the worst case, the memory bandwidth can be reduced by 80% (Bailey, 2015). This can be overcome by careful design of the mapping from the logical image address to the physical memory address.
- Software-based connected component analysis algorithms generally require multiple passes through an image, which requires storing intermediate results in off-chip memory. Direct synthesis using HLS would retain the same sequential multi-pass computational architecture. However, the algorithm can be restructured to operate in a single pass (Bailey and Johnston, 2007; Bailey, 2020) using stream processing (see Section 13.5).

Therefore, while HLS can provide significant advantages for rapid development and design-space exploration, it is no substitute for careful design. It is still essential to consider the hardware being built and treat HLS as an HDL and not as software. However, HLS is maturing, and when used appropriately provides considerable advantages over the more traditional RTL design.

2.3 FPGAs and Image Processing

Since an FPGA implements the logic required by an application by building separate hardware for each function, FPGAs are inherently parallel. This gives them the speed that results from a hardware design while retaining the reprogrammable flexibility of software at a relatively low cost. This makes FPGAs well suited to image processing, particularly for the low- and intermediate-level operations where they can exploit the parallelism inherent in images. FPGAs can readily implement most of the forms of parallelism discussed in Section 1.7.

For a pipelined architecture, separate hardware is built for each image processing operation in the pipeline. In a data synchronous system, data is simply passed from the output of one operation to the input of the next. If the data is not synchronous, appropriate buffers may be incorporated between the operations to manage variations in dataflow or access patterns.

Spatial parallelism may be exploited by building multiple copies of the processing hardware and assigning different image partitions to each of the instances. The extreme case of spatial parallelism consists of building a separate processor for each pixel. Such massive parallelism is not really practical for all but very small images; it does not map well onto FPGAs for realistic sized images, simply because the number of pixels within images exceeds the resources available within an FPGA.

Logical parallelism within an image processing operation is well suited to FPGA implementation, and it is here where many image processing algorithms may be accelerated. This is accomplished by unrolling the inner loops, using parallel hardware rather than performing the operations sequentially.

Streaming feeds image data serially through a single function block. This maps well to a hardware implementation, especially when interfacing directly with a camera or display where the images are naturally streamed. With stream processing, pipelining is required to achieve the required throughput. A complete application implemented using a stream processing pipeline results in a very efficient implementation.

The ability to make significant use of parallelism has tremendous benefits when building embedded vision systems. Performing multiple operations in parallel enables the clock speed to be lowered significantly. A VGA resolution video streamed from a camera at 60 frames per second produces approximately 20 million pixels per second (although the clock frequency is usually higher to account for the blanking periods). On the other hand, 4K video at 60 frames per second requires a clock frequency of 594 MHz. Any significant processing requires many operations to be performed for each pixel, requiring a conventional serial processor to have a much higher clock frequency. A streamed pipelined system implemented on an FPGA can often be operated at the native pixel input (or output) clock frequency. This corresponds to a reduction in clock speed over a serial processor of two orders of magnitude or more. The dynamic power consumption of a system is directly related to the clock frequency, so a slower clock results in a significantly lower power design.

Modern FPGAs also incorporate a hardened CPU, enabling the high-level image processing operations, which are unable to be efficiently parallelised, to be implemented in software on the same chip. If the whole algorithm can be implemented on a single FPGA, the resulting system has a small form factor. Designs with only two or three chips are possible, enabling the whole image processing system to be embedded with the sensor. This enables smart sensors and smart cameras to be built, where the intelligence of the system is built within the camera (Leeser et al., 2004; Mosqueron et al., 2007; Bailey et al., 2012). The result is that vision can then be embedded within many applications as a versatile sensor.

2.3.1 *Choosing an FPGA or Development Board*

The two main FPGA manufacturers in terms of market share are Xilinx (recently acquired by AMD) and Intel (formerly Altera), although there are several others that provide FPGAs.

Most FPGA manufacturers also sell development boards or evaluation boards. For prototyping and algorithm development, such boards can provide an easy entry and save the time and costs of developing the printed circuit board. These boards often consist of an FPGA with a set of peripherals to demonstrate the capabilities of the device within a selection of targeted applications. Most boards provide some form of external memory and some means of interfacing with a host computer (even if it is just for downloading configuration files). Other than this, the specific set of peripherals can vary widely, depending on the target application and market. Boards aimed for education use tend to be of lower cost, often with a smaller FPGA and a wide variety of peripherals enabling them to be used for laboratory instruction.

With the large range of both FPGAs and development boards available, how does one choose the right system? Unfortunately, there is no easy answer to this, because the requirements vary widely depending on the target application, and whether or not a development system or product is being built.

For image processing, the development system should have the largest FPGA that you can afford, with sufficient peripherals to enable a wide range of applications to be considered and programmed. Having a larger FPGA than required in the final product can reduce the compile times during the many design iterations (compile times grow considerably with increased utilisation over about 50%). Some essentials for any image processing application:

- There needs to be some method for getting images into the FPGA system. Depending on the intended input source, there are a number of possibilities. To come from an analogue video camera, some form of codec is necessary to decode the composite video signal into its colour components and to digitise those components. Some systems provide a codec to digitise VGA signals or decoders for high-definition multimedia interface (HDMI), allowing the capture of images produced by another computer system. Common interface standards to connect to digital cameras include USB, Camera Link, and gigabit Ethernet. Most development boards have a USB interface (although actually getting to the video stream is more complex than just the physical interface), but few will have other camera inputs. Some boards provide direct interface to complementary metal oxide semiconductor (CMOS) camera modules, enabling smart cameras to be prototyped.
- The development board must be able to display the results of image processing. While the final implementation may not require image display, it is essential to be able to view the output image while debugging. Either a VGA output or digital video interface (for example HDMI) is suitable.

- The system must have sufficient memory to buffer one or more video frames. While the final application may not require a frame buffer, just capturing and displaying an image will require one unless the camera and display are synchronised. Only the largest FPGAs have sufficient on-chip memory to buffer a whole image, although this memory is probably better used for other purposes. Therefore, the frame buffer is usually off-chip. Dynamic memory has a variable latency, making it less suitable for high-speed random access, although it is suited for stream processing, where the access pattern is known in advance. Static memory has faster access, although it is generally available in smaller blocks. Working with two or more memory banks can often simplify an algorithm implementation by increasing the memory bandwidth. Note that boards with SoC FPGAs will usually have an external memory for the CPU. If this is the only external memory available, it may not be directly accessible from the FPGA fabric but must be accessed via the hard processor system. This can become a bandwidth bottleneck with heavy memory accesses by the FPGA starving the CPU of access.

For systems where the FPGA is installed as a card within a standard PC, the host computer usually manages the image capture and display. The interface between the FPGA and host is usually through PCI Express, enabling high-speed image transfer between FPGA and host. Such development cards usually have hardened PCI Express blocks, reducing the FPGA resources required for the interface. However, depending on the application, the interface may be a bandwidth bottleneck within the system.

When targeting an embedded product, power and size are often the critical constraints. This will involve eliminating any peripherals that are unnecessary in the final application. However, for debugging, it may be useful to have additional peripherals accessible through an expansion port. To minimise the cost of a product, it is desirable to reduce the size of the FPGA used. However, it is important to keep in mind upgrades that may require a larger FPGA. Many families have several different footprints for each size of FPGA. Choosing a footprint that allows a larger FPGA to be substituted will reduce the retooling costs if a larger FPGA is required in later models.

2.4 Summary

FPGAs are composed of a 'sea of gates' consisting of programmable lookup tables and associated grid-like interconnect. Modern FPGAs, however, are heterogeneous and also include hardware multipliers, memory blocks, CPUs, and hardened communications and memory interfaces. FPGAs are programmable in that the specific logic functionality can be configured by the user. Fortunately, a detailed understanding of the architecture of FPGAs is not required; this is taken care of by the programming tools.

The resulting digital hardware is inherently parallel. This makes FPGAs a good platform for accelerating image processing algorithms, especially low- and intermediate-level operations, where parallelism can easily be exploited. The integration of CPUs within SoC FPGAs provides opportunities for implementing in software the high-level operations that are not well suited to FPGA realisation.

HDLs (VHDL and Verilog) are good at describing hardware and are inherently concurrent. Their main limitation is that they require programming at the structural level or RTL, which is quite different from the algorithmic approach to representing image processing algorithms. Consequently, the algorithm is usually developed within a software environment, with a consequent laborious and error-prone task of translating the design from the software into an RTL representation in the HDL. This translation from a software to hardware design can be partially alleviated through HLS, which treats the high-level language (usually some variant of C) as an HDL. The most powerful of these leave the base language largely unchanged and automatically extract parallelism through loop unrolling and pipelining. Often, this is guided by the provision of compiler directives.

HLS gives considerable productivity to writing code for FPGA, by representing the algorithm at a more abstract higher level, with the control structures often implicit in the structure of the algorithm. While there may be parallelism that can be exploited, unless the underlying algorithm is essentially parallel, it is not going to give the most efficient solution. The danger here is that the wrong algorithm is being parallelised, and parallelising serial algorithms is not trivial.

It is important to keep in mind that FPGA-based development is hardware design. Designing for hardware and software requires different skills. Even if a C-based language is used to represent the design, it is important to keep in mind the hardware architecture that is being built or is implied by the algorithm. Not all software algorithms map well to hardware, for example algorithms that rely on pointer arithmetic, and recursion. Operations that require random access to image memory do not integrate well with operations based on stream processing. Under real-time constraints, there is limited access to memory. These and many other aspects of hardware design usually require that the algorithm be transformed rather than simply mapped to hardware. Again, this requires a hardware-oriented mindset.

Image processing algorithm development, however, is usually considered as a software design process. Therefore, the development of efficient FPGA-based image processing systems requires a mix of both hardware and software engineering skills.

References

Ab Rahman, A.A.H., Thavot, R., Mattavelli, M., and Faure, P. (2010). Hardware and software synthesis of image filters from CAL dataflow specification. *Conference on Ph.D. Research in Microelectronics and Electronics (PRIME)*, Berlin, Germany (18–21 July 2010), 4 pages.

Ahmed, E. and Rose, J. (2000). The effect of LUT and cluster size on deep-submicron FPGA performance and density. *International Symposium on Field Programmable Gate Arrays*, Monterey, CA, USA (10–11 February 2000), 3–12. https://doi.org/10.1145/329166.329171.

Al-Dujaili, A. and Fahmy, S.A. (2017). High throughput 2D spatial image filters on FPGAs. *arXiv preprint*, (1710.05154), 8 pages. https://doi.org/10.48550/arXiv.1710.05154.

Alston, I. and Madahar, B. (2002). From C to netlists: hardware engineering for software engineers? *IEE Electronics and Communication Engineering Journal* **14** (4): 165–173. https://doi.org/10.1049/ecej:20020404.

Altera (2009). Generating functionally equivalent FPGAs and ASICs with a single set of RTL and synthesis/timing constraints. White paper, Altera Corporation, USA.

Ashenden, P.J. (2008). *The Designer's Guide to VHDL*, 3e. Burlington, MA: Morgan Kaufmann Publishers.

Bailey, D.G. (2010). Chain coding streamed images through crack run-length encoding. *Image and Vision Computing New Zealand (IVCNZ 2010)*, Queenstown, NZ (8–9 November 2010), 6 pages. https://doi.org/10.1109/IVCNZ.2010.6148812.

Bailey, D.G. (2011). Image border management for FPGA based filters. *6th International Symposium on Electronic Design, Test and Applications*, Queenstown, NZ (17–19 January 2011), 144–149. https://doi.org/10.1109/DELTA.2011.34.

Bailey, D.G. (2015). The advantages and limitations of high level synthesis for FPGA based image processing. *International Conference on Distributed Smart Cameras*, Seville, Spain (8–11 September 2015), 134–139. https://doi.org/10.1145/2789116.2789145.

Bailey, D.G. (2020). History and evolution of single pass connected component analysis. *35th International Conference on Image and Vision Computing New Zealand*, Wellington, NZ, USA (25–27 November 2020), 317–322. https://doi.org/10.1109/IVCNZ51579.2020.9290585.

Bailey, D.G. and Ambikumar, A.S. (2018). Border handling for 2D transpose filter structures on an FPGA. *Journal of Imaging* **4** (12): Article ID 138, 21 pages. https://doi.org/10.3390/jimaging4120138.

Bailey, D.G. and Johnston, C.T. (2007). Single pass connected components analysis. *Image and Vision Computing New Zealand (IVCNZ)*, Hamilton, NZ (5–7 December, 2007), 282–287.

Bailey, D.G., Sen Gupta, G., and Contreras, M. (2012). Intelligent camera for object identification and tracking. *1st International Conference on Robot Intelligence Technology and Applications*, Gwangju, Republic of Korea (16–18 December 2012), *Advances in Intelligent Systems and Computing*, Volume **208**, 1003–1013. https://doi.org/10.1007/978-3-642-37374-9_97.

BDTI (2010). High-Level Synthesis Tools for Xilinx FPGAs. *Technical Report*. Berkeley Design Technology Inc.

Bhasker, J. (1999). *A VHDL Primer*, 3e. New Jersey: Prentice-Hall.

Bhasker, J. (2005). *A Verilog HDL Primer*, 3e. Allentown, PA: Star Galaxy Publishing.

Boland, D. and Constantinides, G.A. (2012). A scalable approach for automated precision analysis. *ACM/SIGDA International Symposium on Field Programmable Gate Arrays*, Monterey, CA (22–24 February 2012), 185–194. https://doi.org/10.1145/2145694.2145726.

Cheah, H.Y. (2016). The iDEA architecture-focused FPGA soft processor. PhD thesis. Singapore: Nanyang Technological University. https://doi.org/10.32657/10356/68990.

Cheah, H.Y., Brosser, F., Fahmy, S.A., and Maskell, D.L. (2014). The iDEA DSP block-based soft processor for FPGAs. *ACM Transactions Reconfigurable Technology and Systems* **7** (3): Article ID 19, 23 pages. https://doi.org/10.1145/2629443.

Clow, J., Tzimpragos, G., Dangwal, D., Guo, S., McMahan, J., and Sherwood, T. (2017). A Pythonic approach for rapid hardware prototyping and instrumentation. *27th International Conference on Field Programmable Logic and Applications (FPL)*, Ghent, Belgium (4–8 September 2017), 7 pages. https://doi.org/10.23919/FPL.2017.8056860.

Cong, J., Liu, B., Neuendorffer, S., Noguera, J., Vissers, K., and Zhang, Z. (2011). High-level synthesis for FPGAs: from prototyping to deployment. *IEEE Transactions on Computer-Aided Design of Integrated Circuits and Systems* **30** (4): 473–491. https://doi.org/10.1109/TCAD.2011.2110592.

Cong, J., Zhang, P., and Zou, Y. (2012). Optimizing memory hierarchy allocation with loop transformations for high-level synthesis. *49th Annual Design Automation Conference (DAC'12)*, San Francisco, CA, USA (3–7 June 2012), 1229–1234. https://doi.org/10.1145/2228360.2228586.

Constantinides, G.A., Cheung, P.Y.K., and Luk, W. (2002). Optimum wordlength allocation. *Symposium on Field-Programmable Custom Computing Machines*, Napa, CA, USA (22–24 April 2002), 219–228. https://doi.org/10.1109/FPGA.2002.1106676.

Coussy, P., Gajski, D.D., Meredith, M., and Takach, A. (2009). An introduction to high-level synthesis. *IEEE Design and Test of Computers* **26** (4): 8–17. https://doi.org/10.1109/MDT.2009.69.

Decaluwe, J. (2004). MyHDL: a Python-based hardware description language. *Linux Journal* **2004** (127): Article ID 5, 12 pages.

DeHon, A. (1999). Balancing interconnect and computation in a reconfigurable computing array. *ACM/SIGDA Seventh International Symposium on Field Programmable Gate Arrays*, Monterey, CA, USA, 69–78. https://doi.org/10.1145/296399.296431.

Dessouky, G., Klaiber, M.J., Bailey, D.G., and Simon, S. (2014). Adaptive dynamic on-chip memory management for FPGA-based reconfigurable architectures. *24th International Conference on Field Programmable Logic and Applications (FPL)*, Munich, Germany (2–4 September 2014), 8 pages. https://doi.org/10.1109/FPL.2014.6927471.

Edwards, S.A. (2006). The challenges of synthesizing hardware from C-like languages. *IEEE Design and Test of Computers* **23** (5): 375–383. https://doi.org/10.1109/MDT.2006.134.

Fingeroff, M. and Bollaert, T. (2010). *High Level Synthesis Blue Book*. USA: Mentor Graphics Corporation.

Francis, R.M. (2013). Exploring Networks-on-Chip for FPGAs. *Technical Report UCAM-CL-TR-828*. Cambridge, UK: University of Cambridge Computer Laboratory.

Friedman, E.G. (2001). Clock distribution networks in synchronous digital integrated circuits. *Proceedings of the IEEE* **89** (5): 665–692. https://doi.org/10.1109/5.929649.

Giamblanco, N.V. and Anderson, J.H. (2019). A dynamic memory allocation library for high-level synthesis. *29th International Conference on Field Programmable Logic and Applications (FPL)*, Barcelona, Spain (8–12 September 2019), 314–320. https://doi.org/10.1109/FPL.2019.00057.

Guo, Z., Buyukkurt, B., and Najjar, W. (2004). Input data reuse in compiling window operations onto reconfigurable hardware. *ACM SIGPLAN Notices* **39** (7): 249–256. https://doi.org/10.1145/998300.997199.

Herbordt, M., VanCourt, T., Gu, Y., Sukhwani, B., Conti, A., Model, J., and DiSabello, D. (2007). Achieving high performance with FPGA-based computing. *IEEE Computer* **40** (3): 50–57. https://doi.org/10.1109/MC.2007.79.

Huang, Q., Lian, R., Canis, A., Choi, J., Xi, R., Calagar, N., Brown, S., and Anderson, J. (2015). The effect of compiler optimizations on high-level synthesis-generated hardware. *ACM Transactions on Reconfigurable Technology and Systems* **8** (3): Article ID 14, 26 pages. https://doi.org/10.1145/2629547.

Hutton, M. (2017). Understanding how the new Intel® HyperFlex™ FPGA architecture enables next-generation high-performance systems. White paper, Intel Corporation, USA.

Hutton, M., Schleicher, J., Lewis, D., Pedersen, B., Yuan, R., Kaptanoglu, S., Baeckler, G., Ratchev, B., Padalia, K., Bourgeault, M., Lee, A., Kim, H., and Saini, R. (2004). Improving FPGA performance and area using an adaptive logic module. *14th International Conference on Field Programmable Logic and Applications*, Antwerp, Belgium (29 August – 1 September 2004), *Lecture Notes in Computer Science*, Volume **3203**, 135–144. https://doi.org/10.1007/b99787.

IEEE 1364-2005 (2006). *IEEE Standard for Verilog Hardware Description Language*. IEEE. https://doi.org/10.1109/IEEESTD.2006.99495.

IEEE 1800-2017 (2018). *IEEE Standard for SystemVerilog–Unified Hardware Design, Specification, and Verification Language*. IEEE. https://doi.org/10.1109/IEEESTD.2018.8299595.

IEEE 1076-2019 (2019). *IEEE Standard for VHDL Language Reference Manual*. IEEE. https://doi.org/10.1109/IEEESTD.2019.8938196.

Khronos (2020). *SYCL™ Specification*, Version 1.2.1, Rev 7. Khronos Group.

Khronos (2021). *OpenCL: open standard for parallel programming of heterogeneous systems*, January 25 2021. Khronos Group. https://www.khronos.org/opencl/ (accessed 18 April 2023).

Kouloheris, J.L. and El Gamal, A. (1991). FPGA performance versus cell granularity. *IEEE Custom Integrated Circuits Conference*, San Diego, CA, USA (12–15 May 1991), Article ID 6.2, 4 pages. https://doi.org/10.1109/CICC.1991.164048.

Krishnan, G. (2005). Flexibility with EasyPath FPGAs. *Xcell Journal* **55**: 96–99.

Kundu, S. and Chattopadhyay, S. (2015). *Network on Chip: The Next Generation of System-on-Chip Integration*. Boca Raton, FL: CRC Press.

Kuon, I. and Rose, J. (2007). Measuring the gap between FPGAs and ASICs. *IEEE Transactions on Computer-Aided Design of Integrated Circuits and Systems* **26** (2): 203–215. https://doi.org/10.1109/TCAD.2006.884574.

Lahti, S., Sjovall, P., Vanne, J., and Hamalainen, T.D. (2019). Are we there yet? A study on the state of high-level synthesis. *IEEE Transactions on Computer-Aided Design of Integrated Circuits and Systems* **38** (5): 898–911. https://doi.org/10.1109/TCAD.2018.2834439.

Lamoureux, J. and Wilton, S.J.E. (2006). FPGA clock network architecture: flexibility vs. area and power. *International Symposium on Field Programmable Gate Arrays*, Monterey, CA, USA (22–24 February 2006), 101–108. https://doi.org/10.1145/1117201.1117216.

Langhammer, M. and Pasca, B. (2015a). Design and implementation of an embedded FPGA floating point DSP block. *IEEE 22nd Symposium on Computer Arithmetic*, Lyon, France (22–24 June 2015), 26–33. https://doi.org/10.1109/ARITH.2015.18.

Langhammer, M. and Pasca, B. (2015b). Floating-point DSP block architecture for FPGAs. *ACM/SIGDA International Symposium on Field-Programmable Gate Arrays*, Monterey, CA, USA (28 February – 2 March, 2015), 117–125. https://doi.org/10.1145/2684746.2689071.

Langhammer, M., Nurvitadhi, E., Pasca, B., and Gribok, S. (2021). Stratix 10 NX architecture and applications. *ACM/SIGDA International Symposium on Field-Programmable Gate Arrays (FPGA '21)*, Virtual (28 February – 2 March, 2021), 57–67. https://doi.org/10.1145/3431920.3439293.

Leeser, M., Miller, S., and Yu, H. (2004). Smart camera based on reconfigurable hardware enables diverse real-time applications. *12th Annual IEEE Symposium on Field-Programmable Custom Computing Machines (FCCM 2004)*, Napa, CA, USA (20–23 April 2004), 147–155. https://doi.org/10.1109/FCCM.2004.53.

Leong, P.H.W. (2008). Recent trends in FPGA architectures and applications. *IEEE International Symposium on Electronic Design, Test and Applications (DELTA 2008)*, Hong Kong (23–25 January 2008), 137–141. https://doi.org/10.1109/DELTA.2008.14.

Li, C., Bi, Y., Benezeth, Y., Ginhac, D., and Yang, F. (2018). High-level synthesis for FPGAs: code optimization strategies for real-time image processing. *Journal of Real-Time Image Processing* **14** (3): 701–712. https://doi.org/10.1007/s11554-017-0722-3.

Liang, T., Zhao, J., Feng, L., Sinha, S., and Zhang, W. (2018). Hi-DMM: high-performance dynamic memory management in high-level synthesis. *IEEE Transactions on Computer-Aided Design of Integrated Circuits and Systems* **37** (11): 2555–2566. https://doi.org/10.1109/TCAD.2018.2857040.

Lim, Y.K. (2012). Algorithmic strategies for FPGA-based vision. Master of Engineering Science thesis. Melbourne, Australia: Monash University. https://doi.org/10.4225/03/583d038fa90c6.

Lim, Y.K., Kleeman, L., and Drummond, T. (2013). Algorithmic methodologies for FPGA-based vision. *Machine Vision and Applications* **24** (6): 1197–1211. https://doi.org/10.1007/s00138-012-0474-9.

Mealy, B. and Tappero, F. (2012). *Free Range VHDL*. freerangefactory.org.

Meeus, W. and Stroobandt, D. (2014). Automating data reuse in high-level synthesis. *Conference on Design, Automation and Test in Europe (DATE '14)*, Dresden, Germany (24–28 March 2014), 4 pages. https://doi.org/10.7873/DATE.2014.311.

Meeus, W., Beeck, K.V., Goedeme, T. Meel, J., and Stroobandt, D. (2012). An overview of today's high-level synthesis tools. *Design Automation for Embedded Systems* **16** (3): 31–51. https://doi.org/10.1007/s10617-012-9096-8.

Mosqueron, R., Dubois, J., and Paindavoine, M. (2007). High-speed smart camera with high resolution. *EURASIP Journal on Embedded Systems* **2007**: Article ID 24163, 19 pages. https://doi.org/10.1155/2007/24163.

Nane, R., Sima, V., Pilato, C., Choi, J., Fort, B., Canis, A., Chen, Y.T., Hsiao, H., Brown, S., Ferrandi, F., Anderson, J., and Bertels, K. (2016). A survey and evaluation of FPGA high-level synthesis tools. *IEEE Transactions on Computer-Aided Design of Integrated Circuits and Systems* **35** (10): 1591–1604. https://doi.org/10.1109/TCAD.2015.2513673.

Ninos, S. and Dollas, A. (2008). Modeling recursion data structures for FPGA-based implementation. *International Conference on Field Programmable Logic and Applications (FPL 2008)*, Heidelberg, Germany (8–10 September 2008), 11–16. https://doi.org/10.1109/FPL.2008.4629900.

Ozer, C. (2014). A dynamic memory manager for FPGA applications. Master of Science thesis. Ankara, Turkey: Middle East Technical University.

Rafi, M. and Din, Nu. (2016). A novel arrangement for efficiently handling image border in FPGA filter implementation. *3rd International Conference on Signal Processing and Integrated Networks (SPIN)*, Noida, India (11–12 February 2016), 163–168. https://doi.org/10.1109/SPIN.2016.7566681.

Reinders, J., Ashbaugh, B., Brodman, J., Kinsner, M., Pennycook, J., and Tian, X. (2021). *Data Parallel C++: Mastering DPC++ for Programming of Heterogeneous Systems using C++ and SYCL*. USA: Apress Open. https://doi.org/10.1007/978-1-4842-5574-2.

Rutenbar, R.A., Baron, M., Daniel, T., Jayaraman, R., Or-Bach, Z., Rose, J., and Sechen, C. (2001). (When) will FPGAs kill ASICs? *38th Annual Design Automation Conference*, Las Vegas, NV, USA (18–22 June 2001), 321–322. https://doi.org/10.1145/378239.378499.

Sanguinetti, J. (2010). Understanding high-level synthesis design's advantages. *EE Times Asia* (26 April 2010), 1–4.

Schafer, B.C. and Wang, Z. (2020). High-level synthesis design space exploration: past, present, and future. *IEEE Transactions on Computer-Aided Design of Integrated Circuits and Systems* **39** (10): 2628–2639. https://doi.org/10.1109/TCAD.2019.2943570.

Singh, S., Rose, J., Chow, P., and Lewis, D. (1992). The effect of logic block architecture on FPGA performance. *IEEE Journal of Solid-State Circuits* **27** (3): 281–287. https://doi.org/10.1109/4.121549.

Sinha, U. (2017). Enabling impactful DSP designs on FPGAs with hardened floating-point implementation. White paper. Intel Corporation, USA.

Siracusa, M., Sozzo, E.D., Rabozzi, M., Tucci, L.D., Williams, S., Sciuto, D., and Santambrogio, M.D. (2022). A comprehensive methodology to optimize FPGA designs via the roofline model. *IEEE Transactions on Computers* **71** (8): 1903–1915. https://doi.org/10.1109/TC.2021.3111761.

Skliarova, I. and Sklyarov, V. (2009). Recursion in reconfigurable computing: a survey of implementation approaches. *International Conference on Field Programmable Logic and Applications (FPL)*, Prague, Czech Republic (31 August – 2 September, 2009), 224–229. https://doi.org/10.1109/FPL.2009.5272304.

Swarbrick, I., Gaitonde, D., Ahmad, S., Gaide, B., and Arbel, Y. (2019). Network-on-chip programmable platform in Versal™ ACAP architecture. *ACM/SIGDA International Symposium on Field-Programmable Gate Arrays*, Seaside, CA, USA (28 February – 2 March 2019), 212–221. https://doi.org/10.1145/3289602.3293908.

Takach, A. (2016). High-level synthesis: status, trends, and future directions. *IEEE Design and Test* **33** (3): 116–124. https://doi.org/10.1109/MDAT.2016.2544850.

Todman, T.J., Constantinides, G.A., Wilton, S.J.E., Mencer, O., Luk, W., and Cheung, P.Y.K. (2005). Reconfigurable computing: architectures and design methods. *IEE Proceedings Computers and Digital Techniques* **152** (2): 193–207. https://doi.org/10.1049/ip-cdt:20045086.

Trimberger, S.M.S. (2015). Three ages of FPGAs: a retrospective on the first thirty years of FPGA technology. *Proceedings of the IEEE* **103** (3): 318–331. https://doi.org/10.1109/JPROC.2015.2392104.

Vallina, F.M. and Alvarez, J.R. (2013). Using OpenCV and Vivado HLS to accelerate embedded vision applications in the Zync SoC. *XCell Journal* **83**: 24–30.

Winterstein, F., Bayliss, S., and Constantinides, G.A. (2013). High-level synthesis of dynamic data structures: a case study using Vivado HLS. *International Conference on Field Programmable Technology*, Kyoto, Japan (9–11 December 2013), 362–365. https://doi.org/10.1109/FPT.2013.6718388.

Wissolik, M., Zacher, D., Torza, A., and Day, B. (2019). Virtex UltraScale+ HBM FPGA: a revolutionary increase in memory performance. White paper. Xilinx Corporation, USA.

Xilinx (2018). *7 Series DSP48E1 Slice*, volume UD479 (v1.10). Xilinx Inc.

Xue, Z. (2019). Dynamic memory management for reconfigurable hardware. PhD thesis. London, UK: Imperial College. https://doi.org/10.25560/74569.

3

Design Process

The process of developing an embedded image processing application involves four steps or stages (Gribbon et al., 2007). The relationship between these is illustrated in Figure 3.1. The problem specification clearly defines the problem in such a way that the success of the proposed solution can be measured. The algorithm development step determines the sequence of image processing operations required to transform the expected input images into the desired result. Architecture selection involves determining the computational structure of the processors that are required to execute the algorithm at both the application and operation levels. Finally, system implementation is the process of mapping the algorithm onto the selected architecture, including construction and testing of the final system.

It has been observed that the successful development of complex algorithms on field-programmable gate arrays (FPGAs) is sensitive to the quality of the implementation (Herbordt et al., 2007). To gain a significant speed improvement over a software implementation, it is necessary that a significant fraction of the algorithm be parallelised. Often, the speedup obtained from implementing an application on an FPGA is disappointing. One of the main factors is that it is not merely sufficient to port an algorithm from software onto an FPGA implementation. To obtain an effective and efficient solution, it is necessary for the computation and the computational architecture to be well matched. Even when using high-level synthesis (HLS), it is still important to consider the overall computational architecture, even though the architecture selection and system implementation stages are partially merged. While algorithm development is usually considered part of the software engineering realm, the design of the processing architecture is firmly in the domain of hardware engineering. Consequently, FPGA-based image processing requires a mix of both skill sets. To fully exploit the resources available on an FPGA, the system implementation stage often requires changes to both the algorithm and the architecture. As a result, the development process is usually iterative, rather than performed as discrete distinct steps as implied by Figure 3.1.

In the description of each of the stages, the important tasks within each step are identified and explained. In particular, the differences between designing for implementation on an FPGA and on a standard software-based system are highlighted.

3.1 Problem Specification

Deriving a detailed specification of the problem or application is arguably the most important step, regardless of whether the resultant system is software or hardware based. Without an adequate specification of the problem, it is impossible to measure how well the problem has been solved. It is impossible to know when a project has been completed when 'completed' has not been defined!

Within image processing, it is common to have a relatively vague description of the problem. For example, a produce inspection problem may be described as inspecting the produce in order to remove blemished and damaged produce. While this may be the aim, such a description is of limited value from an engineering design

Design for Embedded Image Processing on FPGAs, Second Edition. Donald G. Bailey.

Companion Website: www.wiley.com/go/bailey/designforembeddedimageprocessc2e

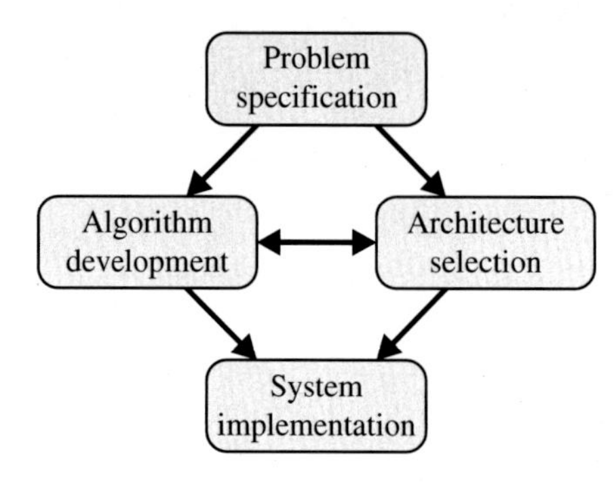

Figure 3.1 The relationship between the four main steps of the design process.

perspective. In an engineering context, determining the problem specification is usually called ***requirements analysis***. To be useful, the problem specification needs to be specific, complete, achievable, and measurable (DAU, 2001).

A problem specification should clearly describe the problem rather than the solution. It should cover what the system needs to do and why it needs to do it but should not cover how the system should be implemented; that is the output of the whole design process. For the specification to be specific, it will need to address at least three areas (DAU, 2001). The first is the system functionality; what the system needs to be able to do. In an image processing application, it needs to specify the desired result of image processing. Secondly, the performance of the system must be addressed; how well must it perform the functions. For real-time image processing, important aspects of this are the maximum allowable latency and minimum throughput (the number of frames that must be processed per second). If the problem involves classification, then for non-trivial problems, it will be inevitable that the system will make misclassifications. If the decision is binary, the allowable failure rate should be specified in terms of both false acceptance and false rejection rates. The third area of consideration is the environment in which the system will operate. Applied image processing consists of more than just the image processing algorithm; it is a systems engineering problem that requires consideration and specification of the complete system (Bailey, 1988). Other important aspects to consider include (Batchelor and Whelan, 1994) lighting, optics, and interfacing with supporting hardware and machinery. The relationships between the image processing system and the rest of the system need to be carefully elucidated and defined.

The problem specification must be complete. It should consider not only the normal operation but also how the system should behave in exceptional circumstances. If the system is interactive, then the interface between the user and the system needs to be clearly defined, at least in terms of functionality. The specification should also not only consider the operation of the system but also address the desired reliability and the required maintenance over the expected life of the system. Again, all aspects of the system should be addressed, not just the image processing components.

The system as defined by the problem specification must be achievable. In a research context, it is essential that the research questions can be answered by the research. For a development project, the final system must be technically achievable at an affordable cost.

Finally, the problem specification must be measurable. This enables the resultant system to be evaluated objectively to verify that it satisfies the specification. Consequently, the requirements should be quantitative and should avoid vague words such as excessive, sufficient, resistant, and similar (DAU, 2001). It is important to determine the constraints on the system. For embedded real-time systems, this includes the frame rate, system latency, size, weight, power, and cost constraints. The resulting set of requirements must be mutually consistent. A distinction should be made between hard constraints (those that are essential for successful operation or completion of the project) and soft constraints (those that are considered desirable but may be relaxed if necessary in the final system). Inevitably, there will be conflicts between the different constraints, particularly with real-time processing. A common trade-off occurs between speed and accuracy (Kehtarnavaz and Gamadia, 2006). These conflicts must be resolved before the development begins.

Successful specification of the problem requires comprehensive knowledge of the problem or task to which image processing is being applied. Such knowledge increases the likelihood that the application-level algorithm will be robust. It is used to select the representative set of sample images that will be used to develop and test the imaging algorithm. During the algorithm development process, it also guides the selection of the image features that are measured. Without such ***problem knowledge***, it is easy to make invalid assumptions about the nature of the task. There is always a danger that a resultant algorithm will work well for the specific cases that were used to develop it but not be sufficiently general or robust to be of practical use. Often, the system developer has limited knowledge of the problem. In such cases, it is essential to have regular feedback and verification from the client or other problem domain expert, particularly during the algorithm-development phase.

Inevitably, the problem specifications evolve through the life cycle of an application. An advantage of FPGAs relative to ASICs is the ability to adapt the design to changing requirements.

3.2 Algorithm Development

The task of image processing algorithm development is to find a sequence of image processing operations that transform the input image into the desired result. Algorithm development is a form of problem-solving activity and therefore usually follows heuristic development principles (Ngan, 1992). Similar to all problem-solving tasks, there is not a unique solution. The usual way of progressing is to find an initial solution and then refine it until it meets the functionality required by the problem specification.

Before the algorithm can be developed, it is necessary to capture a set of sample images. These should be representative of the imaging task that is to be performed and should be captured under conditions that match as closely as possible the conditions under which the system is expected to operate.

In scenarios where it is possible to control the lighting (machine vision for example), then choosing appropriate lighting is essential to simplify the task (Uber, 1986; Batchelor, 1994). In some applications, much of the problem can be solved using structured lighting techniques (such as projecting a grid onto the scene (Will and Pennington, 1972)). Conversely, inadequate lighting can make segmentation more difficult by reducing contrast, illuminating the background or object unevenly, saturating part of the scene, or introducing shadows or specular reflections that may be mistaken for other objects (Bailey, 1988). If the lighting is likely to be variable (for example mobile robotics or video surveillance using natural lighting), then it is essential that the set of sample images include images captured under the full range of lighting conditions that the system is expected to work under.

3.2.1 Algorithm Development Process

Once the sample images have been captured, one or two of these are used to develop the initial sequence of image processing operations. Developing an image processing algorithm is not as straightforward as it sounds. There are many imaging tasks that we, as humans, can perform with ease, that are difficult for computer-based image processing. One classic example is face recognition, something we can do naturally, without thinking. This dichotomy is sometimes called the 'trap of the two-legged existence theorem' (Hunt, 1983); we know that a solution to the problem exists because we, as humans, can perform the task. However, finding that solution (or indeed any solution) in the context of computer vision can be very difficult. While significant advances have been made in understanding how the human brain and visual system works, this does not necessarily lead to algorithmic solutions any more than understanding how a computer works can lead to programming it for a specific task.

At the other extreme lie the deep learning approaches. Through training, they select an appropriate hierarchy of features (LeCun, 2012), which can be used for object detection and classification. Most users of deep learning treat the model as a black box, which is trained by example to optimise the many (often millions) internal weights against the target output. While they can achieve state-of-art performance in many applications, it is often unclear how a particular classification is obtained. Most of this book focuses on 'classical' image processing, with deep learning approaches discussed in Section 14.4.

One of the difficulties of classical image processing is that there is still little underlying theory for developing application-level algorithms. While attempts have been made to formulate the process within a more theoretical framework, for example in terms of mathematical morphology (Vogt, 1986) or statistics (Therrien et al., 1986), these are incomplete. Even within such frameworks, algorithm development remains largely a heuristic (trial and error) process (Bailey and Hodgson, 1988; Ngan, 1992). An operation is chosen from those available and applied to the image. If it performs satisfactorily, it is kept, and an operation is found for the next step. However, if the operation does not achieve the desired result, another is tried in its place. Algorithm development must rely heavily on using the human visual system to evaluate the results of applying each operation. The experience of the developer plays a significant role in this process. The brute-force search for a solution to the algorithm development problem can be significantly streamlined by the application of appropriate heuristics (Ngan, 1992).

This process requires an interactive image processing system that supports such experimentation (Brumfitt, 1984). The algorithm development environment needs to have available many different operations, because it is generally not known in advance which operations will be required in any particular application. It also needs to support the development and integration of new operations, should a particular operation not be supplied with the development environment. It requires an appropriate image display for viewing and evaluating the effects of an operation on an image. The processing time of each operation must be sufficiently short not to distract the developers from their task. It has been suggested that within the development environment, operations should take no longer than about 15 seconds (Cady et al., 1981). It is also desirable to have some form of scripting capability, to avoid the need for re-entering complex sequences of commands when testing the algorithm on different images.

The sequence of operations within an algorithm generally follows that outlined in Section 1.3. However, the particular operations required depend strongly on the image processing task. The order in which the operations are added to the algorithm may not necessarily correspond to the order that they appear in the final algorithm. Attention may focus first on a critical step or initially develop a basic algorithm and then introduce further preprocessing steps to make the algorithm more robust.

3.2.2 *Algorithm Structure*

The remainder of this section will briefly review the operations that may be used at each step within the algorithm, with particular emphasis on how the operations work together to give a robust algorithm. There are many good references that review image processing operations and their functionality, for example Pratt (1978), Jain (1989), Castleman (1996), Russ (2002), Gonzalez et al. (2004), Gonzalez and Woods (2008), Davies (2005), and Solomon and Breckon (2011). The FPGA implementation of many of these operations is considered in more detail in Chapters 7 to 7.

Appropriate preprocessing is essential to make the algorithm robust to the variations encountered within the range of images. The primary goal of preprocessing is to enhance the information or features of interest in the scene while suppressing irrelevant information. However, the selection of particular operation depends strongly on application and the detailed properties and characteristics of the operations that follow in the algorithm.

Physical or environmental constraints may result in non-ideal input images. To compensate for deficiencies in the image capture process, preprocessing can be used: to remove background; to correct for nonlinearity in the image transfer function; to correct for distortion caused by optics or camera angle; and to compensate for deficiencies in illumination. Image filtering may be used to smooth noise or to enhance and detect edges, lines, and other image features. Examples of some of these preprocessing operations are illustrated in Figure 3.2.

One way of suppressing irrelevant information is to normalise the image content. Intensity normalisation, such as contrast expansion and histogram equalisation, can be used to compensate for variations in lighting, reflectivity, or contrast. Position and size normalisation may be used to centre, orient, and scale an image or object to match a template or model. These may require an initial segmentation to first locate the object.

Segmentation is the process of splitting the image into its meaningful components. This can vary from separating objects from the background to separating an image into its constituent parts. Segmentation can be based on any local property of the pixel values, such as intensity, colour, and texture. Two broad approaches to segmentation are based on edges and regions, as shown in Table 3.1. Edge-based methods detect the boundaries between regions, with the regions defined implicitly by the enclosing boundaries. They use edge detection or gradient filters to detect edges and then link the significant edges to form the boundaries. Region-based methods work the other way around. They classify each pixel based on some local property, with the edges determined implicitly by changes in property at the region boundary. They are characterised by filtering (to enhance or detect the property used for segmentation (Randen and Husoy, 1999)) followed by thresholding or statistical

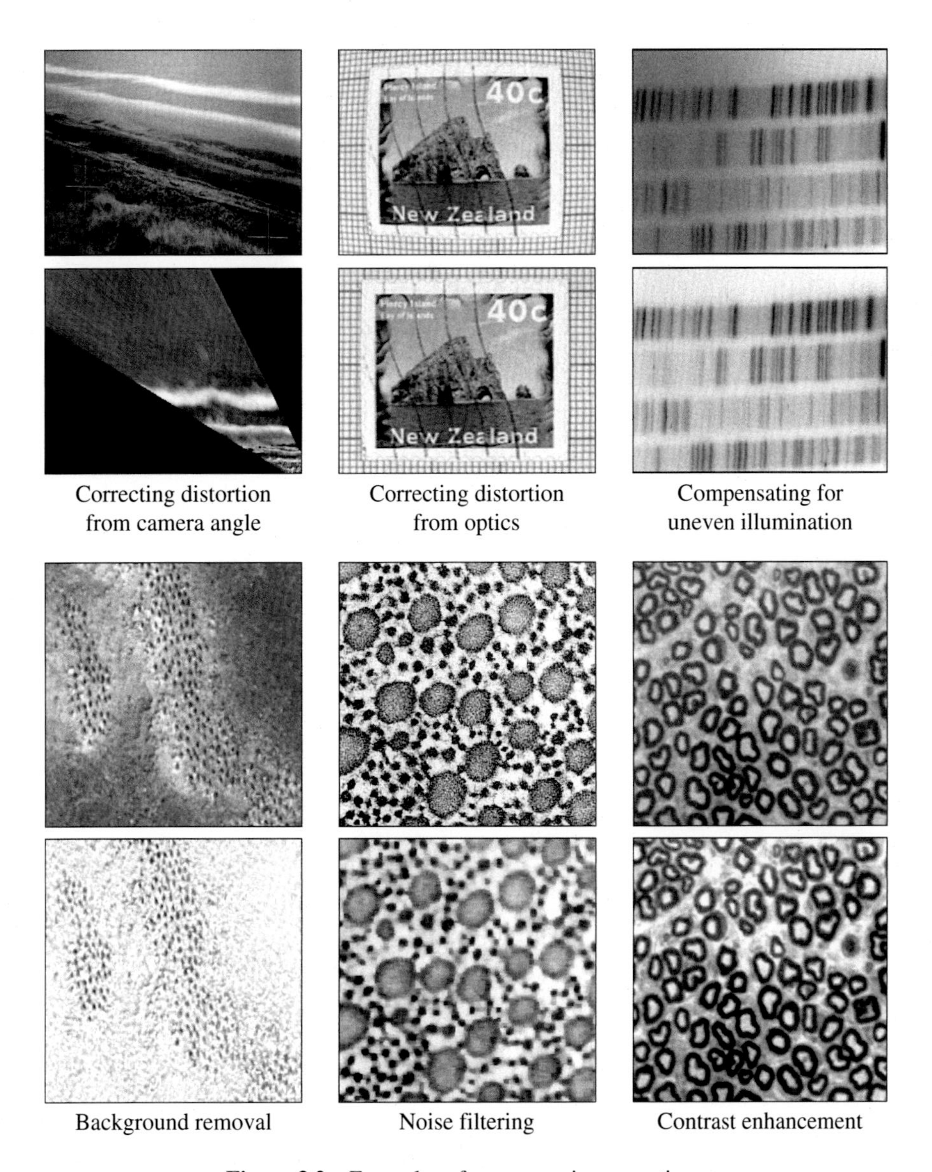

Correcting distortion from camera angle | Correcting distortion from optics | Compensating for uneven illumination

Background removal | Noise filtering | Contrast enhancement

Figure 3.2 Examples of preprocessing operations.

classification methods to assign each pixel to a region. Both edge- and region-based methods can be performed either globally or incrementally (Bailey, 1991) (see Table 3.1). Global methods segment the image by processing every pixel. Incremental methods only operate locally on the part of the image that is being segmented. They are usually initialised with a seed pixel and work by extending the associated edge or region by adding neighbouring pixels until a termination condition is met. Segmentation may also involve using techniques for separating touching objects such as concavity analysis, watershed segmentation, and the Hough transform.

The next stage within the algorithm is to extract any required information from the image. Measurements may be tangible (physical size and intensity) or abstract (for example the fractal dimension to quantify roughness (Brown, 1987) and the average response to a particular filter to characterise texture (Randen and Husoy, 1999)). Most tangible measurements require the imaging system to be calibrated to determine the relationship between pixel measurements and real-world measurements. In image analysis applications, the measurements are the desired output.

Table 3.1 Image segmentation methods.

	Edge-based segmentation	Region-based segmentation
Global processing	Edge detection and linking	Thresholding
Incremental processing	Boundary tracking	Region growing

In other applications, a set of image or object feature measurements is used to characterise or classify the image or object. Boundary features are properties of the edges of the object (such as contrast), while region features are properties of the pixels within the region (colour, texture, and shading). Shape features consider the whole object (for example area, height, width, perimeter, enclosing rectangle, number of corners, convex hull area and perimeter, moments, best-fit ellipse, and Fourier descriptors). Virtually anything that can be measured or can be derived from multiple measurements can be used as a feature. The set of measured features makes up a multi-dimensional ***feature vector***, with each feature corresponding to a dimension.

Classification is the process of recognising or assigning meaning to an object based on the measured feature vector. Each point in feature-space represents a particular combination of feature values. The feature vectors for objects within a class should be similar and fall in the same region within the feature-space. Classification therefore involves clustering or segmentation within the feature-space. The feature-space potentially has a high dimensionality, with many features being strongly correlated. The goal in selecting the best feature set is to choose only as many features as are necessary to unambiguously distinguish the objects. An analysis of the covariance can help to identify significant correlations between the features. Where there are strong correlations, consider replacing one or more of the features by a derived feature. For example, the length and width of an object are often correlated, especially as the scale changes. In this case, more useful information can often be obtained from the aspect ratio (length divided by width) than either the length or width. Unless the size or some other feature is used directly to distinguish between classes, it is often better to use dimensionless derived features such as aspect ratio and compactness. Principal components analysis and related techniques can help to identify the features that account for most of thc variance in the population.

Given a set of feature vectors (for example those derived from the set of sample images), there is a wide selection of techniques used to classify the feature vector to determine the object: decision trees, feature templates, Bayesian classification, neural networks, and support vector machines, among many others. For labelled data (each object has an associated desired classification), these associations form the training data for supervised classification. Training is the process of determining the classification parameters that minimise the classification errors. Some applications have no training data but a large number of samples. Unsupervised classification is the process of determining the classes from analysing the resultant feature vectors. It derives the classification by looking for groups or clusters of feature vectors within the feature space of the data provided. One example of an unsupervised classifier is K-means clustering. The processes of classification and classifier training are described in more detail in Chapter 14.

Finally, the data extracted from the image may require further processing, depending on the application. For example, a Kalman filter may be used for object tracking (Kalman, 1960), or in mobile robotics, the data may be used to update a map or to plan a route through the scene. In many computer vision applications, the data derived from the image is used to update the parameters of an underlying model representing the scene.

Since the initial algorithm is often developed on only a small set of images, the next step is to refine the algorithm using a wider range of images. The algorithm is tested to ensure that it can handle all of the expected conditions in the correct manner. Algorithm refinement often introduces more preprocessing operations to make the complete algorithm more robust. Particular care should be taken to check the algorithm over the full range of expected lighting conditions.

Very few algorithms are perfect. There are two key points where the algorithm is most likely to fail. The first is segmentation failure, where the object is not correctly segmented from the background. Often, a partial segmentation is worse than a complete failure, because the features obtained will generally not reflect those of the complete object. With segmentation failure, it is important to re-evaluate any assumptions that are made about the task or scene to check that they are not the cause of the failure and mitigate these, if necessary. For example, a problem caused by poor contrast or noise may be mitigated by further preprocessing. If necessary, the image capture system may need to be redesigned to obtain better starting images. If possible, the algorithm should detect whether or not the segmentation was successful or at least act appropriately when it encounters errors. The 'garbage in, garbage out' principle applies: if the image quality is very low, the algorithm is less likely to work reliably.

The second point of failure is classification failure. This occurs when the feature vector distributions for two or more classes are not completely separated and are insufficient to discriminate between the classes. This may require measuring an additional feature (or combination of features) that can better discriminate between the overlapping classes. Classification failure is also common when there are insufficient training samples to adequately define the feature vector distributions for each class.

The result of algorithm development is an algorithm that meets the specifications defined during requirements analysis.

3.2.3 FPGA Development Issues

The image processing algorithm cannot easily be developed directly on the FPGA, where the development cycle times are far too long to allow interactive design. Assuming that an implementation of the desired operations is available in a hardware description language, the compilation and place-and-route times required to map them onto an FPGA are prohibitive. Therefore, the algorithm first needs to be developed in a software environment. For this, any image processing algorithm development environment that satisfies the requirements described in Section 3.1 is suitable. Such an environment will usually be distinct from the FPGA development environment unless high-level synthesis is used.

One advantage of separating the algorithm development from the FPGA implementation is that the application-level algorithm can be thoroughly tested before the complex task of mapping the algorithm onto the target hardware. Such testing is often much easier to perform within the software-based image processing environment. Testing the hardware implementation then becomes a matter of verifying that each of the image processing operations has been implemented correctly. If the operations behave correctly, then the whole image processing algorithm will also function in the same way as the software version.

The algorithm development process does not stand alone. The implementation process, as described shortly, is often iterative as the algorithm is modified to make all of the operations architecturally compatible. Whenever any changes are made to the image processing algorithm, it is necessary to retest the algorithm to ensure that it still performs satisfactorily. However, some algorithm testing must be deferred to the final implementation. Evaluating the effectiveness of an image processing algorithm on a continuous, noisy, and real-time video stream is difficult in any environment.

The output from the algorithm development step is the sequence of image processing operations required to transform the input image or images into the desired form. It has been tested (within the software-based image processing environment) and shown to meet the relevant accuracy and robustness specifications for the application.

3.3 Architecture Selection

Once an initial algorithm has been developed, it is necessary to define the implementation architecture. It is important to obtain a good match between the architecture and both the application- and operation-level algorithms. A poor match makes inefficient use of the available resources on the FPGA and can limit the extent to which the algorithm may be accelerated. Consequently, the system may not meet the targeted speed requirements, or the design may require a larger FPGA than is really needed.

The architecture can be considered at two distinct levels. The system-level architecture is concerned primarily with the relationship and connections between the components of the system. Three aspects of the system architecture are of primary importance: the relationship between the FPGA and any other processing elements (whether they be a standard computer, a hardened processor core within an system-on-chip (SoC), or a soft-core processor); the nature of any peripherals, and how they are connected; and the nature and structure of external memory within the system. The computational architecture is concerned primarily with how the computation of the algorithm is performed, or what the loops look like. This includes the forms of parallelism that are being exploited and the split between using hardware and software for implementing the algorithm.

On a software-based image processing system, the computational architecture is fixed, and the system architecture is generally predefined. For an FPGA-based system, however, nothing is provided, and all aspects of the architecture must be developed or selected. One exception to this is perhaps the use of development boards for prototyping, where much of the system-level architecture is fixed by the board vendor.

3.3.1 System Architecture

The biggest factor of the system architecture is the mix between software and hardware computing resources used to implement the design. The spectrum is shown in Figure 3.3 and governs whether a central processing unit (CPU) or FPGA is the central processing element of the system.

In a CPU-centric or ***hosted configuration***, the host may be a server, a desktop machine, or the hardened CPU on a modern SoC platform. The host operating system needs to be augmented to enable it to reconfigure the FPGA (Andrews et al., 2004). This is usually dynamic reconfiguration, allowing the host to change the FPGA configuration, depending on the application (Wigley and Kearney, 2001; Sedcole et al., 2007). The host operating system is also responsible for distributing tasks between the serial processor and the FPGA system (Steiger et al., 2004; Lubbers and Platzner, 2007) and synchronising communications between all of the processes (both hardware and software) (Mignolet et al., 2003; So, 2007). This is where OpenCL and other similar HLS tools have an advantage in that they will provide or automatically build much of the communications infrastructure. In a hosted configuration, the host is responsible for any user interaction and is often responsible for image capture and display, simply providing the data for the FPGA to process.

In an FPGA-centric or ***stand-alone configuration***, all or most of the work is performed on the FPGA. Any software processor may be a soft core implemented using the FPGA's programmable logic or, more commonly now, the hardened CPU within an SoC FPGA. The FPGA is often responsible for any image capture and display. The CPU is usually responsible for any user interaction; if there is no CPU, then any user interaction must be built as hardware modules on the FPGA (Buhler, 2007). Much of the low-level interfacing must be built in hardware. While it is possible to interface directly from the software to the hardware, using a device driver will make access to the hardware more portable and make the interface easier to maintain and manage.

One advantage of using a standard processor within a stand-alone system is that all of the application development tools for that processor can be used. In particular, using the integrated high-level language development and emulation tools for the processor can speed code development and debugging of the software components.

A second advantage is the availability of embedded operating systems such as Linux. Embedded Linux provides all of the features of a standard operating system, subject to the limitations of the memory available to the processor (Williams, 2009). When interfacing with hardware through a Linux device driver, the kernel effectively provides an application programming interface (API) and will provide locking and serialisation

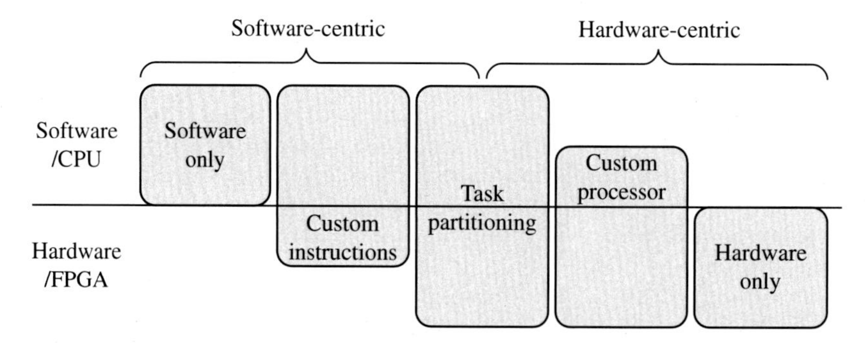

Figure 3.3 Spectrum of hardware and software mix. Note that the CPU can be either external or integrated within the FPGA.

on the device. The availability of a full transmission control protocol and internet protocol (TCP/IP) stack enables the complete system to be networked relatively easily. This can significantly extend the functionality of the system, even if the software processor is not used directly for the application. Having a TCP/IP interface enables:

- an efficient means of communicating images and other data between devices in a distributed system;
- a web interface for control and setting of algorithm parameters and for viewing the results of processing;
- remote debugging and even mounting the file system across the network;
- the ability to remotely update the configuration.

However, one limitation of using a multi-tasking operating system is the variable latency of the software components. Care must be taken when using such systems in a hard real-time context, especially where software latency is critical. In other cases, an embedded Linux can simplify the development and debugging, especially of stand-alone systems.

For debugging and initial development, the stand-alone configuration may have a host computer. This can be used for directly downloading the FPGA configuration file and may provide some support for debugging. However, for normal running, the FPGA configuration is often loaded directly from flash memory, avoiding the need for a host processor.

Also at the system level is the off-chip memory architecture. A key factor to consider is the degree of coupling among the memory, FPGA, and any CPU. Distinction also needs to be made between shared system memory and memory local to the FPGA. For local memory, the technology is also an important consideration. Many high-speed memories, especially those with large capacity, operate most efficiently in burst mode. Dynamic memories typically have several clock cycles latency (which can also vary depending on the current state of the memory and controller) and usually require designing a memory interface (see Section 6.4). Other memory technologies may allow random access on successive clock cycles, although for high-speed operation, they are often pipelined with the address having to be provided several clock cycles before the data is available. This may have implications on the speed of some operation-level algorithms, especially where the memory address is data dependent.

The aspects of interfacing from the FPGA to peripheral devices are covered in more detail in Chapter 6.

3.3.2 Partitioning Between Hardware and Software

Low-level image processing operations that can readily exploit parallelism are ideal for hardware implementation. Such operations are relatively slow when implemented in software simply because of the volume of data that needs to be processed.

However, not all algorithms map well to a hardware implementation. Tasks that have dynamically variable length loops can be clumsy to implement in hardware. Those with complex control sequences (Compton and Hauck, 2002) are usually more efficiently implemented in software. Such tasks are characterised by large amounts of complex code that is primarily sequential in nature. If implemented directly in hardware, this would require significant resources because each operation in the sequence would need to be implemented with a separate hardware block. If there is little scope for sharing the hardware for other operations, much of the hardware will be sitting idle much of the time. The same applies for functions that are only called occasionally, for example once or twice per frame (or less). These tasks or operations can be implemented more easily and more efficiently in software.

Two broad classes of tasks or operations that are best implemented in software are high-level image processing operations (for example object tracking (Arias-Estrada and Rodriguez-Palacios, 2002)) and managing complex communications protocols. High-level image processing operations are characterised by a lower volume of data, more complex control patterns, and often have irregular memory access patterns. Many provide limited opportunities for exploiting parallelism, and often, the computation path is dependent on the data being processed. Similarly, communications such as TCP/IP require a complex protocol stack that can be unwieldy to implement in hardware. While it is possible to implement these tasks directly in hardware (see, for example

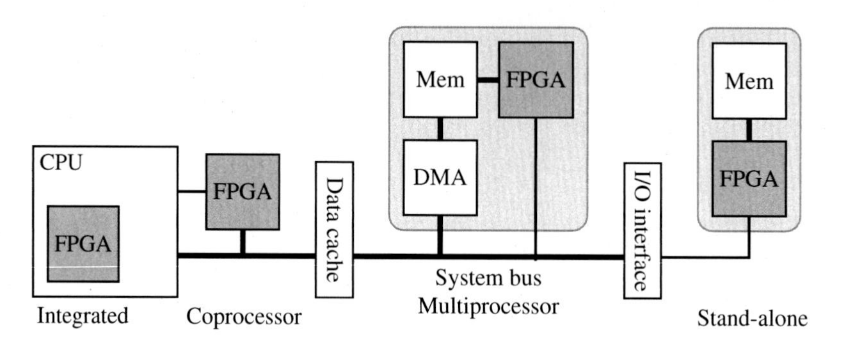

Figure 3.4 Coupling between a host processor and an FPGA. Source: Adapted from (Compton and Hauck, 2002, Fig. 3, p. 177).

(Dollas et al., 2005) for a TCP/IP implementation), it can be more efficient in terms of resource utilisation to implement them in software. Software also has the advantage that it is generally easier to programme.

Therefore, the choice is not just whether to implement the application in either software or hardware. There is a spectrum of partitioning of the application between the two, ranging between full software and full hardware implementations. Compton and Hauck (2002) identified four distinct ways in which programmable logic could be coupled with a host computer system, as illustrated in Figure 3.4. Todman et al. (2005) added a fifth coupling architecture: that of a CPU embedded within the FPGA (either as a hardened CPU or as a soft processor core).

At the instruction level of granularity, the application can be largely implemented in software, with the FPGA providing custom instructions to the CPU. This requires that the programmable logic be built inside the processor and allows instructions to be customised according to the application (Hauck et al., 2004). With modern FPGAs, this requires building a soft-core processor from the logic within the FPGA (the hardened CPU within an SoC is generally unable to efficiently access the FPGA logic in this way). The purpose is to accelerate key instructions within the host processor, while retaining the simplicity of a software processing model. Data is usually passed between the reconfigurable logic and CPU through shared registers (Compton and Hauck, 2002). This also requires building all of the associated tools (optimising compilers and so on, to use the custom instructions). One limitation is the restriction in the amount of data that may be passed to or from the function unit with each custom instruction and the limited time that is available to complete the processing. It is usually necessary to stall the main processor if the custom instruction takes a long time to execute. For image processing, it is impractical to implement the whole algorithm this way. However, this approach can be used to augment the CPU where groups of instructions are clumsy or inefficient when implemented purely in software. This mode is probably too tightly coupled to fully exploit parallelism in low-level image processing operations.

In the next more loosely coupled configuration, an FPGA can be used as a coprocessor (Miyamori and Olukotun, 1998) to accelerate particular types of operations in much the same way that a floating-point coprocessor accelerates floating-point operations within a standard CPU. In this configuration, the FPGA-based processing unit is usually larger than an arithmetic logic unit (ALU) or similar functional unit within the host. The coprocessor operates on its data relatively independently of the main processor. The main processor provides the data to be processed (either directly or indirectly) and then triggers the coprocessor to begin executing. The results are then returned after completion (Compton and Hauck, 2002). The operations performed can run for many clock cycles without intervention from the host processor. This allows the host to continue with other tasks, while waiting for the coprocessor to complete. Note that the FPGA-based coprocessor has access to the data from the host's cache (for example through the accelerator coherency port (ACP) on an SoC). Using dedicated hardware as a coprocessor is most commonly used when the main application is implemented within software, but time critical or computationally dense sections of an algorithm are accelerated through direct hardware implementation. There may also be multiple hardware processors, each developed to accelerate a critical part of the algorithm.

The next arrangement has the FPGA system as an additional processor in a multiprocessor system (Compton and Hauck, 2002). This partitions the application at the task level of granularity, where whole tasks are assigned either to a software or hardware processor. To be effective, this requires the hardware and software tasks to operate relatively independently of each other, only communicating when necessary. The FPGA system sits on the system bus, with DMA (direct memory access) used to transfer data from the host system memory into local memory (either directly or via peripheral component interconnect (PCI) Express). Depending on the processing performed, the results can be transferred back to the host either through DMA or via registers within the FPGA that are mapped into the address space of the host. The host configures the FPGA, indicates where the data to be processed is located, and triggers the FPGA system to begin its execution. When processing is complete, this is signalled back to the host processor. With the modern SoC FPGAs, this approach is the easiest. A profiler can be run on the software version of the algorithm to identify the processing bottlenecks, and the focus of mapping the design to hardware can be placed where the best overall gains can be achieved. If using HLS, the modules being accelerated can be retargeted for hardware realisation. The independent processor arrangement maps well to the image processing pyramid introduced in Figure 1.4, The image capture and low-level image processing operations could be implemented in hardware, with the results being passed to high-level image processing operations implemented in software. However, the performance is strongly influenced by the I/O bandwidth, the amount of local memory available, and its arrangement (Benitez, 2002).

In the stand-alone configuration, the FPGA operates independently of the host computer system, which may not even be connected. Communication with the host is generally sporadic and of relatively low bandwidth. All of the processing is performed by the FPGA, including any embedded CPU it may contain for managing software tasks. If the software tasks are of low complexity, then an alternative to using a full CPU is to design a simplified serial processor. This can be either a lightweight soft core or a completely custom processor. One limitation to using a custom processor is that compilers and other development tools will not be available, and the software for the processor must generally be written in the custom assembly language. At the extreme end, the sequential components of the algorithm could be implemented using a finite state machine. Quite sophisticated, memory-based, finite state machines can be implemented with a small resource footprint (Sklyarov, 2002; Sklyarov and Skliarova, 2008), effectively making them custom processors.

Regardless of the level of partitioning, it is essential to clearly define the interface and communication mechanisms between the hardware and software components. In particular, it is necessary to design the synchronisation and data exchange mechanisms to facilitate the smooth flow of data. When using an SoC device, or a standard embedded soft processor core, these interfaces will generally be defined as part of the core. In the final design, it is important to take into account the communication overhead between the hardware and software portions of the design.

3.3.3 Computational Architecture

The computational architecture defines how the computational aspects of the algorithm are implemented. A particular focus is the form of parallelism being exploited to achieve the algorithm acceleration. Note that with HLS, much of the detailed computational architecture design is automated. However, as indicated earlier, it is still important to consider the high-level design to ensure compatibility between the operations.

3.3.3.1 Stream Processing

The main bottleneck of software-based image processing is that it is memory bound: an operation reads the pixel values from memory, processes them, and writes the results back to memory. Since each operation performs this, the speed of the algorithm will ultimately be limited by the number and speed of memory accesses. Stream processing can overcome this bottleneck by pipelining operations. The input data is read once from memory or streamed from the camera and passed into the first operation. Rather than write the results back to main memory, they are directly passed to the next operation. This will eliminate at least two memory accesses for each stage of the pipeline: writing the results and reading them again as input for the next operation. The benefits of streaming can therefore be significant.

For embedded vision applications, at some stage, the data must initially be read from the camera. At the input, each pixel must be processed as it arrives; otherwise, the data is lost. If possible, as much processing should

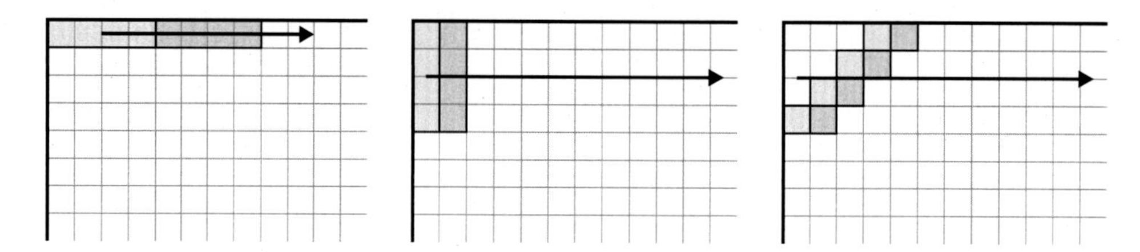

Figure 3.5 Streaming multiple pixels per clock cycle. Left: adjacent pixels on a single row; centre: processing multiple rows in parallel; right: offset rows to account for processing latencies on previous rows.

be performed on this data while it is passing through the FPGA, before it is written to memory. If the whole application can be implemented as a single streamed pipeline, then it may not even be necessary to use a frame buffer to hold the image data. Similarly, if an image is being displayed, the required pixels must be streamed to the output at each clock cycle; otherwise, there will be no data to display, and the output will appear blank for those pixels. Again, much benefit can be gained by performing as much processing on-the-fly as the data is being streamed out.

One of the key characteristics of stream processing is a fixed clock rate. The throughput is usually one clock cycle per pixel, constrained by the input (from a camera) or output (to a display), although higher resolutions or frame rates may require processing multiple pixels per clock cycle. Consequently, the design of stream processing systems is data synchronous, at least on the input and output. If every stage of the pipeline can be operated synchronously, then the design can be simplified greatly.

When streaming multiple pixels per clock cycle, several approaches can be taken, as illustrated in Figure 3.5. The pixel stream usually contains horizontally adjacent pixels, so processing these in parallel is usually the simplest. If the operation has horizontal dependencies, however, such processing may not meet timing constraints, and it may be more effective to stream multiple rows in parallel. Of course, this will require buffering multiple rows of pixels to convert the row parallelism to column parallelism. If there are both horizontal and vertical dependencies, then later rows can be delayed until the data is available, as shown on the right.

Stream processing is well suited to low-level image processing operations such as point operations and local filters where processing can be performed during a raster scan through the image. If an operation, for example a filter, requires data from more than one pixel, it is necessary to design local buffers to cache the necessary data. This is because the synchronous nature of stream processing limits memory access to one read or write per memory port per clock cycle. To satisfy this constraint, the algorithms for some operations may need to be significantly redesigned from that conventionally used in software. The fixed timing constraint arising from the constant throughput of stream processing may be overcome using low-level pipelining to distribute a more complex calculation over several clock cycles.

Stream processing with HLS relies on the compiler to detect the raster scan through the image and infer a stream processing model. From the regular access pattern and data dependencies, appropriate buffers need to be inferred for window filters (Diniz and Park, 2000; Dong et al., 2007).

3.3.3.2 Systolic Arrays and Wavefront Processing

The idea of stream processing may be extended further to systolic arrays (Kung, 1985). A systolic array is a one- or two-dimensional array of processors, with data streamed in and passed between adjacent processors at each clock cycle. It is this latter characteristic that gives systolic arrays their name. Since the communication is regular and local, communication overheads are low. A systolic array differs from pipeline processing in that data may flow in any direction, depending on the type of computation being performed. For many systolic arrays, the processing elements are often relatively simple and perform the same operation at each clock cycle. This makes systolic arrays well suited to FPGA implementation.

The difference between a systolic array and a wavefront array is that systolic arrays are globally synchronous, whereas wavefront arrays are asynchronous. A wavefront array therefore requires handshaking with each data transfer. This means that a wavefront array is self-timed and can operate as fast as the data can be processed. Unfortunately, such asynchronous systems are not well suited to implementation on synchronous FPGAs.

Systolic arrays work best where the processing is recursive with a regular data structure. The systolic array effectively unrolls the recursive loop to exploit concurrency in a way that uses pipelining to make use of the regular data structure. This makes such an architecture well suited for matrix operations (Kung and Leiserson, 1978) and for implementing many low-level image processing operations (Kung and Webb, 1986). They are most suited where the operations are relatively simple and regular, such as filtering (Hwang and Jong, 1990; Diamantaras and Kung, 1997; Torres-Huitzil and Arias-Estrada, 2004), connected component labelling (Nicol, 1995; Ranganathan et al., 1995), and computing the Fourier transform (Nash, 2005). However, a single array is not well matched for implementing a complete application-level image processing algorithm. The streamed nature of the inputs and outputs means that a systolic array processor would fit well with stream processing.

3.3.3.3 Random Access Processing

In contrast with stream processing, random access processing allows pixels to be accessed from anywhere in memory as needed by an operation. Consequently, the data access pattern has no explicit constraints. Random access requires the image to be available in a frame buffer and relaxes the hard timing constraint associated with stream processing. If necessary, several clock cycles may be used to process each pixel accessed or output.

Random access processing is therefore most similar to software processing. Consequently, it is generally easier to map a software algorithm to random access processing than to stream processing. This is the mode implicitly used by HLS, unless it can detect a stream-based access pattern. However, simply porting a software algorithm to hardware using random access processing will generally give disappointing performance. This is because the underlying software algorithm will usually be memory bound, and performance will be limited by memory bandwidth. Where possible, steps must be taken to minimise the number of memory accesses using local buffers. Some of this may be automated by the dependency analysis performed by HLS tools, but some still requires design. It may also be necessary to design an appropriate memory architecture to increase the memory bandwidth. Some of these memory mapping techniques are described in Section 4.2.1.

For complex algorithms, each pixel may take a variable number of clock cycles to process, resulting in an irregular output. This can make it more difficult to synchronise the dataflow between successive operations in the application-level algorithm. This may require synchronisation buffers to manage the uneven data flows. If the inputs and outputs are in a sequential order, then a first-in first-out (FIFO) buffer provides such synchronisation. Otherwise, synchronisation may require a frame buffer.

The clock frequency is generally not constrained for random access. Therefore, the throughput can be increased by maximising the clock frequency using low-level pipelining to reduce the combinatorial delay. Further acceleration can be achieved by examining the dataflow to identify parallel branches and implementing such operations in parallel. This is one of the strengths of HLS, where a compiler directive is used to partially unroll an outer loop to distribute different iterations to separate processors. Each processor will usually have its own local memory to avoid problems with limited bandwidth. However, when considering such parallelism, it is also important to consider not only the operation being parallelised but also the operations before and after it in the processing chain. One overhead to consider is that required to transfer the data from system memory (such as a frame buffer) to local memory and back. If care is not taken, these overheads can remove much of the performance improvement gained from partitioning. Often, the processor will need to be idle during the data transfer phase unless there are sufficient memory resources to enable some form of bank-switching to be used.

3.3.3.4 Massively Parallel Architectures

The extreme case of such data parallelism is to use a massively parallel architecture which has a separate processor for each pixel (or small block of pixels). Algorithms that rely primarily on local information can process whole images in a few clock cycles, regardless of the size of the image. Two limitations of massively parallel architectures are the communication bottleneck and resources required. Communication bottlenecks occur loading data onto and unloading from the processor array and forwarding data between processors for any non-local processing. The large number of processors required limits the processing to relatively small images, even if the processors are relatively simple. As a result of these limitations, massively parallel architectures will not be considered in this book.

3.3.3.5 Hybrid Processing

The best performance can usually be achieved when the whole application can be implemented in a single processing mode, and memory access to a frame buffer is minimised. This may require considerable redesign and development of the operation-level algorithms to make them architecturally compatible. It may also require modifying the application-level algorithm to avoid using operations that do not integrate well with the others.

In some applications, it may not be possible to implement the entire algorithm using stream processing. If it is necessary to combine stream and random access processing, the introduction of one or more frame buffers will usually increase the latency of the algorithm. However, for some operations where each output pixel may depend on data from anywhere within the image, this cannot be avoided. With hybrid processing, frame buffer access may be minimised by maximising the use of stream processing.

3.3.3.6 Computational Architecture Design

The key to an efficient hardware implementation is not to port an existing serial algorithm but to transform the algorithm to produce the computational architecture. There are two stages to this process (before implementation), as shown in Figure 3.6. Note that HLS tools also use these stages as part of the compilation process, so the use of HLS can reduce the need to specifically design the computational architecture.

The first stage is to analyse the algorithm for the underlying algorithmic architecture. In this process, it is important to look at the whole algorithm, not just the individual operations, because gains can be made by considering the interactions between operations, especially if the operations can be made architecturally compatible. Analysing the dataflow is one common method of identifying the underlying computational architecture (Johnston et al., 2006; Sen et al., 2007). Such an analysis not only identifies data dependencies but also potential processing and data bandwidth bottlenecks. In analysing the algorithm, the goal is to identify how the underlying algorithm needs to be transformed to one that can be readily mapped to an efficient hardware implementation.

Several transformations may be applied to the algorithm (Bailey and Johnston, 2010; Bailey, 2011). The first is to convert as many operations as possible to pipelined stream processing to eliminate intervening frame buffers. Any operation that may be implemented by processing the data in a raster-scanned order through the image can be streamed. Some algorithms that are not naturally streamed may be redesigned to use stream processing (for example chain coding, as described in Section 13.3).

The iterations present in many loops may readily be parallelised. Loop unrolling expands loops within an algorithm with multiple copies of the operations within the loop construct. Where there are iteration dependencies, the parallel copies are sequential but allow longer pipelines to be constructed at the expense of additional hardware. With many image processing operations, these innermost loops are inherently parallel (for example weighting the pixels within a linear filter), and it is only the sequential nature of a serial processor that requires a loop. Strip-mining is related to loop unrolling in that it creates multiple parallel copies of the loop body (usually outermost loops). The difference is that each copy operates on a different section of the data. Loop unrolling is one of the most common techniques used by HLS to accelerate an algorithm.

Where there is a series of nested loops, local dependencies both within and between loops can be represented using a polyhedral or polytope model (Lengauer, 1993). The loop indices may be transformed, using an affine

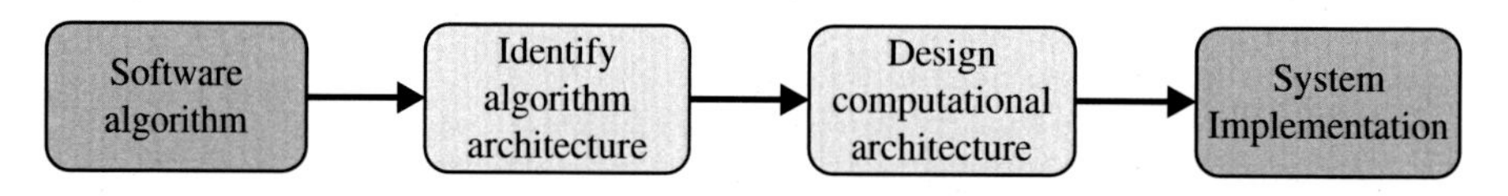

Figure 3.6 Transforming an algorithm to produce the computational architecture.

transformation to re-nest and skew the loops, so that all of the dependencies are between successive iterations of the outermost loop. The polytope representation enables the inner loops to be parallelised with the outermost loop providing the computation schedule. The inner loops can then be allocated to processors, effectively mapping the computation over resources (space) and time (Griebl et al., 1998). When applied to HLS, this can lead to more efficient use of resources and lower latency (Zuo et al., 2013; Campbell et al., 2017).

A less-common transformation is strip-rolling and multiplexing. Sometimes, after strip-mining, only one of the parallel processing units is active in any clock cycle. For example, using stream processing to find the bounding boxes of objects requires operating on multiple objects in parallel. However, a pixel can only belong to one object, so in any clock cycle, only one bounding box processor will be operating. In this case, strip-rolling replaces the multiple processing units with a single processor that is multiplexed between the different instances.

In some algorithms, it is possible to rearrange the order of the operations to simplify the processing complexity or even eliminate some steps. A classic example of this is greyscale morphological filtering followed by thresholding, which is equivalent to thresholding first, followed by binary morphological filtering. The algorithms for binary filtering are significantly simpler than those for greyscale filtering, resulting in considerable hardware savings and often a reduction in latency.

The data can sometimes be coded to reduce the volume and hence the processing speed. Run-length coding is a good example. After the data has been run-length coded, whole runs of pixels can be processed as a group, rather than as individual pixels. For example, Appiah et al. (2008) used run-length coding to accelerate the second pass through the image for connected component labelling.

Gains can often be made by substituting one or more image processing operations with other, functionally similar operations. Substitutions may involve approximating operations with similar but computationally less-expensive operations. Common examples are replacing the L_2 norm in the Sobel filter with the computationally simpler L_1 or L_∞ norm (Abdou and Pratt, 1979) or modifying the coefficients used for colour space conversion to be powers of 2 (Sen Gupta et al., 2004). Note that substituting an operation may change the results of the algorithm. The resulting algorithm will be functionally similar but not necessarily equivalent. In many applications, this does not matter, and the substitution may enable other transformations to be applied to improve the computational efficiency.

The second stage is to design the computational architecture that is implied by the algorithm architecture. This may require further transformation of the algorithm to result in a resource-efficient architecture that can effectively exploit the parallelism available through a hardware implementation. Some of the techniques that may be used for this will be described in more detail in Chapter 4. Operation-specific caching of data ensures that data which is used multiple times will be available when needed. Low-level pipelining and retiming can be used to reduce the combinatorial delay and increase the clock frequency. It is also necessary to select appropriate hardware structures that correspond with data structures and operations used in the software algorithm. HLS will automatically select hardware based on the operations within the algorithm and pipeline the resulting design. The key with HLS is to have an appropriate algorithm for the tools to work on.

The design process can be iterative, with the algorithm and architecture mappings being updated as different design decisions are made. However, with careful design at this stage, the final step of mapping the algorithm onto the architecture to obtain the resultant implementation is straightforward. The speed of HLS can greatly facilitate this stage by allowing rapid exploration of the algorithm to identify hotspots or bottlenecks where the algorithm needs to be transformed. Effort can then be focused on those areas of the algorithm that most benefit from manual redesign.

The key requirement of architecture selection is choosing the most appropriate architecture for the algorithm, both at the computational and system levels. In an embedded application, most of the performance gains come through exploiting parallelism. Power considerations often require maintaining a low clock frequency and minimising the number of memory accesses. Careful design of the image processing algorithm can often allow the low-level vision to operate at the input or output video rate, which is often up to two orders of magnitude slower than current high-end serial processors. Image capture and display almost always operate on streamed data. Therefore, as much processing as possible should be performed on-the-fly as the image is either streamed in or streamed out. This will minimise the number of memory accesses required, and the sequential nature of the accesses will simplify memory system design.

3.4 System Implementation

The final stage within the design process is system implementation. This is where the image processing algorithm is mapped onto the chosen hardware resources as defined by the architecture. The steps here differ significantly from software-based design, where implementation primarily consists of coding the algorithm. Again, it cannot be overemphasised that programming FPGAs is hardware design, not software design. Concurrency and parallelism are implicit in hardware, and the compilers have to build additional control circuitry to make them execute sequential algorithms (Page and Luk, 1991). An FPGA-based implementation requires designing the specific hardware that will perform the required image processing operations. Using an HLS can partially automate this hardware design; however, even when using HLS it is important to consider the hardware implied by the code. Although the languages may look like software, the compilers use the language statements to create the corresponding hardware.

The application-level algorithm is created at a relatively high level of abstraction in terms of individual image processing operations. Some of these operations can be quite complex, and all have their own operation-level algorithms at a lower level of abstraction. Many of these operations can be implemented using several possible algorithms, not all of which necessarily map well to hardware. The operations within the software-based image processing environment will usually have been optimised for serial implementation by software. Simply porting a predominantly serial algorithm onto an FPGA will generally give relatively poor performance. Sequential algorithms will by default run sequentially, usually at a significantly lower clock frequency than on high-end CPUs. Sometimes, the serial algorithm may have a relatively simple transformation to make it suitable for parallel hardware. Many HLS compilers can automatically make such changes to the algorithm to exploit some of the parallelism available, and this is often sufficient to compensate for the lower clock frequency. More often, however, the underlying algorithm may need to be completely redesigned to make it more suitable for FPGA implementation, as described in Section 3.3. In this case, it may be better to first recode the new algorithm within the image processing environment, where it can be tested more easily, rather than attempt to directly develop and test it on the FPGA.

To gain maximum benefit of the parallelism offered by the FPGA, all of the individual operations should be architecturally compatible. If the image processing algorithm has not been developed with the computational architecture in mind, then it may be necessary to modify the algorithm, so that the operations better match the selected architecture. This may require the selection of alternative operations where necessary.

Another necessary task for FPGA-based designs is to develop the interface logic required by peripherals attached to the FPGA. Intellectual property cores may be available for some devices, such as common memory chips and USB ports. For others, these will need to be developed as part of the implementation process. Example interfaces for some specific devices are described in Chapter 6. These hardware interface blocks are the equivalent of device drivers within a conventional operating system. Even if the design uses an embedded CPU running a conventional operating system such as Linux, it will still be necessary to develop the hardware interface between the processor and the external device.

3.4.1 Mapping to FPGA Resources

While it is not necessary to design at the transistor or gate level, it is important to be aware of how the resources of the FPGA are used to implement the design, whether explicitly or implicitly.

Both arithmetic and logic functions are implemented as combinatorial hardware using the configurable logic blocks on the FPGA. Depending on the size of the lookup tables, each logic cell can implement any arbitrary logic function of between three and seven binary inputs, with or without a register on the output. Functions with more inputs are implemented by cascading two or more logic cells.

A simple logic function, such as AND or OR between two N-bit operands may be implemented using N LUTs, one for each bit of the operands. For more complex logic, the number of LUTs required will depend primarily on the number of inputs rather than the actual logic complexity. In many situations, logical inversion may be obtained for free, simply by modifying the contents of the corresponding LUTs.

Addition and subtraction also map onto combinatorial logic. Carry propagation may be implemented using circuits for ripple carry, or more advanced look-ahead carry, or carry-select designs (Hauck et al., 2000). Most

FPGAs have dedicated carry logic that can be used to speed up carry propagation by reducing routing delays and avoiding the need for additional hardware to implement the carry logic. Counters are simply adders (usually adding a constant) combined with a register to hold the current count value.

Relational operators compare the relative values of two numbers. The simplest is testing for equality or inequality, which requires a bitwise comparison using exclusive-NOR, and combining all of the outputs with a single gate to test that all bits in the two numbers are equal. A comparison operation may be implemented efficiently using a subtraction and testing the sign bit. This has longer latency than testing for equality because of the carry propagation. Therefore, if the algorithm involves any loops, tests should be made for equality rather than comparison, if possible, to give the smallest and fastest circuit.

Multiplication can be implemented using a shift-and-add algorithm. If time is not important, this can be implemented serially using not much more logic than an adder. Unrolling the loop results in an array of adders that combines the partial products in parallel. Although, in principle, the length of the carry propagation may be reduced using a Wallace (1964) or Dadda (1965) adder-tree, in practice, this gives a larger and slower implementation than using the FPGA's dedicated carry logic (Rajagopalan and Sutton, 2001). Additional logic is required to perform the carry-save additions, and the structure is not as regular as a simple series of adders, resulting in increased routing delays. If necessary, the parallel multipliers can be pipelined to increase the throughput. Dedicated hardware multipliers found in modern FPGAs are faster and smaller because they do not suffer from FPGA routing delays between stages. However, these multipliers are available only in fixed bit-widths. A single multiplier may readily be used for smaller sized words, but for larger word-widths, either several multipliers must be combined or a narrower multiplier supplemented with programmable fabric logic to make up the width.

In principle, division is only slightly more complex than multiplication, with a combination of shift and subtraction to generate each bit of the quotient. With the standard long division algorithm (for unsigned divisor and dividend), if the result of the test subtraction is positive, the quotient bit is '1', and the result of the test subtraction retained. However, if the test subtraction result is negative, then the corresponding quotient bit is '0', and the test subtraction discarded, restoring the partial remainder to a positive value. This requires that the carry from each subtraction must completely propagate through the partial remainder at each iteration before the sign bit can be determined. The algorithm speed may be improved (Bailey, 2006) using non-restoring division, which allows the partial remainder to go negative. With this scheme, the divisor is added rather than subtracted for negative partial remainders. The number of iterations may be reduced by determining more quotient bits per iteration (using a higher radix). Another approach is to reduce the propagation delay by reducing the length of the critical path. The SRT algorithm (named after Sweeney, Robertson (1958), and Tocher (1958)) achieves this by introducing redundancy in the quotient digit set. This gives an overlap in selecting between adjacent digits, which allows a quotient digit to be determined from an approximation of the partial remainder (hence does not need to wait for the carry to propagate fully). Two disadvantages of the SRT method are that the divisor must be normalised (so that the most significant bit (MSB) is 1) and the quotient digits must be converted to standard binary, both of which require additional logic and time. All of these algorithms may be implemented sequentially to reduce hardware requirements at the expense of throughput or can be pipelined to increase the throughput. There are also many other advanced algorithms (for example (Tatas et al., 2002; Trummer, 2005)). An alternative approach if high-speed multipliers are available is to take the reciprocal of the divisor (using lookup tables or fast iterative methods) and then multiply.

In all of the operations described so far, if one of the operands is always a constant, the hardware can be simplified. This can often reduce both the logic required and the propagation delay of the circuits.

Another logic circuit commonly used in FPGA designs is the multiplexer. These are required to select data from two or more computation paths or when writing to a register or memory from multiple hardware blocks to select the data source (or address) being written. Multiplexers may be generalised to data selection circuits, where a separate select signal is provided for each data source. With such circuits, care must be taken not to simultaneously select multiple sources, as this will combine the data from the multiple sources (using an OR for positive logic or an AND for negative logic with an active low select).

Modern FPGAs also provide memory resources at a range of granularities. At the lowest level, registers may be built up from the flip-flops associated with the logic blocks. In most devices, each logic cell has one flip-flop available on its output, so an N-bit register would require N logic cells. The outputs of these flip-flops are always available for use by other logic, so all of the registers in an array would be available at every clock cycle. An array constructed of registers, if accessed using a variable index, would require one or more multiplexers on the output to select the data from the required register. Similarly, writing to an array also requires a decoder on the input to select the register being written.

The LUTs within the logic cells of some FPGAs can also be configured as fabric memory, providing 16, 32, or 64 memory locations, depending on the available size of the LUTs. These are true memories, with the address decoding and data multiplexing performed using dedicated logic within the logic cell. Note that only one bit per logic cell is accessible for either reading or writing per clock cycle. Some FPGAs allow the inputs to an adjacent logic cell to be configured as a second port, enabling two accesses per clock cycle. Multiple logic cells can be combined in parallel to create arbitrary width memories. In doing so, each parallel memory will share the same address inputs. Fabric RAM is best suited for small, shallow memories because any increase in the depth requires the use of external decoding and multiplexing. Examples where fabric RAM would be used include storing coefficients or parameters that do not have to be accessed simultaneously and short FIFO buffers.

Modern FPGAs also have available larger blocks of dual-port RAM. These generally have flexible word widths from 1 to over 64 bits wide. The size makes them well suited for row buffers for image processing, for larger size FIFO buffers, or for distributed data caching. The RAM can also be used as instruction and data memory for internal CPUs (both hard and soft core).

Large FPGAs have sufficient on-chip memory blocks to hold whole images. However, image frame buffers are usually held off-chip. The number of parallel external RAMs (and ROMs) and their size are limited only by available I/O resources. Off-chip RAM can also be used for other large blocks of intermediate data storage or be used by an internal CPU to hold data and instructions for larger programmes.

A software processor, if used, can be implemented either on the FPGA or as a separate chip. SoC FPGAs provide a hardened CPU on chip, avoiding the need for an external CPU or the use of FPGA resources for building a soft CPU. When using an operating system such as an embedded Linux, an external memory for the CPU is essential.

3.4.2 Algorithm Mapping Issues

As mentioned earlier in this chapter, mapping is not simply a matter of porting the set of operation-level algorithms onto the FPGA, unless they were developed with hardware in mind. The key action is to map what the software is doing, by adapting the algorithm to the hardware. In fact, Lim et al. (2013) argue that it is important to actually design the algorithm to exploit the capabilities available on FPGAs, rather than simply adapting an existing software algorithm.

Again, it is important to keep in mind the different levels of abstraction. The application-level algorithm should be thoroughly tested on a software-based platform before undertaking the time-consuming task of mapping the operations to hardware. In some situations, it may be necessary to substitute different image processing operations to maintain a uniform computational architecture. Generally, though, the application-level algorithm should not change much through the mapping process.

Mapping the individual image processing operations is a different story. Operation-level algorithms may change significantly to exploit parallelism. For some operations, the transformation from a serial to parallel algorithm is reasonably straightforward, while for others a completely new approach, and algorithm, may need to be developed.

During the mapping process, it can be worthwhile to consider not just the operations on their own but also as a sequence. It may be possible to simplify the processing by combining adjacent operations and developing an algorithm for the composite operation. An example of this is illustrated in Section 13.5 with connected component analysis. In other instances, it may be possible to simplify the processing by splitting a single operation into a sequence of simpler operations, for example splitting a separable filter into two simpler filters. In some situations, swapping the order of operations can reduce the hardware requirements; for example, a greyscale morphological filter followed by thresholding requires a more complex filter than its equivalent of thresholding first and then using a binary morphological filter.

A major issue facing the algorithm mapping process for embedded, real-time image processing is meeting the various constraints. The four main constraints facing the system designer are timing, memory bandwidth, resource, and power constraints. Techniques for addressing these are discussed in Chapter 4. A key aspect of the system implementation phase is design-space exploration. Once the base design is working, the next step is to optimise the design to meet the criteria within the problem specification. This can explore aspects such as loop unrolling, hardware replication, and pipelining to accelerate processing speeds where required. Design-space exploration is much simpler when using HLS, as this can often be achieved by adding a few synthesis directives. For register transfer level (RTL)-based designs, this usually requires redeveloping the code for each design-space scenario.

3.5 Testing and Debugging

In any application, it is the fortunate developer who can get the system to function as desired on their first attempt. Inevitably, there are errors in either the design or implementation that require the algorithm to be debugged or tuned. There are four main causes for an algorithm not behaving in the intended manner:

- **Design errors**: This encompasses several issues including faulty logic and the algorithm being unsuitable or not sufficiently robust for the task. A rigorous design process should eliminate algorithm failure before reaching the FPGA implementation stage. However, many development applications follow a less-formal design-as-you-go process, where such design errors can be expected!
- **Implementation errors**: Syntax errors such as spelling mistakes, incorrect operations, and misplaced parentheses will usually (but not always) fail to compile. More subtle errors result from the implementation not matching the design. These include errors in the mapping of operations onto the FPGA, such as not correctly taking into account the latency. Particular problems can relate to conflicts between subsystems working in parallel.
- **Timing errors**: These result from the algorithm being logically correct but being clocked too fast for the propagation delays of the logic. Systems operated at the limit of their speed may fail intermittently as setup and hold times of registers are violated as a result of clock jitter or other noise.
- **Tuning errors**: Here, the algorithm is basically correct, but the selected parameter values do not give the intended behaviour. Tuning involves finding the set of parameters that gives the best performance.

3.5.1 Design

As outlined earlier, the design of an image processing algorithm for a particular task is a heuristic process. It requires considerable experimentation to develop the sequence of image processing operations to achieve the desired result. A software-based image processing system provides the level of user interaction required to experiment with different operations and examine the results. The main issue with algorithm design is how to go about testing a complex algorithm. In all but trivial cases, it is impossible to test the application-level algorithm exhaustively. No algorithm is perfect; all will have flaws or limitations. The primary role of testing is therefore to determine the useable limits of the algorithm and determine its robustness.

One issue with testing image and video processing algorithms is the very large input space. A single low-resolution, 256×256 greyscale (8-bit) image has a total of 524,288 bits. This gives $2^{524,288}$ different images. While only a small fraction of these are 'useful' in any sense of the word, and of those, even fewer may be relevant to a particular task, the number of images likely to be encountered in any application is still impractically large to test exhaustively. By today's standards, a 256×256 image is small. Larger images and image sequences (video) contain even more raw data, compounding the problem.

Testing an image processing algorithm therefore only considers a selection of sample images. In addition to a range of typical cases, it is important to use images which test the boundaries of correct operation. When using image processing to detect defects, it is particularly useful to have test images near the classification boundaries. If necessary, the available images may be modified artificially to simulate the effects of changes in lighting, to adjust the contrast, or to modify the noise characteristics.

Test patterns and noise sources may be used to create test images or modify existing sample images to simulate effects which may occur rarely. For example, uneven lighting may be simulated by adding an intensity wedge to an image. Noise sensitivity may be tested by adding increasing levels of random noise to an image to determine the level at which the algorithm becomes too unreliable to be usable.

Where possible, algorithm failure should be avoided by ensuring that the algorithm works before considering FPGA implementation. However, in some applications, the system must be tested online to obtain a suitably wide range of inputs. Before getting this far, the developer should be fairly certain that the algorithm is not going to change significantly. Most of the adjustments should be parameter tuning rather than wholesale algorithm changes (unless of course the testing determines that the algorithm is not sufficiently robust or is otherwise deficient in some way).

3.5.2 Implementation

The mapping from the initial software representation to a hardware implementation can introduce errors within the algorithm. Verifying each operation-level algorithm is simpler than verifying the complete application-level algorithm in one step. This is because it is easier to test each operation more thoroughly when directly providing input to that operation. When testing as a sequence, some states or combinations may be difficult or impossible to test, given the preceding operations.

Standard testbench techniques can be used to exercise the operation through simulation within the FPGA development environment. For individual operations, most of these tests require a much smaller range of possible inputs than the complete application, so most operations can be tested with relatively small datasets rather than with complete images. The FPGA development platform is also implemented on a standard serial computer, so the simulation of a parallel architecture on a serial computer is usually quite time consuming. This is especially the case for clock-accurate simulations of complex circuits with large datasets (images). Therefore, it is better to test as much of each operation as possible using much smaller datasets. If the algorithm for an operation is complex, or the mapping from software to an RTL hardware design is not straightforward, incremental implementation techniques are recommended. For this, each stage or block of the algorithm is simulated as it is completed, rather than leaving testing of the complete operation to the end. This approach generally enables errors in the implementation to be found earlier and corrected more easily.

Conventional simulators operate from a hardware perspective and usually provide the waveforms of selected signals. Although this can be useful for validating timing and synchronisation, it is less useful for algorithm debugging, which is best achieved with a software style debugger. HLS overcomes many of these problems, because the same source code is compiled to give the hardware implementation.

If each operation behaves the same as its software counterpart, then the complete algorithm should also operate correctly. Testing consists of providing one or more images which exercise each aspect of the algorithm and validating that the correct output is being produced.

3.5.3 Common Implementation Bugs

Experience has shown that there are a number of common bugs which occur during the implementation phase. This list is not exhaustive but serves as a guide for investigating possible causes of malfunctions.

3.5.3.1 Inappropriate Word Length

Hardware allows the bit-width of variables to be adjusted to minimise the hardware resources required. A common cause of errors is to underestimate the number of bits required by an application. Two main consequences of having too few bits are over- and underflow. Overflow occurs when the numbers exceed the available range. With most arithmetic operations, the effect of overflow is for the numbers to wrap around, becoming completely meaningless. During debugging and development, additional logic can be added to detect overflow. This requires adding an additional bit, x_{MSB+1}, to a register. Overflow is then detected as

$$Overflow = \begin{cases} x_{MSB+1}, & \text{for unsigned data,} \\ x_{MSB+1} \oplus x_{MSB}, & \text{for signed data.} \end{cases} \tag{3.1}$$

After thoroughly testing the algorithm, the extra test hardware can be removed in the final implementation. Overflow can be corrected by adding extra bits or by rescaling the numbers.

At the other extreme, underflow occurs when too many bits have been truncated from the least significant end. The effect of underflow is a loss of precision. Underflow occurs primarily when working with fixed-point designs or when using multiplication and division. It also shows when reducing the size of lookup tables to reduce resources. It is harder to build hardware to detect underflow. During the development stages (in software), several designs can be compared with different numbers of least significant bits to check that the errors that arise through truncation are acceptable. Underflow can also be affected by the order in which arithmetic operations are performed, especially when using a floating-point representation. Some applications may require one or more additional least significant guard bits during intermediate steps to reduce the loss of precision.

Both over- and underflow reflect incomplete or inadequate analysis and exploration during the design stages of the project.

3.5.3.2 Wraparound

Closely related to overflow is the wraparound of array indices (unless this is intentional). A similar problem occurs in C programming, especially when using pointers and pointer arithmetic. Apart from the fact that such errors are errors whether in hardware or software, in a hardware system the effects are usually less severe. In software, with its monolithic memory structure, extending past the end of an array will go into following memory and can corrupt completely unrelated variables. In hardware, the effects are usually local. If the array is not a power of 2, then there may be no effect if the hardware past the end of the array does not exist. Otherwise, extending past the end of an array will usually wrap back to the start. The localisation of such errors can make them easier to track and locate in hardware than software. If necessary, additional hardware may be built which detects array-bound errors (and overflow of the index variable).

3.5.3.3 Data Misalignment

Another error relating to data word length is misalignment of data words when performing fixed-point operations. Most hardware description languages have integers as a native type (even if implemented as a bit array) but do not have direct support for fixed-point numbers. It is over to the user to remember the implicit location of the binary point and align the operands appropriately when performing arithmetic operations. The manual processing required can make alignment tedious to check and misalignment bugs difficult to locate. Errors are easily introduced when changing the size or precision of a variable and not updating all instances of its use. This may be overcome to some extent using appropriate parameterised macros to maintain the representation and manage the alignment. The design of an appropriate testbench should detect the presence of any data misalignment errors.

When extending signed numbers (using a two's complement representation), simply padding with zeros will make negative numbers positive. It is necessary to extend the sign bit when extending the most significant bit.

3.5.3.4 Inadequate Arbitration

The parallel nature of FPGA-based designs can result in additional bugs not commonly encountered in software-based systems. One of the most common of these is to write to a register (or access to any other shared resource) simultaneously from multiple parallel threads. The synthesiser will normally build a multiplexer to select the inputs, and if multiple inputs are selected, the result may be a combination of them. Hardware can be added to a design to detect when multiple inputs are selected simultaneously when the multiplexer has been added explicitly to the design (for example using a structural programming style).

Multiplexing errors result from bugs within the control logic, for example failure to build appropriate arbitration logic. Explicitly building in arbitration (see Section 4.3.3) can provide error signals when a parallel thread is blocked from using a resource.

3.5.3.5 Pipeline Synchronisation

Pipelines with multiple parallel data paths can also suffer from synchronisation problems. Temporal misalignment can occur not only within an operation-level algorithm but also within the higher application-level algorithm, making such errors hard to detect. Most hardware description languages treat pipelining as simply another case of parallelism and rely on the developer to implement the correct pipeline control (in particular, priming and flushing). HLS is generally better in this regard.

3.5.3.6 Clock Domain Issues

Synchronisation between clock domains is another complex issue prone to error. Failure to build appropriate synchronisers (see Section 4.1.3) can lead to intermittent failures. Incorrect synchroniser design can lead to hazards being registered in the receiving domain. If throughput is not critical, bidirectional handshaking may be used. The resulting non-uniform data rate on the receiving side of the transfer can cause synchronisation and control issues. For higher throughput (for example sending pixel data), use of a FIFO is essential.

3.5.3.7 Improper RTL Coding Style

Synthesis tools interpret VHDL or Verilog in a particular way, which does not necessarily match the behaviour within simulation tools. Consequently, code that may be legal may not compile to give the intended behaviour. This requires using a 'correct' RTL coding style, particularly with regard to implementation of registers.

3.5.3.8 Incorrect Simulation Models

Only a subset of the VHDL or Verilog is synthesisable; there are many constructs solely used for modelling and simulation. Using such simulation models for synthesis can lead to either compiler errors or, even worse if it compiles, undefined and incorrect behaviour. For intellectual property provided by a third party, it is also important to understand the boundary cases of the provided synthesis models and simulate the design using appropriate simulation models.

3.5.4 Timing

Timing is a critical issue in any high-performance design. ***Timing closure*** is the step of taking a functionally correct design and getting a working implementation that meets the desired operating speed and timing constraints.

Timing closure has generally become more difficult with each successive technology generation. Clock frequencies from video cameras are relatively low, and many early implementations of image processing on FPGA focused on simple image processing operations and small local filters. However, the resolution of digital cameras has increased significantly in recent years. To maintain video frame rates, this has required a corresponding increase in pixel clock frequency by an order of magnitude or more. The complexity of image processing algorithms has also grown. As a result, bigger and faster designs are being implemented on bigger and faster FPGAs, making timing closure more challenging.

There are two aspects of timing closure: one is ensuring that the design on the FPGA can be clocked at the desired rate, and the other is meeting the timing constraints of devices external to the FPGA. As device scales have shrunk, an increasing proportion of the propagation delay between registers on the FPGA comes from the routing (Altera, 2009). With high clock frequencies and large FPGAs, the propagation delay from one side of the chip to the other can exceed the clock period. This can even require pipelining on long routing connections. Therefore, it is essential to keep related logic together to minimise propagation delay. When interfacing with peripherals external to the FPGA, it is important to have the logic placed relatively close to the associated I/O pins. I/O drivers also introduce a significant propagation delay, as do the relatively long tracks on the printed circuit board (PCB) between the FPGA and peripheral. For high-speed devices, such as double data rate (DDR) memory, even the skew between parallel signal lines becomes significant (this is why many high-speed peripherals are going to even higher speed serial connections with the clock embedded with the signal). Fortunately, many of these problems, at least within the FPGA, are managed by the FPGA vendor's tools, which can optimise the placement and routing of designs based on the timing constraints specified by the developer.

The tools also provide a timing report which details the timing characteristics of the design. Such a report is based on a ***static timing analysis***; after the place-and-route phase, the propagation delay through each component and routing wire can be modelled and analysed against the timing constraints. The report will identify any timing violations, the associated critical paths, and determine the maximum operating frequency of a design. However, there are three limitations of such an analysis.

The timing analysis is usually based on the worst case conditions. This usually corresponds with the performance of the slowest device, operating at the maximum temperature and minimum power supply voltage. Consequently, passing under these conditions will guarantee that the propagation delay will be shorter than clock period, taking into account register setup times and any clock skew. With modern high-speed devices, it is also necessary to repeat the analysis under best case conditions (Simpson, 2010) (fastest device, operating at the minimum temperature and maximum power supply voltage) to ensure that register hold times are also met, given the worst case clock uncertainty or skew.

The second limitation is that the timing analysis is only as good as the constraints provided by the user. A path without a timing constraint will not be analysed, and any timing violation will not be reported. This can cause a design to fail. Timing constraints are also used by the place-and-route engine to determine where to focus its effort and optimisations during fitting. Changing any of the source files will change the placement seed, resulting in a different initial placement. Changing the timing constraints will, in general, change the order in which paths are optimised, resulting in a different layout and different timing. It is important to specify realistic timing constraints. If too tight, then the compile times can increase significantly as the place-and-route tools try to satisfy the constraints. Timing constraints are also essential when interfacing with external devices. In setting such constraints, it is also important to take into account the propagation delay of the PCB tracks and clock skew.

The third limitation of static timing analysis is that it can only identify errors within a synchronous design. With multiple clock domains, the clocks in each domain are usually relatively asynchronous. Even when operating at the same frequency, unless one clock is locked to the other, the phase relationship between them is unknown and cannot be relied on when directly transferring data from one domain to the other. When the frequencies are different, the relative phase is always changing, so specifying timing constraints between domains is virtually impossible. Therefore, it is essential that appropriate synchroniser logic be built whenever signals traverse clock domains. Suitable synchronisation logic is described in Section 4.1.3.

Designs that pass timing analysis but have timing errors typically do so for two reasons. The first is that an appropriate time constraint was not set for the path. With multi-phase designs, the setting of suitable constraints is complicated by the fact that the clock might not be enabled on every clock cycle. In such cases, it may be perfectly legitimate for the propagation delay to take up to two or three clock cycles before the register is enabled. Specifying such paths as multi-cycle paths in the constraints file aligns the constraints with reality. (It also avoids unnecessary effort on the part of the place-and-route tools trying to reduce the propagation delay smaller than it needs to be.) The second problem is metastability resulting from asynchronous signals. This is caused by inadequate synchronisation of signals between clock domains. Problems caused by asynchronous signals can be much harder to detect and debug because they are often intermittent.

If your design is failing timing constraints, consider the following checklist:

- **Use realistic timing constraints**: Specifying a higher clock frequency will result in a faster design up to a certain point. This comes at the cost of increased optimisation effort giving larger compile times. Setting the desired clock frequency beyond that which can realistically be achieved will drastically increase the compile times, without achieving timing closure. As the device utilisation increases, the achievable clock frequencies tend to reduce because there is less freedom available in optimising the critical paths.
- **Identify multi-cycle or false paths**: These can give false failure warnings. They also cause the compiler to spend unnecessary effort trying to make these paths conform to the timing constraint.
- **Use low-level pipelining**: Deep logic has a long propagation delay and may need to be pipelined to meet performance targets. Also consider pipelining the control path, as this can also add to the propagation delay,

especially with a series of nested conditional statements. Use retiming to distribute propagation delays more evenly within a pipeline or move control calculations to an earlier clock cycle and pipeline the control signals.

- **Reduce reuse**: Assigning to a register from multiple sources will build a multiplexer on the register input, increasing the number of layers of logic. Propagation delay can be decreased if separate registers are used for each instance if the design allows.
- **Reduce high fanout**: A signal that is used in a large number of places may be subjected to excessive delay, especially if it is distributed over a wide area of the chip. Duplicating the driving registers in different parts of the design can reduce the fanout and increase the locality of the design.
- **Register I/O signals**: When working with high-speed signals, both inputs and outputs should be registered within the I/O blocks. Related signals should use the same I/O banks where possible to reduce differences in propagation delay. Some FPGAs have programmable delays in the I/O blocks to balance small differences in delay.
- **Use modular design with incremental compilation**: If the compile times are excessively long, split the design into modules and use incremental compilation. For this, each module is compiled separately, and the layout within modules is not changed when bringing the modules together under the top layer. The reduced compilation times come at the expense of poorer performance for signals spanning modules (Simpson, 2010). Where possible, module inputs and outputs should be registered, so that the only delay between them is the inter-module routing. Incremental compilation can also result in reduced flexibility in optimising the placement and routing of the top layer of the design. Such techniques can therefore become less effective as the design approaches the capacity of the FPGA.

Timing and timing closure are not just considerations at the end of the implementation but need to be considered at all stages of a design, from the initial specification all the way through design and implementation.

3.5.5 System Debugging

Since it is much harder to debug a failing algorithm in hardware than it is in software, it is important to ensure that the algorithm is working correctly in software before mapping it onto the FPGA. It is essential that any modifications made to the operation-level algorithms during the mapping process be tested in software to ensure that the changes do not affect the correct functioning. However, while it is possible to test the individual operations to check that they function correctly, it is not possible to exhaustively test the complete application-level algorithm. In many real-time video processing applications, every image that is processed is different. Failures, therefore, are not generally the result of the implementation of the algorithm but should be intermittent failures resulting from the algorithm not being able to correctly process some images; in other words, the algorithm is not completely robust.

Unfortunately, the intermittent nature of such algorithm failures makes them hard to detect. Many, if not most, of the images should be processed correctly. Therefore, to investigate the failures, it is necessary to capture the images that cause the algorithm to fail. One way of accomplishing this is to store each image or frame before it is processed. Then, when a failure occurs during testing, the input image that caused the failure will be available for testing. Such testing is actually easier to perform in software, either in the image processing development environment or through simulation of the hardware in the FPGA development environment.

Algorithm debugging in hardware is much more complex. Additional user controls may be required to freeze the input image to enable closer examination of what is happening for a particular input. It may be important to view the image after any of the operations, not just the output. This may require stepping through the algorithm one operation at a time, not only in the forward direction but also stepping backwards through the algorithm to locate the operation that is causing the algorithm to fail. The additional hardware to support these controls is illustrated in Figure 3.7. Video output may be required for debugging, even if the application does not explicitly require it.

Note that if stream processing is used, it is not necessary to have a frame buffer on the output; the output is generated on-the-fly and routed to the display driver. However, non-streamed operations, or non-image-based information, may require a frame buffer to display the results.

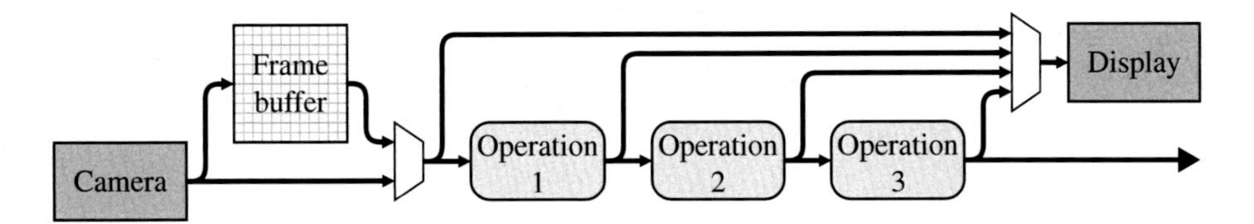

Figure 3.7 Instrumenting an image processing algorithm for debugging. The frame buffer enables the input image to be frozen and fed through the algorithm. The results of any of the steps may be routed to the display (stream processing is assumed here).

At a lower level, debugging the internals of an operation is more difficult. Again, it is emphasised that such debugging should be performed in software or through simulation if possible. The advantage of software debugging is that in software each clock cycle does relatively little compared with hardware where a large number of operations are performed in parallel, and the system state can change considerably (Graham, 2001). Simulation is only as reliable as the simulation model used. If the RTL design is not represented using a proper RTL coding style, it is possible for the simulation to behave differently from the compiled code.

Some methods for debugging a design in hardware are (Graham et al., 2001):

- Testing the algorithm with a low clock frequency will remove potential timing issues from the test. This separates testing for functionality from potential clock rate and propagation delay issues.
- Key signals can be routed to unused I/O ports to make them observable from outside the FPGA. An external oscilloscope or logic analyser is then required to monitor the signals.
- Embedding a logic analyser within the design extends this concept further. Both Xilinx (ChipScope (Xilinx, 2008)) and Intel (SignalTap (Intel, 2018)) provide cores that use memory within the FPGA to record internal signals and later read these recorded signals out to an external companion software package. One limitation of including signal probes and additional signal recording hardware is that these can affect the routing on the FPGA. These affect the timing and can even change the critical paths, possibly masking or creating subtle timing problems that may occur when the modifications are either present or absent (Graham, 2001; Tombs et al., 2004).
- Hardware single stepping can be achieved by gating the clock to stop and restart execution (Tombs et al., 2004). A little extra logic to detect key events can provide hardware breakpoints. Iskander et al. (2010) extended this by controlling the process with an on-chip microprocessor that uses a software style debugging interface for debugging the hardware in place. They used the internal configuration access port (ICAP) of Xilinx FPGAs to read selected hardware registers to provide true hardware debugging.
- Configuration read-back can be used to sample the complete state of a design at one instant. The difficulty with read-back is to associate the logical design with the resources mapped on the FPGA. Signal names, or even the circuit, can be altered for optimisation or other reasons, and with fabric RAM, the address pins may be reordered to improve design routability (Graham, 2001). In spite of these limitations, read-back provides the widest observability of the design (Graham, 2001) although at a very low level. Methods based on read-back require that the clock be stopped, making them impractical for timing verification and in particular for testing of real-time systems.

The models can be validated by directly comparing the results of simulation with hardware execution over the full range of situations that may be encountered (Graham, 2001). This is particularly useful for identifying problems (both functional and timing) associated with device drivers or other interfaces. A hardware structure for testing the operation of an algorithm on an FPGA is given in Figure 3.8. First, the host loads the test image or sequence into a frame buffer. The image is then streamed through the operation under test, saving the results back into the frame buffer. The output is then retrieved by the host and compared with the expected output (produced, for example, by running the software version of the algorithm on the host). Any differences indicate potential errors in the implementation. A lack of differences is only significant if the input image exercises all parts of the algorithm.

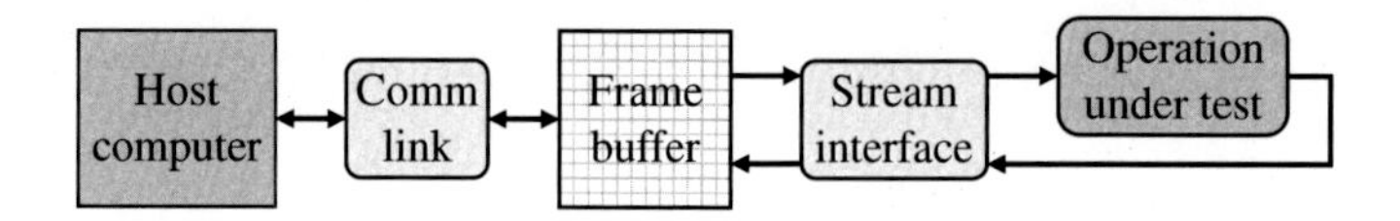

Figure 3.8 Configuration for testing each image processing operation.

3.5.6 Algorithm Tuning

Many application-level algorithms require tuning of parameter values such as threshold levels to optimise their performance. First, a mechanism must be provided for dynamically changing the algorithm parameters. Recompiling the application each time a parameter is changed is very time consuming, so it is not particularly practical. Methods of modifying the parameter through directly manipulating the configuration file (for example using JBits (Guccione et al., 1999) or RapidSmith (Lavin et al., 2010) for Xilinx FPGAs) are not particularly transparent or portable.

Changing parameters dynamically requires building hardware to update the parameters. A standard approach is to provide a set of addressable control and status registers, the basic structure of which is shown in Figure 3.9. These provide a flexible mechanism for directly setting algorithm parameters and can also be used for monitoring the algorithm status. These registers are usually accessed via either a low-speed serial interface or a microprocessor style interface with parallel address and control busses. Intel also provides a mechanism of in-system modification of memories or constants (Intel, 2018). Generally, changes to algorithm parameters should be synchronised to frame boundaries to reduce the occurrence of processing artefacts.

The second key requirement is some form of user interface for adjusting the parameters and visualising the results of the changes. On a hosted system, the host computer can readily provide such a user interface, including displaying the resultant image data; otherwise, it may be beneficial to directly display the image from the FPGA. On a stand-alone system, such a user interface may be implemented through an embedded processor; otherwise, the complete interface will need to be implemented directly in hardware (Buhler, 2007). Note that any tuning parameters will be lost on restarting the application. If they need to be saved, they can be stored in flash memory (or other non-volatile memory) where they can be retrieved by the actual application configuration.

With many applications, it is usually necessary to provide additional information on the context of the parameters being adjusted. This often involves displaying the image or other data structures at intermediate stages within the algorithm, as illustrated in Figure 3.10 for a simple colour tracking application. In this application, the complete algorithm can be implemented using stream processing, without using a frame buffer. However, to set the threshold levels used to detect the coloured targets, it may be necessary to display histograms of red, green, and blue components of the image or of the YCbCr components used for colour thresholding. Since the primary thresholding will be performed in the *Cb–Cr* plane, it is also useful to accumulate and display a two-dimensional *Cb–Cr* histogram of the image. One way of setting the threshold levels is to click on the coloured objects of interest in the input image and use these to initialise or adjust the threshold levels. It is also useful to see the output image after thresholding to see the effects of adjusting the levels, particularly on the boundaries of coloured regions. Similarly, it is also useful to see the effects of the morphological filter in

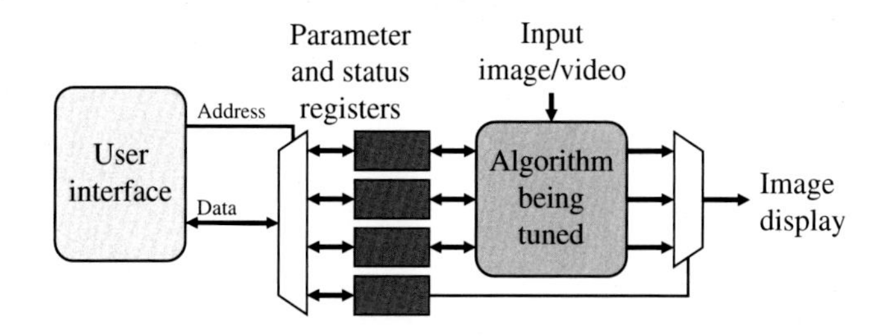

Figure 3.9 Generic structure for algorithm tuning through a set of parameter and status registers.

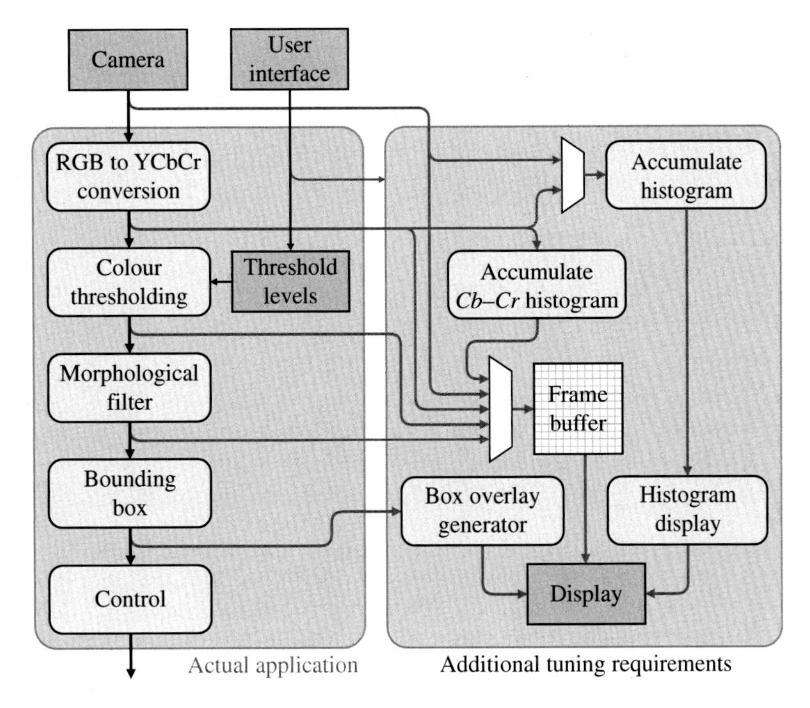

Figure 3.10 Algorithm tuning may require considerable additional hardware to provide appropriate feedback or interaction with the user.

cleaning up the misclassification noise. Finally, it may be useful to include a box generator to overlay the bounding boxes of the detected regions over the top of the original image (or indeed, after any of the other processing stages).

Although this example probably provides more contextual information than is really needed, it does illustrate that the additional logic and resources required to support tuning may actually be more extensive than those of the original application. One way to manage these extra requirements is to use two separate FPGA configuration files: one for tuning, and the other for running the application. The tuning configuration does not need all of the application logic; for example, the final control block in Figure 3.10 may be omitted.

3.5.7 In-field Diagnosis

After the design has been deployed, it may be important to monitor the performance of the system. Depending on the error-handling requirements identified during problem specification, additional actions may be required:

- Error conditions, once detected, can set error flags within status registers, indicating the type and severity of the error. In addition, an error condition may be signalled to the user (either via interrupt or visually via an LED). The status registers can then be examined by the user via the interface described in Figure 3.9 to determine the source and type of error.
- Additional logic may be required to recover from the error state and handle the error. At a minimum, this should reset the logic to a known, stable state.
- If certain errors are anticipated, trace blocks can be designed, which record the history of events leading up to the error. Key signals are recorded in a memory block, usually configured as a circular buffer, so that the window of events prior to the error is recorded. The history can then be downloaded (for example using a mechanism such as in Figure 3.9) and displayed using a logic analyser style of interface for debugging.

- In some cases, no result may be better than an incorrect result. In such cases, the best course of action may be to stop the processing for the current image (or image sequence).
- Images being processed may be streamed to an external memory. In the event of an error, the image may be transferred from memory to more persistent storage for later analysis and debugging.

3.6 Summary

A detailed sequence of steps for the design process is shown in Figure 3.11. The image processing algorithm is first developed and tested using a software-based platform. This is usually based on a rich interactive algorithm development environment.

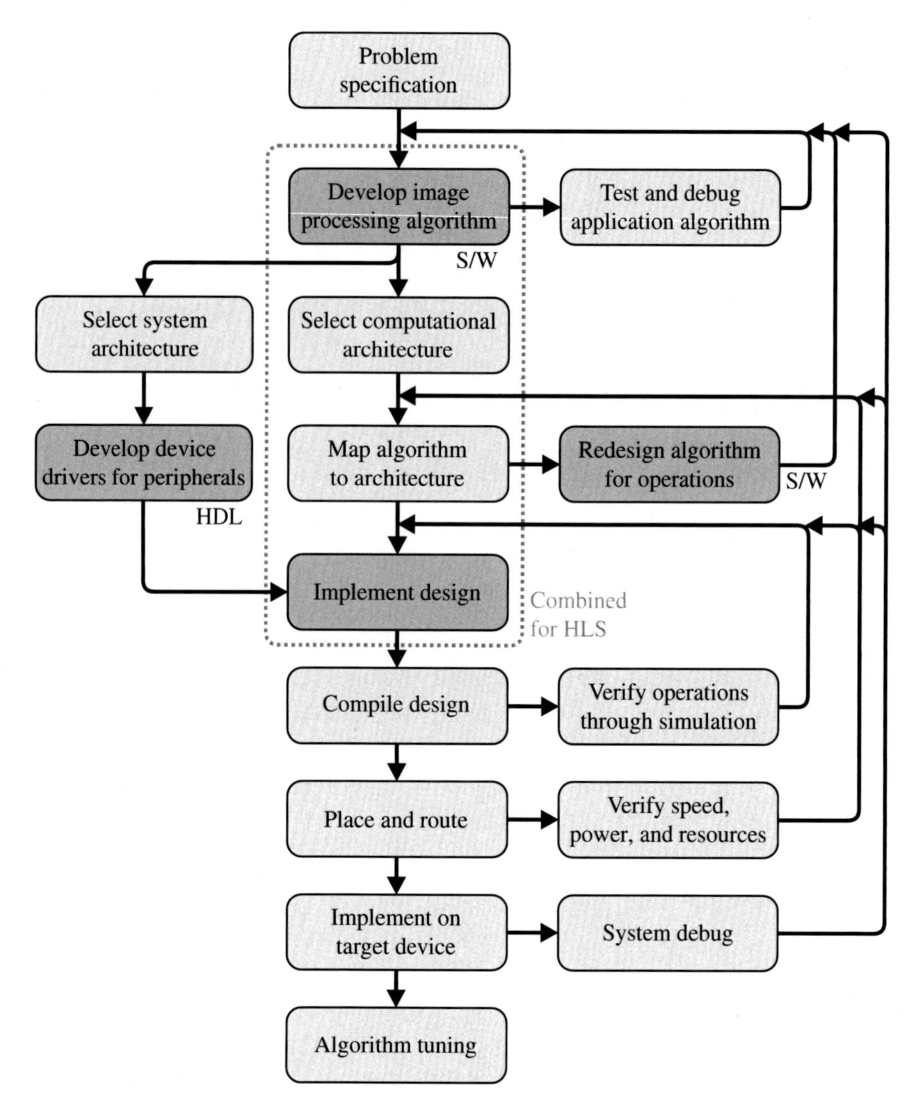

Figure 3.11 Design flow for FPGA-based implementation of image processing algorithms. HDL = hardware description language.

With incremental development, simulation is used to verify that the operation functions correctly as it is implemented. If each operation performs as designed, then the complete application-level algorithm should function as developed and tested within the image processing environment. Note that with HLS, the software used for the initial development can be recompiled for hardware, combining and simplifying many of the steps shown in Figure 3.11.

FPGA vendor tools are used to place-and-route the design onto the FPGA resources. The output from this process will indicate whether there are sufficient resources available on the target FPGA. It also provides timing information based on the actual resources used and can estimate the power used by the design. Feedback from this process is used to modify the algorithms or implementation where necessary to meet timing and power constraints.

The final step is to download the configuration file onto the target device for system testing and algorithm tuning. As part of the debugging process, sections may require re-implementation, remapping, or, in the worst case, redevelopment. It is important that provision for debugging and tuning be considered as part of the design process, rather than something that is added to the implementation at the end.

References

Abdou, I.E. and Pratt, W.K. (1979). Quantitative design and evaluation of edge enhancement / thresholding edge detectors. *Proceedings of the IEEE* **67** (5): 753–763. https://doi.org/10.1109/PROC.1979.11325.

Altera (2009). Timing Closure Methodology for Advanced FPGA Designs. Application note AN-584-1.0. Altera Corporation, USA.

Andrews, D., Niehaus, D., and Ashenden, P. (2004). Programming models for hybrid CPU/FPGA chips. *IEEE Computer* **37** (1): 118–120. https://doi.org/10.1109/MC.2004.1260732.

Appiah, K., Hunter, A., Dickenson, P., and Owens, J. (2008). A run-length based connected component algorithm for FPGA implementation. *International Conference on Field Programmable Technology*, Taipei, Taiwan (8–10 December, 2008), 177–184. https://doi.org/10.1109/FPT.2008.4762381.

Arias-Estrada, M. and Rodriguez-Palacios, E. (2002). An FPGA co-processor for real-time visual tracking. *International Conference on Field Programmable Logic and Applications*, Montpellier, France (2–4 September, 2002), *Lecture Notes in Computer Science*, Volume **2438**, 710–719. https://doi.org/10.1007/3-540-46117-5_73.

Bailey, D.G. (1988). Machine vision: a multi-disciplinary systems engineering problem. *Hybrid Image and Signal Processing*, Orlando, FL, USA (7–8 April 1988), *SPIE*, Volume **939**, 148–155. https://doi.org/10.1117/12.947059.

Bailey, D.G. (1991). Raster based region growing. *6th New Zealand Image Processing Workshop*, Lower Hutt, NZ (29–30 August 1991), 21–26.

Bailey, D.G. (2006). Space efficient division on FPGAs. *Electronics New Zealand Conference (ENZCon'06)*, Christchurch, NZ (13–14 November 2006), 206–211.

Bailey, D.G. (2011). Invited paper: adapting algorithms for hardware implementation. *7th IEEE Workshop on Embedded Computer Vision*, Colorado Springs, CO, USA (20 June 2011), 177–184. https://doi.org/10.1109/CVPRW.2011.5981828.

Bailey, D.G. and Hodgson, R.M. (1988). VIPS - a digital image processing algorithm development environment. *Image and Vision Computing* **6** (3): 176–184. https://doi.org/10.1016/0262-8856(88)90024-8.

Bailey, D.G. and Johnston, C.T. (2010). Algorithm transformation for FPGA implementation. *5th IEEE International Symposium on Electronic Design, Test and Applications (DELTA 2010)*, Ho Chi Minh City, Vietnam (13–15 January 2010), 77–81. https://doi.org/10.1109/DELTA.2010.17.

Batchelor, B.G. (1994). HyperCard lighting advisor. *Machine Vision Applications, Architectures, and Systems Integration III*, Boston, MA, USA (31 October – 2 November 1994), SPIE, Volume **2347**, 180–188. https://doi.org/10.1117/12.188730.

Batchelor, B.G. and Whelan, P.F. (1994). Machine vision systems: proverbs, principles, prejudices and priorities. *Machine Vision Applications, Architectures and Systems Integration III*, Boston, MA, USA (31 October – 2 November 1994), SPIE, Volume **2347**, 374–383. https://doi.org/10.1117/12.188748.

Benitez, D. (2002). Performance of remote FPGA-based coprocessors for image-processing applications. *Euromicro Symposium on Digital System Design*, Dortmund, Germany (4–6 September 2002), 268–275. https://doi.org/10.1109/DSD.2002.1115378.

Brown, S.R. (1987). A note on the description of surface roughness using fractal dimension. *Geophysical Research Letters* **14** (11): 1095–1098. https://doi.org/10.1029/GL014i011p01095.

Brumfitt, P.J. (1984). Environments for image processing algorithm development. *Image and Vision Computing* **2** (4): 198–203. https://doi.org/10.1016/0262-8856(84)90023-4.

Buhler, A. (2007). GateOS: a minimalist windowing environment and operating system for FPGAs. Master of Engineering thesis. Palmerston North, NZ: Massey University. https://doi.org/10179/667.

Cady, F.M., Hodgson, R.M., Pairman, D., Rodgers, M.A., and Atkinson, G.J. (1981). Interactive image processing software for a microcomputer. *IEE Proceedings E: Computers and Digital Techniques* **128** (4): 165–171. https://doi.org/10.1049/ip-e:19810030.

Campbell, K., Zuo, W., and Chen, D. (2017). New advances of high-level synthesis for efficient and reliable hardware design. *Integration* **58**: 189–214. https://doi.org/10.1016/j.vlsi.2016.11.006.

Castleman, K.R. (1996). *Digital Image Processing*, 1e. New Jersey: Prentice-Hall.

Compton, K. and Hauck, S. (2002). Reconfigurable computing: a survey of systems and software. *ACM Computing Surveys* **34** (2): 171–210. https://doi.org/10.1145/508352.508353.

Dadda, L. (1965). Some schemes for parallel multipliers. *Alta Frequenza* **34** (5): 349–356.

DAU (2001). *Systems Engineering Fundamentals*. Fort Belvoir, VA: Defense Acquisition University Press.

Davies, E.R. (2005). *Machine Vision: Theory, Algorithms, Practicalities*, 3e. San Francisco, CA: Morgan Kaufmann.

Diamantaras, K.I. and Kung, S.Y. (1997). A linear systolic array for real-time morphological image processing. *Journal of VLSI Signal Processing* **17** (1): 43–55. https://doi.org/10.1023/A:1007996916499.

Diniz, P.C. and Park, J.S. (2000). Automatic synthesis of data storage and control structures for FPGA-based computing engines. *IEEE Symposium on Field-Programmable Custom Computing Machines*, Napa Valley, CA, USA (17–19 April 2000), 91–100. https://doi.org/10.1109/FPGA.2000.903396.

Dollas, A., Ermis, I., Koidis, I., Zisis, I., and Kachris, C. (2005). An open TCP/IP core for reconfigurable logic. *13th Annual IEEE Symposium on Field-Programmable Custom Computing Machines (FCCM 2005)*, Napa, CA, USA (18–20 April 2005), 297–298. https://doi.org/10.1109/FCCM.2005.20.

Dong, Y., Dou, Y., and Zhou, J. (2007). Optimized generation of memory structure in compiling window operations onto reconfigurable hardware. *3rd International Workshop, Applied Reconfigurable Computing (ARC 2007)*, Mangaratiba, Brazil (27–29 March 2007), *Lecture Notes in Computer Science*, Volume **4419**, 110–121. https://doi.org/10.1007/978-3-540-71431-6_11.

Gonzalez, R.C. and Woods, R.E. (2008). *Digital Image Processing*, 3e. Upper Saddle River, NJ: Pearson Prentice Hall.

Gonzalez, R.C., Woods, R.E., and Eddins, S.L. (2004). *Digital Image Processing using MATLAB*. Upper Saddle River, NJ: Pearson Prentice Hall.

Graham, P.S. (2001). Logical hardware debuggers for FPGA-based systems. PhD thesis. Provo, UT, USA: Brigham Young University.

Graham, P., Nelson, B., and Hutchings, B. (2001). Instrumenting bitstreams for debugging FPGA circuits. *IEEE Symposium on Field Programmable Custom Computing Machines (FCCM'01)*, Rohnert Park, CA, USA (30 April – 2 May 2001), 41–50. https://doi.org/10.1109/FCCM.2001.26.

Gribbon, K.T., Bailey, D.G., and Bainbridge-Smith, A. (2007). Development issues in using FPGAs for image processing. *Image and Vision Computing New Zealand (IVCNZ)*, Hamilton, NZ (5–7 December 2007), 217–222.

Griebl, M., Lengauer, C., and Wetzel, S. (1998). Code generation in the polytope model. *International Conference on Parallel Architectures and Compilation Techniques*, Paris, France (18 October 1998), 106–111. https://doi.org/10.1109/PACT.1998.727179.

Guccione, S., Levi, D., and Sundararajan, P. (1999). JBits: Java based interface for reconfigurable computing. *Military and Aerospace Applications of Programmable Devices and Technologies International Conference*, Laurel, MD, USA (28–30 September 1999), 9 pages.

Hauck, S., Fry, T.W., Hosler, M.M., and Kao, J.P. (2004). The Chimaera reconfigurable functional unit. *IEEE Transactions on VLSI Systems* **12** (2): 206–217. https://doi.org/10.1109/TVLSI.2003.821545.

Hauck, S., Hosler, M.M., and Fry, T.W. (2000). High-performance carry chains for FPGAs. *IEEE Transactions on VLSI Systems* **8** (2): 138–147. https://doi.org/10.1109/92.831434.

Herbordt, M., VanCourt, T., Gu, Y., Sukhwani, B., Conti, A., Model, J., and DiSabello, D. (2007). Achieving high performance with FPGA-based computing. *IEEE Computer* **40** (3): 50–57. https://doi.org/10.1109/MC.2007.79.

Hunt, B.R. (1983). Digital image processing. *Advances in Electronics and Electron Physics* **60**: 161–221. https://doi.org/10.1016/S0065-2539(08)60890-2.

Hwang, J.N. and Jong, J.M. (1990). Systolic architecture for 2-D rank order filtering. *International Conference on Application Specific Array Processors*, Princeton, NJ, USA (5–7 September 1990), 90–99. https://doi.org/10.1109/ASAP.1990.145446.

Intel (2018). *Intel Quartus Prime Standard Edition User Guide: Debug Tools*, Volume UG-20182. Intel Corporation.

Iskander, Y., Craven, S., Chandrasekharan, A., Rajagopalan, S., Subbarayan, G., Frangieh, T., and Patterson, C. (2010). Using partial reconfiguration and high-level models to accelerate FPGA design validation. *International Conference on Field Programmable Technology (FPT 2010)*, Beijing, China (8–10 December 2010), 341–344. https://doi.org/10.1109/FPT.2010.5681432.

Jain, A.K. (1989). *Fundamentals of Image Processing*. Englewood Cliffs, NJ: Prentice Hall.

Johnston, C.T., Bailey, D.G., and Lyons, P. (2006). Towards a visual notation for pipelining in a visual programming language for programming FPGAs. *7th International Conference of the NZ chapter of the ACM's Special Interest Group on Human-Computer Interaction (CHINZ 2006)*, Christchurch, NZ (6–7 July 2006), *ACM International Conference Proceeding Series*, Volume **158**, 1–9. https://doi.org/10.1145/1152760.1152761.

Kalman, R.E. (1960). A new approach to linear filtering and prediction problems. *Journal of Basic Engineering* **82** (1): 35–45.

Kehtarnavaz, N. and Gamadia, M. (2006). *Real-Time Image and Video Processing: From Research to Reality, Synthesis Lectures on Image, Video and Multimedia Processing*. USA: Morgan and Claypool. https://doi.org/10.2200/S00021ED1V01Y200604IVM005.

Kung, S. (1985). VLSI array processors. *IEEE ASSP Magazine* **2** (3): 4–22. https://doi.org/10.1109/MASSP.1985.1163741.

Kung, H.T. and Leiserson, C.E. (1978). Systolic arrays (for VLSI). *Symposium on Sparse Matrix Computations*, Knoxville, TN, USA (2–3 November 1978), 256–282.

Kung, H.T. and Webb, J.A. (1986). Mapping image processing operations onto a linear systolic machine. *Distributed Computing* **1** (4): 246–257. https://doi.org/10.1007/BF01660036.

Lavin, C., Padilla, M., Lundrigan, P., Nelson, B., and Hutchings, B. (2010). Rapid prototyping tools for FPGA designs: RapidSmith. *International Conference on Field Programmable Technology (FPT 2010)*, Beijing, China (8–10 December 2010), 353–356. https://doi.org/10.1109/FPT.2010.5681429.

LeCun, Y. (2012). Learning invariant feature hierarchies. *European Conference on Computer Vision (ECCV 2012) Workshops and Demonstrations*, Florence, Italy (7–13 October 2012), 496–505. https://doi.org/10.1007/978-3-642-33863-2_51.

Lengauer, C. (1993). Loop parallelization in the polytope model. *International Conference on Concurrency Theory (CONCUR'93)*, Hildesheim, Germany (23–26 August 1993), *Lecture Notes in Computer Science*, Volume **715**, 398–416. https://doi.org/10.1007/3-540-57208-2_28.

Lim, Y.K., Kleeman, L., and Drummond, T. (2013). Algorithmic methodologies for FPGA-based vision. *Machine Vision and Applications* **24** (6): 1197–1211. https://doi.org/10.1007/s00138-012-0474-9.

Lubbers, E. and Platzner, M. (2007). ReconOS: an RTOS supporting hard- and software threads. *International Conference on Field Programmable Logic and Applications (FPL 2007)*, Amsterdam, The Netherlands (27–29 August 2007), 441–446. https://doi.org/10.1109/FPL.2007.4380686.

Mignolet, J.Y., Nollet, V., Coene, P., Verkest, D., Vernalde, S., and Lauwereins, R. (2003). Infrastructure for design and management of relocatable tasks in a heterogeneous reconfigurable system-on-chip. *Design, Automation and Test in Europe (DATE'03)*, Munich, Germany (3–7 March 2003), 986–991. https://doi.org/10.1109/DATE.2003.10020.

Miyamori, T. and Olukotun, U. (1998). A quantitative analysis of reconfigurable coprocessors for multimedia applications. *IEEE Symposium on FPGAs for Custom Computing Machines*, Napa Valley, CA, USA (15–17 April 1998), 2–11. https://doi.org/10.1109/FPGA.1998.707876.

Nash, J.G. (2005). Systolic architecture for computing the discrete Fourier transform on FPGAs. *13th Annual IEEE Symposium on Field-Programmable Custom Computing Machines (FCCM 2005)*, Napa, CA, USA (18–20 April 2005), 305–306. https://doi.org/10.1109/FCCM.2005.60.

Ngan, P.M. (1992). The development of a visual language for image processing applications. PhD thesis. Palmerston North, NZ: Massey University. https://doi.org/10179/3007.

Nicol, C.J. (1995). A systolic approach for real time connected component labeling. *Computer Vision and Image Understanding* **61** (1): 17–31. https://doi.org/10.1006/cviu.1995.1002.

Page, I. and Luk, W. (1991). Compiling Occam into field-programmable gate arrays. *Field Programmable Logic and Applications*, Oxford, UK (4–6 September 1991), 271–283.

Pratt, W.K. (1978). *Digital Image Processing*. New York: Wiley.

Rajagopalan, K. and Sutton, P. (2001). A flexible multiplication unit for an FPGA logic block. *IEEE International Symposium on Circuits and Systems (ISCAS 2001)*, Sydney, Australia (6–9 May 2001), Volume **4**, 546–549. https://doi.org/10.1109/ISCAS.2001.922295.

Randen, T. and Husoy, J.H. (1999). Filtering for texture classification: a comparative study. *IEEE Transactions on Pattern Analysis and Machine Intelligence* **21** (4): 291–310. https://doi.org/10.1109/34.761261.

Ranganathan, N., Mehrotra, R., and Subramanian, S. (1995). A high speed systolic architecture for labeling connected components in an image. *IEEE Transactions on Systems, Man and Cybernetics* **25** (3): 415–423. https://doi.org/10.1109/21.364855.

Robertson, J.E. (1958). A new class of digital division methods. *IRE Transactions on Electronic Computers* **7** (3): 218–222. https://doi.org/10.1109/TEC.1958.5222579.

Russ, J.C. (2002). *The Image Processing Handbook*, 4e. Boca Raton, FL: CRC Press.

Sedcole, P., Cheung, P.Y.K., Constantinides, G.A., and Luk, W. (2007). Run-time integration of reconfigurable video processing systems. *IEEE Transactions on VLSI Systems* **15** (9): 1003–1016. https://doi.org/10.1109/TVLSI.2007.902203.

Sen Gupta, G., Bailey, D., and Messom, C. (2004). A new colour-space for efficient and robust segmentation. *Image and Vision Computing New Zealand (IVCNZ'04)*, Akaroa, NZ (21–23 November 2004), 315–320.

Sen, M., Corretjer, I., Haim, F., Saha, S., Schlessman, J., Lv, T., Bhattacharyya, S.S., and Wolf, W. (2007). Dataflow-based mapping of computer vision algorithms onto FPGAs. *EURASIP Journal on Embedded Systems* **2007**: Article ID 49236, 12 pages. https://doi.org/10.1155/2007/49236.

Simpson, P. (2010). *FPGA Design: Best Practices for Team-based Design*. New York: Springer. https://doi.org/10.1007/978-1-4419-6339-0.

Sklyarov, V. (2002). Reconfigurable models of finite state machines and their implementation in FPGAs. *Journal of Systems Architecture* **47** (14–15): 1043–1064. https://doi.org/10.1016/S1383-7621(02)00067-X.

Sklyarov, V. and Skliarova, I. (2008). Design and implementation of parallel hierarchical finite state machines. *2nd International Conference on Communications and Electronics (ICCE 2008)*, Hoi An, Vietnam (4–6 June 2008), 33–38. https://doi.org/10.1109/CCE.2008.4578929.

So, H.K.H. (2007). BORPH: an operating system for FPGA-based reconfigurable computers. PhD thesis. Berkeley, CA, USA: University of California.

Solomon, C. and Breckon, T. (2011). *Fundamentals of Digital Image Processing: A Practical Approach with Examples in Matlab*. Chichester: Wiley-Blackwell. https://doi.org/10.1002/9780470689776.

Steiger, C., Walder, H., and Platzner, M. (2004). Operating systems for reconfigurable embedded platforms: online scheduling of real-time tasks. *IEEE Transactions on Computers* **53** (11): 1393–1407. https://doi.org/10.1109/TC.2004.99.

Tatas, K., Soudris, D.J., Siomos, D., Dasygenis, M., and Thanailakis, A. (2002). A novel division algorithm for parallel and sequential processing. *9th International Conference on Electronics, Circuits and Systems*, Dubrovnik, Croatia (15–18 September 2002), Volume **2**, 553–556. https://doi.org/10.1109/ICECS.2002.1046225.

Therrien, C.W., Quatieri, T.F., and Dudgeon, D.E. (1986). Statistical model-based algorithms for image analysis. *Proceedings of the IEEE* **74** (4): 532–551. https://doi.org/10.1109/PROC.1986.13504.

Tocher, K.D. (1958). Techniques of multiplication and division for automatic binary divider. *Quarterly Journal of Mechanics and Applied Mathematics* **11** (3): 364–384. https://doi.org/10.1093/qjmam/11.3.364.

Todman, T.J., Constantinides, G.A., Wilton, S.J.E., Mencer, O., Luk, W., and Cheung, P.Y.K. (2005). Reconfigurable computing: architectures and design methods. *IEE Proceedings Computers and Digital Techniques* **152** (2): 193–207. https://doi.org/10.1049/ip-cdt:20045086.

Tombs, J., Aguirre Echanove, M.A., Munoz, F., Baena, V., Torralba, A., Fernandez-Leon, A., and Tortosa, F. (2004). The implementation of an FPGA hardware debugger system with minimal system overhead. *14th International Conference on Field Programmable Logic and Application*, Antwerp, Belgium (29 August – 1 September 2004), *Lecture Notes in Computer Science*, Volume **3203**, 1062–1066. https://doi.org/10.1007/b99787.

Torres-Huitzil, C. and Arias-Estrada, M. (2004). Real-time image processing with a compact FPGA-based systolic architecture. *Real-Time Imaging* **10** (3): 177–187. https://doi.org/10.1016/j.rti.2004.06.001.

Trummer, R.K.L. (2005). A high-performance data-dependent hardware integer divider. Masters thesis. Salzburg, Austria: University of Salzburg.

Uber, G.T. (1986). Illumination methods for machine vision. *Optics, Illumination, and Image Sensing for Machine Vision*, Cambridge, MA, USA (30–31 October 1986), SPIE, Volume **728**, 93–102. https://doi.org/10.1117/12.937828.

Vogt, R.C. (1986). Formalised approaches to image algorithm development using mathematical morphology. *Vision'86*, Detroit, MI, USA (3–5 June 1986), 5/17–5/37.

Wallace, C.S. (1964). A suggestion for a fast multiplier. *IEEE Transactions on Electronic Computers* **13** (1): 14–17. https://doi.org/10.1109/PGEC.1964.263830.

Wigley, G. and Kearney, D. (2001). The development of an operating system for reconfigurable computing. *IEEE Symposium on Field-Programmable Custom Computing Machines (FCCM'01)*, Rohnert Park, California, USA (30 April – 2 May 2001), 249–250. https://doi.org/10.1109/FCCM.2001.43.

Will, P.M. and Pennington, K.S. (1972). Grid coding: a novel technique for image processing. *Proceedings of the IEEE* **60** (6): 669–680. https://doi.org/10.1109/PROC.1972.8726.

Williams, J. (2009). *Embedded Linux on Xilinx MicroBlaze, Xilinx University Program and PetaLogix Professor's Workshop Series*. PetaLogix Qld Pty Ltd.

Xilinx (2008). *ChipScope Pro 10.1 Software and Cores User Guide*, Volume UG029. Xilinx Inc.

Zuo, W., Li, P., Chen, D., Pouchet, L., Shunan, Z., and Cong, J. (2013). Improving polyhedral code generation for high-level synthesis. *International Conference on Hardware/Software Codesign and System Synthesis (CODES+ISSS)*, Montreal, Quebec, Canada (29 September – 4 October 2013), 10 pages. https://doi.org/10.1109/CODES-ISSS.2013.6659002.

4

Design Constraints

Chapter 3 described the process of designing an embedded image processing application for implementation on a field-programmable gate array (FPGA). Part of the implementation process involves mapping the algorithm onto the resources of the FPGA. Four types of constraints to this mapping process were identified as timing (limited processing time), memory bandwidth (limited access to data), resource (limited system resources), and power (limited available energy) constraints (Gribbon et al., 2005, 2006). In this chapter, a selection of techniques for overcoming or alleviating these constraints will be described in more detail.

The requirements analysis phase may also identify performance constraints. This chapter concludes with definitions of common performance metrics, both of the implementation and the application.

4.1 Timing Constraints

The data rate requirements of real-time applications impose a strict timing constraint. These are of two types: throughput and latency.

Throughput is the rate at which data must be processed. Stream processing has a strict throughput constraint, usually of one pixel every clock cycle. If this processing rate is not maintained, then data streamed from the camera may be lost, or data streamed to the display may be missing. The clock frequency imposes a timing constraint on the maximum propagation delay between registers within an RTL design. Since the processing per pixel of many image processing operations takes longer than one clock period, low-level pipelining is necessary to satisfy the throughput constraint.

When considering stream processing, it is convenient to distinguish between processes that are source driven, and those that are sink driven, because each has slightly different requirements on the processing structure (Johnston et al., 2008). With a source-driven process, the timing is constrained primarily by the rate at which data is produced by the source. The processing hardware has no direct control over the data arrival; the system needs to respond to the data and control events as they arrive. Conversely, a sink-driven process is controlled primarily by the rate at which data is consumed. An example of this is displaying images and other graphics on a screen.

The other type of timing constraint is latency, which is the maximum permissible time from when the image is captured until the results are output or an action is performed. Latency is particularly important for real-time control applications where image processing is used to provide feedback used to control the state. Delays in the feedback of a control system (caused by the latency of the processing) will limit the response time of the closed loop system and can make control more difficult. In particular, excessive delay can compromise the stability of the system.

Design for Embedded Image Processing on FPGAs, Second Edition. Donald G. Bailey.

Companion Website: www.wiley.com/go/bailey/designforembeddedimageprocessc2e

Whenever processes need to exchange data, they need to be synchronised. This imposes another form of timing constraint. Data transfer is easier when the processes are controlled by the same clock. However, synchronisation is particularly important when the processes are controlled by different (potentially asynchronous) clocks.

4.1.1 Low-level Pipelining

In all non-trivial applications, the propagation delay of the logic required to implement all the image processing operations exceeds the pixel clock period. Pipelining splits the logic into smaller blocks, spread over multiple clock cycles, reducing the propagation delay in any one clock cycle. This allows the design to be clocked faster in order to meet timing constraints.

Consider a simple example. Suppose that the following quadratic must be evaluated for a new pixel value, x, at every clock cycle (a, b, and c are constants):

$$\begin{aligned} y &= ax^2 + bx + c \\ &= (a \times x + b) \times x + c. \end{aligned} \tag{4.1}$$

If implemented within a single clock cycle, as shown in the top panel of Figure 4.1, the propagation delay is that of two multiplications and two additions. If this is too long, the calculations may be spread over two clock cycles, as shown in the middle panel. Register y_1 splits the computation path, allowing the second multiplication and addition to be moved into a second clock cycle. Note that register x_1 is also required to delay the input for the second clock cycle. Without it, the new value of x on the input would be used in the second multiplication, giving the wrong result. The propagation delay per clock cycle is almost halved, although a small additional propagation delay is introduced with the additional register. The two-stage pipeline would enable the clock frequency to be almost doubled. Operating as a pipeline, the throughput is increased as a result of the higher clock frequency. One new calculation begins every clock cycle, even though each calculation now takes two clock cycles. This increase in throughput comes at the expense of a small increase in latency, which is the total time the calculation takes. This is a result of the small increase in total propagation delay.

The registers introduced to construct the pipeline stages may result in a small increase in the number of logic cells required. These registers are implemented using flip-flops within the logic cells. They are typically

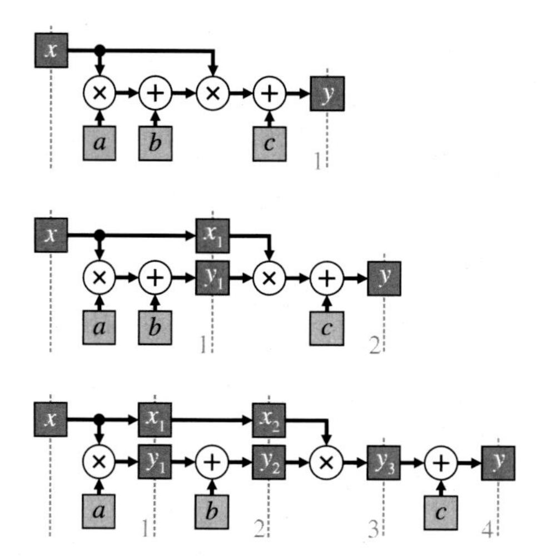

Figure 4.1 Using pipelining for evaluating a quadratic. Top: performing the calculation in one clock cycle; middle: a two-stage pipeline; bottom: spreading the calculation over four clock cycles.

mapped onto unused flip-flops within the logic cells that are already used for implementing the logic. For example, the flip-flops to implement y_1 will most likely be mapped to the flip-flops on the output of the logic cells implementing the adder. However, register x_1 does not correspond to any logic outputs, although it may be mapped to unused flip-flips of other logic within the design.

In the bottom panel of Figure 4.1, the pipeline is extended further, splitting the calculation over four clock cycles. This time, however, the clock period cannot be halved because the propagation delay of the multiplication is significantly longer than that of the addition. Stages 1 and 3 of the pipeline, containing the multipliers, will govern the maximum clock frequency. Consequently, although the maximum throughput can be increased a little, there is a larger increase in the latency because stages 2 and 4 have significant timing slack.

This imbalance in the propagation delay in different stages may be improved through ***retiming*** (Leiserson and Saxe, 1981). This requires moving some of the logic from the multiplication from stages 1 and 3 through registers y_1 and y_3 to stages 2 and 4, respectively. Depending on the logic that is shifted, and how it is shifted, there may be a significant increase in the number of flip-flops required as a result of retiming. Performing such fine-grained retiming manually can be tedious and error prone. Fortunately, retiming can be partially automated and is often provided as an optimisation option within the synthesis tools. Retiming can even move registers within the relatively deep logic created for multipliers and dividers, places that are not easily accessible within the source code. However, automated retiming can make debugging more difficult because it affects both the timing and visibility of intermediate signals.

If retiming is not available, or practical (for example where the output is fed back to the input of the same circuit for the next clock cycle), there are several techniques that can be used to reduce the logic depth. Of the arithmetic operations, division produces the deepest logic, followed by multiplication (unless using a hardened multiplier). Multiplication and division by constants (especially powers of 2) can often be replaced by shifts, or shifts and adds, while modulo operations based on powers of 2 can be replaced by simple bit selection. While long multiplication and division can be implemented efficiently with a loop, if a high throughput is required, the loop can be unrolled and pipelined (Bailey, 2006), with each loop iteration put in a separate pipeline stage. The ripple carry used for wide adders can increase the depth of the logic. Carry propagation can be pipelined using narrower adders and propagating the carry on the next clock cycle. If carried to the extreme, this results in a form of carry-save adder (see (Parhami, 2000) for some of the different adder structures). Greater than and less than comparisons are often based on subtraction. If they can be replaced with equality comparisons, the logic depth can be reduced because the carry chain is no longer required.

It is also important to consider the control flow, and not just the data flow. Quite often, the control flow (conditional statements or detecting the end of loops) is on the critical path. In many cases, it is possible to move some of the control calculation to an earlier clock cycle (retiming) to reduce the critical path and improve timing.

Although a pipeline is a parallel computational structure in that each stage has separate hardware that operates in parallel with all of the other stages, from a dataflow perspective, it is better to think of pipelines as sequential structures (Johnston et al., 2006, 2010). Directly associated with the sequential processing is the ***pipeline latency***. This is the time from when data is input until the corresponding output is produced. This view of latency is primarily a source-driven concept.

From a sink-driven perspective, the latency implies that data must be input to the pipeline at a predetermined time before the output is required. This period is referred to as ***priming*** the pipeline. If the output is synchronous, priming can be scheduled to begin at a preset time before data is required at the output. Unfortunately, this is not an option when data is requested asynchronously by a sink. In this case, it is necessary to prime the pipeline with data and then ***stall*** the pipeline in a primed state until the data is requested at the output. This is discussed in more detail in Section 4.1.2 on synchronisation. Source-driven processes may also be stalled if valid data is not available on every clock cycle.

During priming, the output is invalid, while the valid data propagates through to fill the pipeline. The corresponding period at the end of processing is pipeline ***flushing***, continuing the processing after the last valid data has been input until the corresponding data is output. Note that this use of the term pipeline flushing differs from that used with instruction pipelining of the fetch–decode–execute cycle; there flushing refers to the dumping of the existing, invalid, contents of the pipeline after a branch is taken and repriming with valid data.

The consequence of both priming and flushing is that the clock must be enabled on the pipelined process for more cycles than there are data elements.

4.1.2 Process Synchronisation

Whenever operations work concurrently, it is necessary to coordinate the behaviour and communications between the different processes. This is relatively easy if all of the operations are streamed within a pipeline. However, it is more difficult for random access or hybrid processing modes, especially where the latencies may be different, or where the time taken by the operations may vary. When there is a mix of processing modes, it is often necessary to synchronise the data transfer between operations.

Source-driven processes respond to events as they occur on the input. For video data, the events include new pixels and new-line and -frame events. Based on the event, different parts of the algorithm need to run; for example, when a new frame is started, registers and tables may need to be reset. Processors in a source-driven system are usually triggered by external events. In contrast, sink-driven processes must be scheduled to provide the correct control signals and data on the output at the correct time. Processors in a sink-driven system are often triggered to run at specified times, using internal events.

If all of the external events driving the system are regular, and each operation has a constant processing time or latency, then global scheduling can be used. This can be accomplished using a global counter to keep track of events and the algorithm state as shown in Figure 4.2. The various operations are then scheduled by determining when each operation requires its data and matching the count. An example of global scheduling is processing an image on-the-fly as it is being displayed. Each line is the same length, with regular and synchronous events (the horizontal and vertical sync pulses). This enables the processing for each line to be scheduled to start, so that the pipeline is primed when the data must be available for output to the display. Global scheduling requires all of the operations in the algorithm to be both predictable in their execution time and tightly coupled.

This tight coupling is a disadvantage of global scheduling. For example, if an operation near the start of the pipeline is modified (for example changing the internal pipeline depth as part of design-space exploration), then the schedule for all the downstream operations is affected. This may be overcome by passing control signals along with the data between operations, as illustrated in Figure 4.3. For stream processing with one pixel per clock cycle, and with a horizontal blanking period at the end of each row and a vertical blanking period at the end of each frame, simple *x-valid* (or line valid) and *y-valid* (or frame valid) tags are sufficient to indicate the address of each pixel. It is often convenient to continue using *x-valid* even during the vertical blanking period to assist with counting rows during blanking.

With a continuous pixel stream (no blanking periods), this approach cannot be used because these control signals require at least one blank pixel at the end of each row to indicate the end of row. An alternative is to tag the last pixel in a row (*EOL*) and a frame (*EOF*) to indicate the transition between rows and frames. This approach can also be used with blanking by including an additional data *Valid* tag to show when the data is available, as shown in the bottom panel of Figure 4.3.

These approaches are adequate for source-driven applications. The *Valid* (or *x-valid* ∧ *y-valid*) is used to control the downstream operations or processes by either stalling the process or continuing to run but ignoring the invalid data and marking the corresponding outputs as being invalid. For sink-driven applications, in addition to the *Valid* tag propagated downstream, it is also necessary to have a *Ready* tag propagated upstream, as shown in Figure 4.4. This is asserted to indicate that a processor can accept data on its input; deasserting *Ready*

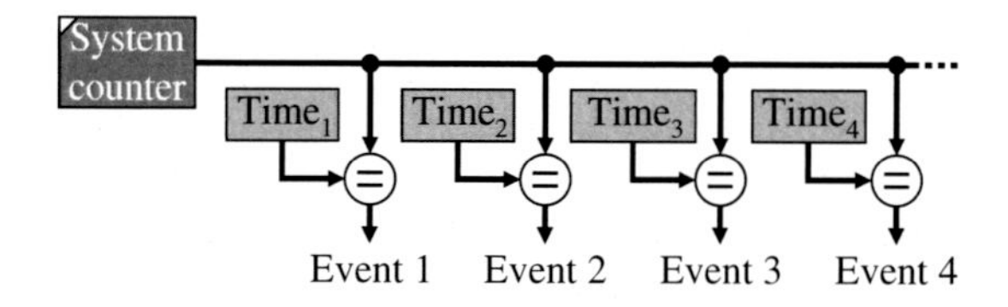

Figure 4.2 Global scheduling derives synchronisation signals from a global counter.

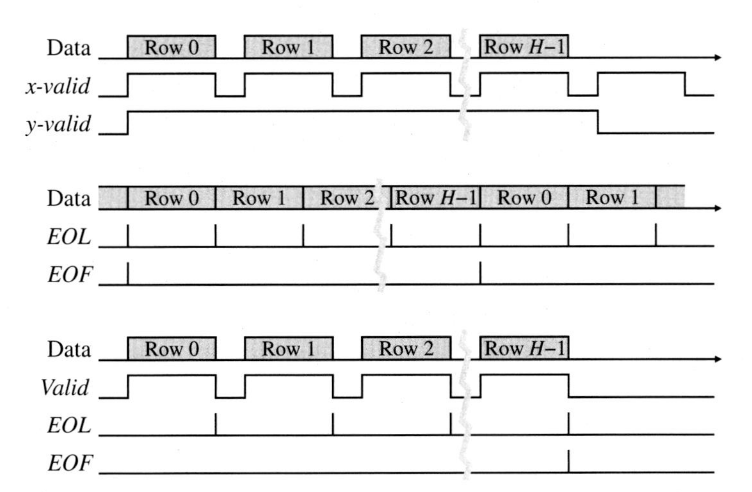

Figure 4.3 Stream control signals. Top: simple stream, with horizontal and vertical blanking; middle: continuous stream, with end-of-line (*EOL*) and end-of-frame (*EOF*) tags on the last pixel; bottom: with blanking, a data *Valid* tag is also required.

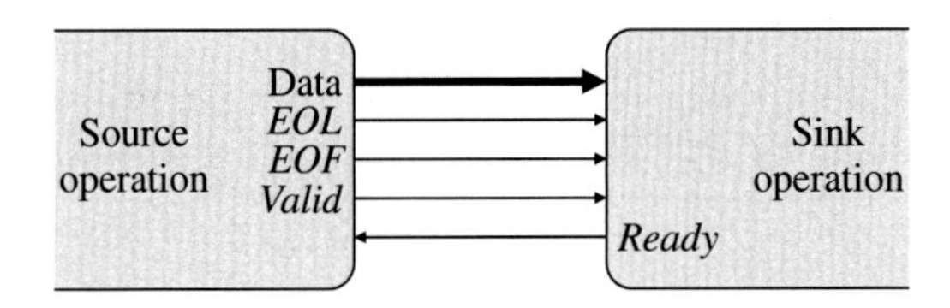

Figure 4.4 Connecting a pixel stream between two operations. *Valid* and *Ready* signals facilitate both source- and sink-driven communication.

requests the upstream process to stop sending data, which may require the upstream process to stall. With such handshaking, data is only transferred when both *Valid* and *Ready* are high, effectively synchronising the source and sink operations. This is also called the communicating sequential processes (CSP) model (Hoare, 1985). With source-driven flow, the downstream operations will stall when data is not available at the input. With sink-driven flow, the upstream operations will automatically prime, and then when the data is not read from the output, the operation will stall waiting for the available data to be transferred. Note that because stalling will block the execution of a process, care must be taken to avoid deadlocks caused by circular dependencies (Coffman et al., 1971).

The AXI4 stream standard (Arm, 2010) follows a similar handshaking mechanism. Rather than an arbitrary data width, it transfers multiple bytes in a transfer, with a *Keep* flag for each byte indicating which bytes contain data. The standard also provides optional destination and identification control signals, allowing multiple data streams to share the same interface. The AXI stream interface is commonly used for connecting stream processes and applications including video applications.

The tight coupling between operations can be relaxed using a first-in first-out (FIFO) buffer between the operations. It allows a source process to continue producing outputs until the FIFO buffer is full before it stalls. Similarly, a sink process will read available data from the FIFO buffer until the buffer is empty before it stalls. In this way, the use of such buffers can simplify the management of pipeline priming and flushing.

The use of FIFO buffers enables operations with variable latency to be combined within a pipeline. In video-based applications, where data is read from a camera or written to a display, the blanking intervals between rows and frames can occupy a significant proportion of the total time. The strict timing of one pixel per clock cycle may be relaxed somewhat by working ahead during the blanking periods (if available), storing the results in a FIFO buffer, which can then be used to smooth out the timing bumps.

4.1.3 Synchronising Between Clock Domains

With some designs, using separate clocks and even different clock frequencies in different parts of a design is unavoidable. This is particularly the case when interfacing with external devices that provide their own clock. It makes sense to use the most natural clock frequency for each device or task, with different sections of the design running in separate clock domains. A problem then arises when communicating or passing data between clock domains, especially when the two clocks are independent, or when receiving asynchronous signals externally from the FPGA.

If the input of a flip-flop is changing when it is clocked (violating the flip-flop's setup or hold times), the flip-flop can enter a metastable state where the output is neither a '0' nor a '1'. The metastable output will eventually resolve to either a '0' or '1', but this can take an indeterminate time. Any logic connected to the output of the metastable flip-flop will be affected, but because the resolution can take an arbitrarily long time, it cannot be guaranteed that the setup times of downstream flip-flops will not also be violated. The probability of staying in the metastable state decreases exponentially with time, so waiting for one clock cycle is usually sufficient to resolve the metastability. A ***synchroniser*** does just this; the output is fed to the input of a second flip-flop, as shown in the circuit in Figure 4.5. The reliability of a synchroniser is expressed as the mean time between failures (Ginosar, 2011):

$$MTBF = \frac{e^{T/\tau}}{T_W f_C f_D}, \tag{4.2}$$

where τ is the settling time constant of the flip-flop, T is the period of the settling window allowed for metastability to resolve; the numerator is the inverse of the probability that any metastability will be resolved within the period T, T_W is the effective width of the metastability window, and is related to the setup and hold times of the flip-flop, f_C is the clock frequency of the synchroniser, and f_D is the frequency of asynchronous data transitions on the clock domain boundary. The denominator of Eq. (4.2) is the rate at which metastable transitions are generated. T is usually chosen so that the $MTBF$ is at least 10 times the expected life of the product (Ginosar, 2003). This is usually satisfied by setting T to be one clock period with a second flip-flop, as in Figure 4.5. If necessary, T can be increased by having more flip-flops in the synchroniser chain.

Note that transitions which occur close to the clock edge may be delayed by a cycle. When transferring a multi-bit word, simply using a separate synchroniser for each bit does not work because the bits may appear in different clock cycles. To guarantee that data is transferred reliably regardless of the relative clock frequencies of the transmitter and receiver, bidirectional synchronised handshaking is required (Ginosar, 2003, 2011). Such a circuit is shown in Figure 4.6. As the data is latched at the sending end, the *Request* line is toggled to indicate that new data is available. This *Request* signal is synchronised using two flip-flops, with a resolution time of one clock period. The XOR gate detects the transition on the *Request* line, which enables the clock on the data latch (the data from the sending clock domain will have been stable for at least one clock cycle and does not require separate synchronisation), and provides a single pulse out on the *Data available* line. An *Acknowledge* signal is sent back to the sending domain where again it is synchronised, and a *Data received* pulse generated. It is only after this signal has been received that new data may be safely latched into the sending register. Otherwise, new data may overwrite the old in the sending data register before the contents have been latched at the receiver.

This whole process takes several clock cycles per data word transferred, limiting the throughput to this rate. If the sending domain is always slower, then the acknowledgement logic may be saved by limiting the input

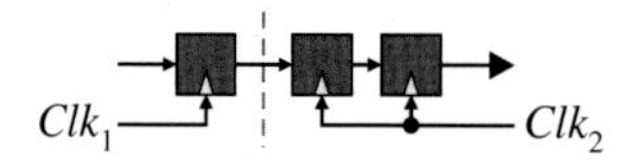

Figure 4.5 Signal synchronisation between clock domains.

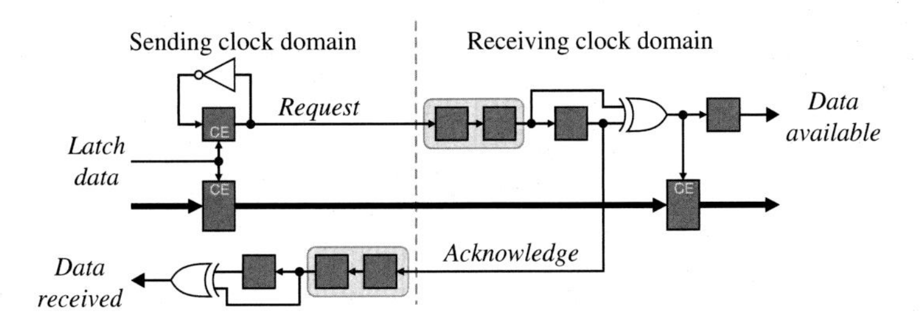

Figure 4.6 Bidirectional handshaking circuit for data synchronisation across clock domains.

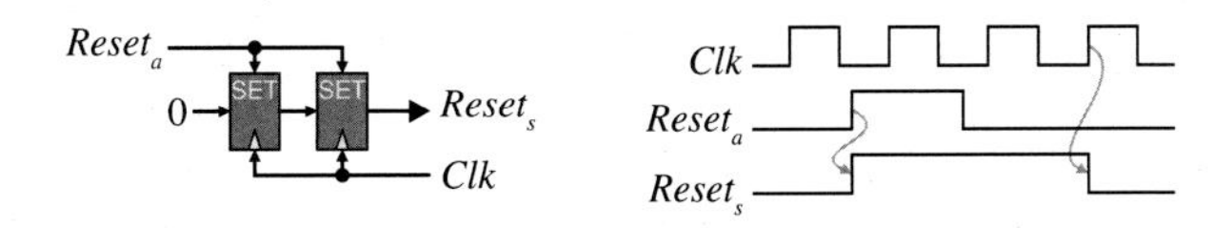

Figure 4.7 Synchronous release of asynchronous resets.

to send every three clock cycles. The danger with this, though, is that if the clock rates are later changed, the synchronisation may no longer be valid (Ginosar, 2003).

To transfer data at a higher rate, it is necessary to use a dual-port RAM as a FIFO buffer (see Section 5.4.1). The RAM has one port in each clock domain and enables data to be queued while the synchronisation signal is sent from one domain to the other. This enables a sustained throughput of one data element per slow clock cycle.

Asynchronous resets also pose synchronisation issues (Cummings and Mills, 2002; Gnusin, 2018). The purpose of a reset signal is to place the circuit into a known state. While asserting a reset asynchronously is not an issue, if the reset signal is released on a clock edge, the flip-flop can become metastable, which is contrary to the purpose of the reset. This is avoided by releasing the reset synchronously as shown in Figure 4.7. The input $Reset_a$ resets the flip-flops asynchronously. When released, the first flip-flop may become metastable, but the second flip-flop is a synchroniser that allows one clock cycle for the metastability to be resolved. The reset output is immediately asserted asynchronously but released in synchronisation with the clock.

4.1.4 I/O Constraints

When interfacing with inputs and outputs, it is essential to specify appropriate timing constraints within the synthesis tools (Gangadharan and Churiwala, 2013). These specify the data signal timing relative to internal or external clocks and account for propagation delays within the external device and also the wiring between the external device and FPGA. A good principle is to register both input and output signals using the flip-flops within the I/O ports to keep the external timing independent of any internal propagation delays.

High-speed interfaces provide their own sets of problems. Careful design and circuit board layout is required to minimise cross-talk, coupling, reflections, and electromagnetic interference (National Semiconductor, 2004). At high speeds, the signals are no longer ideal digital signals and become more analogue. Input sampling appears with an 'eye' pattern, as shown in Figure 4.8. As the clock speed increases, the width of the eye becomes narrower, exacerbating the effects of skew and jitter on the sample timing.

As a consequence, high-speed inputs often have I/O delay elements which can be used to compensate for the effects of skew and ensure that the input is sampled in the centre of the eye. De-skewing the signal requires I/O calibration (Plessas et al., 2011) to set the delays (internal to the high-speed I/O block) for each input line. Initial calibration can be performed during initialisation using a test pattern on the input lines to adjust the I/O delays to optimally sample in the centre of the eye. However, since the skew drifts with changes in device voltage and temperature, very high-speed designs also need dynamic calibration to maintain optimal sampling.

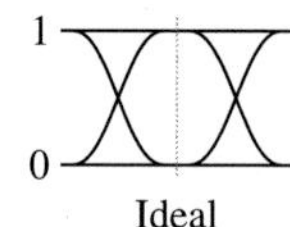

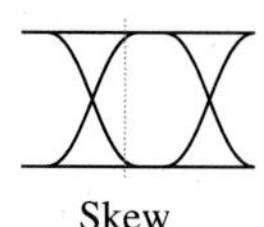

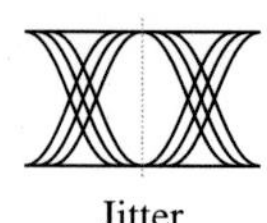

Figure 4.8 Eye patterns from sampling a high-speed signal. The dotted line shows the sample point. Left: ideal case, sampling in the centre of the eye; centre: with signal or clock skew, the sampling moves to the edge of the eye, reducing timing margins; right: with clock jitter, the positions of the transitions become uncertain, reducing the width of the eye.

4.2 Memory Bandwidth Constraints

Closely related to timing constraints are memory bandwidth constraints. Each memory port may only be accessed once per clock cycle. This can be further complicated by pipelined memory architectures, where the address needs to be provided several clock cycles before the data is available. Operations where the next memory location depends on the results of processing at the current location (for example chain coding) exacerbate this problem.

On-chip memory usually has fewer bandwidth issues because such memory consists of a number of relatively small blocks, each of which can be accessed independently, potentially providing a very wide access bandwidth. While high-capacity FPGAs have sufficient on-chip memory to buffer a single lower resolution image, in most applications, it is poor utilisation of this valuable resource to simply use as an image buffer. For this reason, image frames and other large datasets are more usually stored off-chip, where the memory tends to be in larger monolithic blocks. This effectively places large amounts of data behind limited bandwidth and serialised (at the data-word level) connections. Designs which make use of spatial or data-level parallelism can become bandwidth limited if they share access to off-chip memory. To prevent the memory bandwidth from limiting processing speed, it may be necessary to transfer data from off-chip memory into multiple on-chip memory blocks where more bandwidth is available.

The software approach to image processing is to read each pixel required by an operation from memory, apply the operation, and write the results back to memory. A similar process is repeated for every operation that is applied to the image. This continual reading from and writing to memory may be significantly reduced (or even avoided) using a pipelined stream processing architecture. It is essential to process as much of the data as possible while it passes through the FPGA and to limit the data traffic between the FPGA and external memory. This requires designing an appropriate memory architecture to overcome bandwidth problems. Operations may be modified to cache data that is reused. Row buffering is an example of caching, where pixel values are saved for reuse when processing subsequent rows.

4.2.1 Memory Architectures

In this section, several techniques for increasing the memory bandwidth will be described.

4.2.1.1 Parallel Banks

The simplest way to increase memory bandwidth is to have multiple parallel memory systems. If each memory system has separate address and data connections to the FPGA, then each can be accessed independently and in parallel.

The biggest problem occurs when multiple accesses are required to the same data (for example an image in a frame buffer). The crudest approach is to have multiple copies of the image in separate memory banks. Each bank can then be accessed independently to read the required data. For off-chip memory this is not the most memory-efficient architecture, but it can be effective for on-chip caching, allowing multiple processors independent access to the same data.

With many operations, it is often that adjacent pixels need to be accessed simultaneously. By partitioning the address space over multiple banks, for example using one memory bank for odd addresses and one for even addresses, adjacent addresses may be accessed simultaneously. This can be applied to both the row and column addresses, so, for example, four banks could be used to access the four pixels within any 2×2 block (see the top panel of Figure 4.9). Rather than simply partitioning the memory, the addresses may be scrambled to enable a larger number of useful subsets to be accessed simultaneously (Lee, 1988; Kim and Kumar, 1989). For example, the bottom panel of Figure 4.9 enables four adjacent pixels in either a row or a column to be accessed simultaneously. It also allows the main diagonal and the anti-diagonal of a 4×4 block and many (but not all) 2×2 blocks to be accessed concurrently (Kim and Kumar, 1989). It is not possible to have all 4×1, 1×4, and 2×2 blocks simultaneously accessible with only four memory banks. This can be accomplished with five banks, although the address logic is considerably more complex as it requires division and modulo 5 remainders (Park, 1986). In addition to the partitioning and scrambling logic, a series of multiplexers is also required to route the address and data lines to the appropriate memory bank.

4.2.1.2 Packing

A further approach to increasing bandwidth is to increase the word width of the memory. Each memory access will then read or write several pixels. This is effectively connecting multiple banks in parallel but using a single set of address lines common to all banks. Additional circuitry, as shown in Figure 4.10, is usually required to pack the multiple pixels before writing to memory, and to unpack the pixels after reading, which can increase the propagation delay or latency of this scheme. Data packing is effective when pixels that are required simultaneously or on successive clock cycles are packed together. This will be the case when reading or writing streamed images.

4.2.1.3 Bank-Switching

Another common use of multiple memory banks is bank-switching or a ping-pong buffer. Bank-switching is used between two successive image processing operations to decouple the memory access patterns. This enables

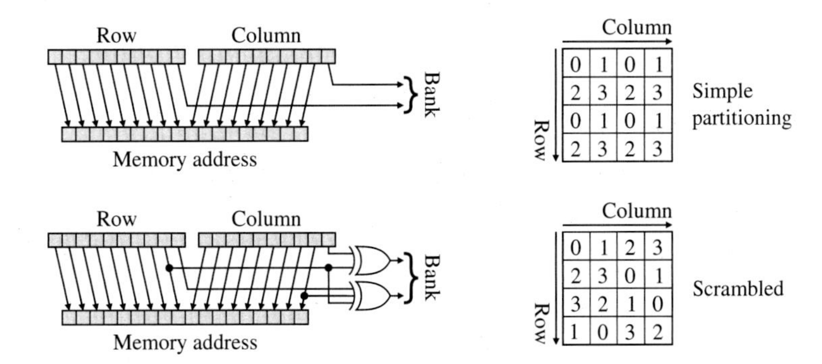

Figure 4.9 Memory partitioning schemes. Top: simple partitioning allows access of 2×2 blocks; bottom: scrambling the address allows row, column, and main diagonal access.

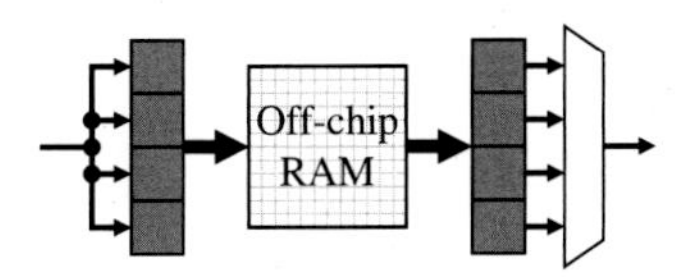

Figure 4.10 Packing multiple pixels per memory location.

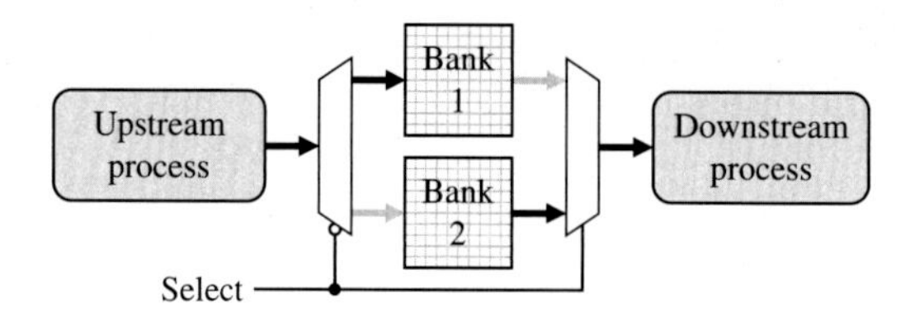

Figure 4.11 Bank-switching or ping-pong buffering.

random access processing operations to be pipelined or to even be integrated within stream processing. It uses two coupled memory banks as shown in Figure 4.11. The upstream process has access to one bank to write its results, while the downstream process accesses data from within a second bank. When the frame is complete, the role of the two banks is reversed (hence, the name ping-pong), making the data just loaded by the upstream process available to the downstream process. Bank-switching therefore adds one frame period to the latency of the algorithm.

If the up- and downstream processes have different frame rates, or are not completely synchronised, then the two processes will want to swap banks at different times. If the banks are switched when one process is not ready, tearing artefacts can result, with part of the frame containing old data, and part new. This problem may be overcome by triple buffering, using a third bank to completely separate the reading and the writing (Khan et al., 2009). This has the obvious implication of increasing the component count of the system and will use a larger number of I/O pins.

4.2.1.4 Multiple Ports

Bandwidth can also be increased using multi-port memory. Here, the increase is gained by allowing two or more concurrent accesses to the same block of memory through separate address and data ports. While this is fine in principle, in practice, it is not used with external memory systems because multi-port memory is more specialised and consequently considerably more expensive.

On-chip memory blocks are usually dual-port. If necessary, a higher number of ports can be emulated using dual-port RAM blocks by having multiple blocks operate in parallel (LaForest et al., 2014).

4.2.1.5 Higher Speed Clock

A more practical approach to mimic multi-port memory is to run the memory at a multiple of the system clock frequency (Manjikian, 2003). The resulting design will allow two or three accesses to memory within a single system clock period. The memory clock should be synchronised with the system clock to avoid synchronisation issues. Double data rate (DDR) memory is one example that allows two data transfers per clock cycle, using both the falling and rising edges of the clock. A higher frequency RAM clock is only practical for systems with a relatively low clock speed or with high-speed memories.

Alternatively, the complete design could be run at the higher clock speed, with the data entering the processing pipeline every several clock cycles. Such approaches are called ***multi-phase*** designs, with several clock cycles or phases associated with the pixel clock. Multi-phase designs not only increase the bandwidth but can also reduce computational hardware by reusing the hardware in different phases.

4.2.2 Caching

A cache memory is a small memory located close to a processor that holds data loaded from an external data source or results that have been previously calculated. It provides rapidly accessible temporary storage for data that is used multiple times. The benefit of a cache is that it has faster access than re-reading the data from external memory (or hard disk) or recomputing the data from original values.

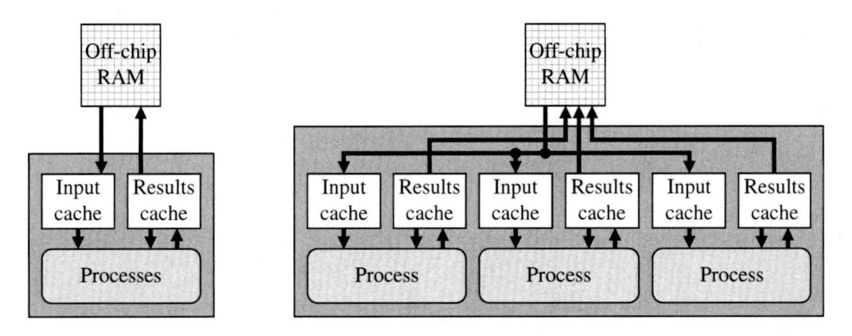

Figure 4.12 Caching as an interface to memory. Left: a single cache is shared between processes; right: data is duplicated in multiple input caches, but each process has a separate results cache.

In software systems, the cache is positioned between the processor and the main memory, as shown in Figure 4.12, with the cache transparent to the processor. Cache misses (data is not available in the cache) require stalling the processor while data is read from memory. This is usually undesirable in FPGA-based systems, especially with stream processing, where such delays can result in loss of data.

In an FPGA, the cache (or local memory) is usually accessed directly, rather than accessing external memory, and the cache controller is responsible for ensuring that the data is available in the cache when required (by preloading the data as necessary). In streaming applications, the cache is usually implemented using a form of FIFO buffer. In this capacity, it can smooth the flow of data between application and off-chip memory, particularly memory with a relatively long latency or access time, or operates in burst mode. On FPGAs, separate caches are often used for input and for holding results to be written back to memory.

The cache can significantly reduce the bandwidth required between the FPGA and memory if the data in the cache is used multiple times (Weinhardt and Luk, 2001). One possible structure for this is an extended FIFO buffer. With a conventional FIFO, only the item at the head of the buffer is available, and when it is read, it is removed from the buffer. However, if the access and deletion operations are decoupled, and the addressing made external so that not just the head of the buffer is accessible, then the extended FIFO can be used as a cache (Sedcole, 2006). With this arrangement, data is loaded into the buffer in logical blocks. Addressing is relative to the head of the buffer, allowing local logical addressing and data reuse. When the processing of a block is completed, the whole block is removed from the head of the FIFO buffer, giving the next block the focus.

With some operations, for example clustering and classifier training algorithms, the task may be accelerated using multiple parallel processors working independently from the same data. Since the block RAMs on the FPGA can all be accessed independently, the cache bandwidth can be increased by having a separate copy of the cached data (Liu et al., 2008) for each process, enabling access to the data without conflict with other processes. Whether there is a single results cache or multiple distributed results caches will depend primarily on how the results are reused by the processes. Figure 4.12 assumes that results are only reused locally, avoiding cache coherency issues with separate caches.

Another arrangement is to have the cache in parallel with the external memory, as shown in Figure 4.13, rather than in series. This allows data to be read from the cache in parallel with the memory, increasing the bandwidth. The cache can be loaded either by copying the data as it is loaded from memory or by the process explicitly saving the data that it may require later into the cache.

4.2.3 Row Buffering

Within image processing, one common form of caching is row buffering. Consider a 3×3 window filter of Figure 4.14, where each output sample is a function of the nine pixel values within the window. Without caching, nine pixels must be read for each window position (each clock cycle for stream processing), with each pixel read nine times as the window is scanned through the image. With stream processing, pixels adjacent horizontally are required in successive clock cycles, so may be buffered and delayed in registers. This reduces the number

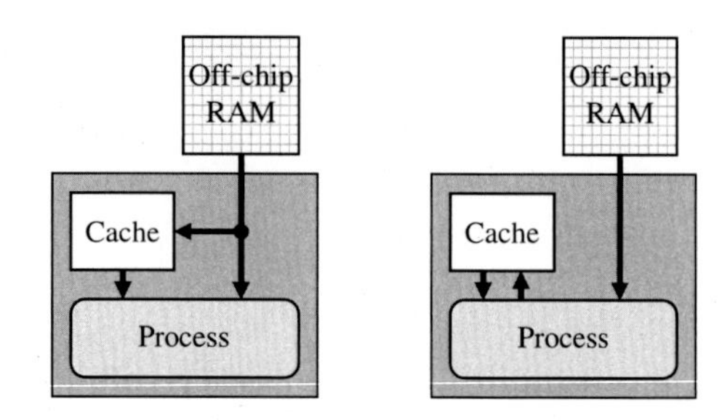

Figure 4.13 Parallel caches.

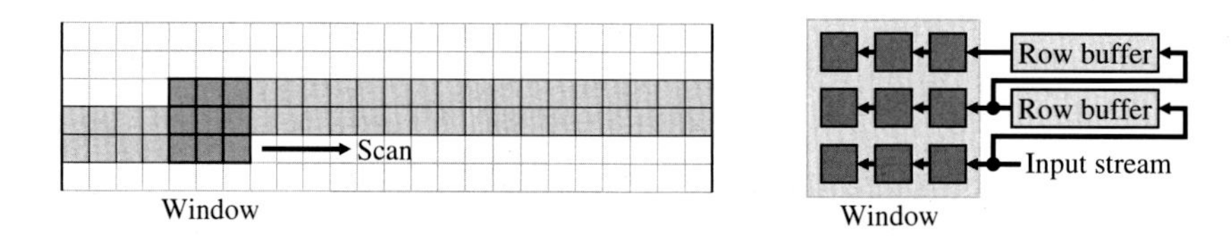

Figure 4.14 Row buffering caches the previous rows, so they do not need to be read in again.

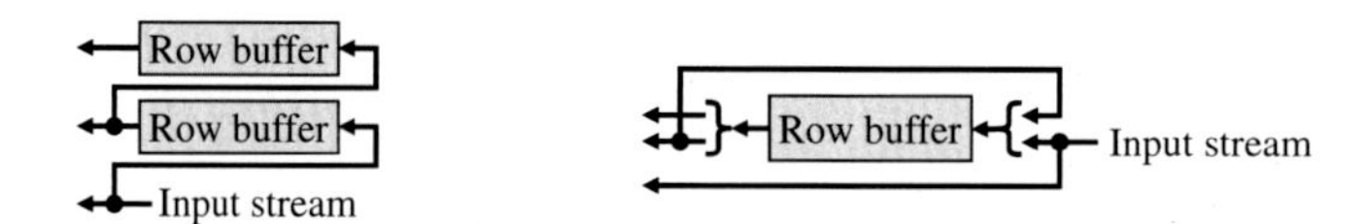

Figure 4.15 Multiple row buffers in parallel can be combined into a single wider row buffer.

of reads to three pixels every clock cycle. A row buffer caches the pixel values of previous rows to avoid having to read these in again. A 3×3 filter spans three rows, the current row and two previous rows; a new pixel is read in on the current row, so two row buffers are required to cache the pixel values for the previous two rows.

Each row buffer effectively delays the input by one row. An obvious implementation of such a digital delay would be to use a W-stage shift register, where W is the width of the image. There are several problems with such an implementation on an FPGA. First, a shift register is made up of a chain of registers, where each bit uses a flip-flop. Each logic cell only has one flip-flop, so a row buffer would use a significant proportion of the FPGA's resources. Some FPGAs allow the logic cells to be configured as shift registers. This would use fewer of the available resources since each cell could provide a delay of between 16 and 64 clock cycles (depending on the lookup table (LUT) size). Second, a shift register is quite power hungry, as multiple bits toggle as the data is shifted along.

A better use of resources is to use a block RAM as a row buffer. Block RAMs cannot be directly configured as shift registers (one exception is Intel's Cyclone and Stratix FPGAs, which can be configured to emulate multi-tap shift registers), although most FPGAs can use them as circular buffers. Rather than have separate address counters, it is often convenient to use the image column index as the address. Note that having multiple buffers in parallel, such as in Figure 4.14, is equivalent to a single buffer but with a wider data path, as shown in Figure 4.15.

4.3 Resource Constraints

The cost of an FPGA generally increases with increasing size (in terms of the number of logic blocks, memory, and other resources). When targeting a commercial application, to minimise the product cost, it is desirable to use an FPGA that is only as large as necessary. However, even the largest FPGA has finite resources. It is therefore important to make efficient use of the available resources. An algorithm that uses fewer resources will generally require less power and have a shorter delay simply because of the shallower logic depth. As the design approaches the capacity of the FPGA, it becomes increasingly harder to route efficiently, and the maximum clock frequency of the design decreases.

A second issue is contention between shared resources (whether shared function blocks, local or off-chip memories). A parallel implementation will have multiple parallel blocks working concurrently. In particular, only one process can access memory, use a shared function, or write to a shared register in any clock cycle. Concurrent access to shared resources must either be scheduled to avoid conflicts, or additional arbitration circuitry must be designed to restrict access to a single process. Arbitration has the side effect of delaying some processes, making it more difficult to determine the exact timing of tasks.

4.3.1 Bit-serial Computation

When resources are constrained, but processing time is not an issue, bit-sequential computation can be used. This performs the opposite of loop unrolling, and spreads the computation over several clock cycles, often calculating one bit of the output per cycle. The hardware for each iteration is effectively reused, significantly reducing resource requirements. Common examples of this are long multiplication and division, and CORDIC operations.

4.3.2 Resource Multiplexing

An expensive resource may be shared between multiple parts of a design using a multiplexer. Three conditions required for this to be practical. First, the cost of the resource must be less than the cost of the multiplexers required to share the resource. For example, sharing an adder is generally not practical because the cost of a separate adder is less than the cost of the multiplexers required to share the adder. However, sharing a complex functional block (such as square root) may well be practical. Second, the resource must only be partially used in each of the tasks using it. If the resource is fully utilised, then sharing a single instance between the multiple tasks will slow the system down and result in variable latency. The third consideration is when the resource is required. If the resource must be used simultaneously in multiple parts of the design, then multiple copies must be used.

When sharing a resource, multiple simultaneous accesses are prohibited. Sharing therefore requires some form of arbitration logic to manage the access to the resource (see Section 4.3.3).

One option when only a few instances of a resource are required at any one time, with instances required by many tasks, is to have a pool of resources, as illustrated in Figure 4.16. However, when the resource pool is shared by a large number of tasks, the multiplexing requirements can become excessive. Alternatives are to connect to the resources via cross-bar switches or busses. Note that the busses are additional resources that must also be shared, because only one task or resource may use a bus at a time.

Sometimes, a careful analysis of a problem can identify potential for resource sharing where it may not be immediately obvious. Consider using stream processing obtaining the bounding boxes of labelled objects (the input is a stream of pixel labels), where each object has a unique label (see Section 13.1). Since the bounding boxes for different objects (labels) may overlap, it is necessary to build the bounding box for each label in parallel. The computational architecture for this has a separate processor for each label, as shown on the left in Figure 4.17. A multiplexer is required on the output to access the data for a particular label for downstream processing. Careful analysis reveals that each pixel has only one label; therefore, a single bounding box processor may be shared, with separate data structures maintained for each label. Rather than multiplexing data registers, an efficient alternative is to use a RAM because the RAM addressing effectively provides the multiplexing for free.

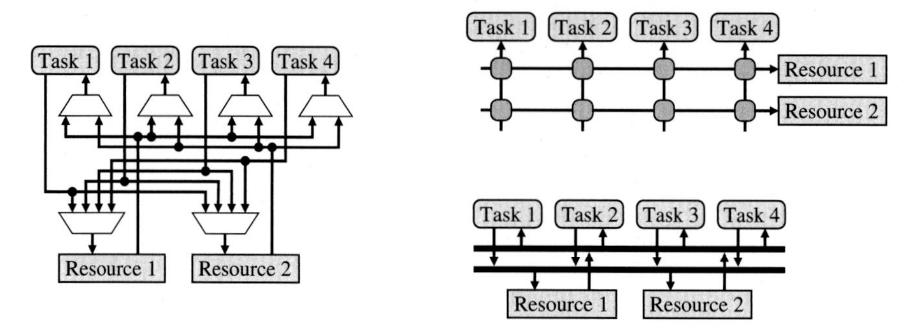

Figure 4.16 Sharing a pool of resources among several tasks. Left: using multiplexers; top right: using cross-bar switches; bottom right: using busses.

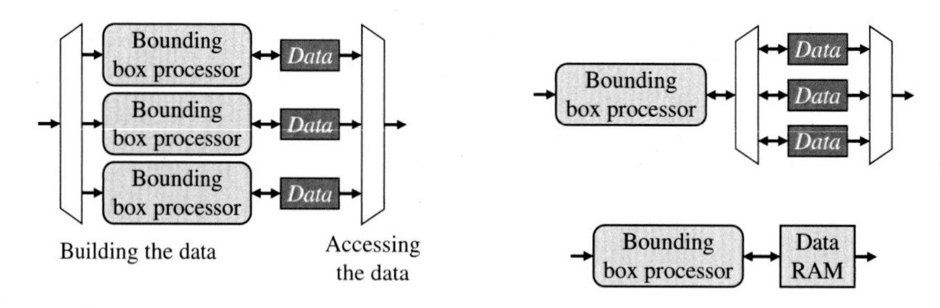

Figure 4.17 Sharing data with a resource. Left: processing each label with a separate bounding box processor; right: sharing a single processor but with separate data registers, which can be efficiently implemented using a RAM.

4.3.2.1 Cross-bar Switches

Ideally, what is required with resource sharing in Figure 4.16 is to reconfigure the connection between the tasks and resources. Unfortunately, in most FPGAs, the routing network is only controlled by the configuration and is not under direct control of the logic. Therefore, the routing must be realised using multiplexers. Each input requires a multiplexer, with the width of the multiplexer dependent on the number of outputs connected. The complexity is therefore proportional to the product of the number of inputs and the number of outputs. This does not scale well as a new input or output is added.

The multiplexers are effectively acting as cross-bar switches. A full switch, enabling any input to connect to any output, requires a full set of multiplexers. However, logic can be reduced significantly using multiple layers of simpler 2×2 cross-bar switches. The reduction in logic (reduced to order $N\log_2 N$ for N connections) comes at the cost of increased latency. A simple butterfly network (as shown in Figure 4.18) requires $\log_2 N$ stages but cannot simultaneously connect every parallel combination (for example if output 4 is connected to input 1, then output 3 cannot connect to input 2 at the same time). A Benes network can connect any permutation but comes at the cost of $2\log_2 N - 1$ stages (Chang and Melhem, 1997).

4.3.2.2 Busses

Busses can also be used to overcome the combinatorial complexity, as shown in the top of Figure 4.19. With a bus, all of the inputs are connected directly to the bus, while the outputs are connected via a tri-state buffer. Enabling an output connects it to the bus, while disabling it disconnects it. The logic scales well, with each new device requiring only a tri-state buffer to connect.

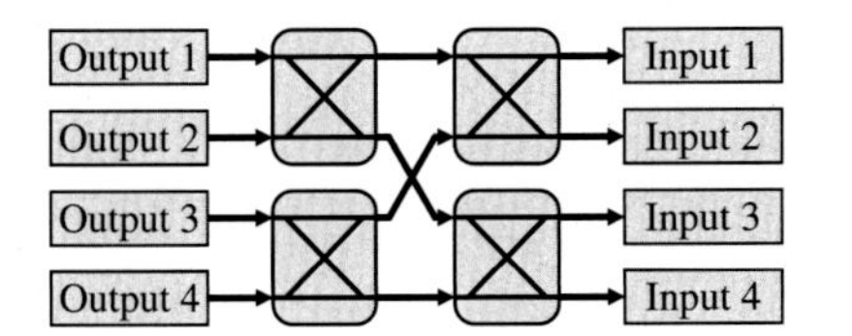

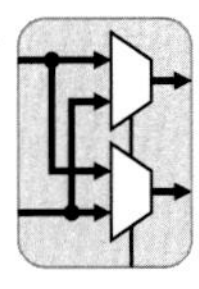

Figure 4.18 Cross-bar switching. Left: butterfly network; right: each cross-bar implemented as a pair of multiplexers.

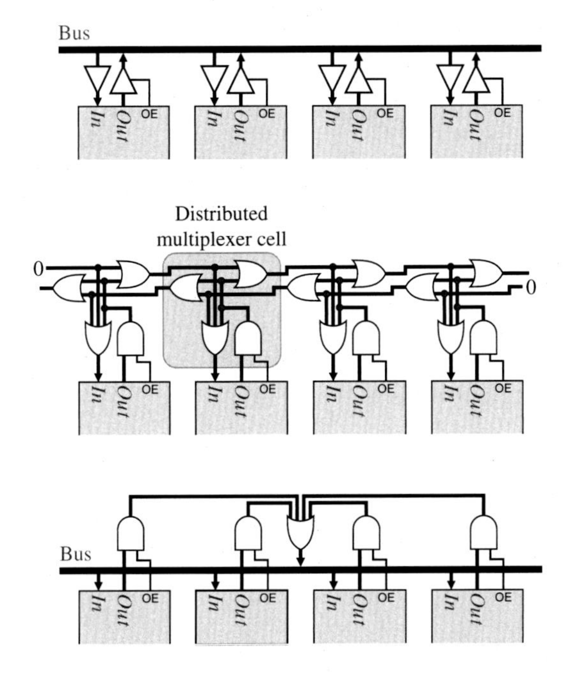

Figure 4.19 Bus structures. Top: connecting to the bus with tri-state buffers; middle: implementing a bus as a distributed multiplexer. Source: Adapted from Sedcole (2006). Bottom: implementing a bus as a shared multiplexer.

This reduction in logic does not come without a cost. First, with multiplexers, each connection is independent of the others. A bus, however, is shared among all of the devices connected to it. While a bus does allow any output to be connected to any input (in fact, it connects to all of the inputs), only one output may be connected at a time. If multiple outputs attempt to connect simultaneously to the bus, this could potentially damage the FPGA (if the outputs were in opposite states). Therefore, an additional hidden cost is the arbitration logic required to control access to the bus.

Unfortunately, not many modern FPGAs have an internal bus structure with the associated tri-state buffers. An equivalent functionality may be constructed using the FPGA's logic resources (Sedcole, 2006), as shown in the middle of Figure 4.19. The bus is replaced by two signals, one propagating to the right, and the other to the left. An enabled output from a device is ORed onto both signals, to connect it to devices in both directions. An input is taken as the OR of both signals. Such an arrangement is effectively a distributed multiplexer; if the expression on the input is expanded, it is simply the OR of the enabled outputs. If multiple outputs are enabled, the signal on all of the inputs may not be what was intended, but it will not damage the FPGA.

The disadvantage of the distributed multiplexer is that it requires the devices to be in a particular physical order on the FPGA. This was not a problem where it was initially proposed, for connecting dynamically reconfigurable functional blocks to a design (Sedcole, 2006). There, the logic for each block connecting to the 'bus' was constrained to a particular region of the FPGA. For general use, a bus may be simply constructed using a shared multiplexer, as shown in the bottom of Figure 4.19.

When a bus is used to communicate between processes, it may be necessary for each process to have a FIFO buffer on its both its input and output (Sedcole, 2006). Appropriately sized buffers reduce the time lost when stalled waiting for the bus. This becomes particularly important with high bus utilisation to cope with time-varying demand while maintaining the required average throughput for each process.

4.3.3 Arbitration

Potential resource conflicts arise whenever a resource is shared between several concurrent blocks. Any task that requires exclusive access to the resource can cause a conflict. Some common examples are:

- **Writing to a shared register**: Only one block may write to a register in any clock cycle. The output of the register, though, is available to multiple blocks simultaneously.
- **Accessing memory, whether reading or writing**: A multi-port memory may only have one access per port. The different ports may write to different memory locations; writing to the same location from multiple ports will result in undefined behaviour.
- **Sending data to a shared function block for processing**: Note that if the block is pipelined, it can simultaneously process several items of data, each at different stages within the pipeline.
- **Writing to a bus**: While a bus may be used to connect to another resource, it is a resource in its own right since only one output may write to the bus at a time.

Since the conflicts may occur between quite different parts of the design that may be executing independently, none of these conflicts can be detected by a compiler. The compiler can only detect potential conflicts through the resource being connected in multiple places, and as a result, it builds the multiplexer needed to connect to the shared resource. The conflicts are runtime events; therefore, they must be detected and resolved as they occur.

With no conflict resolution in place, the default will be for all of the processes to access the resource simultaneously. The multiplexers used to share the resource will prevent any physical damage to the FPGA (although care needs to be taken with busses). However, data from the multiple parallel selections may be combined, and this will generally have undesired results. This is a potential cause of bugs within the algorithm that could be difficult to locate.

With appropriate arbitration, only one process will be granted access to the resource, and the other processes will be denied access. These processes must wait for the resource to become free before continuing. While it may be possible to perform some other task while waiting, this is difficult on an FPGA where the hardware is designed to perform only one task. In some circumstances, it may be possible to queue the access (for example using a FIFO buffer) and continue if processing time is critical. It is more usual, however, to stall or block the process until the resource becomes free.

The simplest form of conflict resolution is scheduled access. With this, conflict is avoided by designing the algorithm such that the accesses are separated or scheduled to occur at different times. In the bank-switch example in Section 4.2.1.3, accesses to each RAM bank are separated in time. While data is being stored in one bank, it is being consumed from the other. At the end of the frame, the control signal is toggled to switch the roles of the banks.

Scheduled access can also be employed at a fine grain using a multi-phase pipeline. Consider the normalisation of the colour components of a YCbCr pixel stream (Johnston et al., 2005) (on the left in Figure 4.20); each of *Cb* and *Cr* are divided by *Y* to normalise for intensity:

$$Cb_n = Cb/Y, \qquad Cr_n = Cr/Y. \tag{4.3}$$

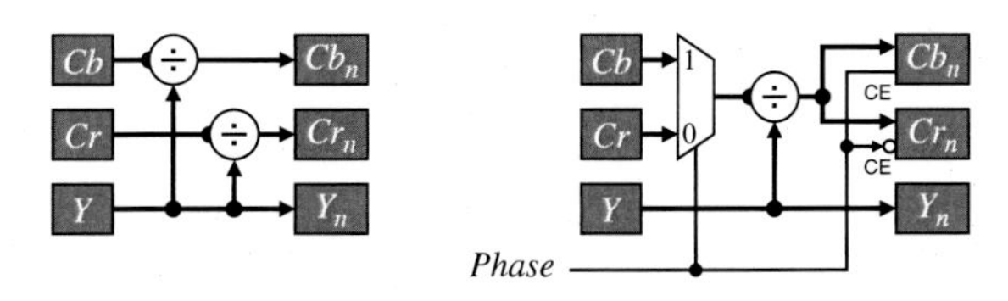

Figure 4.20 Sharing a divider with a two-phase clock.

The divider is a relatively expensive resource, so sharing its access may be desirable. When processing lower resolution images, the pixel clock is often slow compared with the capabilities of modern FPGAs. In such cases, the clock frequency can be doubled, creating a two-phase pipeline. New data is then presented at the input (and is available at the output) on every second clock cycle. However, with two clock cycles available to process every pixel, the divider (and other resources) can now be shared as shown on the right in Figure 4.20. In one phase, Cb is divided by Y and stored in Cb_n, and in the other phase, Cr is similarly processed. If necessary, the divider can be pipelined to meet the tighter timing constraints, since each of Cb and Cr are being processed in separate clock cycles. It is only necessary to use the appropriate phase of clock to latch the results into the corresponding output register.

With stream processing, a shared resource may be available to other processes during the unused horizontal or vertical blanking periods.

However, scheduling requires knowing in advance when the accesses will occur. It is difficult to schedule asynchronous accesses (asynchronous in the sense that they can occur in any clock cycle; the underlying resources must be in a single clock domain). With asynchronous accesses, it is necessary to build an arbiter to resolve the conflict. The type of hardware required depends on the duration of the access. If the duration is only a single clock cycle, for example to provide a clock enable signal for a register, or to select some data, then a simple prioritised acknowledgement may be used. The highest priority request is guaranteed access, while a lower priority request can only gain access if the resource is not being used by a higher priority section. That is,

$$Ack_i = Req_i \wedge \overline{\left(\bigvee_{j=1}^{i-1} Req_j \right)}, \tag{4.4}$$

where a smaller subscript corresponds to a higher priority. With a prioritised access, different parts of the algorithm have different priorities with respect to accessing a resource. The implementation is shown on the left in Figure 4.21. Note that propagation delays of the circuit generating the Req signals mean that there can be hazards on the Ack outputs. If such hazards must be avoided, then the Ack outputs can be registered, delaying the acknowledgement until the following clock cycle.

If the conflict duration is for many clock cycles, then a semaphore is required to lock out other tasks while the granted task has access. This is implemented as a register, Ack_i, on each output, which remembers which task has access. The clock is enabled on the registers only when no request has the semaphore, or when the task with the semaphore releases it by deasserting its request signal (shown on the right in Figure 4.21). With this circuit, once a low-priority task gains the semaphore, it will block a higher priority task. If necessary, the circuit may be modified to enable a higher priority task to pre-empt a lower priority task.

Care must be taken when using semaphores to prevent deadlock. A deadlock may occur when the following four conditions are all true (Coffman et al., 1971):

- Tasks claim exclusive control of the resources they require.
- Tasks hold resources already allocated while waiting for additional resources.
- Resources cannot be forcibly removed by other tasks but must be released by tasks using the resource.
- There is a circular chain of tasks, where each task holds resources being requested by another task.

Deadlock may be prevented by ensuring that any one of these conditions is false. For complex systems, this is perhaps easiest achieved by requiring tasks to request all of the resources they require simultaneously. If not all of the resources are available, then none of the resources are allocated. This requires modification of semaphore

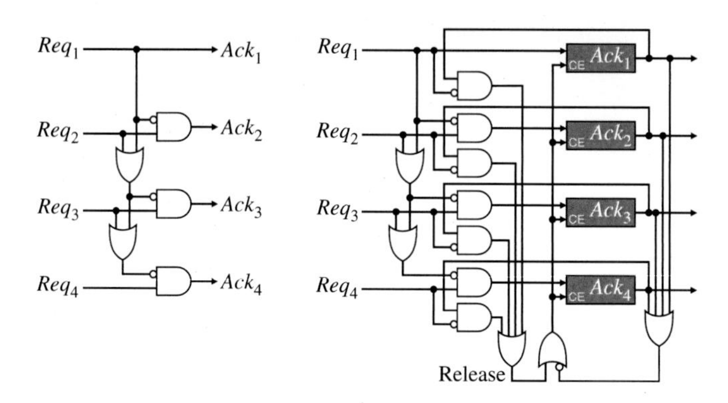

Figure 4.21 Priority-based conflict resolution. Left: a simple prioritised acknowledgement; right: a prioritised multi-cycle semaphore.

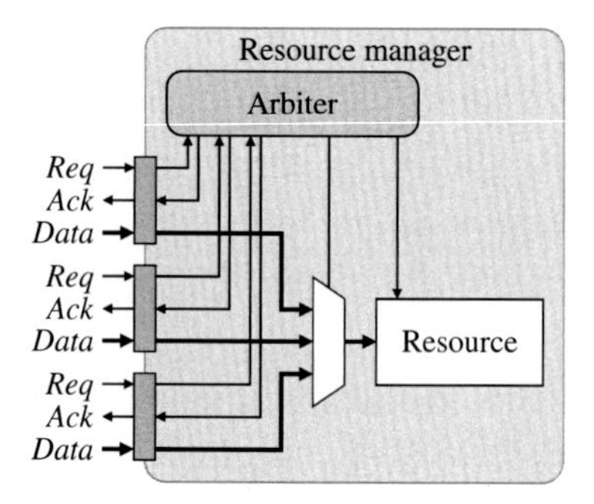

Figure 4.22 Encapsulating a resource and its arbiter within a resource manager simplifies the design.

circuit to couple the requests. Care must also be taken with the priority ordering. For example, consider two tasks that both require the same two resources simultaneously. If the priority order is different for the two resources, then because of the prioritisation, each task will prevent the other from gaining one of the resources, resulting in a deadlock.

4.3.4 *Resource Controllers*

A good design principle is to reduce the coupling as much as possible between parallel sections of an algorithm. This will tend to reduce the number of unnecessary conflicts. However, some coupling between sections is inevitable, especially where expensive resources are shared.

One way of systematically managing potential conflicts is to encapsulate the resource with a resource manager (Vanmeerbeeck et al., 2001) and access the resource via a set of predefined interfaces, as shown in Figure 4.22. The interface should include handshaking signals to indicate that a request to access the resource is granted. Combining the resource with its manager in this way makes the design easier to maintain, as it is only the manager that interacts directly with the resource. It also makes it easier to change the conflict resolution strategy because it is in one place, rather than distributed throughout the code wherever the resource is accessed.

For example, Figure 4.23 shows a possible implementation of the bank-switched memory introduced in Figure 4.11. In this example, the arbitration mechanism is scheduled separate access, so no handshaking is required. The *State* register controls which RAM bank is being written to and which is being read from at any

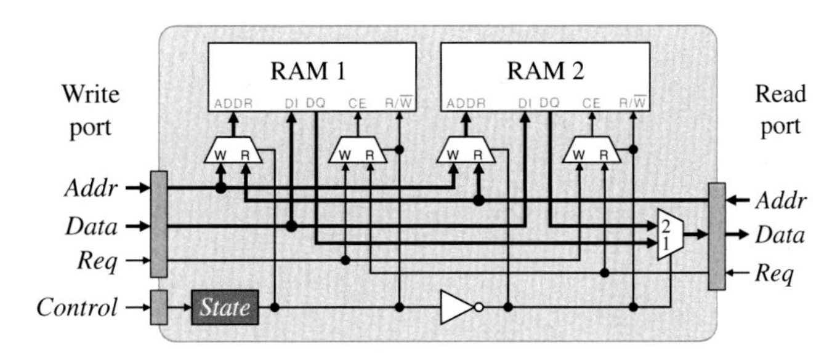

Figure 4.23 Resource manager for bank-switched memories.

one time. The specific banks are hidden from the tasks performing the access. The upstream task just writes to the write port as though it was ordinary memory, and the resource controller directs the request to the appropriate RAM bank. Similarly, when the downstream task reads from the buffer, it is directed to the other bank. The main difference between access via the controller and direct access to a dedicated RAM bank is the slight increase in propagation delay through the multiplexers.

A CSP model (Hoare, 1985) is another approach for managing the communication between the resource controller and the processes that require access to the resource. This effectively uses a channel for each read and write port, with the process blocking until the communication can take place. The blocking is achieved in hardware using a handshake signal to indicate that the data has been transferred.

4.3.5 Reconfigurability

In most applications, a single configuration is loaded onto the FPGA prior to executing the application. This configuration is retained, unchanged, until the application is completed. Such systems are called ***compile-time reconfigurable***, because the functionality of the entire system is determined at compile time and remains unchanged for the duration of the application (Hadley and Hutchings, 1995).

Sometimes, however, there are insufficient resources on the FPGA to hold all of a design. Since the FPGA is programmable, the hardware can be reprogrammed dynamically in the middle of an application. Such systems are termed ***runtime reconfigurable*** (Hadley and Hutchings, 1995). This requires the algorithm to be split into multiple sequential sections, where the only link between the sections is the partially processed data. Since the configuration data for an FPGA includes the initial values of any registers and the contents of on-chip memory, any data that must be maintained from the old configuration to the new must be stored off-chip (Villasenor et al., 1996) to prevent it from being overwritten. Unfortunately, reconfiguring the entire FPGA can take a significant time period; large FPGAs typically have reconfiguration times of tens to hundreds of milliseconds. It is also necessary to take into account reconfiguration overheads, such as fetching the data from external memory, the processor used to perform reconfiguration, and the busses used (Papadimitriou et al., 2011). The reconfiguration latency must be taken into account in any application. With smaller FPGAs, the reconfiguration time may be acceptable (Villasenor et al., 1995, 1996). The latency is usually less important when switching from one application to another, for example between an application used for tuning the image processing algorithm parameters and the application that runs the actual algorithm.

The time required to reconfigure the whole FPGA has spurred research into ***partially reconfigurable*** systems, where only a part of the FPGA is reprogrammed. Partial reconfiguration has two advantages. The first is that the configuration data for part of the FPGA is smaller, so will usually load faster than a complete configuration file. It also has the advantage that the rest of a design can continue to operate, while part of it is being reconfigured. For partial reconfigurability, the bit stream is split into a series of frames, with each frame including an address that specifies its target location within the FPGA (Vipin and Fahmy, 2018). Most of the devices within the Xilinx range (Xilinx, 2018) and newer devices within the Intel range (Intel, 2020) support partial reconfiguration.

Partial reconfiguration requires a modular design, as shown in Figure 4.24, with a static core and a set of dynamic modules (Mesquita et al., 2003). The static core usually deals with interface and control functions

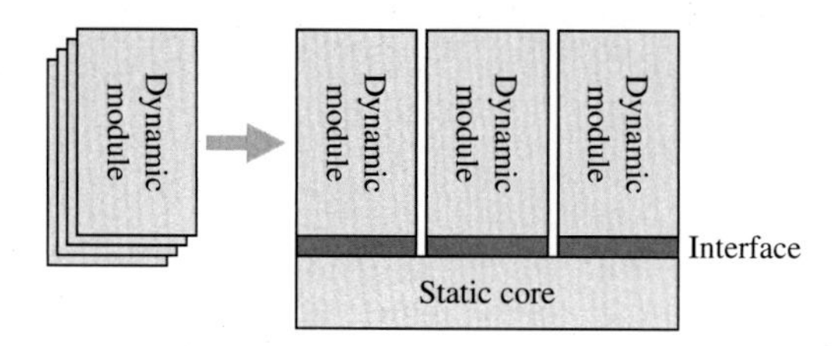

Figure 4.24 Designing for partial reconfigurability. A set of dynamic modules 'plug in' to the static core.

and holds the partial results, while the reconfigurable modules typically provide the computation. The dynamic modules can be loaded or changed as necessary at runtime, a little like overlays or dynamic link libraries in a software system (Horta et al., 2002). Only the portion of the FPGA being loaded needs to be stopped during reconfiguration; the rest of the design can continue executing.

Between the dynamic modules and the static core, a standard interface must be designed to enable each dynamic module to communicate with the core or with other modules (Sedcole, 2006). Common communication mechanisms are to use a shared bus (Hagemeyer et al., 2007) or predefined streaming ports connecting from one module to the next (Sedcole et al., 2003). Both need to be in standard locations to enable them to connect correctly when the module is loaded.

Two types of partial reconfiguration have been considered in the literature. One restricts the reconfigurable modules to being a fixed size that can be mapped to a small number of predefined locations. The other allows modules to be of variable size and be placed more flexibly within the reconfigurable region. The more flexible placement introduces two problems. First, the heterogeneity of resources restricts possible locations that a dynamic module may be loaded on the FPGA. The control algorithm must not only search for a space sufficiently large to hold a module, but the space must also have the required resources in the right place. Second, with the loading and unloading of modules, it is possible for the free space to become fragmented. To overcome this, it is necessary to make the modules dynamically relocatable (Koester et al., 2007). To relocate a module (Dumitriu et al., 2009), it must first be stopped or stalled and then disconnected from the core. The current state (registers and memory) must be read back from the FPGA, before writing the module (including its state) to a new location, where it is reconnected to the core, and restarted.

Partial reconfigurability allows adaptive algorithms. Consider an intelligent surveillance camera tracking vehicles on a busy road. The lighting and conditions in the scene can vary significantly with time, requiring different algorithms for optimal operation (Sedcole et al., 2007). These different sub-algorithms may be designed and implemented as modules. By having the part of the algorithm that monitors the conditions in the static core, it can select the most appropriate modules and load them into the FPGA based on the context.

Simmler et al. (2000) have taken this one step further. Rather than consider switching between tasks that are part of the same application, they propose switching between tasks associated with different applications running on the FPGA. Multi-tasking requires suspending one task and resuming a previously suspended task. The state of a task is represented by the contents of the registers and memory contained within the FPGA (or dynamic module in the case of partial reconfigurability). Therefore, the configuration must be read back from the FPGA to record the state or context. To accurately record the state, the task must be suspended while reading the configuration and while reloading it. However, not all FPGAs support read-back, and therefore, not all will support dynamic task switching.

At present, the tools to support partial dynamic reconfiguration are of quite low level. Each module must be developed with quite tight placement constraints to keep the module within a restricted region and to ensure that the communication interfaces are in defined locations. The system core also requires similar placement constraints to keep it out of the reconfigurable region and to lock the location of the communication interfaces. The partial reconfiguration can be loaded from an external device, or internally from a hard or soft CPU, or even from the programmable logic.

Regardless of whether full or partial reconfigurability is used, runtime reconfigurability can be used to extend the resources available on the FPGA. This can allow a smaller FPGA to be used for the application, reducing both the cost and power requirements.

4.4 Power Constraints

In embedded applications, the power consumption of the system, and in particular of the FPGA, is of importance. The power dissipated by an FPGA can be split into two components: static power and dynamic power. ***Dynamic power*** is the power required for switching signals from '0' to '1' and vice versa. Conversely, ***static power*** is that consumed when not switching, simply by virtue of the device being switched on. Most FPGAs are based on CMOS technology. A CMOS inverter, the most basic CMOS circuit, is shown in Figure 4.25. It consists of a P-channel pull-up transistor and an N-channel pull-down transistor. Only one of the transistors is on at any one time, and since the metal oxide semiconductor (MOS) transistors have a very high off-resistance, CMOS circuits typically have very low static power dissipation.

Any significant power is only dissipated when the output changes state. This results from two main sources (Sarwar, 1997). The first is from charging the capacitance on the output line. The capacitance results from the inputs of any gates connected to the output, combined with the stray capacitance associated with the wiring between the output and any downstream inputs. The power required is given by

$$P = \frac{1}{2}\sum_{n} C_n V_{DD}^2 f_n, \tag{4.5}$$

where the summation is over all of the outputs or nets within a circuit, C_n is the capacitance on each net associated with the output, V_{DD} is the voltage swing (the core supply voltage), and f_n is the frequency of transitions on the output. In a typical FPGA design, 60% of the power is dissipated within the routing, 30% from the logic itself, and 10% from the clock network (Goeders and Wilton, 2012). The power consumption may be reduced by considering each of the terms of Eq. (4.5):

- by minimising the fan-out from each gate (reducing the capacitive load);
- by minimising the use of long wires (line capacitance is proportional to length);
- by reducing the power supply voltage (although the developer has little control over this, it has been reducing with each technology generation);
- by reducing the clock frequency; and
- by limiting the number of outputs that toggle with each clock cycle (including minimising data widths).

Since the core voltage is fixed for a given FPGA family, reducing the clock frequency probably has the most significant effect on reducing the power consumption of a given design. All of the signals toggle in proportion to the clock frequency, so this will reduce not only the power used by the clock network but also the logic and routing. Glitches or hazards caused by differences in propagation delay can result in significant power dissipation (Anderson and Najm, 2004); these can be reduced by keeping the logic depth shallow where possible using pipelining. Minimising the number of clock domains and keeping the logic associated with each clock domain together can also help to limit the power used by the clocking circuitry. Clock gating can be used where the design contains blocks of logic that are used infrequently; stopping the clock for blocks that are not in use

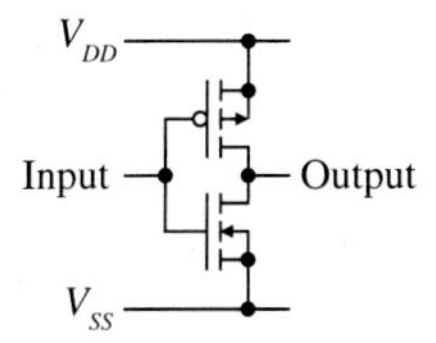

Figure 4.25 CMOS inverter.

can significantly reduce the power (Huda et al., 2009; Oliver et al., 2012). Hardened logic blocks will use less power than implementing the same circuit using programmable fabric, so these should be used when available. There may be limited control over the length of the interconnections, as much of this is managed automatically by the vendor's place-and-route tools. However, these tools are constrained by propagation delay, so in general will try to minimise the length of the interconnections.

Any connections external to the FPGA will often require signals at a higher voltage and will have higher capacitance. Therefore, to reduce power, accesses off-chip should be minimised. This can be achieved by compressing data that must be sent off-chip to reduce the bandwidth and by maximising reuse of the data while it is on-chip (through careful algorithm remapping). In particular, use of external high-speed high-capacity memory should be reduced where possible using on-chip buffering to maximise data reuse. The energy required by an M-bit memory scales with $\sqrt{M}$ (Kadric et al., 2016). In using on-chip memories, there is an optimum level of parallelism that minimises the total power used by matching the size of the local processor storage with the size of the memory blocks available (Kadric et al., 2016). Where possible, using the clock enable signal to disable memories which are not being accessed can reduce power (Tessier et al., 2007).

The second component of dynamic power dissipation results from the fact that as one transistor is switching off, the other is switching on. During the transition, there is a brief interval when both transistors are partially on, and there is a conduction path between V_{DD} and V_{SS}. For fast transition rates, power dissipation from this source is typically much lower than that from charging the load capacitance. The flow-through current can be minimised by keeping the input rise time short, which can be managed by limiting the fan-out from the driving source. This flow-through current is becoming more important with reducing V_{DD} because the threshold voltages of the P- and N-channel transistors are brought closer together, and both transistors are partially on for a wider proportion of the input voltage swing.

Static power dissipation arises from leakage currents from several sources (Unsal and Koren, 2003) of which the sub-threshold leakage is dominant. This is the leakage current through the transistor when the gate voltage is below the switching threshold, and the transistor should be switched off. As the feature size and supply voltage of devices are scaled down, the threshold voltage must also be scaled down to maintain device speed. Unfortunately, the sub-threshold leakage current scales exponentially with decreasing threshold voltage, so, in general, the leakage current contributes to an increasing proportion of the total power dissipation with each technology generation (Kao et al., 2002; Unsal and Koren, 2003). However, more recent technologies, using finFETs, have significantly reduced the leakage current (Abusultan and Khatri, 2014).

Several techniques are used by FPGA manufacturers to mitigate against the sub-threshold leakage current (Kao et al., 2002). One is to use a higher threshold voltage in the sections of the circuit where the speed is not critical, and only use low-threshold devices on the critical path. Another technique is to switch off sections of the circuit which are not being used or at least place them on standby.

Another factor that can influence the design of FPGA systems is the large current burst drawn from the power supply when the device is switched on (Falin and Pham, 2004). This results from the in-rush charging the internal capacitance and the initial activity that takes place as the FPGA is configured. It is essential that the power supplies are designed to provide this peak current, and not just the average required for normal operation. This is of particular concern in low-power, battery-operated devices where the available peak current may be limited, or there is frequent power cycling.

FPGA vendors provide tools for estimating the power requirements of a design. These, however, are dependent on accurate estimates of the toggle frequency and can vary somewhat from the actual power consumed (Meidanis et al., 2011). Others have developed empirical models based on fitting the power of actual designs based on logic and routing resources (see, for example (Anderson and Najm, 2004)).

4.5 Performance Metrics

There are many ways of measuring or evaluating the quality of an FPGA design. The factors of importance will vary depending on the particular application, often with a trade-off between the various criteria discussed here. Part of design-space exploration is to optimise the performance metric of interest.

4.5.1 Speed

There are three key measures of the speed of a design:

- The throughput of a design is a measure of how quickly it can process data. This is often measured in pixels per second or frames per second. Increasing the level of data parallelism can increase the throughput at the expense of increased resources. Pipelining can increase the throughput by increasing the clock frequency. For data-centric applications that require floating-point calculations, the number of floating-point operations per second (flops) is also used.
- The latency of a design is a measure of how quickly it responds to an input. The latency of an operation is the time from an input to its corresponding output. Latency can be reduced by exploiting fine-grained logical parallelism.
- The maximum clock frequency is often used as a proxy for performance, although on its own it is somewhat meaningless. For stream processing at one pixel per clock cycle, it is equivalent to the throughput. Clock frequency can be useful for comparing alternative designs as part of design-space exploration. Methods to increase the clock frequency include using deeper pipelines and moving to a higher speed grade or higher performance FPGA.

These measures can vary significantly depending on whether a high-performance or low-power FPGA is used and also the speed grade of the FPGA. Consequently, a simple comparison, as given in many papers, is often of limited value.

4.5.2 Resources

A design that uses fewer resources for a particular task is more efficient than one which uses more. However, most designs use a mix of logic, memory, and digital signal processing (DSP) blocks; how should these be weighted? When trying to accelerate an application as much as possible through parallelism, often, one of the resources is the limiting factor, making utilisation of that resource the most important.

Different architectures between FPGA vendors, and even between different devices from the same vendor, make it difficult to compare resources used by a design. Even reducing designs to the number of LUTs does not help. Does the device use 4-LUTs, 6-LUTs, or adaptive logic modules (Hutton et al., 2004)? These can affect the number of LUTs used, making comparisons less meaningful. However, in spite of these limitations, the resources used for a design are an important metric, especially when considered during design-space exploration. It must be borne in mind that increased data-level parallelism through loop unrolling will require more resources.

4.5.3 Power

Where there is a limited power envelope, for example in a battery-operated system, the system power is an important metric. Often, average power is considered, but it is also important to consider peak power as this can have an impact on system design. Power can also depend on whether a low-power or high-performance FPGA is used. Vendor tools will estimate the power used by the FPGA but do not consider the whole system. The most reliable measurement comes from monitoring the power supply.

Since performance (speed) often comes at the cost of power, it is often more meaningful to normalise the power by the throughput to give the energy consumed. This is often measured as the energy per frame or energy per operation (Lin and So, 2012).

It is possible to develop systems with extremely low power but are too slow to be practical. Another metric that takes this trade-off into account is the energy-delay product (Lin and So, 2012).

4.5.4 Cost

From a commercial perspective, the system cost is often a key driver. This would imply that the use of low-cost FPGAs is preferred for commodity applications, as long as the performance requirements are met. However, much of the research literature reports results based on high-end FPGAs because these generally give better performance (making the results easier to publish). Consequently, system cost is often not discussed in the literature.

One aspect that also receives little direct attention is the development cost. A rough rule-of-thumb is that the time required to develop an register transfer level (RTL) design for an FPGA is about 10 times longer than that of developing the equivalent in software. This figure is for a single design and does not take into account any significant design-space exploration. Although this is improved considerably through high-level synthesis, a good design can still take two to three times longer than software development.

4.5.5 Application Metrics

So far, the metrics have focused on the performance of the hardware. However, in terms of the application itself, there are several metrics commonly used to assess the quality or accuracy of the design.

To give fast and efficient hardware, many FPGA designs replace floating-point with fixed-point calculations, which can result in approximation error. This is similar in nature to the quantisation error associated with A/D conversion, that of representing continuous values with discrete quantities. Even floating-point numbers have limited precision and will be associated with approximation error. A useful metric is the relative error, RE:

$$RE = \frac{|\hat{x} - x|}{x}, \tag{4.6}$$

where x is the true value, and $\hat{x}$ is its approximate representation.

In image restoration, reconstruction or coding applications, a common metric of the degradation is the mean square error, MSE:

$$MSE = \frac{1}{N}\sum_{p \in I}(\hat{I}(p) - I(p))^2, \tag{4.7}$$

where $I(p)$ are the pixel values of the 'true' image, $\hat{I}(p)$ are the degraded pixel values, and N is the number of pixels in the image. Closely related to this is the RMS error: $RMSE = \sqrt{MSE}$. This is commonly referenced to the maximum possible signal value as peak signal-to-noise ratio, $PSNR$, in decibels:

$$PSNR = 20\log_{10}\frac{2^B - 1}{\sqrt{MSE}}, \tag{4.8}$$

where B is the number of bits per pixel. While a larger PSNR generally indicates better quality, it does not necessarily correlate well with visual quality and is really only valid when comparing different reconstructions of a single image (Huynh-Thu and Ghanbari, 2008).

An alternative metric is the structural similarity index (Wang et al., 2004), which is made up of three components: luminance $l(I,\hat{I})$, contrast $c(I,\hat{I})$, and structure $s(I,\hat{I})$, all calculated over a small window:

$$l(I,\hat{I}) = \frac{2\mu_I\mu_{\hat{I}} + c_1}{\mu_I^2 + \mu_{\hat{I}}^2 + c_1}, \qquad c(I,\hat{I}) = \frac{2\sigma_I\sigma_{\hat{I}} + c_2}{\sigma_I^2 + \sigma_{\hat{I}}^2 + c_2}, \qquad s(I,\hat{I}) = \frac{\sigma_{I\hat{I}}^2 + c_2/2}{\sigma_I\sigma_{\hat{I}} + c_2/2}, \tag{4.9}$$

where μ_I and $\mu_{\hat{I}}$ are the mean values within the window of I and $\hat{I}$, respectively, σ_I^2 and $\sigma_{\hat{I}}^2$ are the variances within the window, $\sigma_{I\hat{I}}^2$ is the window covariance of I and $\hat{I}$, and $c_1 = (0.01(2^B - 1))^2$ and $c_2 = (0.03(2^B - 1))^2$ are small constants to avoid division by zero. These are then combined with a weighted combination (to weight their relative importance),

$$SSIM(I,\hat{I}) = l(I,\hat{I})^\alpha c(I,\hat{I})^\beta s(I,\hat{I})^\gamma, \tag{4.10}$$

to give a structural similarity map (commonly $\alpha = \beta = \gamma = 1$). To obtain a single number, the average value of the SSIM map can be used (Wang et al., 2004).

For object detection or segmentation, it is desired to have a measure of accuracy of detection without having to be exactly correct. One metric which accomplishes this is the intersection-over-union, IoU:

$$IoU = \frac{\text{area}(D \cap G)}{\text{area}(D \cup G)}, \tag{4.11}$$

where D is the set of detected pixels, and G is the set of true ground-truth pixels. This metric can be used either directly or with bounding boxes where these are being detected.

For classification tasks, the accuracy of classification is important. Consider binary classification, where an object is classified as to belonging to a class or not. There are four possible results: true positives (TP), true negatives (TN), false positives (FP), and false negatives (FN). From these, several metrics are commonly derived for evaluating the classifier:

- **Accuracy**: overall performance of the classifier:

$$Accuracy = \frac{TP + TN}{TP + TN + FP + FN}. \tag{4.12}$$

- **Precision**: how accurate the positive predictions are:

$$Precision = \frac{TP}{TP + FP}. \tag{4.13}$$

- **Recall or sensitivity**: proportion of actual positive samples correctly classified:

$$Recall = \frac{TP}{TP + FN}. \tag{4.14}$$

- **Selectivity or specificity**: proportion of actual negative samples correctly classified:

$$Selectivity = \frac{TN}{FP + TN}. \tag{4.15}$$

- **F-measure**: a metric useful for assessing performance when the classes are unbalanced, which is the harmonic mean of precision and recall:

$$F = \frac{2 \times Precision \times Recall}{Precision + Recall} = \frac{2TP}{2TP + FN + FP}. \tag{4.16}$$

Most classifiers have a discrimination threshold, which changes the number of true and false positives as it is changed. This trade-off is plotted as a receiver operating characteristic (ROC) (Fawcett, 2006), which plots the true positive rate (*Recall*) against the false positive rate ($1 - Selectivity$) across the range of threshold settings. The point (0,1) corresponds to a perfect classifier (no false negatives or positives), so the closer the curve passes to this point, the better the classifier. A simple metric of the classifier quality is the area under the ROC curve (AUC), which varies from 0.5 for a completely random classifier up to 1.0 for a perfect classifier.

4.6 Summary

The design process is subject to a number of constraints. These fall into four broad classes:

A real-time application has limited processing time available and is therefore subject to throughput and latency constraints. These may be overcome through low-level pipelining and parallelism. One common issue is synchronising between parallel sections of a design, especially where these are within different clock domains, where appropriate handshaking needs to be designed.

Memory accesses (especially to off-chip memory) have limited bandwidth. While there are several techniques to increase the bandwidth available, one of the most effective is to maximise the use of the data while it is on the FPGA by caching the data in on-chip RAM blocks.

An FPGA has limited available resources. This can lead to sharing expensive resources (including off-chip resources such as memory) between multiple sections of a design. In such cases it is essential to also include arbitration logic, which can be included within a resource controller, exposing a relatively simple and uniform interface.

In power-constrained systems, it is necessary to take a power-aware approach to system design to minimise the power consumption.

There are a number of approaches to evaluating the performance of the resulting system. Depending on the application and environment, these include speed, resources, power, and cost. There are also several metrics for assessing the quality or accuracy of the resulting application.

References

Abusultan, M. and Khatri, S.P. (2014). A comparison of FinFET based FPGA LUT designs. *24th Great Lakes Symposium on VLSI*, Houston, TX, USA (21–23 May 2014), 353–358. https://doi.org/10.1145/2591513.2591596.

Anderson, J.H. and Najm, F.N. (2004). Power estimation techniques for FPGAs. *IEEE Transactions on Very Large Scale Integration (VLSI) Systems* **12** (10): 1015–1027. https://doi.org/10.1109/TVLSI.2004.831478.

Arm (2010). *AMBA 4 AXI4-Stream Protocol Specification*, IHI 0051A. Arm Limited.

Bailey, D.G. (2006). Space efficient division on FPGAs. *Electronics New Zealand Conference (ENZCon'06)*, Christchurch, NZ (13–14 November 2006), 206–211.

Chang, C. and Melhem, R. (1997). Arbitrary size Benes networks. *Parallel Processing Letters* **7** (3): 279–284. https://doi.org/10.1142/s0129626497000292.

Coffman, E.G., Elphick, M., and Shoshani, A. (1971). System deadlocks. *ACM Computing Surveys* **3** (2): 67–78. https://doi.org/10.1145/356586.356588.

Cummings, C.E. and Mills, D. (2002). Synchronous resets? Asynchronous resets? I am so confused! How will I ever know which to use? *Synopsis Users Group Conference (SNUG 2002)*, San Jose, CA, USA (13–15 March 2002), 31 pages.

Dumitriu, V., Marcantonio, D., and Kirischian, L. (2009). Run-time component relocation in partially-reconfigurable FPGAs. *International Conference on Computational Science and Engineering (CSE'09)*, Vancouver, Canada (29–31 August 2009), Volume **2**, 909–914. https://doi.org/10.1109/CSE.2009.493.

Falin, J. and Pham, L. (2004). Tips for successful power-up of today's high-performance FPGAs. *Analog Applications Journal* **3Q 2004**: 11–15.

Fawcett, T. (2006). An introduction to ROC analysis. *Pattern Recognition Letters* **27** (8): 861–874. https://doi.org/10.1016/j.patrec.2005.10.010.

Gangadharan, S. and Churiwala, S. (2013). *Constraining Designs for Synthesis and Timing Analysis: A Practical Guide to Synopsis Design Constraints (SDC)*. New York: Springer. https://doi.org/10.1007/978-1-4614-3269-2.

Ginosar, R. (2003). Fourteen ways to fool your synchronizer. *9th International Symposium on Asynchronous Circuits and Systems*, Vancouver, Canada (12–15 May 2003), 89–96. https://doi.org/10.1109/ASYNC.2003.1199169.

Ginosar, R. (2011). Metastability and synchronizers: a tutorial. *IEEE Design and Test of Computers* **28** (5): 23–35. https://doi.org/10.1109/MDT.2011.113.

Gnusin, A. (2018). Resets and reset domain crossings in ASIC and FPGA designs. White paper. Henderson, NV: Aldec Inc.

Goeders, J.B. and Wilton, S.J.E. (2012). VersaPower: power estimation for diverse FPGA architectures. *International Conference on Field-Programmable Technology*, Seoul, Republic of Korea (10–12 December 2012), 229–234. https://doi.org/10.1109/FPT.2012.6412139.

Gribbon, K.T., Johnston, C.T., and Bailey, D.G. (2005). Design patterns for image processing algorithm development on FPGAs. *IEEE Region 10 Conference (IEEE Tencon'05)*, Melbourne, Australia (21–24 November 2005), 6 pages. https://doi.org/10.1109/TENCON.2005.301109.

Gribbon, K.T., Johnston, C.T., and Bailey, D.G. (2006). Using design patterns to overcome image processing constraints on FPGAs. *3rd IEEE International Workshop on Electronic Design, Test, and Applications (DELTA 2006)*, Kuala Lumpur, Malaysia (17–19 January 2006), 47–53. https://doi.org/10.1109/DELTA.2006.93.

Hadley, J.D. and Hutchings, B.L. (1995). Design methodologies for partially reconfigured systems. *IEEE Symposium on FPGAs for Custom Computing Machines*, Napa Valley, CA, USA (19–21 April 1995), 78–84. https://doi.org/10.1109/FPGA.1995.477412.

Hagemeyer, J., Kettelhoit, B., Koestner, K., and Porrmann, M. (2007). Design of homogeneous communication infrastructures for partially reconfigurable FPGAs. *International Conference on Engineering of Reconfigurable Systems (ERSA'07)*, Las Vegas, NV, USA (25–28 June 2007), 10 pages.

Hoare, C.A.R. (1985). *Communicating Sequential Processes*. London: Prentice-Hall International.

Horta, E.L., Lockwood, J.W., Taylor, D.E., and Parlour, D. (2002). Dynamic hardware plugins in an FPGA with partial run-time reconfiguration. *ACM IEEE Design Automation Conference*, New Orleans, LA, USA (10–14 June 2002), 343–348. https://doi.org/10.1145/513918.514007.

Huda, S., Mallick, M., and Anderson, J.H. (2009). Clock gating architectures for FPGA power reduction. *International Conference on Field Programmable Logic and Applications*, Prague, Czech Republic (31 August – 2 September 2009), 112–118. https://doi.org/10.1109/FPL.2009.5272538.

Hutton, M., Schleicher, J., Lewis, D., Pedersen, B., Yuan, R., Kaptanoglu, S., Baeckler, G., Ratchev, B., Padalia, K., Bourgeault, M., Lee, A., Kim, H., and Saini, R. (2004). Improving FPGA performance and area using an adaptive logic module. *14th International Conference on Field Programmable Logic and Applications*, Antwerp, Belgium (29 August – 1 September 2004), *Lecture Notes in Computer Science*, Volume **3203**, 135–144. https://doi.org/10.1007/b99787.

Huynh-Thu, Q. and Ghanbari, M. (2008). Scope of validity of PSNR in image/video quality assessment. *Electronics Letters* **44** (13): 800–801. https://doi.org/10.1049/el:20080522.

Intel (2020). *Intel Quartus Prime Pro Edition User Guide: Partial Reconfiguration*, Volume UG-20136. Intel Corporation.

Johnston, C.T., Gribbon, K.T., and Bailey, D.G. (2005). FPGA based remote object tracking for real-time control. *International Conference on Sensing Technology*, Palmerston North, NZ (21–23 November 2005), 66–71.

Johnston, C.T., Bailey, D.G., and Lyons, P. (2006). Towards a visual notation for pipelining in a visual programming language for programming FPGAs. *7th International Conference of the NZ Chapter of the ACM's Special Interest Group on Human–Computer Interaction (CHINZ 2006)*, Christchurch, NZ (6–7 July 2006), *ACM International Conference Proceeding Series*, Volume **158**, 1–9. https://doi.org/10.1145/1152760.1152761.

Johnston, C.T., Lyons, P., and Bailey, D.G. (2008). A visual notation for processor and resource scheduling. *IEEE International Symposium on Electronic Design, Test and Applications (DELTA 2008)*, Hong Kong (23–25 January 2008), 296–301. https://doi.org/10.1109/DELTA.2008.76.

Johnston, C.T., Bailey, D., and Lyons, P. (2010). Notations for multiphase pipelines. *5th IEEE International Symposium on Electronic Design, Test and Applications (DELTA 2010)*, Ho Chi Minh City, Vietnam (13–15 January 2010), 212–216. https://doi.org/10.1109/DELTA.2010.29.

Kadric, E., Lakata, D., and Dehon, A. (2016). Impact of parallelism and memory architecture on FPGA communication energy. *ACM Transactions on Reconfigurable Technology and Systems* **9** (4): Article ID 30, 23 pages. https://doi.org/10.1145/2857057.

Kao, J., Narendra, S., and Chandrakasan, A. (2002). Subthreshold leakage modeling and reduction techniques. *IEEE/ACM International Conference on Computer Aided Design*, San Jose, CA, USA (10–14 November 2002), 141–148. https://doi.org/10.1145/774572.774593.

Khan, S., Bailey, D., and Sen Gupta, G. (2009). Simulation of triple buffer scheme (comparison with double buffering scheme). *2nd International Conference on Computer and Electrical Engineering (ICCEE 2009)*, Dubai, United Arab Emirates (28–30 December 2009), Volume **2**, 403–407. https://doi.org/10.1109/ICCEE.2009.226.

Kim, K. and Kumar, V.K.P. (1989). Parallel memory systems for image processing. *IEEE Computer Society Conference on Computer Vision and Pattern Recognition*, San Diego, CA, USA (4–8 June 1989), 654–659. https://doi.org/10.1109/CVPR.1989.37915.

Koester, M., Kalte, H., Porrmann, M., and Rückert, U. (2007). Defragmentation algorithms for partially reconfigurable hardware. *13th International Conference on Very Large Scale Integration of System on Chip*, Perth, Australia (17–19 October 2007), *IFIP Advances in Information and Communication Technology*, Volume **240**, 41–53. https://doi.org/10.1007/978-0-387-73661-7_4.

LaForest, C.E., Li, Z., O'Rourke, T., Liu, M.G., and Steffan, J.G. (2014). Composing multi-ported memories on FPGAs. *ACM Transactions on Reconfigurable Technology and Systems* **7** (3): Article ID 16, 23 pages. https://doi.org/10.1145/2629629.

Lee, D. (1988). Scrambled storage for parallel memory systems. *15th Annual International Symposium on Computer Architecture*, Honolulu, HI, USA (30 May – 2 June 1988), 232–239. https://doi.org/10.1109/ISCA.1988.5233.

Leiserson, C.E. and Saxe, J.B. (1981). Optimizing synchronous systems. *22nd Annual Symposium on Foundations of Computer Science*, Nashville, TN, USA (28–30 October 1981), 23–26. https://doi.org/10.1109/SFCS.1981.34.

Lin, C.Y. and So, H.K.H. (2012). Energy-efficient dataflow computations on FPGAs using application-specific coarse-grain architecture synthesis. *SIGARCH Compututer Architecture News* **40** (5): 58–63. https://doi.org/10.1145/2460216.2460227.

Liu, Q., Constantinides, G.A., Masselos, K., and Cheung, P.Y.K. (2008). Combining data reuse exploitation with data-level parallelization for FPGA targeted hardware compilation: a geometric programming framework. *International Conference on Field Programmable Logic and Applications (FPL 2008)*, Heidelberg, Germany (8–10 September 2008), 179–184. https://doi.org/10.1109/FPL.2008.4629928.

Manjikian, N. (2003). Design issues for prototype implementation of a pipelined superscalar processor in programmable logic. *IEEE Pacific Rim Conference on Communications Computers and Signal Processing (PACRIM 2003)*, Victoria, British Columbia, Canada (28–30 August 2003), Volume **1**, 155–158. https://doi.org/10.1109/PACRIM.2003.1235741.

Meidanis, D., Georgopoulos, K., and Papaefstathiou, I. (2011). FPGA power consumption measurements and estimations under different implementation parameters. *International Conference on Field-Programmable Technology*, New Delhi, India (12–14 December 2011), 6 pages. https://doi.org/10.1109/FPT.2011.6132694.

Mesquita, D., Moraes, F., Palma, J., Moller, L., and Calazans, N. (2003). Remote and partial reconfiguration of FPGAs: tools and trends. *International Parallel and Distributed Processing Symposium*, Nice, France (22–26 April 2003), 8 pages. https://doi.org/10.1109/IPDPS.2003.1213326.

National Semiconductor (2004). *LVDS Owner's Manual: Low-Voltage Differential Signaling*, 3e. Santa Clara, CA: National Semiconductor.

Oliver, J.P., Curto, J., Bouvier, D., Ramos, M., and Boemo, E. (2012). Clock gating and clock enable for FPGA power reduction. *VIII Southern Conference on Programmable Logic*, Bento Goncalves, Spain (20–23 March 2012), 5 pages. https://doi.org/10.1109/SPL.2012.6211782.

Papadimitriou, K., Dollas, A., and Hauck, S. (2011). Performance of partial reconfiguration in FPGA systems: a survey and a cost model. *ACM Transactions on Reconfigurable Technology and Systems* **4** (4): Article ID 36, 24 pages. https://doi.org/10.1145/2068716.2068722.

Parhami, B. (2000). *Computer Arithmetic: Algorithms and Hardware Designs*. New York: Oxford University Press.

Park, J.W. (1986). An efficient memory system for image processing. *IEEE Transactions on Computers* **35** (7): 669–674. https://doi.org/10.1109/TC.1986.1676813.

Plessas, F., Alexandropoulos, A., Koutsomitsos, S., Davrazos, E., and Birbas, M. (2011). Advanced calibration techniques for high-speed source-synchronous interfaces. *IET Computers and Digital Techniques* **5** (5): 366–374. https://doi.org/10.1049/iet-cdt.2010.0143.

Sarwar, A. (1997). CMOS power consumption and C_{pd} calculation. Application note SCAA035B, Texas Instruments, USA.

Sedcole, N.P. (2006). Reconfigurable platform-based design in FPGAs for video image processing. PhD thesis. London: Imperial College.

Sedcole, N.P., Cheung, P.Y.K., Constantinides, G.A., and Luk, W. (2003). A reconfigurable platform for real-time embedded video image processing. *International Conference on Field Programmable Logic and Applications (FPL 2003)*, Lisbon, Portugal (1–3 September 2003), *Lecture Notes in Computer Science*, Volume **2778**, pp. 606–615. https://doi.org/10.1007/b12007.

Sedcole, P., Cheung, P.Y.K., Constantinides, G.A., and Luk, W. (2007). Run-time integration of reconfigurable video processing systems. *IEEE Transactions on VLSI Systems* **15** (9): 1003–1016. https://doi.org/10.1109/TVLSI.2007.902203.

Simmler, H., Levinson, L., and Manner, R. (2000). Multitasking on FPGA coprocessors. *10th International Conference on Field Programmable Logic and Applications*, Villach, Austria (27–30 August 2000), *Lecture Notes in Computer Science*, Volume **1896**, pp. 121–130. https://doi.org/10.1007/3-540-44614-1_13.

Tessier, R., Betz, V., Neto, D., Egier, A., and Gopalsamy, T. (2007). Power-efficient RAM mapping algorithms for FPGA embedded memory blocks. *IEEE Transactions on Computer-Aided Design of Integrated Circuits and Systems* **26** (2): 278–290. https://doi.org/10.1109/TCAD.2006.887924.

Unsal, O.S. and Koren, I. (2003). System-level power-aware design techniques in real-time systems. *Proceedings of the IEEE* **91** (7): 1055–1069. https://doi.org/10.1109/JPROC.2003.814617.

Vanmeerbeeck, G., Schaumont, P., Vernalde, S., Engels, M., and Bolsens, I. (2001). Hardware / software partitioning of embedded system in OCAPI-xl. *International Conference on Hardware Software Codesign*, Copenhagen, Denmark (25–27 April 2001), pp. 30–35. https://doi.org/10.1145/371636.371665.

Villasenor, J., Jones, C., and Schoner, B. (1995). Video communications using rapidly reconfigurable hardware. *IEEE Transactions on Circuits and Systems for Video Technology* **5** (6): 565–567. https://doi.org/10.1109/76.475899.

Villasenor, J., Schoner, B., Chia, K.N., Zapata, C., Kim, H.J., Jones, C., Lansing, S., and Mangione-Smith, B. (1996). Configurable computing solutions for automatic target recognition. *IEEE Symposium on FPGAs for Custom Computing Machines*, Napa Valley, CA (17–19 April 1996), pp. 70–79. https://doi.org/10.1109/FPGA.1996.564749.

Vipin, K. and Fahmy, S.A. (2018). FPGA dynamic and partial reconfiguration: a survey of architectures, methods, and applications. *ACM Computing Surveys* **51** (4): Article ID 72, 39 pages. https://doi.org/10.1145/3193827.

Wang, Z., Bovik, A.C., Sheikh, H.R., and Simoncelli, E.P. (2004). Image quality assessment: from error visibility to structural similarity. *IEEE Transactions on Image Processing* **13** (4): 600–612. https://doi.org/10.1109/TIP.2003.819861.

Weinhardt, M. and Luk, W. (2001). Memory access optimisation for reconfigurable systems. *IEE Proceedings Computers and Digital Techniques* **148** (3): 105–112. https://doi.org/10.1049/ip-cdt:20010514.

Xilinx (2018). *Vivado Design Suite User Guide: Partial Reconfiguration*, Volume UG909 (v2018.1). Xilinx Inc.

5

Computational Techniques

A field-programmable gate array (FPGA) implementation is amenable to a wider range of computational techniques than a software implementation because it can be used to build the specific computational architecture required. Some of the computational considerations and techniques are reviewed in this section. Several of these form the basis for the calculations used within many of the image processing operations described in Chapters 7–14.

5.1 Number Systems

In any image processing computation, the numerical values of the pixels, coordinates, features, and other data must be represented and manipulated. There are many different number systems that can be used. A full discussion of number systems and associated logic circuits for performing basic arithmetic operations is beyond the scope of this book. The interested reader should refer to one of the many good books available on computer arithmetic (for example (Parhami, 2000)).

5.1.1 Binary Integers

Arguably, the most common representation is based on binary, or base-two numbers. As illustrated in Figure 5.1, a positive binary integer, X, is represented by a set of B binary digits, b_i, where each digit weights a successive power of 2. The numerical value of the number is then

$$X = \sum_{i=0}^{B-1} b_i 2^i. \tag{5.1}$$

In software, B is fixed by the underlying computer architecture at 8, 16, 32, or 64 bits. On an FPGA, however, there is no necessity to constrain the representation to these widths. In fact, the hardware requirements can often be significantly reduced by tailoring the width of the numbers to the range of values that need to be represented (Constantinides et al., 2001). For example, the row and column coordinates in a 640×480 image can be represented by 9- and 10-bit integers, respectively.

When representing signed integers, there are three common methods. The ***two's complement*** representation is the format most commonly used by computers. Numbers in the range -2^{B-1} to $2^{B-1} - 1$ are represented as modulo 2^B, which maps the negative numbers into the range 2^{B-1} to $2^B - 1$. This is equivalent to giving the most significant bit a negative weight:

$$X = -b_{B-1}2^{B-1} + \sum_{i=0}^{B-2} b_i 2^i. \tag{5.2}$$

Design for Embedded Image Processing on FPGAs, Second Edition. Donald G. Bailey.

Companion Website: www.wiley.com/go/bailey/designforembeddedimageprocessc2e

$\pm 2^7$	2^6	2^5	2^4	2^3	2^2	2^1	2^0
b_7	b_6	b_5	b_4	b_3	b_2	b_1	b_0

Figure 5.1 Binary number representation. In a two's complement representation, the most significant digit has a negative weight.

Therefore, negative numbers can be identified by a '1' in the most significant bit. Addition and subtraction of two's complement numbers are exactly the same as for unsigned numbers, which is why the two's complement representation is so attractive. Multiplication is a little more complex; it has to take into account that the most significant bit is negative.

The second is the ***sign-magnitude*** representation. This uses a sign bit to indicate whether the number is positive or negative and the remaining $B - 1$ bits to represent the magnitude (absolute value) of the number. The range of numbers represented is from $-(2^{B-1} - 1)$ to $2^{B-1} - 1$. Note that there are two representations for zero: -0 and $+0$. Operations for addition and subtraction depend on the combination of sign bits and the relative magnitudes of the numbers. Multiplication and division are simpler, as the sign and magnitude bits can be operated on separately.

The third representation is ***offset binary***, where an offset or bias is added to all of the numbers:

$$X = \mathit{Offset} + \sum_{i=0}^{B-1} b_i 2^i, \tag{5.3}$$

where *Offset* is usually (but not always) 2^{B-1}. Offset binary is commonly used in image processing to display negative pixel values. It is not possible to display negative intensities, so offsetting zero to mid-grey shows negative values as darker and positive values as lighter. Offset binary is also frequently used with A/D converters to enable them to handle negative voltages. Most A/D converters are unipolar and will only convert positive voltages (or currents). By adding a bias to a bipolar signal, negative voltages can be made positive, enabling them to be successfully converted. When performing arithmetic with offset binary, it is necessary to take into account the offset.

5.1.2 Residue Systems

The residue number system reduces the complexity of integer arithmetic by using modulo arithmetic. A set of k co-prime moduli, $\{M_1, M_2, \ldots M_k\}$, are used, and an integer, X, is represented by its residues or remainders with respect to each modulus: $\{x_1, x_2, \ldots x_k\}$. That is,

$$X = n_i M_i + x_i, \tag{5.4}$$

where each n_i is an integer, and each x_i is a positive integer. Negative values of X may be handled just as easily because modulo arithmetic is used. The representation is unique for

$$0 \le X < \prod_{i=1}^{k} M_i, \tag{5.5}$$

or

$$-\frac{1}{2}\prod_{i=1}^{k} M_i \le X < \frac{1}{2}\prod_{i=1}^{k} M_i. \tag{5.6}$$

Addition, subtraction, and multiplication are performed using the respective modulo operation independently on each of the residues. Replacing a single wide integer with a set of narrower integers and performing the calculations on each residue in parallel can reduce both the propagation delay and logic footprint compared with using wide adders and multipliers. This makes the residue number system attractive for FPGA implementation (Tomczak, 2006).

A disadvantage of the residue number system, however, is that division is impractical. The main application is in digital filtering, which consists of fixed-point multiplications and additions, so the lack of division is not serious.

The Chinese remainder theorem or variants (Wang, 1998) are used to reconstruct a binary number from the residues. A common choice of moduli is $\{2^n - 1, 2^n, 2^n + 1\}$ (Wang et al., 2002), for which the logic required to perform modulo arithmetic is only marginally more complex than for performing standard arithmetic. For this set, there are also efficient techniques for converting between binary and the equivalent residue number representation (Fu et al., 2008).

5.1.3 Redundant Representations

Arithmetic operations are slow compared to simple logic operations because of the propagation delay associated with the carry from one digit to the next. One technique that can be used to speed arithmetic is the use of a redundant representation. A ***redundant number system*** is one that has more digits than the size of the radix (Avizienis, 1961). Standard binary (radix-2) uses the digits $b_i \in \{0, 1\}$. A commonly used redundant system is a ***signed-digit representation*** where $b_i \in \{-1, 0, 1\}$.

With redundancy, carry propagation may be shortened considerably, giving significant speed improvements for additions. A particular signed-digit representation that maximises the number of zeroes in the number is the ***canonical signed-digit*** form (Arno and Wheeler, 1993). This has at least one zero between every nonzero digit, so has at least half of its digits zero. The carry from adding two canonical signed-digit numbers will propagate at most one bit. Minimising the number of digits will also reduce the number of partial products when multiplying numbers (Koc and Johnson, 1994).

The disadvantage of redundant representations is that the additional digits require wider signals (at least two bits are required for the radix-2 signed-digit representation), wider registers for storing the numbers, and usually more logic to process the wider signals. While addition of two canonical signed-digit numbers will give a signed-digit representation, additional logic and propagation delay may be required to convert the result back to canonical form. With any redundant representation, additional logic is required to convert any output back to standard binary.

The signed-digit idea can be extended further, for example to $b_i \in \{-2, -1, 0, 1, 2\}$ (Sam and Gupta, 1990), to further improve the performance of multipliers in particular. The trade-off is between incremental performance improvements and rapidly increasing circuit complexity. An asymmetric signed-digit representation using the digits $b_i \in \{-1, 0, 1, 3\}$ has been proposed for reducing the number of nonzero digits (Kamp et al., 2006). It also has the advantage that each digit only requires two bits to represent; a single 4-LUT (lookup table) is sufficient to perform an arbitrary operation on two digits (Kamp et al., 2006).

Signed digits form the basis of the Booth multiplication algorithm (Booth, 1951), which uses a combination of addition (+1), subtraction (−1), and nothing (0) to reduce the number of additions required. It is also used by non-restoring division algorithms (with quotient digits $b_i \in \{-1, 1\}$) to avoid having to restore the result when a test subtraction gives a negative result (Bailey, 2006). Redundancy is exploited by Sweeney, Robertson and Tocher (SRT) division techniques (Robertson, 1958; Tocher, 1958) by giving an overlap when selecting the correct quotient digit. This overlap means that the quotient digit may be determined from an approximation of the partial remainder and does not need to wait for the carry to propagate fully. Similar techniques have also been applied to shift-and-add methods for the calculation of logarithms and exponentials (Nielsen and Muller, 1996).

5.1.4 Fixed-point Numbers

One limitation of integer representations is their inability to represent real numbers. This may be overcome by scaling the integer by a fraction, so that the integer represents the number of fractions. If the fraction is made a negative power of 2, for example 2^{-k}, then the result is ***fixed-point*** notation. The resulting number is then

$$X_{fixed,k} = X \times 2^{-k} = \pm b_{B-1} 2^{B-1-k} + \sum_{i=0}^{B-2} b_i 2^{i-k}, \tag{5.7}$$

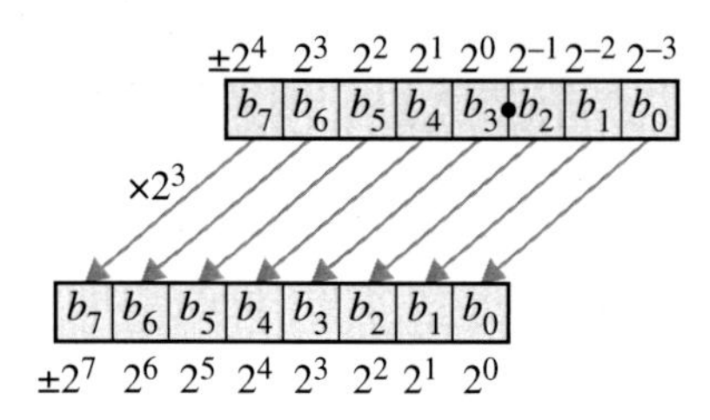

Figure 5.2 Fixed-point number representation with resolution 2^{-3}. Shifting the number 3 bits to the left gives the corresponding integer representation. This number is of type U8.3 or S8.3, depending on whether the weight of the most significant bit is positive or negative, respectively.

where the integer X can be either unsigned or signed (usually two's complement). A convenient shorthand notation for the format of a fixed-point number is UB.k or SB.k where the letter U or S indicates an unsigned or signed B-bit integer, respectively, with k fraction bits. As seen in Figure 5.2, the B-bit integer representation of the fixed-point number corresponds to shifting it k bits to the left.

Most real numbers can only be represented approximately with a fixed-point representation. While increasing the number of fraction bits increases the accuracy of the approximation, many even relatively simple numbers cannot be represented exactly with a finite number of bits. For example, the fraction 1/3 has an infinitely repeating binary representation, 0.01010101010101 … , so any fixed-point (or even floating-point) representation will only ever be an approximation. The absolute error of the representation, or resolution, will be determined by the position of the binary point.

All arithmetic operations are the same as for integers, apart from aligning the binary point for addition and subtraction. This corresponds to an arithmetic shift by a fixed number of bits since the location of the binary point is fixed. Such a shift is effectively free in hardware (it is simply a case of routing the signals appropriately). This makes the use of a fixed-point representation popular for hardware implementation.

A problem with fixed-point numbers is that they have limited dynamic range. The same representation is unable to represent both very small and very large numbers because the fixed location of the binary point fixes the resolution. It is necessary to manually scale large and small numbers to bring them into range to avoid a loss of precision or overflow. This requires a careful analysis of the approximation errors at every stage of the computation to ensure that the final results are meaningful and are not significantly impacted by under- or overflow.

5.1.5 Floating-point Numbers

This fixed resolution may be overcome by allowing the location of the binary point to be variable, which requires another parameter. A ***floating-point*** number therefore has two components: the mantissa or ***significand***, which are the binary digits of the number, and the ***exponent***, which gives the power of 2 that the significand is multiplied by and effectively specifies the position of the binary point.

A ***normalised floating-point*** number has $1.0 \leq$ significand < 2.0. Since the leading digit is always '1', it can be assumed, giving an extra bit of resolution for free. The significand bits therefore represent a binary fraction. Requiring numbers to be normalised maximises the number of significant digits within a number and consequently minimises the relative error in the approximation. The number of bits used for the exponent will determine the dynamic range of numbers (the ratio between the largest and the smallest number) that can be represented. While it could potentially be represented using two's complement, an offset binary representation is commonly used because it enables a simpler unsigned integer comparison to be used for comparing magnitudes. The common representation is shown in Figure 5.3, giving the floating-point number,

$$X_{float} = (-1)^s \times 2^E \times \left(1 + \sum_{j=1}^{B_s} b_{-j} 2^{-j}\right), \tag{5.8}$$

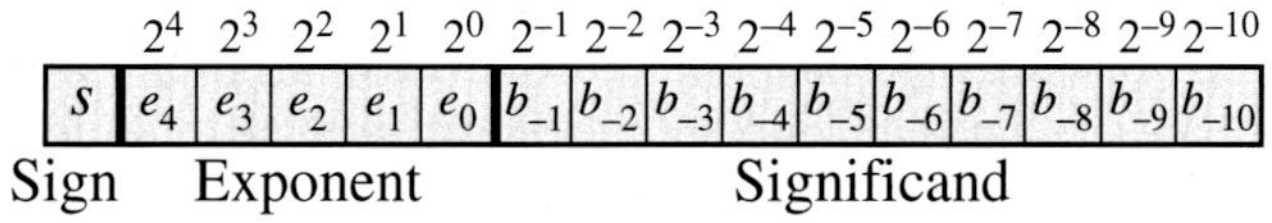

	2^4	2^3	2^2	2^1	2^0	2^{-1}	2^{-2}	2^{-3}	2^{-4}	2^{-5}	2^{-6}	2^{-7}	2^{-8}	2^{-9}	2^{-10}
s	e_4	e_3	e_2	e_1	e_0	b_{-1}	b_{-2}	b_{-3}	b_{-4}	b_{-5}	b_{-6}	b_{-7}	b_{-8}	b_{-9}	b_{-10}
Sign	Exponent					Significand									

Figure 5.3 Floating-point representation.

where

$$E = \sum_{j=0}^{B_e-1} e_j 2^j - (2^{B_e-1} - 1),$$

B_s is the number of bits in the significand, and B_e is the number of bits in the exponent. The term subtracted from the exponent is the standard offset or bias that balances the dynamic range, so that both a number and its reciprocal may be represented.

A disadvantage of the normalised significand is that there is no representation for zero (Goldberg, 1991). It also requires a sign-magnitude representation because the two's complement representation of a negative significand would have a leading '1' regardless of the position of the binary point.

The absolute error of the representation varies depending on the position of the binary point. However, the relative error is constant regardless of the size of the number and is determined by the number of bits in the significand. The relative error of the representation should not be confused with the relative error after an operation. Consider, for example, subtracting two very similar numbers. Many of the significant bits could cancel, giving a much larger relative error for the difference.

The IEEE floating-point standard (IEEE, 2008) enables reproducibility of floating-point calculations between machines by specifying both the representation (how the bits are encoded) and how operations should be performed, so that the same calculation should give the same result regardless of the processor performing the computation. For example, a single-precision (32-bit) floating-point number uses one bit for the sign, 8 bits for the exponent with an offset of 127, and 24 bits (including the implicit leading '1') for the significand. Two exponent values are reserved for special cases. All zeroes are used to represent denormalised numbers (numbers smaller than the smallest normalised number), including zero. All ones are used to represent overflow, infinity, and not-a-number (NaN; for example the result of 0/0 or $\sqrt{-1}$).

When using floating-point calculations on an FPGA, the size of the significand and exponent can be tailored to the accuracy of the calculation being performed, to reduce the hardware required. The consequence of this, however, is that the result will in general be different from performing the same calculation in software using the IEEE standard. As with fixed-point calculations, this requires a careful analysis of the approximation errors to ensure that the results are meaningful.

Multiplication and division are the most straightforward operations for floating-point numbers. For multiplication, the significands are multiplied, and the exponents added. If the resulting significand is greater than two, renormalisation requires shifting it one bit to the right and incrementing the exponent. The product will have more digits than can be represented, so the result must also be rounded to the required number of digits. The sum of the two exponents will include two offsets, so one must be subtracted off. The sign of the output is the exclusive-OR of the input sign bits. Additional logic is required to detect underflow, overflow, and handle other error conditions such as processing infinities and NaNs. Division is similar, except that the significands are divided and the exponents subtracted, and renormalisation may involve a left shift.

Addition and subtraction are more complex. The numbers must first be aligned, so that the exponents are the same. This will involve shifting the significand of the smaller number to the right by the difference in the exponents. When implemented on an FPGA, such a shift is either slow (performed as a series of smaller shifts) or expensive (implemented as a large number of wide multiplexers). If available, a multiplier block may also be used to perform the shift (Gigliotti, 2004). It is necessary to keep an extra bit in the shifted smaller number (a guard bit) to reduce the error introduced by the operation (Goldberg, 1991). The significands are then added or subtracted depending on the operation and signs of the inputs, and the result renormalised. If two similar numbers are subtracted, several of the most significant bits may cancel. Renormalisation therefore requires moving the leftmost one to the most significant bit (for example using a prioritised multiplexer) and adjusting the exponent accordingly. Again, additional logic is required to handle error conditions.

A big advantage of the floating-point system is that it frees programmers from having to think about the magnitude or scale of their data; this is taken care of automatically by the exponent, as long as the number is within range. However, floating-point operations require significantly more resources than the corresponding fixed-point operations. It is not that floating-point operations are that complicated, but managing the exceptions requires a large proportion of the logic, especially if compliance to the IEEE standard is required. For this reason, most developers prefer to use fixed-point operations unless the wider dynamic range of floating-point operations is needed for an application. Where dynamic range is required, one compromise that has been proposed is to use two fixed-point ranges (Ewe et al., 2004, 2005).

5.1.6 Logarithmic Number System

An alternative to the floating-point system is the logarithmic number system, first proposed by Kingsbury and Rayner (1971). This achieves a wide dynamic range by representing a number by its logarithm. A sign-magnitude representation is required because a real logarithm is not defined for negative numbers. The logarithm of zero is also not defined; this requires a separate flag, a special code, or approximating zero by the smallest available number.

The logarithms are usually base-2, although using another base will simply scale the logarithm. The logarithm is represented using fixed-point notation, with B_I bits for the integer component and B_F bits for the fractional component. The integer component will be of the same value as the exponent for a normalised floating-point representation (apart from the offset).

The attraction of the logarithmic number system is that multiplication and division become simple additions and subtractions. Addition and subtraction, however, are more complex. Without loss of generality, assume that $|X| > |Y|$, $L_X = \log_2 X$, and $L_Y = \log_2 Y$. Then,

$$\begin{aligned} \log_2(X \pm Y) &= \log_2 X \left(1 \pm \frac{Y}{X}\right) \\ &= \log_2 X + \log_2 \left(1 \pm 2^{\log_2(Y/X)}\right) \\ &= L_X + \log_2 \left(1 \pm 2^{L_Y - L_X}\right). \end{aligned} \tag{5.9}$$

This requires two functions to represent the second term, one for addition and one for subtraction:

$$\begin{aligned} f_{Add}(L_Y - L_X) &= \log_2 \left(1 + 2^{L_Y - L_X}\right), \\ f_{Sub}(L_Y - L_X) &= \log_2 \left(1 - 2^{L_Y - L_X}\right). \end{aligned} \tag{5.10}$$

These functions can be evaluated either directly (by calculating $2^{L_Y - L_X}$, performing the addition or subtraction, and then taking the logarithm), using table lookup, or by piecewise polynomial approximation (Fu et al., 2006). Since $|X| > |Y|$, then $2^{L_Y - L_X}$ will be between 0.0 and 1.0. Therefore, f_{Add} will be relatively well behaved, although f_{Sub} is more difficult as $\log_2 0$ is not defined. The size of the tables depends on B_F and can be made of reasonable size by cleverly using interpolation (Lewis, 1990).

As would be expected, if the computation involves more multiplications and divisions than additions and subtractions, then the logarithmic number representation has advantages over the floating-point representation in terms of both resource requirements and speed (Haselman, 2005; Haselman et al., 2005).

5.1.7 Posit Numbers

A more recent proposal is the posit number system (Gustafson and Yonemoto, 2017). It achieves better precision (smaller relative error) than the floating-point system for numbers close to 1.0 and a wider dynamic range although this comes at the expense of reduced precision for very large and very small numbers, giving what is

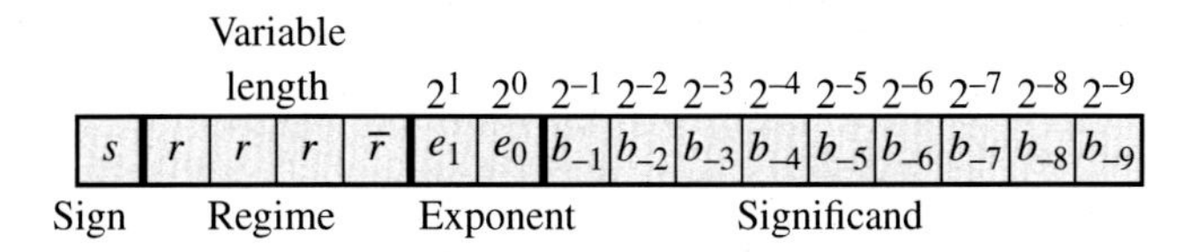

Figure 5.4 Posit number representation. The length of the regime field is variable, with the exponent and significand following (if there are bits left). Here, the exponent size $B_e = 2$.

called a ***tapered accuracy***. It achieves this by effectively having a variable number of bits for the exponent by introducing an extra variable-length regime field.

The representation is illustrated in Figure 5.4 and is parameterised by the total number of bits and the width of the exponent field, B_e. The first bit is a sign bit; if it is a '1', the number is negative, and the two's complement should be taken of the remaining bits before decoding. Next comes the regime field, which is a run of consecutive '0's or '1's, terminated by the opposite bit (or the end of the number). A run of m zeroes corresponds to regime $r = -m$, while a run of m ones gives $r = m - 1$. This gives the regime, R^r, where $R = 2^{2^{B_e}}$. Next comes the exponent, encoded directly in binary: $E = \sum e_i 2^i$. Finally, the significand (if any bits remain) represents a binary fraction, $F = \sum b_{-j} 2^{-j}$, with an implied '1' at the front (similar to floating point). The final value is therefore:

$$X = R^r \times 2^E \times (1 + F). \tag{5.11}$$

There is a single representation of zero (all bits '0') and a single infinity (sign bit '1' with the rest '0').

Chaurasiya et al. (2018) showed that both the resources and latency for processing posit numbers are comparable to that required for floating-point numbers that conform to the IEEE standard. Not having NaN or denormalised numbers simplifies posit numbers. However, if compliance to the standard is not required, then in general, computation and conversion are more expensive for posit than for floating-point numbers, both in terms of hardware and latency (Forget et al., 2019), mainly because of the variable width of the significand. An open source posit core generator has been made available by Jaiswal and So (2019).

The improved precision close to 1.0 gives a more compact representation of numbers within this range. This can potentially allow a reduced word length to reduce the storage (and memory bandwidth) for numbers in this range, for example in neural network and deep learning applications (Langroudi et al., 2018). However, care must be taken with the reduced precision of very large and very small numbers (de Dinechin et al., 2019).

5.2 Elementary Functions

Many image processing algorithms require more than the basic arithmetic operations. Some of the techniques for performing these efficiently in hardware will be outlined in this section.

5.2.1 Square Root

A square root can be calculated using a shift-and-add algorithm. This relies on the relationship

$$(2x + 1)^2 = 4x^2 + 4x + 1 \tag{5.12}$$

to successively add each digit and operates similar to long division, with the quotient changing each step depending on the result so far. The factor of 4 corresponds to a shift of 2 binary places, so the initial number is taken 2 bits at a time. This is illustrated in Figure 5.5 on the left for restoring operation; if the test subtraction makes the partial remainder go negative, the operation is discarded. Non-restoring operations can also be used in calculating the square root (Li and Chu, 1996). This modifies Eq. (5.12) to allow subtraction and addition:

$$(2x \pm 1)^2 = 4x^2 \pm 4x + 1. \tag{5.13}$$

When the partial remainder goes negative, the next remainder digit is -1, and an addition is performed instead of a subtraction. To get a positive remainder, if the final remainder is negative, 1 is subtracted from the result.

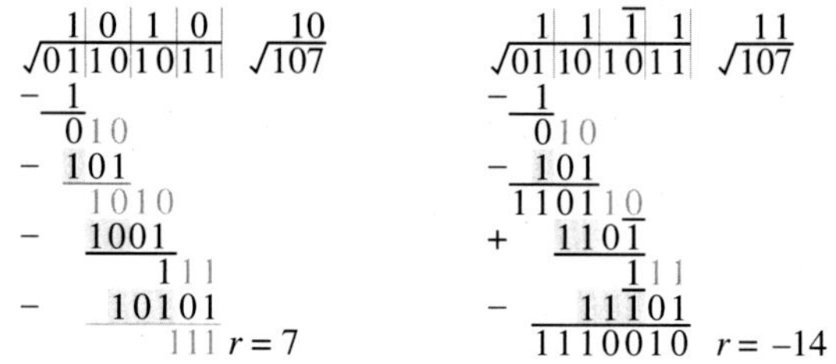

```
   1 0 1 0        10
√01 10 10 11    √107
- 1
  010
- 101
   1010
-  1001
      111
-   10101
        111 r = 7
```

```
   1 1 1̄ 1        11
√01 10 10 11    √107
- 1
  010
- 101
 110110
+   1101
      111
-   11101
 1110010  r = −14
```

Figure 5.5 Square root. Left: restoring operations, when the subtraction goes negative, it is undone; right: non-restoring operations, when the partial remainder goes negative, the next operation is an addition, with the corresponding bit of the square root being -1 (or $\overline{1}$). If the final remainder, R, is negative, 1 needs to be subtracted from the resultant square root.

To round the square root to the nearest integer, if the magnitude of the remainder is greater than the square root, the square root is adjusted by 1. A hardware implementation may be operated serially or unrolled (and pipelined if necessary) for speed. Unrolled, the resulting hardware is actually less than that required for performing a division because the earlier stages require fewer bits.

Other approaches for calculating the square root will be covered in subsequent sections, including using CORDIC (Section 5.2.2), lookup tables (Section 5.2.6), and iterative techniques (Section 5.2.8).

5.2.2 *Trigonometric Functions*

Trigonometric functions can be calculated using CORDIC (from coordinate rotation digital computer) arithmetic (Volder, 1959). It is an iterative shift-and-add technique based on rotating a vector (x, y) by an angle θ_k:

$$\begin{bmatrix} x_{k+1} \\ y_{k+1} \end{bmatrix} = \begin{bmatrix} \cos\theta_k & -\sin\theta_k \\ \sin\theta_k & \cos\theta_k \end{bmatrix} \begin{bmatrix} x_k \\ y_k \end{bmatrix}. \tag{5.14}$$

The trick is choosing the angle and rearranging Eq. (5.14), so that the factors within the matrix are successive negative powers of 2 enabling the multiplications to be performed by simple shifts:

$$\begin{aligned} \begin{bmatrix} x_{k+1} \\ y_{k+1} \end{bmatrix} &= \cos\theta_k \begin{bmatrix} 1 & -d_k\tan\theta_k \\ d_k\tan\theta_k & 1 \end{bmatrix} \begin{bmatrix} x_k \\ y_k \end{bmatrix} \\ &= \frac{1}{\sqrt{1+2^{-2k}}} \begin{bmatrix} 1 & -d_k 2^{-k} \\ d_k 2^{-k} & 1 \end{bmatrix} \begin{bmatrix} x_k \\ y_k \end{bmatrix}, \end{aligned} \tag{5.15}$$

where the direction of rotation, d_k, is $+1$ for an anticlockwise rotation and -1 for a clockwise rotation, and the angle is given by

$$\tan\theta_k = 2^{-k}. \tag{5.16}$$

After a series of K rotations, the total angle rotated is the sum of the angles of the individual rotations:

$$\theta = \sum_{k=0}^{K-1} d_k \tan^{-1}\left(2^{-k}\right). \tag{5.17}$$

The total angle will depend on the particular rotation directions chosen at each iteration; therefore, it is useful to have an additional register (Θ) to accumulate the angle.

The resultant vector is

$$\begin{aligned}\begin{bmatrix} x_K \\ y_K \end{bmatrix} &= \prod_{k=0}^{K-1} \begin{bmatrix} 1 & -d_k 2^{-k} \\ d_k 2^{-k} & 1 \end{bmatrix} \begin{bmatrix} x_0 \\ y_0 \end{bmatrix} \\ &= G \begin{bmatrix} \cos\theta & -\sin\theta \\ \sin\theta & \cos\theta \end{bmatrix} \begin{bmatrix} x_0 \\ y_0 \end{bmatrix},\end{aligned} \tag{5.18}$$

where the scale factor, G, is the gain of the rotation and is constant for a given number of iterations:

$$G = \prod_{k=0}^{K-1} \sqrt{1 + 2^{-2k}} \approx 1.64676. \tag{5.19}$$

The iterations are therefore

$$\begin{aligned} x_{k+1} &= x_k - d_k 2^{-k} y_k, \\ y_{k+1} &= y_k + d_k 2^{-k} x_k, \\ \Theta_{k+1} &= \Theta_k - d_k \tan^{-1}\left(2^{-k}\right), \end{aligned} \tag{5.20}$$

with the corresponding circuit shown in Figure 5.6. On *Load*, the initial values are loaded into the x, y, and Θ registers, and k is reset to 0. Then, with every clock cycle, one iteration of summations is performed and k incremented. Note that it is necessary for the summations and the registers to have an extra $\log_2 N$ guard bits to protect against the accumulation of rounding errors (Walther, 1971). The most expensive part of this circuit is the shifters, which are often implemented using multiplexers. If the FPGA has available spare multipliers, these may be used to perform the barrel shifts (Gigliotti, 2004). A small ROM is required to store the small number of angles given by Eq. (5.16). There are two main modes for choosing the direction of rotation at each step (Volder, 1959): rotation mode and vectoring mode.

Rotation mode starts with an angle in the Θ register and reduces this angle to 0.0 by choosing

$$d_k = \text{sign}(\Theta_k). \tag{5.21}$$

The result after K iterations is

$$\begin{aligned} x_K &= G(x_0 \cos\Theta_0 - y_0 \sin\Theta_0), \\ y_K &= G(x_0 \sin\Theta_0 + y_0 \cos\Theta_0), \\ \Theta_K &= 0. \end{aligned} \tag{5.22}$$

The domain of convergence for the iterations of Eq. (5.20) is

$$|\Theta_0| < \sum_{k=0}^{K-1} \tan^{-1} 2^{-k} \approx 99.88°. \tag{5.23}$$

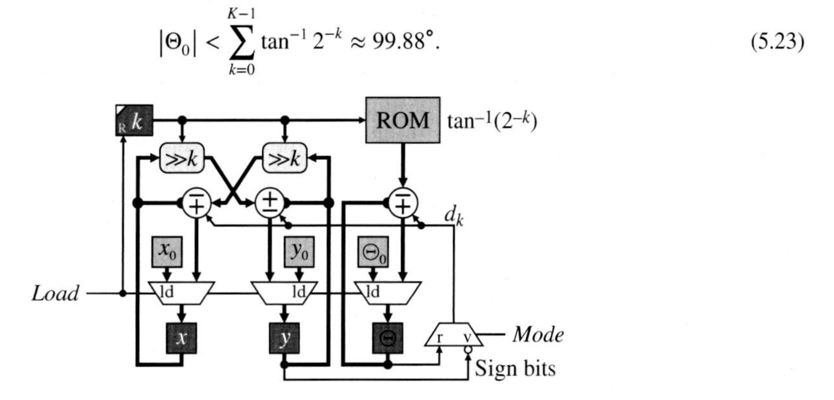

Figure 5.6 Iterative hardware implementation of the CORDIC algorithm.

Starting with $x_0 = \frac{1}{G}$ and $y_0 = 0$, the rotation mode can be used to calculate the sine and cosine of the angle in Θ_0. It may also be used to rotate a vector by an angle, subject to the extension of the magnitude by a factor of G. For angles larger than 90°, the easiest solution is to perform an initial rotation by 180°:

$$\begin{aligned} x_0 &= -x_{-1}, \\ y_0 &= -y_{-1}, \\ \Theta_0 &= \Theta_{-1} \pm 180^\circ. \end{aligned} \tag{5.24}$$

One way of speeding up CORDIC in rotation mode is to consider the Taylor series expansions for sine and cosine for small angles (Walther, 2000):

$$\begin{aligned} \sin\theta &= \theta\left(1 - \frac{1}{6}\theta^2 + \frac{1}{120}\theta^4 + \cdots\right) &&\approx \theta, \\ \cos\theta &= 1 - \frac{1}{2}\theta^2 + \frac{1}{24}\theta^4 + \cdots &&\approx 1.0. \end{aligned} \tag{5.25}$$

After $\frac{N}{2}$ iterations in rotation mode, the angle in the Θ register is such that θ^2 is less than the significance of the least significant guard bit, so it can be ignored. The remaining $\frac{N}{2}$ iterations may be replaced by two small multiplications (assuming that the angle is in radians):

$$\begin{aligned} x_N &= x_{\frac{N}{2}} - y_{\frac{N}{2}}\Theta_{\frac{N}{2}}, \\ y_N &= y_{\frac{N}{2}} + x_{\frac{N}{2}}\Theta_{\frac{N}{2}}. \end{aligned} \tag{5.26}$$

Note that this technique cannot be applied in vectoring mode because it relies on small angles.

The second mode of operation is vectoring mode. It reduces y to zero by choosing

$$d_k = -\text{sign}(y_k), \tag{5.27}$$

which gives the result

$$\begin{aligned} x_K &= G\sqrt{x_0^2 + y_0^2}, \\ y_K &= 0, \\ \Theta_K &= \Theta_0 + \tan^{-1}\left(\frac{y_0}{x_0}\right). \end{aligned} \tag{5.28}$$

This has the same domain of convergence as the rotation mode, given in Eq. (5.23). Therefore, if $x_0 < 0$, it is necessary to bring the vector within range by rotating by 180° (Eq. (5.24)). Vectoring mode effectively converts a vector from rectangular to polar coordinates (apart from the gain term, G). If x_0 and y_0 are both small, then rounding errors can become significant, especially in the calculation of the arctangent (Kota and Cavallaro, 1993). If necessary, this may be solved by renormalising x_k and y_k by scaling by a power of 2 (a left shift).

In both rotation and vectoring modes, each iteration improves the accuracy of the angle by approximately one binary digit.

A vector may be scaled to unit length without needing to perform a division or a square root by coupling two CORDIC units together, one operating in vectoring mode, and the other in rotation mode (Khan et al., 2017). The vectoring mode calculates the angle of the vector, while the same d_k directions control the rotation mode to rotate a unit vector in the opposite direction. Initialising the unit vector to $(\frac{1}{G}, 0)$ compensates for the CORDIC gain.

A third mode of operation has been suggested by Andraka (1998) for determining arcsine or arccosine. For arcsine, it starts with $x_0 = \frac{1}{G}$ and $y_0 = 0$ and rotates in the direction given by

$$d_k = \text{sign}(\sin\theta - y_k). \tag{5.29}$$

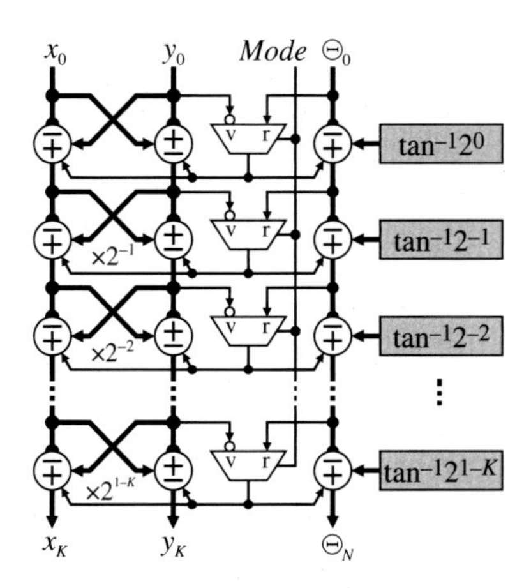

Figure 5.7 Unrolled CORDIC implementation.

The result after converging is

$$\begin{aligned} x_K &= \sqrt{G^2x_0^2 - \sin^2\theta} = \cos\theta, \\ y_K &= \sin\theta, \\ \Theta_K &= \Theta_0 + \sin^{-1}\left(\frac{\sin\theta}{Gx_0}\right) = \Theta_0 + \theta. \end{aligned} \tag{5.30}$$

This mode is effective for $|\sin\theta| < 0.98Gx_0$. Outside this range, the gain of the rotation can cause the wrong direction to be chosen, leading to convergence failure (Andraka, 1998). A similar approach can be used for arccosine.

The implementation in Figure 5.6 is word serial in that each clock cycle performs one iteration at the word level. This iterative loop may be unrolled (Figure 5.7) to use separate hardware for each iteration to perform one complete calculation per clock cycle (Wang et al., 1996; Andraka, 1998). Unrolling enables two simplifications to the design. First, the shifts at each iteration are fixed and can be implemented simply by wiring the appropriate bits to the inputs to the adders. Second, the angles that are added are constants, simplifying the Θ adders.

The propagation delay of the unrolled implementation is limited by the need to propagate the carry all of the way to the sign bit at each iteration before selecting the next operation. However, the design can be readily pipelined with a throughput of one calculation per clock cycle. Two approaches have been developed to reduce the latency, although they only work in rotation mode. The first is to work with the absolute value in the Θ register and detect the sign changes (Dawid and Meyr, 1992), which can be done with redundant arithmetic, working from the most significant bit down, significantly improving the delay. The second approach relies on the fact that the angles added to the Θ register are constant, enabling d_k to be predicted, significantly reducing the latency (Timmermann et al., 1992).

5.2.2.1 Compensated CORDIC

A major limitation of the CORDIC technique is the rotation gain. If necessary, this constant factor may be removed by a multiplication either before or after the CORDIC calculation. Approximate removal (3% error) can be achieved with a single addition by multiplying by $5/8$ (Pettersson and Petersson, 2005). Compensated CORDIC algorithms integrate the multiplication more accurately within the basic CORDIC algorithm in different ways.

Conceptually, the simplest (and the original compensated CORDIC algorithm) is to introduce a scale factor of the form 1 ± 2^{-k} with each iteration (Despain, 1974). The sign is chosen so that the combined gain approaches 1.

This can be generalised in two ways: applying the scale factor only on some iterations, and choosing the terms so that the gain is 2, enabling it to be removed with a simple shift. Since the signs are pre-determined, this results in only a small increase in the propagation delay, although the circuit complexity is considerably increased.

A second approach is to repeat the iterations for some angles to drive the total gain, G, to 2 (Ahmed, 1982). This uses the same hardware as the original CORDIC, with only a little extra logic required to control the iterations. The disadvantage is that it takes significantly longer than the uncompensated algorithm. A hybrid approach is to use a mixture of additional iterations and scale factor terms to reduce the total logic required (Haviland and Tuszynski, 1980). This is probably best applied to the unrolled implementation.

Another approach is to use a double rotation (Villalba et al., 1995). This technique is only applicable to the rotation mode; it performs two rotations in parallel (with different angles) and combines the results. Consider rotating by an angle $(\theta + \beta)$:

$$\begin{aligned} x_K^+ &= G\left(x_0(\cos\theta\cos\beta - \sin\theta\cos\beta) - y_0(\sin\theta\cos\beta + \cos\theta\sin\beta)\right), \\ y_K^+ &= G\left(x_0(\sin\theta\cos\beta + \cos\theta\sin\beta) + y_0(\cos\theta\cos\beta - \sin\theta\cos\beta)\right). \end{aligned} \tag{5.31}$$

Similarly, rotating by an angle $(\theta - \beta)$ gives:

$$\begin{aligned} x_K^- &= G\left(x_0(\cos\theta\cos\beta + \sin\theta\cos\beta) - y_0(\sin\theta\cos\beta - \cos\theta\sin\beta)\right), \\ y_K^- &= G\left(x_0(\sin\theta\cos\beta - \cos\theta\sin\beta) + y_0(\cos\theta\cos\beta + \sin\theta\cos\beta)\right). \end{aligned} \tag{5.32}$$

Then, adding these two results cancels the terms containing $\sin\beta$:

$$\begin{aligned} \frac{1}{2}\left(x_K^+ + x_K^-\right) &= G\left(x_0\cos\theta\cos\beta - y_0\sin\theta\cos\beta\right), \\ \frac{1}{2}\left(y_K^+ + y_K^-\right) &= G\left(x_0\sin\theta\cos\beta - y_0\cos\theta\cos\beta\right). \end{aligned} \tag{5.33}$$

Therefore, by setting $\cos\beta = G^{-1}$, the gain term can be completely cancelled. It takes the same time to perform the double rotation as the original CORDIC apart from the extra addition at the end. However, it does require twice the hardware to perform the two rotations in parallel.

5.2.3 Linear CORDIC

The CORDIC iterations may also be applied in a linear space (Walther, 1971):

$$\begin{aligned} x_{k+1} &= x_k, \\ y_{k+1} &= y_k + d_k 2^{-k} x_k, \\ z_{k+1} &= z_k - d_k 2^{-k}. \end{aligned} \tag{5.34}$$

Again, either rotation or vectoring mode may be used. Rotation mode for linear CORDIC reduces z to zero and in doing so is effectively performing a long multiplication with the result

$$y_K = Y_0 + x_0 z_0. \tag{5.35}$$

Vectoring mode reduces y to zero using Eq. (5.27) and effectively performs a non-restoring division, with the result

$$z_k = z_0 + \frac{y_0}{x_0}. \tag{5.36}$$

Usually, the initial rotation uses $k = 1$ for which Eq. (5.36) will converge provided $y_0 < x_0$. Note that there is no scale factor or gain term associated with linear CORDIC.

5.2.4 Hyperbolic Functions

Walther (1971) generalised CORDIC to operate in a hyperbolic coordinate space. The corresponding iterations are then

$$\begin{aligned} x_{k+1} &= x_k + d_k 2^{-k} y_k, \\ y_{k+1} &= y_k + d_k 2^{-k} x_k, \\ z_{k+1} &= z_k - d_k \tanh^{-1}(2^{-k}). \end{aligned} \tag{5.37}$$

Convergence issues require certain iterations to be repeated ($k = 4, 13, 40, \ldots, k_r, 3k_r + 1, \ldots$) (Walther, 1971). Note too that the iterations start with $k = 1$ because $-1.0 < \tanh x < 1.0$.

Subject to these limitations, operating in rotation mode gives the result

$$\begin{aligned} x_K &= G_h(x_0 \cosh z_0 + y_0 \sinh z_0), \\ y_K &= G_h(x_0 \sinh z_0 + y_0 \cosh z_0), \\ z_K &= 0, \end{aligned} \tag{5.38}$$

where the gain factor (including the repeated iterations) is

$$G_h = \prod_{k=1}^{K-1} \sqrt{1 - 2^{-2k}} \approx 0.82816, \tag{5.39}$$

and the domain of convergence is

$$|z_0| < \sum_{k=1}^{K-1} \tanh^{-1} 2^{-k} \approx 1.118. \tag{5.40}$$

Operating in vectoring mode gives the result

$$\begin{aligned} x_K &= G_h \sqrt{x_0^2 - y_0^2}, \\ y_K &= 0, \\ z_K &= z_0 + \tanh^{-1}\left(\frac{y_0}{x_0}\right). \end{aligned} \tag{5.41}$$

As with circular CORDIC, the gain may be compensated either by appropriate initialisation of x_0 or y_0 (for rotation mode) or by introducing additional, compensating operations. Much of the discussion of compensated CORDIC above may be adapted to hyperbolic coordinates.

Other common functions that may be calculated using hyperbolic coordinates include (Walther, 1971):

$$\exp x = \sinh x + \cosh x, \tag{5.42}$$

$$\ln x = 2\tanh^{-1}\left(\frac{x-1}{x+1}\right), \tag{5.43}$$

$$\sqrt{x} = \sqrt{\left(x + \frac{1}{4}\right)^2 - \left(x - \frac{1}{4}\right)^2}. \tag{5.44}$$

The CORDIC iterations from the different coordinate systems may be combined to use the same hardware (with a multiplexer to select the input to the x register).

5.2.5 Logarithms and Exponentials

In addition to the approach in Section 5.2.4, there is an alternative shift-and-add approach to calculating logarithms and exponentials. Consider the relation (Specker, 1965)

$$\ln x(1 + 2^k) = \ln x + \ln(1 + 2^k). \tag{5.45}$$

The idea is to successively reduce x to 1.0 by a series of multiplications (shift-and-add) and adding the corresponding logarithm terms obtained from a small table. The iteration is given by

$$\begin{aligned} x_{k+1} &= x_k(1 + d_k 2^{-k}), \\ z_{k+1} &= z_k - \ln(1 + d_k 2^{-k}), \end{aligned} \tag{5.46}$$

where

$$d_k = \begin{cases} 1, & x_k + x_k 2^{-k} < 1 \\ 0, & \text{otherwise} \end{cases} \tag{5.47}$$

will converge for $0.4195 < x_0 \leq 1.0$ to

$$\begin{aligned} x_K &= 1, \\ z_K &= z_0 + \ln x_0. \end{aligned} \tag{5.48}$$

Alternatively, if a factor of $(1 - d_k 2^{-k})$ is used, then the domain of convergence is $1.0 \leq x_0 < 3.4627$. For x_0 outside these ranges, it is necessary to prescale the input number to within this range. This is easily accomplished by shifting the input number (multiplying or dividing by a power of 2) and using a small lookup table to provide the logarithm of the corresponding factor. The iteration of Eq. (5.46) may also be readily adapted to taking logarithms in other bases simply by changing the table of constants added to the z register. The algorithm can also be run the other way, by reducing z_k to 0, giving the exponential function.

Muller (1985) showed that this iteration shared the same theory as the CORDIC iteration and placed them within the same framework. Since trigonometric functions can be derived from complex exponentials using the Euler identity,

$$e^{j\theta} = \cos\theta + j\sin\theta, \tag{5.49}$$

the iteration of Eq. (5.46) can be generalised to use complex numbers (Bajard et al., 1994), enabling trigonometric functions to be calculated without the gain factors associated with CORDIC. Another advantage of this implementation is that it allows redundant arithmetic to be used (Nielsen and Muller, 1996) by exploiting the fact that if $d_k \in \{-1, 0, 1\}$, then there is an overlap between the selection ranges for the d_k. This means that only the few most significant bits need to be calculated in order to determine the next digit, reducing the effects of carry propagation delay.

While the CORDIC and other related algorithms based on shift-and-add have relatively simple circuitry, their main limitation is their relatively slow rate of convergence. Each iteration gives only a single bit improvement of the result. A further limitation is that only a few elementary functions may be calculated in this way. More complex or composite functions require the evaluation of many elementary functions.

5.2.6 Lookup Tables

Many functions can be quite expensive to calculate. In some circumstances, an alternative is to pre-calculate the function for the range of expected inputs and store the results in a table. Then, rather than calculate the function, the input is simply looked up in the table, which will return the result.

There are two primary instances when the use of lookup tables can be particularly valuable. The first is when the latency of performing the calculation exceeds the available time constraint. Using a lookup table requires a single clock cycle for the memory access (for non-pipelined on-chip memory) regardless of the complexity of the function. The second is when the complexity of the calculation is such that excessive resources are used on the FPGA. A lookup table only requires a memory sufficiently large to provide an accurate representation of the function.

Let $f(x)$ be the function that is being approximated, and $\tilde{f}(x)$ be the approximation over the domain $a \leq x \leq b$. The error of the approximation is

$$\epsilon(x) = \tilde{f}(x) - f(x). \tag{5.50}$$

A small total squared error between the function and its approximation,

$$SE = \int_a^b \epsilon^2(x)\, dx, \tag{5.51}$$

is one indication of a good approximation. One limitation of the square error is that the absolute error can be large at some points. Therefore, it is usually better to minimise the maximum absolute error (de Dinechin and Tisserand, 2001), given by

$$MaxE = \max_{a \leq x \leq b} |\epsilon(x)|. \tag{5.52}$$

The difference between minimising the squared error and maximum error is illustrated in Figure 5.8, in particular in the first segment.

With lookup tables, there is a trade-off between accuracy and the size of the table. While lookup tables can be very efficient for low to moderate accuracy, they do not scale well, with the table size increasing exponentially with the width of the input. Practical table sizes require reducing the accuracy of the input. This can be achieved simply by dropping the least significant bits. It is not necessary to round the input, as the content of the table can be set to give the best result over the domain implied by the bits that are kept. This is clearly shown in the centre panel of Figure 5.8, where the table entry is set to the middle of the range for a particular bit combination. The error depends on the slope of $f(x)$, with the maximum slope determining the number of bits required on the input to achieve a particular accuracy on the output.

5.2.6.1 Interpolated Tables

Observe that the error function consists of a set of approximately linear segments. This can be exploited using linear interpolation. Rather than each entry in the table being a constant, each entry could be a line segment

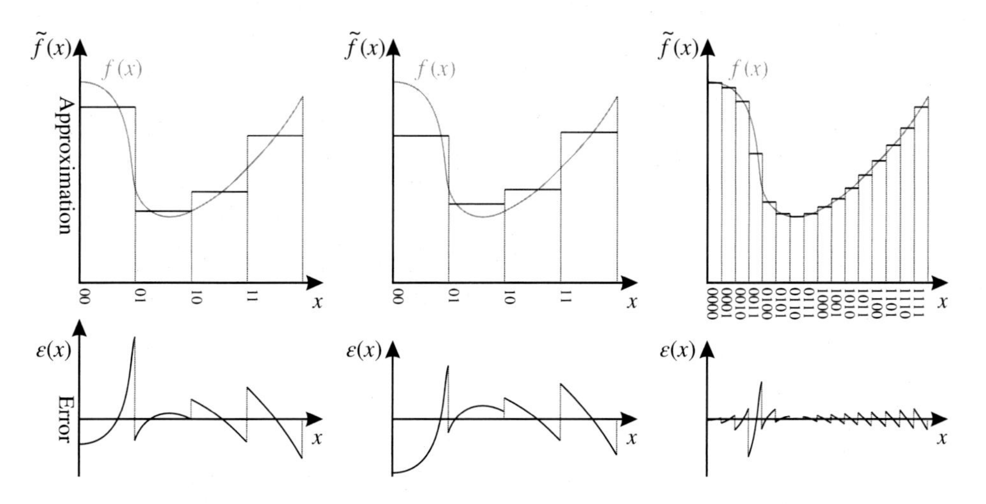

Figure 5.8 Approximating a function by lookup table. Left: minimising the mean square error of the function; centre and right: minimising the maximum absolute error; right: reducing the error by increasing the table size.

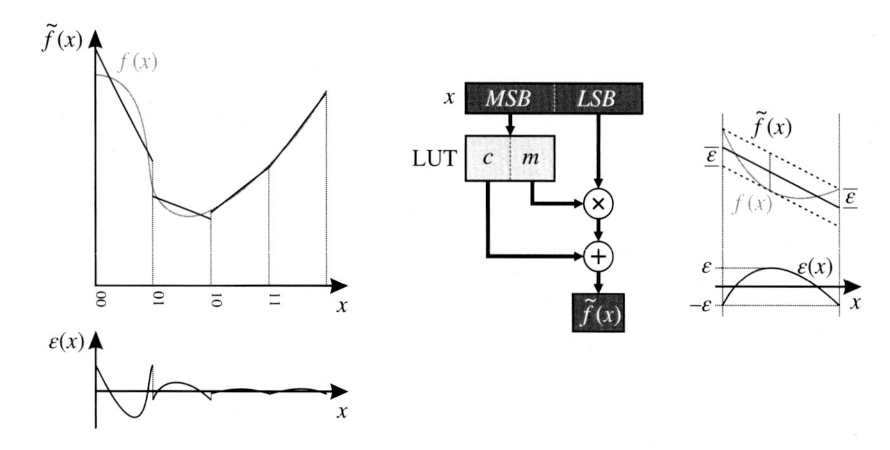

Figure 5.9 Linearly interpolated lookup table. Left: the approximation; centre: the computational architecture for Eq. (5.53); the table contains both the slope and the intercept; right: method of construction that minimises the maximum error.

defined by an intercept, c, and slope, m. The most significant bits are looked up to get the intercept and slope for the corresponding segment, with the least significant bits used to scale the slope to interpolate the values:

$$\tilde{f}(x) = c\left[x|_{MSB}\right] + m\left[x|_{MSB}\right] \times x|_{LSB}. \tag{5.53}$$

The linear segment approximation and the corresponding hardware (Mencer and Luk, 2004) are shown in Figure 5.9. The lookup tables for the intercept and slope have the same addresses; such parallel tables are effectively a single table with a wider data path. Note that it is not necessary for the piecewise linear segments to be continuous; smaller errors can in general be obtained by fitting each segment independently. Additional errors will arise from truncation of the least significant bits after the multiplication and also the finite precision of the slopes stored in the table.

An interpolated table effectively trades the size of the lookup table for increased computational resources and latency. The accuracy of the approximation will depend on the curvature, or the second derivative of $f(x)$. Where the curvature is higher, a smaller step size (more bits used to look up the segment) is required to maintain accuracy.

A variable step size (varying the partition between the most significant bits looked up and the least significant bits used for the interpolation) can significantly reduce the table size (Lee et al., 2003b; Lachowicz and Pfleiderer, 2008). The cost is a little more logic to manage the variable partitioning.

If the approximation is made continuous, as shown in Figure 5.10, then the slope is not needed. The current intercept and the next intercept are looked up using separate ports of a dual-port memory, with the slope calculated from the difference (Gribbon et al., 2003):

$$\tilde{f}(x) = c\left[x|_{MSB}\right] + \left(c\left[x|_{MSB} + 1\right] - c\left[x|_{MSB}\right]\right) \times 2^{-M} \times x|_{LSB}, \tag{5.54}$$

where 2^{-M} is the weight of the least significant bit used to address the lookup table. The 2^{-M} is a shift, which is free in hardware. The approximation error may be slightly larger than if each segment is fitted independently, but this will only occur if the second derivative changes sign, as shown between the first two segments in Figure 5.10.

For higher accuracy, each segment of the table may be represented by a higher order polynomial approximation. This extends Eq. (5.53) to

$$\begin{aligned}\tilde{f}(x) &= \sum_{i=0}^{d} p_i\left[x|_{MSB}\right] \times \left(x|_{LSB}\right)^i \\ &= \left(\left(p_d \times x|_{LSB} + p_{d-1}\right) \times x|_{LSB} + \cdots\right) \times x|_{LSB} + p_0,\end{aligned} \tag{5.55}$$

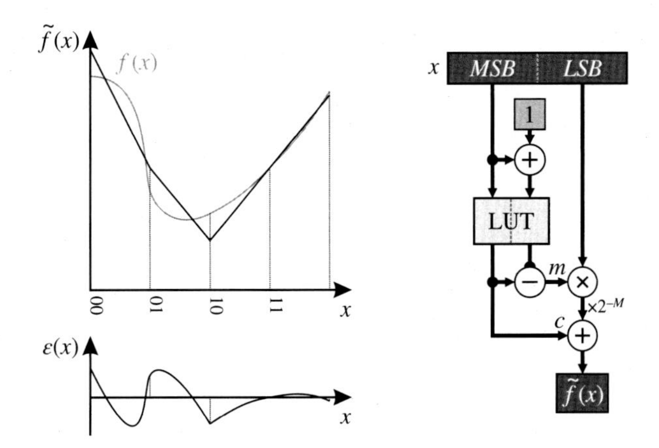

Figure 5.10 Linearly interpolated lookup table using dual-port RAM. The slope is calculated from successive table entries.

where d is the order of the polynomial with coefficients p_i. In practice, to control the accumulation of round-off errors, the polynomial should be evaluated using Horner's method (the second form).

For each segment, the optimum polynomial coefficients may be determined using Remez's exchange algorithm (Lee et al., 2003a). As with a linear approximation, the table size can often be significantly reduced using non-uniform segment widths at the expense of additional decoding logic (Lee et al., 2003a). A higher order polynomial requires fewer segments to fit to a given level of accuracy but requires more multipliers and therefore more resources. It is possible to implement a second-order approximation using only a single multiplier by making careful use of lookup tables (Detrey and de Dinechin, 2004).

5.2.6.2 Bipartite and Multipartite Tables

Observe that in Figure 5.9 the slopes of the last two segments are very similar. This is often the case, especially with smoothly varying functions. The average slope of the last two segments would also give reasonable errors. If the offsets resulting from multiplying the slope by the least significant bits were stored in a table, the multiplication can be avoided. Unfortunately, if the offsets were stored for every segment, the length of the offset table would be the same as looking up every value. However, since the offsets for adjacent segments are similar (because the slopes are similar), then several of these offsets can be shared between segments, giving a significant reduction in the size of the table.

This is the same concept that is used in tables of logarithms (Hassler and Takagi, 1995). Consider a standard four (decimal)-digit table of logarithms. Such a table would require 10^4 or 10,000 entries. Instead, it is arranged as a two-dimensional grid, with 100 rows indexed by the first two digits, and 10 columns indexed by the third digit. A second table, which contains offsets, has the same 100 rows indexed by the first two digits, also with 10 columns, but this time indexed by the fourth digit. The same offsets apply to the whole row of the first table. This bipartite table approach only requires 1000 entries in each table, for a total of 2000 entries. The savings are even greater than they first appear, because the numbers in the second table are smaller, so the bit-width can be narrower.

This approach to bipartite tables (Sarma and Matula, 1995; Hassler and Takagi, 1995) is shown in Figure 5.11. In this example, the eight segments are split into two groups of four for the offset table. Note that at this scale, the bipartite method is not particularly effective for the left group but works well for the right group where the slopes are similar. In general, the input word is split into three approximately equal groups of bits. The most significant two groups are used to index the main table and define the segments. The most and least significant groups are used to give the offsets that apply to the set of segments defined by the first group.

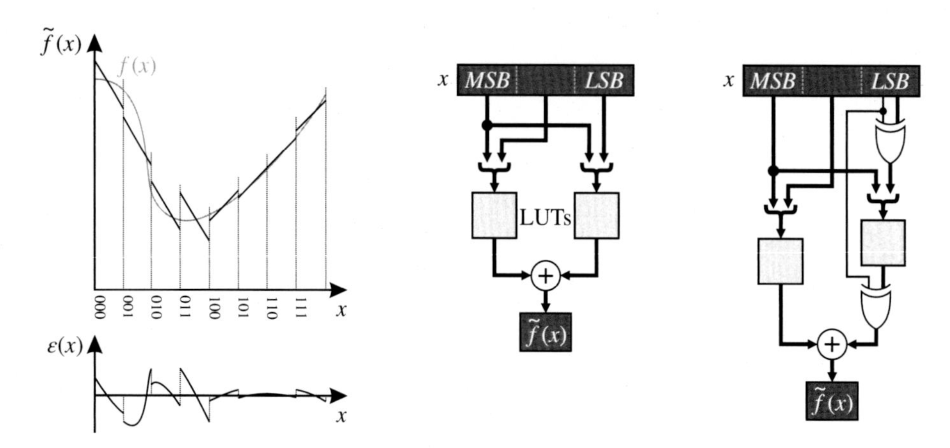

Figure 5.11 Bipartite lookup tables. Left: the approximation; centre: hardware implementation; right: symmetry is exploited to reduce the table size.

Since no multiplication is used, only an addition, bipartite tables are fast and relatively resource efficient. In addition, since the offsets are usually small, the width of the offset table is usually less than that of the main table. If the value in the main table represents the centre of the segment, the entries in the offset table may be made symmetrical. With a little extra logic, this symmetry may be exploited to halve the size of the offset table (Schulte and Stine, 1997). One of the input bits indicates whether the input value is in the left or right part of a segment. It can therefore be used to invert (using exclusive-OR gates) the other address bits within the segment to get the corresponding symmetrical position. The offset from the table is also inverted to give the symmetrical offset, as shown on the right in Figure 5.11.

The bipartite table idea may be extended further to give multiple tables in parallel (Stine and Schulte, 1999; de Dinechin and Tisserand, 2001), which enables the size of the tables to be reduced further, at the cost of an increased number of additions.

Overall, Boullis et al. (2001) and Mencer and Luk (2004) showed that for short inputs (up to about 8–10 bits), a direct table lookup is best. For intermediate size inputs (10–12 bits), bipartite tables require the smallest resources. However, for larger inputs (12–20 bits), interpolated lookup tables are the best compromise.

5.2.7 *Polynomial Approximations*

While the resources to implement a lookup table grow exponentially with output precision, that required for polynomial approximations grow only linearly. Therefore, for high-precision outputs, a polynomial approximation can be considered:

$$\tilde{f}(x) = \sum_{i=0}^{N} p_i x^i. \tag{5.56}$$

A minimax polynomial chooses the polynomial coefficients, p_i, to minimise the maximum approximation error within the domain of interest. Typically, Remez's exchange algorithm is used to find the coefficients. However, such coefficients do not take into account the fact that it is desirable to perform the calculation with limited precision or small multipliers, and simply rounding the coefficients may not give the coefficients with minimum error (Brisebarre et al., 2006). This requires a search for the appropriate coefficients.

An extension of polynomial approximations are rational approximations, where

$$\tilde{f}(x) = \frac{\sum_{i=0}^{N} n_i x^i}{\sum_{i=0}^{D} d_i x^i}. \tag{5.57}$$

For a given level of accuracy, a lower order rational polynomial is required (Koren and Zinaty, 1990) than a regular polynomial. This makes rational polynomial approximations best suited for high precision. However, the rational approximation does require a division operation.

5.2.8 Iterative Techniques

A further approach for high precision is to use an iterative approximation. The shift-and-add techniques described earlier have only linear convergence, which is slow for high-precision outputs. The Newton–Raphson root-finding algorithm,

$$x_{k+1} = x_k - \frac{f(x_k)}{f'(x_k)}, \tag{5.58}$$

has quadratic convergence, so the number of significant bits will approximately double with each iteration (provided the starting approximation is sufficiently close). It works, as illustrated in Figure 5.12, by projecting the slope at x_k to obtain a new estimate closer to the root.

For example, to evaluate the square root of X, an equation is formed for which $\sqrt{X}$ is a root:

$$f(x) = x^2 - X = 0, \tag{5.59}$$

with the corresponding iteration from Eq. (5.58)

$$x_{k+1} = x_k - \frac{x_k^2 - X}{2x_k} = \frac{1}{2}\left(x_k + \frac{X}{x_k}\right). \tag{5.60}$$

Note that the iteration of Eq. (5.60) requires a division, which is relatively expensive. Instead, solving for $\frac{1}{\sqrt{X}}$ and multiplying by x gives $\sqrt{X}$. The corresponding function to solve would be

$$f(x) = x^{-2} - X = 0, \tag{5.61}$$

with the corresponding iteration

$$x_{k+1} = x_k - \frac{x_k^{-2} - X}{-2x_k^{-3}} = \frac{1}{2}x_k\left(3 - x_k^2 X\right). \tag{5.62}$$

Although Eq. (5.62) is more complex than Eq. (5.60), it does not involve a division and can almost certainly be implemented with lower latency.

The number of iterations required to reach a desired level of accuracy depends on the accuracy of the initial estimate. Therefore, other techniques such as lookup tables and other approximations are helpful to give a more accurate initial approximation and reduce the number of iterations (Schwarz and Flynn, 1993). Since the iteration for many functions is a polynomial (for example Eq. (5.62)), the same hardware may be used to form the initial approximation (Habegger et al., 2010).

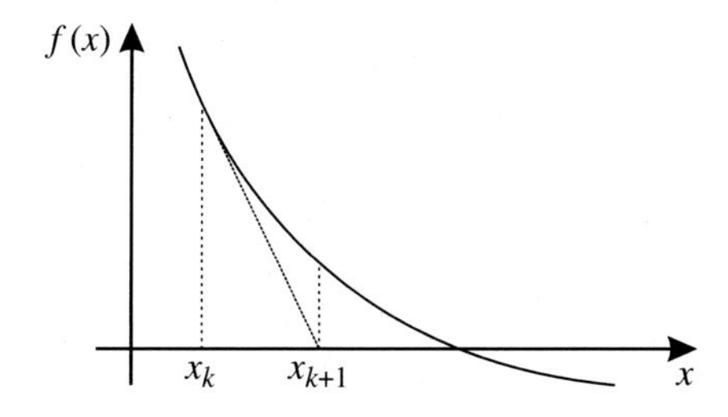

Figure 5.12 Newton–Raphson iteration projects the slope to zero giving the next approximation.

5.3 Other Computation Techniques

Two other techniques that are commonly employed in image processing are incremental update and exploiting separability.

5.3.1 *Incremental Update*

Incremental update reduces the computation by reusing part or all of prior computations. In particular, with stream processing, the known scan sequence enables the calculation to be simplified using the calculation of the previous pixel.

A classic example (Gribbon et al., 2003) is the calculation of x^2 or y^2, where x and y are pixel coordinates. When moving to the next pixel, the identity

$$(x+1)^2 = x^2 + 2x + 1 \tag{5.63}$$

can be used to maintain the value of x^2 without performing a multiplication. The same concept may be extended to calculate higher order polynomials (Chan et al., 1996). A similar technique may be used to determine the input location for an affine transformation incrementally (see Section 10.3.1).

Another example of incremental update is with local filters. As the window moves from one pixel to the next, there is a significant overlap between the corresponding input windows. Many of the pixel values within the window will be the same as the window moves. This may be exploited, for example, by median filters by maintaining a histogram of the values within the window and updating the histogram only for the pixels that change (Garibotto and Lambarelli, 1979; Huang et al., 1979; Fahmy et al., 2005).

5.3.2 *Separability*

Image processing often involves operations that require inputs from two (or more) dimensions. Separable operations are those that can split the 2-D operations into separate operations in each of the X and Y dimensions, which can then be performed either in parallel or sequentially (pipelined). The separate 1-D operations are usually significantly simpler than the composite 2-D operation. The most common application of separability is with filters, but it can also be applied to two-dimensional transformations such as the fast Fourier transform or be used with image warping (Catmull and Smith, 1980).

5.4 Memory Structures

Many algorithms have associated data structures that are stored within memory. On an FPGA, the distributed memory blocks mean that these structures can be independent, enabling the memory-based data structures to be tailored to each algorithm. However, memory is inherently serial. Although internally, data is stored in parallel, with a separate latch (or capacitor for dynamic memory) per bit stored, the addressing and access mechanisms mean that only one memory location can be selected for reading or writing at a time. Multiple data items can therefore only be accessed serially, with the memory bandwidth ultimately limiting the speed of the algorithm (see also Section 4.2).

In this section, the FPGA implementation of several basic data structures will be explored.

5.4.1 *FIFO Buffer*

A first-in first-out (FIFO) buffer behaves as a queue, with the data read out of the buffer in the same order that it was written. Within stream processing, it can be used to relax the strict timing of processing one pixel per clock cycle. Placing a FIFO between operations can decouple the timing of the two operations, so they no longer need to be synchronised. It is commonly used when streaming data to or from external memory, especially dynamic RAM where there may be variable latency depending on the current state of the memory controller. A FIFO

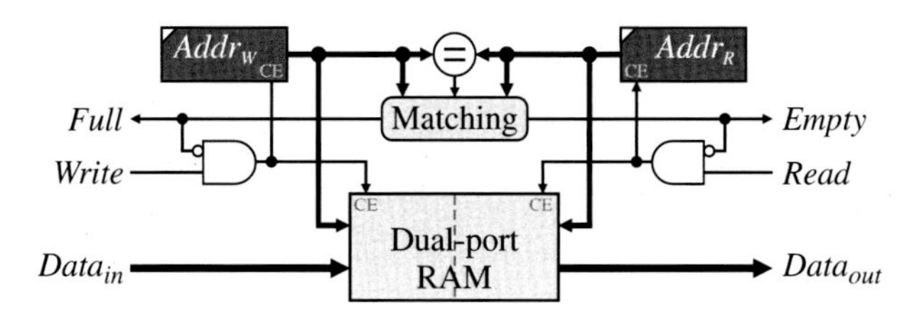

Figure 5.13 Circular memory-based FIFO buffer. The full and empty detection logic is generally not required for use as a row buffer.

can also be used when streaming data between clock domains, so that the data is not subjected to the delays associated with handshaking for every word transferred.

A FIFO buffer may be built from a dual-port memory, as shown in Figure 5.13. Two counters are needed, one for providing the address for writing to the memory and one for reading. The memory is used as a ***circular buffer***, with the first address appearing sequentially after the last address within the memory. For blocks of memory that are a power of 2 long, this is achieved by allowing the address counters to wrap around. If the memory is other than a power of 2 long, then the counter has to be modified to reset to 0 after the last address is reached. A small amount of logic is also required to detect when the buffer is full or empty, although if the FIFO is being used as a row buffer, this is not necessary.

The buffer is empty when the read and write addresses are identical, and the last operation was a read. It is full, if the addresses are identical, and the last operation was a write. There are several approaches for deriving these signals. One is to build a state machine which keeps track of the state. An alternative (for power of 2 buffer sizes) is to extend the counters by one bit and use the most significant bits of the counters to detect full (MSBs are different) or empty (MSBs are the same), as shown on the left panel of Figure 5.14 (Cummings, 2002).

When using the FIFO to cross between clock domains, the comparison for *Full* must be performed in the writing clock domain, and the comparison for *Empty* performed in the reading clock domain. This requires the read and write addresses to be passed to the other clock domain. To minimise delays, the addresses can be converted to Gray-code, so that only one bit changes at a time, enabling a separate synchroniser to be used for each bit. The synchroniser delays the removal of *Empty* when data is added, or *Full* when data is removed, but the data will not over- or underflow because the comparison is performed locally.

Alternatively, the address comparison can be performed asynchronously (Cummings and Alfke, 2002), with the *Full* and *Empty* conditions detected by the counters changing quadrants, as shown in the right panel of Figure 5.14. The *Empty* and *Full* outputs of the matching network need to be synchronised in the appropriate clock domain using the asynchronous set and synchronous clear from Figure 4.7.

In many applications, it is necessary to maintain a column address, even if only for the purposes of reading from or writing to a frame buffer. When using the FIFO as a row buffer, this column address may be used to directly index the buffer rather than using a separate counter. The buffer is then no longer circular but is automatically indexed to the current position in the image.

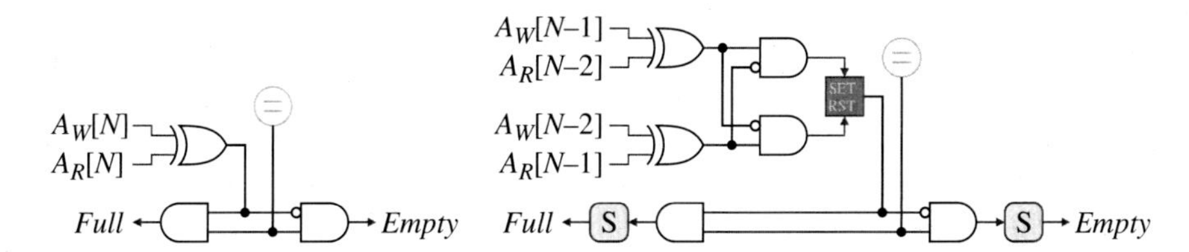

Figure 5.14 Matching network for detecting buffer full and empty. Left: binary counters, with an extra address bit; right: Gray-code counters with quadrant detection; S is an asynchronous set and synchronous clear synchroniser.

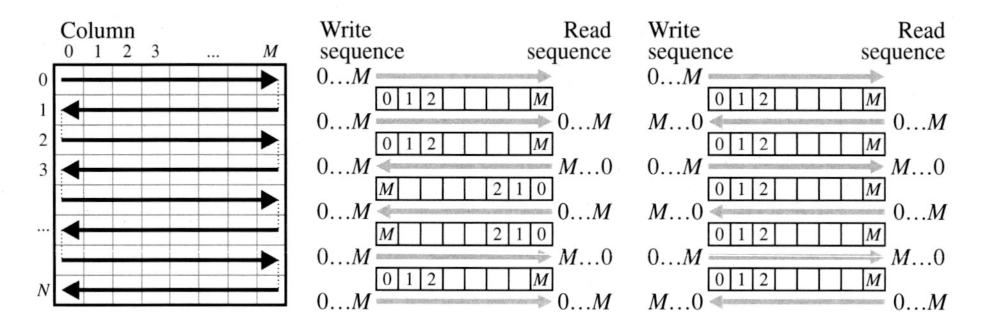

Figure 5.15 Left: zigzag scan sequence. Centre: operation of a zigzag reordering buffer. Source: Adapted from Bailey and Klaiber (2019). Right: operation of a zigzag row buffer. Source: Adapted from Bailey and Klaiber (2019).

Most FPGA vendor tools provide FIFO intellectual property blocks that contain the above structures, simplifying the design in many applications.

5.4.2 Zigzag Buffers

A zigzag reordering buffer converts a standard raster scan stream into a zigzag or serpentine scanned stream, where every second line is scanned from right to left as demonstrated on the left in Figure 5.15. While this can be achieved using double buffering (writing one row to one buffer while reading the previous row from a second buffer), the centre panel of Figure 5.15 shows how this can be accomplished using a single buffer (Bailey and Klaiber, 2019). The trick is to base the indexing on the column number and to read and write in both directions. After the first row, the read and write are to the same address, with the direction reversing every second row. The average latency of the reordering is one row period.

Similarly, a zigzag row buffer provides the pixel value immediately above the current pixel during a zigzag scan. Again, a single buffer can be used as shown on the right in Figure 5.15, reading and writing to the same address, with the direction reversing every row (Bailey, 2012; Bailey and Klaiber, 2019).

5.4.3 Stacks

Stacks are a form of first-in last-out or last-in first-out (LIFO) buffer. Items are ***pushed*** onto a stack by storing them in successive memory locations. When an item is ***popped*** from the stack, the most recently saved item is retrieved first.

A stack can be implemented using a single port memory, combined with a single address counter as the stack pointer, *SP*, as shown in Figure 5.16. It is usual for the stack pointer to address the next available free location.

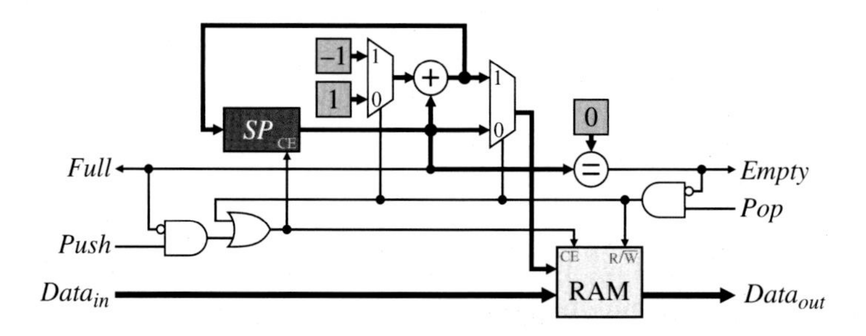

Figure 5.16 Implementation of a stack using a RAM block.

A push writes to the address at the top of stack, as pointed to by *SP*, and increments the stack pointer to point to the next free location. A pop needs to decrement the stack pointer before reading the data on the top of the stack.

In the implementation shown here, a multiplexer is used to either increment or decrement *SP* depending on whether a *Push* or *Pop* operation is used, respectively. A *Push* uses *SP* directly as the RAM address, whereas a *Pop* uses the decremented *SP* as the address. The stack is full after the last memory element is written. Incrementing *SP* will wrap it past the end; if the number of entries is a power of 2, this can be prevented by making *SP* one bit longer than needed to address the memory. The stack is then full when the most significant bit of *SP* is set. The stack is empty if the address is zero; in this condition, a *Pop* is not allowed because there is nothing left on the stack to pop.

5.4.4 Linked Lists

A linked list is useful for maintaining a dynamic collection of data items. Each item is stored in a separate location in memory; however, successive items are not necessarily in sequential memory locations. Associated with each item is a pointer to the next item (this is just the index or memory address of that item), resulting in an implicit ordering of the items within the list.

If ordering is important, linked lists have several advantages over arrays. To insert an item into an array, all the successive items in the array must be moved to new memory locations to make space at the appropriate position in the array. However, with a linked list, the new item can be stored in any free memory location, and the pointers adjusted to put it into the correct place in the sequence. Similarly, when deleting an item from an array, all subsequent items must be shifted back. With a list, it is only necessary to adjust the pointers to skip the item that is removed.

The first item in the list is called the ***head***. The last item has a ***null pointer***, Ø, indicating that there is no next item. In software, 0 is often used as the null pointer because 0 is not usually a valid data address. However, local memory blocks are used on FPGAs, with 0 being a valid data address. This may be resolved either using a separate flag to indicate a null pointer or not using location 0 to hold data. The former approach requires one extra bit per pointer, but it does mean that only one bit needs to be tested to determine if the pointer is null. Reserving location 0 means one less item may be stored in a block of memory. An alternative is to have the end of the list point to itself. This enables items within the list to detect the null pointer, but external pointers to the list would require a separate flag.

One issue with linked lists is keeping track of which memory locations are free. There are two approaches to solving this problem. One is to maintain a register with one bit per memory location to indicate if the corresponding location is free. Associated with this is a priority encoder to translate the first '1' into the corresponding address. While this is suitable for short lists, it can quickly get expensive as the address space gets larger. A second approach is to reuse the linking mechanism to maintain a second list of unused entries. This requires that the memory be initialised with all of the entries linked in the free list before being used.

Most operations on a list require a sequence of several accesses and would require a finite state machine to perform the sequencing. Consider, for example, inserting a new entry after a currently selected entry (see Figure 5.17). Here, *Ptr* points to *D*2, after which a new data item *D*4 will be inserted. *Free* points to the chain of unused entries. Four memory accesses are required:

- *D*2 is read to get the remainder of the list, headed by *D*3.
- *D*2 is then linked to the head of the free list, which will hold the new data, *D*4.
- *F*4 is read to get the remainder of the free list, and *Free* is updated.
- The new data *D*4 is then written in the location of *F*4, with the pointer to the rest of the list, *D*3.

Different operations (initialisation, traversing the list, insertion, and deletion) could be managed by branches within the controlling state machine. Since each operation requires several clock cycles, it is also necessary to include synchronisation logic between the linked list and the rest of the design.

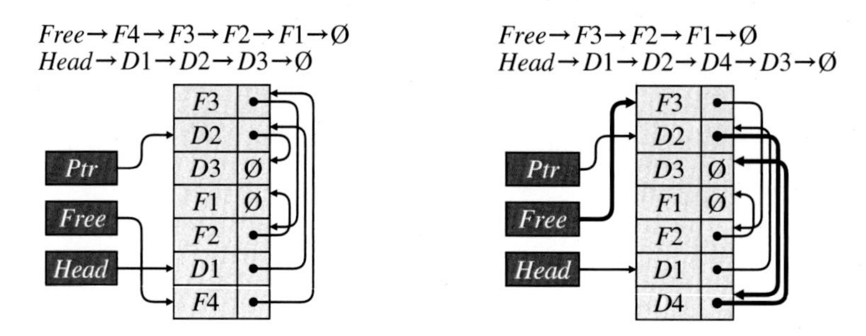

Figure 5.17 Inserting an element into a linked list. Left: initial state; right: final state, changed links are shown in bold.

Note that the sequential ordering requires an insertion or deletion to be either from the head of the list or from a following item. This is because there is no direct reference to previous items in a list (unless a doubly linked list is used, with pointers to both the previous and next items in the list). In addition to requiring wider memory to hold the extra link, a doubly linked list also requires more memory accesses for each operation to maintain the structure.

5.4.5 Trees

Another data structure commonly used in computer vision to represent hierarchically structured data is a tree. An example tree is shown on the left in Figure 5.18. The node at the top of the tree is called the ***root*** node; node *A* is the root in this example. The nodes immediately below a node are called the ***child*** nodes. For example, *D*, *E*, and *F* are children of *B*, and *B* is their parent. Nodes that have no children are ***leaf*** nodes. In this example, *D*, *F*, *G*, *I*, *J*, and *K* are leaf nodes.

In a general tree, a node can have an arbitrary number of children, making it inconvenient for a node to point to all of its children. One general representation of a tree is as a form of two-dimensional linked list. Every node has two pointers: one to its first child (or null if it is a leaf node) and one to its next sibling. In this way, all of the children of a node are contained in a linked list. This approach to representing a tree is shown in the centre of Figure 5.18.

This representation is suitable for navigating from the root of a tree down to a node, but navigating back up the tree is more difficult. This may be overcome by adding a third pointer to each node, linking to its parent. A more efficient alternative in many applications is to use a separate memory structure to remember the context of a node. For example, when performing a depth-first traversal of the tree, parent nodes can be remembered on a stack. As the link from a node to its first child is taken, a pointer to the node can be pushed onto a stack.

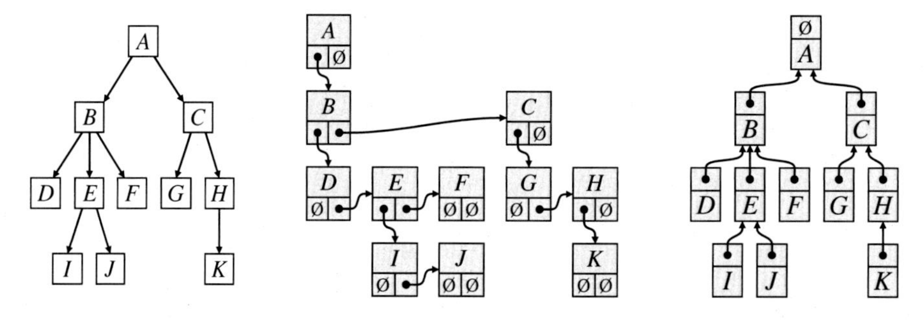

Figure 5.18 A representation of a tree structure. Left: abstract tree; centre: representation with two links per node, one to the first child and one to the next sibling; right: representation with a single link to the parent.

Then, to move back up the tree after reaching a leaf, the most recently visited parent can be popped from the stack. The stack therefore provides the required link from a child back to its parent. If performing a breadth-first search, it is more appropriate to use a FIFO queue. As a node is visited, the pointer to the first child node (if not Ø) is stored in the queue. Then, when the last child is reached, the first entry in the queue is the next node to visit.

Some applications, for example connected component analysis, do not need to traverse the tree but need to be able to find the root node. The corresponding tree structure has each node with only a single link, to its parent, as demonstrated in the right of Figure 5.18.

One particular type of tree that deserves special mention is a ***binary tree***. It has at most two children; so, for a binary tree, it is usual to directly point to the two children rather than using a list. Similarly, a ***quad-tree*** has four children, commonly used for partitioning 2-D space into four disjoint regions.

5.4.6 *Graphs*

Graphs are a generalisation of trees in that any node can link to any other node. Within a graph, the nodes are called vertices, and the links between them are arcs. Arcs can either be undirected (usually indicating an association or similar relationship) or directed (indicating dependence, precedence, or similar relationship). Both vertices and arcs may hold associated data.

Two common representations of graphs are through an adjacency matrix, or links. An adjacency matrix has one row and column for each vertex, with the intersection being '0' or '1' to indicate the presence of an arc between the corresponding vertices, as shown in Figure 5.19. A undirected arc requires two entries, one for each direction, making the adjacency matrix symmetrical. The alternative representation is as a set of linked lists. The vertices are contained with one linked list, with each vertex pointing to a linked list of outgoing arcs (pointing to vertices), as illustrated in Figure 5.19.

The adjacency matrix representation can be stored in either fabric RAM or block RAM and is efficient for a smaller number of vertices when the degree (number of connections) is high. When the degree is low, or when the connectivity is sparse, the linked list can be more efficient. As with linked lists and trees, several clock cycles may be required for each operation on the graph.

5.4.7 *Hash Tables*

A sparse array is an array with a large address space, but only a small proportion of the array has data at any one time. If memory is reserved for the whole array, this can be wasteful of the memory resources available on an FPGA. While the memory requirement of the array may be significantly reduced using a linked list or tree to

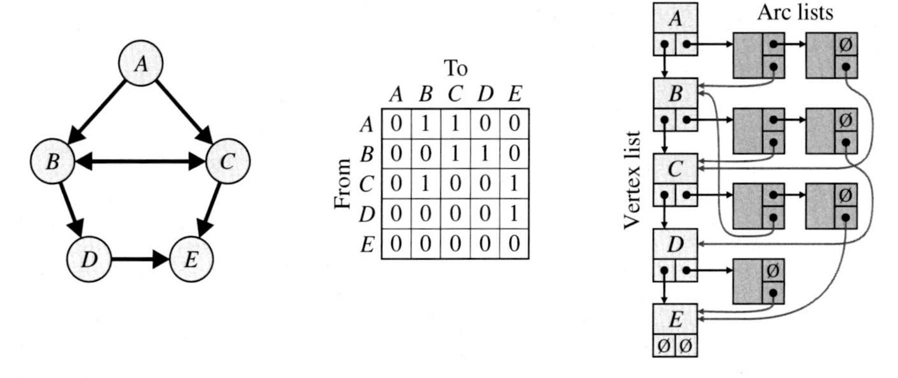

From \ To	A	B	C	D	E
A	0	1	1	0	0
B	0	0	1	1	0
C	0	1	0	0	1
D	0	0	0	0	1
E	0	0	0	0	0

Figure 5.19 A representation of a graph. Left: directed graph; centre: adjacency matrix representation; right: linked lists of nodes and connecting arcs.

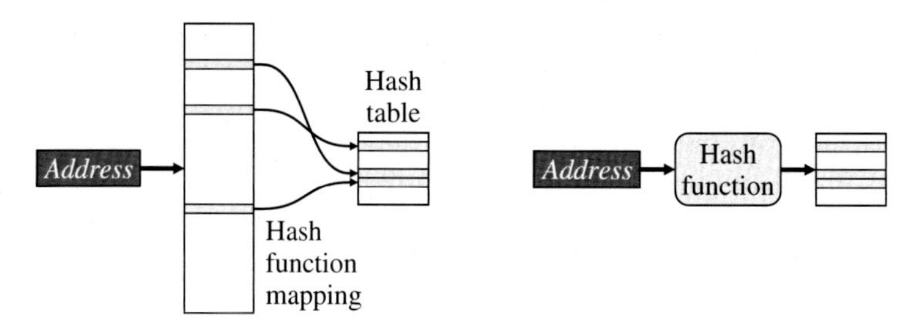

Figure 5.20 Address hashing. Left: conceptual operation of mapping the original sparse address space into a smaller hash table; right: this is accomplished using a hash function.

represent the occupied data elements, these structures incur a significant overhead of having to search through the list or tree to find a particular element.

An alternative approach is to store the data using a hash table. As shown in Figure 5.20, the original address is mapped using a hash function to a much smaller address space, that of the hash table, where the data is actually stored. The reduction in memory comes at the cost of evaluating the hash function. The input addresses do not necessarily need to be memory locations but could be strings or other complex data structures that are reduced using the hash function to provide an index into the hash table.

A good hash function is one that maps the different input addresses that are actually used to different slots within the hash table. If the input addresses can be considered random, then a simple and effective hash function can be to use a subset of the bits of the input address. However, if the set of input addresses are clustered, then a good hash function will distribute these to quite different slots to reduce the chance that multiple input addresses will map to the same slot.

Inevitably, though, there will be address collisions, with multiple input addresses mapping to a common slot. There are two main approaches for resolving such collisions. One is to store multiple elements per slot, using a linked list, for example. This requires additional data structures to hold the list. Another alternative is to store the data in the next available slot (Peterson, 1957). Both methods require a search in the event of collisions. The search is kept small by ensuring that the used slots are randomly distributed, and the hash table is not too full. This approach, however, is sensitive to the fill factor of the hash table (the proportion of slots that are used), and it is generally better to keep the maximum fill factor below about 75%. There are several advanced techniques (Askitis, 2009) that can be used to increase the fill factor, but these generally come at the expense of more complex hash functions.

A hash table can be considered to be a form of associative memory. The sparse addressing of hash tables makes them appropriate for implementing caches. When using a hash table for a cache, the previous entry can be discarded whenever a collision occurs, avoiding the need for searching.

5.5 Summary

This chapter has reviewed a selection of computational techniques from the perspective of an efficient FPGA implementation. First, several number representations were explored, both for integer and real numbers. Then, a diverse selection of techniques were described for implementing elementary functions. Many were based on shift-and-add operations which can be implemented efficiently on an FPGA. Alternatives are lookup tables for low to medium precision, and polynomial approximations and iterative techniques for high precision. Incremental update exploits the predictable patterns associated with stream-based processing, and separability can reduce two-dimensional computations to a sequence of 1-D operations. Finally, a selection of memory-based structures were introduced, and their mapping onto an FPGA outlined. These techniques form the building blocks of many of the image processing algorithms discussed in Chapters 7–14.

References

Ahmed, H.M. (1982). Signal processing algorithms and architectures. PhD thesis. Stanford, CA: Stanford University.

Andraka, R. (1998). A survey of CORDIC algorithms for FPGA based computers. *ACM/SIGDA 6th International Symposium on Field Programmable Gate Arrays*, Monterey, CA, USA (22–25 February 1998), 191–200. https://doi.org/10.1145/275107.275139.

Arno, S. and Wheeler, F.S. (1993). Signed digit representations of minimal Hamming weight. *IEEE Transactions on Computers* **42** (8): 1007–1010. https://doi.org/10.1109/12.238495.

Askitis, N. (2009). Fast and compact hash tables for integer keys. *32nd Australasian Computer Science Conference (ACSC 2009)*, Wellington, NZ (19–23 January 2009), CRPIT, Volume **91**, 101–110.

Avizienis, A. (1961). Signed-digit number representation for fast parallel arithmetic. *IRE Transactions on Electronic Computers* **10** (3): 389–400. https://doi.org/10.1109/TEC.1961.5219227.

Bailey, D.G. (2006). Space efficient division on FPGAs. *Electronics New Zealand Conference (ENZCon'06)*, Christchurch, NZ (13–14 November 2006), 206–211.

Bailey, D.G. (2012). Accelerating the distance transform. *Image and Vision Computing New Zealand*, Dunedin, NZ (26–28 November 2012), 162–167. https://doi.org/10.1145/2425836.2425872.

Bailey, D.G. and Klaiber, M.J. (2019). Zig-zag based single pass connected components analysis. *Journal of Imaging* **5** (4): Article ID 45, 26 pages. https://doi.org/10.3390/jimaging5040045.

Bajard, J.C., Kla, S., and Muller, J.M. (1994). BKM: a new hardware algorithm for complex elementary functions. *IEEE Transactions on Computers* **43** (8): 955–963. https://doi.org/10.1109/12.295857.

Booth, A.D. (1951). A signed binary multiplication technique. *Quarterly Journal of Mechanics and Applied Mathematics* **4** (2): 236–240. https://doi.org/10.1093/qjmam/4.2.236.

Boullis, N., Mencer, O., Luk, W., and Styles, H. (2001). Pipelined function evaluation on FPGAs. *IEEE Symposium on Field-Programmable Custom Computing Machines (FCCM'01)*, Rohnert Park, CA, USA (30 April – 2 May 2001), 304–306. https://doi.org/10.1109/FCCM.2001.37.

Brisebarre, N., Muller, J.M., and Tisserand, A. (2006). Computing machine-efficient polynomial approximations. *ACM Transactions on Mathematical Software* **32** (2): 236–256. https://doi.org/10.1145/1141885.1141890.

Catmull, E. and Smith, A.R. (1980). 3-D transformations of images in scanline order. *ACM SIGGRAPH Computer Graphics* **14** (3): 279–285. https://doi.org/10.1145/965105.807505.

Chan, F.H.Y., Lam, F.K., Li, H.F., and Liu, J.G. (1996). An all adder systolic structure for fast computation of moments. *Journal of VLSI Signal Processing* **12** (2): 159–175. https://doi.org/10.1007/BF00924524.

Chaurasiya, R., Gustafson, J., Shrestha, R., Neudorfer, J., Nambiar, S., Niyogi, K., Merchant, F., and Leupers, R. (2018). Parameterized posit arithmetic hardware generator. *IEEE 36th International Conference on Computer Design (ICCD)*, Orlando, FL, USA (7–10 October 2018), 334–341. https://doi.org/10.1109/ICCD.2018.00057.

Constantinides, G.A., Cheung, P.Y.K., and Luk, W. (2001). The multiple wordlength paradigm. *IEEE Symposium on Field Programmable Custom Computing Machines (FCCM'01)*, Rohnert Park, CA, USA (30 April – 2 May 2001), 51–60. https://doi.org/10.1109/FCCM.2001.46.

Cummings, C.E. (2002). Simulation and synthesis techniques for asynchronous FIFO design. *Synopsis Users Group Conference (SNUG 2002)*, San Jose, CA, USA (13–15 March 2002), 23 pages.

Cummings, C.E. and Alfke, P. (2002). Simulation and synthesis techniques for asynchronous FIFO design with asynchronous pointer comparisons. *Synopsis Users Group Conference (SNUG 2002)*, San Jose, CA, USA (13–15 March 2002), 18 pages.

Dawid, H. and Meyr, H. (1992). VLSI implementation of the CORDIC algorithm using redundant arithmetic. *IEEE International Symposium on Circuits and Systems (ISCAS '92)*, San Diego, CA, USA (10–13 May 1992), Volume **3**, 1089–1092. https://doi.org/10.1109/ISCAS.1992.230290.

de Dinechin, F. and Tisserand, A. (2001). Some improvements on multipartite table methods. *IEEE Symposium on Computer Arithmetic*, Vail, CO, USA (11–13 June 2001), 128–135. https://doi.org/10.1109/ARITH.2001.930112.

de Dinechin, F., Forget, L., Muller, J.M., and Uguen, Y. (2019). Posits: the good, the bad and the ugly. *Conference for Next Generation Arithmetic 2019*, Singapore (2019), Article ID 6, 10 pages. https://doi.org/10.1145/3316279.3316285.

Despain, A.M. (1974). Fourier transform computers using CORDIC iterations. *IEEE Transactions on Computers* **23** (10): 993–1001. https://doi.org/10.1109/T-C.1974.223800.

Detrey, J. and de Dinechin, F. (2004). Second order function approximation using a single multiplication on FPGAs. *14th International Conference on Field Programmable Logic and Application*, Antwerp, Belgium (29 August – 1 September 2004), *Lecture Notes in Computer Science*, Volume **3203**, 221–230. https://doi.org/10.1007/b99787.

Ewe, C.T., Cheung, P.Y.K., and Constantinides, G.A. (2004). Dual fixed-point: an efficient alternative to floating-point computation. *14th International Conference on Field Programmable Logic and Applications*, Antwerp, Belgium (29 August – 1 September 2004), *Lecture Notes in Computer Science*, Volume **3203**, 200–208. https://doi.org/10.1007/b99787.

Ewe, C.T., Cheung, P.Y.K., and Constantinides, G.A. (2005). Error modelling of dual fixed-point arithmetic and its application in field programmable logic. *International Conference on Field Programmable Logic and Applications*, Tampere, Finland (24–26 August 2005), 124–129. https://doi.org/10.1109/FPL.2005.1515710.

Fahmy, S.A., Cheung, P.Y.K., and Luk, W. (2005). Novel FPGA-based implementation of median and weighted median filters for image processing. *International Conference on Field Programmable Logic and Applications*, Tampere, Finland (24–26 August 2005), 142–147. https://doi.org/10.1109/FPL.2005.1515713.

Forget, L., Uguen, Y., and de Dinechin, F. (2019). Hardware cost evaluation of the posit number system. *Compas' 2019-Conference d'informatique en Parallelisme, Architecture et Systeme*, Anglet, France (24–28 June 2019), 1–7.

Fu, H., Mencer, O., and Luk, W. (2006). Comparing floating-point and logarithmic number representations for reconfigurable acceleration. *International Conference on Field Programmable Technology*, Bangkok, Thailand (13–15 December 2006), 337–340. https://doi.org/10.1109/FPT.2006.270342.

Fu, H., Mencer, O., and Luk, W. (2008). Optimizing residue arithmetic on FPGAs. *International Conference on Field Programmable Technology*, Taipei, Taiwan (7–10 December 2008), 41–48. https://doi.org/10.1109/FPT.2008.4762364.

Garibotto, G. and Lambarelli, L. (1979). Fast online implementation of two-dimensional median filtering. *Electronics Letters* **15** (1): 24–25. https://doi.org/10.1049/el:19790018.

Gigliotti, P. (2004). Implementing Barrel Shifters Using Multipliers. Application note XAPP195 (v1.1), Xilinx Inc, USA.

Goldberg, D. (1991). What every computer scientist should know about floating-point arithmetic. *ACM Computing Surveys* **23** (1): 6–48. https://doi.org/10.1145/103162.103163.

Gribbon, K.T., Johnston, C.T., and Bailey, D.G. (2003). A real-time FPGA implementation of a barrel distortion correction algorithm with bilinear interpolation. *Image and Vision Computing New Zealand (IVCNZ'03)*, Palmerston North, NZ (26–28 November 2003), 408–413.

Gustafson, J.L. and Yonemoto, I.T. (2017). Beating floating point at its own game: posit arithmetic. *Supercomputing Frontiers and Innovations* **4** (2): 71–86. https://doi.org/10.14529/jsfi170206.

Habegger, A., Stahel, A., Goette, J., and Jacomet, M. (2010). An efficient hardware implementation of a reciprocal unit. *5th IEEE International Symposium on Electronic Design, Test and Applications (DELTA 2010)*, Ho Chi Minh City, Vietnam (13–15 January 2010), 183–187. https://doi.org/10.1109/DELTA.2010.65.

Haselman, M. (2005). A comparison of floating point and logarithmic number systems for FPGAs. Master of Science thesis. Seattle, WA, USA: University of Washington.

Haselman, M., Beauchamp, M., Wood, A., Hauck, S., Underwood, K., and Hemmert, K.S. (2005). A comparison of floating point and logarithmic number systems for FPGAs. *13th Annual IEEE Symposium on Field-Programmable Custom Computing Machines (FCCM 2005)*, Napa, CA, USA (18–20 April 2005), 181–190. https://doi.org/10.1109/FCCM.2005.6.

Hassler, H. and Takagi, N. (1995). Function evaluation by table look-up and addition. *12th Symposium on Computer Arithmetic*, Bath, UK (19–21 July 1995), 10–16. https://doi.org/10.1109/ARITH.1995.465382.

Haviland, G.L. and Tuszynski, A.A. (1980). A CORDIC arithmetic processor chip. *IEEE Journal of Solid-State Circuits* **15** (1): 4–15. https://doi.org/10.1109/JSSC.1980.1051332.

Huang, T.S., Yang, G.Y., and Tang, G.Y. (1979). A fast two dimensional median filtering algorithm. *IEEE Transactions on Acoustics, Speech and Signal Processing* **27** (1): 13–18. https://doi.org/10.1109/TASSP.1979.1163188.

IEEE 754-2008 (2008). *IEEE Standard for Floating-Point Arithmetic*. IEEE. https://doi.org/10.1109/IEEESTD.2008.4610935.

Jaiswal, M.K. and So, H.K.H. (2019). PACoGen: a hardware posit arithmetic core generator. *IEEE Access* **7**: 74586–74601. https://doi.org/10.1109/access.2019.2920936.

Kamp, W.H.M., McLoughlin, I.V., and Bainbridge-Smith, A. (2006). An exploration of redundant number representations in FPGA. Electronics New Zealand Conference (ENZCon'06), Christchurch, NZ (27–29 November 2006), 63–68.

Khan, T.M., Bailey, D.G., Khan, M.A.U., and Kong, Y. (2017). Efficient hardware implementation for fingerprint image enhancement using anisotropic Gaussian filter. *IEEE Transactions on Image Processing* **26** (5): 2116–2126. https://doi.org/10.1109/TIP.2017.2671781.

Kingsbury, N.G. and Rayner, P.J.W. (1971). Digital filtering using logarithmic arithmetic. *Electronics Letters* **7** (2): 56–58. https://doi.org/10.1049/el:19710039.

Koc, C.K. and Johnson, S. (1994). Multiplication of signed-digit numbers. *Electronics Letters* **30** (11): 840–841. https://doi.org/10.1049/el:19940623.

Koren, I. and Zinaty, O. (1990). Evaluating elementary functions in a numerical coprocessor based on rational approximations. *IEEE Transactions on Computers* **39** (8): 1030–1037. https://doi.org/10.1109/12.57042.

Kota, K. and Cavallaro, J.R. (1993). Numerical accuracy and hardware tradeoffs for CORDIC arithmetic for special-purpose processors. *IEEE Transactions on Computers* **42** (7): 769–779. https://doi.org/10.1109/12.237718.

Lachowicz, S. and Pfleiderer, H.J. (2008). Fast evaluation of the square root and other nonlinear functions in FPGA. *IEEE International Symposium on Electronic Design, Test and Applications (DELTA 2008)*, Hong Kong (23–25 January 2008), 474–477. https://doi.org/10.1109/DELTA.2008.119.

Langroudi, S.H.F., Pandit, T., and Kudithipudi, D. (2018). Deep learning inference on embedded devices: fixed-point vs posit. *1st Workshop on Energy Efficient Machine Learning and Cognitive Computing for Embedded Applications (EMC2)*, Williamsburg, VA, USA (25–25 March 2018), 19–23. https://doi.org/10.1109/EMC2.2018.00012.

Lee, D.U., Luk, W., Villasenor, J., and Cheung, P.Y.K. (2003a). Hierarchical segmentation schemes for function evaluation. *IEEE International Conference on Field-Programmable Technology (FPT)*, Tokyo, Japan (15–17 December 2003), 92–99. https://doi.org/10.1109/FPT.2003.1275736.

Lee, D.U., Luk, W., Villasenor, J., and Cheung, P.Y.K. (2003b). Non-uniform segmentation for hardware function evaluation. *International Conference on Field Programmable Logic and Applications (FPL 2003)*, Lisbon, Portugal (1–3 September 2003), *Lecture Notes in Computer Science*, Volume **2778**, 796–807. https://doi.org/10.1007/b12007.

Lewis, D.M. (1990). An architecture for addition and subtraction of long word length numbers in the logarithmic number system. *IEEE Transactions on Computers* **39** (11): 1325–1336. https://doi.org/10.1109/12.61042.

Li, Y. and Chu, W. (1996). A new non-restoring square root algorithm and its VLSI implementations. *IEEE International Conference on Computer Design: VLSI in Computers and Processors (ICCD '96)*, Austin, TX, USA (7–9 October 1996), 538–544. https://doi.org/10.1109/ICCD.1996.563604.

Mencer, O. and Luk, W. (2004). Parameterized high throughput function evaluation for FPGAs. *Journal of VLSI Signal Processing* **36** (1): 17–25. https://doi.org/10.1023/B:VLSI.0000008067.31043.35.

Muller, J.M. (1985). Discrete basis and computation of elementary functions. *IEEE Transactions on Computers* **34** (9): 857–862. https://doi.org/10.1109/TC.1985.1676643.

Nielsen, A.M. and Muller, J.M. (1996). On-line algorithms for computing exponentials and logarithms. *2nd International Euro-Par Conference*, Lyon, France (26–29 August 1996), *Lecture Notes in Computer Science*, Volume **1124**, 165–174. https://doi.org/10.1007/BFb0024699.

Parhami, B. (2000). *Computer Arithmetic: Algorithms and Hardware Designs*. New York: Oxford University Press.

Peterson, W.W. (1957). Addressing for random-access storage. *IBM Journal of Research and Development* **1** (2): 130–146. https://doi.org/10.1147/rd.12.0130.

Pettersson, N. and Petersson, L. (2005). Online stereo calibration using FPGAs. *IEEE Intelligent Vehicles Symposium*, Las Vegas, NV, USA (6–8 June 2005), 55–60. https://doi.org/10.1109/IVS.2005.1505077.

Robertson, J.E. (1958). A new class of digital division methods. *IRE Transactions on Electronic Computers* **7** (3): 218–222. https://doi.org/10.1109/TEC.1958.5222579.

Sam, H. and Gupta, A. (1990). A generalized multibit recoding of two's complement binary numbers and its proof with application in multiplier implementations. *IEEE Transactions on Computers* **39** (8): 1006–1015. https://doi.org/10.1109/12.57039.

Sarma, D.D. and Matula, D.W. (1995). Faithful bipartite ROM reciprocal tables. *12th Symposium on Computer Arithmetic*, Bath, UK (19–21 July 1995), 17–28. https://doi.org/10.1109/ARITH.1995.465381.

Schulte, M.J. and Stine, J.E. (1997). Symmetric bipartite tables for accurate function approximation. *13th IEEE Symposium on Computer Arithmetic*, Asilomar, CA, USA (6–9 July 1997), 175–183. https://doi.org/10.1109/ARITH.1997.614893.

Schwarz, E.M. and Flynn, M.J. (1993). Hardware starting approximation for the square root operation. *11th Symposium on Computer Arithmetic*, Windsor, Ontario, Canada (29 June – 2 July 1993), 103–111. https://doi.org/10.1109/ARITH.1993.378103.

Specker, W.H. (1965). A class of algorithms for ln(x), exp(x), sin(x), cos(x), $\tan^{-1}(x)$ and $\cot^{-1}(x)$. *IEEE Transactions on Electronic Computers* **14** (1): 85–86. https://doi.org/10.1109/PGEC.1965.264066.

Stine, J.E. and Schulte, M.J. (1999). The symmetric table addition method for accurate function approximation. *Journal of VLSI Signal Processing* **21** (2): 167–177. https://doi.org/10.1023/A:1008004523235.

Timmermann, D., Hahn, H., and Hosticka, B.J. (1992). Low latency time CORDIC algorithms. *IEEE Transactions on Computers* **41** (8): 1010–1015. https://doi.org/10.1109/12.156543.

Tocher, K.D. (1958). Techniques of multiplication and division for automatic binary divider. *Quarterly Journal of Mechanics and Applied Mathematics* **11** (3): 364–384. https://doi.org/10.1093/qjmam/11.3.364.

Tomczak, T. (2006). Residue arithmetic in FPGA matrices. *International Conference on Dependability of Computer Systems*, Szklarska Poreba, Poland (25–27 May 2006), 297–305. https://doi.org/10.1109/DEPCOS-RELCOMEX.2006.43.

Villalba, J., Hidalgo, J.A., Zapata, E.L., Antelo, E., and Bruguera, J.D. (1995). CORDIC architectures with parallel compensation of the scale factor. *International Conference on Application Specific Array Processors*, Strasbourg, France (24–26 July 1995), 258–269. https://doi.org/10.1109/ASAP.1995.522930.

Volder, J.E. (1959). The CORDIC trigonometric computing technique. *IRE Transactions on Electronic Computers* **8** (3): 330–334. https://doi.org/10.1109/TEC.1959.5222693.

Walther, J.S. (1971). A unified algorithm for elementary functions. *Spring Joint Computer Conference*, Atlantic City, NJ, USA (18–20 May 1971), 379–385.

Walther, J.S. (2000). The story of unified CORDIC. *Journal of VLSI Signal Processing* **25** (2): 107–112. https://doi.org/10.1023/A:1008162721424.

Wang, Y. (1998). New Chinese remainder theorems. *32nd Asilomar Conference on Signals, Systems and Computers*, Pacific Grove, CA, USA (1–4 November 1998), Volume **1**, 165–171. https://doi.org/10.1109/ACSSC.1998.750847.

Wang, S., Piuri, V., and Swartzlander, E.E. (1996). A unified view of CORDIC processor design. *IEEE 39th Midwest Symposium on Circuits and Systems*, Ames, IA, USA (18–21 August 1996), Volume **2**, 852–855. https://doi.org/10.1109/MWSCAS.1996.588050.

Wang, Y., Song, X., Aboulhamid, M., and Shen, H. (2002). Adder based residue to binary number converters for ($2^n - 1, 2^n, 2^n + 1$). *IEEE Transactions on Signal Processing* **50** (7): 1772–1779. https://doi.org/10.1109/TSP.2002.1011216.

6

Interfacing

An image processing system never stands alone. This is particularly the case for an embedded vision system, which is usually designed for one specific purpose or task. Therefore, there are usually one or more peripheral devices connected to the field-programmable gate array (FPGA) implementing the system.

Each device must have an appropriate interface within the FPGA. This is responsible for passing any data to, or from, the application logic, including any format conversions required between the peripheral and the algorithm. It is also responsible for providing any control signals required by the peripheral. This may range from a clock signal and simple handshaking through to complex protocol management. Many peripheral devices require initialisation or configuration through the setting of control registers.

The interface logic may be considered to be a device driver, shielding the image processing hardware from the lower level complexities of the physical interface (Bailey et al., 2006). In this context, the term device driver refers primarily to the interface hardware and control signals. This chapter focuses on some of the techniques associated with the design of such interfaces.

It is also useful within this context to consider the interaction between the FPGA system and the user. On a hosted system, many of the user-interface tasks will be performed using the host operating system. However, on a stand-alone system, all of these functions must be performed on the FPGA. One advantage of including an embedded processor is to provide a high-level software environment for managing user interaction. Such tasks tend to be control-centric, making them more difficult to implement directly in hardware. Most user interaction is not time critical, so hardware acceleration is usually not necessary.

Many external devices will have their own natural clock frequency which may be different from the core logic. In such cases, the logic that interacts directly with the peripheral will be in a separate clock domain. The low-level device driver will span clock domains to connect to the rest of the system. For reliable communication between the domains, it is important to build appropriate synchronisation and buffer design, as discussed in more detail in Section 4.1.3.

6.1 Camera Input

In an embedded vision system, it is usually necessary to connect to a camera or other video input. Most modern sensors are complementary metal oxide semiconductor (CMOS) sensors, where each pixel is potentially individually addressable. Although such a random access interface is not usually provided to the user, several common camera operating modes exploit this ability. Some of these modes that are commonly made available to the user are:

- **Windowing**: Rather than read out the entire image each frame, windowing allows a rectangular region of interest to be selected, and only those pixels are read out. This allows the frame rate to be increased by

Design for Embedded Image Processing on FPGAs, Second Edition. Donald G. Bailey.

Companion Website: www.wiley.com/go/bailey/designforembeddedimageprocessc2e

reducing the area of the sensor that is read out. One application of this is object tracking in situations where the object being tracked only occupies a small fraction of the image area.
- **Skipping**: An alternative to windowing for reducing the effective resolution is skipping, where the number of pixels read out is reduced by skipping entire rows or columns. This reduces the pixel count without affecting the field of view.
- **Binning**: Rather than skipping pixels, adjacent pixels can be combined together. This has two advantages over skipping. First, it reduces aliasing around sharp edges because adding adjacent bins is a form of low-pass filtering. Second, since each pixel contributes to the output, binning can improve the sensitivity in low light conditions. Binning often has a lower frame rate because internally multiple pixels must be addressed to produce each output pixel.

There are many different camera communication protocols, and each has its own interfacing requirements. While it is impossible to describe them all in any detail, the broad characteristics of some of the key interfaces will be introduced.

6.1.1 Analogue Video

Although analogue video signals are largely only used in legacy systems, it is instructive to consider their key characteristics which have had a strong influence on digital video.

The 2-D image is scanned in raster format, with the timing as represented in Figure 6.1. Each line is separated by a period without an active video signal called the ***horizontal blanking*** period. During horizontal blanking, a horizontal synchronisation pulse controls the timing. An analogue video signal represents the intensity by an analogue voltage level and implicitly encodes the position with time, as illustrated in Figure 6.2. After the last line, there is a ***vertical blanking*** period consisting of several lines without video data. Vertical blanking also contains a synchronisation pulse to control the vertical timing.

Three techniques are commonly used for representing colour. Component video has three separate signals, one for each of the RGB components. S-video (also called Y/C) has two signals, one for the luminance, and one for the chrominance. The two chrominance components are typically encoded by modulating a colour sub-carrier with quadrature modulation (the SECAM (sequential colour and memory) standard uses frequency modulation instead). Composite video combines the luminance and chrominance together as a single signal.

Television signals consist of two interlaced fields, with the odd and even scan lines within different fields. This effectively reduces the required bandwidth by halving the frame rate without producing the associated annoying flicker. From an image processing perspective, if every field is processed separately, this doubles the frame rate, albeit with reduced vertical resolution and a small vertical offset that alternates between the fields.

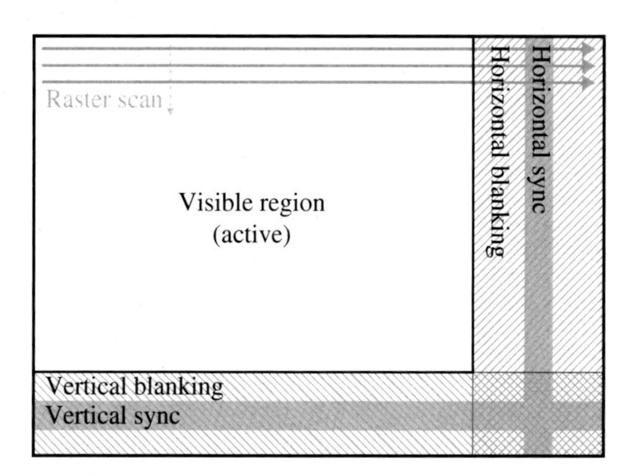

Figure 6.1 Active and blanking regions within a raster-scanned video image.

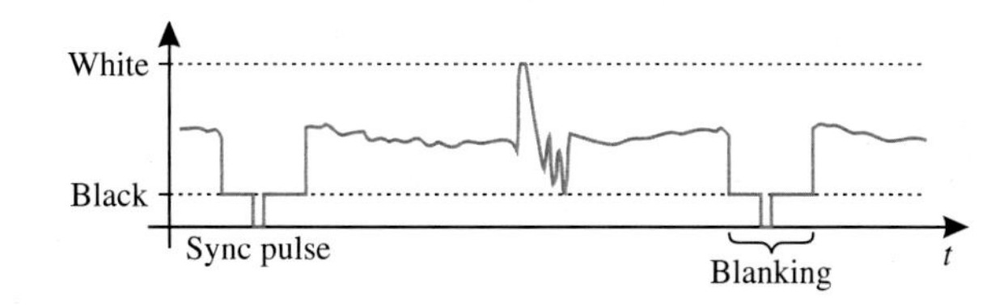

Figure 6.2 Analogue video signal. Left: input scene; right: video signal for the marked line. Source: Photo courtesy of Robyn Bailey.

To process the whole frame, it is necessary to re-interlace the fields. Cameras designed for imaging purposes (rather than consumer video) avoid this by producing a non-interlaced or progressive scan output.

There are also several different analogue video standards. For NTSC (National Television Standards Committee), the signal consists of 525 scan lines (480 containing actual video data) with a frame rate (two fields) of 30 Hz. However, in both PAL (phase alternating line) and SECAM, the signal has 625 lines (576 active) at a frame rate of 25 Hz.

While it is possible to directly convert the incoming composite video signal to digital and perform all of synchronisation and colour separation within an FPGA, this is not recommended because the processing is quite complex and would use considerable resources on the FPGA. The specific processing required also depends on the video standard that is input. Instead, using an analogue video decoder chip will perform all of these functions, and most will automatically detect the signal standard and perform the appropriate processing.

6.1.1.1 Deinterlacing

Deinterlacing converts the image from an interlaced scan to progressive scan, with the output frame rate being double that of the input (so, each field is converted to a full frame). However, each input field has only half of the lines available. Let the nth field be $I_n[x, y]$, where the missing lines are $I_n[x, y] = 0$ for $y \bmod 2 \neq n \bmod 2$. The problem is then to estimate the data within the missing lines. This is an interpolation problem, with solutions involving both spatial and temporal interpolation.

Simple spatial interpolation tends to give lower vertical resolution. This is improved by interleaving data from adjacent fields; however, any motion of objects relative to the camera gives a displacement between adjacent fields, which can result in artefacts. There are many possible approaches to this problem (de Haan and Bellers, 1998), with the more complex methods detecting the motion of objects within the scene and using motion compensated interpolation. One of the simplest methods that gives reasonable results on many images is a simple three-point vertical median filter:

$$Q_n[x, y] = \begin{cases} I_n[x, y], & y \bmod 2 = n \bmod 2, \\ \text{median}(I_n[x, y-1], I_{n-1}[x, y], I_n[x, y+1]), & y \bmod 2 \neq n \bmod 2. \end{cases} \tag{6.1}$$

For static scenes, this effectively applies a median filter only on the missing rows. This will have little effect on edges but may cause a small reduction in fine detail. Where there is motion, the previous field may be incompatible with the current field as a result of the motion of edges. In this case, the pixel from the previous field will be quite different from the adjacent pixels in the current field. The median will therefore select either the pixel above or the pixel below to replace the incompatible pixel.

6.1.2 Direct Digital Interface

Analogue cameras have largely been displaced by digital video cameras, which perform much of the analogue processing internally and directly produce a digital output. Digital cameras allow a much wider range of resolutions and directly provide digital video data and synchronisation signals in parallel. While blanking periods are not strictly necessary for digital cameras, at the sensor level these may be required for the timing of internal camera operations.

In an embedded device, the simplest approach is to interface directly to the digital sensor chip. The digital pixels are streamed from the sensor, with line and frame valid signals. Sensor configuration (shutter speed, triggering, gain control, windowing, and sub-sampling mode) is controlled through a set of registers. Usually, a low-speed serial communication protocol (such as I^2C, see Section 6.3.2, and serial peripheral interface (SPI), see Section 6.3.3) is often used for this.

A disadvantage of direct connection to the sensor is that the signal path between the sensor and FPGA must be kept short to maintain close control over timing and signal integrity. Higher resolution sensors overcome the signal integrity issue by serialising the signals, for example using a MIPI or Camera Link interface, as will be discussed in Sections 6.1.3 and 6.1.4. However, the design and calibration of the high-speed connections is not trivial (Grob, 2010). The alternative to direct connection is to use off-the-shelf cameras with standard interfaces such as USB and Ethernet.

6.1.3 MIPI Camera Serial Interface

The Mobile Industry Processor Interface (MIPI) Alliance has developed several specifications for low-power and high-speed interface of sensors to application processors. For cameras, they have the camera serial interface (CSI-2) which provides a high-speed serial interconnect between the camera sensor and processor that provides a scalable cost-effective solution to addressing the I/O bandwidth issues of high-resolution image sensors (MIPI, 2016).

The CSI-2 specification defines two types of data packet (MIPI, 2019). Short packets are used for synchronisation signals, such as frame-start and frame-end, and optional line-start and line-end signals. Long packets contain the image data, with one image scan line sent per packet. Data packets can be in raw format, RGB, YUV, or even JPEG encoded. The physical layer (PHY) for the communication is sent over either a D-PHY or C-PHY (MIPI, 2016).

The D-PHY (v2.5) provides up to 4.5 Gbps per lane through differential signalling (Icking et al., 2019). It consists of a clock lane and one or more data lanes, as illustrated in Figure 6.3. A full-speed clock is sent on the clock lane, offset by 90°, so that clock transitions occur in the middle of the data bits. The data is sent serially on the data lanes at double data rate (DDR) (one bit per lane for each rising and falling clock edge). If more than one data lane is used, successive bytes are transmitted across the parallel lanes.

The C-PHY uses three wires per lane, with embedded clock and three-phase symbol encoding, with transmission at up to 6 giga-symbols per second (v2.0) (Icking et al., 2019). The three-phase encoding gives approximately 2.28 bits per symbol, giving a significantly higher bandwidth per wire than the D-PHY.

For interfacing with an FPGA, the D-PHY can be implemented through the low-voltage differential signalling (LVDS) ports, with the built-in serialisation and deserialisation (SERDES) blocks used to deserialise the data, at least for lower clock speeds (Microsemi, 2016). Some of the later Xilinx FPGAs have direct support for D-PHY signalling. Alternatively, a D-PHY receiver chip can be used to deserialise the high-speed data and provide the pixel data to the FPGA in parallel. Camera configuration is via an I^2C connection.

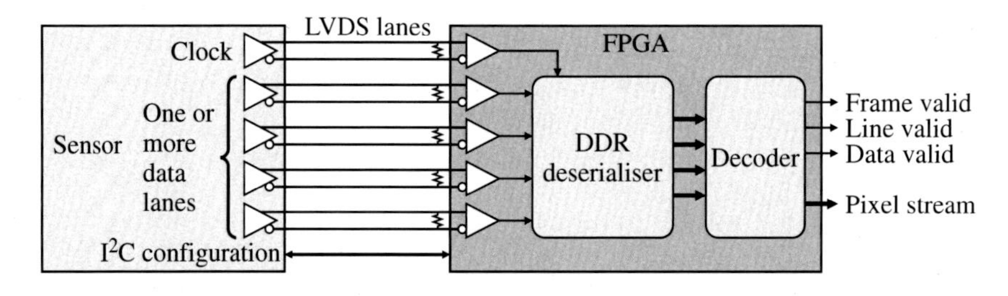

Figure 6.3 MIPI CSI-2 over D-PHY camera connection to FPGA.

6.1.4 Camera Link

While the MIPI interface is at the chip level, a similar interface at the camera level for industrial vision systems is the Camera Link interface. This is based on National Semiconductor's Channel Link technology. A single Channel Link connection provides one-way transmission of 28 data signals and an associated clock over five LVDS pairs. One pair is used for the clock, while the 28 data signals are multiplexed over the remaining four LVDS pairs using a 7 : 1 serialisation of the data on the inputs.

The base configuration for Camera Link (AIA, 2004) has one Channel Link connection, with 24 of the 28 bits allocated for pixel data (three 8-bit pixels or two 12-bit pixels) and the remaining 4 bits contain frame, line, and pixel data valid signals. The pixel clock has a maximum rate of 85 MHz. For higher bandwidth, the medium configuration adds an additional Channel Link for an additional 24 bits. The full configuration adds a third Channel Link, giving a total of 64 bits of pixel data. This may allow, for example, up to eight pixels per pixel clock cycle to be transmitted in parallel from the camera.

In addition, there are four LVDS pairs from the frame grabber to the camera for general-purpose camera control. The use of these is defined by the particular camera manufacturer. Finally, two LVDS pairs are provided for asynchronous serial communication for camera set-up. These are relatively low-speed; the specifications say that a minimum rate of 9600 baud must be supported.

As with the MIPI interface, Camera Link can be interfaced directly to an FPGA using LVDS inputs and SERDES blocks to perform the 7 : 1 deserialisation and demultiplexing (Sawyer, 2018). Alternatively, a Channel Link receiver chip can be used to deserialise the high-speed data to provide parallel pixel data to the FPGA.

6.1.5 USB Cameras

Many consumer cameras have a USB interface. The USB protocol would require the FPGA to act as the host controller. While this can be built in hardware, it would require considerable logic for managing the USB communication protocols (see Section 6.3.4 for interfacing USB with an FPGA). On top of this are the protocols for sending packets to set up the image capture and for receiving and interpreting the video data packets from the camera.

A more effective solution is to have a central processing unit (CPU) (either hard or soft) manage the camera via an appropriate device driver. Many USB-based webcams use the USB video device class (UVC) specification (USB-IF, 2012), so using a Linux operating system enables interface to the camera with a generic UVC driver.

An alternative approach is to use an external USB interface chip to manage the protocol and convert the video to parallel data for interfacing with the FPGA (Abdaoui et al., 2011).

6.1.6 GigE Vision

Higher performance (high speed or large pixel count) devices are tending towards using gigabit Ethernet for data communication. GigE Vision is a camera interface standard developed by the Automated Imaging Association that interfaces to the camera (both video data and control information) using gigabit Ethernet. Because it builds on the back of the Ethernet standard, it can use low-cost standard cables and connectors, allowing high data transfer rates over distances up to 100 m.

The GigE Vision standard has four elements to it. A control protocol defines the communication required to control or configure the camera. A stream protocol defines how the image data is transferred from the camera to the host system. Both of these protocols run over user datagram protocol (UDP). A device discovery mechanism provides a way of identifying which GigE Vision devices are connected and obtaining their Internet addresses. Finally, an extensible markup language (XML) description file provides an online datasheet that specifies the camera controls and image formats available from the device. The structure of this XML datasheet is defined by the GenICam standard (EMVA, 2009).

The design of a GigE Vision-based system is an advanced topic, and a detailed discussion of all of the issues is beyond the scope of this book. Only a brief overview is provided in the following paragraphs.

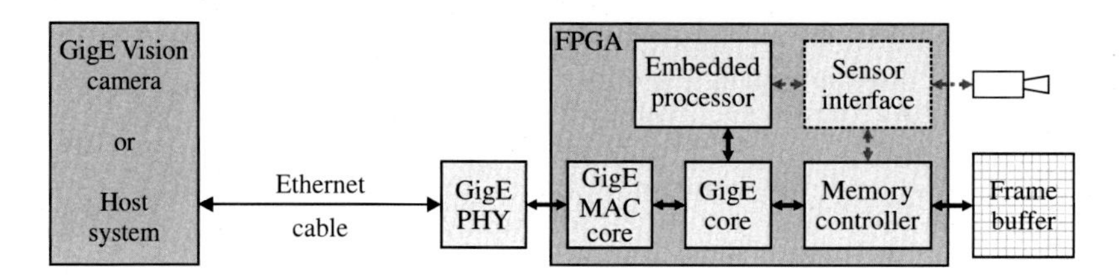

Figure 6.4 Structure of a GigE Vision system.

The complexities of the Ethernet protocol make a pure FPGA implementation difficult. However, in practice, this is simplified by having only one (or few) GigE cameras connected to a system. This enables the packets to be filtered relatively easily by hardware, with the data header packet providing the image size, and the data payload packets providing the image data. The more complex details of managing the protocol stack (see Section 6.3.5) are best handled by an embedded processor.

A reference design is available from Sensor to Image (Sensor to Image, 2016) for a GigE Vision receiver that streams data in from a GigE Vision camera and displays the resultant video on a VGA (video graphics array) output. A more complex, multi-camera design has been developed by Ibraheem et al. (2015), which provides an AXI (advanced extensible interface) stream output for each camera.

The basic structure of a GigE Vision system is shown in Figure 6.4. Unless the FPGA has a built-in gigabit Ethernet PHY, this must be provided by an external chip. The media access controller (MAC) core may be either internal or external to the FPGA. The GigE core manages the low-level networking features, routing the video stream to, or from, the frame buffer memory. The control protocol is routed to the embedded processor, which manages the camera configuration in software.

In embedded vision applications, it would be more practical to build the FPGA system within the camera and perform any processing before sending the results to a host system using GigE Vision. Such a system would also need to interface with the sensor and may require a memory controller for saving images to the frame buffer where necessary. In such a system, the embedded processor also makes available to the GigE Vision side the GenICam-based XML datasheet describing the capabilities of the camera. Not shown in Figure 6.4 is the image processing that adds value to the system!

6.1.7 Camera Processing Pipeline

Associated with image capture, there are several steps within the initial image processing pipeline. These are briefly outlined here, with a more complete discussion given later of the implementation of the particular image processing techniques.

6.1.7.1 Shuttering Issues

The exposure time within a CMOS sensor is the time from when the pixel reset is released until the accumulated charge is transferred out from the sensitive area. There are three common operating modes: global reset, rolling shutter, and global shutter, as illustrated in Figure 6.5.

The simplest approach is global reset, where all pixels are released from reset simultaneously. The consequence is that pixels read out earlier will have a shorter exposure than those read out later. For short exposure times, there can be a significant difference in exposure between the first and the last row read out, resulting in the top of the image being underexposed and the bottom of the image being overexposed. This may be overcome using an external mechanical shutter, or flash illumination.

Many low-cost cameras use an electronic rolling shutter. The reset for each row is successively released, so that the time between release and readout is constant for each row, giving the same exposure. This is common because it uses fewer transistors per sensing photodiode (typically 3 or 4) (Ge, 2012), giving a high fill factor and good sensitivity. However, if any objects within the scene are moving (or if the camera is moving or panning),

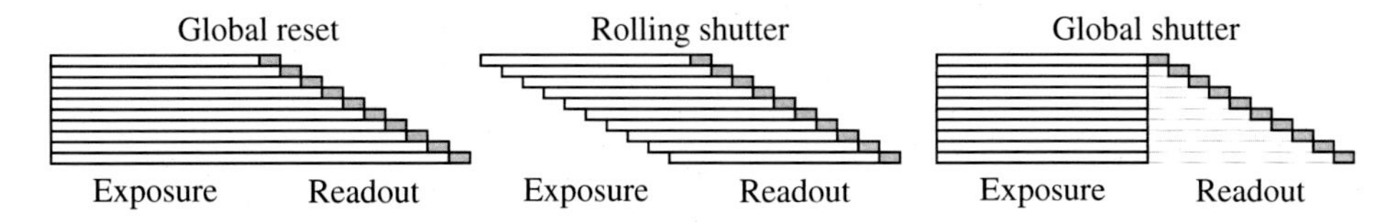

Figure 6.5 Different shuttering arrangements. Left: rolling shutter exposes each row successively before readout; centre: global reset starts the exposure of all rows simultaneously, with the exposure completing on readout; right: global shutter exposes all rows at the same time but requires additional sampling and storage to hold the pixel value until readout.

then since the different rows are exposed at different times, the output image will appear distorted. The nature of this distortion will depend on the relative motion between objects in the scene. Additional effects can be observed with light variations, for example when the scene is illuminated by fluorescent lights (Nicklin et al., 2007) where the periodic variation of light intensity appears as a distinct banding within the image.

Compensation for the rolling shutter requires modelling both the image capture process and the relative motion between objects and the camera (Geyer et al., 2005). Correction is simpler if the scene is stationary, and camera motion can be modelled (Nicklin et al., 2007; Liang et al., 2008). Where multiple images are available, the matching of the corresponding points between images can be used to derive homographies which can be used to compensate for the motion of the camera (Grundmann et al., 2012; Lao and Aider, 2021). In fact, having two sensors with rolling shutters in opposite directions can be used to identify and correct camera motion (Albl et al., 2020).

A global electronic shutter exposes all of the image rows at the same time. This requires that the signal be sampled and held until it is converted, requiring additional transistors per sensing element (typically 5–8) (Ge, 2012). However, with reducing transistor sizes, these have less of an impact on the fill factor. In many applications, the use of a global shutter camera overcomes many of the problems introduced by the use of a rolling shutter.

6.1.7.2 Automatic Focus

Where the camera has an electronically controlled focus, one task is to automatically adjust the focus to achieve the 'best' image. When an image is out of focus, the image becomes blurred. This appears as a reduction of contrast or high-frequency content within an image, and sharp edges become less steep.

These effects can be used to define a focus function, F. This is based on a filter, which can be used to evaluate the focus locally within a particular image to produce a sharpness map, $s(x, y)$. Commonly used filters (see Chapter 9) are the local variance (local contrast), local derivative or difference (edge detection), and second derivative or Laplacian (high frequencies) (Groen et al., 1985). The sharpness map is then summed over a region of interest (ROI, which may be the whole image) to give the focus function. Some commonly used focus functions are the sum of the absolute filter responses,

$$F = \sum_{ROI} |s(x, y)|; \tag{6.2}$$

the sum of squared filter responses,

$$F = \sum_{ROI} |s(x, y)|^2; \tag{6.3}$$

and the number of filter responses that exceed a threshold, thr,

$$F = \sum_{ROI} (|s(x, y)| > thr). \tag{6.4}$$

The chosen focus function is then maximised by adjusting the position of the lens.

For example, Jin et al. (2010) implemented an autofocus algorithm on an FPGA, maximising the number of pixels above an adaptive threshold.

6.1.7.3 Automatic Exposure and White Balance

Correct exposure is important to obtain a good image from a sensor. If the exposure time is too long, then pixels become saturated, and information is lost. Conversely, if the exposure is too short, then all of the pixels are dark, and best use is not made of the available dynamic range.

Measures that are commonly used are to maximise the entropy (which reflects the information available) or to ensure that there are not too many pixels over- or underexposed using the cumulative histogram. These techniques are described more fully in Section 8.1.4.

Note that the exposure can be adjusted through changing either the lens aperture (if such adjustment is available) or the sensor integration time. Maintaining a constant frame rate places an upper limit on the integration time, so beyond that it is necessary to increase the gain within the sensor A/D convertors. However, increasing the gain will also amplify any noise present on the signal.

Closely related to automatic exposure is automatic white balance. Recall from Eq. (1.8) that the colour value output from a sensor depends on the illumination. White balancing compensates for the effects of the spectral content of the light source, giving the appearance under a standard illuminant. This effectively adjusts the relative gains of the red, green, and blue channels based on the estimated illuminant. Approaches for estimating the illuminant are the white-point model (Land and McCann, 1971), which assumes that the highlights correspond to the illumination source; the grey-world model (Finlayson et al., 2001), which assumes that on average the image is achromatic; and the grey-edge model (van de Weijer et al., 2007), which assumes that the average edge difference is achromatic. The details of these algorithms are given in Sections 7.3.5 and 9.7.4.

6.1.7.4 Colour from Raw Images

Most single-chip cameras obtain a colour image by filtering the light falling on each pixel using a mosaic of coloured filters, with a separate filter for each pixel. The most common pattern is the Bayer pattern (Bayer, 1976), as shown on the left in Figure 6.6, which has half of the pixels sensitive to green, one quarter to red, and one quarter to blue. Therefore, to form a full colour image, it is necessary to interpolate the values missing in each of the component images, so that there is an RGB value for each pixel. This interpolation process is called ***demosaicing***, because it removes the mosaic of the colour filter array.

The interpolation filters are spatially dependent because the particular filter function depends on each pixel's position within the mosaic. These vary in complexity from nearest-neighbour interpolation which selects the

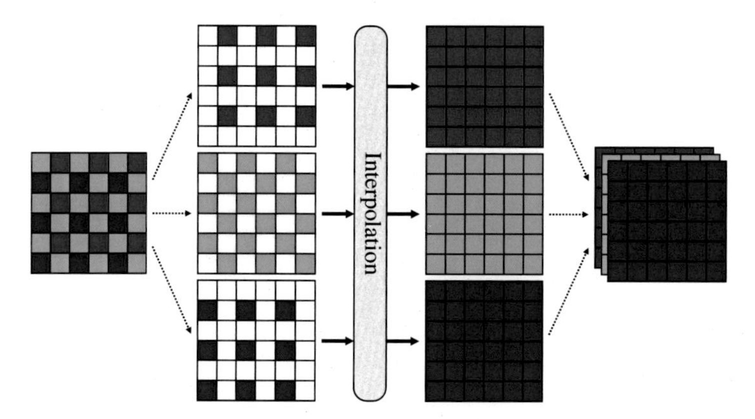

Figure 6.6 Bayer pattern demosaicing. Left: raw image captured with Bayer pattern; centre: interpolating each colour plane; right: full colour image.

nearest available component, through horizontal and vertical linear interpolation, to advanced methods which exploit the correlation between colour channels and perform interpolation along edges. The final result will inevitably be a compromise between interpolation quality and the computation and memory resources required. Several demosaicing algorithms are described in more detail in Section 9.7.3.

6.1.7.5 Gamma Adjustment

The human visual system is not linearly sensitive to light level. Small differences in light level are much more noticeable in dark areas than light areas. This is often represented using a power law (Stevens, 1957) where the perceived brightness, B, depends on the intensity, I, as

$$B = I^{\gamma}, \tag{6.5}$$

where γ is in the range 0.3–0.5. Therefore, to make the best use of the pixel values (to make differences in pixel values perceptually more uniform), the linear values from the sensor are gamma-compressed using Eq. (6.5), typically with $\gamma = 0.45 \approx \frac{1}{2.2}$. Gamma mapping is described in Section 7.1.1.

6.1.7.6 Frame-rate Conversion

Often, the frame rate of a camera is governed by the exposure time, or the application requirements, and may even be variable. However, the frame rate of the display is almost always fixed and may be different from the image capture or processing frame rate. In such situations, it is necessary to change the frame rate of an incoming stream to match that of the output.

The best approach is to use full temporal interpolation. This detects the motion of objects and renders the object in an appropriate position according to the current frame time and object motion. However, motion compensation is quite complex (see Section 12.1.3) and computationally intensive.

If some inter-frame judder can be tolerated, then the easiest approach is to simply output the last complete frame. To avoid intra-frame artefacts when switching between frames, this requires buffering two complete frames, enabling an incoming frame to be stored, while the other frame is being output. If both input and output rates are constant, then two buffers are sufficient, but if either is variable, then a third buffer is required to enable complete frames to be output (Khan et al., 2009).

6.2 Display Output

Just as image capture is important for an embedded image processing system, most imaging systems also need to be able to display images, even if just for debugging. There are two components to displaying an image or indeed any other graphics on an output from an FPGA. The first is the display driver that controls the timing and generates the synchronisation signals for the display device. The second is generating the content, that is controlling what gets displayed where and when.

6.2.1 Display Driver

The basic timing for video signals is introduced in Figure 6.1. The image is sent to a display in a raster-scanned format with a horizontal blanking period between each line. During this blanking period, a synchronisation pulse is used to indicate to the monitor to begin the next line (see Figure 6.7). A similar timing structure is used at the end of each frame to synchronise the monitor for displaying the next frame.

The Video Electronics Standards Association has defined a set of formulae which define the detailed timing for a given resolution and refresh rate known as the coordinated video timing (CVT) standard (VESA, 2013a). The CVT standard gives two sets of timings: one for traditional cathode ray tube (CRT) monitors (where the blanking period is typically ~20% of the row time) and reduced blanking timings for modern flat panel displays that do not require such a long blanking period. The reduction in horizontal blanking period gives a consequent

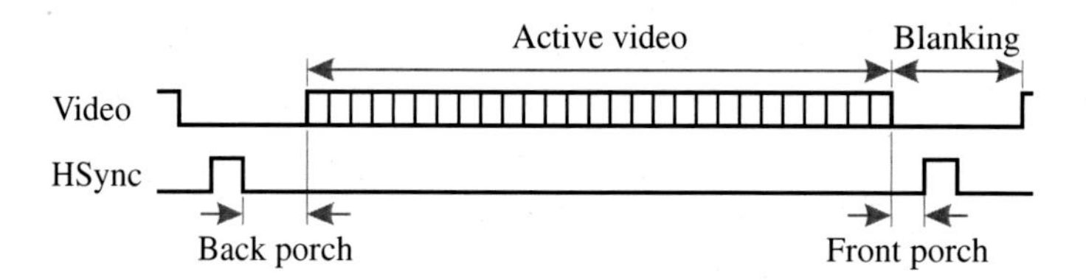

Figure 6.7 Horizontal video timing.

Table 6.1 Timing data for several common screen resolutions for a refresh rate of 60 Hz (VESA, 2013b)

Pixel clock	Visible width	Front porch	Horiz. sync	Back porch	Whole line	Visible height	Front porch	Vert. sync	Back porch	Whole frame	Sync polarity	
MHz			Pixels					Lines			H.	V.
25.175	**640**	16	96	48	800	**480**	10	2	33	525	Neg	Neg
40.00	**800**	40	128	88	1056	**600**	1	4	23	628	Pos	Pos
65.00	**1024**	24	136	160	1344	**768**	3	6	29	806	Neg	Neg
68.25	**1280**	48	32	80	1440	**768**	3	7	12	790	Pos	Neg
79.50	**1280**	64	128	192	1664	**768**	3	7	20	798	Neg	Pos
71.00	**1280**	48	32	80	1440	**800**	3	6	14	823	Pos	Neg
108.00	**1280**	96	112	312	1800	**960**	1	3	36	1000	Pos	Pos
108.00	**1280**	48	112	248	1688	**1024**	1	3	38	1066	Pos	Pos
101.00	**1400**	48	32	80	1560	**1050**	3	4	23	1080	Pos	Neg
108.00	**1600**	24	80	96	1800	**900**	1	3	96	1000	Pos	Pos
162.00	**1600**	64	192	304	2160	**1200**	1	3	46	1250	Pos	Pos
148.50	**1920**	48	44	148	2200	**1080**	4	5	36	1125	Pos	Pos
138.50	**1920**	48	32	80	2080	**1080**	3	5	23	1111	Pos	Neg
154.00	**1920**	48	32	80	2080	**1200**	3	6	26	1235	Pos	Neg
193.25	**1920**	136	200	336	2592	**1200**	3	6	36	1245	Neg	Pos
184.75	**1920**	48	32	80	2080	**1440**	3	4	34	1481	Pos	Neg
156.75	**2048**	48	32	80	2208	**1152**	3	5	25	1185	Pos	Neg
268.50	**2560**	48	32	80	2720	**1600**	3	6	37	1646	Pos	Neg
556.744	**4096**	8	32	40	4176	**2160**	48	8	6	2222	Pos	Neg

Parameters for other refresh rates or resolutions may be obtained from standard tables or from the spreadsheet associated with the CVT standard.
Bold values indicate the visible size of the frame in pixels.

reduction in pixel clock frequency for the same pixel resolution. The polarity of the horizontal and vertical sync pulses indicates whether the CRT timing, the reduced blanking timing, or a legacy timing mode is used. Several common formats are listed in Table 6.1.

The timing can easily be produced by two counters, one for the horizontal timing along each row, and one for the vertical timing, as shown in Figure 6.8. The count is simply compared with the required times of the key events to generate the appropriate timing signals. Note, to reduce the logic, tests for equality are used to set or reset a register. The corresponding 1-bit sync register is set at the start of the signal and cleared at the end. The vertical counter is enabled only when the horizontal counter is reset at the end of each line. The vertical sync pulse is also synchronised with the leading edge of the horizontal sync pulse (VESA, 2013b). In this design, the counters begin at 0 at the start of the addressable active region. However, if any stream processing pipelines require priming, it may be convenient to advance the origin to simplify the programming logic. (This would require adding a comparison for the total, which will no longer be at the end of the blanking period.)

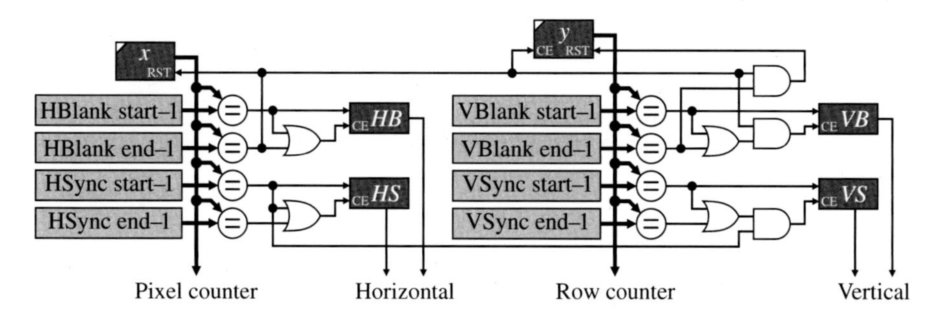

Figure 6.8 Circuit for producing the timing signals for a video display. (CE = clock enable; RST = reset.)

6.2.1.1 VGA Output

Although the VGA resolution is only 640 × 480, the analogue video output connector can work with a range of resolutions, even though it is often called a VGA output. It provides the display with analogue video signals for each of the red, green, and blue components. These require a high-speed D/A converter for each channel; the simplest approach is to use a video D/A chip, which is available from several manufacturers. In addition to the pixel data, separate horizontal and vertical sync signals also need to be sent of the appropriate polarity (see Table 6.1). The sync pulses may be driven directly by the FPGA using 3.3 V TTL (transistor-transistor logic) level signals.

The VGA connector also provides an I^2C channel for communication between the host and the display. This enables the host to determine which resolutions and refresh rates are supported for plug-and-play functionality and to set various display parameters. Generally, an FPGA does not need to use this channel for driving the display as long as the particular resolution and refresh rate are supported.

6.2.1.2 Digital Visual Interface (DVI)

One limitation of the VGA signal is that it is analogue and therefore subject to noise. This becomes more acute at higher resolutions, which requires a higher pixel clock. The digital visual interface (DVI) standard (DDWG, 1999) overcomes this limitation by transmitting the signal digitally.

The RGB pixel values are transmitted serially using a transition-minimised differential signalling (TMDS) scheme. This encodes each 8 bits of input data into a 10-bit sequence that both minimises the number of transitions and averages out the DC level over the long term. The encoded signals are transmitted using a differential pair. As a result of serialisation, the bit clock is 10 times higher than the pixel clock. Since there can be a wide range of clock frequencies (depending on the resolution and refresh rate), a pixel clock is also transmitted on a separate channel. Four additional 10-bit code words are reserved for encoding two control bits per channel, transmitted during the blanking period. These control signals are normally zero; however, the control signals on the blue channel are used to transmit the horizontal and vertical synchronisation signals.

A single-link digital connector contains the clock and three data signals described above plus an I^2C data channel for device identification as described within the VGA section. The maximum pixel clock frequency is 165 MHz. Where the pixel clock exceeds 165 MHz, an additional three data signals provide a dual link, with two pixels transmitted in parallel. For backwards compatibility, analogue video signals can also be provided, enabling a simple adapter to be used with both digital and analogue monitors.

While the 8/10 bit TDMS encoding could potentially be performed on the FPGA (for example using a lookup table), a SERDES block would be required by each channel to meet the required data rates. DVI transmitter chips are available, which remove this logic from the FPGA, requiring just a parallel data output for each channel.

6.2.1.3 High-definition Multimedia Interface (HDMI)

The high-definition multimedia interface (HDMI) specification (HDMI, 2009) extends DVI by adding audio packets, which are sent during the blanking periods. It also adds support for YCbCr encoding, rather than just RGB. However, the HDMI standard interface using RGB mode is compatible with a single-link DVI and just requires a suitable adapter. Later versions of HDMI increase the data rate to enable connection to higher resolution displays. Interface to an HDMI output is easiest through an HDMI transmitter chip.

6.2.2 Display Content

The second part of displaying any output is to produce the required content at the appropriate time. In particular, this requires producing each output pixel value at the correct time for the corresponding object or image to appear on the display.

The conventional approach used in software-based systems is to represent the content to be displayed as an array of data within memory (a frame buffer). Content is then streamed to the display as required. The limitation of this approach is that the data to be displayed must be calculated and written to a memory. Then, the contents of that memory are periodically read and streamed out to the display. To reduce flicker effects when updating the contents, double buffering is often used. A copy is made of the display image, which is then updated independently of what is being displayed, and then rapidly written back to the frame buffer to change the display smoothly. This approach spends a significant fraction of the time reading from and writing to memory. Arbitration logic is also required to avoid access conflicts with the process that is actually displaying the image.

Using random access processing on an FPGA, similar to its software counterpart, requires a frame buffer to hold the image being displayed. Generally bank switching is also used to give a clean display and meet the bandwidth requirements of both the image generation and display driver processes.

However, an FPGA provides an alternative approach using stream processing. By synchronising the image processing with the display driver, it is possible to perform much of the processing on the data as it is being displayed. This is easiest when the processing has a fixed latency; processing is started in advance of when it is required. Since the display driver produces regular, predictable signals, these can be used to trigger the processing to begin at precisely the right time. The fixed latency requirement may be relaxed if necessary by introducing a first-in first-out (FIFO) buffer to take up any slack. In such cases, the blanking periods at the end of each line and each frame may be used to catch up and refill the FIFO buffer.

Even within such a rigid processing regime, considerable flexibility is possible. For example, the display can be segmented, with different images or stages within the process displayed in each section.

6.2.2.1 Character Generation

It may be useful to have textual or other annotation overlaid on the display. This may be useful for labelling objects, regions of interest, or displaying textual debugging information. While text could be displayed in a frame buffer, this is inefficient in terms of memory usage and makes dynamic update more difficult.

A more efficient approach is to represent the text directly as ASCII codes and interpret these on-the-fly to produce the bitmaps that need to be displayed. Character generation logic is simplified if character locations are restricted to a grid, but more general positioning can be accommodated. The process is outlined in Figure 6.9. The display driver produces the position of the current pixel being displayed. On the leading edge of each character to be displayed, the annotation processor outputs the ASCII code and row within the character to the character generator. The line of pixels is read from the character-generator bitmap and is ORed into the output shift register, combining it with any other characters that are also being produced. The bits are shifted out sequentially, producing the character glyph to be overlaid onto the display.

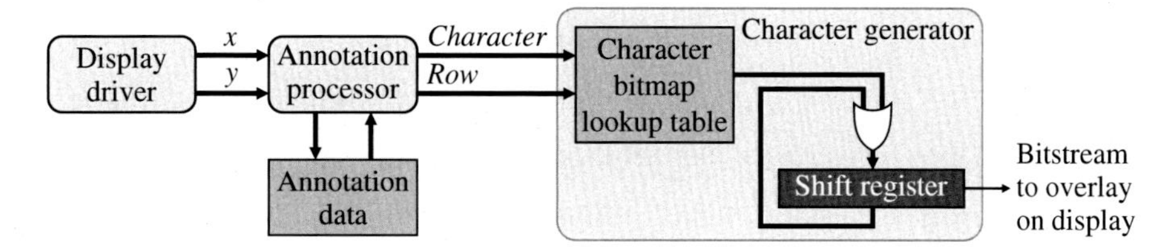

Figure 6.9 Character generation process.

6.3 Serial Communication

Many peripheral devices are configured using some form of serial communications. This section briefly describes the most common of these and associated FPGA implementation issues.

6.3.1 RS-232

A common serial interface standard is RS-232. While in many applications it has been superseded by USB, it still remains popular because of its simplicity and is a standard peripheral in many microcontroller chips. While technically RS-232 refers to the particular electrical signalling standard, it is often used synonymously for asynchronous serial communication, and this is primarily the sense used here. To convert from normal logic levels to RS-232 signalling levels where required, an appropriate interface chip is the easiest solution. The need for a common ground limits the length of an RS-232 cable. In an industrial setting, the differential signalling of RS-422 (or RS-485 for multi-drop) is more robust.

Asynchronous communication is so-named because it does not transmit a clock signal. Before transmission, the data line is held in an idle state (a logic '1'). Transmission is started by sending a start bit (logic '0'); the falling edge is used to synchronise the receiver to the transmitter. Following the start bit, 5–8 data bits are sent, followed by an optional parity bit, and finally, one to two stop bits (logic '1') are used to enable the next start bit to be detected. The following timing and data framing parameters must be agreed between the transmitter and receiver before transmission:

- the clock frequency;
- the number of data bits and the order in which they are sent (most significant bit [MSB] or least significant bit [LSB] first);
- whether to use a parity bit, and if so, whether odd or even parity is used;
- the number of stop bits (the minimum time before a new transmission may begin).

A half-duplex connection uses a single data line for both transmitting and receiving. This requires hardware handshaking to control the direction of the data transfer. The handshaking is asymmetric, using separate signal lines. One end is designated as a data terminal, and it signals its desire to transmit by setting the request-to-send (RTS) signal low. This is acknowledged by a reply on the clear-to-send (CTS) line. Full-duplex connection uses separate independent data lines for transmitting and receiving.

On an FPGA, the communication can be managed by a simple state machine to step through the bits. The transmitter in Figure 6.10 works as follows. From the idle state, data is loaded into the shift register; the transmitter sends a start bit; and the parity counter, P, is either set or cleared, depending on whether odd or even parity is required. The data bits are then shifted out in turn, while a 1-bit counter (the exclusive-OR gate) determines the parity. After the data bits, the parity bit is transmitted (if required) followed by the stop bits, and the transmitter goes back to the idle state.

The receiver in Figure 6.10 is a little more complex. A synchroniser is used on the asynchronous input to prevent metastability problems. A falling-edge detector detects the start of the start bit and resets the sampling counter to the middle of its period, so that the data is sampled in the middle of a bit. The parity counter is also

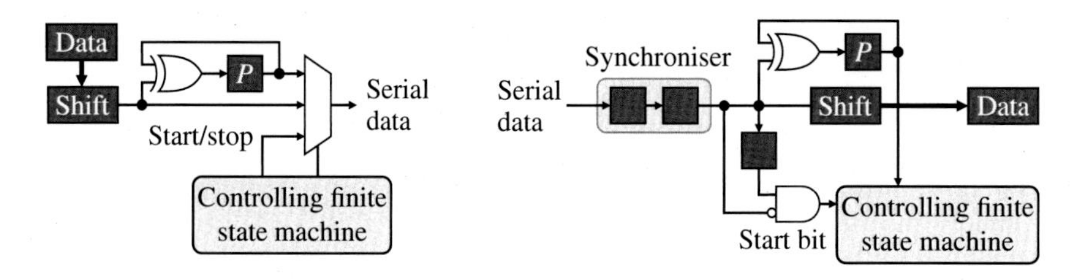

Figure 6.10 Skeleton implementation for an RS-232 driver. Left: transmitter; right: receiver.

reset at this stage. Successive data bits are shifted into the shift register, and the parity counter is checked after the parity bit is received. If there are no errors, the received word is transferred from the shift resister to the output data register, and the receiver goes back to the idle state waiting for the next start bit.

Data transfer is quite robust in that the transmitter and receiver clocks can be different by about 5% before getting framing errors. The implementation of asynchronous communication with a predefined set of parameters is straightforward. However, if the ability to change the parameters at runtime is necessary, then the system becomes a little more complex.

6.3.2 I^2C

Many devices and interface chips are configured using an Inter IC (I^2C) bus (NXP, 2014) or related communications protocol (for example SMBus and PMBus used within computer systems). This is a two-wire serial bus, with a data line (*SDA*) and clock line (*SCL*). Many devices may be connected on the single bus, including multiple bus masters, with each device on the bus individually addressable. The bus supports multiple data rates, from 100 kbit/s up to 3.4 Mbit/s, with the speed limited by the slowest device on the bus. When connecting to peripherals from an FPGA, it will be assumed here that the FPGA is the single bus master, and that it will be programmed to run at the speed of the slowest peripheral. (To handle the more complex case of multiple bus masters, consult the I^2C specifications (NXP, 2014).)

The bus lines normally use a passive pull-up, which can be achieved on the FPGA by always outputting a '0' and using the output enable to disable the output for a high signal, as illustrated in Figure 6.11. When operating with a single master, and if none of the devices will stretch the clock, then active pull-up can be used on the clock line. The clock signal is always generated by the master, which also initiates and terminates any data transfer. Normally, the data signal is only allowed to change while the clock signal is low. The exceptions to this are indicating a start condition (*S*) or a stop condition (*P*). Each byte transferred requires nine clock cycles: eight to transfer data bits (MSB first) and one to acknowledge.

A transfer is initiated by the master (the FPGA) signalling a start condition: pulling *SDA* low while the clock is high. The first byte in any data transfer is always sent by the bus master and consists of the 7-bit address of the slave device and an $R/\overline{W}$ bit. This is then acknowledged by the addressed slave pulling *SDA* low for the ninth clock cycle. The remainder of the transfer depends on whether the master signalled a read or a write (see Figure 6.12). When the FPGA is writing, each byte transferred is acknowledged by the slave, unless there is some error condition (for example it did not understand a command that was sent). At the end of the

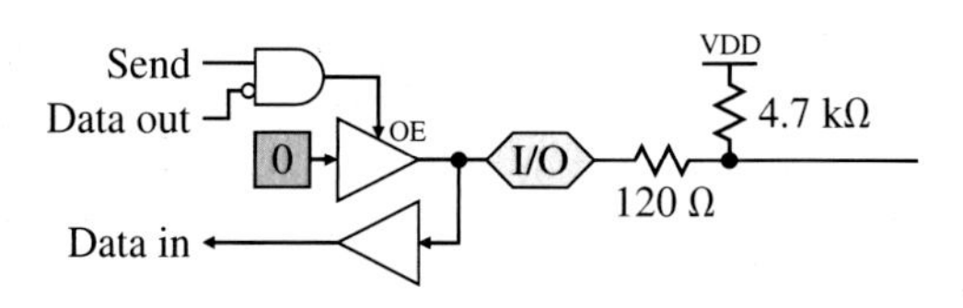

Figure 6.11 Implementation of an open-drain or passive pull-up data bus using an external pull-up resistor.

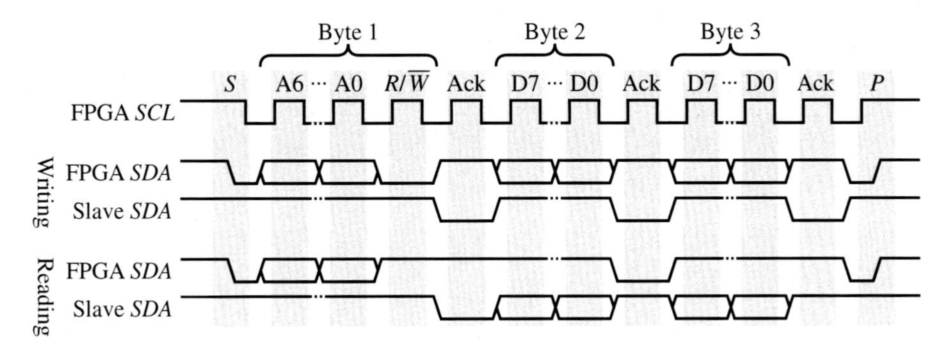

Figure 6.12 I^2C communication sequence. Top: clock; middle: writing to the device; bottom: reading.

transfer, the master signals a stop condition (a rising edge on *SDA* after the clock has gone high) or another start condition if it wishes to make another transfer. When reading from the device, after the first address byte, the slave begins transmitting data which is acknowledged by the FPGA. After the last byte, the FPGA sends a negative acknowledgement followed by the stop or repeated start condition.

There are four conditions that can lead to a negative acknowledgement (NXP, 2014), indicating that the transfer should be aborted:

- There was no slave on the bus with the specified address. Since there will be no response, the default will be for the bus to stay high.
- The slave device is not ready to communicate or cannot receive any more data from the master.
- The data sent to the slave was invalid, out of range, or otherwise not understood by the slave.
- When the master is receiving data from a slave, to indicate that no more data should be transmitted.

Many I^2C devices are based on registers in that data is written to, or read from, a numbered register on the device. In such cases, a common protocol is for the first data byte (or word for 16-bit addresses) of a write transfer to set an internal register pointer, with subsequent data bytes (or words) written to the registers. After each register is accessed, the internal pointer increments to point to the next register. Reading from a register is a little more involved. First, the FPGA needs to send a write to the slave, to set the internal register address. Then, a repeated start condition is used to terminate the first transaction and begin a read transfer to read the contents of the register addressed in the first message.

An FPGA implementation of I^2C is made easier by the FPGA being the bus master and providing the clock signal. The clock signal can be produced simply by dividing down the clock within the main clock domain. A finite state machine may be used to control the driver, using a circuit similar to Figure 6.10 (obviously, the parity generator and checker is not required). The control is a little more complex as a result of slave addressing. However, the driver can be kept reasonably simple by separating the driver from the data interpretation.

Initialising a peripheral often consists of sending data to several registers. The data for this can be stored in a small ROM and then streamed to the peripheral device through the I^2C core.

6.3.3 Serial Peripheral Interface (SPI)

Another serial bus architecture is the SPI. It is sometimes used with sensors and is commonly used as an interface to serial memories. It is both simpler and faster than I^2C, although it has no formal protocols, and there are several variations. The basic structure of the interface is as shown in Figure 6.13. The master selects the slave device, and then data is communicated in synchronisation with the clock (*SCLK*). Each slave has its own separate slave select ($\overline{SS}$) pin, although the clock and data lines can be shared. The data transfer is full duplex, with *MoSi* being the master output/slave input, and *SoMi* the slave output/master input, although both lines do not always carry meaningful data simultaneously. The meaning of the data transferred, and even the size of the

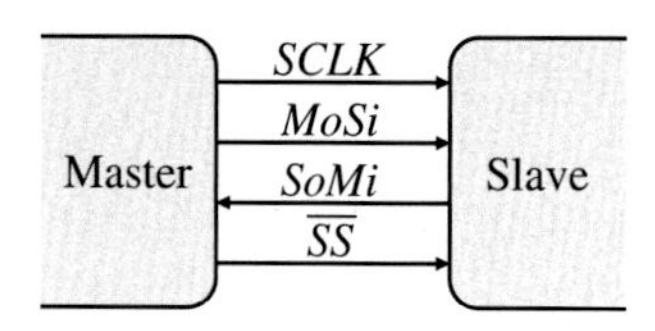

Figure 6.13 Serial peripheral interface.

transfer, is dependent on the particular device. Therefore, it is necessary to consult with the manufacturer's documentation on the specific commands and transfer protocol used in order to design an appropriate driver for the peripheral.

6.3.4 Universal Serial Bus (USB)

Interfacing with devices using USB is quite complex (Axelson, 2009). USB is based on a tiered star topology, with a single host and up to 127 devices. All communication is initiated from the host, which is responsible for detecting when a device is connected, assigning an address, determining the device's functionality, and initiating data transfers. A device must participate in the enumeration and configuration processes and respond appropriately (by sending or receiving data) when addressed.

An embedded FPGA system is most likely to be a USB device, where it is a peripheral providing data to a host computer. At a minimum, this requires an external PHY chip to manage the physical signalling. However, because of the complexity of the protocols, it is generally not worth the effort interfacing hardware directly with a USB connection. The alternative is to use a CPU (either soft core or hardened) on the FPGA to manage the communication and protocols. For simple serial communication, the simplest approach is to connect to an integrated USB chip that enables asynchronous serial data to be streamed directly to it. Many such devices have a built-in microcontroller which manages the details of the USB protocols, with the data interfaced directly through a serial connection.

The FPGA may also be a USB host, with one or more USB peripherals connected to it, for example a USB camera. Implementing a host controller is even more complex than a device controller. The simplest approach is to use an embedded CPU and interface via appropriate software drivers. If necessary, the data packets can be intercepted directly by the FPGA and interpreted under the direction of the CPU.

6.3.5 Ethernet

The use of Ethernet for point-to-point communication is relatively straightforward, although writing a driver for interfacing between the rest of the design and the Ethernet is not for the faint-hearted. Fortunately, logic cores are available to do most of the hard work. The communication interface requires a PHY for managing the physical signalling and a MAC core for encapsulating the data with the required Ethernet headers and controlling data transmission and reception.

Using an Ethernet connection to connect to the Internet adds a whole new layer of complexity. Communication on the Internet is based on the TCP/IP. This is a suite or stack of protocols (Roberts, 1996), where the Internet protocol (IP) encapsulates data transmitted with a header that describes the particular protocol, which is then transmitted (as payload) over Ethernet. Several of the core protocols of the Internet protocol suite are:

- The transmission control protocol (TCP) is one of the most commonly used protocols for the transmission of data on the Internet. It establishes and maintains a virtual stream connection between the two end points, on which reliable bidirectional transfer of data may take place. The protocol has mechanisms for detecting dropped or corrupted packets and automatically retransmitting the data until it is received correctly.
- The user datagram protocol (UDP) provides a lightweight mechanism for the transmission of data packets from one end point to another. While there is a checksum for checking data integrity, there is no mechanism for

ensuring reliable delivery. UDP is used for time-sensitive applications where the overhead of retransmitting lost or missing data is unacceptable, for example streamed media.
- The Internet control message protocol (ICMP) works behind the scenes to communicate status and error messages between hosts and routers in an Internet-based network.
- The address resolution protocol (ARP) maps an IP address to an Ethernet address to enable the packet to be sent to the correct destination on an Ethernet-based network.

On top of UDP and TCP are the application layer protocols, which specify how the actual data content is transferred for a particular application, such as file transfer, mail, news, domain name resolution, and web (HTTP).

While all of these component protocols could be implemented in hardware (see, for example (Dollas et al., 2005; Sidler et al., 2015; Ruiz et al., 2019)), the complexity of the layered architecture can quickly become unwieldy. Consequently, the Internet protocol is probably best managed by a software-based protocol stack, with the Ethernet MAC implemented in hardware using the FPGA logic.

Equipping an embedded system with a full TCP/IP stack provides a generic mechanism for remote communication. A web interface allows remote control and setting of algorithm parameters. The results, including images, can be readily displayed directly through a web interface. An Internet connection also allows the FPGA configuration to be remotely updated, providing a ready path for upgrades.

6.3.6 PCI Express

Peripheral component interconnect express (PCIe) is the current standard for connecting to peripherals within a high-speed computer system (Wilen et al., 2003). Previous generations of the peripheral component interconnect (PCI) bus were parallel, with 32 or 64 data lines. However, as the bus speed is increased, increased crosstalk from the coupling between adjacent signal lines and increased skew between data lines resulting from small differences in path length limited the effective I/O bandwidth. In switching to a serial connection with the clock embedded within the data, PCI Express solves the skew problem (although a much higher clock rate is required to maintain the bandwidth) and uses differential signalling to significantly reduce the effects of crosstalk. PCIe is a point-to-point connection with separate transmit and receive, avoiding overheads associated with bus arbitration and contention. Each transmit and receive pair is called a lane. Version 1.0 had a raw transfer rate of 2.5 Gbit/s, with the latest version (6.0) supporting 64 Gbit/s. Connections requiring higher bandwidth can run multiple lanes in parallel (up to 32 are supported), with successive bytes sent over successive lanes.

Early versions of PCIe used 10-bit symbols for each 8-bit data byte (TDMS, to minimise the number of transitions and average the DC level). Later versions reduce the overhead using 130-bit symbols for each 128 bits. At the link layer, additional overheads are introduced for CRC (cyclic redundancy check) error detection and packet acknowledgements. Each data packet is kept in the transmit buffer until acknowledgement is received, enabling packets signalled as corrupt to be automatically resent.

Higher layers of the protocol create virtual links between the components that share the underlying physical link. This is implemented through a transaction layer, where each transaction consists of a request followed by a response. Between the request and response, the link is available for other traffic. Flow control uses a credit-based scheme where the receiver advertises the space available within the receive buffer as credit. The transmitter keeps track of its remaining credit and must stop transmitting before the buffer is full. As the receiver processes data and removes it from the buffer, it restores credit to the transmitting device.

PCIe is most commonly used to interface between an FPGA board and its host computer system, enabling the rapid transfer of data in high-performance computing applications. Recent FPGAs have hardened PCIe ports, with logic cores available for interfacing between the port and the application.

6.4 Off-chip Memory

In many applications, it is necessary to use memory external to the FPGA for frame buffers and other large memory blocks. External memory is also required when running an embedded processor with an operating

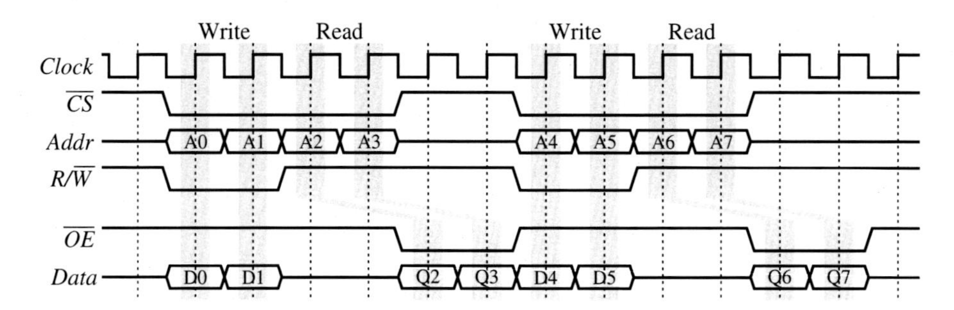

Figure 6.14 Timing for a typical pipelined memory with two clock cycles latency. Two cycles are lost when a write follows a read due to bus turnaround.

system such as Linux. There are three main types of memory in widespread use: static RAM, dynamic RAM, and flash memory (although there are several other technologies under development). The issues involved in interfacing with these will be described in turn.

6.4.1 Static RAM

Static memory stores each data bit using a latch. Each memory bit therefore requires a minimum of four transistors, with an additional two transistors used for reading and writing. This is considerably larger than the one transistor used by dynamic memory, making it more expensive to manufacture. Its advantage is its speed, making it commonly used for intermediate memory sizes (in the low megabyte range).

Almost all high-speed synchronous memories are pipelined. Therefore, when reading, the data is available several clock cycles after the address is provided, although the memory has a throughput of one memory access per clock cycle. This access latency must be built into the design of the application. This makes external memories a little more difficult to use than internal block RAMs.

Usually when writing, both the address and data must be applied at the same time. Consequently, when following a read by a write, there is a dead time to avoid bus contention while the read operation completes before the data pins are available for providing the data to be written (see Figure 6.14). This bus turnaround time must be taken into account in the timing budget and reduces the effective bandwidth available. While such memories may be suitable for reading or writing large blocks, such as might be encountered when streaming an image to or from a frame buffer, they are significantly less efficient when reads and writes are alternated, for example in two-phase designs.

Zero bus turnaround (ZBT) memories have no dead period between writes and reads. This is accomplished by providing the data to be written after the address, as shown in Figure 6.15. The ability to switch between reads and writes from one cycle to the next makes ZBT memories significantly easier to interface to an application. The data to be written may be delayed within the FPGA using a pipeline.

6.4.2 Dynamic RAM

Dynamic random access memory (DRAM), because of its lower cost, is well suited for larger volumes of memory. It uses a single transistor per memory cell, with the data stored as charge on a capacitor. Since the charge leaks off the capacitor, it must be refreshed at least once every 64 ms to prevent the data from being lost. To achieve this with large memories, DRAMs are structured so that a complete row (within the 2-D memory array) is refreshed at a time. This leads to a paged structure (see Figure 6.16), where the memory is divided into banks, and within each bank a row is selected first, and then the column within that row is read from or written to.

DRAMs are significantly harder to interface to than static memory for two reasons. First, they must be refreshed periodically by accessing every row at least once every 64 ms. Second, the paged address structure

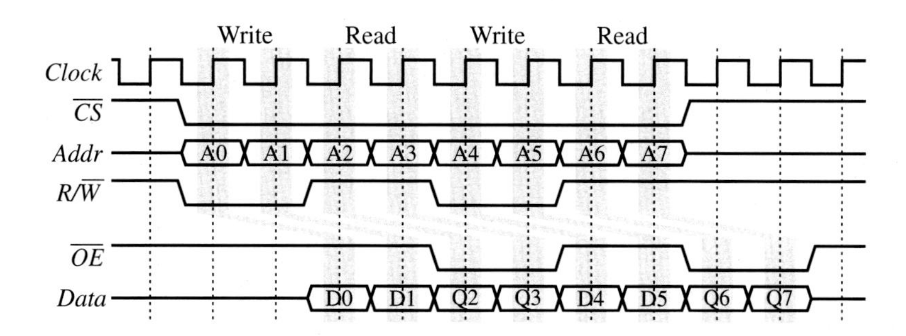

Figure 6.15 Typical timing for a pipelined ZBT static memory with two clock cycles latency.

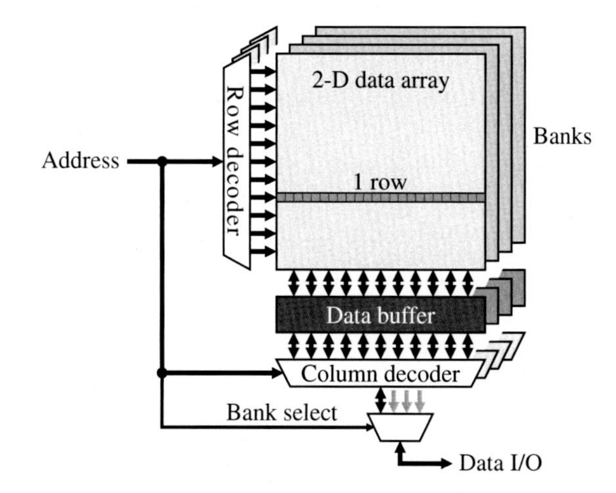

Figure 6.16 Page structure of a synchronous DRAM. Within each bank, only one row of data is accessible at a time.

requires the row address before the column address, with both sets of address lines sharing the same pins. Interface to a chip is controlled through a sequence of commands which are provided through a set of control pins. A typical command sequence is as follows:

- An activate command uses the address lines to select the memory row to use. Activating a row reads the row into the column sense amplifiers which effectively buffer the data, enabling the data within that row to be accessed. This is also called opening the row and typically takes several clock cycles.
- Once activated, read and write commands may be made to that row. With a read or write command, the address lines specify the column that is to be read or written. A read or write has several clock cycles latency; however, these are pipelined, enabling one access per clock cycle to data on that row. When writing, the data must be provided with the address, so there is a bus turnaround delay when switching from reads to writes (see Figure 6.14). Operating in burst mode enables several successive memory locations to be read or written with a single command. This is exploited with DDR memories, giving two reads or writes per clock cycle, one on each edge of the clock.
- Before selecting another memory row, it is necessary to close the row with a pre-charge command. Closing the row writes the data buffer back, refreshing the charge on the capacitors within that row. It also pre-charges the sense amplifiers to an intermediate state ready to sense the next row.
- A refresh command refreshes one row of memory. An internal counter is used to refresh the rows in sequence. The refresh command internally activates a row followed by a pre-charge to refresh the charge on

the capacitors. While it is possible to do this manually, having a refresh command simplifies the memory controller.

There is considerable latency, especially when switching from accessing one row to another. To help alleviate this, most DRAM chips have multiple banks of memory internally. Interleaving commands to different banks may be used to hide the activation and pre-charge overheads by performing reads and writes to one bank, while activating rows or precharging within another bank. However, such interleaving does not eliminate the bus turnaround, which uses an increasing proportion of the bandwidth for higher speed memories (Ecco and Ernst, 2017).

Reading or writing from a dynamic memory is therefore not a simple matter of providing the address and data. The sequence of operations to access a particular memory and ensure that the memory is kept refreshed requires a memory controller to hide the low-level details from the application. Many modern FPGAs incorporate hardened memory controllers to simplify access. Otherwise, it is necessary to build the controller from the fabric of the FPGA. The latency of DRAM is variable, depending on the memory state, so data buffering is usually required to maintain a smooth flow of data to or from the memory.

A common issue, particularly with high-speed DDR memories, is ensuring that any signal skew in the parallel data and address lines does not violate setup and hold times.

6.4.2.1 Frame Buffers

External memory is almost always required for building a frame buffer. If random access processing is required, then static memory is most efficient because it avoids the overhead of opening and closing rows. ZBT memories are designed with no dead period enabling true low-latency random access, making them easier to interface to an application. This simplifies the design of the associated memory controller (see, for example (Vanegas et al., 2010)).

Dynamic memory is more commonly used for frame buffers because it is less expensive for large volumes of data. Bus turnaround makes them poorly suited for alternating reads and writes. They have their highest bandwidth when operated in burst mode, with a series of reads or writes on the same row. Therefore, when working with video streams, it is most efficient to read or write a whole image row at a time. The bank address bits are positioned immediately above the column address bits, so that the opening and closing overheads can be hidden when accessing sequential addresses. Since the access pattern is known, each row can be requested from the frame buffer one row in advance and stored on the FPGA within a row buffer, from which it can be streamed to the application. This minimises bus turnaround and hides the variable latency of the DRAM. Similarly, when writing, a row of data can be accumulated into a row buffer and written to the DRAM as a long sequence of writes.

Note that if regular access to each DRAM row can be guaranteed (for example continually outputting the frame buffer to a display), then separate memory refresh operations are not required; simply accessing an entry on each DRAM row is sufficient to refresh the row.

Streamed access by columns (for example with a 2-D fast Fourier transform (FFT)) is generally less practical because of the large activation and pre-charge latencies when switching memory rows. These can partially be hidden by storing successive rows within separate banks, so accesses can be interleaved. This, however, requires sufficiently long bursts within each memory row, so there is enough bandwidth on the command bus for interleaving. This may be overcome to some extent by creating a mapping between logical memory and physical memory to increase the number of sequential accesses to the same page (Bailey, 2015). Mapping some of the image row address bits into the DRAM column address space, as demonstrated in Figure 6.17, ensures that multiple successive row accesses are within the same DRAM row. Combining some image column bits to select the bank also ensures that when traversing an image row, not all accesses are to the same bank. Note that a good mapping is dependent on both the image size and the DRAM row size.

If random access to DRAM is required, then the application needs to be tolerant of the variable latency and potentially reduced memory bandwidth.

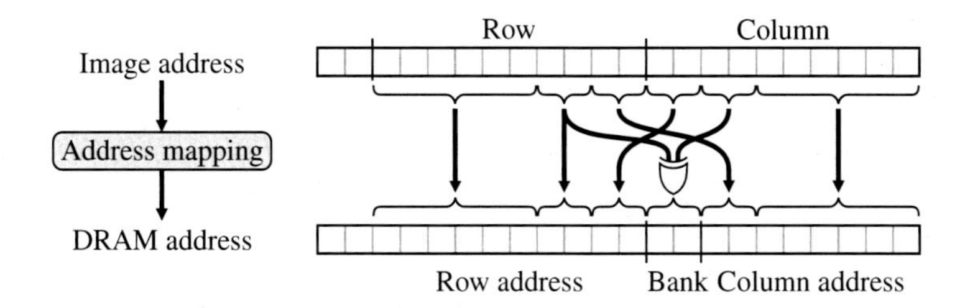

Figure 6.17 Address mapping to enable interleaving accesses between banks. Source: Adapted from Bailey (2015).

6.4.3 *Flash Memory*

Flash memory is non-volatile, making it useful for storing data required by an application while the FPGA is switched off, or when switching from one configuration file to another. The basic mechanism is to have charge stored on a floating gate transistor, which results in a change in transistor threshold voltage, enabling the different states to be identified (Pavan et al., 1997). Programming consists of injecting charge onto the floating gate, while erasing removes the charge. There are two main types of flash memory, based on how these transistors are arranged: NOR flash and NAND flash.

NOR flash is similar to regular memory, with address and data lines allowing individual transistors to be read or written using a random access memory pattern. Read access times are reasonably fast, although writes are very slow in comparison with static or dynamic memories, and this must be taken into account by the application. Erasure is on a block-by-block basis and is also very slow. Erasing a block sets all the bits to '1's, and writing can only change '1's to '0's. To change a '0' to a '1' requires erasing the complete block, losing any data contained within it.

NAND flash arranges the transistors differently to enable a higher packing density (resulting in lower cost). The consequence is that NAND flash has page- or sector-based structure, making it better suited for sequential access than random access. Programming individual bytes and random access are difficult, but programming and erasing blocks is significantly faster than NOR flash (Micron, 2010). Internally, reading copies the whole block from non-volatile storage into buffer register (typically 25 μs), from which the data can then be streamed out (typically 25 ns per byte or word). Similarly, writing streams the data into the buffer register (25 ns per transfer) from which the whole block is programmed at once (200–500 μs). Since NAND flash memories are oriented to sequential access, the data pins are usually multiplexed to carry commands and the page address.

NAND flash requires error checking and correction to ensure data integrity. Each block usually has a number of spare bytes that can be used for storing error correction codes. However, it is up to the controller to implement any error checking and correction. It is also important to check the status after a program or erase operation to confirm that the operation has completed successfully. If not successful, such blocks should be marked as bad and not be used.

6.5 Processors

Partitioning an application between software and hardware implementation requires data communication between the CPU and the hardware. This section discusses the interface between the FPGA logic and an embedded hardened Arm CPU within an system-on-chip (SoC) FPGA.

6.5.1 *AXI Interface*

One of the most common interconnects, especially for ARM processors, is the AXI bus (Arm, 2021). This establishes communication between manager and subordinate (master and slave) interfaces, where the manager

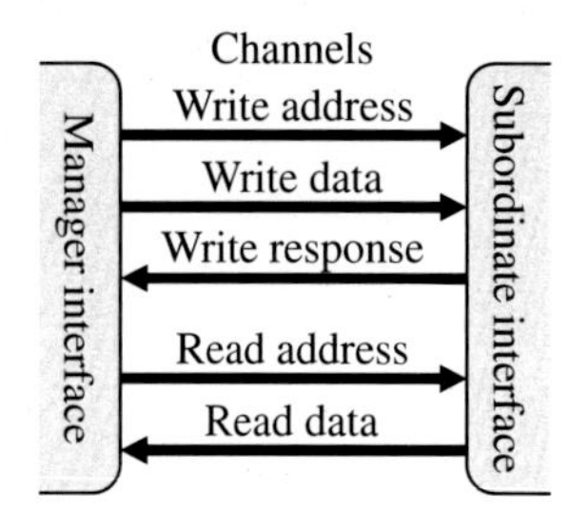

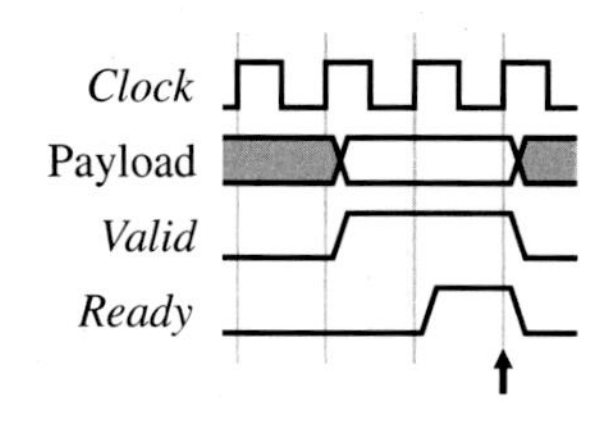

Figure 6.18 AXI interface. Left: the interconnect consists of five channels, three for writing and two for reading; the direction of the arrow corresponds to the payload transfer direction. right: handshake mechanism; the payload is only transferred when *Valid* and *Ready* are both asserted.

provides the address (and control information) on an address channel, with the data transferred on a data channel. For write transactions, there is an additional write response channel, as illustrated in Figure 6.18. The transfer of data occurs on the rising edge of the clock when both the sender (*Valid*) and receiver (*Ready*) are ready. A transaction consists of an address and control transfer on the address channel, the transfer of a burst of data words on the data channel, and for write transactions, an acknowledgement on the write response channel. The burst length can be up to 256 words, specified as part of the control.

Multiple transactions can be underway at a time; a manager does not need to wait for one transaction to complete before issuing another. The transfers have an associated *ID* signal which is used to distinguish the transactions.

The AXI coherency extensions (ACE) protocol (Arm, 2021) provides a framework for maintaining correctness of data which is shared across caches (for example within manager ports), enabling maximum reuse of cached data. The protocol extends the AXI communication with additional signalling to enable the validity of cached data to be determined.

The accelerator coherency port (ACP) is an AXI slave port that enables an external manager to access data from the cache rather than directly from external memory, effectively putting the manager interface on the processor side of the cache. This may be useful for fine-grained low-level tasks, where the FPGA is accelerating particular functions or providing custom instructions.

For transfer of small amounts of data, for example setting control registers or reading status registers on the FPGA, the AXI-Lite protocol can be used. This simplifies the transfers, with only one transaction allowed at a time, with a burst length of 1.

For large volumes of continuous data (for example a video stream), the AXI stream protocol (Arm, 2010) may be simpler to use. A stream is a unidirectional flow of data from a master to a slave. There is no explicit addressing of data within a stream, although *ID* and *Destination* signals can be used for routing streams to appropriate destinations. Bursts are replaced by packets, which can be arbitrary size (with the last transfer of a packet marked by a *Last* signal).

Vendor tools provide most of the interfacing IP, removing the need to design the AXI interface for each peripheral built on the FPGA. Generally, all that is required is the setting of a few parameters.

6.5.2 Avalon Bus

Intel uses its proprietary Avalon bus in a similar manner for connecting components within an Intel FPGA (Intel, 2020). This is primarily used for connecting peripherals to a Nios-II soft-core processor but can also be used for interfacing with hardened Arm cores. The memory mapped and streaming interfaces have similar functionality to the corresponding AXI interfaces, although with different signalling. Again, the development tools provide most of the interfacing IP, simplifying the process of connecting custom peripherals.

6.5.3 Operating Systems

When developing code for the software processor, two main approaches can be taken: bare metal and having the code run under an operating system.

Bare-metal programming is where the software interacts directly with the hardware, without any underlying operating system. This usually consists of a main while-loop for tasks that are not time critical, with interrupts used for responding directly to time-critical events. Bare-metal software often has lower latency because it is dealing directly with the hardware, without the overheads of an operating system. However, it requires direct knowledge of the hardware that it is interacting with and becomes more complex when working with multiple tasks. Bare-metal programming is appropriate for small, simple systems with few resources.

The alternative is to run the code under an operating system, whether some form of real-time operating system (RTOS) or embedded Linux. Interaction with the hardware is made through device drivers which handle many of the low-level details, including data transfer protocols. The operating system can provide a whole array of supporting services, including scheduling, memory management, and access to higher level resources, such as USB and networking. It is appropriate for more complex code, with more sophisticated user interfaces (for example a web interface).

An RTOS provides real-time scheduling, with controlled latency, which is important if the code is performing or controlling a real-time task. However, it does not necessarily include services such as networking, and a small RTOS may provide only a limited file system. A more general operating system, such as Linux, will not necessarily meet real-time constraints. However, Linux has vast selection of other software and services that will run on the system.

6.5.4 Implications for System Design

Having available a CPU (whether soft core or hardened) has an impact on the overall system design. Software-based device drivers give access to many more peripherals much more easily than would be available if they had to be interfaced directly to hardware (although perhaps at the cost of latency). Even relatively trivial tasks such as RS-232 communication and I^2C configuration of peripherals can be offloaded from hardware to software. Within an embedded image processing application, hardware can be used for the low-level vision tasks, with the high-level tasks implemented more efficiently in software.

6.6 Summary

This chapter has reviewed interfacing issues particularly related to embedded vision applications.

A camera or input sensor is perhaps the key peripheral in a vision system. The most straightforward approach is to interface directly with the sensor chip. This gives the lowest latency and avoids many of the complications of interfacing with a higher level communications protocol, such as USB and Ethernet.

Image display is another important capability, even if it is just used for debugging and system tuning. A simple display is relatively easy to generate, although an analogue video output will require external digital to analogue converters. A DVI or HDMI output keeps the system all-digital, although will require more logic on the FPGA and high-speed SERDES interfaces to produce the serial signals (unless an external PHY is used).

Several serial communication protocols have been outlined. The control channel for many devices (including cameras and displays) is usually over a relatively low-speed serial link. The slower speed makes the design easier, enabling simple interfaces to be built with relatively few resources. Higher speed connections such as gigabit Ethernet and PCI Express are significantly more complex and should be built using intellectual property blocks. While both will be less common in embedded vision applications, PCI Express is commonly used to interface between a host computer system and an FPGA in a high-performance reconfigurable computing platform.

Issues with interfacing with external memories have been considered briefly. The main complication is the memory latency with pipelined access. This is particularly acute with dynamic memory which requires that a page be selected before it is accessed. Careful design of the system, and in particular the memory controller, is required to prevent the acceleration gained elsewhere in the algorithm from being eroded due to memory bandwidth limitations.

Finally, the interface to an embedded processor was briefly outlined. Much of this is simplified through the use of a standard interface, such as an AXI interface or similar, with much of the actual interface detail handled by the development tools.

References

Abdaoui, A., Gurram, K., Singh, M., Errandani, A., Chatelet, E., Doumar, A., and Elfouly, T. (2011). Video acquisition between USB 2.0 CMOS camera and embedded FPGA system. *5th International Conference on Signal Processing and Communication Systems (ICSPCS)*, Honolulu, HI, USA (12–14 December 2011), 5 pages. https://doi.org/10.1109/ICSPCS.2011.6140863.

AIA (2004). *Camera Link Specifications*. Automated Imaging Association.

Albl, C., Kukelova, Z., Larsson, V., Polic, M., Pajdla, T., and Schindler, K. (2020). From two rolling shutters to one global shutter. *IEEE/CVF Conference on Computer Vision and Pattern Recognition (CVPR)*, Seattle, WA, USA (13–19 June 2020), 2502–2510. https://doi.org/10.1109/CVPR42600.2020.00258.

Arm (2010). *AMBA 4 AXI4-Stream Protocol Specification, IHI 0051A*. Arm Limited.

Arm (2021). *AMBA AXI and ACE Protocol Specification, IHI 0022Hc*. Arm Limited.

Axelson, J.L. (2009). *USB Complete: The Developer's Guide*. 4e. Madison, WI: Lakeview Research LLC.

Bailey, D.G. (2015). The advantages and limitations of high level synthesis for FPGA based image processing. *International Conference on Distributed Smart Cameras*, Seville, Spain (8–11 September 2015), 134–139. https://doi.org/10.1145/2789116.2789145.

Bailey, D.G., Gribbon, K., and Johnston, C. (2006). GATOS: a windowing operating system for FPGAs. *3rd IEEE International Workshop on Electronic Design, Test, and Applications (DELTA 2006)*, Kuala Lumpur, Malaysia (17–19 January 2006), 405–409. https://doi.org/10.1109/DELTA.2006.51.

Bayer, B.E. (1976). Color imaging array. United States of America Patent 3971065.

DDWG (1999). *Digital Visual Interface (DVI)*. Digital Display Working Group.

de Haan, G. and Bellers, E.B. (1998). Deinterlacing - an overview. *Proceedings of the IEEE* **86** (9): 1839–1857. https://doi.org/10.1109/5.705528.

Dollas, A., Ermis, I., Koidis, I., Zisis, I., and Kachris, C. (2005). An open TCP/IP core for reconfigurable logic. *13th Annual IEEE Symposium on Field-Programmable Custom Computing Machines (FCCM 2005)*, Napa, CA, USA (18–20 April 2005), 297–298. https://doi.org/10.1109/FCCM.2005.20.

Ecco, L. and Ernst, R. (2017). Tackling the bus turnaround overhead in real-time SDRAM controllers. *IEEE Transactions on Computers* **66** (11): 1961–1974. https://doi.org/10.1109/TC.2017.2714672.

EMVA (2009). GenICam Standard 2.0. European Machine Vision Association.

Finlayson, G.D., Hordley, S.D., and Hubel, P.M. (2001). Color by correlation: a simple, unifying framework for color constancy. *IEEE Transactions on Pattern Analysis and Machine Intelligence* **20** (11): 1209–1221. https://doi.org/10.1109/34.969113.

Ge, X. (2012). The design of a global shutter CMOS image sensor in 110nm technology. Master of Science thesis. Delft, The Netherlands: Delft University of Technology.

Geyer, C., Meingast, M., and Sastry, S. (2005). Geometric models of rolling-shutter cameras. *6th Workshop on Omnidirectional Vision, Camera Networks and Non-classical Cameras (OMNIVIS05)*, Beijing, China (21 October 2005), 12–19.

Grob, T. (2010). Implementation of a FPGA-based interface to a high speed image sensor. Masters thesis. Stuttgart, Germany: University of Stuttgart.

Groen, F.C.A., Young, I.T., and Ligthart, G. (1985). A comparison of different focus functions for use in autofocus algorithms. *Cytometry* **6** (2): 81–91. https://doi.org/10.1002/cyto.990060202.

Grundmann, M., Kwatra, V., Castro, D., and Essa, I. (2012). Calibration-free rolling shutter removal. *IEEE International Conference on Computational Photography (ICCP)*, Seattle, WA, USA (28–29 April 2012), 8 pages. https://doi.org/10.1109/ICCPhot.2012.6215213.

HDMI (2009). *High-Definition Multimedia Interface Specification Version 1.4*. HDMI Licensing, LLC.

Ibraheem, O.W., Irwansyah, A., Hagemeyer, J., Porrmann, M., and Rueckert, U. (2015). A resource-efficient multi-camera GigE vision IP core for embedded vision processing platforms. *International Conference on ReConFigurable Computing and FPGAs (ReConFig)*, Mexico City, Mexico (7–9 December 2015), 6 pages. https://doi.org/10.1109/ReConFig.2015.7393282.

Icking, H., Nagpal, R.K., and Wiley, G. (2019). A look at MIPI's two new PHY versions, 30 March 2021. MIPI Alliance. https://resources.mipi.org/blog/a-look-at-mipis-two-new-phy-versions (accessed 24 April 2023).

Intel (2020). *Avalon Interface Specifications, Volume MNL-AVABUSREF*. Intel Corporation.

Jin, S., Cho, J., Kwon, K.H., and Jeon, J.W. (2010). A dedicated hardware architecture for real-time auto-focusing using an FPGA. *Machine Vision and Applications* **21** (5): 727–734. https://doi.org/10.1007/s00138-009-0190-2.

Khan, S., Bailey, D., and Sen Gupta, G. (2009). Simulation of triple buffer scheme (comparison with double buffering scheme). *2nd International Conference on Computer and Electrical Engineering (ICCEE 2009)*, Dubai, United Arab Emirates (28–30 December 2009), Volume **2**, 403–407. https://doi.org/10.1109/ICCEE.2009.226.

Land, E.H. and McCann, J.J. (1971). Lightness and retinex theory. *Journal of the Optical Society of America* **61** (1): 1–11. https://doi.org/10.1364/JOSA.61.000001.

Lao, Y. and Aider, O.A. (2021). Rolling shutter homography and its applications. *IEEE Transactions on Pattern Analysis and Machine Intelligence* **43** (8): 2780–2793. https://doi.org/10.1109/TPAMI.2020.2977644.

Liang, C.K., Chang, L.W., and Chen, H.H. (2008). Analysis and compensation of rolling shutter effect. *IEEE Transactions on Image Processing* **17** (8): 1323–1330. https://doi.org/10.1109/TIP.2008.925384.

Micron (2010). NAND Flash 101: An Introduction to NAND Flash and How to Design It Into your Next Product. Technical note. USA: Micron Technology, Inc.

Microsemi (2016). Building MIPI CSI-2 applications using SmartFusion2 and IGLOO2 FPGAs. Application note AC460 (Revision 1.0). USA: Microsemi Corporation.

MIPI (2016). *Evolving CSI-2 Specification*. Technology brief. MIPI Alliance Inc.

MIPI (2019). *MIPI Alliance Specification for Camera Serial Interface 2 (CSI-2), v3.0*. MIPI Alliance Inc.

Nicklin, S.P., Fisher, R.D., and Middleton, R.H. (2007). Rolling shutter image compensation. In: *RoboCup 2006: Robot Soccer World Cup X*, Lecture Notes in Computer Science, vol. **4434**, 402–409. https://doi.org/10.1007/978-3-540-74024-7_39.

NXP (2014). *I^2C-Bus Specification and User Manual, UM10204 Rev. 6*. NXP Semiconductors.

Pavan, P., Bez, R., Olivo, P., and Zanoni, E. (1997). Flash memory cells-an overview. *Proceedings of the IEEE* **85** (8): 1248–1271. https://doi.org/10.1109/5.622505.

Roberts, D. (1996). *Internet Protocols Handbook*. Scottsdale, AZ: Coriolis Group Books.

Ruiz, M., Sidler, D., Sutter, G., Alonso, G., and Lopez-Buedo, S. (2019). Limago: an FPGA-based open-source 100 GbE TCP/IP stack. *29th International Conference on Field Programmable Logic and Applications (FPL)*, Barcelona, Spain (8–12 September 2019), 286–292. https://doi.org/10.1109/FPL.2019.00053.

Sawyer, N. (2018). LVDS source synchronous 7:1 serialization and deserialization using clock multiplication. Application note XAPP585 (v1.1.2). USA: Xilinx Inc.

Sensor to Image (2016). GigE Vision Reference Design, volume A-1.0.0. Sensor to Image GmbH.

Sidler, D., Alonso, G., Blott, M., Karras, K., Vissers, K., and Carley, R. (2015). Scalable 10Gbps TCP/IP stack architecture for reconfigurable hardware. *IEEE 23rd Annual International Symposium on Field-Programmable Custom Computing Machines*, Vancouver, Canada (2–6 May 2015), 36–43. https://doi.org/10.1109/FCCM.2015.12.

Stevens, S.S. (1957). On the psychophysical law. *Psychological Review* **64** (3): 153–181. https://doi.org/10.1037/h0046162.

USB-IF (2012). Universal Serial Bus Device Class Definition for Video Devices, Revision 1.5. USB Implementers Forum Inc. https://www.usb.org (accessed 24 April 2023).

Vanegas, M., Tomasi, M., Diaz, J., and Ros, E. (2010). Multi-port abstraction layer for FPGA intensive memory exploitation applications. *Journal of Systems Architecture* **56** (9): 442–451. https://doi.org/10.1016/j.sysarc.2010.05.007.

van de Weijer, J., Gevers, T., and Gijsenij, A. (2007). Edge-based color constancy. *IEEE Transactions on Image Processing* **16** (9): 2207–2214. https://doi.org/10.1109/TIP.2007.901808.

VESA (2013a). Coordinated Video Timings Standard, Version 1.2. Video Electronics Standards Association.

VESA (2013b). VESA and Industry Standards and Guidelines for Computer Display Monitor Timing (DMT), Version 1.0, Revision 13. Video Electronics Standards Association.

Wilen, A., Schade, J.P., and Thornburg, R. (2003). *Introduction to PCI Express: A Hardware and Software Developer's Guide*. USA: Intel Press.

7

Point Operations

Beginning with this chapter, the focus will be on how specific image processing operations may be implemented on an field-programmable gate array (FPGA). Chapters 7–11 will describe preprocessing and other low-level operations. Chapters 12–14 will discuss intermediate-level operations.

7.1 Point Operations on a Single Image

The simplest class of image processing operations is that of point operations. They are so named because the output value for a pixel depends only on the corresponding pixel value of the input image:

$$Q[x, y] = f(I[x, y]), \tag{7.1}$$

where $f(\cdot)$ is some arbitrary function. Point operations may therefore be represented by a mapping, or transfer function, as shown in Figure 7.1.

The simplest hardware implementation is to use stream processing to pass each input pixel through a hardware block implementing the function, as illustrated in the right panel of Figure 7.1. Since each pixel is processed independently, point operations can also be easily parallelised by partitioning the image over multiple processors.

In spite of their simplicity, point operations have wide use in terms of contrast enhancement, segmentation, colour filtering, change detection, masking, and many other applications.

7.1.1 Contrast and Brightness Adjustment

One interpretation of the mapping function is to consider its effect on the brightness and contrast of an image. To make an image brighter, the output pixel value needs to be increased. This may be accomplished by adding a constant, as illustrated in Figure 7.2. Similarly, an image may be made darker by decreasing the pixel value, for example by subtracting a constant. In practice, adjusting the brightness is more complicated than this because the human visual system is nonlinear and does not consider each point in an image in isolation but relative to its context within the overall image (as demonstrated by the optical illusion in Figure 7.3).

The contrast of an image is affected or adjusted by the slope of the mapping function. A slope greater than 1 increases the contrast (see Figure 7.4), and a slope less than 1 corresponds to a contrast reduction. This may be accomplished by multiplying by a constant greater than or less than 1, respectively.

A relatively simple point operation for adjusting both the brightness and contrast is

$$Q = aI + b = a(I + b'), \tag{7.2}$$

Design for Embedded Image Processing on FPGAs, Second Edition. Donald G. Bailey.

Companion Website: www.wiley.com/go/bailey/designforembeddedimageprocessc2e

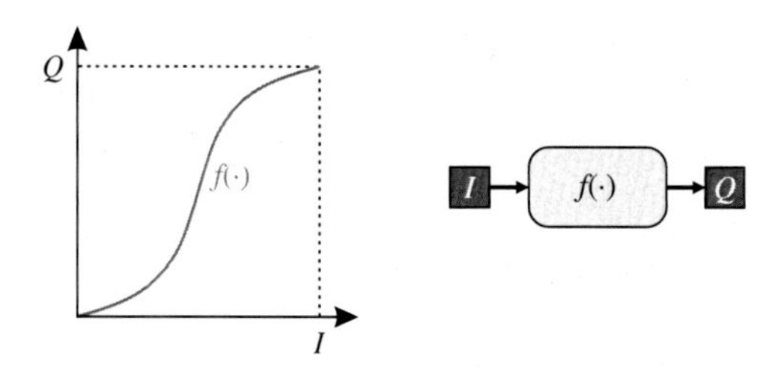

Figure 7.1 Left: a point operation maps the input pixel value, I, to the output value, Q, via an arbitrary mapping function, $f(\cdot)$; right: implementation of a point operation using stream processing.

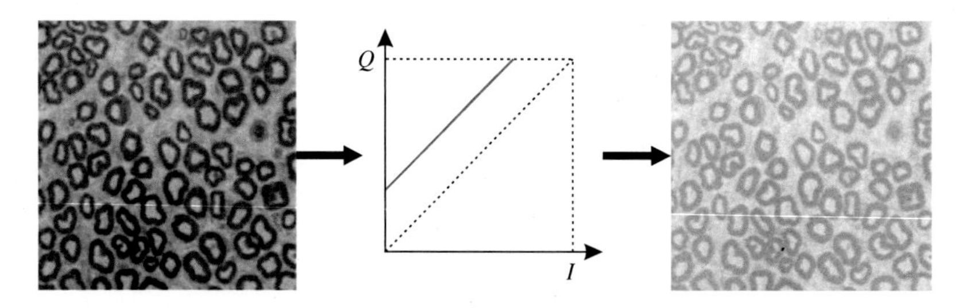

Figure 7.2 Increasing the brightness of an image by adding a constant.

Figure 7.3 Brightness illusion. The band through the centre has a constant pixel value.

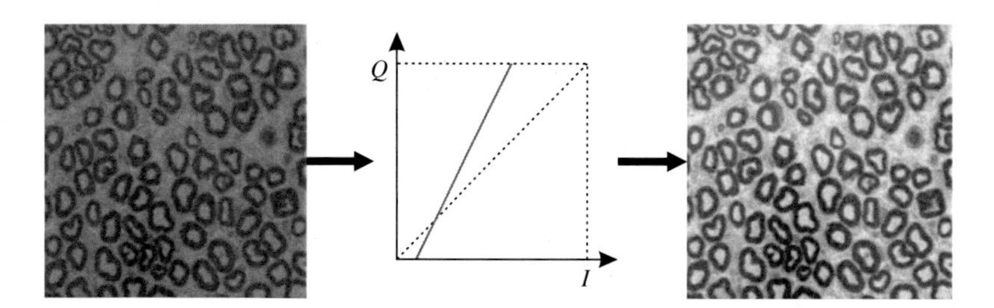

Figure 7.4 Increasing the contrast of an image.

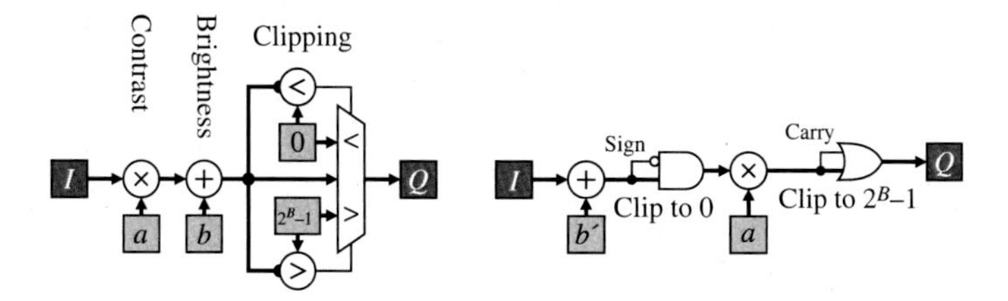

Figure 7.5 Left: Schematic representation of a simple contrast enhancement operation; right: simplifying the circuit using overflow.

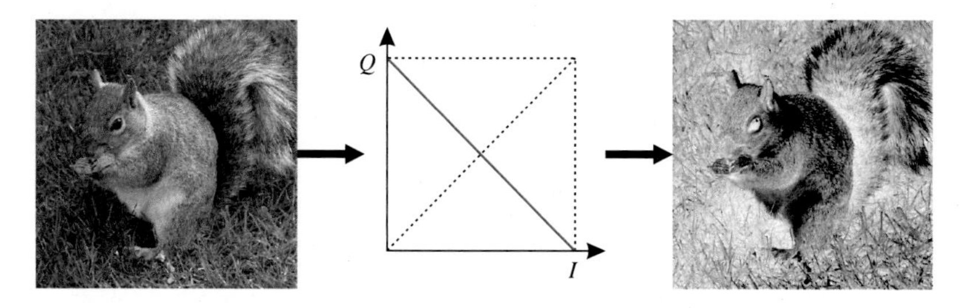

Figure 7.6 Inverting an image. Source: Photo courtesy of Robyn Bailey.

where a and b are arbitrary constants that control the brightness and contrast. One issue is what to do when the value for Q exceeds the range of representable values (with B bits per pixel, for $Q < 0$ or $Q > 2^{B-1}$). The computation default of ignoring overflow would cause values outside this range to wrap around, usually leading to undesired results. More sensible would be for the output to saturate or clip to the limits. Clipping requires building hardware to detect when the limits are exceeded and adjust the output accordingly, as shown schematically on the left in Figure 7.5. Note that either the brightness or contrast operation may be performed first, depending on the form of Eq. (7.2).

This circuit may be simplified to the right panel of Figure 7.5 by relying on the fact that the maximum pixel value is $2^B - 1$. With positive gain, only the offset can cause the result to go negative. Using the second form of Eq. (7.2), the sign bit of the addition can be used to clip the output to 0. Similarly, the carry from the multiplication can be used to clip to the maximum by setting all the bits to '1'.

A common contrast enhancement method is to derive the gain and offset from the minimum and maximum values available in the input image. The contrast is then maximised by stretching the pixel values to occupy the full range of available pixel values. This obviously requires that the whole image be read to determine the extreme pixel values, with consequent timing issues. Such issues will be discussed in Section 8.1.3 with histogram equalisation, another adaptive contrast enhancement technique.

A contrast reversal is obtained if the slope of the transfer function is negative. For example, to invert the image, a gain of −1 is used, as shown in Figure 7.6, where

$$Q = (2^B - 1) - I = \bar{I}. \tag{7.3}$$

Logically, this may be obtained by taking the one's complement (inverting each bit) of the input.

The transformation does not necessarily have to be linear. The same principle applies with a nonlinear mapping: raising or lowering the output will increase or decrease the brightness of the corresponding pixels; slopes greater than 1.0 correspond to a local increase in contrast, while slopes less than 1.0 correspond to a decrease in contrast; a negative slope will result in contrast inversion. A nonlinear mapping allows different effects at different intensities.

A common mapping is the gamma transformation. This is easier to represent when considering normalised values (where $Q_N = \frac{Q}{2^B-1}$ and $I_N = \frac{I}{2^B-1}$ represent fractions between zero and one):

$$Q_N = I_N^{\gamma}. \tag{7.4}$$

The gamma curve was designed primarily to compensate for nonlinearities in the transfer function at various stages within the imaging system. For example, the relationship between the drive voltage of a cathode ray tube

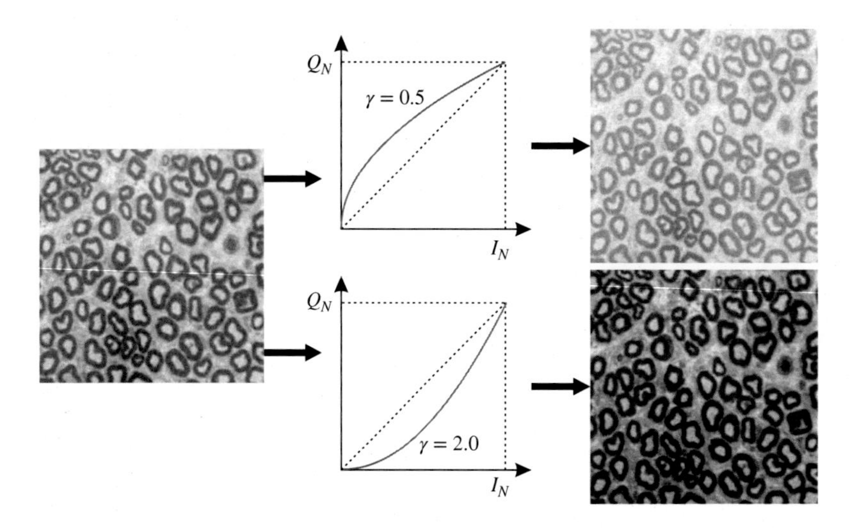

Figure 7.7 The effects of a gamma transformation.

(CRT) and the output intensity is not linear but is instead a power law. Therefore, a compensating gamma curve is required to make the contrast appear more natural on the display.

The effects of the gamma transformation can be seen in Figure 7.7. When γ is less than 1.0, the curve moves up, increasing the overall brightness of the image. From the slope of the curve, the contrast is increased for smaller pixel values at the expense of the lighter pixels, which have reduced contrast. For γ greater than 1.0, the opposite occurs. The image becomes darker, and the contrast is enhanced in the lighter areas at the expense of the darker areas.

A sigmoid curve is also sometimes used to selectively enhance the contrast of the mid-tones at the expense of the highlights and shadows (Braun and Fairchild, 1999).

7.1.2 Global Thresholding and Contouring

Thresholding classifies each pixel into one of two (or more) classes based on its pixel value and is commonly used to segment between object and background. In its basic form, thresholding compares each pixel in the image with a threshold level, *thr*,

$$Q = \begin{cases} 1, & I \geq thr, \\ 0, & I < thr, \end{cases} \tag{7.5}$$

and assigns a label to the corresponding output pixel (usually white or black), as shown in Figure 7.8. As the output is binary, it only requires a single bit of storage per pixel. Therefore, if an image can be successfully

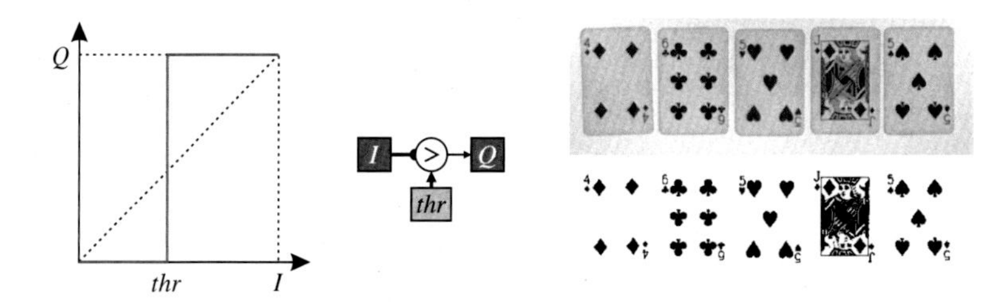

Figure 7.8 Simple global thresholding.

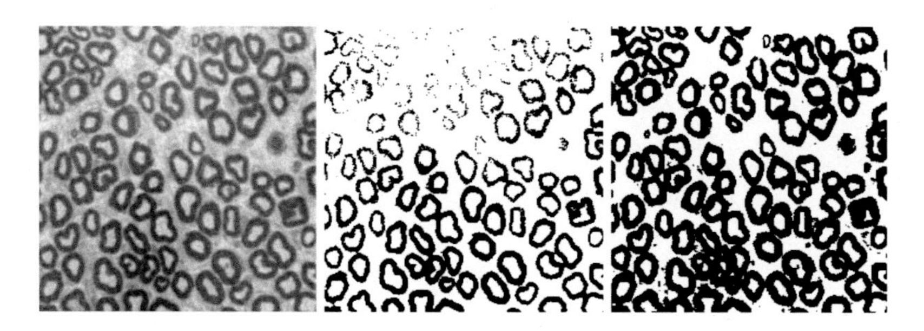

Figure 7.9 Global thresholding is not always appropriate. Left: input image; centre: optimum threshold for the bottom of the image; right: optimum threshold for the top of the image.

thresholded as it is streamed from the camera, the memory requirements for buffering the image are significantly reduced.

An appropriate threshold level can often be selected by analysing the statistics of the image, for example as contained in a histogram of pixel values. Several such threshold selection methods will be described in Section 8.1.5 under histogram processing.

Sometimes, however, a single global threshold does not suit the whole image. Consider the example in Figure 7.9. The optimum threshold for one part of the image is not the optimum for other parts. Using a global threshold to process such images will involve a compromise in the quality of the thresholding of some parts of the image. A point operation is unsuitable to process such images because to give good results over the whole image the operation needs to take into account not only the input pixel value but also the local context. Such images require adaptive thresholding, a form local filtering, and will be considered further in Section 9.3.6.

Simple thresholding may be generalised to operate with more than one threshold level.

$$Q = \begin{cases} 2, & thr_2 \le I, \\ 1, & thr_1 \le I < thr_2, \\ 0, & I < thr_1. \end{cases} \tag{7.6}$$

An example of this is shown in Figure 7.10. Here, two thresholds have been used to separate the image into three regions: the background, the cards, and the pips. Each output pixel is assigned a label based on its relationship to the two threshold levels.

In this example, the darkest region is never adjacent to the lightest region. This is not the case in Figure 7.11, where the interior of the white bubbles is black. There is also a gradation in contrast between the white and black giving a ring of pixels within the bubbles detected as the intermediate background level. However, even if there was a sharp change in contrast between the black and white, many of the boundary pixels are still likely to have intermediate levels. This is because images are area sampled; the pixel value is proportional to the average light falling within the area of the pixel. Therefore, if the boundary falls within a pixel, as illustrated on the right

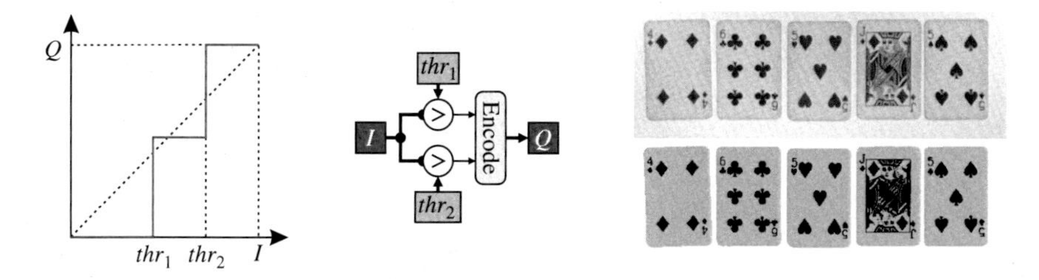

Figure 7.10 Multi-level thresholding.

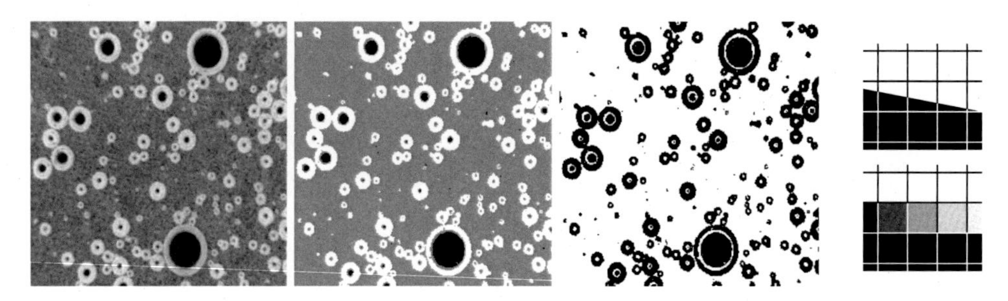

Figure 7.11 Misclassification problem with multi-level thresholding. Left: original image; centre-left: classifying the three ranges; centre-right: the background level labelled with white to show the misclassifications within each bubble; right: edge pixels have an intermediate value leading to potential misclassification.

in Figure 7.11, part of the pixel will be white and part black, resulting in intermediate pixel values. Any blurring or lens defocus will exacerbate this effect.

Consequently, it is inevitable that some of the boundary pixels will fall between the thresholds with multi-level thresholding and will be assigned an incorrect label. Such misclassifications require the information from the context (filtering) as preprocessing either before the thresholding or after thresholding to reclassify the incorrectly labelled pixels.

Another form of multi-level thresholding is contouring. This selects pixel values within a range, as seen in the centre-right panel of Figure 7.11, and requires a minor modification to Eq. (7.6):

$$Q = \begin{cases} 0, & thr_2 \leq I, \\ 1, & thr_1 \leq I < thr_2, \\ 0, & I < thr_1. \end{cases} \tag{7.7}$$

Contouring is so named because in an image with slowly varying pixel values, the pixels selected appear like contour lines on a map. Although this looks like edge pixels are being detected, contouring makes a poor edge detector because it lacks the context required to give connectivity.

7.1.3 Lookup Table Implementation

Most of the point operations considered so far are relatively simple mappings and can be implemented directly using logic. More complex mappings may require pipelining to improve the speed of deep logic.

Since the output value depends only on the input, the mapping for any point operation can be pre-calculated and stored in a lookup table. The input pixel value is then used to index into the table to find the corresponding output. An example of this is shown in Figure 7.12.

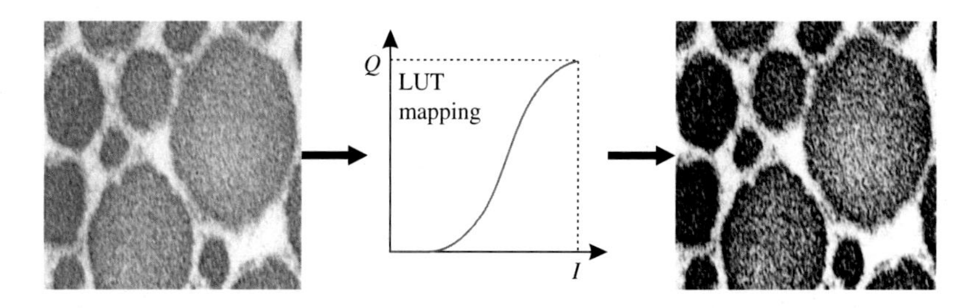

Figure 7.12 Performing an arbitrary mapping using a lookup table.

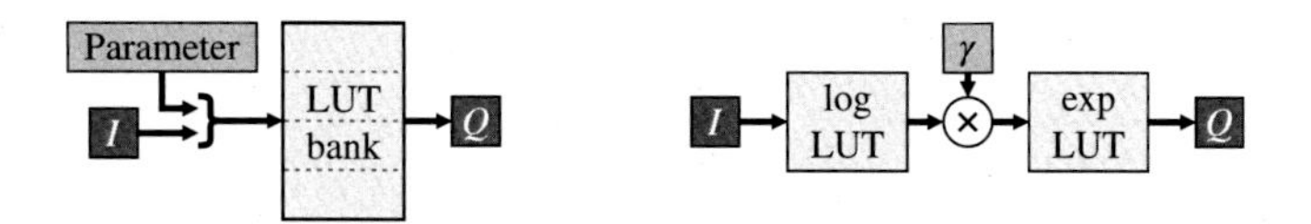

Figure 7.13 Parameterised lookup tables. Left: using the parameter as an address; right: combining tables with logic (implementing the gamma mapping for arbitrary γ).

The biggest advantage of a lookup table implementation is the constant access time, regardless of the complexity of the function. The size of the table depends on the width of the input stream, with each extra bit on the input doubling the size of the table.

A disadvantage of using a lookup table is that once the table has been set, the mapping is fixed. If the mapping is parameterised (for example contrast enhancement or thresholding), then it is also necessary to build logic to construct the lookup table before the frame is processed. Constructing the table will use the same (or similar) logic as implementing the point operation directly. The only difference is that timing is less critical. In a system with a software coprocessor, the table could also be populated by software rather than hardware.

If the parameter can be reduced to a small number of predefined values, then one possibility is to use the parameter to select between several preset lookup tables (as shown on the left in Figure 7.13). This is achieved by concatenating the parameter with the input pixel value to form the table address. Another possibility for more complex parameterised functions is to combine one or more lookup tables with logic, as shown for the gamma mapping on the right in Figure 7.13.

A sequence of two or more point operations is in itself a point operation. For example, contrast enhancement followed by thresholding is equivalent to thresholding with a different threshold level. This implies that a complex sequence of point operations may be replaced by a single lookup table that implements the whole sequence in a single clock cycle.

7.2 Point Operations on Multiple Images

Point operations can be applied not only to single images but also between multiple images. For this, Eq. (7.1) is extended to

$$Q[x,y] = f(I_1[x,y], I_2[x,y], \ldots, I_k[x,y]). \tag{7.8}$$

The corresponding pixels of the images are combined by an arbitrary function, $f(\cdot)$, to give the output pixel value at the corresponding pixel location.

Consider the case of two or more images derived from the same input image, as shown in Figure 7.14. Stream processing requires that the streams for each of the images be synchronised. If the processing branches on the inputs to the point operation have different latencies, then it will be necessary to introduce a delay within the faster branches to synchronise the streams with the slowest branch. The delay may follow the faster process (as with Process 2), or be inserted before it (as with Process 3), or even distributed throughout the process.

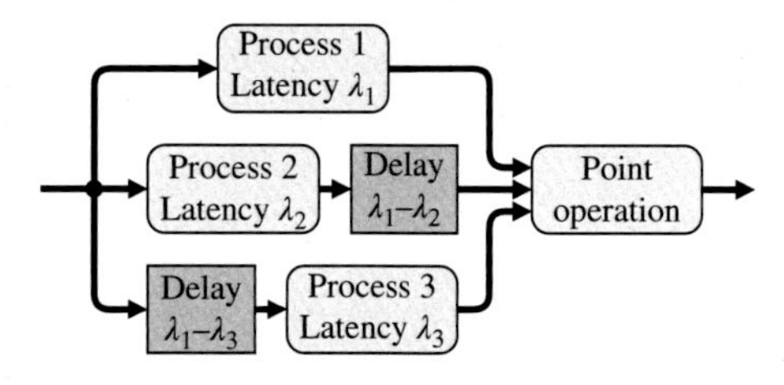

Figure 7.14 Point operation applied to processed versions of the same input image. A delay is inserted into the faster branches to equalise the latency.

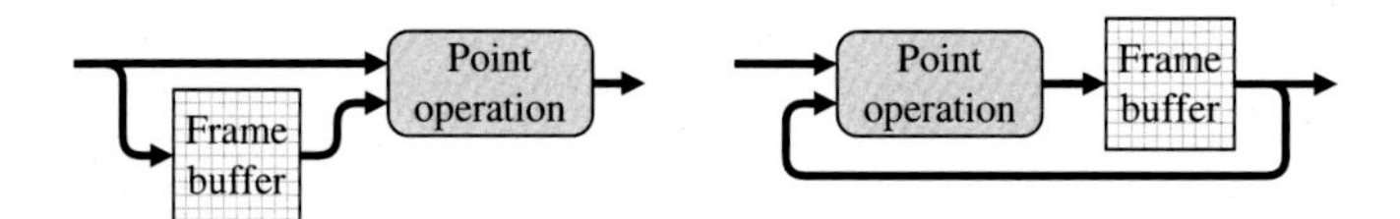

Figure 7.15 Two-image point operation applied to successive images. A frame buffer is required to hold the image from the previous frame. Left: sequential access; right: recursive access.

However, keeping the delays together at the start may enable the shift registers or a first-in first-out (FIFO) buffer to be shared.

A similar situation occurs when combining images derived from different, but synchronised, sensors.

If the images are captured at different times, then one or more frame buffers are required to hold the previous frames, as shown in Figure 7.15. Sequential access operates directly on the input image and one or more previous images. Recursive access feeds the output image back to be combined in some way with the incoming input image. In both cases, the frame buffer is used for both reading and writing. Some of the techniques of Section 4.2.1 may need to be used to enable the multiple parallel accesses.

Just as single-image point operations may be implemented using a lookup table, lookup tables may also be used to implement multiple-image point operations. The table input or address is formed by concatenating the individual inputs. Unfortunately, the memory requirements of a table grow exponentially with the number of address lines. This makes multiple-input point operation lookup tables impractical for all but the most complex operations, or where the latency is critical. As with single-input lookup tables, the latency of multiple-input LUTs is constant at one clock cycle.

While there are potentially many multi-image point operations, this section focuses on the more common operations.

7.2.1 Image Averaging

Real images inevitably contain noise, both from the imaging process and also from quantisation when creating a digital image. Let the captured image be the sum of the ideal, noise-free image and noise:

$$I[x,y] = \tilde{I}[x,y] + n[x,y]. \tag{7.9}$$

By definition, the noise-free image is the same from frame to frame. Let the noise have zero mean and be independent from frame to frame:

$$E(n_i[x,y]n_j[x,y]) = 0 \text{ for } i \neq j, \tag{7.10}$$

where $E(\cdot)$ is the expectation operator.

If several images of the same scene are averaged temporally (weights, $\sum_i w_i = 1$),

$$\begin{aligned} Q[x,y] &= \sum_i w_i I_i[x,y] \\ &= \sum_i w_i \tilde{I}[x,y] + \sum_i w_i n[x,y] \\ &= \tilde{I}[x,y] + \hat{n}[x,y]. \end{aligned} \tag{7.11}$$

The image content is the same from frame to frame, so will reinforce when added. However, since the individual noise images are independent, the noise in one frame will partially cancel the noise in other frames. The noise variance will therefore decrease and is given by (Mitra, 1998)

$$\sigma_{\hat{n}}^2 = \sigma_n^2 \sum_i w_i^2. \tag{7.12}$$

If averaging a constant number of images, N, the greatest noise reduction is given when the weights are all equal ($w_i = \frac{1}{N}$). The signal-to-noise ratio (SNR, defined as the ratio of the energy in the signal to the energy in the noise) of the output image is then

$$SNR_Q = \frac{\sigma_{\tilde{I}}^2}{\sigma_{\hat{n}}^2} = N\frac{\sigma_{\tilde{I}}^2}{\sigma_n^2} = N \times SNR_I. \tag{7.13}$$

Since the SNR depends on the noise variance, the noise amplitude will therefore decrease by $\sqrt{N}$.

Averaging multiple frames cannot be used to reduce quantisation noise when the noise from other sources (before quantisation) is significantly less than one pixel value. The quantisation noise will be similar from one frame to the next, violating Eq. (7.10). The reduction given by Eq. (7.12) no longer applies. However, if the noise standard deviation from other sources is larger than one pixel value, then this effectively makes the quantisation noise independent, enabling it to be reduced by averaging.

The main problem with averaging multiple frames is the requirement to store the previous $N - 1$ frames in order to perform the averaging. This limitation may be overcome by reducing the output frame rate by a factor of N; images are accumulated until N have been summed, and then the accumulator is reset to begin accumulation of the next set.

To overcome this limitation, one technique is to calculate the mean recursively, giving an exponentially decreasing sequence of weights:

$$\begin{aligned} Q_i[x,y] &= (1-\alpha)Q_{i-1}[x,y] + \alpha I_i[x,y] \\ &= Q_{i-1}[x,y] + \alpha(I_i[x,y] - Q_{i-1}[x,y]) \\ &= \sum_{n=0}^{\infty}(\alpha(1-\alpha)^i)I_{i-n}[x,y]. \end{aligned} \tag{7.14}$$

The output noise variance, from Eq. (7.12), is then given by

$$\sigma_{\hat{n}}^2 = \sigma_n^2 \sum_{i=0}^{\infty} \alpha^2(1-\alpha)^{2i} = \sigma_n^2 \frac{\alpha}{2-\alpha}, \tag{7.15}$$

with the resulting SNR

$$SNR_Q = \left(\frac{2}{\alpha} - 1\right)\frac{\sigma_{\tilde{I}}^2}{\sigma_n^2}. \tag{7.16}$$

By comparing Eq. (7.16) with Eq. (7.13), the weight to give an equivalent noise smoothing to averaging N images is

$$\alpha = \frac{2}{N+1}. \tag{7.17}$$

The assumption with this analysis is that no additional noise is introduced by the averaging process. In practice, multiplication by α will require truncating the result. This will introduce additional quantisation noise with each iteration, limiting the improvement in SNR. The total quantisation noise variance is

$$\sigma_{\hat{q}}^2 = \sigma_q^2 \sum_{i=0}^{\infty} (1-\alpha)^{2i} = \sigma_q^2 \frac{1}{\alpha(2-\alpha)}, \tag{7.18}$$

where σ_q^2 is the variance of the noise introduced with each quantisation. To reduce the effects of quantisation noise, it is necessary to maintain G additional guard bits on the accumulator image:

$$G > \log_2 \frac{1}{\alpha(2-\alpha)} \approx -1 - \log_2 \alpha. \tag{7.19}$$

An example implementation of Eq. (7.14) is shown in Figure 7.16. Since the frame buffer will almost certainly be off-chip memory, one of the schemes of Section 4.2.1 will be needed to enable the accumulated image

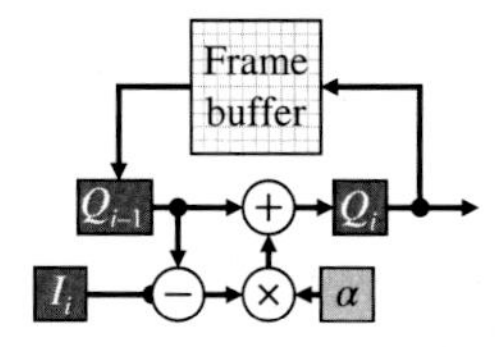

Figure 7.16 An implementation of recursive weighted image averaging using Eq. (7.14).

to be both read and written for each pixel. For practical reasons, α can be made a power of 2, enabling the multiplication to be implemented with a fixed binary shift. If $\alpha = 2^{-k}$, then the effective noise averaging window, from Eq. (7.17), is

$$N = \frac{2}{\alpha} - 1 = 2^{k+1} - 1. \tag{7.20}$$

7.2.2 Image Subtraction

The main application of image subtraction is to detect changes within a scene. This is illustrated in Figure 7.17; where there is no change, the difference will be small, with any changes in the image resulting in larger differences. This detects not only the addition or removal of objects but also subtle shifts in the position of objects, which result in differences at intensity edges within the image.

When subtracting images taken at different times, it is important that the lighting remains constant or the images be normalised. This is particularly difficult with outdoor scenes, where there is little control over the lighting. One approach to manage this problem is to take differences only between images closely spaced in time where it is assumed that conditions do not change significantly from frame to frame. Another is to maintain a dynamic estimate of the background that adapts as conditions change.

Assuming that the object and background have different pixel values, inter-frame differencing can detect both the addition and removal of objects. For moving objects, there will be differences at both the new and old locations of the object, as seen clearly in Figure 7.18. Significant differences are detected by thresholding the absolute difference. Double differencing (Kameda and Minoh, 1996) then detects regions that are common between successive difference images. It therefore detects objects that are both arriving in the first difference and leaving in the second.

Although the double difference requires data from three frames, a recursive architecture can be used, as illustrated in Figure 7.19. The saved pixel value is augmented with one extra bit (Δ) to indicate that the pixel was different to the previous frame. This saved difference bit is then combined with the difference bit for the current two frames to give the detected output pixels.

Figure 7.17 Image subtraction to detect change. Left: original scene; centre: the changed scene; right: the difference image (offset to represent zero difference as a mid-grey to enable positive and negative differences to be represented).

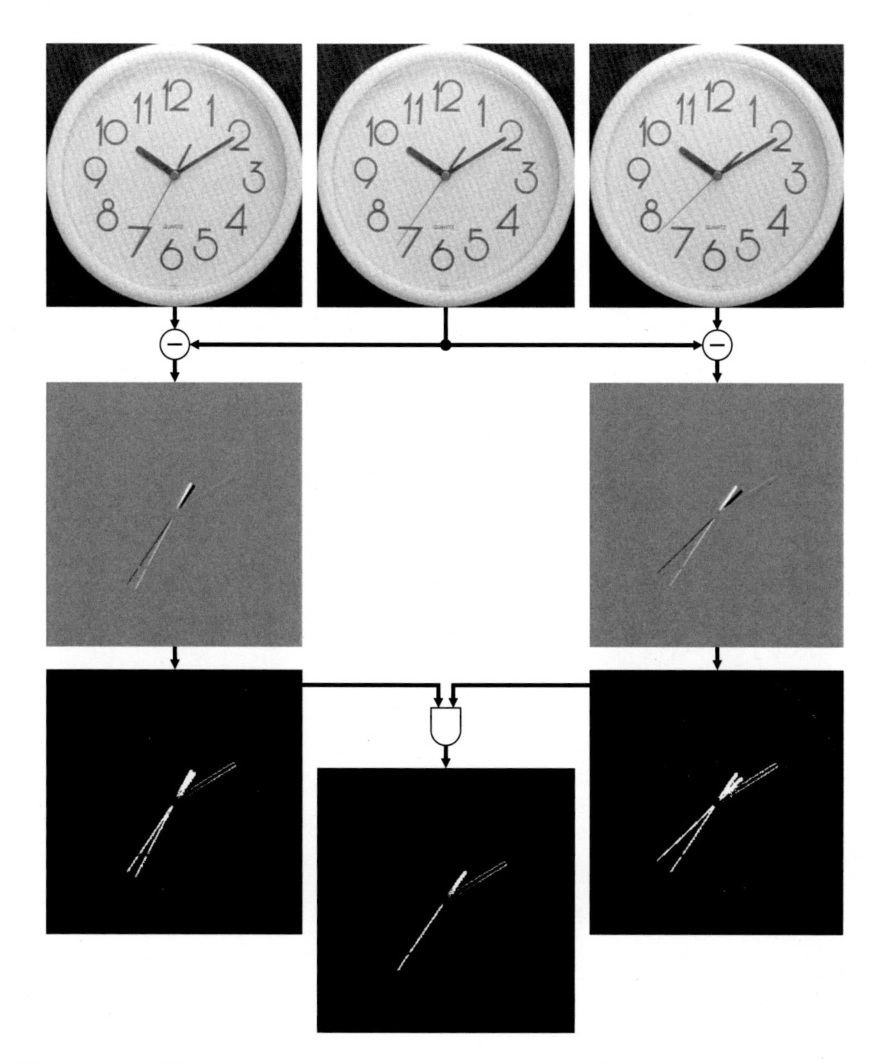

Figure 7.18 Double difference approach. Top row: three successive input images; middle row: differences between successive images; bottom row: left and right: absolute differences above the threshold; centre: the double difference, corresponding to the position in the centre frame.

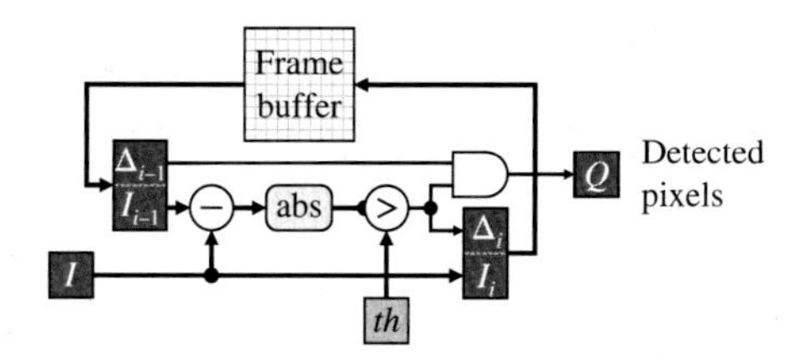

Figure 7.19 Implementing double differencing.

For more slowly moving objects (that obscure the background for several successive images), it is necessary that the object has sufficient texture to create differences in both difference frames. Double differencing has been applied to vehicle detection using an FPGA by Cucchiara et al. (1999). With multiple vehicles, all travelling at the same speed, it is still possible to get ghost false positives (Abdelli and Choi, 2017), although these can be reduced using more frames.

7.2.3 Background Modelling

The alternative is to construct a model of the static background and take the difference between each successive image and the background model. To account for changing conditions, it is necessary for the background image (or model) to be adaptive. Such background models can vary significantly in complexity (McIvor et al., 2001; Piccardi, 2004).

7.2.3.1 Mean and Variance

The simplest model is to represent the background image by the mean (see, for example (Heikkila and Silven, 1999)). It can be estimated efficiently using the recursive update of Eq. (7.14), where α controls how quickly changes to the scene are updated into the background. Larger values of α result in faster adaption of the background but may result in artefacts such as trails behind moving objects as they are partially assimilated into the background (Heikkila and Silven, 1999).

However, such a simple model is unable to cope with regions of the image that are naturally variable, for example leaves fluttering in the wind. By estimating the variance associated with each pixel, pixels that vary significantly may either be masked out or only be detected as foreground object pixels if the difference is significant (for example it exceeds k standard deviations, where k is typically 2.5 or 3) (Orwell et al., 1999). Let B_i be the background model estimated using Eq. (7.14). A recursive estimate of the variance may be derived as

$$\sigma_i^2[x,y] = (1-\alpha)\sigma_{i-1}^2[x,y] + \alpha(I_i[x,y] - B_i[x,y])^2. \tag{7.21}$$

Guo et al. (2015) suggested having a minimum difference threshold, d_{min}, to avoid detecting noise in an unchanging background, with the output detected as

$$Q = \begin{cases} 1, & |I_i[x,y] - B_i[x,y]| \geq \max(d_{min}, k\sigma_i[x,y]), \\ 0, & \text{otherwise}. \end{cases} \tag{7.22}$$

Calculating the standard deviation requires taking the square root of the variance. This can be avoided by comparing the squared difference directly with the variance:

$$Q = \begin{cases} 1, & (I_i[x,y] - B_i[x,y])^2 \geq \max(d_{min}^2, k^2\sigma_i^2[x,y]), \\ 0, & \text{otherwise}. \end{cases} \tag{7.23}$$

A simpler alternative is to replace the standard deviation with the mean absolute deviation:

$$d_i[x,y] = (1-\alpha)d_{i-1}[x,y] + \alpha|I_i[x,y] - B_i[x,y]|, \tag{7.24}$$

$$Q = \begin{cases} 1, & |I_i[x,y] - B_i[x,y]| \geq \max(d_{min}, kd_i[x,y]), \\ 0, & \text{otherwise}. \end{cases} \tag{7.25}$$

This scheme is illustrated in Figure 7.20. All pixels that differ from the mean by more than k deviations are detected as foreground pixels. As before, it is desirable to have α to be a power of 2 to reduce the logic. Setting $k = 2.5$ or $k = 3$ can also be implemented with a shift-and-add algorithm.

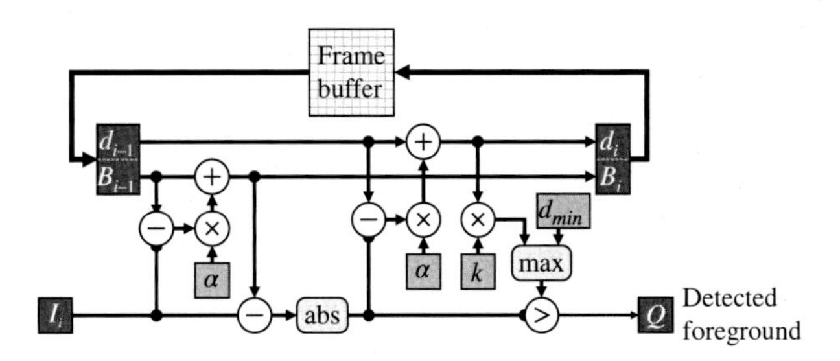

Figure 7.20 Detecting foreground pixels using the mean and mean absolute deviation.

7.2.3.2 Gaussian Mixture Model

A limitation of the mean and variance approach is that the large variance (or mean absolute deviation) is often the result of two or more different distributions being represented by a pixel. For example, with a flag waving in the wind, a pixel may alternate between the background and one of the colours on the flag. More sophisticated models account for the multimodal pixel value distributions resulting from such effects by representing each pixel by a weighted mixture of Gaussian distributions (Stauffer and Grimson, 1999). The mixture and Gaussian parameters are then updated as each image is acquired.

First, the Gaussian within the mixture that represents the current pixel is determined (starting with the Gaussian that has the most weight). If the pixel value is within k (typically 2.5) standard deviations of the mean, the pixel value is incorporated into that Gaussian, using Eqs. (7.14) and (7.21). The weights for all of the Gaussians in the mixture are updated using a similar weighted update, increasing the weight of the matched Gaussian and decreasing the weights of the others. If none of the Gaussians in the mixture matches the current pixel, the Gaussian with the least weight is discarded and replaced by a new Gaussian with a large standard deviation and low weight. The most probable Gaussians are considered background, and the lowest weight Gaussians represent new events, so are considered to be foreground object pixels.

A significant advantage of the mixture model over a single Gaussian occurs when an object stops long enough to become part of the background and then moves again. A single Gaussian will drift to the new value and then drift away again, resulting in an object being detected for a considerable time after it moves. With the mixture model, however, the mean does not drift, but a new distribution is begun for the object. When this receives sufficient weight, it is considered background. The model for the original background is retained, so when the object moves again, the pixels are correctly classified as background.

Typically, three to five Gaussians are used to model most images (McIvor et al., 2001). An FPGA implementation would be a relatively straightforward extension of Figure 7.20. Having more terms for each pixel requires more storage and logic to maintain the model, requiring a wide bandwidth to external memory, especially when processing colour images. Alternatively, the frame rate can be reduced to enable sequential memory locations to be used for the component distributions. Appiah and Hunter (2005) implemented a simpler version of multimodal background modelling using fixed width Gaussians. Their implementation considered both greyscale and colour images. Genovese and Napoli (2014) implemented a Gaussian mixture model with three Gaussians for real-time processing of high-definition video. Rather than model the distribution with Gaussians, Juvonen et al. (2007) directly used a coarse histogram of the distribution for each pixel, using the histogram peak as the background value.

Other related statistical background modelling methods are reviewed by Bouwmans (2011).

7.2.3.3 Minimum and Maximum

If it is desired to detect dark objects moving against a light background, then the maximum of successive images may provide a good estimate of the background (Shilton and Bailey, 2006). Sufficient images must be

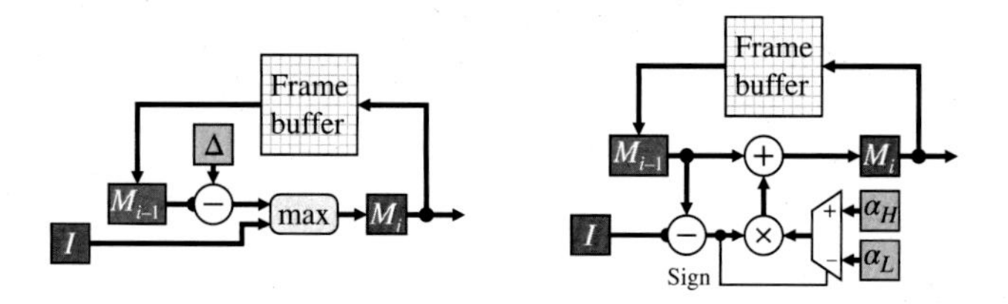

Figure 7.21 Estimating a light background using a recursive maximum. Left: the maximum has a fixed decay, from Eq. (7.28); right: using an adaptive mean, from Eq. (7.30).

combined, so that for each pixel, at least one image within the set contains a background pixel. Similarly, light objects against a dark background can use the minimum. These may be implemented directly with the recursive architecture:

$$M_i[x,y] = \max(M_{i-1}[x,y], I_i[x,y]), \tag{7.26}$$

$$m_i[x,y] = \min(m_{i-1}[x,y], I_i[x,y]). \tag{7.27}$$

Two limitations of these are that any outlier will automatically become part of the background, and if conditions change, the background is unable to adapt. These may be overcome by building a fixed linear decay into expression:

$$M_i[x,y] = \max(M_{i-1}[x,y] - \Delta, I_i[x,y]), \tag{7.28}$$

$$m_i[x,y] = \min(m_{i-1}[x,y] + \Delta, I_i[x,y]), \tag{7.29}$$

where Δ is related to the expected noise level. An implementation of this using the recursive architecture is shown on the left in Figure 7.21.

An alternative approach taken by Benetti et al. (2018) is to adapt the recursive mean to Eq. (7.14), depending on whether the input pixel is above or below the current maximum:

$$M_i[x,y] = \begin{cases} M_{i-1}[x,y] + \alpha_H(I_i[x,y] - M_{i-1}[x,y]), & I_i[x,y] \geq M_{i-1}[x,y], \\ M_{i-1}[x,y] + \alpha_L(I_i[x,y] - M_{i-1}[x,y]), & \text{otherwise}, \end{cases} \tag{7.30}$$

where $\alpha_H > \alpha_L$. The higher value of α_H causes the model to adapt quickly to values exceeding the maximum, while the lower α_L allows the maximum to slowly decay towards the input. Swapping α_H and α_L will approximate the adaptive minimum rather than the maximum.

7.2.3.4 Median

An alternative to the temporal mean is the median of the previous several images (Cutler and Davis, 1998). The median discards outliers, and with a step scene change (for example from a moving object) the background will come from the majority distribution. A problem with using the median, however, is that multiple images must be stored to enable its calculation on a pixel-by-pixel basis. The memory storage and consequent bandwidth issues restrict the usefulness of median calculation to short time windows for real-time operation.

However, since the median minimises the minimum absolute difference to the samples, it can be approximated by simple incrementing (Bylander and Rosen, 1997)

$$Med_i[x,y] = \begin{cases} Med_{i-1}[x,y] - \Delta, & I[x,y] < Med_{i-1}[x,y], \\ Med_{i-1}[x,y], & I[x,y] = Med_{i-1}[x,y], \\ Med_{i-1}[x,y] + \Delta, & I[x,y] > Med_{i-1}[x,y], \end{cases} \tag{7.31}$$

where Δ is chosen to balance the accuracy (a larger Δ) and the drift of the estimate with noise in the input (implies a smaller Δ is best). The median can be initialised to the first frame ($Med_0[x, y] = I_0[x, y]$). If the median is not stationary (the statistics change with time), then Eq. (7.31) can adapt to the changing median. For a maximum rate of change of the median of β per frame, the optimal value is $\Delta = 3\beta$ (Bylander and Rosen, 1997).

7.2.4 Intensity Scaling

Multiplication or division can be used to selectively enhance the contrast within an image. One image is the input image being enhanced, with the other image representing the pixel-dependent gain. As described earlier, a gain greater than 1 will increase the contrast, and a gain less than 1 will reduce the contrast. The gain image is usually obtained by preprocessing the input image in some way.

One application of pixel-by-pixel division is correcting for non-uniform pixel sensitivity within the image sensor, vignetting caused by the lens, or uneven illumination. The basic principle is to characterise the response by capturing a reference image of a uniform field or plain, non-textured background. Any variation in the pixel value is a result of deficiencies in the capture process, whether caused by variations in sensor gain or illumination. An image of a scene can then be corrected by dividing by the reference image (Aikens et al., 1989):

$$Q[x, y] = k\frac{I[x, y]}{Ref[x, y]}, \tag{7.32}$$

where the constant k controls the dynamic range of the output image and is chosen so that the full range of pixel values is used. If the input and reference images are captured under identical conditions, then k is typically set to $2^B - 1$ for B-bit images. After calibration, the reference image does not change, so $k/Ref[x, y]$ can be precalculated, enabling a multiplication to be used rather than a division in Eq. (7.32).

For best accuracy, when capturing the reference image, the exposure should be as large as possible without actually causing any pixels to saturate. In practice, a uniform illumination field is difficult to achieve (Schoonees and Palmer, 2009), but a similar reference image may be obtained by processing a sequence of images of a moving background containing a modulation pattern. The processing enables the variations caused by the imaging process to be separated from the variations in the illumination (Schoonees and Palmer, 2009).

Note that any errors or noise present in the reference image will be introduced into the processed image through Eq. (7.32). It is therefore important to minimise the noise. This may be accomplished if necessary by averaging several images or applying an appropriate noise-smoothing filter.

Once calibrated, the reference image needs to be stored, so that it can be made available as needed for processing. A full resolution image would likely require an external frame buffer. However, when correcting vignetting or lighting, the reference image is generally slowly varying and can readily be compressed. A simple compression scheme would be to down-sample the reference image and reconstruct the reference using interpolation (see Section 10.1.2). Down-sampling by a factor of 8 or 16 would give a significant data reduction, enabling the reference to be stored directly on the FPGA.

When considering only vignetting, an alternative approach is to model the intensity falloff with radius (Goldman and Chen, 2005). The model can then be used to perform the correction, rather than requiring an explicit reference image.

7.2.5 Masking

Masking is commonly used to select a region of interest to process in an image, while ignoring irrelevant regions. Preprocessing is used to construct a binary mask image, which is then used to select the regions of interest (see Figure 7.22). The background can easily be set to either black or white.

Alternatively, the mask could be used to multiplex between multiple images. One application is image compositing, creating a single image from a series of offset images, for example when making a panorama from a series of panned images. To ensure a good result, it is necessary to correct for lens and other distortions and to ensure accurate registration in the region of overlap. Where the scene geometry is known, this may be accomplished by rectifying the images. If the images are subject to vignetting, the seams between the images can be

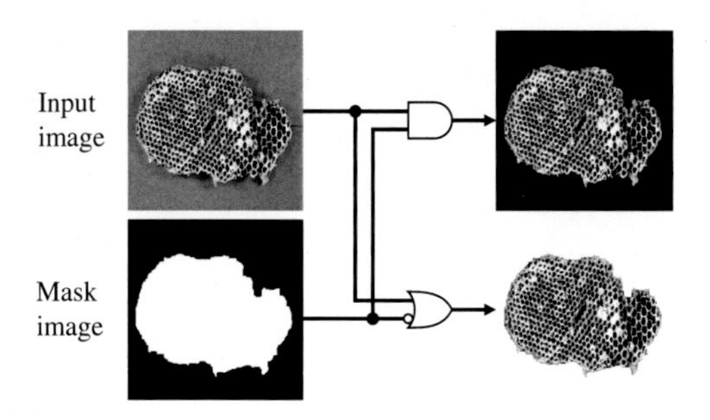

Figure 7.22 Masking to set the background to black (AND gate) or white (OR gate).

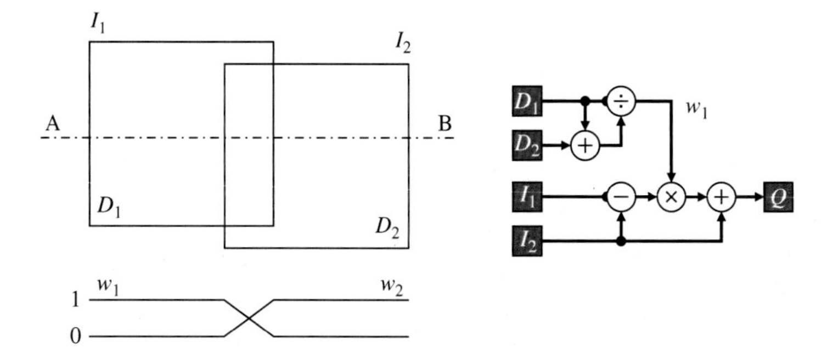

Figure 7.23 Merging of images in the overlap region. Left: the weights corresponding to the input images are shown for the cross-section A–B. Right: schematic for image merging, given registered images and distance weighted masks.

visible. Correcting for vignetting (Goldman and Chen, 2005) can significantly improve the results by making the pixel values of on each side of the join similar. Rather than simply switch from one image to the other at the border between the images, the seam resulting from mismatched pixel values between the images may be reduced by merging the images in the overlap region. Note that this requires good registration to avoid ghosting. Consider two frames, I_1 and I_2, as illustrated in Figure 7.23. In the region of overlap, the two frames are merged with a smooth transition from one to the other. The weights applied to each of the images, w_1 and w_2, respectively, depend on the width of overlap, which may vary from image to image, and also with the position within the image.

One relatively simple technique is to apply the distance transform (see Section 13.6) to each of the mask images. This labels each pixel within the mask image with the distance to the nearest edge of the mask. The weights may then be calculated from the distance-transformed masks, D_1 and D_2:

$$w_1 = \frac{D_1}{D_1 + D_2}, \quad w_2 = \frac{D_2}{D_1 + D_2} = 1 - w_1. \tag{7.33}$$

The output image is then given by

$$Q = w_1 I_1 + w_2 I_2 = I_2 + w_1(I_1 - I_2). \tag{7.34}$$

A streamed implementation of Eqs. (7.33) and (7.34) is shown on the right in Figure 7.23, assuming synchronised input streams. Equation (7.34) is factorised to reduce the computation to a single division to calculate the weight and a single multiplication to combine the two images.

7.2.6 *High Dynamic Range (HDR) Imaging*

In applications where there is a very wide dynamic range in intensities it can be difficult to capture a good image using a single exposure without a specially designed sensor. Sensor approaches include multiple gains for each pixel, enabling an extended range readout (Takayanagi et al., 2018), or multiple sensing elements per pixel, either with different sensitivities or exposures (Stampoglis, 2019) or using a mosaic of neutral density filters (Nayar and Mitsunaga, 2000).

7.2.6.1 HDR Image Fusion

Alternatively, multiple images from a regular sensor may be captured spanning a range of exposures. It is then necessary to combine the information from these images to give a single high dynamic range (HDR) image (Mann and Picard, 1995). (Note that any motion with the scene will result in motion artefacts (ghosting) when the images are fused (Bouderbane et al., 2016).) It is useful to consider each image on a logarithmic intensity scale, because a change in exposure (a multiplicative factor) corresponds to a shift on a logarithmic scale, as illustrated in Figure 7.24. The relative scale of the images (or offsets on the logarithmic axis) must be estimated, either through calibration or by analysing the corresponding pixels in adjacent exposures (Mann and Picard, 1995). The range of useful pixel values in each image is limited on the upper end by saturation of pixels which exceed the intensity range and on the lower end by noise. Any nonlinearity in the exposure curve may be found by fitting a curve to the inverse exposure function (Debevec and Malik, 2008). The goal of image fusion is therefore to estimate the true intensity at each pixel by selecting data from the corresponding input images.

Rather than simply select pixels from each exposure, a weighted average of the input images may be used, with the offset images weighted based on local image quality. These would give more weight to mid-range pixel values, rather than the extremes, which may be more affected by saturation or noise, as indicated in Figure 7.24. This would give

$$Q[x,y] = \frac{\sum_i w_i(I_i[x,y]) S_i I_i[x,y]}{\sum_i w_i(I_i[x,y])}, \tag{7.35}$$

where w_i is the image-dependent weight, and S_i is the relative scale of the ith image. The resultant HDR output image may then be processed in the normal way, although with more bits required to represent each pixel.

Mertens et al. (2007) extended this idea and used three weights, to represent contrast, saturation, and well-exposedness. Contrast used a Laplacian filter to detect edges and textured regions. Saturation used the standard deviation of the RGB values for each pixel to weight regions of good colour. The well-exposedness weight was based on a Gaussian curve centred at 0.5 intensity (with $\sigma = 0.2$), with the results for each colour channel multiplied. To avoid halo artefacts, the fusion can be performed using a combined Laplacian (for the images) and Gaussian (for the weights) pyramid (Mertens et al., 2007).

To overcome the problem of ghosting, the most thorough approach is to use optical flow to estimate motion between the different exposures and combine the corresponding pixels. However, this is unrealistic for real-time

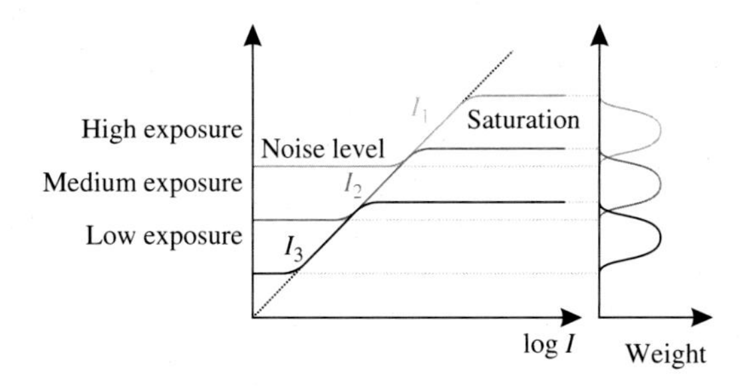

Figure 7.24 A change in exposure is a shift on a logarithmic scale. The goal is to estimate the intensity at each pixel (along the diagonal line), for example by weighting the individual input images.

FPGA implementation. One approach is to simply remove from the exposure sequence any pixels which violate the known exposure ordering sequence (Sidibe et al., 2009). The ordering criterion is quite loose, so Bouderbane et al. (2016) compared the expected pixel values based on the known exposures and gave a low weight to anomalous pixels, although this still has faint residual ghosts. Nosko et al. (2020) similarly detected anomalous pixels based on the known exposure and gave zero weight to pixels which are more than 20% different from expected.

7.2.6.2 Tone Mapping Operators

After fusion, the resulting dynamic range can exceed the dynamic range of displays. This requires compressing the dynamic range to match the capabilities of the output device. When applied to video, temporal coherence is important to avoid flicker effects resulting from the mapping changing from frame to frame (Eilertsen et al., 2013). Two approaches are commonly used (DiCarlo and Wandell, 2000; Ou et al., 2022): tone reproduction curves, which compress the contrast by map individual pixels, and filter-based approaches, which maintain local contrast but compress global contrast. This section discusses tone reproduction curves, which are point operations, with filtering approaches discussed in Section 9.3.7.

A tone reproduction curve is a nonlinear monotonic mapping of pixel values. Two commonly used compression curves (Ou et al., 2022) are power function (the gamma transformation of Eq. (7.4) with $\gamma < 1$) and a logarithmic mapping:

$$Q = k\log(I + \delta), \tag{7.36}$$

where k is a scale factor chosen to match the maximum values of the range, and the $+\delta$ avoids an undefined mapping for zero pixel values. A related function proposed by Reinhard et al. (2002) is

$$Q_N = \frac{I}{I + \bar{I}/g_N}, \tag{7.37}$$

where

$$\bar{I} = \exp\left(\frac{1}{N}\sum_{x,y}\log(I[x,y] + \delta)\right), \tag{7.38}$$

and g_N is the target level for mid-grey (typically $g_N = 0.18$). The logarithmic mapping tends to over-emphasise the contrast within darker areas at the expense of lighter areas, whereas the gamma transformation and Reinhard's mapping can balance this better (through appropriate choice of γ or g_N).

Histogram equalisation (see Section 8.1.3) is an alternative that dynamically adjusts the mapping for light and dark areas (Bailey, 2012; Ou et al., 2022). It enhances the contrast for pixel values where there are many pixels at the expense of pixel value ranges where there are fewer pixels. Rather than apply the mapping globally, it can also be applied locally as adaptive histogram equalisation (see Section 9.5).

Simply averaging the individual images taken at different exposures performs both fusion and tone mapping (Bailey, 2012). This works because the saturation of the images at high intensities means that the mapping function is effectively piecewise linear, with a similar shape to the gamma mapping.

7.2.6.3 Systems

Mann et al. (2012) implemented a real-time system which captured images four stops apart (a factor of 16 exposure). Since the exposure ratio was fixed, the weighted average fusion of two images could be pre-calculated, along with the tone mapping of the result, using a gamma transformation with $\gamma = 0.2$. The pre-calculated fusion and mapping was stored in a triangular $256 \times 256 \times 3$ (for colour) lookup table, enabling a single lookup of the input pixel values to give the mapped output.

Lapray et al. (2013) used three exposures, which were automatically calculated and optimised using histogram analysis (see Section 8.1.4). Each new frame from the camera is combined with the previous two frames which had been saved to a frame buffer. The images are transformed into logarithmic space and offset by the logarithm of the exposure time. Fusion is based on a weighted average, with weights dropping linearly from mid-grey, with the final result converted back to a linear scale, using the Reinhard's tone mapping of Eq. (7.37).

Nosko et al. (2020) similarly used three exposures, although provided an output at one-third the capture frame rate. The images were directly fused, with demosaicing performed on the resulting HDR image. Bilateral filtering was used for the tone mapping (see Section 9.3.7).

7.3 Colour

Colour images are a logical extension to greyscale images. The main difference is that each pixel consists of a vector of components rather than a scalar. We can define a vector-based image as **I**, where

$$\mathbf{I}[x,y] = \{I_1[x,y], I_2[x,y], I_3[x,y]\} = \begin{bmatrix} I_1[x,y] \\ I_2[x,y] \\ I_3[x,y] \end{bmatrix}, \quad (7.39)$$

with component images I_1, I_2, and I_3. As described in Section 1.2.2, a colour image is formed by capturing images with sensors that are selectively sensitive to the red, green, and blue wavelengths of the spectrum that approximately match the sensitivity of the cones within the human visual system. Each colour pixel is then a three-dimensional vector, with red, green, and blue components, $\mathbf{I}_{RGB} = \{R, G, B\}$.

7.3.1 False Colour

False colour is so named because the colours seen in the output image do not reflect the true underlying colour of the scene. The purpose is to make features that may not be apparent or visible in the original image more apparent to a human viewer. Therefore, it is seldom used for automatic image processing.

A greyscale image can be mapped onto a colour image, where each pixel value is assigned a separate colour. This is implemented using a lookup table to produce each of the red, green, and blue components, as shown in Figure 7.25. Its usefulness relies on the ability of the human visual system to more readily distinguish colours than shades of grey, particularly when the local contrast varies (see, for example Figure 7.3). An appropriate pseudocolour both enhances the contrast and can facilitate the manual selection of an appropriate threshold level.

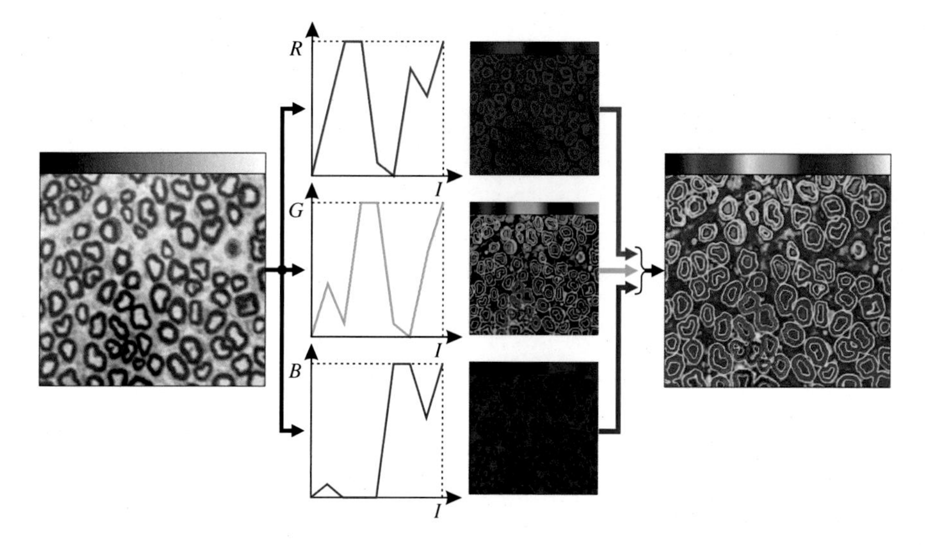

Figure 7.25 Pseudocolour or false colour mapping of a greyscale image using lookup tables.

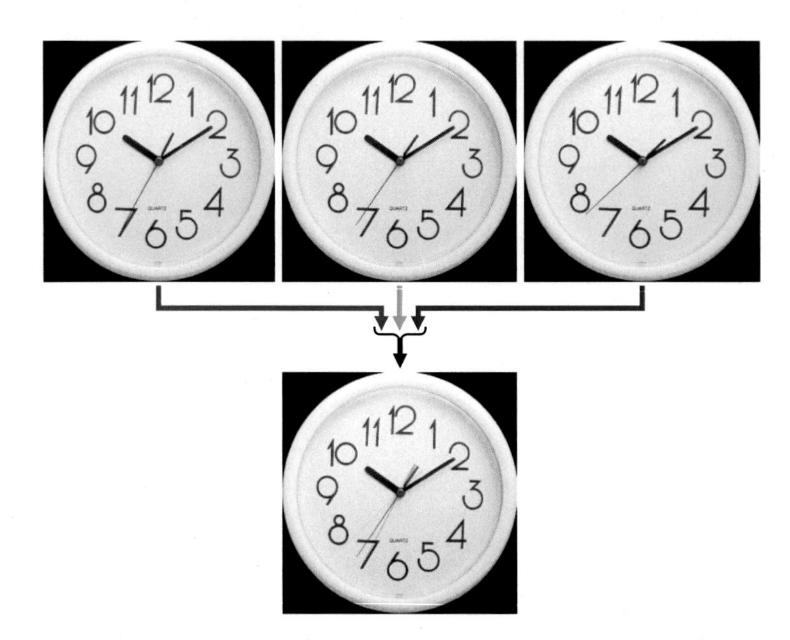

Figure 7.26 Temporal false colouring. Images taken at different times are assigned to different channels, with the resultant output showing coloured regions where there are temporal differences.

Another application of false colour is to map the non-visible components of a multi-spectral image into the visible part of the spectrum. For example, in remote sensing, different types of vegetation have different spectral signatures that are most obvious in the near infrared, red, and green components (Duncan and Leeson, 1999). Therefore, a mapping commonly used within remote sensing is to map these onto the red, green, and blue components, respectively, of the output image. The resulting colours are not the same as seen by the unaided human eye; however, mapping the infrared band into the visible range does allow subtle distinctions to be more readily seen and distinguished. Similarly, false colour can be used to provide a visualisation of the key components of a hyperspectral image by mapping selected bands to RGB components.

A related technique is to combine different images of the same scene as the different components of a colour image. One example is to verify the correct registration of images taken from different viewpoints (Reulke et al., 2008). Registered components should align, so any misalignment will show as coloured fringes. A similar application is to combine images taken at different times as different components, as shown in Figure 7.26. Common regions will have equal red, green, and blue components, making them appear grey. However, any differences will cause an imbalance in the components, resulting in a coloured output. In the example here (7.26), dark objects appear in their complimentary colour.

7.3.2 Colour Space Conversion

While the RGB colour space is native to most display devices, it is not always the most natural to work in. Colour space conversion involves transforming one vector representation into another that makes subsequent analysis or processing easier.

7.3.2.1 RGB

The RGB colour space is referred to as ***additive***, because a colour is made by adding particular levels of red, green, and blue light. Figure 7.27 illustrates this with the RGB components of a colour image. Traditionally, 24 bits are used, with 8 bits for each of the RGB components, although HDR images can use 12–16 bits for each component.

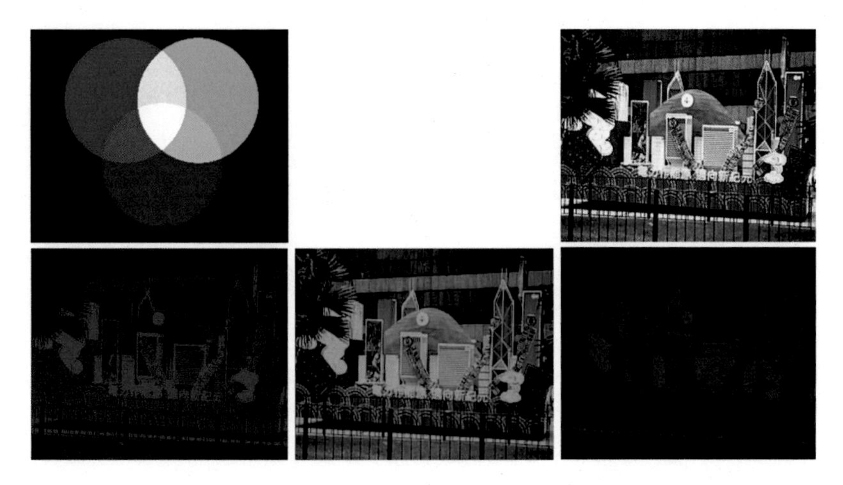

Figure 7.27 RGB colour space. Top left: combining red, green, and blue primary colours; bottom: the red, green, and blue components of the colour image on the top right.

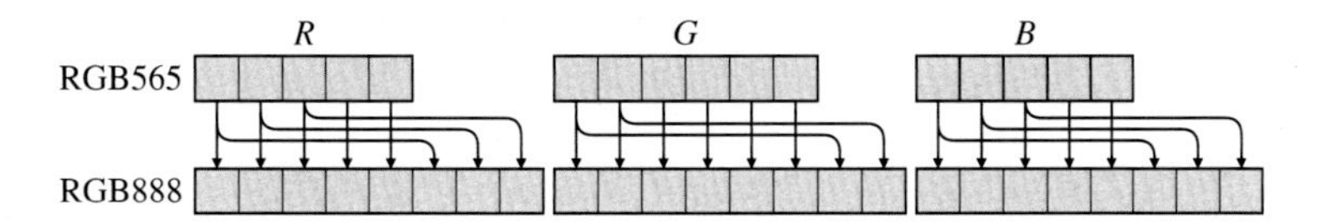

Figure 7.28 Converting from RGB565 to RGB888.

One variation of note is 16-bit RGB565, where the entire colour vector is contained within 16 bits, usually as a result of bandwidth limitations. With RGB565, 5 bits are allocated for each of the R and B components, and 6 bits are allocated for G (to take into account the increased sensitivity of the human visual system to green). To convert from RGB565 to 24-bit RGB, rather than append 0s, it is better to append the two (for G) or three (for R and B) most significant bits of the component, as shown in Figure 7.28.

The RGB colour components (whether from capture or display) are device dependent. In a scene captured by a camera, the colour vector for each pixel depends not only on the colour in the scene and the illumination but also on the spectral response of the filters used to measure each component. Similarly, the actual colour produced on a display will depend on the spectral content of the red, green, and blue light sources within the display. Therefore, there are many different RGB colour spaces depending on the particular wavelengths (or spectral mix) used for each of the red, green, and blue primaries. To combat this, a standard device-independent colour space was defined: sRGB (IEC, 1999). It defines the specific colours of the three primaries used for red, green, and blue and a nonlinear gamma-like mapping between linear RGB and a nonlinear RGB that approximates the response of CRT-based displays.

Conversion from a device-dependent RGB to sRGB requires first multiplying the device-dependent colour vector by a 3×3 matrix that depends on the red, green, and blue spectral characteristics of the device. This transformation, determined by calibration, is sometimes called the ***colour profile*** of the device. The result of the transformation represents a vector of normalised (between 0.0 and 1.0) linear RGB values. Any component outside this range is outside the gamut of sRGB and is clipped to within this range. Each component, C_N, is then mapped from linear RGB to sRGB values by (IEC, 1999):

$$C_{NsRGB} = \begin{cases} 12.92C_{NRGB}, & C_{NRGB} \leq 0.0031308, \\ 1.055C_{NRGB}^{1/2.4} - 0.055, & C_{NRGB} > 0.0031308. \end{cases} \tag{7.40}$$

The final values can then be represented as 8-bit quantities by scaling by 255.

Unless specified otherwise, images are assumed to be in sRGB colour space. However, many image processing operations assume that a linear space is being used. Eq. (7.40) can readily be inverted to remove the gamma mapping and return to linear components:

$$C_{NRGB} = \begin{cases} \dfrac{C_{NsRGB}}{12.92}, & C_{NsRGB} \leq 0.04045, \\ \left(\dfrac{C_{NsRGB} + 0.055}{1.055}\right)^{2.4}, & C_{NsRGB} > 0.04045. \end{cases} \tag{7.41}$$

The gamma mapping (and its inverse) is best implemented using lookup tables.

7.3.2.2 CMY and CMYK

In printing, rather than actively producing light, the image begins with the white paper, and colour is produced by filtering or blocking some of the colour. The ***subtractive*** primaries are cyan, magenta, and yellow and consist of appropriate inks, dyes, or filters. Consider the yellow dye; it will allow the red and green spectral components to pass through, but will attenuate the blue, making that part of the scene appear yellow. The more yellow, the more the blue is attenuated, and the yellower the scene will appear. Similarly, the magenta dye attenuates the green component, and the cyan dye attenuates the red component. A CMY image is therefore formed from the cyan, magenta, and yellow components: $\mathbf{I}_{CMY} = \{C, M, Y\}$.

Therefore, as illustrated in Figure 7.29, the colours of a scene may be produced by mixing different quantities of yellow, magenta, and cyan dyes to give the required spectral content at each point. Normalised CMY components are approximately given by

$$C_N \approx 1 - R_N, \qquad M_N \approx 1 - G_N, \qquad Y_N \approx 1 - B_N. \tag{7.42}$$

The exact relationship depends on the spectral content of the RGB components and the spectral transmissivity of the CMY components. While the simple conversion of Eq. (7.42) works reasonably well with lighter, unsaturated colours, it becomes less accurate for darker and more saturated colours where one (or more) of the CMY components is larger. There are two reasons for this. First, the attenuation is not linear with the amount of dye used but is exponential. This makes the relationship approximately linear for lower levels of CMY components but deviates more for the higher levels. Consequently, a particular spectral component cannot be completely removed, making it difficult to produce fully saturated colours and dark colours. The second reason

Figure 7.29 CMY colour space. Top left: combining yellow, magenta, and cyan secondary colours; bottom: the yellow, magenta, and cyan components of the colour image on the top right.

is that equal levels of yellow, magenta, and cyan seldom result in a flat spectral response. Both of these factors make black appear as a muddy colour, often with a colour cast.

In printing, this problem is overcome with the addition of a black dye, resulting in the CMYK colour space. Equal amounts of yellow, magenta, and cyan dyes are replaced by the appropriate amount of black dye. This changes the approximate conversion from Eq. (7.42) to

$$\begin{aligned} K_N &\approx 1 - \max(R_N, G_N, B_N), \\ C_N &\approx (1 - K_N - R_N)/(1 - K_N), \\ M_N &\approx (1 - K_N - G_N)/(1 - K_N), \\ Y_N &\approx (1 - K_N - B_N)/(1 - K_N). \end{aligned} \tag{7.43}$$

Other than for printing, the CMY or CMYK colour spaces are not often used for image processing.

7.3.2.3 YUV and YCbCr

When colour television was introduced, backward compatibility with existing black and white television was desired. The luminance signal, Y, is a combination of RGB components, weighted to reflect the relative sensitivities of the human visual system. The colour (chrominance) is provided by two difference signals, $B - Y$ and $R - Y$. Assuming normalised RGB, the image can be represented in YUV colour space, $\mathbf{I}_{YUV} = \{Y, U, V\}$, with the components given by

$$\begin{aligned} Y &= 0.299R + 0.587G + 0.114B, \\ U &= 0.492(B - Y), \\ V &= 0.877(R - Y), \end{aligned} \tag{7.44}$$

or in matrix form as

$$\begin{bmatrix} Y \\ U \\ V \end{bmatrix} = \begin{bmatrix} 0.299 & 0.587 & 0.114 \\ -0.147 & -0.289 & 0.436 \\ 0.615 & -0.515 & -0.100 \end{bmatrix} \begin{bmatrix} R \\ G \\ B \end{bmatrix}. \tag{7.45}$$

Strictly speaking, the YUV colour space is an analogue representation. The corresponding digital representation is called YCbCr, with $\mathbf{I}_{YCbCr} = \{Y, Cb, Cr\}$. There are two commonly used YCbCr formats: the first scales the values by less than 255 to give 8-bit quantities with headroom and footroom and offsets the chrominance components to enable an unsigned representation:

$$\begin{bmatrix} Y \\ Cb \\ Cr \end{bmatrix} = \begin{bmatrix} 65.481 & 128.553 & 24.966 \\ -37.797 & -74.203 & 112.0 \\ 112.0 & -93.786 & -18.214 \end{bmatrix} \begin{bmatrix} R_N \\ G_N \\ B_N \end{bmatrix} + \begin{bmatrix} 16 \\ 128 \\ 128 \end{bmatrix}. \tag{7.46}$$

The second scales the chrominance components to use the full range, from −0.5 to 0.5 (for example as used by the JPEG 2000 standard (ISO, 2000)). This is used where the headroom and footroom are not considered as important as maximising the dynamic range:

$$\begin{bmatrix} Y \\ Cb \\ Cr \end{bmatrix} = \begin{bmatrix} 0.299 & 0.587 & 0.114 \\ -0.169 & -0.331 & 0.500 \\ 0.500 & -0.419 & -0.081 \end{bmatrix} \begin{bmatrix} R \\ G \\ B \end{bmatrix}. \tag{7.47}$$

Optionally, the Cb and Cr are converted from twos-complement to offset binary (by adding 2^{B-1}, or toggling the most significant bit). The YCbCr components of a colour image are illustrated in Figure 7.30.

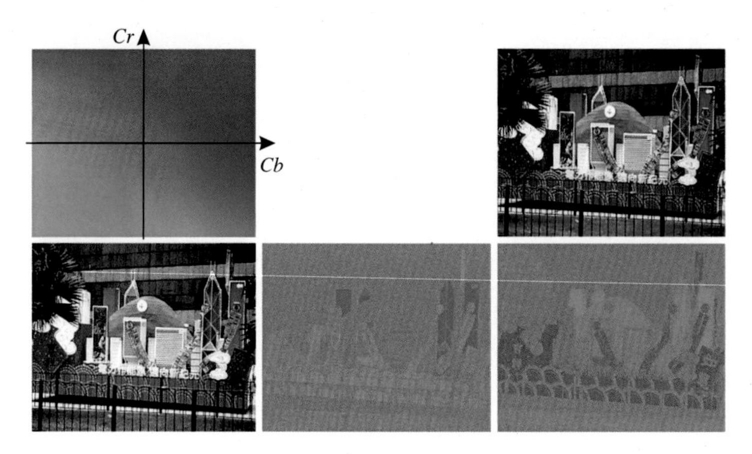

Figure 7.30 YCbCr colour space. Top left: the *Cb–Cr* colour plane at mid-luminance; bottom: the luminance and chrominance components of the colour image on the top right. (The chrominance components are displayed using offset binary to enable negative values to be displayed.)

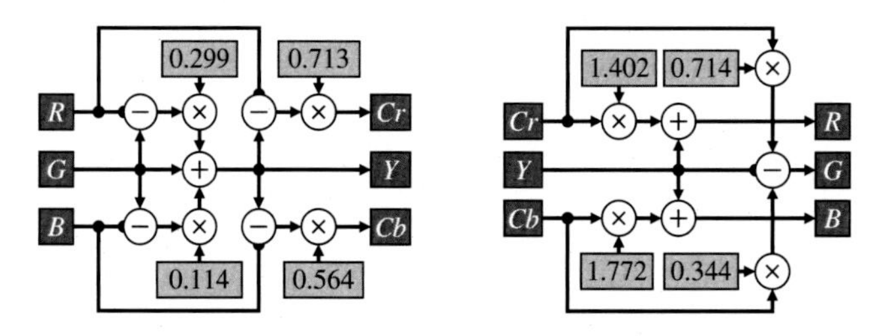

Figure 7.31 Left: conversion from RGB to YCbCr using Eq. (7.48); right: conversion from YCbCr to RGB using Eq. (7.49).

The implementation of Eq. (7.47) will be considered here. Using the form of Eq. (7.44), this matrix can be factorised to use four multiplications:

$$\begin{aligned} Y &= 0.299(R-G)+0.114(B-G)+G, \\ Cb &= \frac{0.5}{(1-0.114)}(B-Y) = 0.564(B-Y), \\ Cr &= \frac{0.5}{(1-0.299)}(R-Y) = 0.713(R-Y), \end{aligned} \tag{7.48}$$

with the implementation shown in Figure 7.31. Since the multiplications are by constants, they may be implemented efficiently using a few additions. The inverse similarly requires four nontrivial multiplications:

$$\begin{bmatrix} R \\ G \\ B \end{bmatrix} = \begin{bmatrix} 1 & 0 & 1.402 \\ 1 & -0.344 & -0.714 \\ 1 & 1.772 & 0 \end{bmatrix} \begin{bmatrix} Y \\ Cb \\ Cr \end{bmatrix}. \tag{7.49}$$

From an image processing perspective, the advantage of using YCbCr over RGB is the reduction in correlation between the channels. This is particularly useful for image compression (Chapter 12). It also helps with colour thresholding and colour enhancement, as described later in this section.

A limitation of the YCbCr colour space is that the transformation is a (distorted) rotation of the RGB coordinates. The scale factors are chosen to ensure that every RGB combination has a legal YCbCr representation; however, the converse is not true. Some combinations of YCbCr fall outside the legal range of RGB values. This implies that more bits must be kept in the YCbCr representation if data is not to be lost and the transform is to be reversed. Note that this cannot be avoided when using rotation-based transformation of the colour space.

A second issue is that the weights for the Y component (which are based on human perception of luminance) require fixed-point (or floating-point) multiplications to perform the transformation. For image processing, weighting based on human perception is often less relevant and can be relaxed.

7.3.2.4 Related Colour Spaces

One obvious simplification is to restrict the coefficients to powers of 2. There are several ways of accomplishing this. One is to adjust the weights of the Y component to be powers of 2:

$$\begin{aligned} Y &= \frac{1}{4}(R-G)+\frac{1}{4}(B-G)+G, \\ Cb &= \frac{1}{2}(B-Y), \\ Cr &= \frac{1}{2}(R-Y). \end{aligned} \tag{7.50}$$

However, this does not use the full dynamic range available for Cb and Cr (with these ranging from $-\frac{3}{8}$ to $\frac{3}{8}$). Using the full range would require scaling the chrominance terms by $\frac{4}{3}$.

Closely related to this is the reversible colour transform (RCT) used by lossless coding in JPEG 2000 (ISO, 2000). This has

$$\begin{aligned} Y &= \left\lfloor \frac{R+2G+B}{4} \right\rfloor, \\ Cb &= B-G, \\ Cr &= R-G, \end{aligned} \tag{7.51}$$

where $\lfloor \cdot \rfloor$ represents truncation. Although information appears to be lost by the truncation, it is retained in the other two terms and can be recovered exactly (hence, reversible colour transform). The Cb and Cr terms require an extra bit to prevent wrap around. The reverse transformation is

$$\begin{aligned} G &= Y-\left\lfloor \frac{Cr+Cb}{4} \right\rfloor, \\ B &= Cb+G, \\ R &= Cr+G. \end{aligned} \tag{7.52}$$

None of these transforms are orthogonal. One YCbCr-like orthogonal transformation (Sen Gupta and Bailey, 2008) is

$$\begin{bmatrix} Y \\ C_1 \\ C_2 \end{bmatrix} = \begin{bmatrix} 1 & 1 & 1 \\ 1 & -2 & 1 \\ 1 & 0 & -1 \end{bmatrix} \begin{bmatrix} R \\ G \\ B \end{bmatrix}, \tag{7.53}$$

although the inverse transformation requires a division by 3. It is not possible to have an orthogonal transformation that uses powers of 2 for both the forward and inverse transformations. Closely related to this is the YCoCg colour space that is nearly orthogonal and is also easily inverted (Malvar and Sullivan, 2003):

$$\begin{bmatrix} Y \\ Co \\ Cg \end{bmatrix} = \begin{bmatrix} \frac{1}{4} & \frac{1}{2} & \frac{1}{4} \\ \frac{1}{2} & 0 & -\frac{1}{2} \\ -\frac{1}{4} & \frac{1}{2} & -\frac{1}{4} \end{bmatrix} \begin{bmatrix} R \\ G \\ B \end{bmatrix}. \tag{7.54}$$

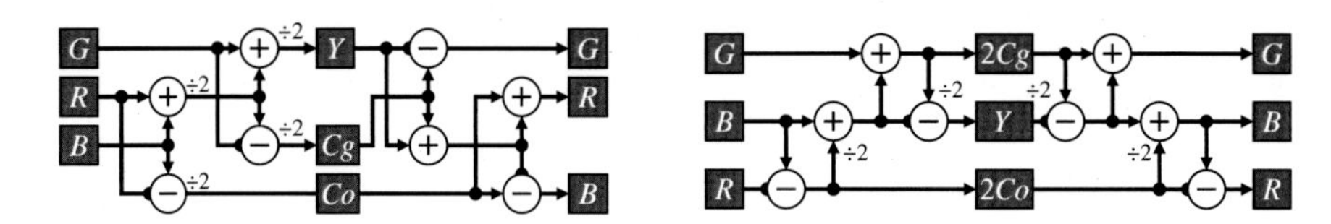

Figure 7.32 Left: a multiplierless YCoCg transformation and its inverse; right: implemented using lifting (an extra bit is kept in *Co* and *Cg* to make it reversible).

As shown in Figure 7.32, this may be further factorised to reduce the complete transformations to four additions, with the division by 2 performed for free by a shift. Its inverse is similarly simple:

$$\begin{bmatrix} R \\ G \\ B \end{bmatrix} = \begin{bmatrix} 1 & 1 & 1 \\ 1 & 0 & 1 \\ 1 & -1 & -1 \end{bmatrix} \begin{bmatrix} Y \\ Co \\ Cg \end{bmatrix}. \tag{7.55}$$

The YCoCg can also be implemented using lifting as shown on the right in Figure 7.32, making it reversible (immune to truncation errors) (Malvar and Sullivan, 2003).

7.3.2.5 HSV and HLS

The RGB or YCbCr colour spaces do not reflect the psychological way in which we interpret or think about colour. When we interpret a colour, we are less sensitive to the intensity but consider a colour in terms of its hue and the strength of the colour (saturation). Therefore, it is reasonable to have a colour space with hue, saturation, and intensity as the three components. Two such colour spaces are $\mathbf{I}_{HSV} = \{H, S, V\}$ (hue, saturation, and value) and $\mathbf{I}_{HLS} = \{H, L, S\}$ (hue, lightness, and saturation) (Foley and Van Dam, 1982).

The hue, saturation and value (HSV) colour space is typically represented as a cone as shown in the left panel of Figure 7.33. The hue represents the angle around the cone, resulting in the colour wheel on the right panel of Figure 7.33. By definition, the black–white axis up the centre of the cone has a hue of 0. The remainder of the colour wheel is split into three sectors based on which component is the maximum, and the proportion between the other two components is used to give the angle.

Mathematically, the hue can be defined as

$$H = \begin{cases} 0, & R = G = B, \\ \dfrac{(G-B) \times 60°}{\max(R,G,B) - \min(R,G,B)} \bmod 360°, & R \geq G, B, \\ \dfrac{(B-R) \times 60°}{\max(R,G,B) - \min(R,G,B)} + 120°, & G \geq R, B, \\ \dfrac{(R-G) \times 60°}{\max(R,G,B) - \min(R,G,B)} + 240°, & B \geq R, G. \end{cases} \tag{7.56}$$

With a binary number system, the use of degrees (or even radians) is inconvenient. One alternative is to normalise the angle, so that a complete cycle goes from 0.0 to 1.0, which maximises the hue resolution for a given number of bits and avoids the need to explicitly manage wraparound. This may be preferable when performing many manipulations on the hue. Another alternative is to represent 60° by a power of 2 (for example 64) to simplify the multiplication. This also makes it easier to convert back to RGB.

The value component, V, is similar to Y, except that equal weight is given to the RGB components. V represents the height up the cone, as shown in the left panel of Figure 7.33, and is usually represented normalised between 0.0 and 1.0:

$$V_N = \max(R_N, G_N, B_N). \tag{7.57}$$

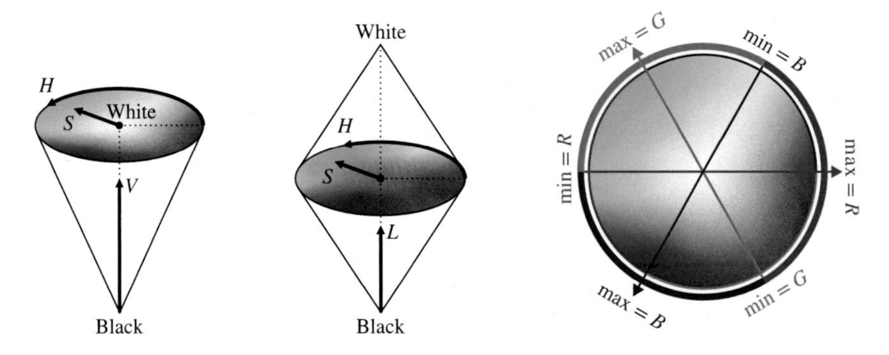

Figure 7.33 HSV and HLS colour spaces. Left: the HSV cone; centre: the HLS bicone; right: the hue colour wheel showing the relationship with RGB components.

Figure 7.34 HSV and HLS colour spaces. Left: original colour image; top row: the HSV hue, saturation, and value components; bottom row: the HLS hue, saturation, and lightness components.

The saturation represents the strength of the colour and is represented by the radius of the cone relative to the maximum radius for a given value:

$$S_{HSV} = \begin{cases} 0, & \max(R,G,B) = 0, \\ \dfrac{\max(R,G,B) - \min(R,G,B)}{\max(R,G,B)}, & \text{otherwise.} \end{cases} \tag{7.58}$$

Normalisation of S means that only V changes as the intensity of the image is scaled. The HSV components of a sample image are shown in the top row of Figure 7.34. Note that the hue is meaningless for greys because even a small amount of noise on the components will determine the hue.

The hue, lightness, and saturation (HLS) representation differs from the HSV colour space in that it is a bicone rather than a cone, as seen in the centre panel of Figure 7.33. The main advantage of HLS over HSV is that it is symmetric with respect to black and white. The definition of hue is the same for both representations, given by Eq. (7.56), but the lightness and saturation differ from the HSV space. Again, the lightness and saturation are usually represented as normalised values:

$$L_N = \frac{1}{2}(\max(R_N, G_N, B_N) + \min(R_N, G_N, B_N)), \tag{7.59}$$

$$S_{HLS} = \begin{cases} 0, & L_N = 0, \\ \dfrac{\max(R_N, G_N, B_N) - \min(R_N, G_N, B_N)}{2L_N}, & L_N < \dfrac{1}{2}, \\ \dfrac{\max(R_N, G_N, B_N) - \min(R_N, G_N, B_N)}{2 - 2L_N}, & L_N \geq \dfrac{1}{2}. \end{cases} \tag{7.60}$$

The HLS components of the sample image are shown in the bottom row of Figure 7.34.

Equation (7.59) places fully saturated primary colours at half lightness, at the intersection of the two cones. In general, this makes L less than V, although this has no real consequence from an image processing perspective.

The major limitation of HLS is that the saturation is constant with scaling lightness only in the lower cone ($L_N \leq \frac{1}{2}$). This gives the anomaly that pastel colours can appear as fully saturated. In particular, this can be seen in the lighter areas of Figure 7.34, where the saturation is high.

For these reasons, only the implementation of HSV will be considered here. The left panel of Figure 7.35 shows an implementation of the conversion from RGB to HSV. The minimum and maximum components are determined from the sign bits of the differences between the input colour channels. These differences need to be calculated anyway to give the numerators of Eq. (7.56), so this information is effectively for free. Apart from the multiplexers to select the minimum and maximum components, the main complexity is the two dividers for normalising the hue and the saturation. The offset added is based on the sector and is given by

$$Offset = \begin{cases} 128, & \text{when } G_{max}, \\ 256, & \text{when } B_{max}, \\ 384, & \text{when } R_{max} \wedge G_{min}, \end{cases} \tag{7.61}$$

with the hue taking a value between 0 and 383. Note that for the full hue range to be a power of 2, the final H needs to be multiplied by $\frac{2}{3}$.

To convert HSV back to RGB (assuming H is in the range 0 to 6×2^B), the three most significant bits of the hue represent the sector and hence can be used to select the appropriate values for the RGB components:

$$\mathbf{I}_{RGB} = \begin{cases} \{V, V - VS(1 - H_{LSB}), V - VS\}, & H_{MSB} = 0 \text{ (red to yellow)}, \\ \{V - VSH_{LSB}, V, V - VS\}, & H_{MSB} = 1 \text{ (yellow to green)}, \\ \{V - VS, V, V - VS(1 - H_{LSB})\}, & H_{MSB} = 2 \text{ (green to cyan)}, \\ \{V - VS, V - VSH_{LSB}, V\}, & H_{MSB} = 3 \text{ (cyan to blue)}, \\ \{V - VS(1 - H_{LSB}), V - VS, V\}, & H_{MSB} = 4 \text{ (blue to magenta)}, \\ \{V, V - VS, V - VSH_{LSB}\}, & H_{MSB} = 5 \text{ (magenta to red)}, \end{cases} \tag{7.62}$$

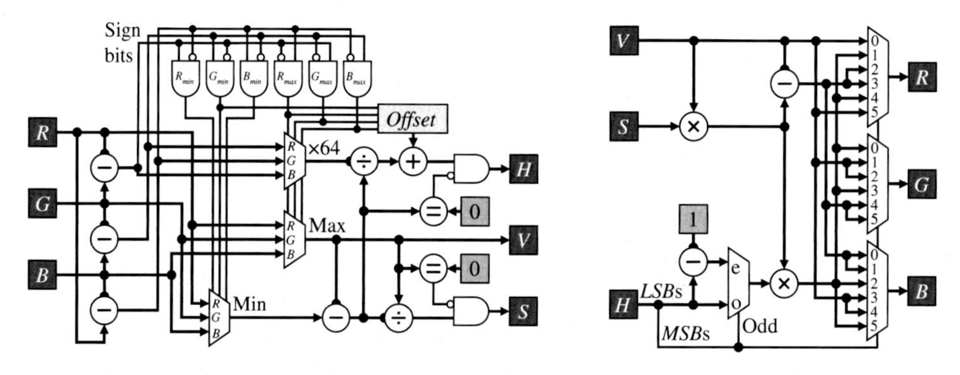

Figure 7.35 Left: conversion from RGB to HSV, with H in the range 0–383. right: conversion from HSV back to RGB.

where the least significant bits of hue are considered as a fraction. Equation (7.62) is implemented in the right panel of Figure 7.35. If the full range is used for H, then the H needs to be multiplied by 1.5 before selecting the sector bits.

A form of hue and saturation may also be derived from the YCbCr colour space by converting the chrominance components into polar coordinates (Punchihewa et al., 2005). This may be readily implemented using a CORDIC (coordinate rotation digital computer) transformation:

$$\hat{H} = \tan^{-1}\frac{Cb}{Cr}, \qquad \hat{S} = \frac{\sqrt{Cb^2 + Cr^2}}{Y}, \tag{7.63}$$

where the scaling of the saturation by Y makes it independent of intensity (it could alternatively be scaled by the maximum RGB component).

The HSV colour space is useful in image processing for colour detection and enhancement. The advantage gained is the intensity independence of hue and saturation, enabling more robust segmentation. One limitation, however, is the need to have a correct white balance. The hue (and also the saturation to a lesser degree) is affected by any colour cast, especially for colours with low saturation. Section 7.3.5 considers correcting the white balance.

In some applications, only one part of the colour wheel is of interest. For example, many fruits vary in colour from green to yellow or red. In this case, the sectors between green and red are of particular interest, whereas hues outside this range are less important. In these cases, a simplified substitute for hue may be used. For example, when analysing the colour of limes, colours in the range from green (chlorophyll dominates) to yellow (carotenoids dominate) are important (Bunnik et al., 2006); these may be captured by a ratio of the components:

$$H_{G-Y} = \frac{R}{G}. \tag{7.64}$$

This measure ranges from 0.0 for pure green to 1.0 for yellow, enabling the health of the fruit to be assessed. Such a measure is less suitable for reds (such as would be encountered when grading tomatoes, where the red-pigmented lycopene dominates), because the red component can be much larger than the green. A more useful hue measure in this case is (Bunnik et al., 2006)

$$H_{G-R} = \frac{R-G}{R+G}, \tag{7.65}$$

which ranges from -1.0 for pure greens through 0.0 for yellow to 1.0 for pure reds. Note that both Eqs. (7.64) and (7.65) require the correct scaling or white balance between the red and green channels to accurately reflect the colour.

In other applications, other colour distinctions may enable similar simplifications to be used.

7.3.2.6 CIE *XYZ* and *xyY*

In 1931, the International Commission on Illumination (CIE) standardised the *XYZ* device-independent colour space, $\mathbf{I}_{XYZ} = \{X, Y, Z\}$. The X, Y, and Z are imaginary (in the sense that they do not correspond to any real illumination) tristimulus values chosen such that the Y corresponds with perceived intensity, and all visible colours are represented by a combination of positive values of *XYZ* (Hoffmann, 2000).

A 3-D colour space is difficult to represent graphically in two dimensions. However, since the underlying colour does not change with intensity, a normalised colour space $\mathbf{I}_{xyY} = \{x, y, Y\}$ may be derived:

$$x = \frac{X}{X+Y+Z}, \qquad y = \frac{Y}{X+Y+Z}, \tag{7.66}$$

which then allows the colour or chromaticity to be represented by the normalised x–y components, resulting in the chromaticity diagram shown on the left in Figure 7.36. Given two points on the chromaticity diagram corresponding to two colour sources, a linear combination of those sources will lie on a straight line between those points. Therefore, any three points, corresponding, for example, to red, green, and blue light sources,

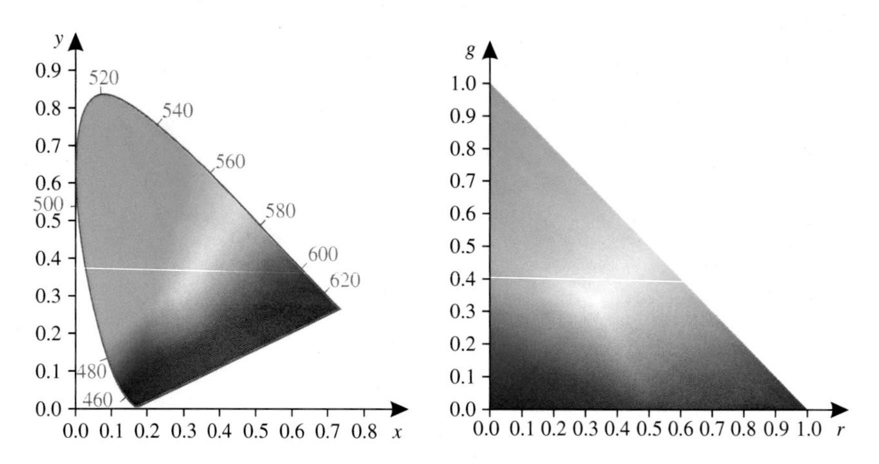

Figure 7.36 Chromaticity diagrams. Left: x–y chromaticity; the numbers are wavelengths of monochromatic light in nm. right: device-dependent r–g chromaticity.

Table 7.1 Chromaticity values for HDTV and sRGB.

	x	y	$z = 1 - x - y$
Red	0.6400	0.3300	0.0300
Green	0.3000	0.6000	0.1000
Blue	0.1500	0.0600	0.7900
White point	0.3127	0.3290	0.3583

can be mixed to produce only the colours within the triangle defined by those three points. To go outside the triangle would require one (or more) of the sources to have a negative intensity, which is clearly impossible. Such a triangle, therefore, defines the gamut of colours that can be produced by those primaries, for example on a display.

The X and Z values may be recovered from

$$X = \frac{Y}{y}x, \qquad Z = \frac{Y}{y}(1 - x - y). \tag{7.67}$$

With any linear tristimulus colour space, the component values may be transformed into any other linear colour space simply by a matrix multiplication (Hoffmann, 2000). The matrix depends, of course, on the spectral characteristics of the particular components used. For example, the primaries used in high-definition television (HDTV) have the chromaticity values listed in Table 7.1. The white point defines the chromaticity coordinates of white, defined as the colour observed when all three RGB components are equal. The corresponding transformation matrix and its inverse can then be derived as (Hoffmann, 2000)

$$\begin{bmatrix} R \\ G \\ B \end{bmatrix} = \begin{bmatrix} 3.2410 & -1.5374 & -0.4984 \\ -0.9692 & 1.8760 & 0.0416 \\ 0.0556 & -0.2040 & 1.00570 \end{bmatrix} \begin{bmatrix} X \\ Y \\ Z \end{bmatrix}, \tag{7.68}$$

$$\begin{bmatrix} X \\ Y \\ Z \end{bmatrix} = \begin{bmatrix} 0.4124 & 0.3576 & 0.1805 \\ 0.2126 & 0.7152 & 0.0722 \\ 0.0193 & 0.1192 & 0.9505 \end{bmatrix} \begin{bmatrix} R \\ G \\ B \end{bmatrix}. \tag{7.69}$$

Note that these apply to the linear device-independent RGB components before applying the gamma mapping of Eq. (7.40) to give sRGB.

If device independence is not a requirement, a device-dependent chromaticity may also be formed. This normalises the RGB components in a similar manner to Eq. (7.66):

$$r = \frac{R}{R+G+B}, \qquad g = \frac{G}{R+G+B}, \tag{7.70}$$

with the device-dependent r–g chromaticity shown on the right in Figure 7.36. Normalisation removes the intensity, enabling the r–g chromaticity to be used for thresholding similar to YCbCr or HSV.

7.3.2.7 CIE $L^*a^*b^*$ and CIE $L^*u^*v^*$

One limitation of the colour spaces described so far is that they are not perceptually uniform. In estimating colour differences or colour errors, it is useful if the distance between two colour vectors is a measure of the perceived colour difference. Several colour spaces have been derived from the CIE XYZ space that are perceptually more uniform. Two considered here are the CIE $L^*a^*b^*$ and CIE $L^*u^*v^*$ colour spaces.

The $L^*a^*b^*$ space is defined relative to a reference white point $\{X_n, Y_n, Z_n\}$ and introduces the following nonlinearity to make the space more uniform:

$$f(C) = \begin{cases} C^{1/3}, & C > \left(\frac{6}{29}\right)^3, \\ \frac{1}{3}\left(\frac{29}{6}\right)^2 C + \frac{4}{29}, & \text{otherwise.} \end{cases} \tag{7.71}$$

The conversion is to $\mathbf{I}_{L^*a^*b^*} = \{L^*, a^*, b^*\}$ is then

$$\begin{aligned} L^* &= 116 f(Y/Y_n) - 16, \\ a^* &= 500(f(X/X_n) - f(Y/Y_n)), \\ b^* &= 200(f(Y/Y_n) - f(Z/Z_n)), \end{aligned} \tag{7.72}$$

resulting in the lightness, L^*, ranging from 0 to 100. A different range may be arranged by appropriate scaling. The a^*-axis goes from green (negative) to red or magenta (positive), and the b^*-axis goes from blue (negative) to yellow (positive). The ranges of a^* and b^* are often clipped to -128 to 127, giving an 8-bit integer value. The function, $f(\cdot)$, may be implemented using a lookup table, although to convert one pixel per clock cycle, three will be required, one for each component. If the reference white point does not change, the division can be combined into the tables.

The inverse conversion,

$$\begin{aligned} f(Y/Y_n) &= \frac{L^* + 16}{116}, \\ f(X/X_n) &= \frac{a^*}{500} - f(Y/Y_n), \\ f(Z/Z_n) &= f(Y/Y_n) - \frac{b^*}{200}, \end{aligned} \tag{7.73}$$

requires three divisions before inverting Eq. (7.71):

$$C = \begin{cases} f(C)^3, & f(C) > \frac{6}{29}, \\ 3\left(\frac{6}{29}\right)^2 \left(f(C) - \frac{4}{29}\right), & \text{otherwise,} \end{cases} \tag{7.74}$$

which is again best implemented as a lookup table. The resultant components then need to be scaled by the reference white point components, which can again be combined into the lookup table if constant.

The $L^*u^*v^*$ colour space was designed to be a little easier to calculate than $L^*a^*b^*$, although on an FPGA there may be little difference because of the normalising divisions. The lightness is the same as Eq. (7.72), with the two chrominance components given by

$$u^* = 13L^*(u' - u'_n), \qquad v^* = 13L^*(v' - v'_n), \tag{7.75}$$

where

$$u' = \frac{4X}{X + 15Y + 3Z}, \qquad v' = \frac{9Y}{X + 15Y + 3Z}, \tag{7.76}$$

and u'_n and v'_n are the (u', v') chromaticities of the white point. The resulting u^* and v^* chrominance components range from -100 to 100. Again, the components may be scaled to give a more convenient range.

Conversion from $L^*u^*v^*$ back to XYZ is

$$u' = \frac{u^*}{13L^*} + u'_n, \qquad v' = \frac{v^*}{13L^*} + v'_n, \tag{7.77}$$

$$\begin{aligned} Y &= \begin{cases} Y_n\left(\frac{3}{29}\right)^3 L^*, & L^* \le 8, \\ Y_n\left(\frac{L^* + 16}{116}\right)^3, & L^* > 8, \end{cases} \\ X &= Y\frac{9u'}{4v'}, \\ Z &= Y\frac{12 - 3u' - 20v'}{4v'}. \end{aligned} \tag{7.78}$$

The colour difference between two points in either the $L^*a^*b^*$ or $L^*u^*v^*$ colour space may be determined by calculating the Euclidean distance between the corresponding vectors.

7.3.3 Colour Thresholding

Colour thresholding, similar to scalar thresholding, assigns a label to each pixel. With colour images, the purpose is to detect which pixels belong to each of a set of colours of interest. The output of colour thresholding is an image of labels, with each label corresponding to a colour class.

A computationally simple approach is to associate a rectangular box in the colour coordinates with a colour class. This corresponds to using a pair of thresholds for each component to define the boundaries of the box along that component, extending contouring (Eq. (7.7)) to each of the three channels.

RGB space is generally not appropriate for this unless the illumination is fixed. This is because the red, green, and blue components all scale with illumination. As the intensity changes, points will move diagonally in RGB space, requiring the box to be large. Consequently, only a few very different colours can be detected, and there is poor discrimination between colours.

Converting to YCbCr gives improvement because rectangular boxes aligned with the YCbCr axes will be diagonal in RGB space. However, the chrominance components still scale with the luminance, although this scaling is away from the central grey axis. The specific YCbCr transformation is not usually important, enabling the simpler variations of Eq. (7.53) or (7.54) to be used. Box-based colour thresholding can be implemented as in Figure 7.37. This circuit needs to be repeated for each detected colour class. The comparisons can easily be realised using a lookup table for each channel; multiple classes may be segmented simultaneously by having a separate bit-plane associated with each colour class. Better discrimination may be obtained by normalising the chrominance components by the luminance (or the maximum of the red, green, and blue components (Johnston et al., 2005b)).

Alternatively, the thresholding may be performed directly in the HSV colour space where the hue and saturation are independent of the luminance. Note that the HLS colour space is usually less suitable than HSV because the saturation changes with lightness when the lightness is greater than 50%.

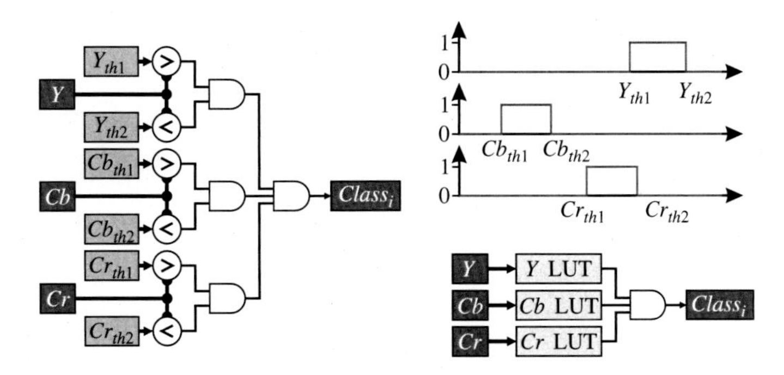

Figure 7.37 Colour thresholding using a box in YCbCr colour space. Left: comparisons required for each colour class; right: lookup table approach, with one lookup table per colour channel.

The set of pixel values corresponding to a colour class do not necessarily need to fall within a rectangular box aligned with the component axes. However, the logic becomes increasingly more complex to handle arbitrary shaped regions. One approach is to use a single, large, lookup table taking as input the concatenated components of the colour input vector. The mapping can then be arbitrarily complex and can include any necessary colour space conversions. Unfortunately, such a lookup table is prohibitively large ($2^{24} = 16{,}777{,}216$ entries for 24-bit colour) and would need to be implemented in external memory. The table size may be reduced using fewer bits of the input at the cost of poorer discrimination.

A composite approach can also be used. For example, Johnston et al. (2005a) used two lookup tables, one for each of *Cb* and *Cr*, and also combined a normalisation division by *Y* (with a reduced number of bits) into those tables (see Section 15.1 for more details).

7.3.4 Colour Enhancement

When performing colour enhancement, it is often easier to work in HSV or HLS space. Since the hue represents the basic colour, generally it is left unchanged (unless correcting colour errors as described in Section 7.3.5). Colours can be made move vivid by increasing the saturation. This can be achieved by multiplying the saturation components by a constant or using a gamma enhancement (with $\gamma < 1$). Overall, the contrast of the image may be enhanced by enhancing the lightness or value component.

If working in YCbCr colour space, the saturation may be increased by multiplying the chrominance components by a constant. If the colour is represented in terms of its luminance and chrominance vectors:

$$\mathbf{Colour} = \mathbf{Luminance} + \mathbf{Chrominance}, \tag{7.79}$$

then increasing the chrominance gives

$$\begin{aligned}\mathbf{Enhanced} &= \mathbf{Luminance} + (1+\alpha)\mathbf{Chrominance} \\ &= (1+\alpha)\mathbf{Colour} - \alpha\mathbf{Luminance}.\end{aligned} \tag{7.80}$$

Using the definition of *Y* from Eq. (7.50) or (7.54) enables the saturation to be enhanced directly in RGB colour space:

$$\begin{bmatrix} R \\ G \\ B \end{bmatrix}_{Enhanced} = (1+\alpha)\begin{bmatrix} R \\ G \\ B \end{bmatrix} - \alpha\frac{R+2G+B}{4} = \begin{bmatrix} 1+\frac{3\alpha}{4} & -\frac{\alpha}{2} & -\frac{\alpha}{4} \\ -\frac{\alpha}{4} & 1+\frac{\alpha}{2} & -\frac{\alpha}{4} \\ -\frac{\alpha}{4} & -\frac{\alpha}{2} & 1+\frac{3\alpha}{4} \end{bmatrix}\begin{bmatrix} R \\ G \\ B \end{bmatrix}. \tag{7.81}$$

Any resulting values outside the range 0 to $2^B - 1$ need to be clipped.

7.3.5 White Balance

From Eq. (1.8), the colour of each pixel observed in an image depends not only on the spectral reflectivity of the corresponding object point but also on the illumination. The interaction of the illumination spectrum and the reflectance spectrum within the spectral characteristics of the sensor reduces the perceived value to a single number (or three numbers for colour). Consequently, the colours of objects within a scene illuminated by incandescent light can be quite different from those illuminated by LED light. For outdoor scenes, the spectral characteristics of the illumination depend on whether the object is illuminated by sunlight, or in shadow, or whether the sky is clear or overcast, or even on the time of day. There are many more complex illumination estimation methods, which will not be considered here. Several of these are discussed by Funt et al. (1998) and Finlayson et al. (2001).

The colour constancy problem is that of estimating how the scene would appear given canonical illumination (Hordley, 2006). Calibrating the camera for the illumination is essential if the imaging system is used to make meaningful colour measurements (Akkaynak et al., 2014). Unfortunately, even with full knowledge of the actual illumination spectrum, estimating the response under a canonical illumination is ill-posed, and in many circumstances, prior knowledge of the illumination is not even available. However, under the assumption that the illumination spectrum is relatively smooth and broad, reasonable approximations can be made to remove the effects of the illumination spectrum from an image. Colour correction involves transforming the RGB value of each pixel to achieve this.

Assuming a linear mixing model (Eq. (1.8) is linear, although the human visual system is not), this requires determining the components of a 3×3 linear transformation matrix (Akkaynak et al., 2014). However, given that the spectral distribution at the sensor is the product of the reflectivity and illumination distributions, this is often simplified to a diagonal matrix, independently scaling the R, G, and B components (Finlayson et al., 1994). Note that if any of the components are saturated, or there are other nonlinearities, this will result in an additional colour change that cannot easily be corrected.

7.3.5.1 Grey-world Model

Perhaps the simplest model is to assume that on average the image is achromatic (Finlayson et al., 2001). This implies that the average image colour should be grey. The effective illumination colour is then given by the mean of the red, green, and blue components of the image. Correction is performed by dividing by the mean value and scaling to ensure that all pixel values are within the range of allowed values:

$$\hat{R} = k\frac{R}{\mu_R}, \quad \hat{G} = k\frac{G}{\mu_G}, \quad \hat{B} = k\frac{B}{\mu_B}, \tag{7.82}$$

where $\hat{R}\hat{G}\hat{B}$ are the corrected colour components, and the scaling factor, k, that maximises the output contrast is given by

$$k = (2^B - 1)\min\left(\frac{\mu_R}{\max\limits_{x,y}(R)}, \frac{\mu_G}{\max\limits_{x,y}(G)}, \frac{\mu_B}{\max\limits_{x,y}(B)}\right), \tag{7.83}$$

with the result shown in the centre panel of Figure 7.38. Alternatively, the scaling factor can be set to $k = 2^{B-1}$ (mapping the mean to mid-grey) (Funt et al., 1998), although this has the danger that one or more components may saturate. Or, if the image has been exposed correctly, only the red and blue channels need to be adjusted, with $k = \mu_G$.

Grey-world white balancing can also be performed in the YCbCr colour space. A grey average should give $\mu_{Cb} = 0$ and $\mu_{Cr} = 0$, with the blue and red channels adjusted, respectively, to achieve this.

The major limitation of the grey-world model is that if the image contains a dominant colour, the estimate of the illumination colour will be biased towards that colour (Gershon et al., 1987). As a result, a colour cast complementary to the dominant colour will be introduced into the image. An advantage of working in the YCbCr colour space is that the bias induced by strong (saturated) colours may be reduced by only averaging the Cb and Cr values below a threshold.

Figure 7.38 Simple colour correction. Left: original image captured under incandescent lights, resulting in a yellowish-red cast; centre: correcting assuming the average is grey, using Eq. (7.82); right: correcting assuming the brightest pixel is white, using Eq. (7.84).

7.3.5.2 White-point Model

Another approach for estimating the illuminant is to assume that the highlights correspond to the colour of the light source. This assumes that the brightest point in the image is white, and the white-point model scales each component by this value to make it achromatic (Land and McCann, 1971; Funt et al., 1998). In practice, this assumption is relaxed by finding the maximum value of each RGB component and scaling each component by its corresponding maximum:

$$\hat{R} = (2^B - 1)\frac{R}{\max\limits_{x,y}(R)}, \quad \hat{G} = (2^B - 1)\frac{G}{\max\limits_{x,y}(G)}, \quad \hat{B} = (2^B - 1)\frac{B}{\max\limits_{x,y}(B)}. \tag{7.84}$$

The results of this scheme are shown in the right panel of Figure 7.38.

One limitation of this approach is that it is easy for the assumptions to be violated. If the white pixel is the brightest in the image, there is a strong possibility that one or more of the channels will be saturated, underestimating the true value of that component. In many cases, the brightest point in an image is the result of specular reflection with the resulting pixel affected by the colour of the reflecting surface. This is mitigated to some extent by treating each colour channel independently (and not requiring the maximum of each channel to come from the same pixel).

Basing the correction on a single pixel will inevitably introduce noise into the estimate of the illumination colour, affecting the accuracy of any correction. This may be mitigated using a noise-smoothing filter before finding the maximum value.

The brightest object in the image may not be white, and if it is white, it may not be pure white. If the colour is slightly off-white, making it white will introduce a cast. This limitation is harder to overcome, because it is impossible to distinguish between an off-white surface and slightly coloured illumination.

7.3.5.3 White-patch Model

The white-point model may be extended by deliberately placing a white patch within the image. If the illumination and camera settings do not change with time, this may even be captured off-line as a calibration image. The white patch can then be detected automatically, or by having the patch in a known location, or by manually selecting the region of interest. Let $\mu_{(white)}$ be the mean of the pixels within the white patch, then the correction may be performed as

$$\hat{R} = (2^B - 1)\frac{R}{\mu_{R(white)}}, \quad \hat{G} = (2^B - 1)\frac{G}{\mu_{G(white)}}, \quad \hat{B} = (2^B - 1)\frac{B}{\mu_{B(white)}}. \tag{7.85}$$

If the white patch is not the brightest point within the image (for example, a light grey is used instead to prevent any of the components of the calibration patch from saturating), then the $2^B - 1$ in Eq. (7.85) may be replaced by k similar to Eq. (7.83).

Figure 7.39 Correcting using black, white, and grey patches. Left: original image with the patches marked; centre: stretching each channel to correct for black and white, using Eq. (7.86); right: adjusting the gamma of the red and blue channels using Eq. (7.87) to make the grey patch achromatic.

This will ensure that the white patch will appear achromatic. However, the camera may also introduce a DC offset in the output (for example as an offset in the amplifier or A/D converter, or not accurately estimating the dark current or black-level). The colour correction may be improved by introducing a black patch into the model and setting the black patch a pixel value of 0. Let $\mu_{(black)}$ be the mean of the pixels within the black patch. Colour correction using both patches is therefore

$$\begin{aligned}\hat{R} &= (2^B - 1)\frac{R - \mu_{R(black)}}{\mu_{R(white)} - \mu_{R(black)}},\\ \hat{G} &= (2^B - 1)\frac{G - \mu_{G(black)}}{\mu_{G(white)} - \mu_{G(black)}},\\ \hat{R} &= (2^B - 1)\frac{B - \mu_{B(black)}}{\mu_{B(white)} - \mu_{B(black)}}.\end{aligned} \tag{7.86}$$

The results of this correction are shown in the centre panel of Figure 7.39.

The colour correction is making a brightness and contrast adjustment to each of the channels, as described in Section 7.1.1. If the calibration is performed online, then the best approach is to place the black and white patches at the top of the image, so the mean values may be used to correct the remainder of the image. Otherwise, the whole image needs to be stored in a frame buffer to enable the whole image to be corrected. If the illumination is changing slowly, then it may be sufficient to use the channel gains and offsets calculated from the previous frame.

A more sophisticated model adds a third mid-grey patch, with mean value $\mu_{(grey)}$. This allows slight variations in gamma to be corrected between the channels. The green channel is left unchanged, but a gamma correction is applied to the red and blue channels to make the components equal for the grey patch. Considering just the red component (a similar equation results for blue):

$$\hat{R} = (2^B - 1)\left(\frac{R - \mu_{R(black)}}{\mu_{R(white)} - \mu_{R(black)}}\right)^{\gamma_R}, \tag{7.87}$$

where

$$\gamma_R = \frac{\ln\left(\frac{\mu_{G(grey)} - \mu_{G(black)}}{\mu_{G(white)} - \mu_{G(black)}}\right)}{\ln\left(\frac{\mu_{R(grey)} - \mu_{R(black)}}{\mu_{R(white)} - \mu_{R(black)}}\right)}. \tag{7.88}$$

The results of adjusting the gamma are shown in the right panel of Figure 7.39.

Correcting for grey requires the addition of two gamma correction blocks, one for each of red and blue, as shown in Figure 7.40. The circuit for these is complicated by the fact that the gamma values are not constant but are a result of calibration. If implemented online, the calculation of the gamma is also required. However,

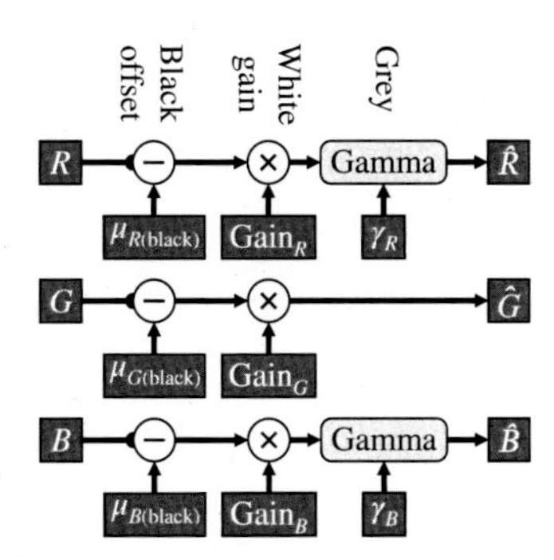

Figure 7.40 Implementing colour correction: the various parameters are stored in registers. Overflow detection and clipping has been omitted for clarity.

since this in only performed once per frame, it is probably best implemented in software rather than directly in hardware. In that case, the software can calculate the values for a lookup table implementation of not only the gamma correction blocks but also for the whole mapping.

7.3.5.4 Filter-based Models

As well as point operations, there are also approaches based on filtering the image, including the grey-edge and the centre-surround models. These approaches will be described in Section 9.7.4.

7.4 Multi-spectral and Hyperspectral Imaging

Sensors are not restricted to using only three spectral channels from only the visible part of the electromagnetic spectrum. Including the near infrared has been found useful in many applications, including land use and vegetation classification in remote sensing (Ehlers, 1991), and produce grading (Duncan and Leeson, 1999) for detecting blemishes that may not be as apparent in the visible spectrum.

The difference between multi- and hyperspectral imaging is that multi-spectral images only have a few relatively broad frequency bands that are not necessarily contiguous. Hyperspectral images effectively capture a continuous spectrum with a typical spectral resolution of 5–10 nm. Rather than the three channels of a colour image, a hyperspectral image can have up to several hundred channels or bands, depending on the spectral resolution and wavelength range. Since the underlying image data has three dimensions (two spatial and one spectral), it is often referred to as a ***hyperspectral cube***.

Having finer spectral resolution enables the detailed spectral signature to be obtained for each pixel. This overcomes the issues of metamerism associated with conventional RGB or RGBI (with infrared) images. Having a detailed spectrum opens many more application areas including mineral exploration (Bedini, 2017); agriculture (Teke et al., 2013); food processing (Feng and Sun, 2012; Liu et al., 2017); medicine (Lu and Fei, 2014); and material and waste sorting (Tatzer et al., 2005).

7.4.1 Hyperspectral Image Acquisition

There are three main approaches to capturing the hyperspectral cube, as illustrated in Figure 7.41:

Spectral scanning: The traditional approach to capturing hyperspectral images was to capture images of each band sequentially, using a series of filters or a tuneable optical bandpass filter to select the wavelength. This can give very high spectral resolution but comes at the cost of the time to capture the image (one frame per spectral band) and relies on the subject being stationary during the capture.

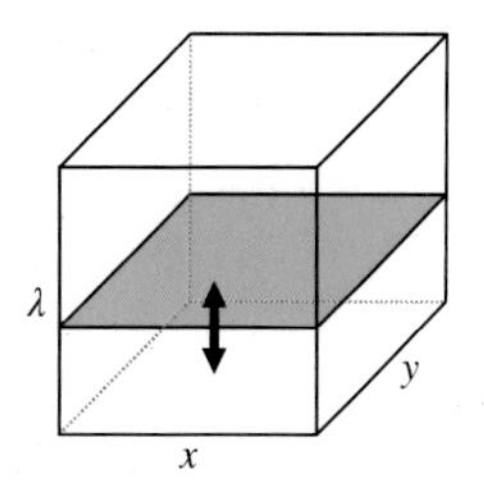

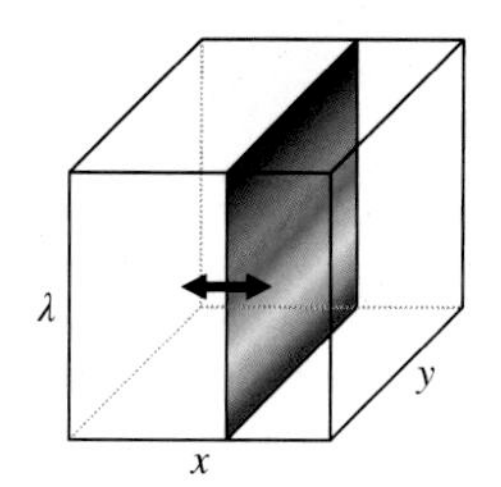

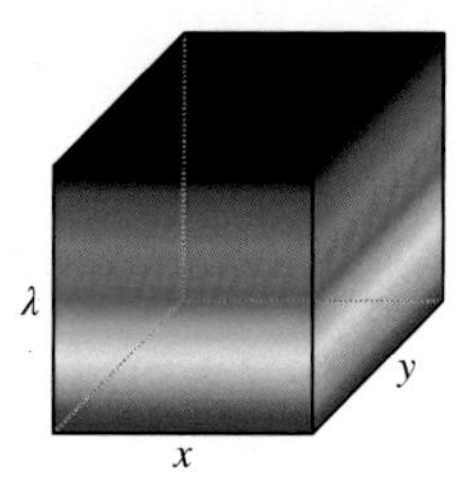

Figure 7.41 Hyperspectral cube acquisition approaches. Left: spectral scanning; centre: spatial scanning; right: snapshot (non-scanning).

Spatial scanning: A 2-D sensor captures a single line within the image, with the second sensor dimension sampling the spectrum. Such cameras typically use a prism or diffraction grating to spread the spectrum. This gives good spectral resolution but requires spatial scanning to capture the x spatial dimension. Either the sensor can be scanned (for example the motion of a satellite or aircraft), or the subject scanned (for example on a moving conveyor belt).

Snapshot: Captures the whole cube at once, trading off spectral and spatial resolution for temporal resolution. Typically only 16–36 spectral bands are captured, with each pixel capturing only a single band. The sensor is either divided into tiles, with each tile containing a different spectral band, or a mosaic of filters is used over individual pixels. With the mosaic, each band is spatially offset from the others and may be interpolated to give higher spatial resolution (Mihoubi et al., 2017) in much the same way as demosaicing the Bayer pattern in RGB images.

7.4.2 Processing Steps

The huge volume of data contained within the hyperspectral cube, and the demand for real-time processing in many applications, makes hyperspectral imaging a good candidate for FPGA-based acceleration.

Where scanning approaches have been used for acquisition, it may be necessary to register the individual images. The strong correlation between adjacent bands makes image alignment straightforward for spectral scanning approaches. Spatial scanning is a little more difficult; any non-uniform motion can result in successive scans not being parallel (such motion generally requires additional sensors to detect). However, any motion in the y-direction can be detected by correlation, enabling individual lines to be aligned. Several image registration techniques are discussed in Section 10.4.

As the sensor measures the incoming light, the data requires radiometric calibration to convert the radiance measurements to reflectance (or absorbance in the case of transmission imaging) (Feng and Sun, 2012; Lu and Fei, 2014). This serves the same purpose as white balancing for a colour sensor, to remove the effects of illumination and calibrate the sensor. Calibration requires capturing an image with no light (the dark current), I_{dark}, and a broad spectrum diffuse white target, I_{white}. The reflectance, ρ, is then

$$\rho = \frac{I - I_{dark}}{I_{white} - I_{dark}}. \tag{7.89}$$

For absorbance, A, a logarithmic scale is more commonly used to give the optical density:

$$A = -\log \frac{I - I_{dark}}{I_{reference} - I_{dark}}, \tag{7.90}$$

where $I_{reference}$ is a reference image (without any subject).

There is a large amount of redundancy within the spectra, which are usually relatively smooth except at particular wavelengths where there are resonances or energy absorbed as a result of chemical bonds. To reduce the processing, a data reduction step is often used (Feng and Sun, 2012; Lu and Fei, 2014). The goal is to find a subspace suitable for classification. Common operations here are principal components analysis for unsupervised selection or linear discriminant analysis for supervised selection. Note that the principal components carry no intrinsic meaning, and consequently, classification performance can suffer, particularly around the edges of regions (Rodarmel and Shan, 2002). Often, individual pixels are not from a single class, so the spectrum for each pixel should be considered as a mixture of endpoints (Keshava and Mustard, 2002). Several techniques can be used to identify these endpoints and determine the proportions of each endpoint associated with each pixel.

In many applications, a few distinct frequency bands carry much of the important information. Selecting only those bands is another important data reduction technique. Bands of interest can include the location of peaks and valleys in the spectrum or peaks in the spectral derivative (where the changes are greatest). Clustering can aid in band selection by grouping bands into clusters and selecting one band from each cluster (Wang et al., 2018).

Each pixel can be assigned one of several classes based on the hyperspectral signature (or a reduced dimensionality version). For this, several different classification methods have been used.

Clustering can also be used in an unsupervised manner to identify patterns of interest within a dataset (Zhao et al., 2019; Le Moan and Cariou, 2018). However, these can suffer from the curse of dimensionality (in high dimensions, the volume becomes so huge that the data is sparse, and distance measures become less meaningful, making similarity search more difficult). This sparsity also poses challenges for density estimation without large volumes of data (Le Moan and Cariou, 2018).

7.5 Summary

Point operations are some of the simplest to implement on an FPGA because each output pixel value depends only on the corresponding input pixel value. This enables them to be implemented using any processing mode. With no dependencies between pixels, point operations can readily be parallelised by building multiple processing units. It is also common to pipeline point operations.

Point operations on a single image are primarily concerned with adjusting the brightness or contrast (or equivalently adjusting the colour balance or contrast of a colour image). Thresholding is another common point operation used for segmenting an image on the basis of intensity or colour. Any point operation may be directly implemented using a lookup table, giving constant processing time regardless of the complexity of the operation.

Point operations may also be applied between multiple images. Virtually any operation may be applied between the pixels of the images being combined. Most commonly, these are images captured at different times, requiring a frame buffer or other storage. A common architecture is recursive processing, where a newly acquired image is combined with one or more images held in a frame buffer, with the result written back to the frame buffer. Applications include noise filtering, background modelling, change detection, and image fusion.

Basic colour processing can be thought of as a point operation, where each pixel comprises a vector. Common operations here are colour space conversion, colour enhancement, and thresholding (colour-based classification). Colour can be generalised to hyperspectral imaging, where each pixel contains a spectral distribution. The high dimensionality of such data (each pixel has tens or hundreds of spectral samples) poses particular challenges for real-time processing.

The simplicity of point operations is also a limitation. Frequently, global data is required to set appropriate parameters (for example for contrast stretching or selecting an appropriate threshold level). The use of histograms to capture and derive some of these parameters is described in Chapter 8. A point operation considers each pixel in isolation and does not take context into account. Local filters, where the output value is dependent on the pixels within a local neighbourhood, are described in Chapter 9.

References

Abdelli, A. and Choi, H.J. (2017). A four-frames differencing technique for moving objects detection in wide area surveillance. *IEEE International Conference on Big Data and Smart Computing (BigComp)*, Jeju, Republic of Korea (13–16 February 2017), 210–214. https://doi.org/10.1109/BIGCOMP.2017.7881701.

Aikens, R.S., Agard, D.A., and Sedat, J.W. (1989). Solid-state imagers for microscopy. In: *Methods in Cell Biology*, Chapter 16, vol. **29** (ed. Y.L. Wang and D.L. Taylor), 291–313. New York: Academic Press.

Akkaynak, D., Treibitz, T., Xiao, B., Gurkan, U.A., Allen, J.J., Demirci, U., and Hanlon, R.T. (2014). Use of commercial off-the-shelf digital cameras for scientific data acquisition and scene-specific color calibration. *Journal of the Optical Society of America A* **31** (2): 312–321. https://doi.org/10.1364/JOSAA.31.000312.

Appiah, K. and Hunter, A. (2005). A single-chip FPGA implementation of real-time adaptive background model. *IEEE International Conference on Field-Programmable Technology*, Singapore (11–14 December 2005), 95–102. https://doi.org/10.1109/FPT.2005.1568531.

Bailey, D.G. (2012). Streamed high dynamic range imaging. *International Conference on Field Programmable Technology (ICFPT)*, Seoul, Republic of Korea (10–12 December 2012), 305–308. https://doi.org/10.1109/FPT.2012.6412153.

Bedini, E. (2017). The use of hyperspectral remote sensing for mineral exploration: a review. *Journal of Hyperspectral Remote Sensing* **7** (4): 189–211. https://doi.org/10.29150/jhrs.v7.4.p189-211.

Benetti, M., Gottardi, M., Mayr, T., and Passerone, R. (2018). A low-power vision system with adaptive background subtraction and image segmentation for unusual event detection. *IEEE Transactions on Circuits and Systems I: Regular Papers* **65** (11): 3842–3853. https://doi.org/10.1109/TCSI.2018.2857562.

Bouderbane, M., Dubois, J., Heyrman, B., Lapray, P.J., and Ginhac, D. (2016). Ghost removing for HDR real-time video stream generation. *Real-Time Image and Video Processing 2016*, Brussels, Belgium (7 April 2016). SPIE, Volume **9897**, Article ID 98970F, 6 pages. https://doi.org/10.1117/12.2230313.

Bouwmans, T. (2011). Recent advanced statistical background modeling for foreground detection - a systematic survey. *Recent Patents on Computer Science* **4** (3): 147–176.

Braun, G.J. and Fairchild, M.D. (1999). Image lightness rescaling using sigmoidal contrast enhancement functions. *Journal of Electronic Imaging* **8** (4): 380–393. https://doi.org/10.1117/1.482706.

Bunnik, H.M.W., Bailey, D.G., and Mawson, A.J. (2006). Objective colour measurement of tomatoes and limes. *Image and Vision Computing New Zealand (IVCNZ'06)*, Great Barrier Island, NZ (27–29 November 2006), 263–268.

Bylander, T. and Rosen, B. (1997). A perceptron-like online algorithm for tracking the median. *International Conference on Neural Networks (ICNN'97)*, Houston, TX, USA (12 June 1997), Volume **4**, 2219–2224. https://doi.org/10.1109/ICNN.1997.614292.

Cucchiara, R., Onfiani, P., Prati, A., and Scarabottolo, N. (1999). Segmentation of moving objects at frame rate: a dedicated hardware solution. *7th International Conference on Image Processing and its Applications*, Manchester, UK (13–15 July 1999), 138–142. https://doi.org/10.1049/cp:19990297.

Cutler, R. and Davis, L. (1998). View-based detection and analysis of periodic motion. *14th International Conference on Pattern Recognition*, Brisbane, Australia (16–20 August 1998), Volume **1**, 495–500. https://doi.org/10.1109/ICPR.1998.711189.

Debevec, P.E. and Malik, J. (2008). Recovering high dynamic range radiance maps from photographs. *SIGGRAPH '08: Special Interest Group on Computer Graphics and Interactive Techniques Conference*, Los Angeles, CA, USA (11–15 August 2008), Article ID 31, 10 pages. https://doi.org/10.1145/1401132.1401174.

DiCarlo, J.M. and Wandell, B.A. (2000). Rendering high dynamic range images. *Sensors and Camera Systems for Scientific, Industrial, and Digital Photography Applications*, San Jose, CA, USA (24–26 January 2000). SPIE, Volume **3965**, 392–401. https://doi.org/10.1117/12.385456.

Duncan, D.B. and Leeson, G. (1999). Cost effective real-time multi-spectral digital video imaging. *Sensors, Cameras, and Systems for Scientific/Industrial Applications*, San Jose, CA, USA (25–26 January 1999). SPIE, Volume **3649**, 100–108. https://doi.org/10.1117/12.347065.

Ehlers, M. (1991). Multisensor image fusion techniques in remote sensing. *ISPRS Journal of Photogrammetry and Remote Sensing* **46** (1): 19–30. https://doi.org/10.1016/0924-2716(91)90003-E.

Eilertsen, G., Wanat, R., Mantiuk, R.K., and Unger, J. (2013). Evaluation of tone mapping operators for HDR-video. *Computer Graphics Forum* **32** (7): 275–284. https://doi.org/10.1111/cgf.12235.

Feng, Y.Z. and Sun, D.W. (2012). Application of hyperspectral imaging in food safety inspection and control: a review. *Critical Reviews in Food Science and Nutrition* **52** (11): 1039–1058. https://doi.org/10.1080/10408398.2011.651542.

Finlayson, G.D., Drew, M.S., and Funt, B.V. (1994). Color constancy: generalized diagonal transforms suffice. *Journal of the Optical Society of America A* **11** (11): 3011–3019. https://doi.org/10.1364/JOSAA.11.003011.

Finlayson, G.D., Hordley, S.D., and Hubel, P.M. (2001). Color by correlation: a simple, unifying framework for color constancy. *IEEE Transactions on Pattern Analysis and Machine Intelligence* **20** (11): 1209–1221. https://doi.org/10.1109/34.969113.

Foley, J.D. and Van Dam, A. (1982). *Fundamentals of Interactive Computer Graphics*. Reading, MA: Addison-Wesley.

Funt, B., Barnard, K., and Martin, L. (1998). Is machine colour constancy good enough? In: *5th European Conference on Computer Vision (ECCV'98)*, Freiburg, Germany (2–6 June 1998), *Lecture Notes in Computer Science*, Volume **1406**, 445–459. https://doi.org/10.1007/BFb0055655.

Genovese, M. and Napoli, E. (2014). ASIC and FPGA implementation of the Gaussian mixture model algorithm for real-time segmentation of high definition video. *IEEE Transactions on Very Large Scale Integration (VLSI) Systems* **22** (3): 537–547. https://doi.org/10.1109/TVLSI.2013.2249295.

Gershon, R., Jepson, A.D., and Tsotsos, J.K. (1987). From [R,G,B] to surface reflectance: computing color constant descriptors in images. *International Joint Conference on Artificial Intelligence*, Milan, Italy (23–29 August 1987), Volume **2**, 755–758.

Goldman, D.B. and Chen, J.H. (2005). Vignette and exposure calibration and compensation. *10th IEEE International Conference on Computer Vision (ICCV 2005)*, Beijing, China (17–21 October 2005), Volume **1**, 899–906. https://doi.org/10.1109/ICCV.2005.249.

Guo, G., Kaye, M.E., and Zhang, Y. (2015). Enhancement of Gaussian background modelling algorithm for moving object detection and its implementation on FPGA. *IEEE 28th Canadian Conference on Electrical and Computer Engineering (CCECE)*, Halifax, Nova Scotia, Canada (3–6 May 2015), 118–122. https://doi.org/10.1109/CCECE.2015.7129171.

Heikkila, J. and Silven, O. (1999). A real-time system for monitoring of cyclists and pedestrians. *2nd IEEE Workshop on Visual Surveillance*, Fort Collins, CO, USA (26 June 1999), 74–81. https://doi.org/10.1109/VS.1999.780271.

Hoffmann, G. (2000). *CIE colour space*. http://www.fho-emden.de/hoffmann/ciexyz29082000.pdf (accessed 1 May 2023).

Hordley, S.D. (2006). Scene illuminant estimation: past, present, and future. *Color Research and Application* **31** (4): 303–314. https://doi.org/10.1002/col.20226.

IEC 61966-2-1:1999 (1999). *Multimedia Systems and Equipment - Colour Measurement and Management - Part 2-1: Colour Management - Default RGB Colour Space - sRGB*. International Electrotechnical Commission.

ISO/IEC 15444-1:2000 (2000). *JPEG 2000 Image Coding System - Part 1: Core Coding System*. International Standards Organisation.

Johnston, C.T., Bailey, D.G., and Gribbon, K.T. (2005a). Optimisation of a colour segmentation and tracking algorithm for real-time FPGA implementation. *Image and Vision Computing New Zealand (IVCNZ'05)*, Dunedin, NZ (28–29 November 2005), 422–427.

Johnston, C.T., Gribbon, K.T., and Bailey, D.G. (2005b). FPGA based remote object tracking for real-time control. *International Conference on Sensing Technology*, Palmerston North, NZ (21–23 November 2005), 66–71.

Juvonen, M.P.T., Coutinho, J.G.F., and Luk, W. (2007). Hardware architectures for adaptive background modelling. *3rd Southern Conference on Programmable Logic*, Mar del Plata, Argentina (28–26 February 2007), 149–154. https://doi.org/10.1109/SPL.2007.371739.

Kameda, Y. and Minoh, M. (1996). A human motion estimation method using 3-successive video frames. *International Conference on Virtual Systems and Multimedia (VSMM'96)*, Gifu, Japan (18–20 September 1996), 135–140.

Keshava, N. and Mustard, J.F. (2002). Spectral unmixing. *IEEE Signal Processing Magazine* **19** (1): 44–57. https://doi.org/10.1109/79.974727.

Land, E.H. and McCann, J.J. (1971). Lightness and retinex theory. *Journal of the Optical Society of America* **61** (1): 1–11. https://doi.org/10.1364/JOSA.61.000001.

Lapray, P.J., Heyrman, B., Rosse, M., and Ginhac, D. (2013). A 1.3 megapixel FPGA-based smart camera for high dynamic range real time video. *7th International Conference on Distributed Smart Cameras (ICDSC)*, Palm Springs, CA, USA (29 October – 1 November 2013), 6 pages. https://doi.org/10.1109/ICDSC.2013.6778230.

Le Moan, S. and Cariou, C. (2018). Parameter-free density estimation for hyperspectral image clustering. *International Conference on Image and Vision Computing New Zealand (IVCNZ)*, Auckland, NZ (19–21 November 2018), 6 pages. https://doi.org/10.1109/IVCNZ.2018.8634706.

Liu, Y., Pu, H., and Sun, D.W. (2017). Hyperspectral imaging technique for evaluating food quality and safety during various processes: a review of recent applications. *Trends in Food Science and Technology* **69**: 25–35. https://doi.org/10.1016/j.tifs.2017.08.013.

Lu, G. and Fei, B. (2014). Medical hyperspectral imaging: a review. *Journal of Biomedical Optics* **19** (1): Article ID 10901, 23 pages. https://doi.org/10.1117/1.JBO.19.1.010901.

Malvar, H.S. and Sullivan, G.J. (2003). Transform, Scaling and Color Space Impact of Professional Extensions. *Technical report JVT-H031r1*. Joint Video Team (JVT) of ISO/IEC MPEG and ITU-T VCEG.

Mann, S. and Picard, R.W. (1995). On being 'undigital' with digital cameras: extending dynamic range by combining differently exposed pictures. *Proceedings of IST's 48th Annual Conference*, Cambridge, MA, USA (7–11 May 1995), 422–428.

Mann, S., Lo, R.C.H., Ovtcharov, K., Gu, S., Dai, D., Ngan, C., and Ai, T. (2012). Realtime HDR (High Dynamic Range) video for eyetap wearable computers, FPGA-based seeing aids, and glasseyes (EyeTaps). *25th IEEE Canadian Conference on Electrical and Computer Engineering (CCECE)*, Montreal, Quebec, Canada (29 April – 2 May 2012), 6 pages. https://doi.org/10.1109/CCECE.2012.6335012.

McIvor, A., Zang, Q., and Klette, R. (2001). The background subtraction problem for video surveillance systems. In: *International Workshop on Robot Vision (RobVis 2001)*, Auckland, NZ (16–18 February 2001), *Lecture Notes in Computer Science*, Volume **1998**, 176–183. https://doi.org/10.1007/3-540-44690-7_22.

Mertens, T., Kautz, J., and Reeth, F.V. (2007). Exposure fusion. *15th Pacific Conference on Computer Graphics and Applications (PG'07)*, Maui, HI, USA (29 October – 2 November 2007), 382–390. https://doi.org/10.1109/PG.2007.17.

Mihoubi, S., Losson, O., Mathon, B., and Macaire, L. (2017). Multispectral demosaicing using pseudo-panchromatic image. *IEEE Transactions on Computational Imaging* **3** (4): 982–995. https://doi.org/10.1109/TCI.2017.2691553.

Mitra, S.K. (1998). *Digital Signal Processing: A Computer-based Approach*. Singapore: McGraw-Hill.

Nayar, S.K. and Mitsunaga, T. (2000). High dynamic range imaging: spatially varying pixel exposures. *IEEE Conference on Computer Vision and Pattern Recognition*, Hilton Head Island, South Carolina, USA (13–15 June 2000), Volume **1**, 472–479. https://doi.org/10.1109/CVPR.2000.855857.

Nosko, S., Musil, M., Zemcik, P., and Juranek, R. (2020). Color HDR video processing architecture for smart camera. *Journal of Real-Time Image Processing* **17** (3): 555–566. https://doi.org/10.1007/s11554-018-0810-z.

Orwell, J., Remagnino, P., and Jones, G.A. (1999). Multi-camera colour tracking. *2nd IEEE Workshop on Visual Surveillance*, Fort Collins, CO, USA (26 June 1999), 14–21. https://doi.org/10.1109/VS.1999.780264.

Ou, Y., Ambalathankandy, P., Takamaeda, S., Motomura, M., Asai, T., and Ikebe, M. (2022). Real-time tone mapping: a survey and cross-implementation hardware benchmark. *IEEE Transactions on Circuits and Systems for Video Technology* **32** (5): 2666–2686. https://doi.org/10.1109/TCSVT.2021.3060143.

Piccardi, M. (2004). Background subtraction techniques: a review. *IEEE International Conference on Systems, Man and Cybernetics*, The Hague, The Netherlands (10–13 October 2004), Volume **4**, 3099–3104. https://doi.org/10.1109/ICSMC.2004.1400815.

Punchihewa, A., Bailey, D.G., and Hodgson, R.M. (2005). Colour reproduction performance of JPEG and JPEG2000 codecs. *8th International Symposium on DSP and Communication Systems, (DSPCS'2005) and 4th Workshop on the Internet, Telecommunications and Signal Processing, (WITSP'2005)*, Noosa Heads, Australia (19–21 December 2005), 312–317.

Reinhard, E., Stark, M., Shirley, P., and Ferwerda, J. (2002). Photographic tone reproduction for digital images. *29th Annual Conference on Computer Graphics and Interactive Techniques*, San Antonio, TX, USA (21–26 July 2002), 267–276. https://doi.org/10.1145/566570.566575.

Reulke, R., Meysel, F., and Bauer, S. (2008). Situation analysis and atypical event detection with multiple cameras and multi-object tracking. In: *2nd International Workshop on Robot Vision*, Auckland, NZ (18–20 February 2008), *Lecture Notes in Computer Science*, Volume **4931**, 234–247. https://doi.org/10.1007/978-3-540-78157-8_18.

Rodarmel, C. and Shan, J. (2002). Principal component analysis for hyperspectral image classification. *Surveying and Land Information Science* **62** (2): 115–122.

Schoonees, J.A. and Palmer, T. (2009). Camera shading calibration using a spatially modulated field. *Image and Vision Computing New Zealand (IVCNZ 2009)*, Wellington, NZ (23–25 November 2009), 191–196. https://doi.org/10.1109/IVCNZ.2009.5378412.

Sen Gupta, G. and Bailey, D. (2008). Discrete YUV look-up tables for fast colour segmentation for robotic applications. *IEEE Canadian Conference on Electrical and Computer Engineering (CCECE 2008)*, Niagara Falls, Canada (4–7 May 2008), 963–968. https://doi.org/10.1109/CCECE.2008.4564679.

Shilton, A. and Bailey, D. (2006). Drouge tracking by image processing for study of laboratory scale pond hydraulics. *Flow Measurement and Instrumentation* **17** (1): 69–74. https://doi.org/10.1016/j.flowmeasinst.2005.04.002.

Sidibe, D., Puech, W., and Strauss, O. (2009). Ghost detection and removal in high dynamic range images. *17th European Signal Processing Conference*, Glasgow, Scotland (24–28 August 2009), 2240–2244.

Stampoglis, P. (2019). The design of a high speed CMOS image sensor: featuring global shutter, high dynamic range and flexible exposure control in 110nm technology. Master of Science thesis. Delft, The Netherlands: Delft University of Technology.

Stauffer, C. and Grimson, W.E.L. (1999). Adaptive background mixture models for real-time tracking. *IEEE Computer Society Conference on Computer Vision and Pattern Recognition*, Fort Collins, CO, USA (23–25 June 1999), Volume **2**, 246–252. https://doi.org/10.1109/CVPR.1999.784637.

Takayanagi, I., Yoshimura, N., Mori, K., Matsuo, S., Tanaka, S., Abe, H., Yasuda, N., Ishikawa, K., Okura, S., Ohsawa, S., and Otaka, T. (2018). An over 90 dB intra-scene single-exposure dynamic range CMOS image sensor using a 3.0 micrometre triple-gain pixel fabricated in a standard BSI process. *Sensors* **18** (1): Article ID 203, 11 pages. https://doi.org/10.3390/s18010203.

Tatzer, P., Wolf, M., and Panner, T. (2005). Industrial application for inline material sorting using hyperspectral imaging in the NIR range. *Real-Time Imaging* **11** (2): 99–107. https://doi.org/10.1016/j.rti.2005.04.003.

Teke, M., Deveci, H.S., Haliloglu, O., Gurbuz, S.Z., and Sakarya, U. (2013). A short survey of hyperspectral remote sensing applications in agriculture. *6th International Conference on Recent Advances in Space Technologies (RAST)*, Istanbul, Turkey (12–14 June 2013), 171–176. https://doi.org/10.1109/RAST.2013.6581194.

Wang, Q., Zhang, F., and Li, X. (2018). Optimal clustering framework for hyperspectral band selection. *IEEE Transactions on Geoscience and Remote Sensing* **56** (10): 5910–5922. https://doi.org/10.1109/TGRS.2018.2828161.

Zhao, Y., Yuan, Y., and Wang, Q. (2019). Fast spectral clustering for unsupervised hyperspectral image classification. *Remote Sensing* **11** (4): Article ID 399, 21 pages. https://doi.org/10.3390/rs11040399.

8

Histogram Operations

As indicated in Chapter 7, the parameters for many point operations can be derived from the histogram of pixel values of the image. This chapter is divided into two parts: the first considers greyscale histograms and some of their applications, and the second extends this to multidimensional histograms.

8.1 Greyscale Histogram

The histogram of a greyscale image, as shown in Figure 8.1, gives the count of the number of pixels in an image as a function of pixel value:

$$H[i] = \sum_{x,y} \begin{cases} 1, & I[x,y] = i, \\ 0, & \text{otherwise.} \end{cases} \tag{8.1}$$

Closely related is the ***cumulative histogram***, which counts the number of pixels less than or equal to a pixel value:

$$S_0H[i] = \sum_{x,y} \begin{cases} 1, & I[x,y] \leq i \\ 0, & \text{otherwise} \end{cases} = \sum_{n=0}^{i} H[n]. \tag{8.2}$$

The cumulative histogram is always monotonically increasing as is shown in Figure 8.1, with the total number of pixels in the image, $N_P = S_0H[2^B - 1]$.

There are two main steps associated with using histograms for image processing. The first step is to build the histogram, and the second is to extract data from the histogram and use it for processing the image. Building the histogram will be described in Section 8.1.1, whereas the applications are discussed in Sections 8.1.2, 8.1.3, 8.1.4, 8.1.5, 8.1.6.

8.1.1 Building the Histogram

To build the histogram, it is necessary to accumulate the counts for each pixel value. Since each pixel must be visited once, building the histogram is ideally suited to stream processing. An array of counters, one for each pixel value, may be used for this purpose. The input pixel is used to enable the clock of the corresponding counter as shown on the left in Figure 8.2. A minor change to the decoder would also enable the cumulative histogram to be built (Fahmy et al., 2005). To read the array, the output counters need to be indexed using a multiplexer.

A disadvantage of this approach is that the decoding and output multiplexing require quite a lot of logic. The decoder of the input pixel stream may be reused as part of the output multiplexer; however, the number of histogram bins means that a large amount of logic cannot be avoided. This approach is therefore best suited if only a few counters are required (for example after thresholding), or if processing requires the individual counters to be accessed in parallel.

Design for Embedded Image Processing on FPGAs, Second Edition. Donald G. Bailey.

Companion Website: www.wiley.com/go/bailey/designforembeddedimageprocessc2e

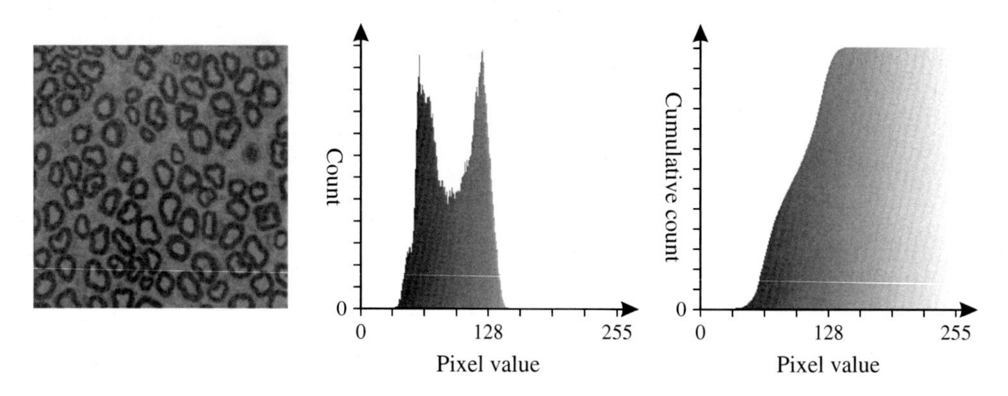

Figure 8.1 Left: an image; centre: its histogram; right: its cumulative histogram.

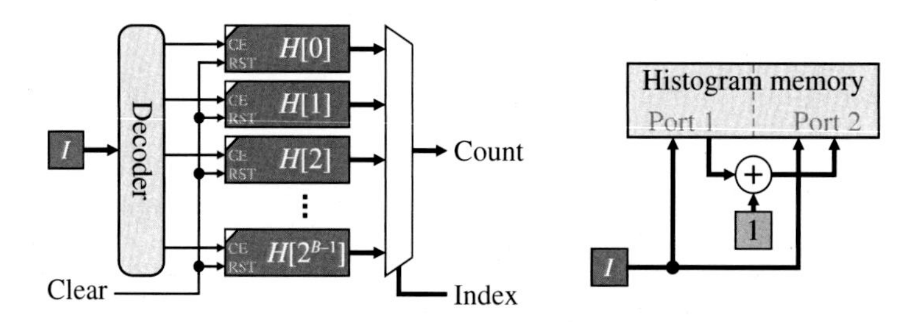

Figure 8.2 Histogram accumulation: left: built from parallel counters; right: built from memory. (CE = clock enable; RST = reset.)

Since each pixel can have only one pixel value, for data streamed at one pixel per clock cycle, only one counter is incremented in any clock cycle. This implies that the accumulator registers may be implemented in memory. Incrementing an accumulator requires reading the associated memory location, adding one, and writing the sum back to memory. This requires a dual-port memory, with one port used to read the memory, and one port to write the result, as shown conceptually on the right in Figure 8.2. Memory is effectively a multiplexed register bank, with the multiplexers implicit in the memory addressing, so significantly fewer logic resources are required.

The memory-based approach does have its disadvantages. All of the accumulators cannot be reset in parallel at the start of building the histogram; it requires sequentially cycling through the memory locations to set them to zero. It also cannot be used directly for building the cumulative histogram; that also requires a cycle through the memory locations after building the histogram:

$$S_0H[i] = \begin{cases} H[0], & i = 0, \\ S_0H[i-1] + H[i], & 0 < i \leq 2^{B-1}. \end{cases} \tag{8.3}$$

The circuit of Figure 8.2 requires the data to be read near the start of the clock cycle, with the incremented value written at the end of the clock cycle. This may be readily achieved with asynchronous memory by delaying the write pulse. However, with synchronous memory, the value is read into a memory output register, with the incremented value being written in the following clock cycle. If the subsequent pixel has the same value, a read-before-write hazard results, with the old value read before the updated value is written. This may be solved with appropriate data forwarding by saving the value being written back to memory in a register (H^+) as shown in Figure 8.3.

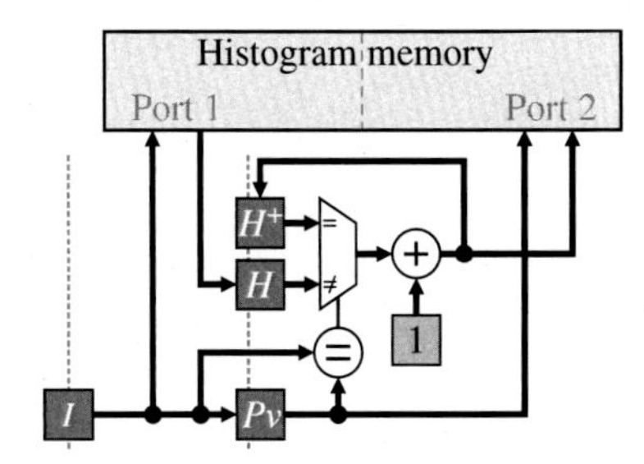

Figure 8.3 Pipelined histogram accumulation with data forwarding to handle read-before-write hazards.

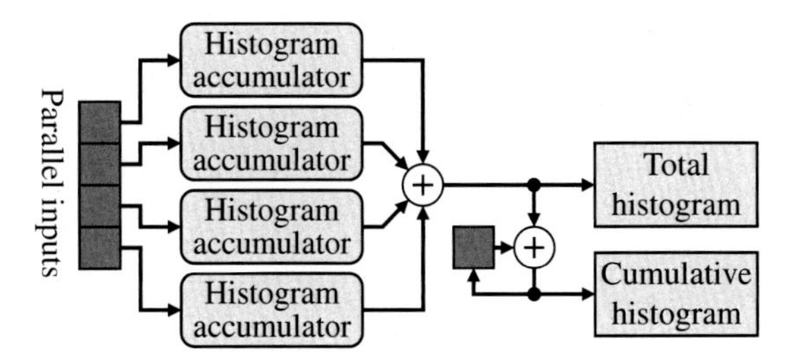

Figure 8.4 Partitioning the accumulation over several parallel processors.

To increase the throughput (either a multiple-pixel stream or through partitioning the image), it is necessary to use several parallel accumulators (Figure 8.4). A final pass through the accumulators is then required to combine the results from each partition to give the histogram for the complete image. If required, the cumulative histogram can also be built during the transfer process.

The circuits that reset the memory and extract data from histogram must connect to the same histogram memory. These require multiplexing the address and data lines to the appropriate processors. These multiplexers will be assumed in the following sections.

8.1.2 Data Gathering

Histograms may be used to gather data from the objects within the image.

8.1.2.1 Area

Consider an image of a single object, and that object can be separated from the background by thresholding. The area of the object is simply given by the number of pixels detected by thresholding. This may be obtained from the histogram of the corresponding binary image; such a histogram accumulator requires only a single counter (compare with Figure 8.2). However, if the threshold level is obtained by processing the histogram, then the area may be found directly from the histogram without necessarily thresholding the input image:

$$\text{Area}(i_{min} \leq I \leq i_{max}) = \sum_{i=i_{min}}^{i_{max}} H[i] = S_0H[i_{max}] - S_0H[i_{min} - 1]. \tag{8.4}$$

This idea may be generalised to measuring the area of many objects within an image. If the image is processed to assign a unique label to the pixels associated with each object (connected component labelling; see Section 13.4), then each histogram bin will be associated with a different label, giving the area of each object.

8.1.2.2 Mean and Variance

While the mean and variance may be calculated directly from the image, they may also be extracted from the histogram. If the statistics are taken from a restricted range of pixel values that is chosen dynamically, it is more efficient to obtain the histogram as an intermediate step.

The mean pixel value of an image is given by

$$\mu_I = \frac{\sum_{x,y} I[x,y]}{\sum_{x,y} 1} = \frac{\sum_i iH[i]}{\sum_i H[i]}, \tag{8.5}$$

with the variance as

$$\sigma_I^2 = \frac{\sum_{x,y} (I[x,y] - \mu_I)^2}{\sum_{x,y} 1} = \frac{\sum_i (i-\mu_I)^2 H[i]}{\sum_i H[i]} = \frac{\sum_i i^2 H[i]}{\sum_i H[i]} - \mu_I^2. \tag{8.6}$$

The last form of Eq. (8.6) allows the variance to be calculated in parallel with the mean, rather than having to calculate the mean first. Often, the mean and variance within a restricted range of pixel values are required rather than over the full range. In this case, the summations of Eqs. (8.5) and (8.6) are performed over that range (i_{min} to i_{max}).

A direct implementation of these summations is shown in Figure 8.5. The *Start* signal selects i_{min} into counter i and resets the summations to zero. On subsequent clock cycles, $H[i]$ is read and added to the appropriate accumulators. If the delay through the two multiplications is too long, then this may be pipelined. When i reaches i_{max}, the final summation is performed, and then divisions are used to calculate the mean and variance. The *Done* output goes high to indicate the completion of the calculation.

While the divisions on the output cannot easily be avoided, it is possible to share the hardware for a single division between the two calculations. The division can also be pipelined if necessary. If the standard deviation is required (rather than the variance), then a square root operation (see Section 5.2.1) must be applied to the variance output.

The multiplications associated with the summations may be eliminated using incremental update. First, consider the summation:

$$\sum_i (K-i)H[i] = K\sum_i H[i] - \sum_i iH[i], \tag{8.7}$$

where K is an arbitrary constant. Therefore,

$$\mu_I = K - \frac{\sum_i (K-i)H[i]}{\sum_i H[i]}. \tag{8.8}$$

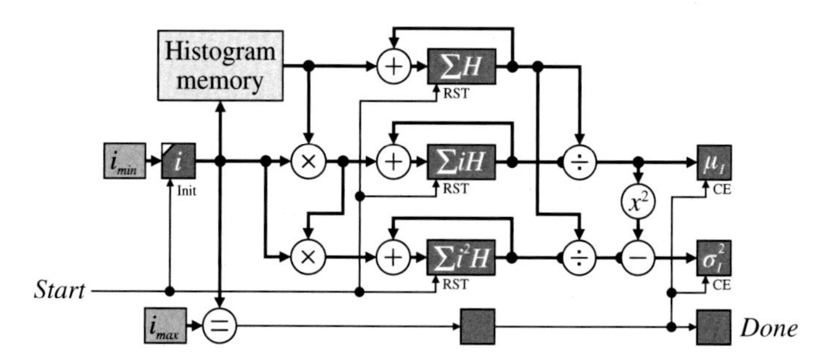

Figure 8.5 Calculating the mean and variance from a range of values within the histogram. (CE = clock enable; Init = initialisation; RST = reset.)

Similarly,

$$\frac{\sum_i (K-i)^2 H[i]}{\sum_i H[i]} = \frac{\sum_i i^2 H[i]}{\sum_i H[i]} - 2K\frac{\sum_i iH[i]}{\sum_i H[i]} + K^2\frac{\sum_i H[i]}{\sum_i H[i]}$$
$$= \sigma_I^2 + \mu_I^2 - 2K\mu_I + K^2$$
$$= \sigma_i^2 + (\mu_I - K)^2. \tag{8.9}$$

Therefore,

$$\sigma_I^2 = \frac{\sum_i (K-i)^2 H[i]}{\sum_i H[i]} - \left(\frac{\sum_i (K-i)H[i]}{\sum_i H[i]}\right)^2. \tag{8.10}$$

Next, consider using incremental calculation for Eqs. (8.7) and (8.9):

$$\sum_i^{K}(K-i)H[i] = \sum_i^{K-1}(K-i)H[i]$$
$$= \sum_i^{K-1}((K-1)-i)H[i] + \sum_i^{K-1}H[i]. \tag{8.11}$$

From

$$\sum_i^{K-1}((K-i)-1)^2 H[i] = \sum_i^{K-1}(K-i)^2 H[i] - 2\sum_i^{K-1}(K-i)H[i] + \sum_i^{K-1}H[i], \tag{8.12}$$

$$\sum_i^{K}(K-i)^2 H[i] = \sum_i^{K-1}(K-i)^2 H[i]$$
$$= \sum_i^{K-1}((K-i)-1)^2 H[i] + 2\sum_i^{K-1}(K-i)H[i] - \sum_i^{K-1}H[i]$$
$$= \sum_i^{K-1}((K-1)-i)^2 H[i] + 2\sum_i^{K-1}((K-1)-i)H[i] + \sum_i^{K-1}H[i]. \tag{8.13}$$

Therefore, starting K with i_{min} and incrementing until i_{max} allows the summations to be performed without any multiplications (the factor of 2 in Eq. (8.13) is a left shift), giving the circuit in Figure 8.6.

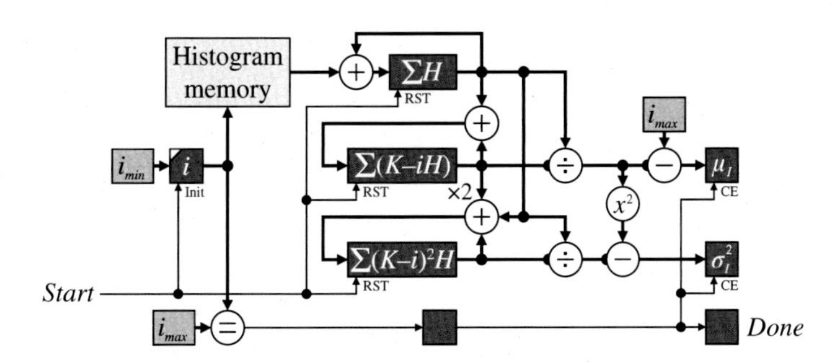

Figure 8.6 Using incremental calculations for the mean and variance. (CE = clock enable; Init = initialisation; RST = reset.)

The circuits of Figures 8.5 and 8.6 are suitable if only a single calculation is required. However, if the mean and variance are required for several different ranges, then the repeated summation can become a time bottleneck. If timing becomes a problem, then this may be overcome by building sum tables that contain cumulative moments of the histogram:

$$S_0H[j] = \sum_{i=0}^{j} H[i], \qquad S_1H[j] = \sum_{i=0}^{j} iH[i], \qquad S_2H[j] = \sum_{i=0}^{j} i^2H[i]. \tag{8.14}$$

These tables require a single pass through the histogram to construct, and after that, the mean and variance of any range may be calculated with two accesses to each table:

$$\mu_I(i_{min} \leq I \leq i_{max}) = \frac{S_1H[i_{max}] - S_1H[i_{min} - 1]}{S_0H[i_{max}] - S_0H[i_{min} - 1]}, \tag{8.15}$$

$$\sigma_I^2(i_{min} \leq I \leq i_{max}) = \frac{S_2H[i_{max}] - S_2H[i_{min} - 1]}{S_0H[i_{max}] - S_0H[i_{min} - 1]} - \left(\frac{S_1H[i_{max}] - S_1H[i_{min} - 1]}{S_0H[i_{max}] - S_0H[i_{min} - 1]}\right)^2. \tag{8.16}$$

Higher order moments (skew and kurtosis) may also be calculated in a similar manner by extension of the above circuit.

8.1.2.3 Range

The statistical range of pixel values within an image is probably best calculated directly from the image as it is streamed. However, if the histogram is already available for other processing, then it is a simple manner to extract the range from the histogram. The minimum pixel value, I_{min}, is the smallest index that has a non-zero count, and the maximum pixel value, I_{max}, is the largest. A common operation using the range is contrast expansion using Eq. (7.2) with

$$a = \frac{2^B - 1}{I_{max} - I_{min}}, \qquad b' = I_{min}. \tag{8.17}$$

8.1.2.4 Median

Calculating the median of an image is easiest using the cumulative histogram of the image. The image median is defined as the bin that contains the 50th percentile. Ideally, an inverse cumulative histogram could be used; this maps from the cumulative count back to the corresponding pixel value. However, this can be derived from the cumulative histogram:

$$med_I = \min_i \left\{ S_0H[i] \geq \frac{N_P}{2} \right\}. \tag{8.18}$$

Rather than scan through the cumulative histogram to find the median, Eq. (8.18) may be solved with a binary search, with each successive iteration giving one bit of the median.

The image median may be generalised to determine the median value within a range of pixel values:

$$med_I(i_{min} \leq I \leq i_{max}) = \min_i \left\{ S_0H[i] \geq \frac{S_0H[i_{max}] + S_0H[i_{min} - 1]}{2} \right\}. \tag{8.19}$$

The architecture of Figure 8.2 may be adapted to calculate the median (Fahmy et al., 2005). The decoder enables the clock of all of the counters less than or equal to the pixel value, building the cumulative histogram. As the outputs of all these counters are available, they can be compared with the median count in parallel, with a priority encoder to find the minimum output that has a '1', as required by Eq. (8.18), and returns the index corresponding to the median.

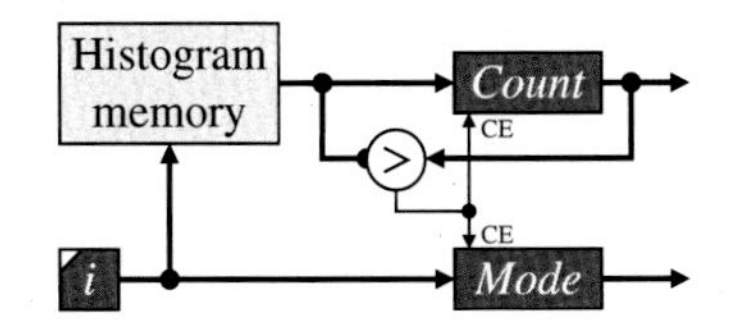

Figure 8.7 Finding the mode of a histogram. (CE = clock enable.)

8.1.2.5 Mode

Another statistic that requires the histogram is the mode. The mode is defined as the index of the highest peak within the histogram. This requires a single scan through the histogram, for example as in Figure 8.7.

There are two limitations to this approach for finding the mode. If the histogram is multi-modal, only the global mode is found. The global mode (with the highest peak) is not necessarily the most significant peak. Finding multiple peaks requires detecting the significant valleys and finding the mode between successive pairs of valleys. Valley detection is described in Section 8.1.5 on threshold selection.

A second limitation is that histograms tend to be noisy. This is clearly seen in Figure 8.1. Consequently, the mode may not necessarily be representative of the true location of the peak. The accuracy of the peak location may be improved through filtering the histogram before finding the mode. A one-dimensional linear smoothing filter may be pipelined between reading the histogram values and finding the maximum. Suitable filters are described in detail in Chapter 9. Note that care must be taken when designing the filter response to avoid bias being introduced if the peak is skewed.

8.1.2.6 Entropy

The entropy of an image provides a measure of the information content within an image. It is given by

$$H_I = -\sum_{i=0}^{2^B-1} p(i)\log_2 p(i), \tag{8.20}$$

where $p(i)$ is the probability of pixel value i in the image. It represents the number of bits per pixel to represent the image when considering each pixel independently (see Section 12.1.6).

The entropy can be calculated directly from the histogram of the image by normalising the counts by the number of pixels to give a probability:

$$\begin{aligned} H_I &= -\sum_{i=0}^{2^B-1} \frac{H[i]}{N_P}\log_2\frac{H[i]}{N_P} \\ &= \log_2 N_P - \frac{1}{N_P}\sum_{i=0}^{2^B-1} H[i]\log_2 H[i]. \end{aligned} \tag{8.21}$$

A priority encoder may be used to find the position of the most significant bit of the count in $H[i]$, enabling a barrel-shift to normalise the count into the range 0.5–1.0. The logarithm of the fraction may be calculated using a lookup table or the shift-and-add method described in Section 5.2.5.

8.1.3 Histogram Equalisation

Histogram equalisation is a contrast enhancement or contrast normalisation technique. It uses the histogram to construct a monotonic mapping that results in a flattened output histogram. From an information theoretic perspective, histogram equalisation attempts to maximise the entropy of an image by making each pixel value in the output equally probable.

The mapping for performing histogram equalisation is the normalised cumulative histogram. Intuitively, if the input bin count is greater than the average, the slope of the mapping will be greater than 1, and conversely if less than the average. Normalisation requires scaling the cumulative histogram by the number of pixels, so that the output maps to the maximum pixel value:

$$Q = \frac{2^B - 1}{N_P} S_0 H[I]. \tag{8.22}$$

Note that the scaling only needs to be applied once to the cumulative histogram to obtain the mapping, rather than to every pixel.

From a contrast enhancement perspective, the assumption behind histogram equalisation is that the peaks of the input histogram contain the information of interest, and the contrast between the peaks is of lesser importance. To make the output histogram flat, it is necessary to reduce the average height of the peaks by spreading the pixel values out. This increases the contrast of these regions. The valleys between the peaks are compressed to raise the average value, effectively reducing the contrast of these features.

This is seen in Figure 8.8 where the regions with counts greater than the average are enhanced at the expense of those below the average. In this image, the effect has been to enhance the contrast within the background material; this is probably not the intended result as it would have been more useful to enhance the contrast between the regions. However, in many applications, histogram equalisation provides a useful contrast enhancement.

Note that the main peak of the output histogram is exactly the same height as that of the input. This is a consequence of using a point operation to perform the mapping, which requires all pixels with one value in the input to map to the same value in the output. The average is obtained by spacing consecutive input values

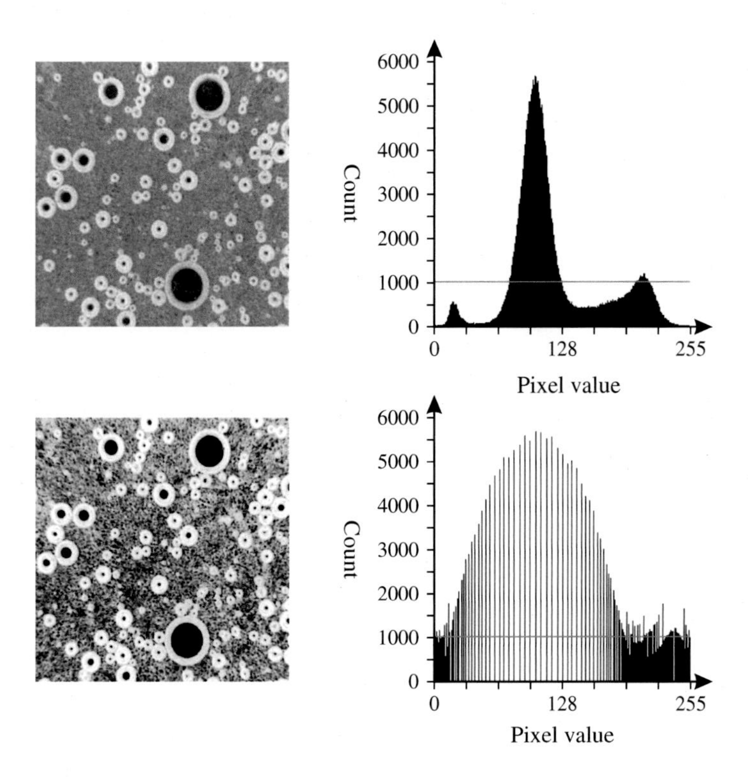

Figure 8.8 Histogram equalisation. Top: input image with its histogram; bottom: the image and histogram after histogram equalisation. The grey line on the histograms shows the average ‘flat’ level.

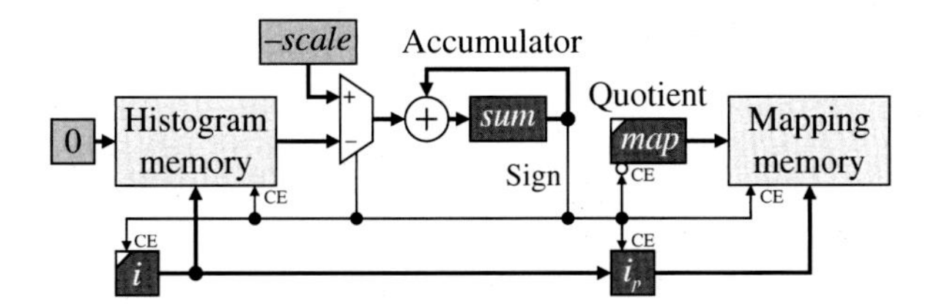

Figure 8.9 Using repeated subtraction to avoid division for calculating the histogram equalisation mapping. (CE = clock enable.)

to more separated values in the output. The opposite occurs for bins with counts below the average. Several consecutive input pixel values may map to the same output pixel.

A flatter output histogram may be obtained using local information to distribute the pixels with an input value with a large count over several output pixel values (Hummel, 1975, 1977) although such operations are technically filters rather than point operations.

The division operation in Eq. (8.22) is the most expensive part of histogram equalisation, although it can readily be pipelined. Alternatively, since N_p is constant, the division can be performed by multiplying by $\frac{1}{N_p}$. The division may also be performed by repeated subtraction. This takes advantage of the fact that the cumulative histogram is monotonic to extend the calculations performed at lower pixel values to higher values. Therefore, the repeated subtractions calculate the mapping at the same time as the total is accumulating. The architecture for this is shown in Figure 8.9. The counters *map* and *i* are initialised to 0, and the *sum* is initialised to $-scale$ (or $-scale/2$ for rounding) where

$$scale = \frac{N_P}{2^B - 1}. \tag{8.23}$$

If the accumulated sum, which represents the current remainder, is negative, then the count from the next histogram bin is accumulated. However, if the accumulated sum is positive, the scale factor is subtracted from the sum, and the quotient incremented. When the index i is incremented, the previous value is also saved as i_p. As a histogram bin is accumulated, the current quotient, *map*, is saved at the index i_p. For 8-bit images, *map* will be incremented at most 255 times, and 257 clock cycles are required to read the input histogram and write the mapping into the lookup table.

Several control issues are associated with histogram equalisation (or any histogram processing):

- The circuitry for building the histogram (for example from Figure 8.3) must be combined with that for calculating the mapping used to perform the histogram equalisation (Figure 8.9). This requires multiplexing the address and data lines of the histogram memory.
- Similarly, the mapping lookup table for performing the histogram equalisation must be shared between the circuit for building the mapping (Figure 8.9) and applying the mapping to the image. Each of these processes only requires a single port, so use of a dual-port memory would simplify the sharing. Otherwise, the map address lines would also require an appropriate multiplexer, with additional control required to provide the memory read or write signal.
- It is necessary to reset the histogram accumulators to zero before the next frame. This may be accomplished while building the map, using the histogram memory write port, as shown in Figure 8.9.
- The registers for building the map (in Figure 8.9) may be initialised to appropriate values while the histogram is being accumulated.
- Constructing the histogram equalisation map requires data from the complete image, so it is necessary to buffer the image before the mapping is applied. In many instances, the image contents (and consequently the mapping) change relatively slowly. In this case, the mapping built from the previous frame can be applied while building the histogram for the current frame (McCollum et al., 1988). Although the histogram equalisation might not be quite correct, there is a significant savings in both latency and resources (the frame buffer is not required). These two options are shown in Figure 8.10.

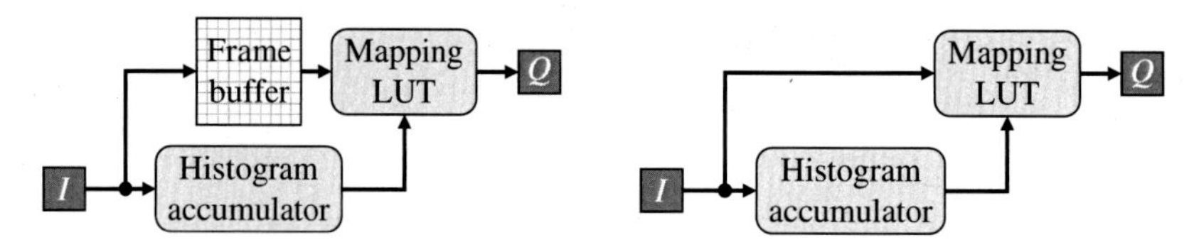

Figure 8.10 Using the histogram for histogram equalisation. Left: buffering the frame during histogram accumulation; right: directly applying table from the last image to the current image.

8.1.3.1 Histogram Equalisation Variants

Histogram equalisation has several limitations, with several modifications devised to overcome them.

The average pixel value within the image can change significantly with histogram equalisation. Several variations designed to improve this (Kaur et al., 2011). The simplest is perhaps bi-histogram equalisation, which divides the histogram into two partitions by the mean and performs histogram equalisation on each partition. Alternatively, the histogram can be divided into multiple sections based on inflection points, and histogram equalisation performed on each section (Chiang et al., 2013).

A second issue is the over-enhancement at some pixel values, usually around peaks in the histogram. Since the height of a peak governs the slope of the cumulative histogram, the contrast can be increased significantly around such peaks. Such over-enhancement may be avoided by limiting the slope of the mapping, giving contrast-limited histogram equalisation. This is achieved by clipping the histogram and redistributing the clipped counts to the other bins. Since this redistribution can result in further bins exceeding the limit, this redistribution procedure is usually recursive. Schatz (2013) approximates this by constraining the redistribution to a single pass.

The mapping for histogram equalisation is determined by the global statistics of the image. When the area of interest occupies only a small proportion of the image, the enhancement will be dominated by the statistics of the background, which are often of little relevance to the area of interest. To avoid this, histogram equalisation can be applied locally, within a window, effectively making it a filter. Adaptive histogram equalisation is discussed in Section 9.5.

8.1.3.2 Successive Mean Quantisation Transform

Related to histogram equalisation is the successive mean quantisation transform (Nilsson et al., 2005b). This works using a series of means to derive successive threshold levels that define successive bits of the output pixel value. Initially, the whole range of pixel values is defined as a single partition. The mean value of the pixels within a partition is used to split the partition into two at the next level, with the corresponding output bit defined by the partition to which a pixel is assigned. This partitioning process is illustrated in Figure 8.11. Sum tables may be used to calculate successive means efficiently using Eq. (8.15).

It has been demonstrated that in many applications, the successive mean quantisation transform provides a subjectively better enhancement than histogram equalisation, because it preserves the gross structure of the histogram and does not over-enhance the image (Nilsson et al., 2005a,c). In the context of the successive mean quantisation transform, histogram equalisation corresponds to successively partitioning using the median rather than the mean.

8.1.3.3 Histogram Shaping

The procedure for histogram equalisation may readily be adapted to perform other histogram transformations, for example hyperbolisation (Frei, 1977). It is even possible to derive a target histogram from one image and apply a mapping to another image such that the histogram of the output image approximates that of the target. The basic procedure for this is illustrated graphically in Figure 8.12. To derive a monotonic mapping, it is

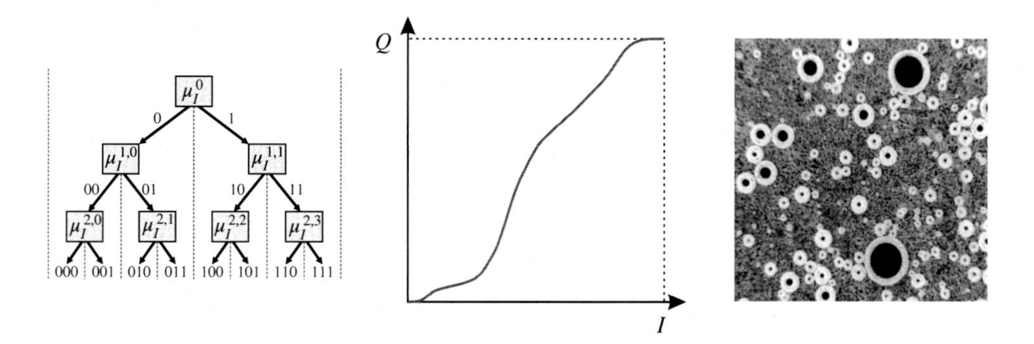

Figure 8.11 Successive mean quantisation transform. Left: determining the output code or pixel value. Source: Adapted from (Nilsson et al., 2005b, Fig.2, p. IV-430) © 2005 IEEE, reproduced with permission. Centre: the resultant mapping; right: applying the mapping to the input image.

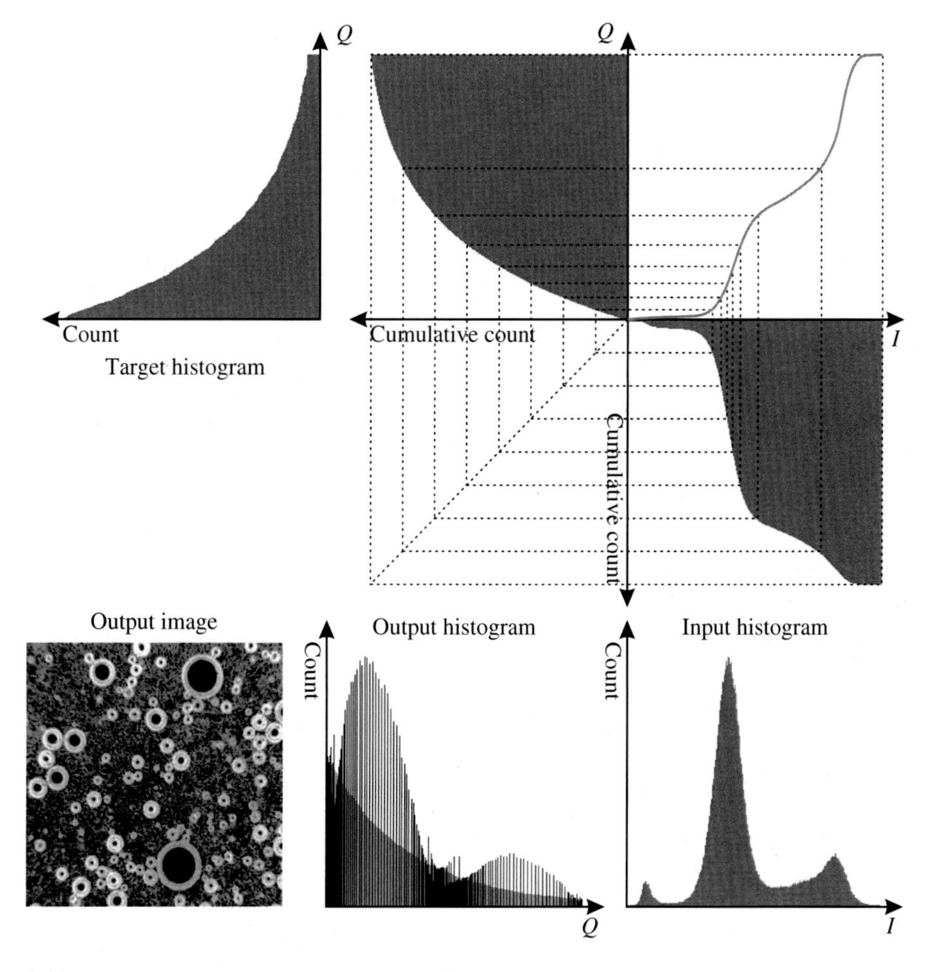

Figure 8.12 Determining the mapping to produce an arbitrary output histogram shape by matching cumulative histograms. Top: target histogram and cumulative histogram; right: input histogram and cumulative histogram; top-right: building the mapping; bottom: output image and output histogram (with target histogram in the background).

necessary for all of the pixels less than or equal to a pixel value in the input image to map to values that are less than or equal to the corresponding output value. This may be achieved by mapping the cumulative histogram of the input to the cumulative histogram of the target. As with histogram equalisation, the output histogram approximates the target on average, as shown in the bottom of Figure 8.12.

The circuit of Figure 8.9 may be adapted to calculate this mapping. Rather than subtract off the constant *scale*, the value to be subtracted may be obtained from the target histogram, $H_T[map]$ (assuming that both images have the same area).

8.1.4 Automatic Exposure

Related to contrast enhancement is automatic exposure. This involves using data obtained from the captured image to modify the image capture process to optimise the resultant images. In sensors with electronic shuttering, this may require adjusting the control signals that determine the exposure time. It may also involve automatically adjusting the aperture or even controlling the intensity of the light source.

The histogram of the captured image can provide useful data for gauging the quality of the exposure. If the maximum pixel value within the image is significantly less than $2^B - 1$, then the image is not making the best use of the available dynamic range. This may be improved by increasing the exposure. On the other hand, if the image contains a large number of pixels saturated at 2^{B-1}, then important detail may be lost. This may be corrected by reducing the exposure.

Although the maximum pixel value may be determined without needing a histogram, there are two limitations of using the maximum pixel value to assess exposure. First, the maximum value is more likely to be an outlier as a result of noise. In many images, the maximum value may result from specular reflections or other highlights that are not significant from the perspective of image content. Second, the maximum pixel value may actually reach $2^B - 1$ without being saturated.

An alternative approach to maximising the exposure without significant saturation is to use the histogram to find the pixel value of the 99th percentile (Bailey et al., 2004) (or even higher, depending on the expected image content). If the histogram is flat, the 99th percentile should have a value of about 252 for an 8-bit image, so the optimisation criterion would be to adjust the exposure to achieve this value. However, most histograms are non-uniform, and in many applications there may be variability from one frame to the next that must be tolerated.

This may be managed by having a target acceptable range for the given percentile. Consider the image in Figure 8.13. The histogram clearly shows that some pixels are saturated (from the peak at 255). These are the specular reflections and highlights, and no information of importance is lost if these are saturated. However,

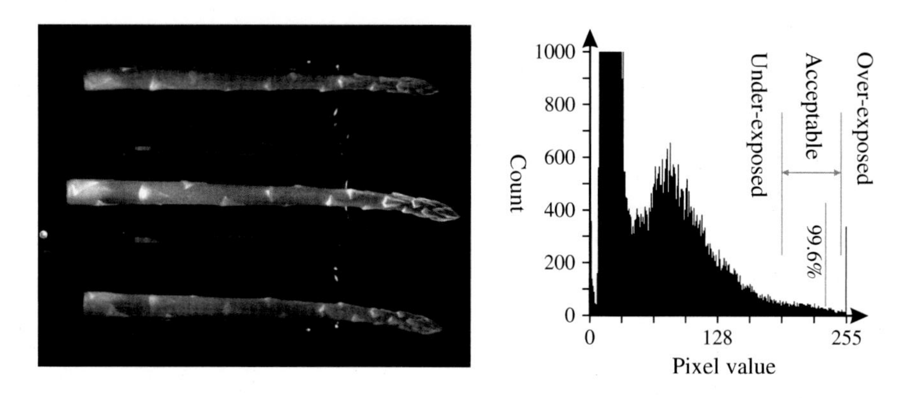

Figure 8.13 Using the histogram for exposure. Left: example image; right: histogram with its 99.6th percentile shown, along with the acceptable bounds. The large dead-band (acceptable range) allows for variation from one image to the next.

the 99.6th percentile (allowing for the equivalent of two rows worth of pixels to be saturated) is within the allowed band of 192–250, indicating that the exposure for this image is acceptable. In this application, the large dead-band allows for the considerable variability in the produce from one image to the next, while ensuring that an adequate exposure is maintained.

When using the cumulative histogram to determine the exposure, it is faster to accumulate the histogram counts starting from the maximum pixel value and counting down, rather than counting up from 0. Alternatively, rather than building the whole histogram, the two cumulative histogram bins corresponding to the upper and lower limits of the dead band may be determined directly by counting the pixels which exceed those two limits. Note that changing the camera exposure will not affect the current image. It may also be best to adjust the exposure incrementally over several images rather than in a single step, especially if the object being imaged changes from one frame to the next.

To balance between over- and underexposure, Popadic et al. (2017) aimed for an exposure such that the count of the 16 (out of 256) brightest histogram bins was equal to the 16 darkest histogram bins. Alternative approaches are to set the mean value to mid-grey (2^{B-1}) (Ning et al., 2015) or to maximise the entropy (hence information content) within the image (Ning et al., 2015; Rahman et al., 2011). Note that maximising the entropy is harder to achieve, especially if the image is changing from frame to frame.

8.1.5 Threshold Selection

The purpose of thresholding is to segment the image into two (or more) classes based on pixel value. Where the pixel values of each class are well separated, the resulting histogram of pixel values is multi-modal. This implies that the distributions of each of the classes, and hence the best places to threshold the image, may be determined by analysing the histogram.

The most common case is separating objects from a background, where there are two classes: object and background (see Figure 8.14). This will be the focus of this section. The resulting histogram is often bimodal, with peaks representing the distributions of object and background pixel values. If the peaks are clearly separated and do not overlap, choosing an appropriate threshold between the peaks is relatively trivial. In most cases, however, the peaks are broad, with significant overlap between the distributions. The 'best' threshold then depends on the segmentation criterion, and a clean image may not even be possible without significant preprocessing.

The image may be filtered to improve the histogram obtained. Noise smoothing may result in narrower peaks, making them more distinct and easier to detect and separate. Applying an edge enhancement filter prior to accumulating the histogram will reduce the number of pixel values in the valley between the peaks, making the result less sensitive to the actual threshold level used (Bailey, 1995).

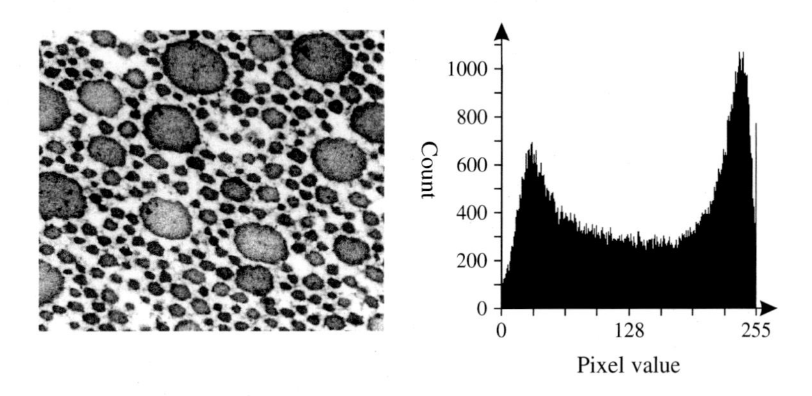

Figure 8.14 An image of objects against a background and its corresponding histogram.

If the number of either object or background pixels is significantly smaller than the other, the valley between the peaks can be indistinct. Obtaining the histogram only of pixels near edges will result in approximately equal numbers of object and background pixels (Weszka et al., 1974), making threshold selection a little easier.

Only global threshold selection based on data extracted from the histogram is considered in this section. Adaptive thresholding chooses a different threshold at each location based on local statistics, effectively making it a form of filter. Adaptive thresholding will be examined in Section 9.3.6.

Many of the threshold selection techniques described in this section would be easier to implement in software, using an embedded processor, than directly in hardware. The timing is not particularly critical because the threshold could be calculated during the vertical blanking period between frames.

8.1.5.1 Simple Parametric Approaches

If the proportion of object pixels is known in advance, then the corresponding threshold level may be determined from the inverse cumulative histogram (Doyle, 1962).

The threshold can be determined by scanning through histogram accumulating the counts until the desired proportion is reached:

$$T = \max_t \{S_0 H[t] < N_{obj}\} = \max_t \left\{ \sum_{i=0}^{t} H[i] < N_{obj} \right\}. \tag{8.24}$$

Alternatively, if the cumulative histogram is available, Eq. (8.24) can be solved using a binary search.

Usually, however, the number of object pixels is not known in advance but is determined as a result of thresholding. In this case, a simple approach is to set the threshold to a preset number of standard deviations from the mean:

$$T = \mu_I + \alpha\sigma_I, \tag{8.25}$$

where α is chosen empirically based on the knowledge of the image being thresholded.

8.1.5.2 Valley Finding Approaches

A good threshold is generally in the valley between the peaks. In general, however, both the peaks and the valley between them are noisy, so simply choosing the deepest valley can result in considerable variation from one image to the next.

One approach to this problem is to smooth the histogram to make the valley between the peaks clearer. This was first proposed by Prewitt and Mendelsohn (1966). The histogram is repeatedly filtered with a narrow filter until there are only two local maxima with a single local minimum in the valley in between. The threshold is this local minimum, which will be the only pixel value in the histogram that satisfies

$$H[T-1] > H[T] \le H[T+1]. \tag{8.26}$$

A disadvantage of this approach is the repeated filtering required to obtain the necessary smoothness.

The global valley approach of Davies (2008) identifies the valleys by the height difference relative to the peaks on either side. For a given pixel value, t, the maximum height of the histogram peaks on left and right sides of t are, respectively:

$$H_L[t] = \max_{i<t} H[i], \quad H_R[t] = \max_{i>t} H[i]. \tag{8.27}$$

The pixel value of the deepest valley (the threshold level) is then given by

$$T = \arg\max_t \{ \sqrt{\text{clip}(H_L[t] - H[t])\text{clip}(H_R[t] - H[t])} \}, \tag{8.28}$$

where

$$\text{clip}(x) = \begin{cases} 0, & x \le 0, \\ x, & x > 0. \end{cases} \tag{8.29}$$

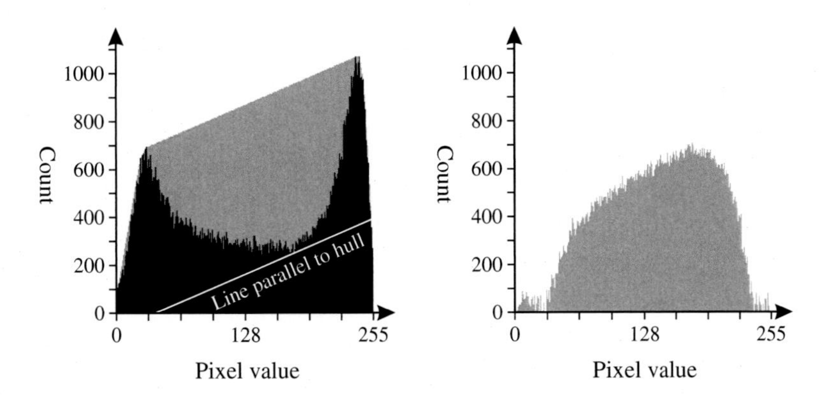

Figure 8.15 Convex hull of a histogram. Left: the convex hull superimposed on the histogram (note that the artificial peaks due to clipping at 0 and 255 have been suppressed). The line parallel to the top of the hull shows how the threshold is biased towards the higher peak; right: the difference between the hull and the histogram.

The geometric mean in Eq. (8.28) is better than the arithmetic mean because it prevents pedestals at the ends of the distribution (Davies, 2008). Only two passes are required through the histogram to evaluate the criterion function: a reverse pass to incrementally calculate H_R and a forward pass to incrementally calculate both H_L and Eq. (8.28). Note that in finding the threshold, the square root operation of Eq. (8.28) may be omitted.

Rosenfeld and de la Torre (1983) proposed an approach to finding the valley by finding the deepest pixel below the convex hull of the histogram, as illustrated in Figure 8.15. The largest difference will correspond to the pixel value in the valley that has a tangent parallel with the hull. The slope of the hull biases the threshold location towards the higher peak, which is desirable, as it enables even a small bump on the side of a unimodal histogram to be detected.

The convex hull of the histogram is the smallest extension that has no concavities. The convex hull may be represented by the set of convex vertices, which correspond to the set of counts in the histogram that are unaffected by the extension process. The slope of the convex hull between the vertices decreases monotonically with increasing pixel value. This property is used to find the set of convex vertices. Consider three consecutive candidate vertices: $(i_1, H[i_1])$, $(i_2, H[i_2])$, and $(i_3, H[i_3])$. If

$$\frac{H[i_2] - H[i_1]}{i_2 - i_1} < \frac{H[i_3] - H[i_2]}{i_3 - i_2} \tag{8.30}$$

or, equivalently,

$$(H[i_2] - H[i_1])(i_3 - i_2) < (H[i_3] - H[i_2])(i_2 - i_1), \tag{8.31}$$

then the vertex at i_2 is concave and is eliminated. Finding the convex vertices requires at most 2^{B+1} steps in a single scan through the histogram (for each pixel value added as a vertex, prior entries are eliminated until the set of vertices is convex; each pixel value is added once and removed at most once from the set). A final pass is used to reconstruct the convex hull from the vertices, subtract off the corresponding histogram count, and find the maximum (see the right panel in Figure 8.15).

8.1.5.3 Modelling Approaches

An alternative approach is to model the shape of the histogram and derive the threshold from the model.

A common assumption is that the shape of each of the modes is a Gaussian distribution, modelling the histogram as a sum of Gaussians (a Gaussian mixture model):

$$H[i] = \sum_j A_j e^{-\frac{(i-\mu_j)^2}{2\sigma_j^2}}. \tag{8.32}$$

In the case where the histogram is bimodal, the histogram is separated into two parts, each modelled with a Gaussian distribution. The minimum error will correspond to the crossover point between the two distributions, which is given by (Kittler and Illingworth, 1986)

$$T = \arg\min_t(p_1 \ln\sigma_1 + p_2 \ln\sigma_2 - p_1 \ln p_1 - p_2 \ln p_2), \tag{8.33}$$

where p_1 and p_2 are the proportions of pixels on each side of the threshold (in each of the two peaks), and σ_1 and σ_2 are their respective standard deviations:

$$p_1(t) = \frac{\sum_{i=0}^{t} H[i]}{N_P} = \frac{S_0H[t]}{N_P}, \quad p_2(t) = 1 - p_1(t) \tag{8.34}$$

and

$$\sigma_1(t) = \sigma_I(0 \le I \le t), \quad \sigma_2(t) = \sigma_I(t < I \le 2^B - 1). \tag{8.35}$$

Equation (8.33) may be evaluated with a single pass through the cumulative histogram, although several clock cycles may be needed for each pixel value to evaluate the logarithms. Note that the division by N_P is a constant scale factor and can be left out. Using the variance instead of the standard deviation avoids the need for square roots:

$$\ln\sigma = \frac{1}{2}\ln\sigma^2. \tag{8.36}$$

Rather than fitting to the histogram, an alternative is to make the distributions on each side of the threshold as compact as possible. This leads to the popular approach by Otsu (1979) of minimising the intra-class variance or, equivalently, maximising the interclass variance:

$$T = \arg\min_t\{p_1\sigma_1^2 + p_2\sigma_2^2\} = \arg\max_t\{p_1p_2(\mu_2 - \mu_1)^2\}, \tag{8.37}$$

where μ_1 and μ_2 are the mean pixel values on each side of the threshold:

$$\mu_1(t) = \mu_I(0 \le I \le t), \quad \mu_2(t) = \mu_I(t < I \le 2^B - 1). \tag{8.38}$$

Since the criterion function of Eq. (8.37) can have multiple local extrema (Lee and Park, 1990), iterative methods may converge to a local optimum rather than the global optimum. Therefore, it is necessary to search by scanning through the histogram.

The interclass variance criterion function may be evaluated directly from the cumulative histogram and summed first moment:

$$p_1p_2(\mu_2 - \mu_1)^2 = \frac{S_0H[t]}{N_P}\frac{(N_P - S_0H[t])}{N_P}\left(\frac{S_1H[2^B-1] - S_1H[t]}{N_P - S_0H[t]} - \frac{S_1H[t]}{S_0H[t]}\right)^2. \tag{8.39}$$

For maximisation, the constant factor N_P^2 in the denominator of the first two terms may be left out, and Eq. (8.39) rearranged to reduce the number of division operations:

$$p_1p_2(\mu_2 - \mu_1)^2 \propto \frac{(S_0H[t]S_1H[2^B-1] - S_1H[t]N_P)^2}{S_0H[t](N_P - S_0H[t])}. \tag{8.40}$$

If the sum of all pixels is known ($S_1H[2^B-1] = \sum I(x,y)$ and can be accumulated while building the histogram), then Eq. (8.40) can be evaluated in a single pass during the incremental accumulation of the sum tables (Eq. (8.14)).

Otsu's method works best with a bimodal distribution, with a clear distinct valley between the modes (Lee and Park, 1990). Performance can deteriorate when there is a significant imbalance between the number of object and background pixels, or when the object and background are not well separated, or when there is significant noise (Lee and Park, 1990).

An alternative approach that gives similar results to Otsu's method for many images is to iteratively refine the threshold to midway between the means of the object and background classes (Ridler and Calvard, 1978; Trussell, 1979):

$$T_k = \frac{\mu_1+\mu_2}{2} = \frac{\mu_I(0\le i\le T_{k-1}) + \mu_I(T_{k-1}<i\le 2^B-1)}{2}$$
$$= \frac{1}{2}\left(\frac{S_1H[T_{k-1}]}{S_0H[T_{k-1}]} + \frac{S_1H[2^B-1]-S_1H[T_{k-1}]}{N_P - S_0H[T_{k-1}]}\right), \tag{8.41}$$

with the initial threshold given from the global mean

$$T_0 = \mu_I. \tag{8.42}$$

Either a single iteration can be used ($T = T_1$), or the iteration of Eq. (8.41) is repeated until it converges ($T_k = T_{k-1}$), although this generally occurs within four iterations (Trussell, 1979). Note that the resultant threshold level is dependent on the initial value. The initialisation of Eq. (8.42) works well if there are approximately equal numbers of object and background pixels, or the object and background are well separated. An alternative initialisation if the distribution is very unbalanced is the mid-range of the distribution:

$$T_0 = \frac{\min(I)+\max(I)}{2}. \tag{8.43}$$

The attractiveness of this iterative method is its simplicity and speed, and with good contrast images it gives good results. This method has been implemented on an field-programmable gate array (FPGA) by Anghelescu (2021).

Another approach to thresholding is to use the total entropy of two regions of the histogram as the criterion function (Kapura et al., 1985). The threshold is then chosen to maximise the information between the object and background distributions:

$$T = \arg\max_t \left\{ -\sum_{i=0}^{t} \frac{H[i]}{S_0H[t]} \ln \frac{H[i]}{S_0H[t]} - \sum_{i=t+1}^{2^B-1} \frac{H[i]}{N_P - S_0H[t]} \ln \frac{H[i]}{N_P - S_0H[t]} \right\}$$
$$= \arg\max_t \left\{ \ln(S_0H[t](N_P - S_0H[t])) - \frac{\sum_{i=0}^{t} H[i]\ln H[i]}{S_0H[t]} - \frac{\sum_{i=t+1}^{2^B-1} H[i]\ln H[i]}{N_P - S_0H[t]} \right\}. \tag{8.44}$$

Finding an optimum threshold using this criterion function can be assisted by building a sum table of

$$S_{ln}H[j] = \sum_{i=0}^{j} H[i]\ln H[i], \tag{8.45}$$

which requires a single pass through the histogram. This avoids having to perform the summation in Eq. (8.44) for each candidate threshold level, t, enabling the optimum to be found in a second pass. Maximising the entropy has a tendency to push the threshold towards the middle of the distribution, so this method works best when there is a clear distinction between the two distributions.

8.1.6 Histogram Similarity

Histogram similarity is one method of object classification. The intensity histograms of a set of candidate model objects are maintained in a database. The histogram of the test object is obtained and compared with the histograms of the model objects, with the object classified based on which model has the most similar histogram. Unlike most other methods of object classification, the object does not need to be completely separated from the background, as long as it dominates the content of the image.

The histogram may require some normalisation prior to matching (usually limited to contrast stretching) to account for varying illumination intensity. Normalisation such as histogram equalisation, or shaping, is obviously not appropriate. Often, the resolution of the histogram is reduced (the bin size is increased) by combining adjacent bins. This reduces the sensitivity to small changes in illumination and problems resulting from gaps in the histogram that result from normalisation. It also helps to speed the search for a match.

Let all the model histograms be normalised to correspond to the same area image, and let $M_j[i]$ be the histogram of the jth model. The similarity between the test histogram, $H[i]$, and the model can be found from their intersection (Swain and Ballard, 1991)

$$Match_j = \sum_i \min(H[i], M_j[i]). \tag{8.46}$$

This measure will be a maximum when the histograms are identical and will decrease as the histograms become more different. Therefore, the input image is classified as the object that has the largest match.

All spatial relationships between the pixel values are lost in the process of accumulating the histogram. Consequently, many quite different images can have the same histogram. Classification based on histogram similarity will therefore only be effective when the histograms of all of the models are distinctly different. However, the method has the advantage that the histogram has a considerably smaller volume of data than that of the whole image, significantly reducing comparison times. This is helped further by reducing the number of bins within the histograms. Where speed is important, the test histogram may be compared with a number of model histograms in parallel.

To reduce problems with changes in illumination, Geninatti and Boemo (2019) used the cross-correlation between histograms to detect similarity between successive frames in a video:

$$S_t(k) = \frac{\sum_i H_t(i)H_{t-1}(i+k)}{\sqrt{\sum_i H_t^2(i)}\sqrt{\sum_i H_{t-1}^2(i+k)}}. \tag{8.47}$$

The maximum correlation, $\max_k(s_t(k))$, is then a measure of similarity; a low value indicates a scene change.

8.2 Multidimensional Histograms

Just as pixel-value histograms are useful for gathering data from greyscale images, multidimensional histograms can be used to accumulate data from colour or other vector valued images. The simplest approach is simply to accumulate a 1-D histogram of each channel. While this is useful for some purposes, for example gathering the data required for colour balancing, it cannot capture the relationships between channels that give each pixel its colour. This requires a true multidimensional histogram.

A multidimensional histogram has one axis or dimension for each channel in the input image. For example, the histogram of an RGB image would require three dimensions. The histogram gives a count of the pixels within the image that have a particular vector value. For an RGB image, Eq. (8.1) would be extended to give

$$H[r,g,b] = \sum_{x,y} \begin{cases} 1, & \mathbf{I}_{RGB}[x,y] = \{r,g,b\}, \\ 0, & \text{otherwise.} \end{cases} \tag{8.48}$$

Data is accumulated in exactly the same way as for a greyscale histogram. The larger address space of a multidimensional histogram requires a significantly larger memory, which may not necessarily fit within the FPGA. The size of the histogram memory can become prohibitively large, especially for three and higher dimensions. This problem may be overcome by reducing the number of bins by applying a coarse quantisation to each component.

Indexing into the multidimensional array is performed in a similar manner to that used for software:

$$H[x, y, z] = H[n_x n_y z + n_x y + x] = H[(zn_y + y) \times n_x + x], \tag{8.49}$$

where n_x and n_y are the number of bins used for the x and y components, respectively. Indexing may be simplified if the range of each component is a power of 2, as the memory address may then be formed by concatenating the vector components. The quantisation to accomplish this may be readily implemented by truncating the least significant bits of each vector component.

Two histogram memory accesses are required for each pixel: one to read the previous count, and the other to write the updated count. This may be accomplished either with dual-port memory, in the same manner as 1-D histograms (see Figure 8.3), or by running the accumulator memory at twice the pixel clock frequency.

One limitation with using multidimensional histograms is the time required to both initialise them and to extract the data. Throughput issues may be overcome using pipelining. Bandwidth issues may require multiple parallel banks to initialise and process the histograms in a realistic time. Once the histogram data has been collected, the histograms can be used and processed as images in their own right if necessary.

8.2.1 Triangular Arrays

When processing colour images, intensity independence is often desired. Removing the intensity through colour space conversion gives two chrominance dimensions. If chromaticity is used (either x–y or r–g), the resulting 2-D histogram accumulator is triangular, almost halving its size. To achieve these savings, it is necessary to set the number of quantisation bins per channel to $n = 2^k - 1$, giving

$$N_E = \frac{1}{2}n(n+1) = 2^{2k-1} - 2^{k-1} \tag{8.50}$$

bins within the accumulator histogram. The non-power-of-2 size for n requires scaling the input before truncating the least significant bits, which may be implemented by a single subtraction:

$$x = \left\lfloor x_{in}\frac{n}{2^b} \right\rfloor = \left\lfloor x_{in}\frac{2^k - 1}{2^b} \right\rfloor = \left\lfloor \frac{x_{in} - 2^{-k}x_{in}}{2^{b-k}} \right\rfloor . \tag{8.51}$$

Two approaches may be used to pack the data into the smaller array. The first successively shifts each row back in memory to remove the unused entries:

$$\begin{aligned} H[x, y] &= H\left[x + ny - \frac{1}{2}y(y-1)\right] \\ &= H\left[x + 2^k y - \frac{1}{2}y(y+1)\right] . \end{aligned} \tag{8.52}$$

The multiplication may be avoided using a small lookup table on the y address, as shown on the left in Figure 8.16.

The second approach is to maintain the spacing but fold the upper addresses into the unused space at the end of each row:

$$\begin{aligned} H[x, y] &= \begin{cases} H[x + (n+1)y], & y \leq \frac{n}{2} \\ H[(n-x) + (n+1)(n-y)], & y > \frac{n}{2} \end{cases} \\ &= \begin{cases} H[x + 2^k y], & y_{MSB} = \text{‘0’}, \\ H[\overline{x} + 2^k \overline{y}], & y_{MSB} = \text{‘1’}. \end{cases} \end{aligned} \tag{8.53}$$

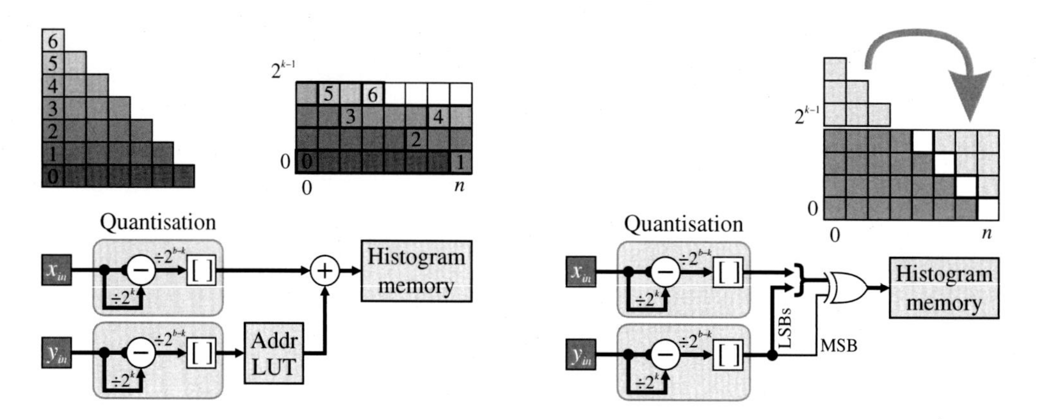

Figure 8.16 Packed addressing schemes for a triangular 2-D histogram with $n = 2^k - 1$. Left: packing the rows sequentially. The numbered memory locations represent the addresses stored in the Addr LUT (lookup table). Right: folding the upper addresses into the unused space. (MSB = most significant bit.)

Note that $(n - x)$ and $(n - y)$ are the one's complements of x and y, respectively ($\overline{x}$ and $\overline{y}$), enabling them to be implemented with an exclusive-OR gate as shown in Figure 8.16.

8.2.2 Multidimensional Statistics

8.2.2.1 Mean and Covariance

The scalar mean and variance extend to vector valued quantities as the mean vector and covariance matrix:

$$\boldsymbol{\mu}_\mathbf{I} = \frac{\sum\limits_{x,y} \mathbf{I}[x,y]}{\sum\limits_{x,y} 1} = \begin{bmatrix} \mu_{I_1} \\ \mu_{I_2} \\ \mu_{I_3} \end{bmatrix}, \tag{8.54}$$

$$\boldsymbol{\Sigma}_\mathbf{I} = \frac{\sum\limits_{x,y} (\mathbf{I} - \boldsymbol{\mu}_\mathbf{I})(\mathbf{I} - \boldsymbol{\mu}_\mathbf{I})^\mathsf{T}}{\sum\limits_{x,y} 1} = \begin{bmatrix} \sigma^2_{I_1} & \sigma^2_{I_1,I_2} & \sigma^2_{I_1,I_3} \\ \sigma^2_{I_1,I_2} & \sigma^2_{I_2} & \sigma^2_{I_2,I_3} \\ \sigma^2_{I_1,I_3} & \sigma^2_{I_2,I_3} & \sigma^2_{I_3} \end{bmatrix}, \tag{8.55}$$

where the superscript $^\mathsf{T}$ denotes the vector transpose (for complex data, the complex conjugate of the transpose is used). Each covariance term is therefore

$$\sigma^2_{I_i,I_j} = \frac{\sum\limits_{x,y} (I_i - \mu_{I_i})(I_j - \mu_{I_j})}{\sum\limits_{x,y} 1} = \frac{\sum\limits_{x,y} I_i I_j}{\sum\limits_{x,y} 1} - \mu_{I_i}\mu_{I_j}. \tag{8.56}$$

The diagonal elements of $\boldsymbol{\Sigma}_\mathbf{I}$ correspond to the variances of the individual components. The off-diagonal elements measure how the corresponding components vary with respect to each other, which reflects the correlation between that pair of components. The covariance matrix characterises the spread of distribution within the multidimensional space. If the distribution is Gaussian, then the shape will be ellipsoidal, with the orientation and extent of the ellipsoid defined by the covariance matrix.

In one dimension, the distance of a point, x, from the centre of the distribution may be normalised by the number of standard deviations:

$$d = \frac{x - \mu}{\sigma} = \sqrt{(x - \mu)^2(\sigma^2)^{-1}}. \tag{8.57}$$

The Mahalanobis distance (Mahalanobis, 1936) generalises this to measure the distance of a vector, $\mathbf{x}$, from the centre of a distribution, $\mathbf{X}$:

$$d_{Mahalanobis} = \sqrt{(\mathbf{x} - \boldsymbol{\mu}_{\mathbf{X}})^{\top}\boldsymbol{\Sigma}_{\mathbf{X}}^{-1}(\mathbf{x} - \boldsymbol{\mu}_{\mathbf{X}})}. \tag{8.58}$$

When calculating global statistics for the image, the data for the mean and covariance may be accumulated directly on the input as it is streamed in. For a colour image (with three components or dimensions), the mean therefore requires three accumulators, one for each dimension. If the denominator is not fixed (for example extracting statistics from a region of the image), an additional accumulator is required for the denominator. From the symmetry in the covariance matrix, only six accumulators are required for the numerator of Eq. (8.56).

Alternatively, a multidimensional histogram may be used as an intermediary, in a similar manner to that described for 1-D histograms in Section 8.1.2.2. While using a histogram as an intermediate step is efficient in 1-D, in two and higher dimensions their utility becomes marginal as the histogram can have more entries than the original image. While this may be overcome using coarse quantisation, this will also reduce the accuracy of the statistics derived from the histograms. Therefore, gathering data from a multidimensional histogram is usually only practical if many separate calculations must be performed for different regions within the multidimensional data space (for example to obtain the mean and covariance for each of a set of colour classes).

8.2.2.2 Principal Components Analysis

Principal components analysis (PCA) performs a rotation of the multidimensional space to align the axes with the distribution. This alignment is defined by the covariance matrix being diagonal (the off-diagonal covariances are all zero), and the axes are ordered in terms of decreasing variance. This means that the first principal component is in the direction that accounts for the most variation within the data. The dimensionality of the dataset may be reduced by discarding the lower order components. It can be shown that for a given number of dimensions or axes retained, this will minimise the error of the approximation in a least squares sense.

The principal component axes correspond to the eigenvectors of the covariance matrix, which are orthogonal since the covariance matrix is symmetric. Therefore, the matrix $\mathbf{E}$ formed from the unit eigenvectors of the covariance matrix will represent a pure rotation of the multidimensional space about the centre (or mean vector). The corresponding eigenvalues represent the variance in the direction of each eigenvector. This is related by

$$\boldsymbol{\Sigma}_{PCA} = \mathbf{E}^{\top}\boldsymbol{\Sigma}_{\mathbf{I}}\mathbf{E}, \tag{8.59}$$

where the $\boldsymbol{\Sigma}_{PCA}$ is the diagonal covariance matrix of the transformed data. The transformation is performed by projecting each pixel onto the new axes:

$$\mathbf{I}_{PCA} = \mathbf{E}^{\top}\mathbf{I}. \tag{8.60}$$

The complexity of determining the eigenvectors from the covariance matrix implies that this step is probably best performed in software rather than hardware. Of course, the arithmetic logic unit (ALU) for the processor may be enhanced with dedicated instructions to accelerate the matrix operations. Once the transformation matrix, $\mathbf{E}$, has been determined, the projection of Eq. (8.60) may be readily implemented in hardware.

From an image processing perspective, PCA has two main applications. When applied across the components of an image, it determines the linear combinations of components that account for most of the variation within the image. This may be used, for example, to reduce the number of channels of a multi-spectral or hyperspectral image. It can also be used for image compression to concentrate or compact the energy in the signal; fewer bits are required for the components that have lower variance. In this regard, the principal components of an RGB image are often similar to the YUV components (unless the image is dominated by a strong colour).

The second application of PCA is in object recognition. In this case, rather than consider the components as the axes, the pixel locations are the axes; each image may then be considered a point in this very high-dimensional space. When considering a set of images of related objects, each image will result in a separate point, with the resulting distribution representing the variability from one image to another. The dimensionality of this high-dimensional space may be significantly reduced by PCA. Each axis in the new space

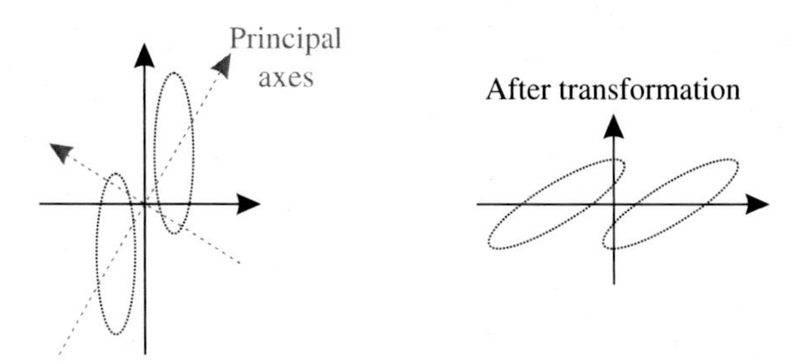

Figure 8.17 Principal components counter-example. Left: the data on the original axes, with the principal component axes overlaid. The ellipses represent different data classes. Right: after PCA transformation; the original axes were better for class separation than the PCA axes.

corresponds to an eigen-image (the high-dimensional eigenvector of the covariance matrix). The variation between the set of images may be reduced to a small number of components, representing the projections onto, hence the weights of, each eigen-image. These weights may then be used for classifying the objects. One classic example of this approach is face recognition (Turk and Pentland, 1991), where the eigen-images are often referred to as ***eigen-faces***.

It must be noted that the principal components are only independent axes with respect to the underlying data only if the whole dataset is Gaussian distributed. This may be the case if it is performed on a cluster of data points within the dataset (for example a particular colour class) but is generally not the case for whole images which consist of multiple objects or classes. In this case, PCA will merely decorrelate the axes. These axes may not necessarily have any useful meaning; in particular, they may not necessarily be useful features for performing classification (see, for example (Cheriyadat and Bruce, 2003)). The 2-D (contrived) example of Figure 8.17 is such a counter-example that illustrates a case where the original axes were better for separating the two classes than the principal axes.

8.2.2.3 Vector Median

Other statistical measures, such as the median and range, are based on an ordering of the data. Since vectors have no natural ordering (Barnett, 1976), the standard definitions no longer apply. The simple approach of taking the median or range of each component gives an approximation but is not the true result. One approach is to convert the vector to a scalar, which can then be used to define an ordering. Such conversions enable a reduced ordering (Barnett, 1976), with the results determined by the particular conversion used. A limitation of these approaches is that many quite different vectors would be considered equal, with no ordering between them. One way of overcoming this for integer components is to interleave the bits of each of the components (Chanussot et al., 1999). This keeps each distinct vector unique, although the ordering is still dependent on interleaving order.

It is possible to define a true median (Astola et al., 1988, 1990) for vector data. The median of a set of vectors, $\mathbf{X}$, is the vector within the set, $\mathbf{x}$, that minimises the distances to all of the other points in the set. That is,

$$\mathbf{x}_{med} = \underset{\mathbf{x}_j \in \mathbf{X}}{\operatorname{argmin}} \sum_i \|\mathbf{x}_j - \mathbf{x}_i\|. \tag{8.61}$$

The distances may be measured using either Euclidean (L_2 norm) or Manhattan (L_1 norm) distances (Astola et al., 1990). The Manhattan distance is easier to calculate in hardware, although the median vector will then depend on the particular coordinates used. The Euclidean distance can be calculated using CORDIC (coordinate rotation digital computer) arithmetic (Section 5.2.2). Calculating the median requires measuring the distance between every pair of points within the set, making the computational complexity of order $O(N^2)$ for N points. This is impractical for an image; however, for smaller N, such as filtering, this is more feasible.

One limitation of this definition of the vector median is that it is not unique. Consider a set of points distributed symmetrically (for example in two dimensions, evenly around the circumference of a circle as shown

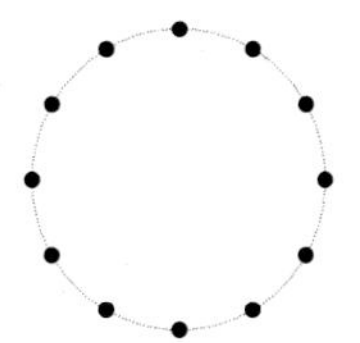

Figure 8.18 Vector median of Eq. (8.61) is not unique.

in Figure 8.18). Each one of the points is the median as defined by Eq. (8.61). This non-uniqueness results from requiring the median to be one of the input samples. Davies (2000) argues that this constraint is unnecessary and results in selecting a point that is not representative of the data as a whole. Relaxing this constraint gives the ***geometric median***:

$$\mathbf{x}_{med} = \underset{\mathbf{x}_j \in \mathbb{R}^n}{\text{argmin}} \sum_i \|\mathbf{x}_j - \mathbf{x}_i\|_2, \tag{8.62}$$

where n is the dimensionality of the vector. The geometric median is unique (Ostresh, 1978) and may be found iteratively by Weiszfeld's algorithm (Ostresh, 1978; Chandrasekaran and Tamir, 1989):

$$\mathbf{x}_{r+1} = \frac{\sum_i \frac{\mathbf{x}_i}{\|\mathbf{x}_i - \mathbf{x}_r\|_2}}{\sum_i \frac{1}{\|\mathbf{x}_i - \mathbf{x}_r\|_2}}. \tag{8.63}$$

Each iteration of Eq. (8.63) requires summing over the set of vectors. Since successive iterations must be performed sequentially, the only scope for parallelism is to partition the set of input vectors over multiple processors. Calculating the inverse Euclidean distance is also relatively expensive but is amenable to pipelining using a CORDIC processor. The resulting weight may be shared by the numerator and denominator. The iteration of Eq. (8.63) converges approximately linearly, although it can get stuck if one of the intermediate points is one of the input vectors (Chandrasekaran and Tamir, 1989).

8.2.2.4 Range

For multidimensional data, the range can be defined as the distance between the two points that are furthermost apart (Barnett, 1976):

$$range = \max_{\mathbf{x}_i, \mathbf{x}_j \in \mathbf{X}} \|\mathbf{x}_i - \mathbf{x}_j\|. \tag{8.64}$$

Again the distance between every pair of points must be calculated, giving complexity of order $O(N^2)$. A simpler approximation is to combine the ranges of each component. This assumes that the distribution is rectangular and aligned with the axes, so will tend to over-estimate the true range.

8.2.3 Colour Segmentation

Multidimensional histograms provide a convenient visualisation for segmentation, particularly when looking at colour images. Segmentation is equivalent to finding clusters of similar colours within the dataset or equivalently finding peaks within the multidimensional histogram.

When considering colour segmentation or colour thresholding, using 3-D histograms is both hard to visualise and expensive in terms of resources. In this case, pairs of components are usually considered; this projects the 3-D histogram onto a 2-D plane. If colour is of primary interest, then the colour can be converted to YCbCr or HSV (hue, saturation, and value), and the intensity channel ignored. The corresponding 2-D histograms are then taken of either *Cb–Cr* or *H–S*, respectively. Alternatively, the components can be normalised, and the *x–y* or *r–g* histogram taken.

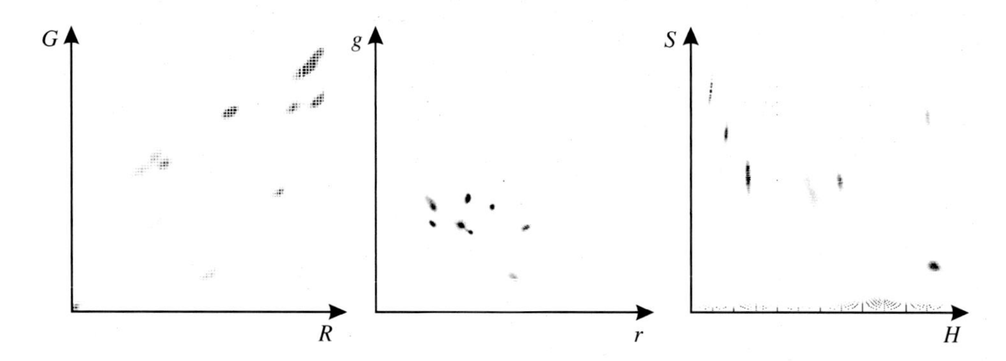

Figure 8.19 2-D histograms of the colour corrected image from Figure 7.39. Left: R–G histogram; centre: r–g histogram; right: H–S histogram. For key features of each histogram, refer to the text.

Several of these are compared in Figure 8.19. Note the predominantly diagonal structure in the R–G histogram in the left panel. This is even more exacerbated if the lighting varies across the scene. The grid structure is caused from the stretching of each component during colour correction.

Normalising the coordinates by dividing the sum (as in Eq. (7.70)) gives the r–g histogram. This is a triangular histogram with nothing above the line $r + g = 1$. The intensity dependence has been removed, resulting in tighter groups. The dense group in the centre corresponds to the white background. The isolated 'random' points result from the black region where a small change in one of the components due to noise drastically changes the proportions of the components. Black, grey, and white regions must be distinguished based on luminance rather than colour. The faint lines connecting each cluster with the background result from mixing around the edges of each coloured patch; the edge pixels consist partly of the coloured object and partly of the background. Although present in all of the histograms, these edges are more visible in the r–g or Cb–Cr histograms.

The third example is the H–S histogram on the right. This too removes the intensity dependence. The best separation is achieved after proper colour balancing, which has quite a strong effect on both the hue and saturation. The pattern of points at low saturation results from noise affecting both the hue and saturation of grey and white pixels. Similarly, the scattered points at higher saturation result from noise affecting the black pixels.

The colour histogram can then be used as the basis for segmentation. This process is illustrated in Figure 8.20 with the Co–Cg histogram of the colour corrected image of Figure 7.39. Since each coloured patch results in a distinct cluster, simply thresholding the 2-D histogram is effective in this case. Each connected component in the histogram is assigned a unique label (see Section 13.4), represented here by a colour. This labelled 2-D

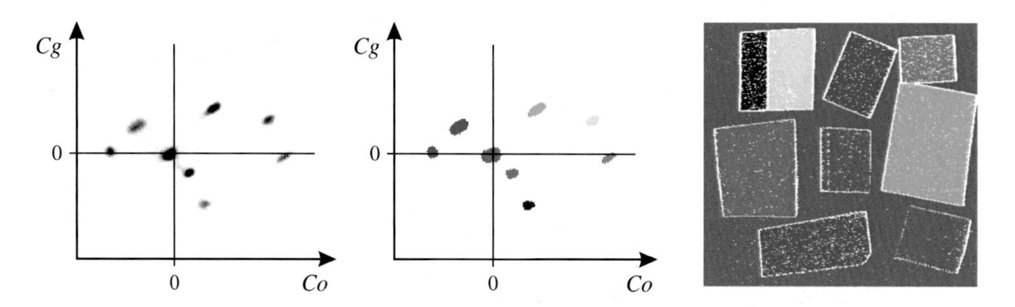

Figure 8.20 Using a 2-D histogram for colour segmentation. Left: Co–Cg histogram using Eq. (7.54); centre: after thresholding and labelling, used as a 2-D lookup table; right: labelled image.

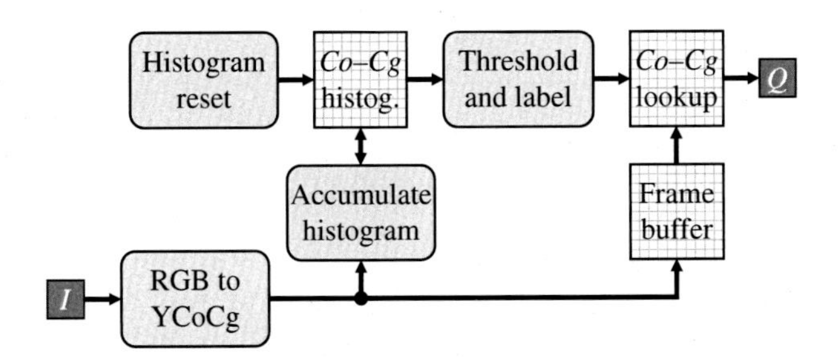

Figure 8.21 Block diagram of a simple colour segmentation scheme.

histogram can then be used as a lookup table on the original *Co* and *Cg* components to give the labelled, segmented image.

In the example here, the white pixels represent unclassified colour values. The corresponding points within the histogram were below the threshold. The number of unclassified pixels within a patch may be reduced by expanding the labelled regions within the 2-D lookup table (for example using a morphological filter) and filtering the output image to remove isolated unclassified pixels. Note the large number of unclassified pixels around the borders of each patch. These result from edge pixels being a mixture of the patch and background colours. They therefore fall partway between the corresponding clusters in the histogram. The number of such pixels may be reduced using an appropriate edge enhancement filter prior to classification. The grey and black patches cannot be distinguished from the white background in this classification because the Y component was not used for segmentation. The projection onto the *Co–Cg* plane collapses these regions into a single cluster.

A block diagram for implementing such colour segmentation is shown in Figure 8.21. If the labelling is determined offline, then the *Co* and *Cg* components may be directly looked up to produce the label. However, the advantage of using the histogram is that the segmentation can be adaptive. The lookup table is also not necessarily required to perform the classification; any of the fixed threshold methods described in Section 7.3.3 could be used, with the histogram data used to adapt the threshold levels.

8.2.3.1 Bayesian Classifier

Once common application of multidimensional histograms is for detecting skin-coloured regions within images (Vezhnevets et al., 2003; Kakumanua et al., 2007). Histograms can represent the colour distributions of skin and non-skin regions, with the initial distributions established through training on examples of skin and general background. Let $H_j[\mathbf{I}]$ be the colour histogram for class C_j. Normalised, this gives the likelihood of observing a particular colour, $\mathbf{I}$, given the class, that is

$$p(\mathbf{I} \mid C_j) = \frac{H_j[\mathbf{I}]}{\sum_{\mathbf{I}} H_j[\mathbf{I}]}. \tag{8.65}$$

Then, assuming that the training samples are provided in the expected proportion of observation, the Bayesian classifier uses Bayes' rule to select the most probable class, given the observed colour, $\mathbf{I}$:

$$C = \underset{j}{\operatorname{argmax}}\{p(C_j \mid \mathbf{I})\} = \underset{j}{\operatorname{argmax}}\left\{\frac{p(\mathbf{I} \mid C_j)p(C_j)}{p(\mathbf{I})}\right\} = \underset{j}{\operatorname{argmax}}\{p(\mathbf{I} \mid C_j)p(C_j)\}. \tag{8.66}$$

The last step discards the denominator since it is constant for all classes, so will not affect which one is the maximum. Substituting the histogram data from Eq. (8.65) and replacing $p(C_j)$ with the number of samples in

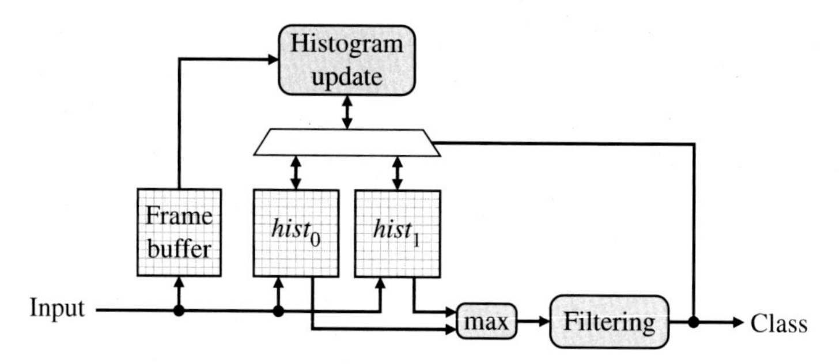

Figure 8.22 Adaptive Bayesian colour segmentation, for example for skin detection.

the class $\sum H_j[\mathbf{I}]$ (ignoring a constant scale factor) gives

$$C = \underset{j}{\operatorname{argmax}} \left\{ \frac{H_j[\mathbf{I}]}{\sum_{\mathbf{I}} H_j[\mathbf{I}]} \sum_{\mathbf{I}} H_j[\mathbf{I}] \right\} = \underset{j}{\operatorname{argmax}} \{H_j[\mathbf{I}]\}. \tag{8.67}$$

In other words, the particular class that has the maximum histogram entry for the observed colour is the most likely class. Alternatively, if the cost of false positives and false negatives differs, then the histograms may be weighted accordingly before selecting the maximum (Chai and Bouzerdoum, 2000).

The process can then be made adaptive, using the classification results to further train the classifier. For this to provide new data, rather than simply to reinforce the existing classification, the output has to be filtered to remove as many misclassifications (both false positives and false negatives) as possible before being fed back. This is shown in block diagram form (without the control logic) in Figure 8.22.

Using only the chrominance components will make the colour segmentation independent of the light intensity. However, it has also been shown (Kakumanua et al., 2007) that this will also decrease the discrimination ability, and that better accuracy is obtained with a full 3-D colour space.

Toledo et al. (2006) have implemented on an FPGA a histogram-based skin colour detector using a combination of 3-D probability maps (histograms) with 32 bins per RGB and YCbCr channel and a combination of 2-D histograms using *I*–*Q* and *U*–*V* components. They used a software processor to monitor the results and adapt the segmentation parameters to improve the tracking performance.

8.2.4 Colour Indexing

Histogram similarity, as described in Section 8.1.6, can readily be extended to identification of colour images. This histogram in Eq. (8.46) is simply replaced with a multidimensional histogram. The main application of histogram similarity is searching for similar images within an image database. The colour histogram contains much important information that can be used to distinguish different colour images, is invariant to translation and rotation, and is relatively insensitive to the orientation of the object, and occlusion. If the colour histogram has a reduced number of bins, then the histograms can be much faster to compare than the images themselves. Reasonable results can be achieved with as few as 200 colour bins (Swain and Ballard, 1991).

One limitation of simple colour indexing is that the search is linear with the size of the database. However, the scaling factor may be improved in a number of ways. The simplest is to reduce the number of bins by combining the counts of groups of adjacent bins. Since

$$\min(A, A') + \min(B, B') \le \min(A + B, A' + B'), \tag{8.68}$$

the reduced resolution will always give a higher match score than the higher resolution histogram. Therefore, database entries with a low match score may be eliminated at a lower cost.

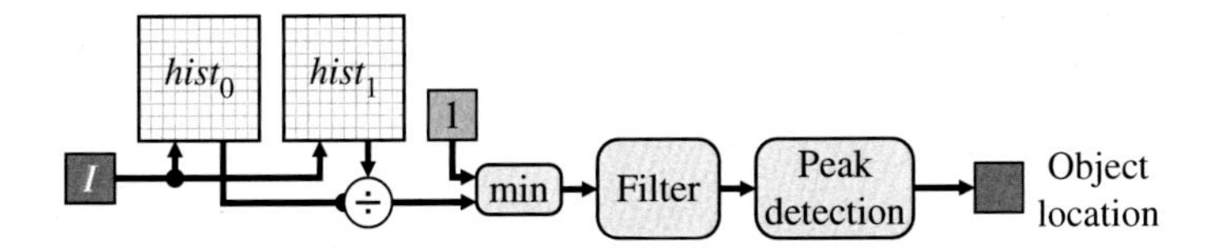

Figure 8.23 Object location through histogram back-projection.

For many histograms, most of the total count is concentrated in a few bins, corresponding to the dominant colours in the image. The match from these bins has the greatest influence on the match score. An approximate match score may be produced quickly by comparing only those bins with the database histograms. The search speed may be further improved if the database is sorted based on the count in each bin (Swain and Ballard, 1991). This requires a separate database index for each bin, but this index only needs to include those database histograms for which that bin is significant. In this way, large sections of the database that will give a poor match are not even searched. Only those that are likely to give a good match are compared, and even then, only a small number of bins need to be matched. Swain and Ballard (1991) demonstrated that as few as 10 bins actually need to be compared to find the best match, significantly improving the search time.

Once a match is found, the next step is to locate the actual object within the image. This can be accomplished through histogram back-projection (Swain and Ballard, 1991). The basic idea for this is illustrated in Figure 8.23. Each pixel in the input image is looked up in both the model histogram from the database, H_M, and the histogram for the image, H_I. The value of the back-projected pixel is then

$$bp[x,y] = \min\left(\frac{H_M[I[x,y]]}{H_I[I[x,y]]}, 1\right). \tag{8.69}$$

This effectively assigns a high value to the pixels that have a strong colour match with the model, and a low value to the pixels that are not significant in the model. The back-projected image is then smoothed using a suitable filter, and the location of the maximum detected. Since the largest peak does not necessarily correspond to the object (sometimes it will be the second or third largest peak), an alternative is to threshold the filtered image to detect candidate locations.

8.2.5 *Texture Analysis*

Visual texture is characterised by spatial patterns within the pixel values of an image. Since it has a spatial aspect, simple first-order statistics are unable to characterise a texture properly. For example, many quite different textures have identical grey-level histograms. It is necessary to capture the spatial relationships using second-order statistics. One common method for texture analysis and classification is based on second-order conditional distribution functions, which can be captured in a co-occurrence matrix. This is essentially a 2-D histogram, where the first component is the input greyscale image, and the second component is the input image offset by distance d in direction θ:

$$H_{d,\theta}[a,b] = \sum_{x,y} \begin{cases} 1, & (I[x,y]=a) \wedge (I[x+\Delta x, y+\Delta y]=b), \\ 0, & \text{otherwise,} \end{cases} \tag{8.70}$$

where $\|(\Delta x, \Delta y)\| = d$ and $\angle(\Delta x, \Delta y) = \theta$. Typically, the Chebyshev or chessboard distance is used, and the directions are limited to 45° increments to simplify calculation.

The spatial grey-level dependence method (SGLDM) (Conners and Harlow, 1980) of texture analysis extracts several features from the co-occurrence matrix. Haralick et al. (1973) defines 14 features, of which 5 are commonly used (Conners and Harlow, 1980):

energy:

$$E(H_{d,\theta}) = \frac{\sum_{i,j} (H_{d,\theta}[i,j])^2}{\left(\sum_{i,j} H_{d,\theta}[i,j]\right)^2}, \tag{8.71}$$

entropy:

$$H(H_{d,\theta}) = -\frac{\sum_{i,j} H_{d,\theta}[i,j] \log_2 H_{d,\theta}[i,j]}{\sum_{i,j} H_{d,\theta}[i,j]}, \tag{8.72}$$

correlation:

$$C(H_{d,\theta}) = \frac{\sum_{i,j} (i - \mu_i)(j - \mu_j) H_{d,\theta}[i,j]}{\sigma_i \sigma_j}, \tag{8.73}$$

local homogeneity:

$$L(H_{d,\theta}) = \frac{\sum_{i,j} \dfrac{H_{d,\theta}[i,j]}{1 + (i-j)^2}}{\sum_{i,j} H_{d,\theta}[i,j]}, \tag{8.74}$$

inertia or contrast:

$$I(H_{d,\theta}) = \frac{\sum_{i,j} (i-j)^2 H_{d,\theta}[i,j]}{\sum_{i,j} H_{d,\theta}[i,j]}, \tag{8.75}$$

where μ_i, μ_j, σ_i, and σ_j are the means and standard deviations of the i and j components, respectively, of the SGLDM. To obtain meaningful results for comparison or classification, the counts within the two-dimensional histogram must be normalised to give probabilities.

Several features are used, over a range of angles, and for a selection of distances that characterise the textures of interest. Some of the features are shown for a sample texture in Figure 8.24, along with the associated SGLDMs displayed as images. The 1-D histogram is also shown for reference; it is indistinguishable from a Gaussian distribution and is unable to capture any spatial relationships.

If the whole image contains a single texture, the above analysis is suitable. However, for texture segmentation, the co-occurrence matrices must be calculated from smaller regions or windows within the image.

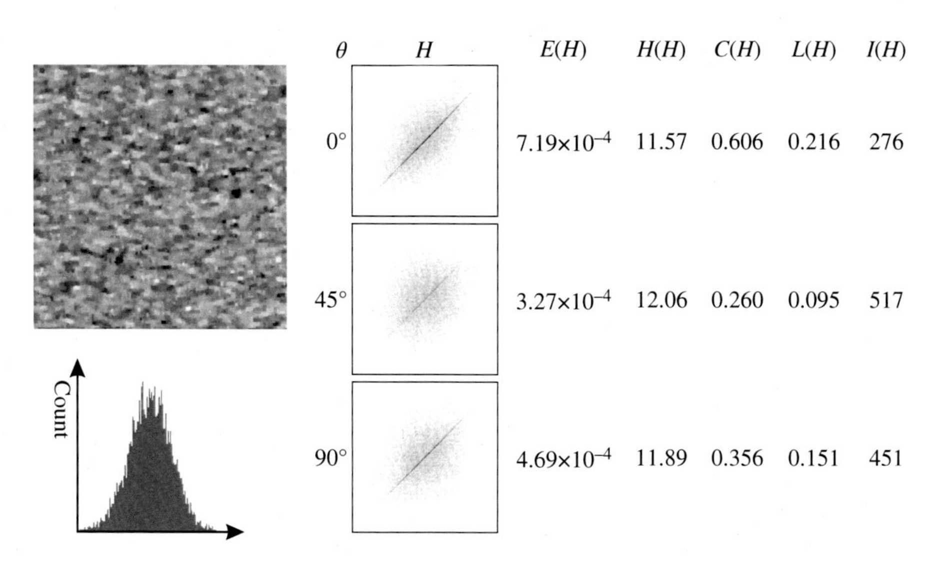

Figure 8.24 A sample texture with its histogram and some associated SGLDM features for $d = 3$.

The co-occurrence matrix is more accurate when formed from a large number of pixels. However, for small windows, the measures can become meaningless unless the number of bins in the histogram is reduced accordingly. This has an advantage from an implementation perspective, since the memory required to store the matrix is smaller, and calculating the texture features is faster. On an FPGA, the different features may be calculated in parallel as the matrix is scanned. If several matrices are used (with different separations and orientations), the features from each matrix may also be calculated in parallel. Tahir et al. (2003a) did exactly this; they built 16 co-occurrence matrices in parallel with distance ranges of $d = \{1, 2, 3, 4\}$ and angles $\theta = \{0°, 45°, 90°, 135°\}$, with 32 pixel value bins in each dimension of their 2-D histogram. They also used the FPGA to calculate seven texture features from each matrix in parallel (Tahir et al., 2003b).

8.3 Summary

Histograms provide a useful analytic tool for extraction of data from images. By capturing the distribution of pixel values, several statistical measures can be derived from an image. These are useful either directly as features or as data that can be used for contrast enhancement, threshold selection, or exposure adjustment. Histogram similarity can also provide an efficient method for object identification or classification.

When applied to colour images, multidimensional histograms can provide colour statistics and be used for colour segmentation and colour object identification and classification. The covariance matrix may be used for dimensionality reduction via PCA.

Spatial relationships captured by a two-dimensional co-occurrence matrix can be used for deriving textural features from an image or image patch.

References

Anghelescu, P. (2021). Automatic thresholding technique using reconfigurable hardware. *Cybernetics and Systems* **52** (2): 145–168. https://doi.org/10.1080/01969722.2020.1804220.

Astola, J., Haavisto, P., Heinonen, P., and Neuvo, Y. (1988). Median type filters for color signals. *IEEE International Symposium on Circuits and Systems*, Espoo, Finland (7–9 June 1988), Volume 2, 1753–1756. https://doi.org/10.1109/ISCAS.1988.15274.

Astola, J., Haavisto, P., and Neuvo, Y. (1990). Vector median filters. *Proceedings of the IEEE* **78** (4): 678–689. https://doi.org/10.1109/5.54807.

Bailey, D.G. (1995). Pixel calibration techniques. *New Zealand Image and Vision Computing '95 Workshop*, Lincoln, NZ (28–29 August 1995), 37–42.

Bailey, D.G., Mercer, K.A., Plaw, C., Ball, R., and Barraclough, H. (2004). High speed weight estimation by image analysis. *New Zealand National Conference on Non Destructive Testing*, Palmerston North, NZ (27–29 July 2004), 89–96.

Barnett, V. (1976). Ordering of multivariate data. *Journal of the Royal Statistical Society Series A - Statistics in Society* **139** (3): 318–355. https://doi.org/10.2307/2344839.

Chai, D. and Bouzerdoum, A. (2000). A Bayesian approach to skin color classification in YCbCr color space. *IEEE Region 10 Conference (TENCON 2000)*, Kuala Lumpur, Malaysia (24–27 September, 2000), Volume 2, 421–424. https://doi.org/10.1109/TENCON.2000.888774.

Chandrasekaran, R. and Tamir, A. (1989). Open questions concerning Weiszfeld's algorithm for the Fermat-Weber location problem. *Mathematical Programming* **44** (1–3): 293–295. https://doi.org/10.1007/BF01587094.

Chanussot, J., Paindavoine, M., and Lambert, P. (1999). Real time vector median like filter: FPGA design and application to color image filtering. *IEEE International Conference on Image Processing (ICIP'99)*, Kobe, Japan (24–28 October 1999), Volume 2, 414–418. https://doi.org/10.1109/ICIP.1999.822929.

Cheriyadat, A. and Bruce, L.M. (2003). Why principal component analysis is not an appropriate feature extraction method for hyperspectral data. *IEEE International Geoscience and Remote Sensing Symposium (IGARSS 2003)*, Toulouse, France (21–25 July 2003), Volume 6, 3420–3422. https://doi.org/10.1109/IGARSS.2003.1294808.

Chiang, S.E., Sun, C.C., and Lin, C.Y. (2013). Fast contrast enhancement based on a novel dynamic histogram equalization algorithm. *Asia-Pacific Workshop on FPGA Applications*, Nanjing, China (1–3 November 2013), 5 pages.

Conners, R.W. and Harlow, C.A. (1980). A theoretical comparison of texture algorithms. *IEEE Transactions on Pattern Analysis and Machine Intelligence* **2** (3): 204–223. https://doi.org/10.1109/TPAMI.1980.4767008.

Davies, E.R. (2000). Accuracy of multichannel median filter. *Electronics Letters* **36** (25): 2068–2069. https://doi.org/10.1049/el:20001465.

Davies, E.R. (2008). Stable bi-level and multi-level thresholding of images using a new global transformation. *IET Computer Vision* **2** (2): 60–74. https://doi.org/10.1049/iet-cvi:20070071.

Doyle, W. (1962). Operations useful for similarity invariant pattern recognition. *Journal of the Association for Computing Machinery* **9** (2): 259–267. https://doi.org/10.1145/321119.321123.

Fahmy, S.A., Cheung, P.Y.K., and Luk, W. (2005). Novel FPGA-based implementation of median and weighted median filters for image processing. *International Conference on Field Programmable Logic and Applications*, Tampere, Finland (24–26 August 2005), 142–147. https://doi.org/10.1109/FPL.2005.1515713.

Frei, W. (1977). Image enhancement by histogram hyperbolisation. *Computer Graphics and Image Processing* **6** (3): 286–294. https://doi.org/10.1016/S0146-664X(77)80030-0.

Geninatti, S. and Boemo, E. (2019). A proposal of two histogram circuits to calculate similarities between video frames using FPGAs. *X Southern Conference on Programmable Logic (SPL)*, Buenos Aires, Argentina (10–12 April 2019), 103–108. https://doi.org/10.1109/SPL.2019.8714375.

Haralick, R.M., Shanmugam, K., and Dinstein, I. (1973). Textural features for image classification. *IEEE Transactions on Systems, Man and Cybernetics* **3** (6): 610–621. https://doi.org/10.1109/TSMC.1973.4309314.

Hummel, R.A. (1975). Histogram modification techniques. *Computer Graphics and Image Processing* **4** (3): 209–224. https://doi.org/10.1016/0146-664X(75)90009-X.

Hummel, R.A. (1977). Image enhancement by histogram transformation. *Computer Graphics and Image Processing* **6** (2): 184–195. https://doi.org/10.1016/S0146-664X(77)80011-7.

Kakumanua, P., Makrogiannisa, S., and Bourbakis, N. (2007). A survey of skin-color modeling and detection methods. *Pattern Recognition* **40** (3): 1106–1122. https://doi.org/10.1016/j.patcog.2006.06.010.

Kapura, J.N., Sahoob, P.K., and Wong, A.K.C. (1985). A new method for gray-level picture thresholding using the entropy of the histogram. *Computer Vision, Graphics, and Image Processing* **29** (3): 273–285. https://doi.org/10.1016/0734-189X(85)90125-2.

Kaur, M., Kaur, J., and Kaur, J. (2011). Survey of contrast enhancement techniques based on histogram equalization. *International Journal of Advanced Computer Science and Applications* **2** (7): 137–141. https://doi.org/10.14569/IJACSA.2011.020721.

Kittler, J. and Illingworth, J. (1986). Minimum error thresholding. *Pattern Recognition* **19** (1): 41–47. https://doi.org/10.1016/0031-3203(86)90030-0.

Lee, H. and Park, R.H. (1990). Comments on "An optimal multiple threshold scheme for image segmentation". *IEEE Transactions on Systems, Man and Cybernetics* **20** (3): 741–742. https://doi.org/10.1109/21.57290.

Mahalanobis, P.C. (1936). On the generalised distance in statistics. *Proceedings of the National Institute of Sciences of India* **2** (1): 49–55.

McCollum, A.J., Bowman, C.C., Daniels, P.A., and Batchelor, B.G. (1988). A histogram modification unit for real-time image enhancement. *Computer Vision, Graphics and Image Processing* **42** (3): 387–398. https://doi.org/10.1016/S0734-189X(88)80047-1.

Nilsson, M., Dahl, M., and Claesson, I. (2005a). Gray-scale image enhancement using the SMQT. *IEEE International Conference on Image Processing (ICIP '05)*, Genoa, Italy (11–14 September 2005), Volume 1, 933–936. https://doi.org/10.1109/ICIP.2005.1529905.

Nilsson, M., Dahl, M., and Claesson, I. (2005b). The successive mean quantization transform. *IEEE International Conference on Acoustics, Speech and Signal Processing (ICASSP '05)*, Philadelphia, PA, USA (18–23 March 2005), Volume 4, 429–432. https://doi.org/10.1109/ICASSP.2005.1416037.

Nilsson, M., Sattar, F., Chng, H.K., and Claesson, I. (2005c). Automatic enhancement and subjective evaluation of dental X-ray images using the SMQT. *5th International Conference on Information, Communications and Signal Processing*, Bangkok, Thailand (6–9 December 2005), 1448–1451. https://doi.org/10.1109/ICICS.2005.1689298.

Ning, J., Lu, T., Liu, L., Guo, L., and Jin, X. (2015). The optimization and design of the auto-exposure algorithm based on image entropy. *8th International Congress on Image and Signal Processing (CISP)*, Shenyang, China (14–16 October 2015), 1020–1025. https://doi.org/10.1109/CISP.2015.7408029.

Ostresh, L.M. (1978). On the convergence of a class of iterative methods for solving the Weber location problem. *Operations Research* **26** (4): 597–609. https://doi.org/10.1287/opre.26.4.597.

Otsu, N. (1979). A threshold selection method from gray-level histograms. *IEEE Transactions on Systems, Man and Cybernetics* **9** (1): 62–66. https://doi.org/10.1109/TSMC.1979.4310076.

Popadic, I., Todorovic, B.M., and Reljin, I. (2017). Method for HDR-like imaging using industrial digital cameras. *Multimedia Tools and Applications* **76** (10): 12801–12817. https://doi.org/10.1007/s11042-016-3692-8.

Prewitt, J.M.S. and Mendelsohn, M.L. (1966). The analysis of cell images. *Annals of the New York Academy of Sciences* **128** (3): 1035–1053. https://doi.org/10.1111/j.1749-6632.1965.tb11715.x.

Rahman, M., Kehtarnavaz, N., and Razlighi, Q. (2011). Using image entropy maximum for auto exposure. *Journal of Electronic Imaging* **20** (1): Article ID 013007, 10 pages. https://doi.org/10.1117/1.3534855.

Ridler, T.W. and Calvard, S. (1978). Picture thresholding using an iterative selection method. *IEEE Transactions on Systems, Man and Cybernetics* **8** (8): 630–632. https://doi.org/10.1109/TSMC.1978.4310039.

Rosenfeld, A. and de la Torre, P. (1983). Histogram concavity analysis as an aid in threshold selection. *IEEE Transactions on Systems, Man and Cybernetics* **13** (3): 231–235. https://doi.org/10.1109/TSMC.1983.6313118.

Schatz, V. (2013). Low-latency histogram equalization for infrared image sequences: a hardware implementation. *Journal of Real-Time Image Processing* **8** (2): 193–206. https://doi.org/10.1007/s11554-011-0204-y.

Swain, M.J. and Ballard, D.H. (1991). Colour indexing. *International Journal of Computer Vision* **7** (1): 11–32. https://doi.org/10.1007/BF00130487.

Tahir, M.A., Bouridane, A., Kurugollu, F., and Amira, A. (2003a). An FPGA based coprocessor for calculating grey level co-occurrence matrix. *IEEE International Symposium on Micro-NanoMechatronics and Human Science*, Cairo, Egypt (27–30 December 2003), Volume 2, 868–871. https://doi.org/10.1109/MWSCAS.2003.1562424.

Tahir, M.A., Roula, M.A., Bouridane, A., Kurugollu, F., and Amira, A. (2003b). An FPGA based co-processor for GLCM texture features measurement. *10th IEEE International Conference on Electronics, Circuits and Systems*, Sharjah, United Arab Emirates (14–17 December 2003), volume 3, 1006–1009. https://doi.org/10.1109/ICECS.2003.1301679.

Toledo, F., Martinez, J.J., Garrigos, J., Ferrandez, J., and Rodellar, V. (2006). Skin color detection for real time mobile applications. *International Conference on Field Programmable Logic and Applications (FPL'06)*, Madrid, Spain (28–30 August 2006), 721–724. https://doi.org/10.1109/FPL.2006.311299.

Trussell, H.J. (1979). Comments on "Picture thresholding using an iterative selection method". *IEEE Transactions on Systems, Man and Cybernetics* **9** (5): 311–311. https://doi.org/10.1109/TSMC.1979.4310204.

Turk, M.A. and Pentland, A.P. (1991). Face recognition using eigenfaces. *IEEE Conference on Computer Vision and Pattern Recognition (CVPR'91)*, Maui, Hawaii, USA (3–6 June 1991), 586–591. https://doi.org/10.1109/CVPR.1991.139758.

Vezhnevets, V., Sazonov, V., and Andreeva, A. (2003). A survey on pixel-based skin color detection techniques. *Graphicon-2003*, Moscow, Russia (3–10 September 2003), 85–92.

Weszka, J.S., Nagel, R.N., and Rosenfeld, A. (1974). A threshold selection technique. *IEEE Transactions on Computers* **23** (12): 1322–1326. https://doi.org/10.1109/T-C.1974.223858.

9

Local Filters

Local filters extend point operations by having the output be some function of the pixel values within a local neighbourhood or window:

$$Q[x,y] = f(I[x,y], \ldots, I[x+\Delta x, y+\Delta y]), \quad (\Delta x, \Delta y) \in \mathbf{W}, \tag{9.1}$$

where $\mathbf{W}$ is the window or local neighbourhood centred on $I[x,y]$, as illustrated in Figure 9.1. The window can be of any shape or size, but is usually square, $W \times W$ pixels in size, with W odd, so that the window centre is well defined. As the window is scanned through the input image, each possible position generates an output pixel according to Eq. (9.1). The filter function, $f(\cdot)$, determines the type of filter. Since the output depends not only on the input pixel but also its local context, filters can be used for noise removal or reduction, edge detection, edge enhancement, line detection, and feature detection.

9.1 Window Caching

The software approach to filtering has both the input and output images stored in frame buffers. For each output pixel, the pixels within the input window are retrieved from memory, with the filtered result written back. In this form, the filter is ultimately limited by the bandwidth of memory holding the images. Any acceleration of filtering must exploit the fact that each pixel is used in multiple windows. With stream processing, this is accomplished through caching pixels as they are read in to enable them to be reused in later window positions.

Pixel caching for stream processing of $W \times W$ filters requires a series of $W - 1$ row buffers. These are effectively first-in first-out (FIFO) buffers, as described in Section 5.4.1, although the memory may be addressed directly by column number. Scanning the window through the image is equivalent to streaming the image through the window. The row buffers can either be placed in parallel with the window or, since the window consists of a set of shift registers, in series with the window as shown in Figure 9.2. Computationally they are equivalent. Parallel row buffers need to be slightly longer (the full width of the image) but have the advantage that they are kept independent of the window and filter.

Although the internal memory of most field-programmable gate arrays (FPGAs) is dual-port, it is possible to cache the input rows using only single-port memory. This requires W buffers, with the input being written to successive buffers in round robin fashion. The window is then formed from the input, and the other row buffers as shown in Figure 9.3. The state is easiest implemented as a ring counter, directly controlling the multiplexers and the read/write signals to the row buffers.

If the resources are limited, and the row buffer is too long for the available memory, then other stream access patterns are possible, as illustrated in Figure 9.4. The image may be scanned down the columns rather than across the rows. Alternatively, the image may be partitioned and processed as a series of vertical tiles (Sedcole, 2006), although this would introduce a small overhead, because to process the complete image the vertical tiles must overlap by $W - 1$ pixels.

Design for Embedded Image Processing on FPGAs, Second Edition. Donald G. Bailey.

Companion Website: www.wiley.com/go/bailey/designforembeddedimageprocessc2e

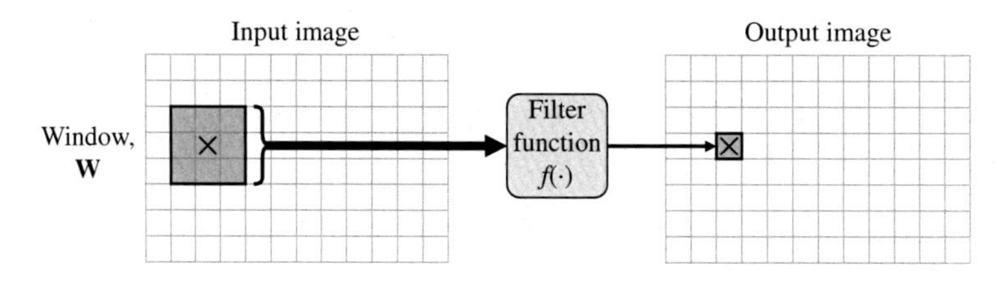

Figure 9.1 A local window filter. The shaded pixels represent the input window located at × that produces the filtered value for the corresponding location in the output image. Each possible window position generates the corresponding pixel value in the output image.

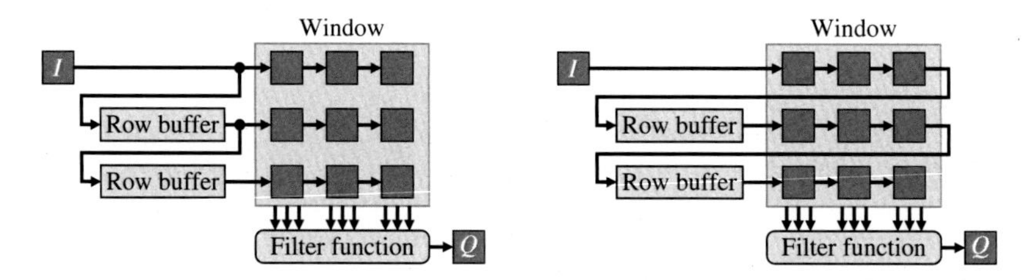

Figure 9.2 Row buffers. Left: in parallel with window; right: in series with window.

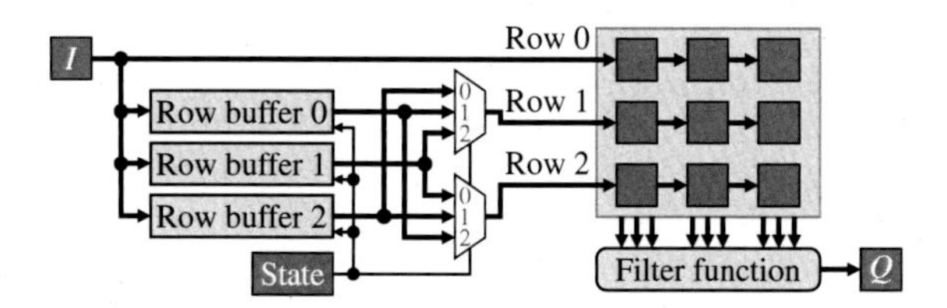

Figure 9.3 Filtering with single-port row buffers.

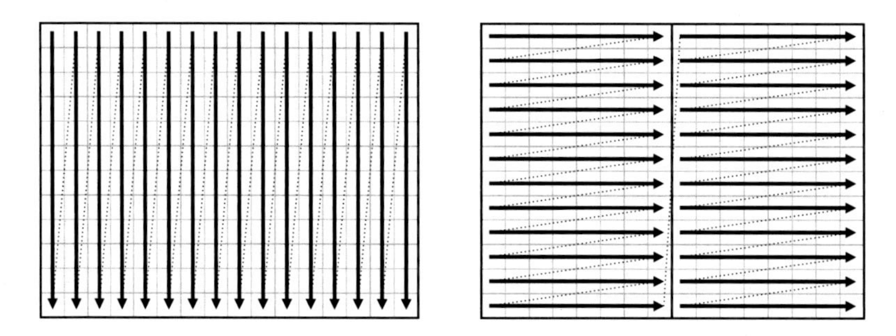

Figure 9.4 Alternative stream access patterns: Left: stream by column; right: tiling into vertical strips.

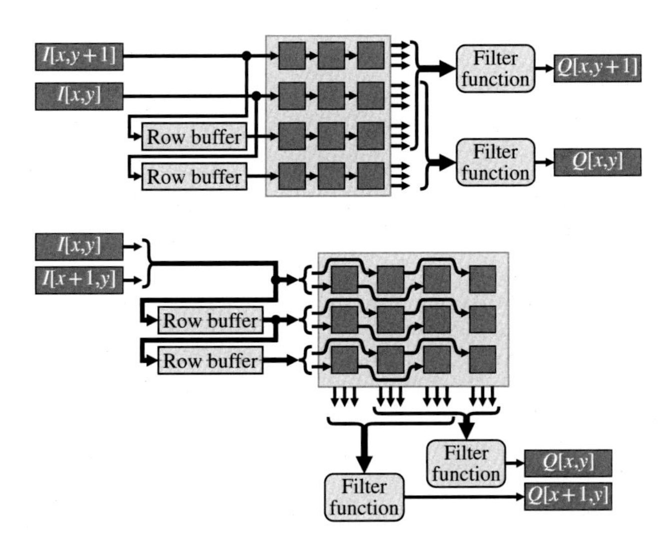

Figure 9.5 Parallel window processing through partial loop unrolling. Top: vertical unrolling; bottom: horizontal unrolling.

If additional resources and memory bandwidth are available, multiple filters may operate in parallel. The overlap between adjacent windows may be exploited by processing multiple adjacent rows or columns (Draper et al., 2003). As illustrated in Figure 9.5, partially unrolling either the vertical or horizontal scan loop through the image will input multiple adjacent pixels per clock cycle. For an unroll factor of k (processing k pixels in parallel), the combined window size is $W \times (W + k - 1)$ or $(W + k - 1) \times W$. Significant memory resources are saved because the number of row buffers is unchanged. For vertical unrolling, the buffers are arranged with a pitch of k rows. For horizontal unrolling, k pixels are processed each clock cycle, making the buffers k times wider but also shorter by a factor of k. Parallel implementation requires k copies of the filter function. However, with some filters, the overlap between windows may further enable some of the filter function logic to be shared (Lucke and Parhi, 1992).

9.1.1 *Border Handling*

One problem with local filters is managing what happens when the window is not completely within the input image. Several approaches can be considered for handling such border windows (Bailey, 2011; Rafi and Din, 2016; Al-Dujaili and Fahmy, 2017; Ozkan et al., 2017; Bailey and Ambikumar, 2018), most of which involve estimating $(W - 1)/2$ pixel values beyond each border to fill the window and produce a full-sized image at the output (see Figure 9.6):

Do nothing: The input stream is simply wrapped. The pixels at the end of a row are immediately followed by the pixels at the start of the next row. Similarly, the last row of one frame is immediately followed by the first

C	C	C	C	C	C	C
C	C	C	C	C	C	C
C	C	C	C	C	C	C
C	C	C	0,0	0,1	0,2	0,3
C	C	C	1,0	1,1	1,2	1,3
C	C	C	2,0	2,1	2,2	2,3
C	C	C	3,0	3,1	3,2	3,3

0,0	0,0	0,0	0,0	0,1	0,2	0,3
0,0	0,0	0,0	0,0	0,1	0,2	0,3
0,0	0,0	0,0	0,0	0,1	0,2	0,3
0,0	0,0	0,0	0,0	0,1	0,2	0,3
1,0	1,0	1,0	1,0	1,1	1,2	1,3
2,0	2,0	2,0	2,0	2,1	2,2	2,3
3,0	3,0	3,0	3,0	3,1	3,2	3,3

1,1	1,0	1,1	1,0	1,1	1,2	1,3
0,1	0,0	0,1	0,0	0,1	0,2	0,3
1,1	1,0	1,1	1,0	1,1	1,2	1,3
0,1	0,0	0,1	0,0	0,1	0,2	0,3
1,1	1,0	1,1	1,0	1,1	1,2	1,3
2,1	2,0	2,1	2,0	2,1	2,2	2,3
3,1	3,0	3,1	3,0	3,1	3,2	3,3

2,2	2,1	2,0	2,0	2,1	2,2	2,3
1,2	1,1	1,0	1,0	1,1	1,2	1,3
0,2	0,1	0,0	0,0	0,1	0,2	0,3
0,2	0,1	0,0	0,0	0,1	0,2	0,3
1,2	1,1	1,0	1,0	1,1	1,2	1,3
2,2	2,1	2,0	2,0	2,1	2,2	2,3
3,2	3,1	3,0	3,0	3,1	3,2	3,3

3,3	3,2	3,1	3,0	3,1	3,2	3,3
2,3	2,2	2,1	2,0	2,1	2,2	2,3
1,3	1,2	1,1	1,0	1,1	1,2	1,3
0,3	0,2	0,1	0,0	0,1	0,2	0,3
1,3	1,2	1,1	1,0	1,1	1,2	1,3
2,3	2,2	2,1	2,0	2,1	2,2	2,3
3,3	3,2	3,1	3,0	3,1	3,2	3,3

Figure 9.6 Practical edge extension schemes illustrated for the top-left corner of an image. Left: constant extension; left-centre: edge pixel duplication; centre: two-phase duplication; right-centre: mirroring with duplication; right: mirroring without duplication.

row of the next frame. Since the opposite edges of an image are unrelated in most applications, the border pixels are likely to be invalid. If this is an issue, the invalid pixels can be discarded, effectively shrinking the output image by $(W - 1) \times (W - 1)$ pixels.

Constant extension: A predefined value is used for the pixels outside the image. This could be black (0) or white (for example with morphological filters (Ramirez et al., 2010)), or the average pixel value from the image, depending on the type of filter and the expected scene.

Duplication or clamping: The nearest pixel value within the image is used (nearest-neighbour extrapolation); see Figure 9.6. This approach has been used with median and rank filters (for example by (Choo and Verma, 2008)).

Two-phase duplication: The outermost two rows are alternated; see Figure 9.6. This is required when duplicating the borders of raw Bayer patterned images (Bailey and Li, 2016).

Mirroring with duplication: The rows and columns just inside the border are reflected to the outside of the image (effectively duplicating the outside row, also called half-sample symmetry); see Figure 9.6. Mirroring is commonly used with wavelet filters (and adaptive histogram equalisation (Kurak, 1991)).

Mirroring without duplication: Similar to the previous case, but the outside pixels are not duplicated (also called whole-sample symmetry) (McDonnell, 1981; Benkrid et al., 2003); see Figure 9.6.

Periodic extension: Extends the image by periodically tiling the image horizontally and vertically. This is implicit with the discrete Fourier transform (Bracewell, 2000). However, the storage requirements make periodic extension impractical for stream processing on FPGA.

Modified filter function: The filter function could be modified to work with the smaller, truncated window when part of the window extends beyond the boundary of the image. For many filters, this is not an option, and where it is, the logic requirements can grow rapidly to manage the different cases.

In software, border processing is usually managed with additional code to provide appropriate processing of windows which extend past the edge of the image. Naively mapping such approaches to hardware can result in significant additional logic for processing the border windows. For some solutions, the logic required to handle the boundary cases can be just as much or more than the regular window logic.

Hardware can be simplified by separating the border handling from both the row buffering and filter function. The extensions are separable in the sense that extensions in the horizontal and vertical directions are independent (Bailey, 2011; Bailey and Ambikumar, 2018), allowing borders to be managed by a small amount of additional control logic. This consists of separate vertical and horizontal routing networks (basically a set of multiplexers controlled by the row or column index) to route the incoming pixels to the correct row and column of the window (Chakrabarti, 1999; Rafi and Din, 2016) as shown in Figure 9.7.

If the delay introduced by the routing multiplexers presents a timing issue, then the routing can be incorporated as part of the window shift register chain as shown in Figure 9.8. With no horizontal blanking (the pixels of one row immediately follow the pixels of the previous row), additional registers are required to hold the data while the window for the previous row is flushed. Then, all of the window registers can be primed in parallel for the start of the next row. If the horizontal blanking period is longer than the horizontal priming period $\left(\frac{W-1}{2}\right)$, then this can be relaxed (for example Kumar and Kanhe (2022) used a last-in first-out (LIFO)

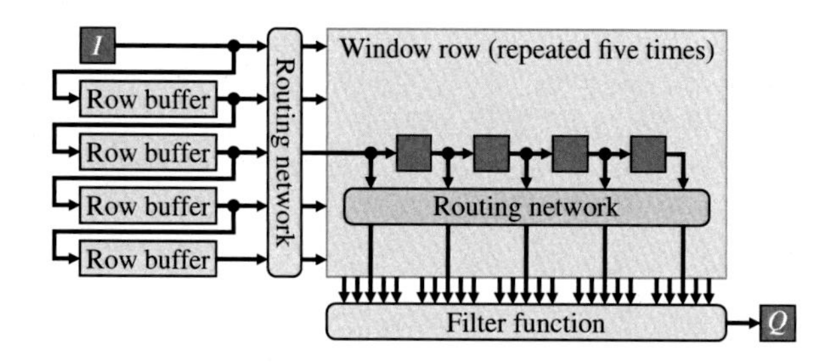

Figure 9.7 Border management through row and column routing networks.

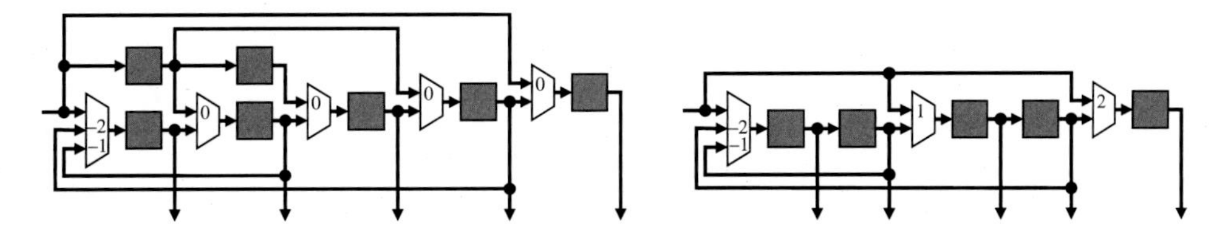

Figure 9.8 Moving the column routing to before the window registers to improve timing. Shown here is mirroring without duplication. Multiplexer labels −1 and −2 correspond to the first and second pixels after the last valid pixel in a row. Left: no delay between rows; right: additional horizontal blanking is available between rows, enabling the design to be simplified. Source: Adapted from Bailey (2011).

buffer for mirroring). Figure 9.8 illustrates mirroring without duplication; similar networks can be devised for the other border extension schemes (Bailey, 2011).

While a similar procedure could be used for transitioning from the bottom of one image to the top of the next, this would require additional (relatively expensive) row buffers to perform the counterpart of the preload registers along the row. Therefore, row routing is best handled using a routing network.

A similar routing network approach can also be applied to transpose filter structures (Bailey and Ambikumar, 2018), with a similarly light overhead, while retaining the clock speed improvements resulting from the pipelining due to the transpose filter structures (described in more detail in Section 9.2.1.1).

9.1.2 *Filter Latency*

The latency of a window filter is measured from when a pixel is input to when the corresponding pixel is output. For a filter, it comprises two components: the time required to gather the data to form the window, t_W, and the latency of the filter function that calculates the output pixel value, t_f:

$$t_{Filter} = t_W + t_f. \tag{9.2}$$

While it is necessary to load a whole window of data before the filter function can calculate the output, the output usually corresponds to the centre of the window (see Figure 9.1). Therefore, the window latency is the time required to load the pixels after (and including) the centre pixel. Assuming that the image is streamed one pixel per clock cycle, let the time required to stream a complete image row (including any blanking period if present) be R. This makes

$$t_W = R\frac{W-1}{2} + \frac{W+1}{2} = \frac{(R+1)(W-1)}{2} + 1 \tag{9.3}$$

for odd window sizes. If W is even, this needs to be adjusted based on the defined location of the output pixel relative to the window. As shown in Section 9.1.1, t_W is unaffected by appropriate border handling, apart from any pipelining introduced by the routing network.

If the system is sink-driven, then it is necessary to start the filtering at least t_{Filter} before the first data is required on the output. It is also necessary to continue to clock the filter for this many clock cycles after the last data is input to flush the contents of the filter (for example during vertical blanking).

9.2 Linear Filters

For a linear filter, the filter function is a weighted sum of the pixel values within the window:

$$Q[x,y] = \sum_{i,j\in \mathbf{W}} w[i,j]I[x+i,y+j]. \tag{9.4}$$

The particular set of weights is called the filter ***kernel***, with the filter function determined by the kernel used. Linear filtering performs a 2-D convolution with the flipped kernel, $w[-i,-j]$. Since the Fourier transform of a convolution is the product of the respective Fourier transforms (Bracewell, 2000), the operation of linear filters can also be considered from their effects in the frequency domain. This will be explored further in Section 11.1.

In signal processing terminology, the filters described by Eq. (9.4) are ***finite impulse response*** filters. Recursive, ***infinite impulse response*** filters can also be used; these use the previously calculated outputs in addition to the inputs to calculate the current output:

$$Q[x,y] = \sum_{i,j \in \mathbf{W}} w_q[i,j]Q[x+i,y+j] + \sum_{i,j \in \mathbf{W}} w[i,j]I[x+i,y+j]. \tag{9.5}$$

By necessity, w_q must be 0 for the pixels that have not been calculated yet. Since recursive filters use previous outputs, care must be taken in their implementation because the clock frequency can be limited by the propagation delay of the feedback loop. Recursive filters are not often used in image processing, because their nonlinear phase response results in a directional smearing of the image contents. To avoid this, it is necessary to apply each filter twice, once with a top-to-bottom, left-to-right scan pattern, and again with a bottom-to-top, right-to-left scan pattern. The two passes make such filters difficult to pipeline, although for a given frequency response, infinite impulse response filters require a significantly smaller window (Mitra, 1998). Some finite impulse response filters, such as the box filter described in Section 9.2.2.1, can also be implemented efficiently using recursion. Apart from this, only finite impulse response filters will be discussed further in this section.

9.2.1 Filter Techniques

The obvious implementation of a linear filter is illustrated in Figure 9.9, especially if the FPGA has plentiful multiplication or digital signal processing (DSP) blocks. Each pixel within the window is multiplied in parallel by the corresponding filter weight and then added; the DSP blocks of many FPGAs can perform multiply-and-accumulate, with the addition being wider to account for the increased word width from the multiplication. Alternatively, rather than adding the terms sequentially, an adder-tree can be used to minimise the propagation delay (Chu, 2006).

9.2.1.1 Pipelining and Transpose Structure

The propagation delay through the multiplication and adders in Figure 9.9 may exceed the system clock cycle and require pipelining. Since each input pixel contributes to several output pixels, rather than delaying the

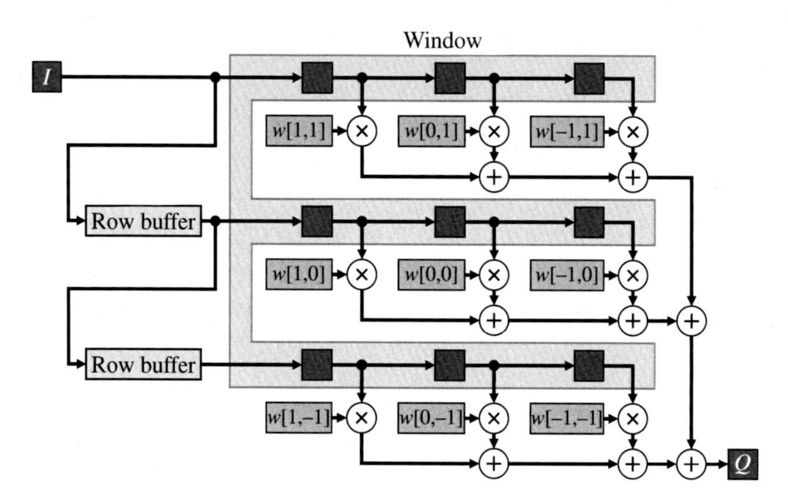

Figure 9.9 Direct implementation of linear filtering. Note that the bottom right window position is the oldest pixel and corresponds to the top left pixel within the image in the image. The filter weights are shown here as constants, but could also be programmable, and stored in a set of registers.

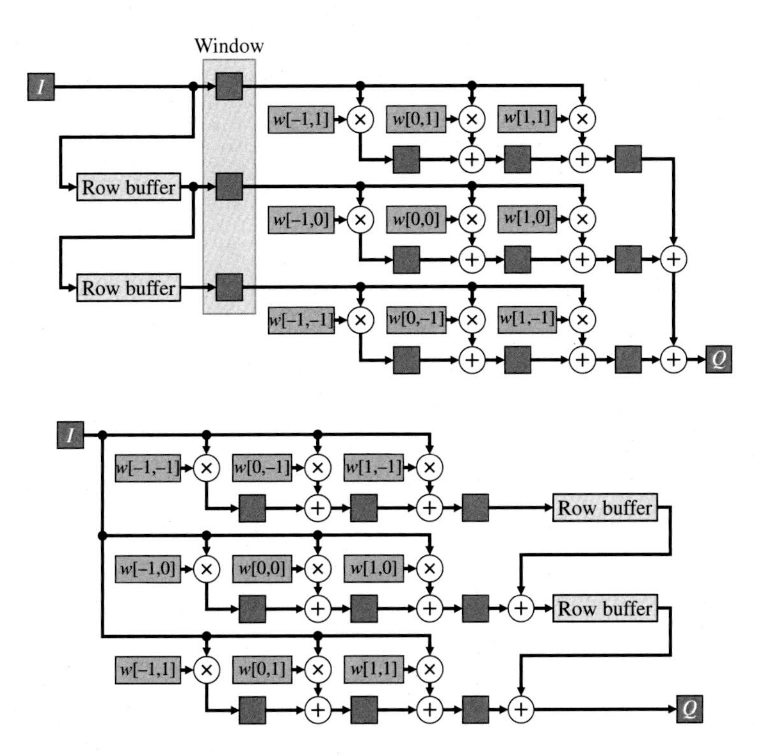

Figure 9.10 Pipelined linear filter using a transpose filter structure. Top: transpose structure for each row; bottom: transpose structure for the whole window. The transpose structure reverses to order of coefficients compared to Figure 9.9.

inputs and accumulating the output, the transpose filter structure performs all the multiplication with the input and delays the product terms (Mitra, 1998). The transpose filter structure does this by feeding data through the filter backwards (swapping the input and the output) and swapping summing junctions and pick-off points. This is applied in Figure 9.10. The advantage of the transpose structure is that the output is automatically pipelined. When transposing the whole window, it is necessary to cache the partial sums using row buffers until the remainder of the window appears. Depending on the filter coefficients, these partial sums may require a few guard bits to prevent accumulation of rounding errors. The row buffers therefore may need to be a few bits wider than the input pixel stream.

The transpose filter structures require a little more effort to manage the image boundaries, although border management can also be incorporated into the filter structure (Benkrid et al., 2003; Bailey and Ambikumar, 2018). As illustrated in Figure 9.11, this consists of a combination network to route the product terms corresponding to pixels outside the image to be combined with the appropriate output pixel. The combination network requires an additional adder and a multiplexer to select the combination for the border pixels (Bailey and Ambikumar, 2018).

9.2.1.2 Decomposition and Separability

Many filters can be decomposed into a cascade of simpler filters. The kernel of a composite linear filter is the convolution of the kernels of the constituent filters:

$$w[i,j] = w_1[i,j] \otimes w_2[i,j] = \sum_{x,y} w_1[x,y]w_2[i-x,j-y]. \tag{9.6}$$

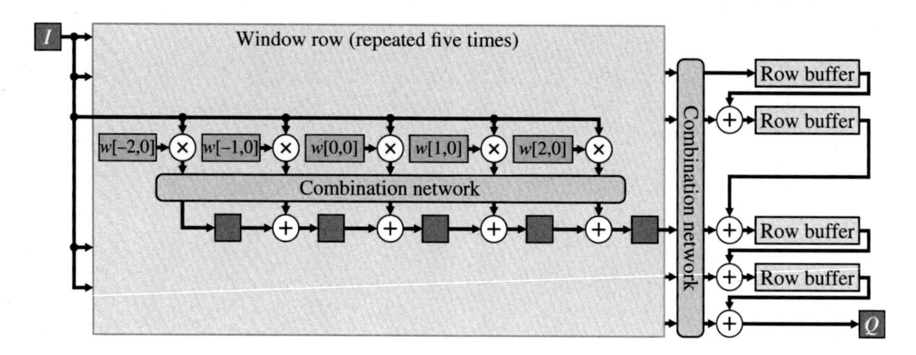

Figure 9.11 Image border management with transpose filter structures. Source: Adapted from Bailey and Ambikumar (2018).

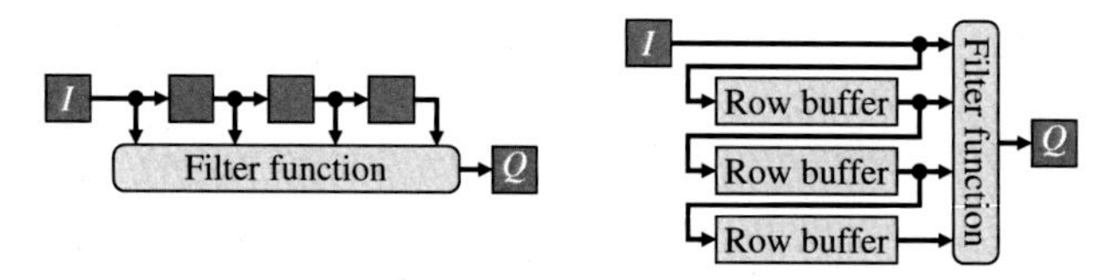

Figure 9.12 Converting a row filter to a column filter by replacing pixel delays with row buffers. Left: row filter; right: column filter.

If the 2-D filter can be decomposed into a cascade of two 1-D filters, the filter is said to be separable. A separable linear filter requires

$$w[i,j] = w_x[i]w_y[j]. \tag{9.7}$$

Separability will reduce the number of both multiplications and associated additions for implementing the filter from W^2 to $2W$. A 1-D row filter can be converted into a column filter by replacing the pixel delays within the window by row buffers, as shown in Figure 9.12.

While not all filters are directly separable, even arbitrary 2-D filters may be decomposed as a sum of separable filters (Lu et al., 1990; Bouganis et al., 2005)

$$w[i,j] = \sum_k w_{x,k}[i]w_{y,k}[j] \tag{9.8}$$

through singular value decomposition of the filter kernel. The vectors associated with the k largest singular values will account for most of the information within the filter. Such a decomposition can give savings if $k \leq W/2$, which will be the case if the original filter is symmetrical.

When implementing such a set of parallel separable filters, the relatively expensive row buffers can be shared between the column filters. Similarly, the window registers for the row filters can be shared using the transpose filter structure for the row filters (Bailey, 2010; Bailey and Ambikumar, 2018).

Other decompositions and approximations are also possible, especially for large windows. For example, Kawada and Maruyama (2007) show that large circularly symmetric filters may be approximated by octagons and rely on the fact that the pixels with a common coefficient lie on a set of lines. As the window scans, the total for each octagon edge is updated to reflect the changes from one window position to the next.

9.2.1.3 Reducing the Logic

Many filters are symmetric; therefore, many of their coefficients share the same value. If using the direct implementation of Figure 9.9, the corresponding input pixel values can be added prior to the multiplication.

Alternatively, since each input pixel is multiplied by a number of different coefficients, the input pixel value can be multiplied by each unique coefficient once, with the results cached until needed (Porikli, 2008b). This fits well with the transposed implementation of Figure 9.10. In both cases, the number of multipliers is reduced to the number of unique coefficients.

When implementing constant coefficient filters, if any of the coefficients is a power of 2, the corresponding multiplication is a trivial shift of the corresponding pixel value. If multiplier blocks are in short supply, the multiplications may be implemented with shift-and-add algorithms. For any given coefficient, the smallest number of shifts and adds (or subtracts) is obtained using the canonical signed-digit representation of the coefficients. Further optimisations may be obtained by identifying common sub-expressions and factorising them out (Hartley, 1991). For example, $165 = 10100101_2$ requires three adders for a canonical signed-digit multiplication but only two by factorising as $101_2 \times 100001_2$. Many techniques have been developed to find the optimal factorisation and arrangement of terms (see, for example (Potkonjak et al., 1996; Martinez-Peiro et al., 2002; Al-Hasani et al., 2011; Hore and Yadid-Pecht, 2019)). However, the increasing prevalence of high-speed pipelined multipliers within FPGAs has reduced the need for such decompositions. The optimised hardware multipliers are hard to outperform with the relatively slow adder logic of the FPGA fabric (Zoss et al., 2011).

If using a relatively slow clock rate, and hardware is at a premium, then significant hardware savings can often be made by doubling or tripling the clock rate but keeping the same data throughput. The result is a multi-phase design which enables expensive hardware (multiply-and-accumulate in the case of linear filters) to be reused in different phases of the clock. A three-phase system is shown in Figure 9.13. A zero is fed back in phase zero to reset the accumulation for a new output pixel.

9.2.2 *Noise Smoothing*

One of the most common filtering operations is to smooth noise. The basic principle behind noise smoothing is to use the central limit theorem to reduce the noise variance. With only one image, it is not possible to average with time as is used in Section 7.2.1. Instead, adjacent pixels are averaged. To avoid changing the overall contrast, a noise-smoothing filter requires unity average gain:

$$\sum_{i,j \in \mathbf{W}} w_{smooth}[i,j] = 1.0. \tag{9.9}$$

Most images have their energy concentrated in the low frequencies, with the high frequencies containing information relating to fine details and edges (rapidly changing pixel values). Random (white) noise has a uniform frequency distribution. Therefore, a low-pass filter will reduce the noise while having the least impact on the energy content of the image. Unfortunately, attenuating the noise also attenuates the high-frequency content of the image, resulting in a loss of fine detail and a blurring of edges.

With linear filters, their effect on the noise and image content can be analysed separately. Assuming that the noise is independent for each pixel, then from Eq. (7.12) the best improvement in noise variance for a given

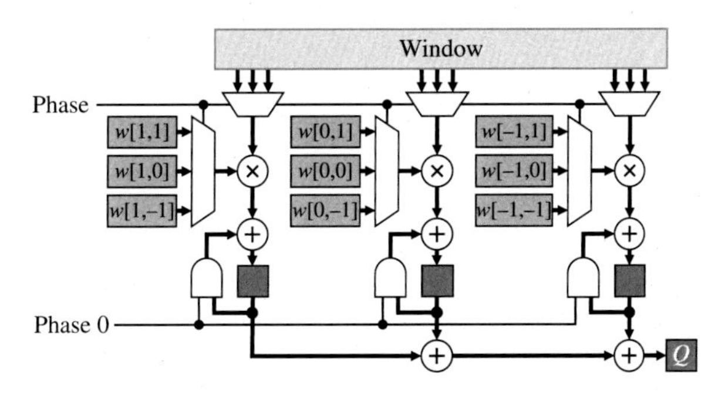

Figure 9.13 Using a multi-phase design with a higher clock speed can reduce the number of multipliers.

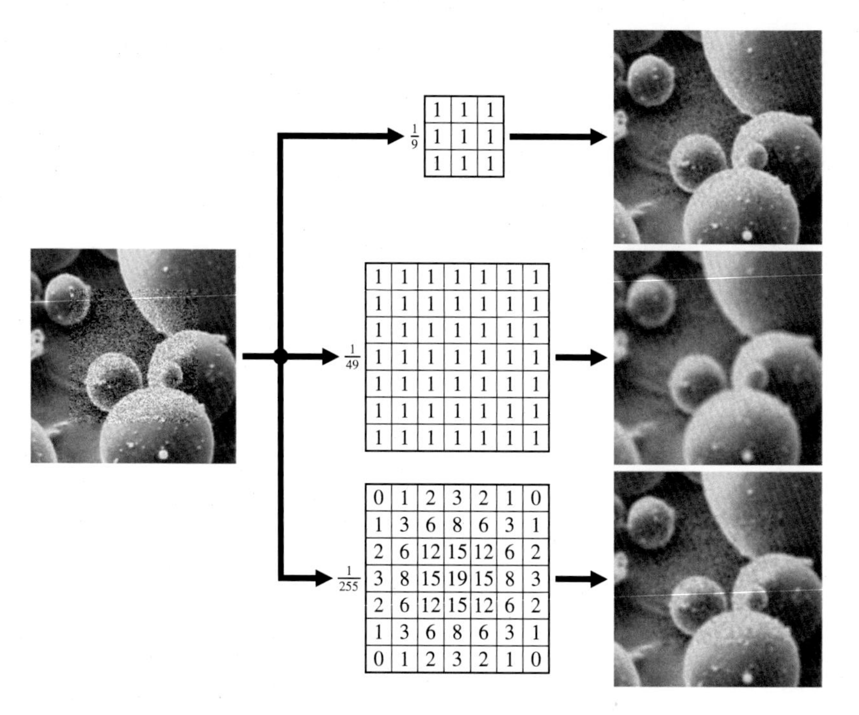

Figure 9.14 Noise smoothing applied to an image with added noise (*PSNR* = 15.3 dB). Top: a 3×3 box filter (noise *PSNR* = 24.7 dB, distortion *PSNR* = 34.8 dB, and total *PSNR* = 24.2 dB); centre: a 7×7 box filter (noise *PSNR* = 31.3 dB, distortion *PSNR* = 26.1 dB, and total *PSNR* = 24.7 dB); bottom: a Gaussian filter with $\sigma = 1.5$ (noise *PSNR* = 29.0 dB, distortion *PSNR* = 29.8 dB, and total *PSNR* = 26.0 dB).

size window will result if the weights are all equal (a box filter (McDonnell, 1981)). A $W \times W$ box filter will reduce the noise variance by W^2. A larger width window will therefore result in more noise smoothing. These noise suppression effects can be clearly seen in Figure 9.14.

Unfortunately, the blurring effects will become worse with increased noise smoothing (as seen in the distortion *PSNR* in Figure 9.14). So while filtering the noise will improve the signal-to-noise ratio, the distortion introduced through filtering reduces the signal-to-noise ratio. This trade-off between loss of fine details and noise suppression is also demonstrated in Figure 9.14. The optimum (from a least squares error) is reached when these two effects balance and is given by a Wiener filter.

9.2.2.1 Box Filter

The rectangular window of a box filter is separable, allowing the filter to be implemented as a cascade of two 1-D filters. Additionally, the equal weights make the filter amenable to a recursive implementation. Let $S[x]$ be the sum of pixel values within a 1-D window:

$$S[x] = \sum_{i=-\frac{W-1}{2}}^{\frac{W-1}{2}} I[x+i]. \tag{9.10}$$

Then,

$$S[x+1] = S[x] + I\left[x+1+\frac{W-1}{2}\right] - I\left[x-\frac{W-1}{2}\right], \tag{9.11}$$

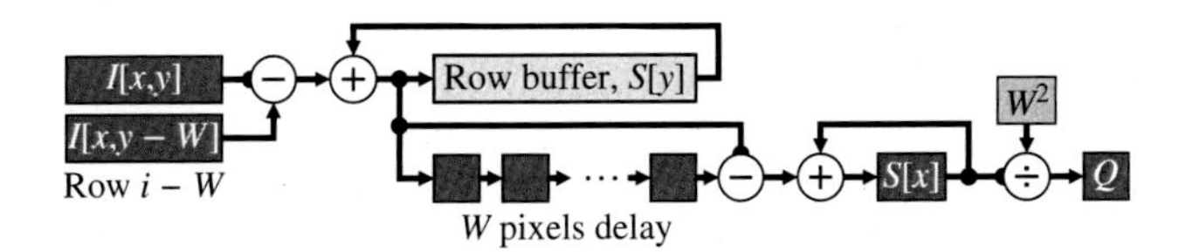

Figure 9.15 Efficient implementation of a 2-D $W \times W$ box average filter.

which reduces the 1-D filter to a single addition and subtraction regardless of the size of the window. An input pixel is then only required twice: when it enters, and when it leaves the window.

In two dimensions, this is applied vertically, as a column filter ($S[y]$), and then horizontally, as a row filter ($S[x]$), as is illustrated in Figure 9.15. For large windows, a large number of row buffers is required to cache the input for its second access. Therefore, if the input stream is from a frame buffer which has sufficient bandwidth to stream two rows simultaneously, the row buffering requirements can be reduced significantly, with only the current column sum buffered.

The Fourier transform of a uniform filter is a sinc function, so the frequency response of a box filter has poor side-lobe performance at high frequencies. These side-lobes mean that significant high-frequency noise still remains after filtering. This is apparent in the fine textured block in the centre image in Figure 9.14. A better low-pass filter can be obtained by rolling off the weights towards the edge of the kernel. While the noise variance will be higher than from a box filter (for a given window size), there is better attenuation of the high-frequency noise and fewer filtering artefacts.

9.2.2.2 Gaussian Filter

One such commonly used low-pass filter has Gaussian weights:

$$w_{Gaussian}[i,j] = \frac{1}{2\pi\sigma^2} \exp\left(-\frac{i^2+j^2}{2\sigma^2}\right), \tag{9.12}$$

where the standard deviation, σ, is the scale or equivalent radius of the filter. Note that the Gaussian weights never actually go to zero, although with quantisation they are zero beyond a certain radius. A practical implementation will often further truncate the weights to within a finite square kernel. The filter width, W, should generally be greater than about 4σ, and if accuracy is important, then $W \geq 6\sigma$. With truncation, the kernel coefficients may require rescaling to make the total equal to 1.0.

The Gaussian filter kernel is separable, enabling the logic to be significantly reduced for large window sizes. It is also symmetric, reducing the number of multiplications required for a 1-D kernel by a factor of 2. A cascade of two Gaussian filters in series gives a wider Gaussian filter. If the standard deviations of the two constituent filters are σ_1 and σ_2, then the combination has

$$\sigma = \sqrt{\sigma_1^2 + \sigma_2^2}. \tag{9.13}$$

The central limit theorem enables a Gaussian filter to be approximated by a sequence of rectangular box filters. For a sequence of k box filters of width W, the resultant approximate Gaussian has

$$\sigma = \sqrt{\frac{k}{12}(W^2 - 1)}. \tag{9.14}$$

For many applications, $k = 3$ or $k = 4$ will provide a sufficiently close approximation to a Gaussian filter.

Applying a series of Gaussian filters with successively larger standard deviations will remove features of increasing size or scale from the image, giving a scale-space representation of an image (Lindeberg, 1994). This is used, for example, in the scale invariant feature transform (see Section 10.4.1.1). Down-sampling successive lower resolution images gives a ***Gaussian pyramid***.

9.2.3 *Edge Detection*

Edge detection is another common application of filtering. At the edges of objects, there is often a change in pixel value, reflecting the contrast between an object and the background or between two objects. Since the edge is characterised by a difference in pixel values, a derivative filter can be used to detect edges. Alternatively, since the edges contain high-frequency information, a high-pass filter can be used to detect edges of all orientations. The main limitation of high-pass filters is their strong sensitivity to noise (see the top filter of Figure 9.16). Note that for an edge detection filter

$$\sum_{i,j\in \mathbf{W}} w_{edge}[i,j] = 0, \tag{9.15}$$

so that the response is zero in areas of uniform pixel value.

9.2.3.1 First Derivative

The first derivative of pixel values gives the intensity gradient within an image. The gradient is a vector, which is usually determined by calculating the gradient in two orthogonal directions:

$$\nabla I = \begin{bmatrix} \dfrac{\partial I}{\partial x} \\ \dfrac{\partial I}{\partial y} \end{bmatrix} = \begin{bmatrix} I_x \\ I_y \end{bmatrix}. \tag{9.16}$$

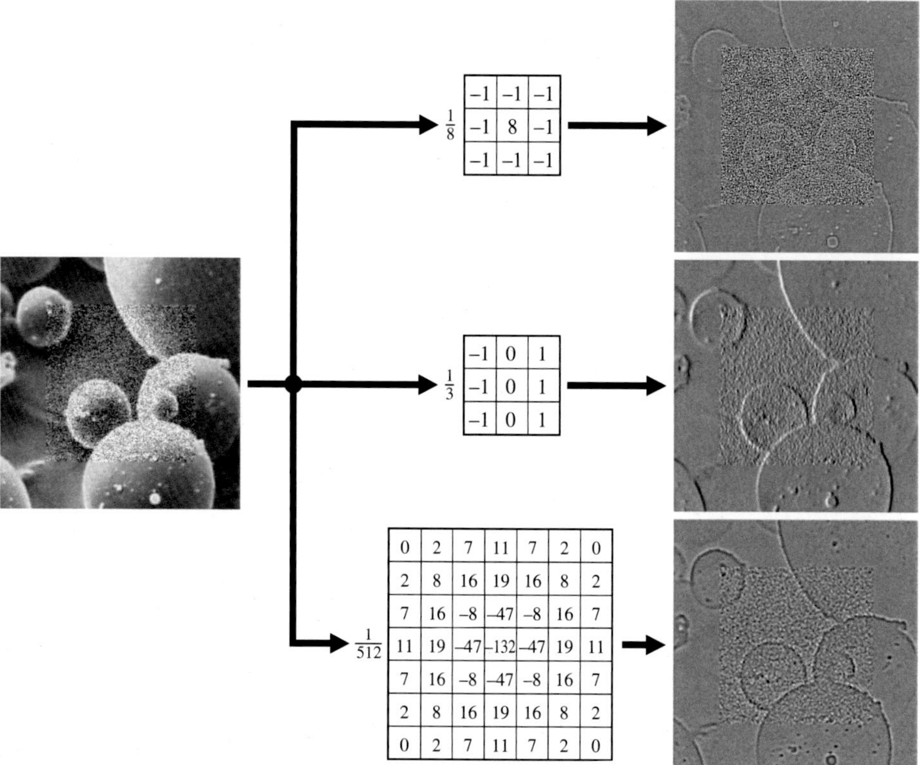

Figure 9.16 Linear edge detection filters; the output pixel values are offset to show negative values. Top: high-pass filter; centre: vertical edge detection using a Prewitt kernel. Source: Prewitt (1970)/with permission of ELSEVIER. Bottom: Laplacian of Gaussian filter, with $\sigma = 1.0$.

Table 9.1 Filter coefficients for optimal smoothing filters and derivative filters.

Filter	−3	−2	−1	0	1	2	3
Smoothing			0.229879	0.540242	0.229879		
Derivative			−0.425287	0.000000	0.425287		
Smoothing		0.037659	0.249153	0.426375	0.249153	0.037659	
Derivative		−0.104550	−0.292315	0.000000	0.292315	0.104550	
Smoothing	0.004711	0.069321	0.245410	0.361117	0.245410	0.069321	0.004711
Derivative	−0.018708	−0.125376	−0.193091	0.000000	0.192091	0.125376	0.018708

Source: Adapted from Farid and Simoncelli (2004).

The vector of the first partial derivatives is also sometimes called a Jacobian, $\mathbf{J}(I)$. On a discrete sample grid, the gradient is usually approximated in one dimension by taking the difference $w_D = \frac{1}{2}[-1\ 1]$ (which offsets the gradient by half a pixel) or the central difference $w_D = [-1\ 0\ 1]$. More accurate approximations can be obtained using longer filter kernels (Fornberg, 1988).

Differentiation is sensitive to noise, so the Prewitt filter in the middle of Figure 9.16 averages vertically (to give noise smoothing) while using a horizontal central difference to detect vertical edges. If accuracy in estimating the orientation of the gradient is required, Farid and Simoncelli (2004) argue that it is important to match the derivative filter with the smoothing filter. They derived the optimal smoothing filters and their associated derivatives, and these are listed in Table 9.1. In two dimensions, these filters are applied separably, using fixed-point arithmetic.

The noise sensitivity can also be improved significantly by first filtering the image with a Gaussian filter (for example as used in the Canny filter); however, there is a trade-off between accurate detection (improves with smoothing) and accurate localisation (deteriorates with smoothing) (Canny, 1986). With linear filters, it does not matter whether the derivatives are taken before or after smoothing. Indeed they can even be combined with the smoothing to give a single derivative of the Gaussian filter: in the x-direction

$$w_{Gaussian_x}[i,j] = \frac{i}{2\pi\sigma^4}\exp\left(-\frac{i^2+j^2}{2\sigma^2}\right) = \frac{i}{\sigma^2}w_{Gaussian}, \tag{9.17}$$

and similarly in the y-direction. With the smoothed image, the peaks of the first derivative will correspond to the best location of the edges. (Finding the magnitude of the gradient vector and peaks are nonlinear operations and will be discussed in Sections 9.3.1 and 9.3.3, respectively.)

9.2.3.2 Second Derivative

For edge detection, the position of maximum slope has a peak in the first derivative and consequently a zero-crossing of the second derivative. Note that finding the zero-crossings requires a nonlinear filter, which will be described in Section 9.3.4. Just as the first derivative measures gradient, the second derivative measures curvature, which is usually represented by the Hessian matrix:

$$\mathbf{H}(I) = \nabla\nabla I = \begin{bmatrix} \dfrac{\partial^2 I}{\partial x^2} & \dfrac{\partial^2 I}{\partial x \partial y} \\ \dfrac{\partial^2 I}{\partial x \partial y} & \dfrac{\partial^2 I}{\partial y^2} \end{bmatrix} = \begin{bmatrix} I_{xx} & I_{xy} \\ I_{xy} & I_{yy} \end{bmatrix}. \tag{9.18}$$

The eigenvectors of the Hessian give the principal directions of curvature, with the corresponding curvature values given by the eigenvalues. A scalar second derivative filter is the Laplacian,

$$\nabla^2 I = I_{xx} + I_{yy}. \tag{9.19}$$

The filter kernels for the different second derivative filters are shown in Figure 9.17.

0	0	0
1	-2	1
0	0	0

0	1	0
0	-2	0
0	1	0

$\frac{1}{4}$

1	0	-1
0	0	0
-1	0	1

0	1	0
1	-4	1
0	1	0

Figure 9.17 Second derivative filter kernels. From the left: I_{xx}, I_{yy}, I_{xy}, Laplacian.

In detecting edges, the noise response of the Laplacian may be tempered with a Gaussian filter. Consider the Laplacian of the Gaussian (LoG) filter (Marr and Hildreth, 1980):

$$w_{LoG}[i,j] = \frac{i^2 + j^2 - 2\sigma^2}{2\pi\sigma^6} \exp\left(-\frac{i^2 + j^2}{2\sigma^2}\right) = \frac{i^2 + j^2 - 2\sigma^2}{\sigma^4} w_{Gaussian}. \tag{9.20}$$

It is effectively a band-pass filter, where the standard deviation, σ, controls the centre frequency, or the scale from a scale-space perspective, at which the edges are detected. The bottom image of Figure 9.16 shows the effects of a Laplacian of Gaussian filter. Since the LoG filter is not separable, it may be efficiently implemented as a separable Gaussian followed by a Laplacian filter.

A filter with a similar response to the LoG filter is the difference of Gaussians (or DoG) filter. It is formed by subtracting the output of two Gaussian filters with different standard deviations:

$$w_{DoG} = \frac{1}{2\pi k^2\sigma^2} \exp\left(-\frac{i^2 + j^2}{2k^2\sigma^2}\right) - \frac{1}{2\pi\sigma^2} \exp\left(-\frac{i^2 + j^2}{2\sigma^2}\right), \tag{9.21}$$

where k is the ratio of standard deviations of the two Gaussians. The row buffers for the two parallel vertical filters may be shared, significantly saving resources. The DoG filter closely approximates the LoG filter when $k = 1.6$ (Marr and Hildreth, 1980). A ***Laplacian pyramid*** is a scale-space representation formed by taking the differences between the levels of a Gaussian pyramid (Lindeberg, 1994).

9.2.4 Edge Enhancement

An edge enhancement filter sharpens edges by increasing the gradient at the edges. In this sense it is the opposite of edge blurring, where the gradient is decreased by attenuating the high-frequency content. Edge enhancement works by boosting the high-frequency content of the image. One such filter is shown in Figure 9.18. Again note that

$$\sum_{i,j \in \mathbf{W}} w_{enhancement}[i,j] = 1 \tag{9.22}$$

to maintain the global contrast of the image. A major limitation of the high-frequency gain is that any noise within the image will also be amplified. This cannot be easily avoided with a linear filter, so linear edge enhancement filters should only be applied to relatively noise-free images. Over-enhancement of edges can result in ringing, which will become more severe as the enhancement is increased.

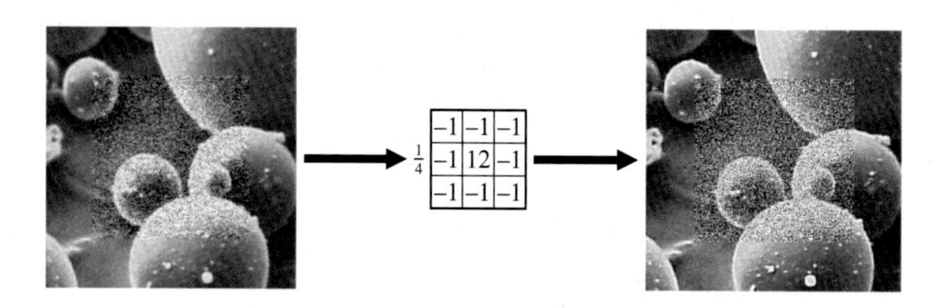

Figure 9.18 Linear edge enhancement filter.

9.3 Nonlinear Filters

Restricting $f(\cdot)$ in Eq. (9.1) to be a linear combination is quite limiting (Bailey et al., 1984). In particular, linear filters have difficulty distinguishing between legitimate changes in pixel value (for example at an edge) from undesirable changes resulting from noise. Consequently, linear filters will blur edges while attempting to reduce noise or be sensitive to noise while detecting edges or lines within the image.

Linear filters may be modified to improve their characteristics, for example by making the filter weights adaptive or dependent on the image content. The simplest of these are ***trimmed filters***, which omit some pixels within the window from the calculation. While the mean possesses good noise reduction properties, it is sensitive to outliers. A trimmed mean will discard the outliers by discarding the extreme pixels and calculate the mean of the remainder (Bednar and Watt, 1984). With a heavy-tailed noise distribution, this can give an improvement in signal-to-noise ratio by smoothing noise without significantly blurring edges. By selecting only the pixel values similar to the central pixel value, pixels primarily from one side of the edge are averaged, with a corresponding reduction in blur.

Gated filters take this one step further. They select between two (or more) different filters to produce an output based on a gating function, which is a function of the pixel values within the window (another filter). The gating is determined independently for each output pixel, as represented in Figure 9.19.

One example of a gated filter is edge-sensitive noise smoothing. By detecting the gradient within the window, and only smoothing perpendicular to the gradient direction (parallel to any edges), reduces noise without significantly blurring edges. This is a simple case of a steerable filter, where the orientation of the filter is selected based on the orientation of the structure within the window.

Another example of a gated filter is a noise-smoothing edge enhancement filter. If an edge is present (the gating function), an edge enhancement filter is applied; otherwise, a noise-smoothing filter can be selected. A more complex example would be to perform a statistical test to determine if the pixel values in a window come from a single Gaussian distribution (Gasteratos et al., 2006). If so, the mean can be used to smooth the noise; otherwise, the original pixel value is retained to prevent degrading the edges (which will generally have a mixed distribution).

While there are many useful nonlinear filters, only a few of the most common will be described here.

9.3.1 *Gradient Magnitude*

Nonlinear combinations of linear filters can also be useful. The most common of these is the Sobel edge detection filter (Abdou and Pratt, 1979). The outputs of two linear gradient filters (horizontal gradient I_x and vertical gradient I_y) are combined to give the 2-D gradient magnitude:

$$Q = \sqrt{I_x^2 + I_y^2}. \tag{9.23}$$

Both the horizontal and vertical Sobel filters are separable and may be decomposed to reduce the logic required, as shown in Figure 9.20. Equation (9.23) may be conveniently calculated using a CORDIC (coordinate rotation digital computer) unit (see Section 5.2.2), especially if the edge orientation is also required. However, if only the edge strength is required, two simpler alternatives are commonly used (Abdou and Pratt, 1979):

$$Q = \max\left(|I_x|, |I_y|\right) \tag{9.24}$$

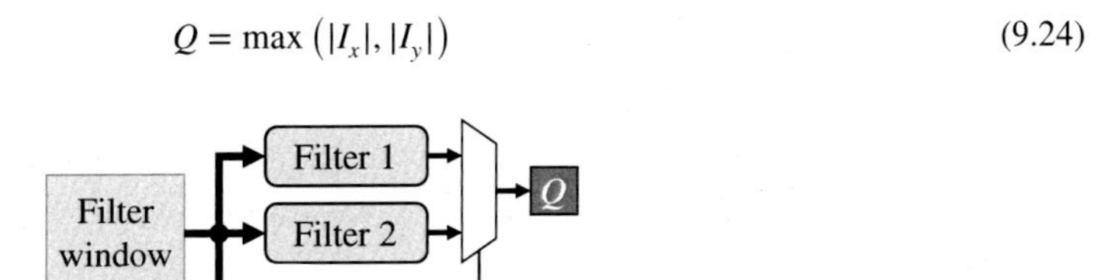

Figure 9.19 Gated filter.

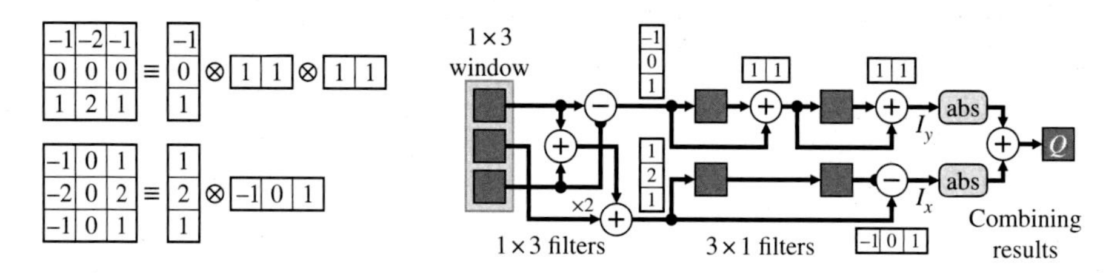

Figure 9.20 Sobel filter. Left: decomposition of component linear filters; right: efficient separable implementation using Eq. (9.25) to combine the results.

or

$$Q = |I_x| + |I_y|. \tag{9.25}$$

9.3.2 Edge Orientation

As indicated above, the orientation of edges, or intensity gradients, may be found by taking the arctangent of the gradient vector (for example using a CORDIC unit):

$$\theta = \tan^{-1}\left(\frac{I_y}{I_x}\right). \tag{9.26}$$

In general, using the Sobel filter to calculate I_x and I_y gives reasonably accurate results (Abdou and Pratt, 1979; Kittler, 1983), although best accuracy can be obtained using a matched filter and derivative filter (Farid and Simoncelli, 2004) (see Section 9.2.3.1).

If necessary, the effects of noise may be reduced by applying a Gaussian or other noise-smoothing filter before detecting the gradients. This also smooths the edge, making the gradient detection using a discrete filter more accurate. This process is illustrated in Figure 9.21.

In many applications, the output angle is required in terms of the neighbouring pixels. Rather than calculate the arctangent and quantise the result, the orientation may be determined directly. Figure 9.22 shows some of the possibilities. To distinguish between a horizontal or vertical gradient, a threshold of 45° may be evaluated by comparing the magnitudes as shown on the left. If necessary, the sign bits of the gradients may be used to distinguish between the four compass directions. The logic for this is most efficiently implemented using a small lookup table to give the two bits out. If an eight-neighbourhood is desired, the threshold angle is 22.5°. A sufficiently close approximation in many circumstances is

$$Horizontal = \left|I_x\right| > 2\left|I_y\right|, \tag{9.27}$$

which corresponds to a threshold of 26.6°. This will give a small bias to horizontal and vertical gradients over the diagonals. Again, the sign bits are used to resolve the particular direction and which quadrant the diagonals are in.

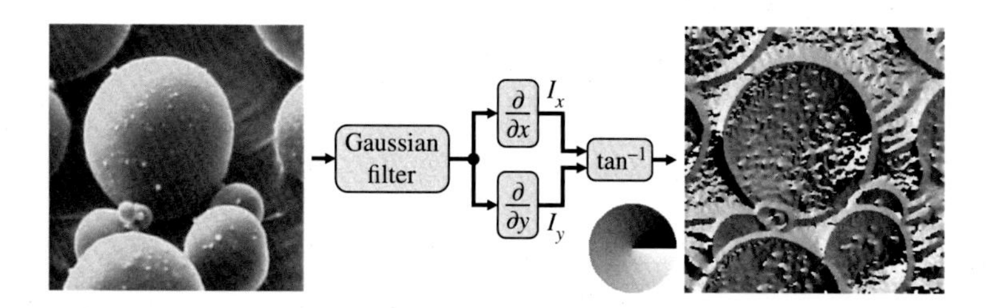

Figure 9.21 Edge orientation. The circle shows the mapping between angle and output pixel value.

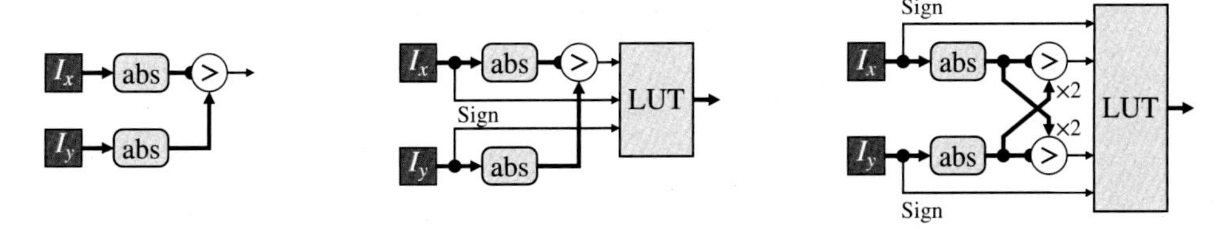

Figure 9.22 Circuits for detecting limited orientations. Left: distinguishing horizontal and vertical edges; centre: gradient in four directions (0°, ±90°, 180°); right: including diagonal directions (0°, ±45°, ±90°, ±135°, 180°). (LUT = lookup table.)

9.3.3 Peak Detection and Non-maximal Suppression

Peaks consist of pixels which are the local maximum (see Section 9.6.2.1) within a window. Non-maximal suppression simply sets the output pixel to 0 if it is not a local maximum, as shown in Figure 9.23. The delay is to account for the latency of the maximum filter but can often make use of the resources within the maximum filter (especially the row buffers).

Its use in edge detection is a little more complex. Consider the Canny edge detector (Canny, 1986), which first smooths the image, then detects the gradient, and suppresses points that are not a local maximum along the direction of the steepest gradient. Since the Gaussian smoothing will also spread the edge, the detector defines the position of the maximum gradient as the location of the edge. Non-maximal suppression therefore thins the edges, keeping only those points where the edge is the strongest. Since the edge strength can vary along the edge, to maintain edge continuity, it is necessary to find the maximum perpendicular to the edge.

The direction of the steepest gradient may be determined as horizontal, vertical, or one of the two diagonals using either a CORDIC block or the scheme described in Section 9.3.2. If the gradient at a point is larger than at both points on either side in the direction of the gradient, then the magnitude is kept. Otherwise, it is not a local maximum and is reset to zero. One circuit for implementing this is presented in Figure 9.24. The orientation must be delayed by one row to compensate for the latency of getting the magnitude on the row below the current pixel. The orientation is used to select the two adjacent magnitudes for comparison. If either of the neighbours is greater, this masks the output. Note that the Canny filter also performs hysteresis thresholding, which cannot be implemented as a simple filter (see Section 13.4.5 for a description of hysteresis thresholding).

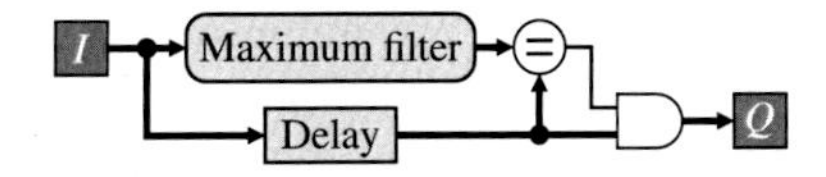

Figure 9.23 General non-maximal suppression.

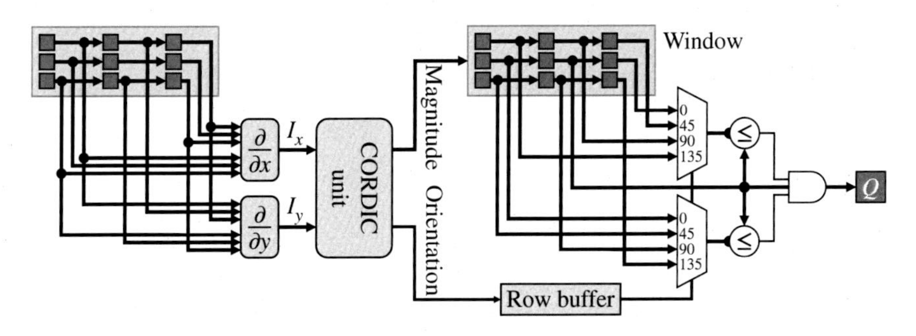

Figure 9.24 Non-maximal suppression for an edge detector.

9.3.4 Zero-crossing Detection

A zero-crossing detector is required to detect the edge locations from the output of Laplacian of Gaussian or difference of Gaussian filters. Whenever two adjacent pixels have opposite signs, there must be a zero-crossing somewhere in the space between them (only very occasionally does the zero-crossing occur precisely at a pixel location). Arbitrarily, the pixel on the positive side of the edge is defined as being the location of the zero-crossing. Referring to Figure 9.25, if the centre pixel is positive, and any of the four adjacent pixels is negative, then the centre pixel is a zero-crossing. Using only the sign bit also handles the case when the centre pixel is exactly zero.

The zero-crossings will form closed contours within a 2-D image. However, many of the detected crossings will correspond to insignificant or unimportant features. A cleaner edge image may be obtained by requiring the adjacent pixels to not only be of opposite signs but also be significantly different. This thresholding is shown in the right panel of Figure 9.25, although comes at the cost of potential missing edge points.

9.3.5 Bilateral Filter

When using a Gaussian filter to smooth noise, edges are also blurred because the pixels from both sides of an edge are averaged together. The idea behind trimmed filters can be extended to give the bilateral filter where the filter weights depend not only on the spatial position but also on the intensity difference (Tomasi and Manduchi, 1998):

$$Q[x,y] = \frac{\sum_{i,j \in \mathbf{W}} w_d[i,j]\ w_r[i,j]\ I[x+i,y+j]}{\sum_{i,j \in \mathbf{W}} w_d[i,j]\ w_r[i,j]}, \tag{9.28}$$

where

$$w_d[i,j] = \exp\left(-\frac{i^2+j^2}{2\sigma_d^2}\right) \tag{9.29}$$

is the spatial or domain weight which weights each pixel according to its distance from the window centre (the same as for the Gaussian filter). The second term,

$$w_r[i,j] = \exp\left(-\frac{\|I[x+i,y+j]-I[x,y]\|^2}{2\sigma_r^2}\right), \tag{9.30}$$

is the intensity or range weight, based on the similarity of the pixel value in the window to the input pixel. For colour images, using the $L^*a^*b^*$ colour space is suggested (Tomasi and Manduchi, 1998) because it is approximately perceptually uniform, and the Euclidean distance is meaningful.

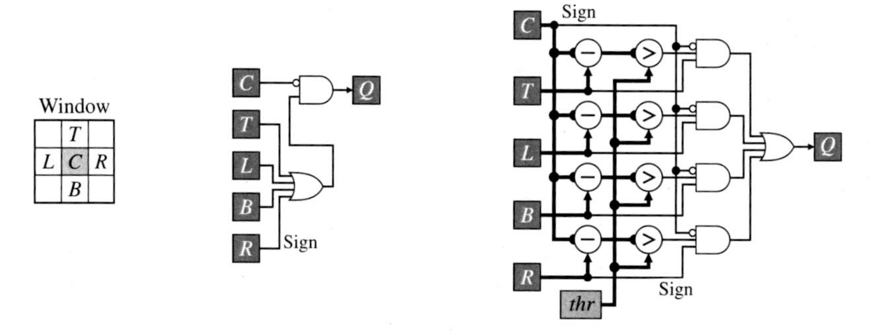

Figure 9.25 Zero-crossing detection filter. Left: definition of pixel locations within the window; centre: all zero-crossings; right: those above a threshold.

Since the range weight depends on the pixel values within the window, it is necessary to accumulate the weights to normalise the result (the denominator in Eq. (9.28)). These factors make the bilateral filter more complex to implement. For this reason, many implementations only work with small (3×3 or 5×5) spatial windows. For example, Kutty et al. (2014) use a 3×3 window. The absolute difference with the centre pixel value is looked up in a table to get the combined $w_r w_d$ coefficient (the symmetry of w_d enables the use of dual-port memory, requiring only one RAM for every two window pixels). The coefficients are multiplied by the window pixels, and both the products and coefficients are added using an adder-tree before a final division. Gabiger-Rose et al. (2014) work with a larger 5×5 window and use a lookup table to derive only the range weights. To reduce the logic, the internal filter runs at four times the pixel clock. For the domain filter, a separable Gaussian filter is applied to both the range weighted window pixels and the weights in parallel before the final division. Johnson et al. (2019) extend this approach for windows up to 15×15.

Several approximations can be used to reduce the overall complexity. These include using a piece-wise approximation of the range function (rather than using a lookup table) and replacing the Gaussian domain filter with a box filter. Although the bilateral filter is not separable, Pham and van Vliet (2005) claim that using two 1-D bilateral filters still gives good noise removal.

A different approach that scales well for large window sizes is to use a histogram of the pixels within the window (Porikli, 2008a). This enables the range filter to be calculated in constant time, regardless of the window size, although the running histogram requires considerable resources (see Figure 9.37).

9.3.6 Adaptive Thresholding

Often when thresholding using a constant global threshold (see Section 7.1.2), it is difficult to find a single threshold suitable for segmenting the whole image. Adaptive thresholding is a form of nonlinear filter where the threshold level, $thr[x, y]$, is made to depend on the local context:

$$Q[x,y] = \begin{cases} 1, & I[x,y] \geq thr[x,y], \\ 0, & I[x,y] < thr[x,y]. \end{cases} \tag{9.31}$$

To distinguish between object and background, the threshold level should be somewhere between the pixel value distributions of these regions. The simplest estimate therefore is an average of the pixel values within the window, provided the window is larger than the object size. (A Gaussian filter gives a smoother threshold, although the box average is simpler to calculate.) This approach works best around the edges of the object, where thresholding is most critical, but for larger objects or empty background regions may result in misclassifications. These may be reduced by offsetting the average. This is illustrated on the top path in Figure 9.26 for the image of Figure 7.9 (where global thresholding did not work).

An alternative (equivalent) way to look at adaptive thresholding is to consider the filter as a preprocessing step that adjusts the image enabling a global threshold to be effective:

$$\hat{I}[x,y] = I[x,y] - thr[x,y], \tag{9.32}$$

$$Q[x,y] = \begin{cases} 1, & \hat{I}[x,y] \geq 0, \\ 0, & \hat{I}[x,y] < 0. \end{cases} \tag{9.33}$$

An example of this is to use morphological filtering to estimate the background level, which can then be subtracted from the image enabling a global threshold to be used (the bottom path in Figure 9.26).

9.3.6.1 Error Diffusion

When producing images for human output, one problem of thresholding is a loss of detail. A similar problem occurs when quantising an image to a few grey-levels, where the spatial correlation of the quantisation error results in contouring artefacts. What is desired is for the local average level to be maintained, while reducing the number of quantisation levels.

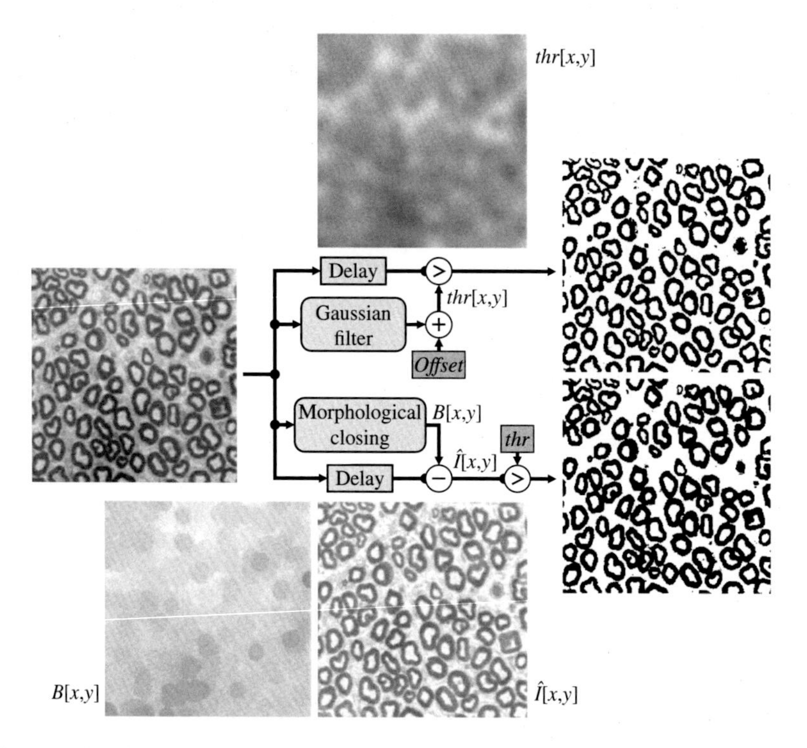

Figure 9.26 Adaptive thresholding. Top: using a Gaussian filter ($\sigma = 8$ and *Offset* = 16) to calculate the threshold level $thr[x, y]$; bottom: estimating the background using a 19×19 circular morphological closing to remove the objects ($B[x, y]$) and subtracting to remove the background variation ($\hat{I}[x, y]$).

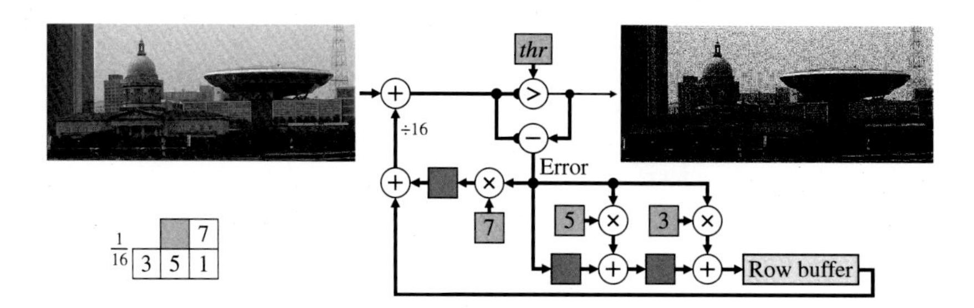

Figure 9.27 Floyd–Steinberg binary error diffusion. Source: Photo courtesy of Robyn Bailey.

Floyd and Steinberg (1975) devised such an algorithm for displaying greyscale images on a binary display. The basic principle is that after thresholding or quantisation, the quantisation error (the difference between the input and the output) is propagated on to the neighbouring four pixels that have not yet been processed. A possible implementation of this is shown in Figure 9.27. A row buffer holds the accumulated error for the next row, where it is added to the incoming pixel stream. The small integer multiplications can be implemented with a shift-and-add algorithm. Error diffusion is closely related to quantisation with noise shaping (Kite et al., 2000).

Shiau and Fan (1996) modified the window slightly and used power-of-2 distribution weights to reduce the computation and also some of the pattern artefacts resulting from the error diffusion. Ostromoukhov (2001) eliminated the pattern artefacts altogether using a zigzag or serpentine scan and having a different set of distribution weights for each pixel value.

The threshold level is normally set at mid-grey. Thresholding is then equivalent to selecting the most significant bit of the input. However, the threshold level is actually arbitrary (Knuth, 1987); regardless of the threshold level, the errors will average out to zero, giving the desired average grey-level on the output. Modulating the threshold level (making it dependent on the input) can be used to enhance edges (Eschbach and Knox, 1991), with the recommended level given as the one's complement of the input:

$$thr = \overline{I[x, y]}. \tag{9.34}$$

Although described here for a binary output, the same principles can be applied with any number of quantisation levels by propagating the quantisation error to reduce contouring effects. When applied to colour images, error diffusion may be applied to each component independently. With a binary output for each component, the outputs are in general uncorrelated between the channels. Threshold modulation may be used to force either correlation or anti-correlation between the channels (Bailey, 1997). A simple modulation based on intensity (applying Eq. (9.34) to the average of the RGB components) is sufficient to force resynchronisation.

9.3.7 High Dynamic Range Tone Mapping

Simple pixel-based global tone mapping methods for high dynamic range imaging were introduced in Section 7.2.6. However, in compressing the global dynamic range, the local dynamic range is also compressed, making the images look flat. This may be overcome using local tone mapping methods.

One approach to this is to use a local image scaling operation to increase the brightness within darker areas, while reducing the brightness of highlights (Bailey, 2012). A low-pass filter, $LPF(\cdot)$, such as a Gaussian or box filter, is used to estimate the average local brightness:

$$Q[x, y] = I[x, y]s(LPF(I)), \tag{9.35}$$

where $s(\cdot)$ is a scaling function that decreases monotonically for increasing brightness. Such a filter maintains local contrast since adjacent pixels are given similar weights since the output of the low-pass filter is slowly varying. However, one limitation of such a simple approach is a halo effect around large step edges (DiCarlo and Wandell, 2000), caused from the other side of the edge influencing the local average brightness. To overcome this, a trimmed filter or bilateral filter may be used.

A more common approach is to separate the image into base and detail components (Ou et al., 2022) as illustrated in Figure 9.28. The base component is usually estimated by a low-pass filter, for example a Gaussian filter, although using a bilateral filter (Nosko et al., 2020) can avoid halo effects. If necessary, multiple levels or scales of detail components can be used (DiCarlo and Wandell, 2000). Then, each of the components is compressed using a separate tone map (for example a logarithmic or power mapping). The base layer accounts for most of the wide dynamic range, so its dynamic range is compressed most strongly. The detail layer is compressed less and enables the local contrast to be maintained.

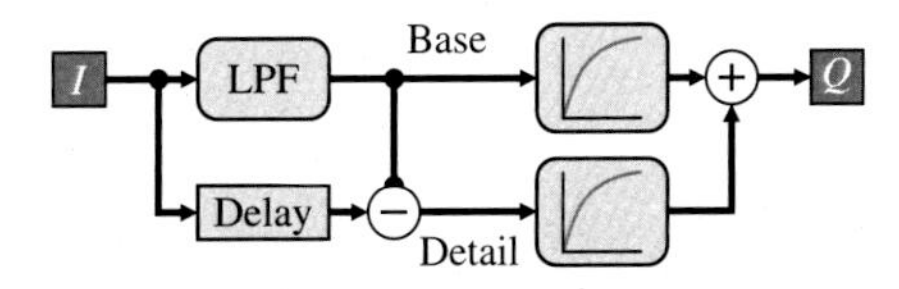

Figure 9.28 Tone mapping to reduce dynamic range. The low-pass filter separates the base image from the details, and a separate tone mapping is applied to each component.

9.4 Rank Filters

An important class of nonlinear filters are rank filters (Heygster, 1982; Hodgson et al., 1985). These sort the pixel values within the window and output the value from a selected rank position in the list, as illustrated in Figure 9.29. They are a generalisation of the minimum, maximum (Nakagawa and Rosenfeld, 1978), and median filters. Obviously, if only a single, constant rank position is required, then only that sorted output needs to be produced by the sorting network, and the multiplexer is not required.

Rank filters, or combinations of rank filters, can be used in many image processing operations. Using ranks close to the median will discard outliers giving them good noise-smoothing properties, especially for heavy-tailed distributions. However, the median filter is not as effective at removing additive Gaussian noise as a box filter of the same size. For impulsive noise, this is demonstrated in Figure 9.30 by almost completely removing the salt-and-pepper noise from a corrupted image. Since the median filter can corrupt non-noisy pixels, a switching median filter (Sun and Neuvo, 1994) is sometimes used. This is a gated filter that uses an impulse detector to select either the median or keep the original pixel value if the pixel is not corrupted. The recursive median filter performs better than the median (Qiu, 1996), although a real-time implementation on FPGA suffers problems with the propagation delay of the sorting network in the feedback loop. Peng et al. (2022) implemented a recursive median by dividing the window into two parts to minimise the propagation delay.

Straight edges are not affected by median filters, although if the edge is curved, its position can shift, affecting the accuracy of any measurements made on the object (Davies, 1999). In the absence of noise, median filters

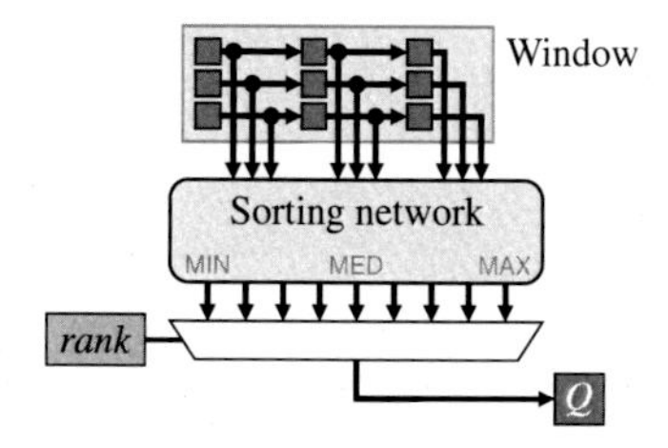

Figure 9.29 Rank filter.

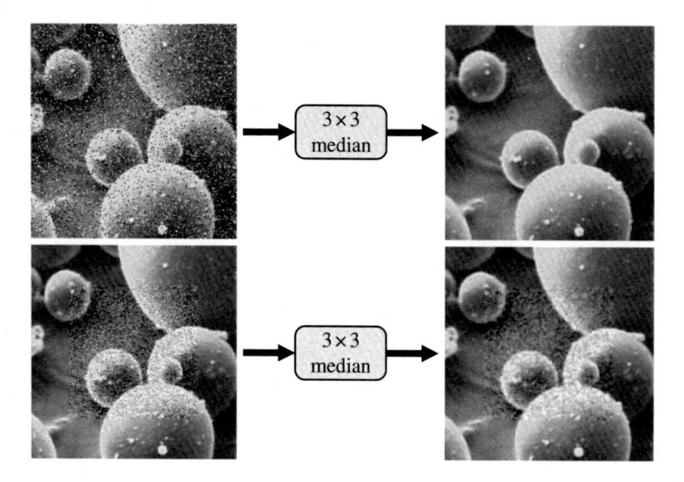

Figure 9.30 Median filter. Top: salt-and-pepper noise with 12% pixels corrupted (*PSNR* = 23.7 dB), after filtering *PSNR* = 34.0 dB; bottom: additive Gaussian noise (*PSNR* = 15.3 dB), after filtering *PSNR* = 22.1 dB.

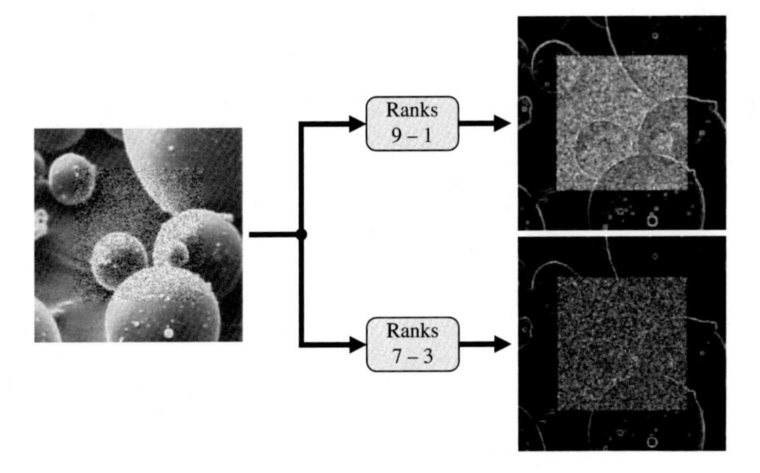

Figure 9.31 Subrange filters for edge detection.

do not blur edges (Nodes and Gallagher, 1982). However, any noise will cause edges to be slightly blurred with the level of blurring dependent on the noise level (Justusson, 1981).

To detect edges, the difference between two rank values can be used (Bailey and Hodgson, 1985), effectively using the range (or subrange) of pixel values within the window. Ranks above the median will select pixels from the light side of the edge, while ranks below the median will select from the dark side of the edge. Their difference will therefore indicate edge height. Using rank values closer to the extremes will give a stronger response but is also more sensitive to noise. Rank values closer to the median are less noise sensitive but will give a weaker response. These effects are clearly seen in Figure 9.31. Generally a 3×3 window is used for edge detection; a larger window will result in thick responses at the edges.

A gated rank filter can be used for enhancing edges (Bailey, 1990). Consider a blurred edge as shown on the left in Figure 9.32. When the centre of the window is over the edge, there are several pixels on each side of the edge within the window. These will be selected by rank values near the extremes. Let the pixel values selected by the rank filters to represent the dark and light sides of the edge be D and L, respectively. An edge enhancement filter classifies the centre pixel as being either light or dark depending on which is closer:

$$Q = \begin{cases} D, & C - D < L - C, \\ L, & \text{otherwise}. \end{cases} \tag{9.36}$$

Selecting the extreme values for D and L will be more sensitive to noise and outliers and will tend to give a slight over-enhancement. Ringing is avoided because one of the pixel values within the window is always selected for

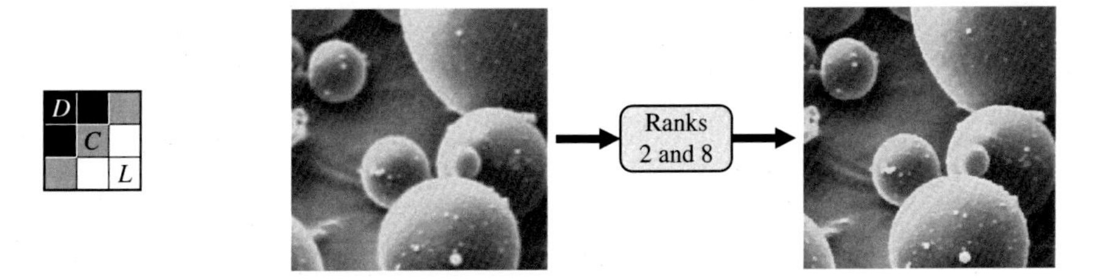

Figure 9.32 Rank-based edge enhancement filter. Left: pixels on either side of a blurred edge; right: edge enhancement using ranks 2 and 8.

output. Selecting rank values closer to the median will give some edge enhancement, while at the same time suppressing noise.

The rank-based edge enhancement filter is particularly effective immediately prior to multi-level thresholding and can significantly reduce the problems described in Section 7.1.2 (Figure 7.11). It also makes threshold selection easier because the classification process of Eq. (9.36) depopulates the spread between the histogram peaks that results from pixels that are a mixture of two regions. The thresholding results are therefore less sensitive to the exact threshold level used. For best enhancement, the size of the window should be twice the width of the blur, so that when the window is centred over a blurred edge, there will be representatives of each side of the edge within the window to be selected.

Any monotonically increasing transformation of the pixel values will not change their rank order. Rank filters are therefore commutative with monotonic transformations. Note that the results of rank-based edge detection or edge enhancement will be affected, because the output is a function of multiple selected values, and the relative differences will in general be affected by the transformation.

9.4.1 Sorting Networks

Key to the implementation of rank filters is the sorting network that takes the pixel values within the window and sorts them into ranked order. Several techniques have been developed for implementing this, both in software and in hardware.

Perhaps the simplest and most obvious approach is to use a bubble sorting odd–even transposition network as shown in Figure 9.33. However, this is only suitable for small windows because the logic requirements grow with the square of the window area. Running the filter with a higher clock rate (using a multi-phase design) can reduce the hardware requirements by sharing hardware between clock cycles.

An improvement can be gained by exploiting the overlap between successive window positions. This may be achieved by sorting a column once as the new pixel comes into the window and then merging the sorted columns (Chakrabarti and Wang, 1994; Waltz et al., 1998). One arrangement of this is shown in Figure 9.34. Note that for three inputs it is both faster and more efficient to perform the comparisons in parallel and use a single larger multiplexer than it is to use a series of three two-input compare-and-swap blocks. In the second layer of compare-and-sorts, since the inputs are successively delayed with each window position, the middle comparator can also be removed and replaced with the delayed output of the first comparator. This reduces the number of comparators from 36 required for a bubble sort to 18. Again, this can be further reduced if not

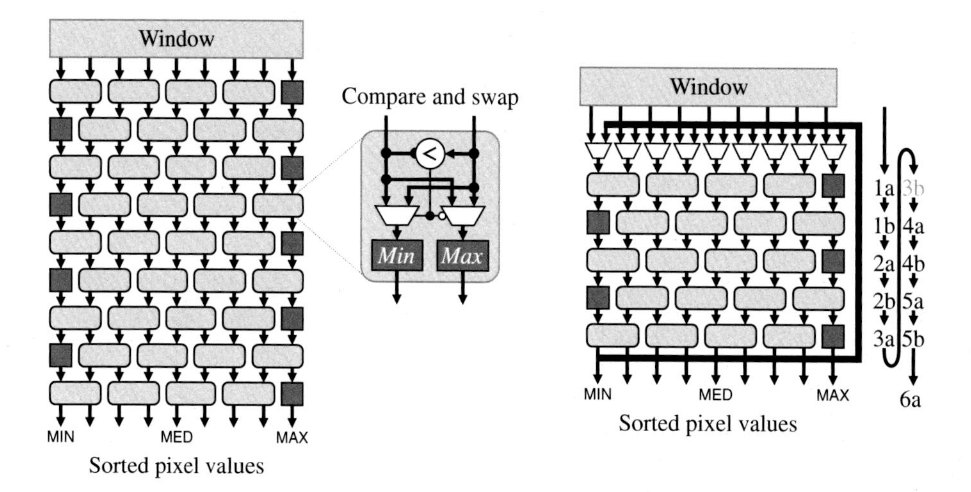

Figure 9.33 Bubble sorting using an odd–even transposition network. Left: each stage is pipelined; right: two-phase design.

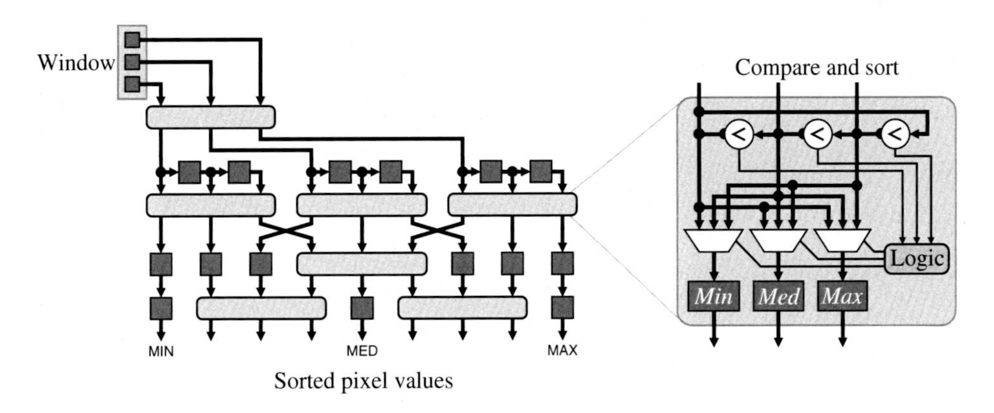

Figure 9.34 Sorting columns first, followed by a merging.

every rank value is required (for example minimum, maximum, or median filtering). A similar analysis may be performed for larger window sizes (Kumar et al. (2017) applied it for a 5×5 median). Although savings are gained by sorting the columns only once, the logic requirements still increase approximately with the square of the window size. Similarly, multiple adjacent windows may be calculated in parallel, reusing the calculations. For example, Chen et al. (2019) used this approach to calculate a 2×2 block of medians using a 3×3 window.

The concept of 2-sorting (compare-and-swap) and 3-sorting (shown in Figure 9.34) can be generalised to single-stage N-sorting (Kent and Pattichis, 2021). The key principle is to compare each pair of inputs to derive a set of control signals. These are then added to give a count for each input which determines which output it gets routed to. Finally, a multiplexer (effectively a cross-bar switch) routes each input to the corresponding output in a single stage.

Rather than sort the actual pixel values, each pixel in the window can have a corresponding rank register which counts the number of other pixels in the window greater than that pixel (Swenson and Dimond, 1999). Then as pixels enter and exit the window, it is necessary to compare these with each of the existing window pixels to determine their rank and to adjust the ranks accordingly. For one-dimensional filtering, a systolic array may be used to efficiently compare each pixel with every other pixel in the window and assign a rank value to each sample. In two dimensions, it is necessary to remove and insert a whole new column at each step (Hwang and Jong, 1990), giving an output every W cycles.

An alternative approach is to use the threshold decomposition property of rank filtering (Fitch et al., 1985), as is illustrated in Figure 9.35. Rank filtering is commutative with thresholding, so if the image is thresholded first, a binary rank filter may be used. Therefore, a grey-level rank filter may be implemented by decomposing the input using every possible threshold, performing a binary rank filter, and then reconstructing the output by adding the results.

A simple implementation of a binary rank filter is to count the number of '1's within the window and compare this with the rank value, as shown in Figure 9.36. The counter here is parallel; it is effectively an adder network,

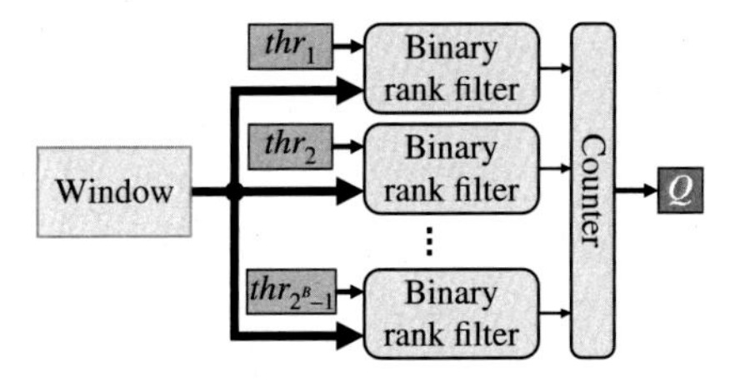

Figure 9.35 Threshold decomposition: the window values are thresholded to enable separate binary filters and reconstructed by combining the outputs.

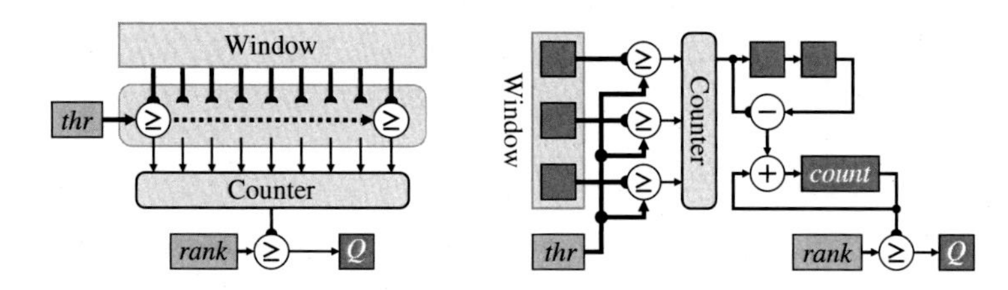

Figure 9.36 Binary rank filter for threshold decomposition. Left: thresholding to create binary values, and counting; right: an efficient implementation (similar to box filtering) exploiting commonality between adjacent window positions.

adding the inputs to give the count. As with other implementations, it is possible to exploit the significant window overlap for successive pixels. An implementation of this is shown on the right in Figure 9.36. The column counts are maintained, with the new column added and the old column subtracted as the window moves. This reduces the size of the counter from W^2 to W inputs.

Threshold decomposition requires $2^B - 1$ parallel binary rank filters. This is expensive for small windows but becomes more efficient for larger window sizes. This principle of threshold decomposition may also be generalised to stack filters (Wendt et al., 1986).

A closer look at the left panel of Figure 9.36 reveals that the count produced is actually the bin from cumulative histogram of the window corresponding to the threshold, *thr*. An early software approach to median filtering maintained a running histogram of the window, updating it as the data moved into and out of the window (Garibotto and Lambarelli, 1979; Huang et al., 1979). In hardware, the trick used for box filtering can be applied to the running histogram as shown in Figure 9.37. By maintaining column histograms within the window, as the window moves to the next pixel, the histogram for the new column is updated and added to the window histogram. The column that has just left the window is subtracted from the window histogram. The column histograms must be cached in a row buffer until the next line. Since the histograms must be added and subtracted in parallel, they must be stored in registers, although a block RAM may be used for the window delay and histogram row buffer. The resources required to build the running histogram are independent of the window size. However, significant memory resources are required for the histogram row buffer and for registers for the column and window histograms. This approach is therefore suitable for large window sizes, although it does require a larger modern FPGA to provide the required on-chip memory.

Goel et al. (2019) simplified the process by only building the histogram using the upper four bits of the input pixels. Although this only gives an approximate result, it significantly reduces the memory resources. The histogram was rebuilt for every pixel within a 3×3 window every clock cycle.

Given the local histogram, it is necessary to search the histogram to find the corresponding rank value. This can be made faster using a cumulative histogram (which can be built in a similar manner to Figure 9.37).

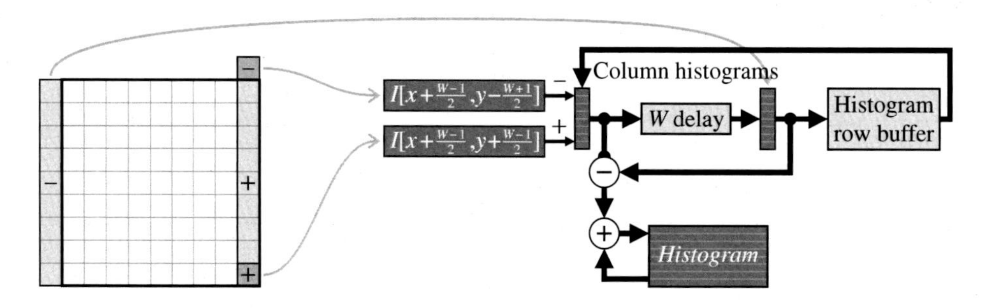

Figure 9.37 Maintaining a running histogram of a rectangular window. One pixel is added to and one removed from a column, and one column is added and one removed for each window position.

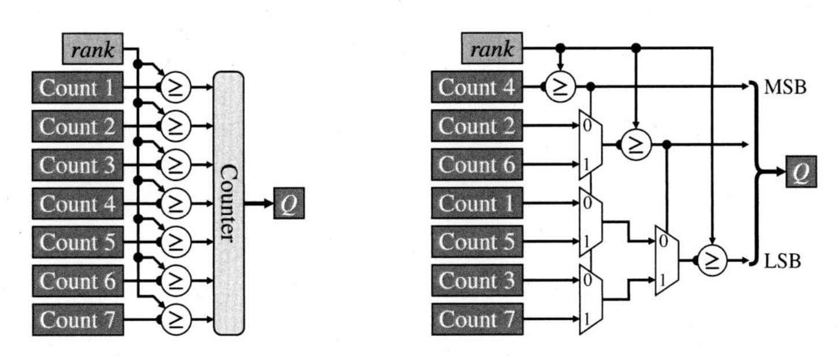

Figure 9.38 Searching the cumulative histogram for a rank value (illustrated for 3-bit data). Left: parallel approach of Figure 9.35; right: binary search. (MSB = most significant bit; LSB = least significant bit.)

This approach was used by Fahmy et al. (2005) for 1-D windows. An alternative to the direct output shown in Figure 9.35 is to use a binary search of the cumulative histogram (Harber and Neudeck, 1985). This is illustrated in Figure 9.38 for 3-bit data words. In the right panel, the cumulative histogram counts are arranged in a bit-reversed order to minimise the interconnect length. The binary search can easily be extended to wider words, using pipelining if necessary to meet timing constraints. The binary search circuitry needs to be duplicated for each rank value required at the output.

This binary search can be applied directly to the original data using a bit voting approach. The idea is to reduce the logic by performing a binary sequence of threshold decompositions, with each decomposition yielding one bit of the output (Ataman et al., 1980; Danielsson, 1981). The threshold decomposition is trivial, selecting the most significant bit of those remaining, while passing on the remaining bits to the next stage. If the most significant bit of the input differs from the desired rank output, that input pixel value can never be selected for the rank output. This is achieved by setting the remaining bits to '0' or '1' accordingly to prevent them from being selected in subsequent stages. The circuit for this scheme is shown in Figure 9.39. A scheme similar to this has been implemented on an FPGA by Benkrid et al. (2002), Prokin and Prokin (2010) for median filtering, and Choo and Verma (2008) for rank filtering. Note that only a single rank value is output; to produce additional rank values, the sorting network must be duplicated.

While the median and other rank filters are not separable, a separable median filter can be defined that applies a 1-D median filter first to the rows, then to the columns of the result, or vice versa (Shamos, 1978; Narendra, 1981). Although this is not the same as the true 2-D median, it is a sufficiently close approximation in most image processing applications. It also gives a significant savings in terms of computational resources. Note that in general a different result will be obtained when performing the row or column median first.

An alternative approach is to reformulate the 1-D median and apply it in two dimensions as a multi-shell filter (Li, 1989; Li and Ramsingh, 1995). Multi-shell filters give good performance at removing isolated noise points while preserving fine linear features.

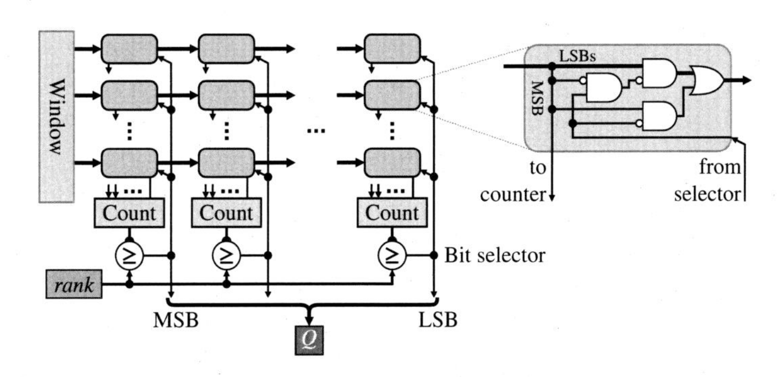

Figure 9.39 Bit sequential rank sorter. (MSB = most significant bit; LSB = least significant bit.)

9.5 Adaptive Histogram Equalisation

Closely related to rank filtering is adaptive histogram equalisation for contrast enhancement. Global histogram equalisation, as described in Section 8.1.3, is ineffective if there is a wide variation in background level, making the global histogram approximately uniform. Adaptive histogram equalisation overcomes this by performing the histogram equalisation locally using a window, with only the centre pixel value being transformed, using separate mapping based on local context (Hummel, 1977). This expands the contrast locally while making the image more uniform globally. To reduce the computational complexity, many software implementations divide the image into a series of overlapping tiles and either use a single transformation for all of the pixels within each tile or interpolate the transformations between tiles and apply the interpolated transformation to each pixel (Pizer et al., 1987).

Since histogram equalisation uses the scaled cumulative histogram within the window, the output is effectively the scaled rank position of central pixel value within the window (Pizer et al., 1987). For small windows, the threshold decomposition of Figure 9.36 can be modified to calculate the cumulative histogram bin for the current centre pixel value (see the left panel in Figure 9.40). Unfortunately, because the threshold varies with each pixel it is not possible to exploit the overlap as the window is scanned.

For larger window sizes, a running histogram can be used to calculate the cumulative histogram bin for the centre pixel by summing the counts for the bins less than the centre. This is then normalised to give the output pixel value (Kokufuta and Maruyama, 2009), as represented on the right in Figure 9.40.

This approach of local contrast enhancement can also be extended to other variants, such as contrast limited adaptive histogram equalisation. This may be achieved either through dividing the image into tiles and interpolating the mapping between tiles (Reza, 2004) or by separately processing each pixel (Kokufuta and Maruyama, 2010).

9.6 Morphological Filters

Based on set theory, mathematical morphology defines a set of transformations on a set based on a structuring element (Serra, 1986). In terms of image processing filters, the set can be defined by the pixel values within the image, and the structuring element may be thought of as an arbitrary window. As implied by the name, morphological filters will filter an input image on the basis of the shape or morphology of objects within the image. The structuring element defines the shape of the filter and defines or controls what is being filtered. The basic principles of morphological filtering are easiest understood in terms of binary images. These principles are later extended to filtering greyscale images.

9.6.1 Binary Morphology

A binary image considers each pixel to belong to one of two classes: object and background. Let the object pixels be represented by '1', and the background by '0'. The structuring element, **S**, consists of a set of vectors or offsets. It can therefore be considered as another binary image or shape and has the same function as the local window for other filters.

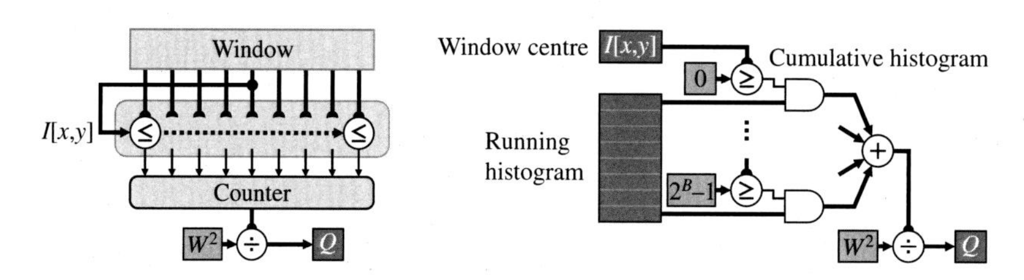

Figure 9.40 Adaptive histogram equalisation. Left: using window thresholding, suitable for small windows; right: using a running histogram (see Figure 9.37), suitable for large windows.

The two basic operations of morphological filtering are erosion and dilation. With ***erosion***, an object pixel is kept only if the structuring element placed at that pixel fits completely within the object. Erosion is therefore a logical AND of the pixels within the window:

$$Q_{erosion}[x,y] = I \ominus S = \bigwedge_{i,j \in \mathbf{S}} I[x+i, y+j]. \tag{9.37}$$

It is called an erosion, because the object size becomes smaller as a result of the processing.

With ***dilation***, each input pixel is replaced by the shape of the structuring element within the output image. This is equivalent to outputting an object pixel if the flipped structuring element overlaps an object pixel in the input. Dilation is therefore a logical OR of the flipped window pixels:

$$Q_{dilation}[x,y] = I \oplus S = \bigvee_{i,j \in \mathbf{S}} I[x-i, y-j]. \tag{9.38}$$

Note that the flipping of the window (rotation by 180° in two dimensions) will only affect asymmetric windows. In contrast with erosion, dilation causes the object to expand, and the background to become smaller. Erosion and dilation are duals in that a dilation of the object is equivalent to an erosion of the background and vice versa.

Many other morphological operations may be defined in terms of erosion and dilation. Two of the most commonly used morphological filters are opening and closing. An ***opening*** is defined as an erosion followed by a dilation with the same structuring element:

$$I \circ S = (I \ominus S) \oplus S. \tag{9.39}$$

The erosion will remove features smaller than the size of the structuring element, while the dilation will restore the remaining objects back to their former size. Therefore, the remaining object points are only where the structuring element fits completely within the object.

The converse of this is a ***closing***, which is a dilation followed by an erosion:

$$I \bullet S = (I \oplus S) \ominus S. \tag{9.40}$$

Background regions are kept only if the structuring element completely fits within them. It is called a closing because any holes or gaps within the object smaller than the structuring element are closed as a result of the processing. These four morphological filters are illustrated in Figure 9.41.

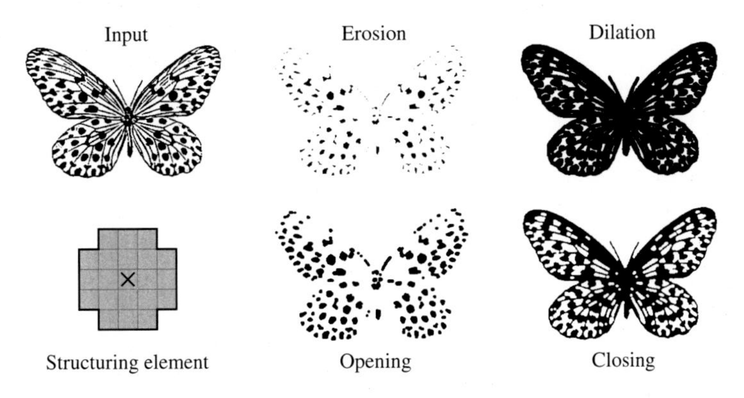

Figure 9.41 Morphological filtering. In this example, black pixels are considered object, with a white background. Source: Photo courtesy of Robyn Bailey.

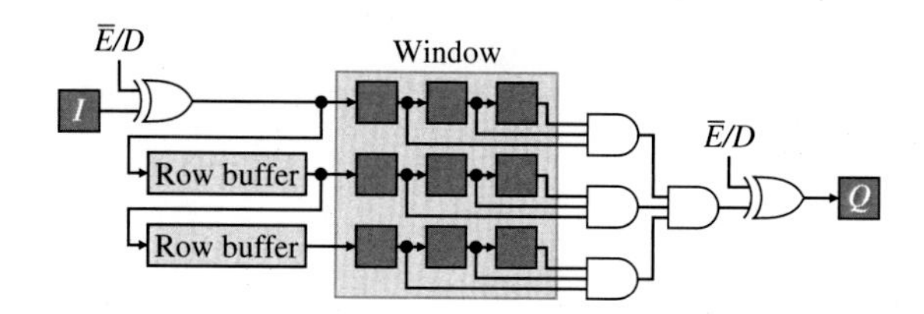

Figure 9.42 Circuit for erosion and dilation. The control signal selects between erosion and dilation.

9.6.1.1 Implementation

The relatively simple processing and modest storage required by morphological filters has made them one of the most commonly implemented image processing filters, especially in the early days when FPGAs were quite small and the available resources limited what could be accomplished. The direct implementation of erosion and dilation is relatively trivial, as seen in Figure 9.42. Duality enables both to be implemented with a single circuit, with a control signal used to complement the input and output.

Opening and closing may be implemented as a pipeline of erosion followed by dilation or dilation followed by erosion, respectively.

9.6.1.2 Decomposition

Just as linear filters could be decomposed as a sequence of simpler smaller filters, morphological filters can also be decomposed. A series decomposition makes use of the fact that dilation is associative, allowing a complex structuring element to be made up of a sequence of simpler structuring elements:

$$(I \oplus S_1) \oplus S_2 = I \oplus (S_1 \oplus S_2), \tag{9.41}$$

$$(I \ominus S_1) \ominus S_2 = I \ominus (S_1 \oplus S_2). \tag{9.42}$$

A series decomposition of the diameter 5 circular structuring element is presented on the left in Figure 9.43. The most obvious application of the chain rule is to make the operations for rectangular structuring elements separable, with independent filters in each of the horizontal and vertical directions.

A parallel decomposition is based on combining filters in parallel. For dilations:

$$I \oplus (S_1 \vee S_2) = (I \oplus S_1) \vee (I \oplus S_2), \tag{9.43}$$

And for erosions:

$$I \ominus (S_1 \vee S_2) = (I \ominus S_1) \wedge (I \ominus S_2). \tag{9.44}$$

A parallel decomposition is illustrated on the right in Figure 9.43. A rectangular window is often used for the parallel decomposition to exploit separability (Bailey, 2010). The row buffers can be shared for the parallel column filters, and using a transpose structure for the row filters allows the registers to be shared for those. This approach is demonstrated in Figure 9.44 for a 5 × 5 circular erosion filter.

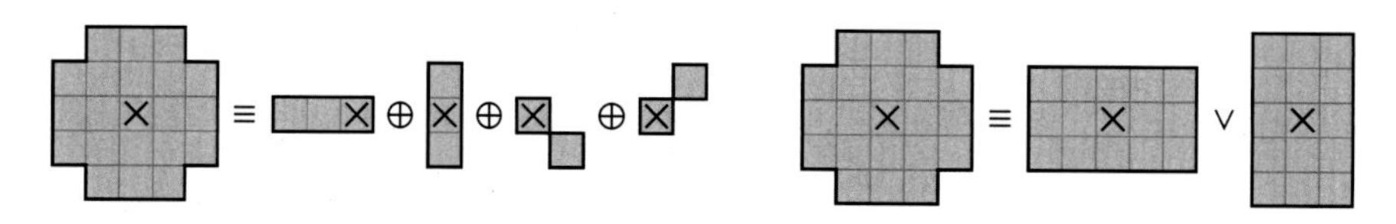

Figure 9.43 Examples of structuring element decompositions. Left: series decomposition using dilation; right: parallel decomposition of dilation using OR.

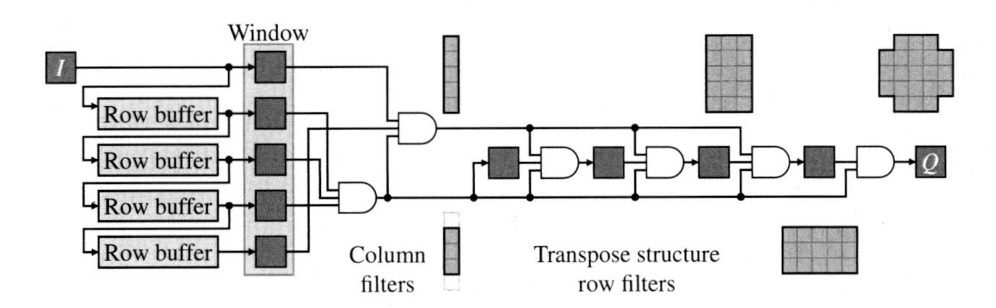

Figure 9.44 Parallel separable filter decomposition for a 5×5 circular erosion, sharing row buffers and register chain. (A dilation filter would replace all the AND gates with OR gates.)

9.6.1.3 State Machine Implementation

An alternative approach is to implement the filter directly as a finite state machine (Waltz, 1994b). First, consider a 1-D erosion filter. The pattern of '1's and '0's in the window can be considered as the state, since this information must be maintained to perform the filtering. If the structuring element is continuous, a useful representation of the state is the count of successive '1's within the window. The output is a '1' only when all of the inputs within the window are '1's; therefore, whenever a '0' is encountered, the count is reset to 0, and when the count reaches the window width, it saturates. An example state transition diagram and the corresponding implementation are given in Figure 9.45. This approach may be extended to arbitrary patterns by maintaining a more complex state machine. The state is effectively a compressed representation of the elements within the window.

The finite state machine approach can also be extended to two dimensions by creating two state machines, one for operating on the rows and one for the columns (Waltz, 1994b; Waltz and Garnaoui, 1994), as illustrated in Figure 9.46. It is not necessary for the structuring element to be separable in the normal sense of separability; the row state machine passes the pattern information representing the current state of the row to the column state machine.

First, the required patterns on each of the rows of the structuring element are identified. The row state machine is then designed to identify the row patterns in parallel as the data shifts in. Combinations of the row patterns are then encoded in the output to the column state machine. The column state machine then identifies the correct sequence of row patterns as each column shifts in. In this example, the row machine has only seven states, and the column machine six states. The finite state machines may therefore be implemented using small lookup tables.

The technique may be extended to filters using several different structuring elements in parallel. The corresponding state machines are coded to detect all of the row patterns for all of the filters in parallel and similarly detect the outputs corresponding to all of the different row pattern combinations in parallel.

The extension to dilation is trivial; the pattern is made up of '0's rather than '1's. It may also be extended to binary template matching (Waltz, 1994a), where the row patterns consist of a mixture of '0's and '1's (and don't cares). Binary correlation is implemented in a similar manner (Waltz, 1995), although it counts the number of

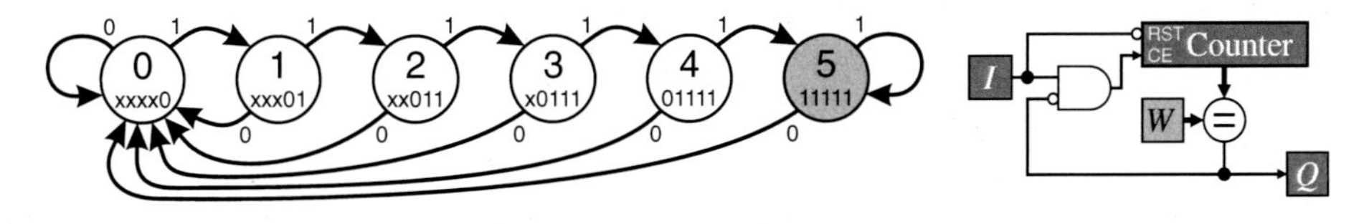

Figure 9.45 Left: horizontal 1×5 erosion based on a state machine implementation. The shaded state outputs a '1', and the other states output a '0'; right: implementation of a $1 \times W$ horizontal erosion. (CE = clock enable; RST = reset.)

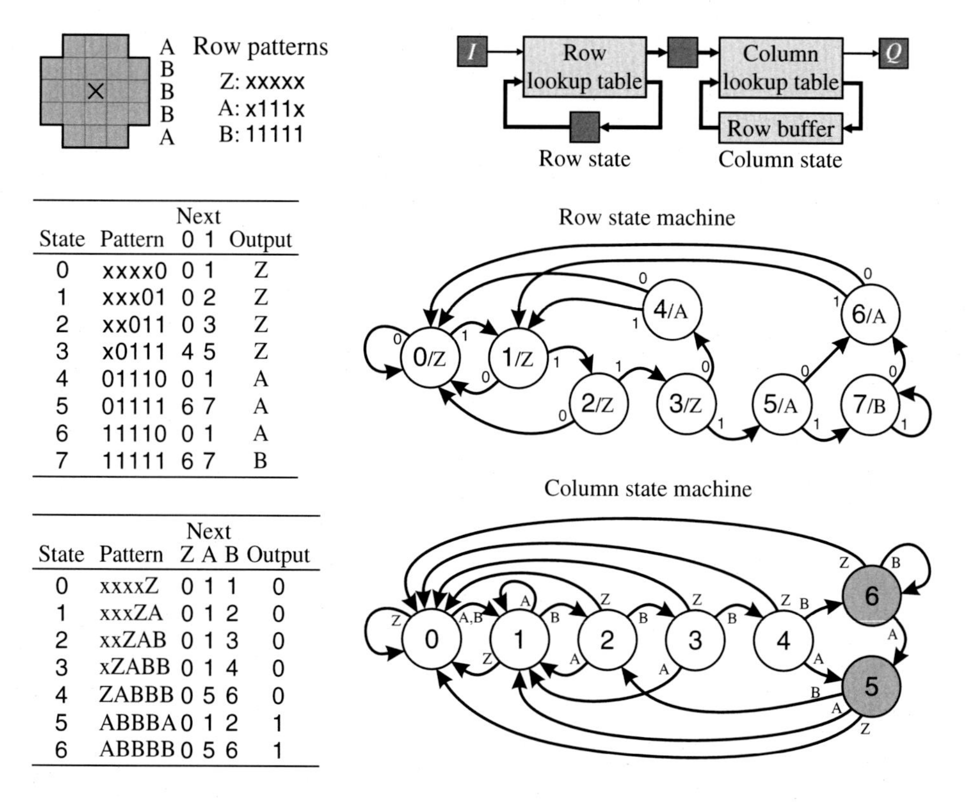

State	Pattern	Next 0	Next 1	Output
0	xxxx0	0	1	Z
1	xxx01	0	2	Z
2	xx011	0	3	Z
3	x0111	4	5	Z
4	01110	0	1	A
5	01111	6	7	A
6	11110	0	1	A
7	11111	6	7	B

State	Pattern	Next Z	Next A	Next B	Output
0	xxxxZ	0	1	1	0
1	xxxZA	0	1	2	0
2	xxZAB	0	1	3	0
3	xZABB	0	1	4	0
4	ZABBB	0	5	6	0
5	ABBBA	0	1	2	1
6	ABBBB	0	5	6	1

Figure 9.46 Separable finite state machines for erosion with a non-separable window.

pixels that match the template. The state machines for correlation are more complex because they must also account for mismatched patterns with appropriate scores.

9.6.2 Greyscale Morphology

The concepts of binary morphology may be extended to greyscale images using threshold decomposition. Reconstructing the greyscale image leads to selecting the minimum pixel value within the structuring element for erosion and the maximum pixel value for dilation. An alternative viewpoint is to treat each pixel value as a fuzzy representation between '0' and '1' and use fuzzy AND and OR in Eqs. (9.37) and (9.38) (Goetcherian, 1980). Either way, the resulting representations are

$$I \ominus S = \min_{i,j \in \mathbf{S}} \{I[x+i, y+j]\} \tag{9.45}$$

and

$$I \oplus S = \max_{i,j \in \mathbf{S}} \{I[x-i, y-j]\}, \tag{9.46}$$

with opening and closing defined as combinations of erosion and dilation as before. Examples of greyscale morphological operations are illustrated in Figure 9.47. Opening removes lighter blobs from the image that are smaller than the structuring element, whereas closing removes darker blobs.

9.6.2.1 Decomposition

Any of the decompositions described for binary morphological filters also apply to their greyscale counterparts. Minimum and maximum filters with rectangular windows are separable, providing an efficient implementation

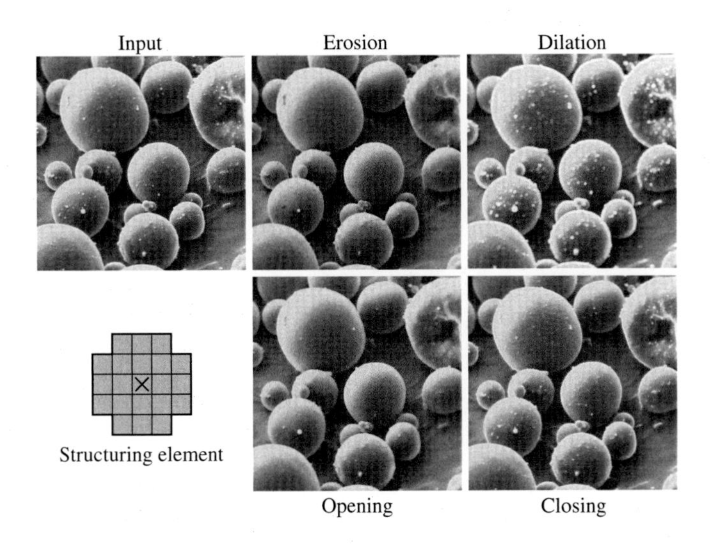

Figure 9.47 Greyscale morphological filtering.

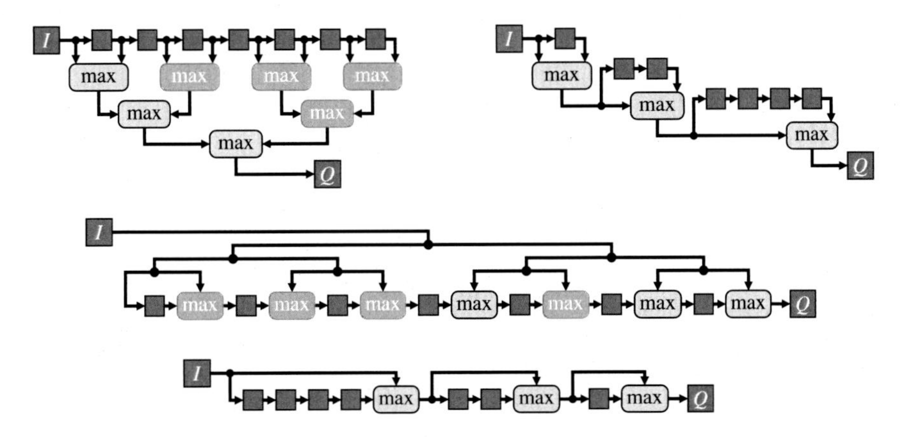

Figure 9.48 Reducing the number of operations in a maximum filter. Top left: a tree structure requires $W-1$ operations; top right: exploiting reuse reduces this to $\log_2 W$ operations; middle: transpose tree structure requires $W-1$ operations; bottom: exploiting reuse in the transpose structure.

for large window sizes. For the constituent 1-D filters, the overlap in window positions as the window moves enables the filter to be further decomposed using $\lceil \log_2 W \rceil$ operations (Bailey, 2010). The basic principle is shown in Figure 9.48 for $W = 8$. To find the maximum directly requires $W-1$ two-input maximum operations, arranged in a tree to minimise propagation delay as shown in the top left panel. For streamed inputs, the shaded operations have already been calculated and can be buffered from before, as shown in the top right panel. The circuit can readily be modified for window lengths that are not a power of 2, simply by reducing the number of delays in the last stage. The middle and bottom panels show a similar reuse transformation applied to the transpose filter structure (swapping input with output and maximum operations with pick-off points). To implement a column filter, the sample delays are replaced by row buffers in the manner of Figure 9.12.

Circular structuring elements may be decomposed using a parallel decomposition, with the corresponding rectangular windows implemented separably (Bailey, 2010). Figure 9.49 demonstrates this for the 5×5 circular structuring element. The row buffers and delays are reused for both filters by merging the two filters.

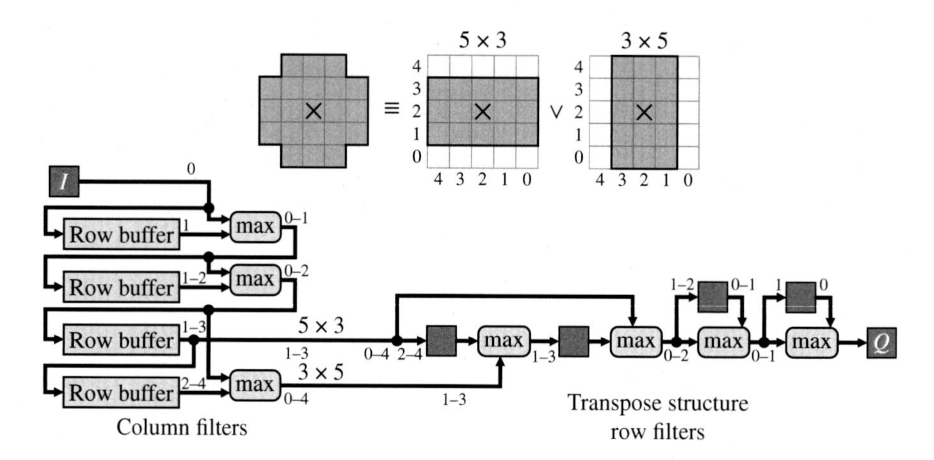

Figure 9.49 Efficient circular window implementation of greyscale dilation. Top: parallel decomposition of the structuring element; bottom: the separable implementation of both components. A transpose filter structure is used for the row filters. The numbers with the column filter represent the number of rows delay from the input. The numbers in the row filter represent the number of pixel delays to the output.

This requires that a transpose structure be used for the row filters. If necessary, additional pipeline registers may be added to reduce the delay through the chain of maximum operations. Similar structures can be applied for more filters in parallel for larger structuring elements (Bailey, 2010).

9.6.2.2 Greyscale Structuring Elements

The binary structuring elements of Eqs. (9.45) and (9.46) may be further extended to use greyscale structuring elements:

$$I \ominus S = \min_{i,j \in S} \{I[x+i, y+j] - S[i,j]\} \tag{9.47}$$

and

$$I \oplus S = \max_{i,j \in S} \{I[x-i, y-j] + S[i,j]\}\,. \tag{9.48}$$

The introduction of the extra component makes efficient implementation more difficult, because it is difficult to reuse results of previous calculations in all but special cases of the structuring element.

9.7 Colour Filtering

Filtering colour images carries its own set of problems. Linear filters may be applied independently to each component of an RGB image. When using nonlinear filters, it is important to operate on the colour components consistently. This usually means treating each colour vector value as a single unit and operating on it as such. If a trimmed filter removes a pixel in one component, all components for that pixel should be trimmed. A gated filter should apply the same gating to all channels. Applying a nonlinear filter independently to each component can lead to colour fringing artefacts when the different components are treated differently.

Problems can also be encountered when filtering hue images, because the hue wraps around. For example, the average of 12° and 352° is not 182° as obtained by direct numerical averaging but 2°. If the dominant hue is red (close to 0°), the hues may be offset by 180° before filtering and adjusted back again afterwards. The alternative is to convert from hue and saturation polar coordinates to rectangular coordinates before filtering and convert back to polar after filtering.

9.7.1 Colour Morphology and Vector Median

Vectors, being multidimensional, do not have a natural ordering. Therefore, ranking (including minimum and maximum) is not defined for colour images. This problem is discussed in more detail in Section 8.2.2.3. Simply applying the corresponding greyscale rank or morphological filters independently on each channel can give colour fringing where the relative contrast in different channels is reversed (for example on a border between red and green).

One approach (Comer and Delp, 1999) is to define a vector to scalar transformation that may be used to define a reduced ordering, enabling the pixel values to be sorted and the minimum or maximum pixel to be selected (Barnett, 1976). A simple transformation is to use the vector magnitude. For colour images, one common strategy is to use the luminance component (the Y from YUV or YCbCr, the V from hue, saturation, and value (HSV), or L^* from $L^*a^*b^*$ or $L^*u^*v^*$) as the scalar value for ordering. In fact, any weighted combination of the RGB components could be used (Comer and Delp, 1999). When filtering in this way, it is important to keep the components together through any selection process to avoid colour fringing artefacts. In general, different transformations will give different results.

The median is, however, defined for colour data (Astola et al., 1988, 1990) (see Section 8.2.2.3), and can be used for noise smoothing. One problem is the search through all of the points to find the median. When filtering streamed images, this must be performed within a pixel clock cycle. To do this efficiently, it is necessary to reuse the large proportion of data that is common from one window position to the next. Even so, efficient search algorithms such as those proposed by Barni (1997) are difficult to map to a hardware implementation. For small windows, the median may be calculated directly, although for larger windows, the separable median may be more appropriate.

Many variations of the vector median filter have been proposed in the literature. The simplest is to interleave the bits from each of the components and perform a scalar median filter (Chanussot et al., 1999) with the wider data word. Several gated vector median filters have been proposed (Celebi and Aslandogan, 2008) that remove noise without losing too many fine details.

9.7.2 Edge Enhancement

The rank-based edge enhancement filter can readily be adapted to colour images. The key is to first identify when a pixel is on a blurred edge and then identify which of the two regions the edge pixel is most associated with. Consider three pixels across a boundary between two different coloured regions, as illustrated on the left in Figure 9.50. If point **C** is on a blurred edge between regions **A** and **B**, the colour will be a mixture of the colours of **A** and **B**:

$$\mathbf{C} = \alpha\mathbf{A} + (1 - \alpha)\mathbf{B}, \quad 0 < \alpha < 1. \tag{9.49}$$

In this case, since **C** is between **A** and **B**, the central pixel **C** will be the vector median of the three points. In this case, from Eq. (8.61),

$$\|\mathbf{C} - \mathbf{A}\| + \|\mathbf{C} - \mathbf{B}\| < \begin{cases} \|\mathbf{A} - \mathbf{B}\| + \|\mathbf{A} - \mathbf{C}\| \\ \|\mathbf{B} - \mathbf{A}\| + \|\mathbf{B} - \mathbf{C}\| \end{cases} \tag{9.50}$$

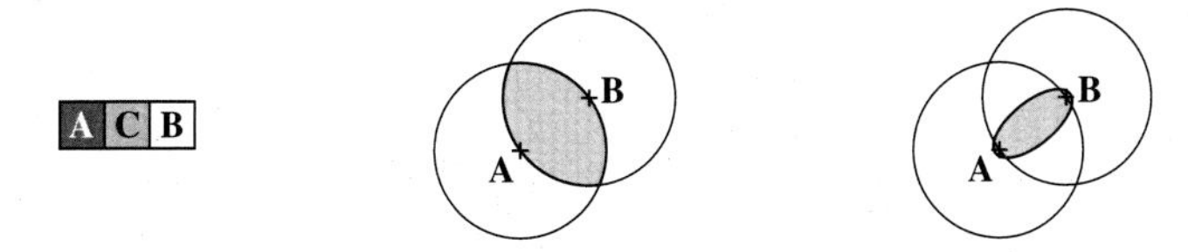

Figure 9.50 Constructs for colour edge enhancement. Left: three pixels, with **C** on the border between regions **A** and **B**; centre: point **C** would be the vector median in the shaded region; right: a tighter criterion defined by Eq. (9.53).

or

$$\|\mathbf{A}-\mathbf{B}\| > \begin{cases} \|\mathbf{C}-\mathbf{B}\| \\ \|\mathbf{C}-\mathbf{A}\| . \end{cases} \tag{9.51}$$

If this is used as the criterion for detecting an edge pixel as proposed by Tang et al. (1994), a point **C** anywhere in the shaded region in the centre panel of Figure 9.50 would be considered an edge pixel and enhanced. In three dimensions, this is a discus-shaped region, which can include points that deviate considerably from Eq. (9.49).

If **C** is indeed a mixture of the two colours, then

$$\|\mathbf{C}-\mathbf{B}\| + \|\mathbf{C}-\mathbf{A}\| \approx \|\mathbf{A}-\mathbf{B}\| . \tag{9.52}$$

An alternative edge criterion, allowing for some small deviation from the line as a result of noise, is (Gribbon et al., 2004)

$$\|\mathbf{C}-\mathbf{B}\| + \|\mathbf{C}-\mathbf{A}\| < \|\mathbf{A}-\mathbf{B}\| + T, \tag{9.53}$$

which is an ellipsoidal region, illustrated by the shaded region in the right panel of Figure 9.50. T can either be fixed, or it may be proportional to the distance between **A** and **B**. In Figure 9.50, T is $\|\mathbf{A}-\mathbf{B}\| /8$, which can easily be calculated using a simple shift.

If **C** is an edge pixel, as determined by either Eq. (9.51) or (9.53), then the edge pixel may be enhanced by selecting the closest:

$$\mathbf{Q} = \begin{cases} \mathbf{A}, & \|\mathbf{C}-\mathbf{A}\| < \|\mathbf{C}-\mathbf{B}\|, \\ \mathbf{B}, & \|\mathbf{C}-\mathbf{A}\| \geq \|\mathbf{C}-\mathbf{B}\|, \end{cases} \tag{9.54}$$

otherwise it is left unchanged.

In extending to two dimensions, using a larger 2-D window presents a problem because of the difficulty in determining which two pixels in the window are the opposite sides of the edge (the **A** and **B** above). One alternative is to apply the 1-D enhancement filter to the rows and columns (Tang et al., 1994). An alternative is to check if the edge is primarily horizontal or vertical and perform the enhancement perpendicular to the edge (Gribbon et al., 2004). This approach is shown in Figure 9.51 using the edge detection criterion of Eq. (9.53).

The distances may be measured using either L_1 or L_2 norms. The L_1 norm is simpler to calculate, although it is not isotropic. The L_2 norm is isotropic, although for colours it requires three multiplications, two sums, and a square root. The alternative is to use CORDIC arithmetic, although two compensated CORDIC units are required for a three-component colour space. Since the norms do not need to be exact, an approximation to the L_2 norm may be obtained by taking a linear combination of the ordered components (Barni et al., 2000). For three components, the relative weighting of the three components that gives the minimum error is (Barni et al., 2000) $1.0 : (\sqrt{2}-1) : (\sqrt{3}-\sqrt{2})$. A simple approximation that is reasonably close is $1.0 : 0.5 : 0.25$.

For larger edge blur, the filter can be extended by considering points two or three pixels apart (Tang et al., 1994). If noise is an issue, the filter may be preceded by an appropriate noise-smoothing filter.

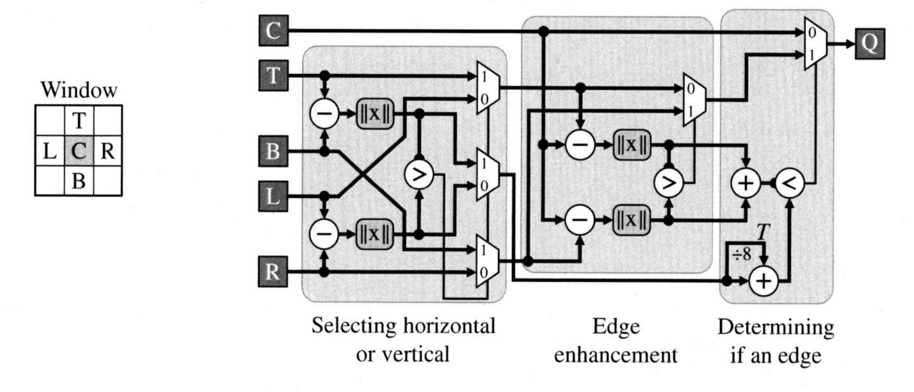

Figure 9.51 Colour edge enhancement using Eq. (9.53).

9.7.3 Bayer Pattern Demosaicing

Single-chip cameras capture only a single colour channel at each pixel. The most common colour filter pattern is the Bayer pattern. The resulting mosaic must be interpolated to obtain a full colour image.

The simplest form of filtering is nearest-neighbour interpolation. For the red and blue components, this duplicates the available component within each 2×2 block. The missing green pixels may be duplicated either from above or to the left. While nearest-neighbour interpolation is simple, it does not really improve the resolution, with the image appearing blocky, particularly along edges.

An improvement may be gained using linear interpolation. The missing red and blue components with two horizontal or vertical neighbours are obtained by averaging the adjacent pixels. Those in the middle of a block are obtained by averaging the four neighbouring values. This requires a 3×3 window as shown in Figure 9.52. The vertical average is reused to reduce the computation required. The divide by 2 from the averaging is free in hardware.

Using linear interpolation can result in blurring of edges and fine detail. Since a different filter is applied to each colour channel, this can result in colour fringes around sharp edges and fine lines. To reduce the blurring and consequent colour artefacts, it is necessary to detect the orientation of any features within the image and perform any interpolation in the direction along the edges rather than perpendicular to them. Locating edges is complicated by the fact that only one colour component is available for each pixel.

To obtain an accurate estimate of the position of the edge, a much larger window and significantly more computation is required. There are many more complex algorithms described in the literature; a few examples will be mentioned here. The ratio between the colour components at each pixel can be iteratively adjusted to make them consistent on each side of an edge (Kimmel, 1999). While this method handles edges reasonably well, the underlying assumption is that there are several consistent samples on each side of an edge. This is not the case with textured regions and fine detail, which result in colour fringing effects. This approach of enforcing consistency between colour channels is taken further by Gunturk et al. (2002), where the image is first decomposed into low- and high-frequency bands using a form of wavelet transform before applying consistency constraints within each band.

An alternative to interpolation is extrapolation. The missing value is estimated by extrapolating in each of the four directions and using a classifier to select the best result (Randhawa and Li, 2005). However, while this approach gives excellent results, it requires a larger window with multiple stages of filtering (Bailey et al., 2015).

The final result is inevitably a compromise between interpolation quality and the computation and memory resources required.

One relatively simple algorithm that provides significantly better results than simple interpolation with only a modest increase in computation is that by Hsia (2004). An edge-directed weighting is used for the green channel, with simple interpolation used for the red and blue channels. The results are then enhanced using a local contrast-based gain to reduce blurring across edges.

First, the edge-directed weighting requires the horizontal and vertical edge strengths to be estimated:

$$\Delta H = |G[x-1,y] - G[x+1,y]|, \qquad \Delta V = |G[x,y-1] - G[x,y+1]|. \tag{9.55}$$

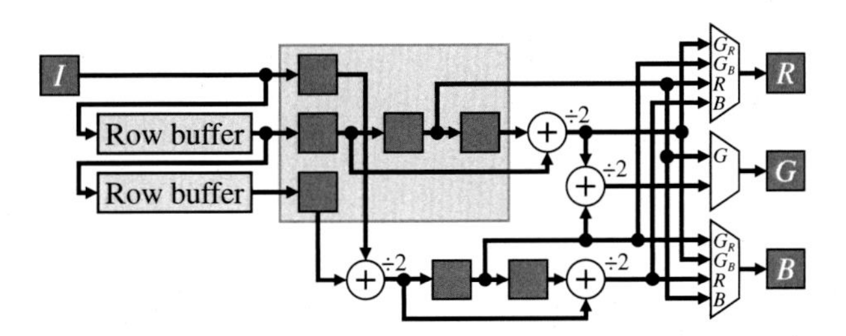

Figure 9.52 Bayer pattern demosaicing using linear interpolation. Within the multiplexers, G_R and G_B mean the green pixel on the red and blue rows, respectively.

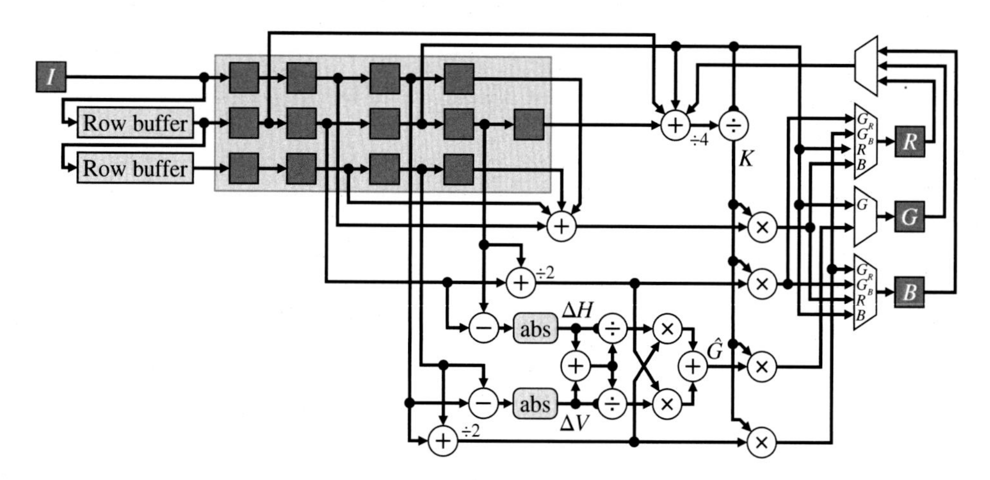

Figure 9.53 Direct implementation of Hsia's demosaicing scheme using Eqs. (9.56) and (9.57).

Then, these are used to weight the horizontal and vertical averages:

$$\hat{G} = \frac{\Delta V}{\Delta V + \Delta H}\frac{(G[x-1,y] + G[x+1,y])}{2} + \frac{\Delta H}{\Delta V + \Delta H}\frac{(G[x,y-1] + G[x,y+1])}{2}. \tag{9.56}$$

If the denominator is 0, then the two weights are made equal at 0.5.

The interpolated output values are scaled by a local contrast-based gain, which is calculated horizontally to reduce the number of row buffers required (Hsia, 2004). The gain term, K, is

$$K[x,y] = \frac{4I[x,y]}{I[x-2,y] + Q[x-1,y] + I[x,y] + I[x+2,y]}, \tag{9.57}$$

where $Q[x-1,y]$ is the previously calculated output for the colour channel corresponding to the input pixels. The final value is the product of Eqs. (9.56) and (9.57). A similar gain term is used for the interpolated red and blue channels.

A direct implementation of this scheme is shown in Figure 9.53. The outputs are fed back through a multiplexer to select $Q[x-1,y]$ for the channel that determines the gain term. The effective window size is 5×3 requiring only two row buffers as with the simple bilinear interpolation.

One limitation of this circuit from a real-time implementation is the feedback term used to calculate the local gain. The propagation delay through the division, multiplication, and two multiplexers cannot be pipelined and will therefore limit the maximum pixel clock rate. A potential alternative for K is

$$K_3[x,y] = \frac{3I[x,y]}{I[x-2,y] + I[x,y] + I[x+2,y]}, \tag{9.58}$$

which drops the feedback term. The multiplication by 3 may be performed by a single addition either before or after the division.

The circuit of Figure 9.53 may be simplified by reusing parts of the filter and factorising Eq. (9.56) to give Figure 9.54. The number of multiplications by K was reduced from four to two by observing that only two of the results will be used for any particular output. An additional stage of pipelining has been added to perform the division and multiplications in separate clock cycles, and higher speed designs could further pipeline the division.

9.7.4 *White Balancing*

Previously, Section 7.3.5 looked at global white balancing, considering each pixel independently. The assumption was that the illumination is constant for the whole scene, enabling it to be estimated and corrected globally.

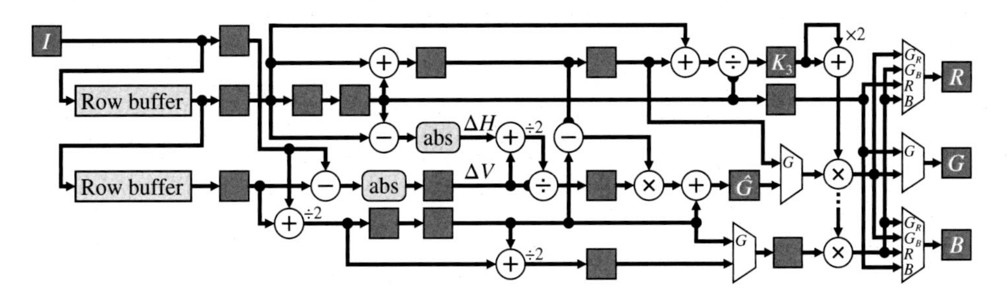

Figure 9.54 Simplified logic for Hsia's demosaicing, using Eq. (9.58).

However, many scenes have a variation in illumination (even if it is just a variation in intensity due to shading or position of the light source). Land and McCann (1971) showed that the human visual system is not sensitive to slow spatial variations, but to step edges, and developed the retinex theory.

To account for variation in illumination, the illumination needs to be estimated locally (using a low-pass filter). This leads to the centre-surround approach to white balancing. Since the received signal from the camera is a product of illumination and reflectance, Misaka and Okuno (2020) took the logarithm of the signal to convert the product to a sum. The local illumination can be estimated by a wide Gaussian window filter; this needs to be wide enough that the average represents the colour of the light. Subtracting this average will remove the illumination component and leave the reflectance.

Park et al. (2020) used a similar approach to enhance the contrast within an image.

9.7.4.1 Grey-edge Model

The grey-edge model is based on the hypothesis that the average edge difference (or image derivative) is achromatic (van de Weijer et al., 2007). This requires filtering the image, first with a Gaussian filter to reduce noise and then a derivative filter (for example the Sobel filter). The mean in each of the three colour channels provides an estimate of the illumination, which is then used to calculate gains for each of the RGB channels in a similar way to Eq. (7.82) from the grey world model.

Tan et al. (2015) implemented a simplified version of this, using only horizontal filtering (to avoid the need for row buffers). The Gaussian filter was replaced with a box average followed by down-sampling. A simple horizontal difference was used to estimate the derivative, which was then used for adjusting the RGB gains.

9.8 Summary

Local filters extend point operations by making the output depend not only in the corresponding input pixel value but also its local context (a window into the input image surrounding the corresponding input pixel). To implement filters efficiently on an FPGA, it is necessary to cache the input values as they are loaded, so that each pixel is only loaded once. The regular access pattern of filters enables the cache to be implemented with relatively simple row buffers built from the block RAMs within the FPGA. When combined with pipelining of the filter function, such caching allows one pixel to be processed every clock cycle, making local filters ideal for stream processing.

Linear filters are arguably the most widely used class of local filters. Linear filters output a linear combination of the input pixels within the window. Linearity enables such filters also to be considered in terms of their effect on different spatial frequencies; this aspect will be explored more fully in Chapter 11.

One disadvantage of linear filters is their limited ability to distinguish between signals of interest and noise. Many nonlinear filters have been developed to address this problem, of which only a small selection has been reviewed in this chapter. Two important classes of nonlinear filters considered in some detail are rank filters and morphological filters. Several efficient structures for implementing these filters have been described in some detail.

Filtering colour images poses its own problems. Linear filters can be applied separately to each component, but for nonlinear filters it is important to treat the colour vector as a single entity. This is particularly so for rank and morphological filters, where edges within the image can be shifted as a result of filtering. To prevent colour artefacts, the edges within all components must be moved similarly.

Local filters are an essential part of any image processing application. They are commonly used for preprocessing to improve the robustness of the algorithm, by filtering out irrelevant details from the image, or for enhancing information of importance. Therefore, the techniques and principles described in this chapter are indispensable for accelerating any embedded image processing application.

References

Abdou, I.E. and Pratt, W.K. (1979). Quantitative design and evaluation of edge enhancement / thresholding edge detectors. *Proceedings of the IEEE* **67** (5): 753–763. https://doi.org/10.1109/PROC.1979.11325.

Al-Dujaili, A. and Fahmy, S.A. (2017). High throughput 2D spatial image filters on FPGAs. *arXiv preprint*, (1710.05154), 8 pages. https://doi.org/10.48550/arXiv.1710.05154.

Al-Hasani, F., Hayes, M., and Bainbridge-Smith, A. (2011). A new sub-expression elimination algorithm using zero dominant representations. *6th International Symposium on Electronic Design, Test and Applications*, Queenstown, NZ (17–19 January 2011), 45–50. https://doi.org/10.1109/DELTA.2011.18.

Astola, J., Haavisto, P., Heinonen, P., and Neuvo, Y. (1988). Median type filters for color signals. *IEEE International Symposium on Circuits and Systems*, Espoo, Finland (7–9 June 1988), Volume **2**, 1753–1756. https://doi.org/10.1109/ISCAS.1988.15274.

Astola, J., Haavisto, P., and Neuvo, Y. (1990). Vector median filters. *Proceedings of the IEEE* **78** (4): 678–689. https://doi.org/10.1109/5.54807.

Ataman, E., Aatre, V.K., and Wong, K.M. (1980). A fast method for real-time median filtering. *IEEE Transactions on Acoustics, Speech and Signal Processing* **28** (4): 415–420. https://doi.org/10.1109/TASSP.1980.1163426.

Bailey, D.G. (1990). A rank based edge enhancement filter. *5th New Zealand Image Processing Workshop*, Palmerston North, NZ (9–10 August 1990), 42–47.

Bailey, D.G. (1997). Colour plane synchronisation in colour error diffusion. *IEEE International Conference on Image Processing*, Santa Barbara, CA, USA (26–29 October 1997), Volume **1**, 818–821. https://doi.org/10.1109/ICIP.1997.648089.

Bailey, D.G. (2010). Efficient implementation of greyscale morphological filters. *International Conference on Field Programmable Technology (FPT 2010)*, Beijing, China (8–10 December 2010), 421–424. https://doi.org/10.1109/FPT.2010.5681450.

Bailey, D.G. (2011). Image border management for FPGA based filters. *6th International Symposium on Electronic Design, Test and Applications*, Queenstown, NZ (17–19 January 2011), 144–149. https://doi.org/10.1109/DELTA.2011.34.

Bailey, D.G. (2012). Streamed high dynamic range imaging. *International Conference on Field Programmable Technology (ICFPT)*, Seoul, Republic of Korea (10–12 December 2012), 305–308. https://doi.org/10.1109/FPT.2012.6412153.

Bailey, D.G. and Ambikumar, A.S. (2018). Border handling for 2D transpose filter structures on an FPGA. *Journal of Imaging* **4** (12): Article ID 138, 21 pages. https://doi.org/10.3390/jimaging4120138.

Bailey, D.G. and Hodgson, R.M. (1985). Range filters: local intensity subrange filters and their properties. *Image and Vision Computing* **3** (3): 99–110. https://doi.org/10.1016/0262-8856(85)90058-7.

Bailey, D. and Li, J.S.J. (2016). FPGA based multi-shell filter for hot pixel removal within colour filter array demosaicing. *Image and Vision Computing New Zealand (IVCNZ)*, Palmerston North, NZ (21–22 November 2016), 196–201. https://doi.org/10.1109/IVCNZ.2016.7804450.

Bailey, D.G., Hodgson, R.M., and McNeill, S.J. (1984). Local filters in digital image processing. *National Electronics Conference (NELCON)*, Christchurch, NZ (22–24 August 1984), Volume **21**, 95–100.

Bailey, D.G., Randhawa, S., and Li, J.S.J. (2015). Advanced Bayer demosaicing on FPGAs. *International Conference on Field Programmable Technology (FPT)*, Queenstown, NZ (7–9 December 2015), 216–219. https://doi.org/10.1109/FPT.2015.7393154.

Barnett, V. (1976). Ordering of multivariate data. *Journal of the Royal Statistical Society Series A - Statistics in Society* **139** (3): 318–355. https://doi.org/10.2307/2344839.

Barni, M. (1997). A fast algorithm for 1-norm vector median filtering. *IEEE Transactions on Image Processing* **6** (10): 1452–1455. https://doi.org/10.1109/83.624972.

Barni, M., Buti, F., Bartolini, F., and Cappellini, V. (2000). A quasi-Euclidean norm to speed up vector median filtering. *IEEE Transactions on Image Processing* **9** (10): 1704–1709. https://doi.org/10.1109/83.869182.

Bednar, J.B. and Watt, T.L. (1984). Alpha trimmed means and their relationship to median filters. *IEEE Transactions on Acoustics, Speech and Signal Processing* **32** (1): 145–153. https://doi.org/10.1109/TASSP.1984.1164279.

Benkrid, K., Crookes, D., and Benkrid, A. (2002). Design and implementation of a novel algorithm for general purpose median filtering on FPGAs. *IEEE International Symposium on Circuits and Systems (ISCAS 2002)*, Phoenix, AZ, USA (26–29 May 2002), Volume **4**, 425–428. https://doi.org/10.1109/ISCAS.2002.1010482.

Benkrid, A., Benkrid, K., and Crookes, D. (2003). A novel FIR filter architecture for efficient signal boundary handling on Xilinx VIRTEX FPGAs. *11th Annual IEEE Symposium on Field-Programmable Custom Computing Machines (FCCM 2003)*, Napa, CA, USA (9–11 April 2003), 273–275. https://doi.org/10.1109/FPGA.2003.1227267.

Bouganis, C.S., Constantinides, G.A., and Cheung, P.Y.K. (2005). A novel 2D filter design methodology for heterogeneous devices. *13th Annual IEEE Symposium on Field-Programmable Custom Computing Machines (FCCM 2005)*, Napa, CA, USA (18–20 April 2005), 13–22. https://doi.org/10.1109/FCCM.2005.10.

Bracewell, R.N. (2000). *The Fourier Transform and its Applications*. 3e. New York: McGraw Hill.

Canny, J. (1986). A computational approach to edge detection. *IEEE Transactions on Pattern Analysis and Machine Intelligence* **8** (6): 679–698. https://doi.org/10.1109/TPAMI.1986.4767851.

Celebi, M.E. and Aslandogan, Y.A. (2008). Robust switching vector median filter for impulsive noise removal. *Journal of Electronic Imaging* **17** (4): Article ID 043006, 9 pages. https://doi.org/10.1117/1.2991415.

Chakrabarti, C. (1999). A DWT-based encoder architecture for symmetrically extended images. *IEEE International Symposium on Circuits and Systems (ISCAS '99)*, Orlando, FL, USA (30 May – 2 June 1999), Volume **4**, 123–126. https://doi.org/10.1109/ISCAS.1999.779957.

Chakrabarti, C. and Wang, L.Y. (1994). Novel sorting network-based architectures for rank order filters. *IEEE Transactions on VLSI Systems* **2** (4): 502–507. https://doi.org/10.1109/92.335027.

Chanussot, J., Paindavoine, M., and Lambert, P. (1999). Real time vector median like filter: FPGA design and application to color image filtering. *IEEE International Conference on Image Processing (ICIP'99)*, Kobe, Japan (24–28 October 1999), Volume **2**, 414–418. https://doi.org/10.1109/ICIP.1999.822929.

Chen, W., Chen, P., Hsiao, Y., and Lin, S. (2019). A low-cost design of 2D median filter. *IEEE Access* **7**: 150623–150629. https://doi.org/10.1109/ACCESS.2019.2948020.

Choo, C. and Verma, P. (2008). A real-time bit-serial rank filter implementation using Xilinx FPGA. *Real-Time Image Processing 2008*, San Jose, CA, USA (28–29 January 2008). SPIE, Volume **6811**, Article ID 68110F, 8 pages. https://doi.org/10.1117/12.765789.

Chu, P.P. (2006). *RTL Hardware Design Using VHDL*. Hoboken, NJ: Wiley. https://doi.org/10.1002/0471786411.

Comer, M. and Delp, E. (1999). Morphological operations for color image processing. *Journal of Electronic Imaging* **8** (3): 279–289. https://doi.org/10.1117/1.482677.

Danielsson, P.E. (1981). Getting the median faster. *Computer Graphics and Image Processing* **17** (1): 71–78. https://doi.org/10.1016/S0146-664X(81)80010-X.

Davies, E.R. (1999). High precision discrete model of median shifts. *7th International Conference on Image Processing and its Applications*, Manchester, UK (13–15 July 1999), Volume **1**, 197–201. https://doi.org/10.1049/cp:19990310.

DiCarlo, J.M. and Wandell, B.A. (2000). Rendering high dynamic range images. *Sensors and Camera Systems for Scientific, Industrial, and Digital Photography Applications*, San Jose, CA, USA (24–26 January 2000). SPIE, Volume **3965**, 392–401. https://doi.org/10.1117/12.385456.

Draper, B.A., Beveridge, J.R., Bohm, A.P.W., Ross, C., and Chawathe, M. (2003). Accelerated image processing on FPGAs. *IEEE Transactions on Image Processing* **12** (12): 1543–1551. https://doi.org/10.1109/TIP.2003.819226.

Eschbach, R. and Knox, K.T. (1991). Error-diffusion algorithm with edge enhancement. *Journal of the Optical Society of America A* **8** (12): 1844–1850. https://doi.org/10.1364/JOSAA.8.001844.

Fahmy, S.A., Cheung, P.Y.K., and Luk, W. (2005). Novel FPGA-based implementation of median and weighted median filters for image processing. *International Conference on Field Programmable Logic and Applications*, Tampere, Finland (24–26 August 2005), 142–147. https://doi.org/10.1109/FPL.2005.1515713.

Farid, H. and Simoncelli, E.P. (2004). Differentiation of discrete multidimensional signals. *IEEE Transactions on Image Processing* **13** (4): 496–508. https://doi.org/10.1109/TIP.2004.823819.

Fitch, J.P., Coyle, E.J., and Gallagher, N.C. (1985). Threshold decomposition of multidimensional ranked order operations. *IEEE Transactions on Circuits and Systems* **32** (5): 445–450. https://doi.org/10.1109/TCS.1985.1085740.

Floyd, R.W. and Steinberg, L. (1975). An adaptive algorithm for spatial grey scale. *Society for Information Display Symposium, Digest of Technical Papers*, 36–37.

Fornberg, B. (1988). Generation of finite difference formulas on arbitrarily spaced grids. *Mathematics of Computation* **51**: 699–706. https://doi.org/10.1090/S0025-5718-1988-0935077-0.

Gabiger-Rose, A., Kube, M., Weigel, R., and Rose, R. (2014). An FPGA-based fully synchronized design of a bilateral filter for real-time image denoising. *IEEE Transactions on Industrial Electronics* **61** (8): 4093–4104. https://doi.org/10.1109/TIE.2013.2284133.

Garibotto, G. and Lambarelli, L. (1979). Fast online implementation of two-dimensional median filtering. *Electronics Letters* **15** (1): 24–25. https://doi.org/10.1049/el:19790018.

Gasteratos, I., Gasteratos, A., and Andreadis, I. (2006). An algorithm for adaptive mean filtering and its hardware implementation. *Journal of VLSI Signal Processing* **44** (1–2): 63–78. https://doi.org/10.1007/s11265-006-5920-3.

Goel, A., Ahmad, M.O., and Swamy, M.N.S. (2019). Design of a 2D median filter with a high throughput FPGA implementation. *62nd International Midwest Symposium on Circuits and Systems (MWSCAS)*, Dallas, TX, USA (4–7 August 2019), 1073–1076. https://doi.org/10.1109/MWSCAS.2019.8885009.

Goetcherian, V. (1980). From binary to grey tone imaging processing using fuzzy logic concepts. *Pattern Recognition* **12** (1): 7–15. https://doi.org/10.1016/0031-3203(80)90049-7.

Gribbon, K.T., Bailey, D.G., and Johnston, C.T. (2004). Colour edge enhancement. *Image and Vision Computing New Zealand (IVCNZ'04)*, Akaroa, NZ (21–23 November 2004), 291–296.

Gunturk, B.K., Altunbasak, Y., and Mersereau, R.M. (2002). Color plane interpolation using alternating projections. *IEEE Transactions on Image Processing* **11** (9): 997–1013. https://doi.org/10.1109/TIP.2002.801121.

Harber, R. and Neudeck, S.B.G. (1985). VLSI implementation of a fast rank order filtering algorithm. *IEEE International Conference on Acoustics, Speech, and Signal Processing (ICASSP '85)*, Tampa, FL, USA (26–29 March 1985), 1396–1399. https://doi.org/10.1109/ICASSP.1985.1168229.

Hartley, R. (1991). Optimization of canonic signed digit multipliers for filter design. *IEEE International Symposium on Circuits and Systems*, Singapore (11–14 June 1991), Volume **4**, 1992–1995. https://doi.org/10.1109/ISCAS.1991.176054.

Heygster, G. (1982). Rank filters in digital image processing. *Computer Graphics and Image Processing* **19** (2): 148–164. https://doi.org/10.1016/0146-664X(82)90105-8.

Hodgson, R.M., Bailey, D.G., Naylor, M.J., Ng, A.L.M., and McNeill, S.J. (1985). Properties, implementations and applications of rank filters. *Image and Vision Computing* **3** (1): 3–14. https://doi.org/10.1016/0262-8856(85)90037-X.

Hore, A. and Yadid-Pecht, O. (2019). On the design of optimal 2D filters for efficient hardware implementations of image processing algorithms by using power-of-two terms. *Journal of Real-Time Image Processing* **16** (2): 429–457. https://doi.org/10.1007/s11554-015-0550-2.

Hsia, S.C. (2004). Fast high-quality color-filter-array interpolation method for digital camera systems. *Journal of Electronic Imaging* **13** (1): 244–247. https://doi.org/10.1117/1.1631443.

Huang, T.S., Yang, G.Y., and Tang, G.Y. (1979). A fast two dimensional median filtering algorithm. *IEEE Transactions on Acoustics, Speech and Signal Processing* **27** (1): 13–18. https://doi.org/10.1109/TASSP.1979.1163188.

Hummel, R.A. (1977). Image enhancement by histogram transformation. *Computer Graphics and Image Processing* **6** (2): 184–195. https://doi.org/10.1016/S0146-664X(77)80011-7.

Hwang, J.N. and Jong, J.M. (1990). Systolic architecture for 2-D rank order filtering. *International Conference on Application Specific Array Processors*, Princeton, NJ, USA (5–7 September 1990), 90–99. https://doi.org/10.1109/ASAP.1990.145446.

Johnson, B., Moncy, J.K., and Rani, J.S. (2019). Self adaptable high throughput reconfigurable bilateral filter architectures for real-time image de-noising. *Journal of Real-Time Image Processing* **16**: 1745–1764. https://doi.org/10.1007/s11554-017-0684-5.

Justusson, B.I. (1981). Median filtering: statistical properties. In: *Two-Dimensional Digital Signal Processing II: Transforms and Median Filters, Topics in Applied Physics*, vol. **43** (ed. T. Huang), 161–196. Berlin, Germany: Springer-Verlag. https://doi.org/10.1007/BFb0057597.

Kawada, S. and Maruyama, T. (2007). An approach for applying large filters on large images using FPGA. *International Conference on Field Programmable Technology*, Kitakyushu, Japan (12–14 December 2007), 201–208. https://doi.org/10.1109/FPT.2007.4439250.

Kent, R.B. and Pattichis, M.S. (2021). Design, implementation, and analysis of high-speed single-stage N-sorters and N-filters. *IEEE Access* **9**: 2576–2591. https://doi.org/10.1109/ACCESS.2020.3047594.

Kimmel, R. (1999). Demosaicing: image reconstruction from color CCD samples. *IEEE Transactions on Image Processing* **8** (9): 1221–1228. https://doi.org/10.1109/83.784434.

Kite, T.D., Evans, B.L., and Bovik, A.C. (2000). Modeling and quality assessment of halftoning by error diffusion. *IEEE Transactions on Image Processing* **9** (5): 909–922. https://doi.org/10.1109/83.841536.

Kittler, J. (1983). On the accuracy of the Sobel edge detector. *Image and Vision Computing* **1** (1): 37–42. https://doi.org/10.1016/0262-8856(83)90006-9.

Knuth, D.E. (1987). Digital halftones by dot diffusion. *ACM Transactions on Graphics* **6** (4): 245–273. https://doi.org/10.1145/35039.35040.

Kokufuta, K. and Maruyama, T. (2009). Real-time processing of local contrast enhancement on FPGA. *International Conference on Field Programmable Logic and Applications (FPL)*, Prague, Czech Republic (31 August – 2 September 2009), 288–293. https://doi.org/10.1109/FPL.2009.5272284.

Kokufuta, K. and Maruyama, T. (2010). Real-time processing of contrast limited adaptive histogram equalization on FPGA. *International Conference on Field Programmable Logic and Applications*, Milan, Italy (31 August – 2 September 2010), 155–158. https://doi.org/10.1109/FPL.2010.37.

Kumar, K.P. and Kanhe, A. (2022). FPGA architecture to perform symmetric extension on signals for handling border discontinuities in FIR filtering. *Computers and Electrical Engineering* **103**: Article ID 108307, 13 pages. https://doi.org/10.1016/j.compeleceng.2022.108307.

Kumar, V., Asati, A., and Gupta, A. (2017). Low-latency median filter core for hardware implementation of 5 × 5 median filtering. *IET Image Processing* **11** (10): 927–934. https://doi.org/10.1049/iet-ipr.2016.0737.

Kurak, C.W. (1991). Adaptive histogram equalization: a parallel implementation. *4th Annual IEEE Symposium Computer-Based Medical Systems*, Baltimore, MD, USA (12–14 May 1991), 192–199. https://doi.org/10.1109/CBMS.1991.128965.

Kutty, J.S.S., Boussaid, F., and Amira, A. (2014). A high speed configurable FPGA architecture for bilateral filtering. *IEEE International Conference on Image Processing (ICIP)*, Paris, France (27–30 October 2014), 1248–1252. https://doi.org/10.1109/ICIP.2014.7025249.

Land, E.H. and McCann, J.J. (1971). Lightness and retinex theory. *Journal of the Optical Society of America* **61** (1): 1–11. https://doi.org/10.1364/JOSA.61.000001.

Li, J.S.J. (1989). A class of multi-shell min/max median filters. *IEEE International Symposium on Circuits and Systems*, Portland, OR, USA (8–11 May 1989), 421–424. https://doi.org/10.1109/ISCAS.1989.100380.

Li, J.S.J. and Ramsingh, A. (1995). The relationship of the multi-shell to multistage and standard median filters. *IEEE Transactions on Image Processing* **4** (8): 1165–1169. https://doi.org/10.1109/83.403424.

Lindeberg, T. (1994). Scale-space theory: a basic tool for analysing structures at different scales. *Journal of Applied Statistics* **21** (2): 225–270. https://doi.org/10.1080/757582976.

Lu, W.S., Wang, H.P., and Antoniou, A. (1990). Design of two-dimensional FIR digital filters by using the singular-value decomposition. *IEEE Transactions on Circuits and Systems* **37** (1): 35–46. https://doi.org/10.1109/31.45689.

Lucke, L.E. and Parhi, K.K. (1992). Parallel structures for rank order and stack filters. *IEEE International Conference on Acoustics, Speech and Signal Processing (ICASSP-92)*, San Francisco, CA, USA (23–26 March 1992), Volume **5**, 645–648. https://doi.org/10.1109/ICASSP.1992.226513.

Marr, D. and Hildreth, E. (1980). Theory of edge detection. *Proceedings of the Royal Society of London, Series B, Biological Sciences* **207** (1167): 187–217. https://doi.org/10.1098/rspb.1980.0020.

Martinez-Peiro, M., Boemo, E.I., and Wanhammar, L. (2002). Design of high-speed multiplierless filters using a nonrecursive signed common subexpression algorithm. *IEEE Transactions on Circuits and Systems II: Analog and Digital Signal Processing* **49** (3): 196–203. https://doi.org/10.1109/TCSII.2002.1013866.

McDonnell, M.J. (1981). Box filtering techniques. *Computer Graphics and Image Processing* **17** (1): 65–70. https://doi.org/10.1016/S0146-664X(81)80009-3.

Misaka, K. and Okuno, H. (2020). FPGA implementation of an algorithm that enables color constancy. *IEEE/SICE International Symposium on System Integration (SII)*, Honolulu, HI, USA (12–15 January 2020), 991–995. https://doi.org/10.1109/SII46433.2020.9025949.

Mitra, S.K. (1998). *Digital Signal Processing: A Computer-based Approach*. Singapore: McGraw-Hill.

Nakagawa, Y. and Rosenfeld, A. (1978). A note on the use of local MIN and MAX operations in digital picture processing. *IEEE Transactions on Systems, Man and Cybernetics* **8** (8): 632–635. https://doi.org/10.1109/TSMC.1978.4310040.

Narendra, P.M. (1981). A separable median filter for image noise smoothing. *IEEE Transactions on Pattern Analysis and Machine Intelligence* **3** (1): 20–29. https://doi.org/10.1109/TPAMI.1981.4767047.

Nodes, T.A. and Gallagher, N.C. (1982). Median filters: some modifications and their properties. *IEEE Transactions on Acoustics, Speech and Signal Processing* **30** (5): 739–746. https://doi.org/10.1109/TASSP.1982.1163951.

Nosko, S., Musil, M., Zemcik, P., and Juranek, R. (2020). Color HDR video processing architecture for smart camera. *Journal of Real-Time Image Processing* **17** (3): 555–566. https://doi.org/10.1007/s11554-018-0810-z.

Ostromoukhov, V. (2001). A simple and efficient error-diffusion algorithm. *28th Annual Conference on Computer Graphics and Interactive Techniques*, Los Angeles, CA, USA (12–17 August 2001), 567–572. https://doi.org/10.1145/383259.383326.

Ou, Y., Ambalathankandy, P., Takamaeda, S., Motomura, M., Asai, T., and Ikebe, M. (2022). Real-time tone mapping: a survey and cross-implementation hardware benchmark. *IEEE Transactions on Circuits and Systems for Video Technology* **32** (5): 2666–2686. https://doi.org/10.1109/TCSVT.2021.3060143.

Ozkan, M.A., Reiche, O., Hannig, F., and Teich, J. (2017). Hardware design and analysis of efficient loop coarsening and border handling for image processing. *IEEE 28th International Conference on Application-specific Systems, Architectures and Processors (ASAP)*, Seattle, WA, USA (10–12 July 2017), 155–163. https://doi.org/10.1109/ASAP.2017.7995273.

Park, J.W., Lee, H., Kim, B., Kang, D., Jin, S.O., Kim, H., and Lee, H. (2020). A low-cost and high-throughput FPGA implementation of the retinex algorithm for real-time video enhancement. *IEEE Transactions on Very Large Scale Integration (VLSI) Systems* **28** (1): 101–114. https://doi.org/10.1109/TVLSI.2019.2936260.

Peng, B., Zhou, Y., Li, Q., Lin, M., Weng, J., and Zeng, Q. (2022). FPGA implementation of low-latency recursive median filter. *International Conference on Field Programmable Technology*, Hong Kong (5–9 December 2022), 312–318. https://doi.org/10.1109/ICFPT56656.2022.9974273.

Pham, T.Q. and van Vliet, L.J. (2005). Separable bilateral filtering for fast video preprocessing. *IEEE International Conference on Multimedia and Expo*, Amsterdam, The Netherlands (6 July 2005), 4 pages. https://doi.org/10.1109/ICME.2005.1521458.

Pizer, S.M., Amburn, E.P., Austin, J.D., Cromartie, R., Geselowitz, A., Greer, T., ter Haar Romeny, B., Zimmerman, J.B., and Zuiderveld, K. (1987). Adaptive histogram equalization and its variations. *Computer Vision, Graphics, and Image Processing* **39** (3): 355–368. https://doi.org/10.1016/S0734-189X(87)80186-X.

Porikli, F. (2008a). Constant time O(1) bilateral filtering. *IEEE Conference on Computer Vision and Pattern Recognition (CVPR 2008)*, Anchorage, AK, USA (23–28 June 2008), 8 pages. https://doi.org/10.1109/CVPR.2008.4587843.

Porikli, F. (2008b). Reshuffling: a fast algorithm for filtering with arbitrary kernels. *Real-Time Image Processing 2008*, San Jose, CA, USA (28–29 January 2008). SPIE, Volume **6811**, Article ID 68110M, 10 pages. https://doi.org/10.1117/12.772114.

Potkonjak, M., Srivastava, M.B., and Chandrakasan, A.P. (1996). Multiple constant multiplications: efficient and versatile framework and algorithms for exploring common subexpression elimination. *IEEE Transactions on Computer-Aided Design of Integrated Circuits and Systems* **15** (2): 151–165. https://doi.org/10.1109/43.486662.

Prewitt, J.M.S. (1970). Object enhancement and extraction. In: *Picture Processing and Psychopictorics* (ed. B. Lipkin and A. Rosenfeld), 75–149. New York: Academic Press.

Prokin, D. and Prokin, M. (2010). Low hardware complexity pipelined rank filter. *IEEE Transactions on Circuits and Systems II: Express Briefs* **57** (6): 446–450. https://doi.org/10.1109/TCSII.2010.2048371.

Qiu, G. (1996). An improved recursive median filtering scheme for image processing. *IEEE Transactions on Image Processing* **5** (4): 646–648. https://doi.org/10.1109/83.491340.

Rafi, M. and Din, N.u. (2016). A novel arrangement for efficiently handling image border in FPGA filter implementation. *3rd International Conference on Signal Processing and Integrated Networks (SPIN)*, Noida, India (11–12 February 2016), 163–168. https://doi.org/10.1109/SPIN.2016.7566681.

Ramirez, J.M., Flores, E.M., Martinez-Carballido, J., Enriquez, R., Alarcon-Aquino, V., and Baez-Lopez, D. (2010). An FPGA-based architecture for linear and morphological image filtering. *20th International Conference on Electronics, Communications and Computer (CONIELECOMP)*, Cholula, Mexico (22–24 February 2010), 90–95. https://doi.org/10.1109/CONIELECOMP.2010.5440788.

Randhawa, S. and Li, J.S. (2005). CFA demosaicking with improved colour edge preservation. *IEEE Region 10 Conference (IEEE Tencon'05)*, Melbourne, Australia (21–24 November 2005), 5 pages. https://doi.org/10.1109/TENCON.2005.301070.

Reza, A.M. (2004). Realization of the contrast limited adaptive histogram equalization (CLAHE) for real-time image enhancement. *Journal of VLSI Signal Processing Systems for Signal, Image and Video Technology* **38** (1): 35–44. https://doi.org/10.1023/B:VLSI.0000028532.53893.82.

Sedcole, N.P. (2006). Reconfigurable platform-based design in FPGAs for video image processing. PhD thesis. London, UK: Imperial College.

Serra, J. (1986). Introduction to mathematical morphology. *Computer Vision, Graphics and Image Processing* **35** (3): 283–305. https://doi.org/10.1016/0734-189X(86)90002-2.

Shamos, M.I. (1978). Robust picture processing operators and their implementation as circuits. *ARPA Image Understanding Workshop*, Pittsburgh, PA, USA (November 1978), 127–129.

Shiau, J.N. and Fan, Z. (1996). Set of easily implementable coefficients in error diffusion with reduced worm artifacts. *Color Imaging: Device-Independent Color, Color Hard Copy, and Graphic Arts*, San Jose, CA, USA (29 January – 1 February 1996). SPIE, Volume **2658**, 222–225. https://doi.org/10.1117/12.236968.

Sun, T. and Neuvo, Y. (1994). Detail-preserving median based filters in image processing. *Pattern Recognition Letters* **15** (4): 341–347. https://doi.org/10.1016/0167-8655(94)90082-5.

Swenson, R.L. and Dimond, K.R. (1999). A hardware FPGA implementation of a 2D median filter using a novel rank adjustment technique. *7th International Conference on Image Processing and its Applications*, Manchester, UK (13–15 July 1999), Volume **465**, 103–106. https://doi.org/10.1049/cp:19990290.

Tan, X., Lai, S., Wang, B., Zhang, M., and Xiong, Z. (2015). A simple gray-edge automatic white balance method with FPGA implementation. *Journal of Real-Time Image Processing* **10** (2): 201–217. https://doi.org/10.1007/s11554-012-0318-x.

Tang, K., Astola, J., and Neuvo, Y. (1994). Multichannel edge enhancement in color image processing. *IEEE Transactions on Circuits and Systems for Video Technology* **4** (5): 468–479. https://doi.org/10.1109/76.322994.

Tomasi, C. and Manduchi, R. (1998). Bilateral filtering for gray and color images. *6th International Conference on Computer Vision*, Bombay, India (4–7 January 1998), 839–846. https://doi.org/10.1109/ICCV.1998.710815.

van de Weijer, J., Gevers, T., and Gijsenij, A. (2007). Edge-based color constancy. *IEEE Transactions on Image Processing* **16** (9): 2207–2214. https://doi.org/10.1109/TIP.2007.901808.

Waltz, F.M. (1994a). Application of SKIPSM to binary template matching. *Machine Vision Applications, Architectures, and Systems Integration III*, Boston, MA, USA (31 October – 2 November 1994). SPIE, Volume **2347**, 417–427. https://doi.org/10.1117/12.188752.

Waltz, F.M. (1994b). Separated-kernel image processing using finite-state machines (SKIPSM). *Machine Vision Applications, Architectures, and Systems Integration III*, Boston, MA, USA (31 October – 2 November 1994). SPIE, Volume **2347**, 386–395. https://doi.org/10.1117/12.188749.

Waltz, F.M. (1995). Application of SKIPSM to binary correlation. *Machine Vision Applications, Architectures, and Systems Integration IV*, Philadelphia, PA, USA (23–24 October 1995). SPIE, Volume **2597**, 82–91. https://doi.org/10.1117/12.223967.

Waltz, F.M. and Garnaoui, H.H. (1994). Application of SKIPSM to binary morphology. *Machine Vision Applications, Architectures, and Systems Integration III*, Boston, MA, USA (31 October – 2 November 1994). SPIE, Volume **2347**, 396–407. https://doi.org/10.1117/12.188750.

Waltz, F.M., Hack, R., and Batchelor, B.G. (1998). Fast efficient algorithms for 3x3 ranked filters using finite-state machines. *Machine Vision Systems for Inspection and Metrology VII*, Boston, MA, USA (4–5 November 1998). SPIE, Volume **3521**, 278–287. https://doi.org/10.1117/12.326970.

Wendt, P.D., Coyle, E.J., and Gallagher, N.C. (1986). Stack filters. *IEEE Transactions on Acoustics, Speech and Signal Processing* **34** (4): 898–911. https://doi.org/10.1109/TASSP.1986.1164871.

Zoss, R., Habegger, A., Bandi, V., Goette, J., and Jacomet, M. (2011). Comparing signal processing hardware-synthesis methods based on the Matlab tool-chain. *6th International Symposium on Electronic Design, Test and Applications*, Queenstown, NZ (17–19 January 2011), 281–286. https://doi.org/10.1109/DELTA.2011.58.

10

Geometric Transformations

The next class of operations to be considered is that of geometric transformations. These redefine the geometric arrangement of the pixels within an image (Wolberg, 1990). Examples include zooming, rotating, and perspective transformation. Such transformations are typically used to correct spatial distortions resulting from the imaging process, to register an image to a predefined coordinate system (for example registering an aerial image to a particular map projection, or rectifying a stereo pair of images so that the rows corresponds to epipolar lines), or to register one image with another.

Unlike point operations and local filters, the output pixel does not, in general, come from the same input pixel location. This means that some form of buffering is required to manage the delays resulting from the changed geometry. The simplest approach is to hold the input image or the output image (or both) in a frame buffer and using random access processing. It is usually quite difficult to implement geometric transformations with streaming for both the input and output.

The mapping between input and output pixels may be defined in two different ways, as illustrated in Figure 10.1. The ***forward mapping*** defines the output pixel coordinates, (u, v), as a function, $m_f(\cdot)$, of the input coordinates:

$$Q[u, v] = Q[m_{fu}(x, y), m_{fv}(x, y)] = I[x, y]. \tag{10.1}$$

Since the mapping is specified in terms of input pixel locations, the forward mapping can process a streamed input (for example from a camera). However, as a result of the mapping, sequential input addresses do not necessarily correspond to sequential output addresses.

Conversely, the ***reverse mapping*** defines the input pixel coordinates as a function, $m_r(\cdot)$, of the output coordinates:

$$Q[u, v] = I[m_{rx}(u, v), m_{ry}(u, v)] = I[x, y]. \tag{10.2}$$

It can process sequential output addresses, enabling it to produce a streamed output (for example to a display). Since the mapping specifies where each output pixel comes from, it will, in general require random access for the input image.

The basic architectures for implementing forward and reverse mappings are compared in Figure 10.2. For streamed access of the input (forward mapping) or output (reverse mapping), the address will be provided by appropriate counters synchronised with the corresponding stream. While the input to the mappings will usually be integer image coordinates, in general the mapped addresses will not be. Consequently, the implementation will be more complex than that implied by Figure 10.2. The issues associated with the forward and reverse mappings are different, so will be discussed in more detail separately.

Design for Embedded Image Processing on FPGAs, Second Edition. Donald G. Bailey.

Companion Website: www.wiley.com/go/bailey/designforembeddedimageprocessc2e

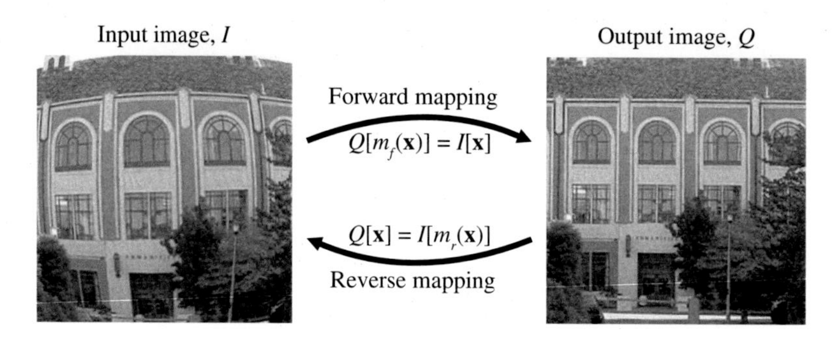

Figure 10.1 Forward and reverse mappings for geometric transformation.

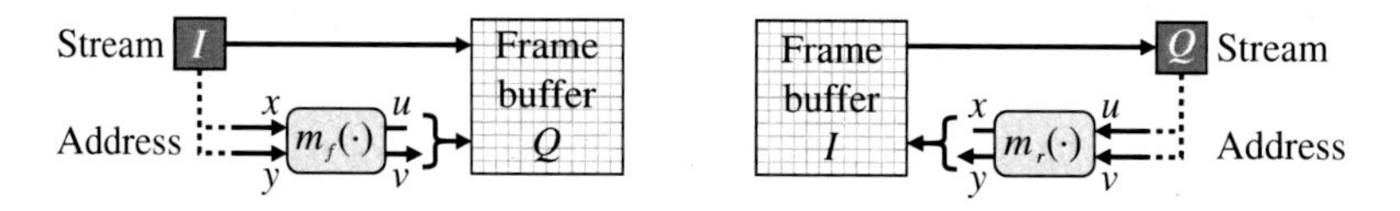

Figure 10.2 Basic architectures for geometric transformation. Left: forward mapping; right: reverse mapping.

10.1 Reverse Mapping

The reverse mapping is the most similar to software, where the output image is scanned, and the output pixel obtained from the mapped point in the input image. This makes the reverse mapping well suited for producing a streamed output for display or passing to downstream operations.

Two factors complicate this process. First, when the magnification is less than 1, it is necessary to filter the input image to reduce the effects of aliasing. Second, the mapped coordinates do not necessarily fall exactly on the input pixel grid. This requires some form of interpolation to calculate the pixel value to output. Methods for managing these two aspects will be described in more detail in Sections 10.1.1 and 10.1.2, respectively.

Although an arbitrary warp requires the complete input image to be available for the reverse mapping, if the transformation keeps the image mostly the same (for example correcting lens distortion or rotating by a small angle), then the transformation can begin before the whole image is buffered. Sufficient image rows must be buffered to account for the maximum shift in the vertical direction between the input and output images. While a new input row is being streamed in, the output is calculated and streamed from the row buffers, as shown in Figure 10.3. Address calculations (to select the appropriate row buffer) are simplified by having the number of row buffers be a power of 2, although this may be overcome using a small lookup table to map addresses to row buffers. Reducing the number of row buffers required can significantly reduce both the resource requirements and the latency (Akeila and Morris, 2008).

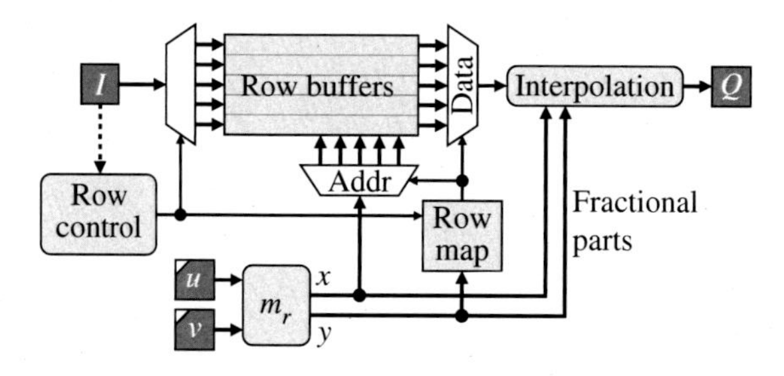

Figure 10.3 Reverse mapping with a partial frame buffer (using row buffers).

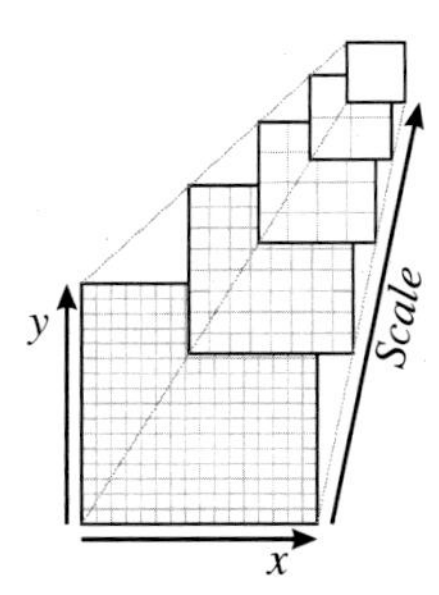

Figure 10.4 A multi-resolution pyramid is formed by filtering and down-sampling by 2 at each level.

10.1.1 Anti-alias Filtering

When down-sampling an image, it is necessary to reduce aliasing artefacts by filtering the image to remove spatial frequency content higher than twice the new sample frequency. If the magnification factor is approximately constant, then a simple low-pass filter (as described in Chapter 9) may be used. However, if the magnification varies significantly with position (for example with a perspective transformation), then the filter must be spatially variant, with potentially large windows where the magnification is small.

Generally, there is insufficient time to filter the image while producing the output, so it may be necessary to pre-filter the image before the geometric transformation. There are two main approaches to filtering that will allow for a range of magnifications. One is to use a pyramidal data structure with a series of filtered images at a range of resolutions, and the other is to use an integral image.

An ***image pyramid*** is a multi-resolution data structure used to represent an image at a range of scales. Each level of the pyramid is formed by successively filtering and down-sampling the layer below it by a factor of 2, as illustrated in Figure 10.4. Typically, each pixel in the pyramid is the average of the four pixels below it, although using a Gaussian filter to derive the down-sampled images (Gaussian pyramid) is better at removing the high frequencies. A pyramid can be constructed in a single pass through the image and requires 33% more storage than the original image.

The pyramid can be used for filtering for geometric transformation in several ways. The simplest is to determine which level corresponds to the desired level of filtering based on the magnification and select the nearest pixel within that level (Wolberg, 1990) (nearest-neighbour interpolation). This limits the filter window to be square, with size and position determined by pyramid structure. This may be improved by interpolating within the pyramid (Williams, 1983) in the x, y, and *Scale* directions (trilinear interpolation). Such interpolation requires reading eight locations in the pyramid, requiring innovative mapping and caching techniques to achieve in a single clock cycle.

An alternative approach is to use an ***integral image*** or summed area table (Crow, 1984). An integral image is one where each pixel contains the sum of all of the pixels above and to the left of it:

$$S[x,y] = \sum_{i\leq x} \sum_{j\leq y} I[i,j]. \tag{10.3}$$

Integral images can be formed efficiently by separating the column and row summations:

$$\begin{aligned} C[x,y] &= C[x,y-1] + I[x,y], \\ S[x,y] &= S[x-1,y] + C[x,y], \end{aligned} \tag{10.4}$$

as shown on the left in Figure 10.5. In the first row of the image, the previous column sum is initialised as 0 (rather than feeding back from the row buffer), and in the first column, the sum table for the row is initialised to 0. If necessary, this can be readily parallelised to process multiple pixels each clock cycle by processing multiple rows in parallel, and delaying each subsequent row by one clock to remove data dependencies (see the right panel in Figure 3.5) (Ehsan et al., 2009), or using a systolic array (De la Cruz, 2011).

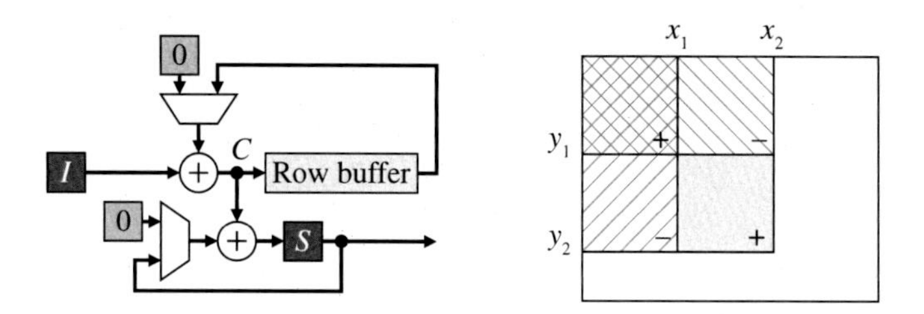

Figure 10.5 Left: recursive streamed implementation of an integral image; right: using an integral image to find the sum of pixel values within an arbitrary rectangular area.

The sum over any arbitrary rectangular window may then be calculated in constant time (with four accesses to the integral image as shown on the right in Figure 10.5):

$$\sum_{i=x_1}^{x_2} \sum_{j=y_1}^{y_2} I[i,j] = S[x_2, y_2] - S[x_1 - 1, y_2] - S[x_2, y_1 - 1] + S[x_1 - 1, y_1 - 1], \tag{10.5}$$

regardless of the size and position of the rectangular window. Filtering requires dividing the sum from Eq. (10.5) by the area to give the average output pixel value. This can be seen as an extension of box filtering as described in Section 9.2.2.1 to allow filtering with a variable window size. If required, integral images may be generalised to approximate Gaussian filtering (Derpanis et al., 2007).

In the context of filtering for geometric transformation, summed area tables have an advantage over pyramid structures in that they give more flexibility to the shape and size of the window. Note that with magnifications close to and greater than 1, filtering is not necessary, and the input pixel values can simply be interpolated to give the required output.

10.1.2 Interpolation

The mapped coordinates are usually not on integer locations. Estimating the output pixel value requires estimating the underlying continuous image, $C(x, y)$, and resampling this at the desired locations. $C(x, y)$ is formed by convolving the input image samples with a continuous interpolation kernel, $K(x, y)$:

$$C(x, y) = \sum_{i,j} I[i,j] K(x - i, y - j), \tag{10.6}$$

which is illustrated in one dimension in Figure 10.6.

Band-limited images sampled at greater than twice the maximum frequency may be reconstructed exactly using a sinc interpolation kernel. Unfortunately, the sinc kernel has infinite extent and decays slowly. Instead, simpler kernels with limited extent (or ***region of support***) that approximate the sinc are used. A smaller support reduces the bandwidth required to access the input pixels.

Sampling the continuous image is equivalent to weighting the corresponding pixel values by a sampled interpolation kernel:

$$I[u, v] = C(x, y)|_{x=u, y=v} = \sum_{i,j} I[i,j] K(u - i, v - j). \tag{10.7}$$

Resampling is therefore filtering the input image with the weights given by the sampled interpolation kernel. Note that the kernel indices correspond to the fractional part of the desired sample location. There are two important differences with the filtering described in Section 9.2. First, the fractional part of the input location (and consequently the filter weights) changes from one output pixel to the next. Therefore, some of the optimisations based on filtering with constant weights can no longer be applied. Second, the path through the input image corresponding to an output image row will not, in general, follow a raster scan but will follow

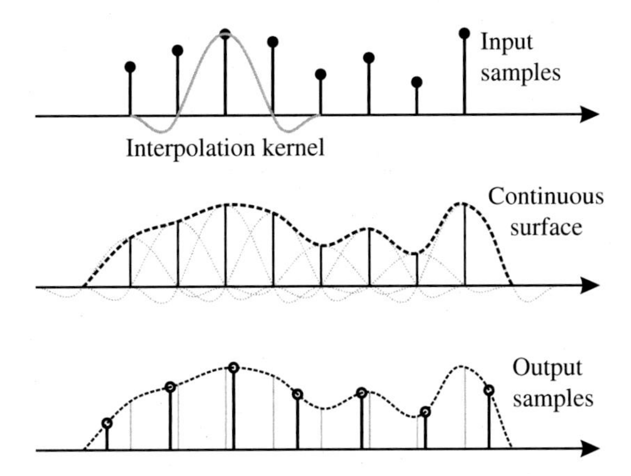

Figure 10.6 The interpolation process illustrated in one dimension: Top: the input samples are convolved with a continuous interpolation kernel; middle: the convolution gives a continuous image; bottom: this is then resampled at the desired locations to give the output samples.

an arbitrary curve depending on the geometric transformation. In many cases, this makes the simple, regular, caching arrangements for local filters unsuited for interpolation, requiring more complex caching arrangements to be devised to manage the curved path through the input image.

Nearest-neighbour interpolation is the simplest form of interpolation; it simply selects the nearest pixel to the desired location. The filter kernel is effectively a rectangular function, the size of one pixel. This requires rounding the coordinates into the input image to the nearest integer, which is achieved by adding 0.5, and selecting the integer component. Incorporating this addition into the mapping function means that the fractional component can simply be truncated from the calculated coordinates. The corresponding pixel in the input image can then be read directly from the frame buffer.

Other interpolation schemes combine the values of neighbouring pixels to estimate the value at a fractional position. The input pixel location is split into its integer and fractional components:

$$(x, y) = (x_i + x_f, y_i + y_f) = (x_i, y_i) + (x_f, y_f). \tag{10.8}$$

The integer parts of the desired location, (x_i, y_i), are used to select the input pixel values, which are then weighted by values derived from the fractional components, (x_f, y_f), as defined in Eq. (10.7).

As with image filtering, the logic also needs to appropriately handle boundary conditions. When the output pixel does not fall within the input image, the interpolation is not required, and an empty pixel (however this is defined for the application) is provided to the output. If one or more of the neighbouring pixels are outside the image, then it is necessary to provide appropriate values as input to the interpolation depending on the image extension scheme used.

10.1.2.1 Bilinear Interpolation

Perhaps the most commonly used interpolation method is bilinear interpolation. It combines the values of the four nearest pixels using separable linear interpolation:

$$\begin{aligned} I[x, y] = {} & I[x_i, y_i](1 - x_f)(1 - y_f) + I[x_i + 1, y_i]x_f(1 - y_f) \\ & + I[x_i, y_i + 1](1 - x_f)y_f + I[x_i + 1, y_i + 1]x_f y_f. \end{aligned} \tag{10.9}$$

An alternative interpretation of Eq. (10.9) is demonstrated on the right in Figure 10.7. Each pixel location is represented by a unit square, with the weights given by the area of overlap between the output pixel and the

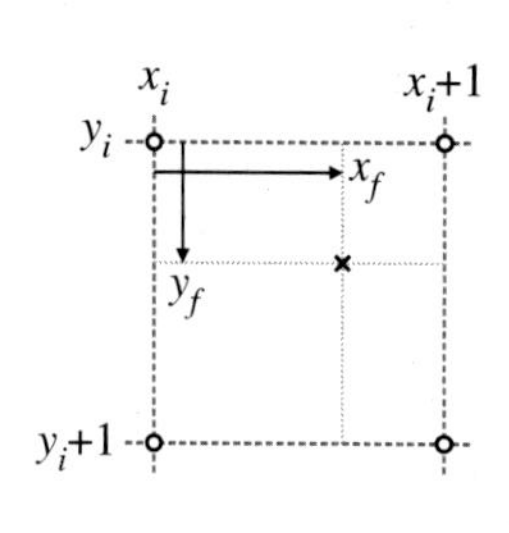

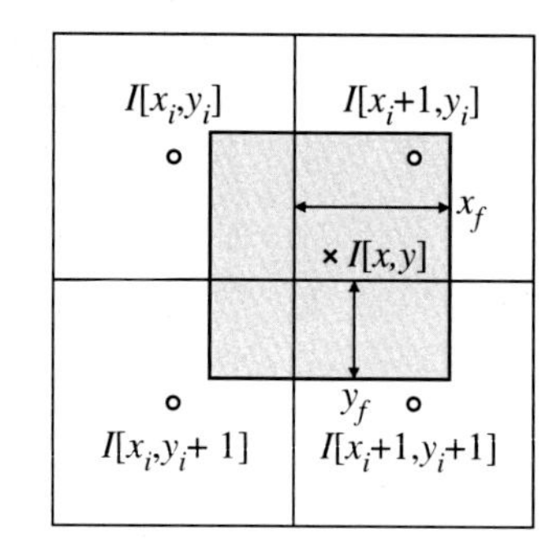

Figure 10.7 Left: coordinate definitions for interpolation; subscripts i and f denote integer and fractional parts of the respective coordinates; right: bilinear interpolation.

available input pixels. Although Eq. (10.9) appears to require eight multiplications, exploiting separability and careful factorisation can reduce this to three:

$$\begin{aligned} I_{y_i} &= I[x_i, y_i] + x_f(I[x_i + 1, y_i] - I[x_i, y_i]), \\ I_{y_{i+1}} &= I[x_i, y_i + 1] + x_f(I[x_i + 1, y_i + 1] - I[x_i, y_i + 1]), \\ I[x, y] &= I_{y_i} + y_f(I_{y_{i+1}} - I_{y_i}). \end{aligned} \tag{10.10}$$

While a random access implementation (such as performed in software) is trivial, accessing all four input pixels within a single clock cycle requires careful cache design. If the whole image is available on chip, it will be stored in block RAMs. By ensuring that successive rows are stored in separate RAMs, a dual-port memory can simultaneously provide both input pixels on each line, hence all four pixels, in a single clock cycle. If only a single port is available, the data must be partitioned with odd and even addresses in separate banks, so that all four input pixels may be accessed simultaneously. In most applications, however, there is insufficient on-chip memory available to hold the whole image, and the image is held in an external frame buffer with only one access per clock cycle.

In this case, memory bandwidth becomes the bottleneck. Each pixel should only be read from external memory once and be held in an on-chip cache for use in subsequent windows. Two main approaches can be used: one is to load the input pixels as necessary, and the other is to preload the cache and always access pixels directly from there. Since the path through the input image does not, in general, follow a raster scan, it is necessary to cache not only the pixel values but also their coordinates. To avoid the complexity of an associative memory, the row address can be implicit in which buffer the data is stored in. Consider a B row cache consisting of B separate block RAMs. The particular block RAM used to cache a pixel may be determined by taking the row address modulo B and using this to index the cache blocks. This is obviously simplified if B is a power of 2.

Loading input pixels on demand requires that successive output rows and pixels be mapped sufficiently closely in the input image that the reuse from one location to the next is such that the required input pixels are available (Gribbon and Bailey, 2004). This generally requires that the magnification be greater than 1; otherwise, multiple external memory accesses are required to obtain the input data. If the output stream has horizontal blanking periods at the end of each row, this requirement may be partially relaxed through using a first-in first-out (FIFO) buffer on the output to smooth the flow. For a bilinear interpolation, a new frame may be initialised by loading the first row twice; the first time loads data into the cache but does not produce output, while the second pass produces the first row of output. This is similar to directly priming the row buffers for a regular filter. The requirements are relaxed somewhat in a multi-phase design, because more clock cycles are available to gather the required input pixel values.

Preloading the cache is a little more complex as it requires calculating in advance which pixels are going to be required. The mapping is used to determine which pixels to load into the cache, with the mapping buffered for use when all of the pixels are available in the cache (see Figure 10.8). Alternatively, the cache control can

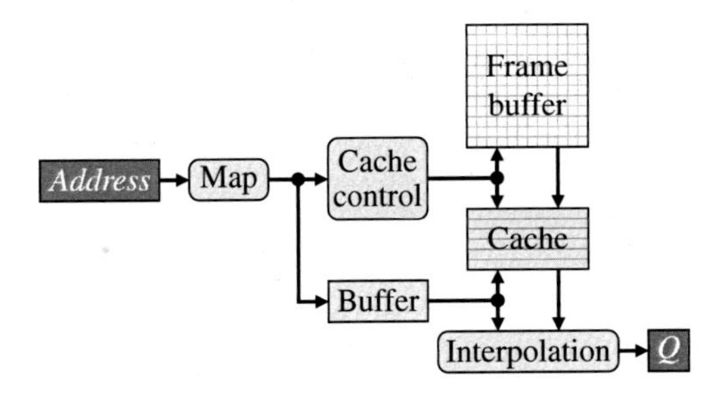

Figure 10.8 Preloading the cache; the calculated mapping is buffered for when the data is available.

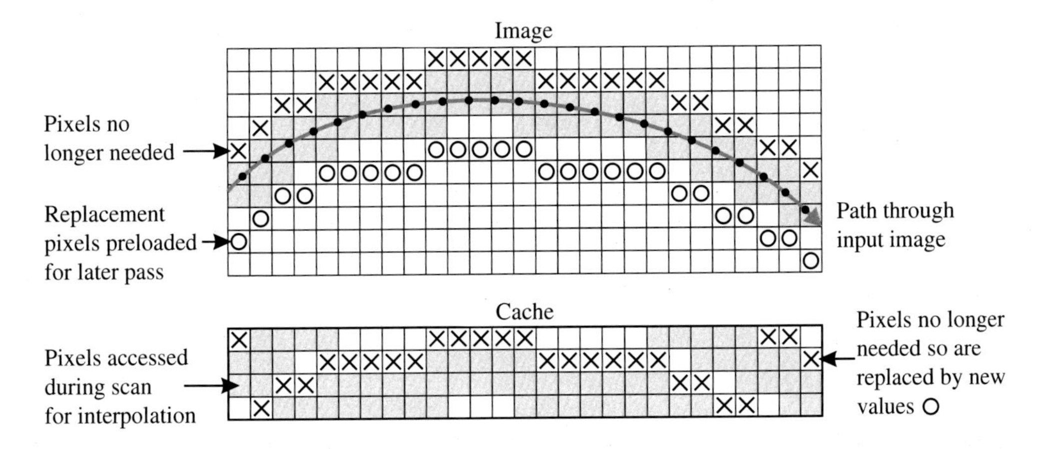

Figure 10.9 Preloading pixels which are no longer required.

monitor when data is no longer required and automatically replace the expired values with new data. This relies on the progression of the scan through the image in successive rows, so that when a pixel is not used in a column, it is replaced by the next corresponding pixel from the frame buffer. This process is illustrated in Figure 10.9.

10.1.2.2 Bicubic Interpolation

While bilinear interpolation gives reasonable results in many applications, a smoother result may be obtained through bicubic interpolation. Bicubic interpolation uses a separable piecewise cubic interpolation kernel with the constraints that it is continuous and has a continuous first derivative. These constraints result in the following continuous 1-D kernel (Keys, 1981):

$$K_{Keys}(x) = \begin{cases} (a+2)|x|^3 - (a+3)|x|^2 + 1, & 0 \leq |x| < 1, \\ a|x|^3 - 5a|x|^2 + 8a|x| - 4a, & 1 \leq |x| < 2, \\ 0, & 2 \leq |x|, \end{cases} \tag{10.11}$$

where a is a free parameter. Selecting $a = -0.5$ makes the interpolated function agree with the Taylor series approximation to a sinc in as many terms as possible (Keys, 1981). The resulting 1-D interpolation has a region of support of four samples:

$$I[x] = \sum_{k=-1}^{2} w_k I[x_i + k], \tag{10.12}$$

where the weights are derived by substituting the fractional part of the desired pixel location into Eq. (10.11):

$$
\begin{aligned}
w_{-1} &= -\frac{1}{2}x_f^3 + x_f^2 - \frac{1}{2}x_f, \\
w_0 &= \frac{3}{2}x_f^3 - \frac{5}{2}x_f^2 + 1, \\
w_1 &= -\frac{3}{2}x_f^3 + 2x_f^2 + \frac{1}{2}x_f, \\
w_2 &= \frac{1}{2}x_f^3 - \frac{1}{2}x_f^2.
\end{aligned} \tag{10.13}
$$

These weights may be either calculated on-the-fly or determined through a set of small lookup tables. Since the weights sum to 1, Eq. (10.12) may be refactored to reduce the number of multiplications at the expense of additional subtractions (see Figure 10.10):

$$
\begin{aligned}
I[x] &= I[x_i+2]\left(1 - \sum_{k=-1}^{1} w_k\right) + \sum_{k=-1}^{1} w_k I[x_i+k] \\
&= I[x_i+2] + \sum_{k=-1}^{1} w_k (I[x_i+k] - I[x_i+2]).
\end{aligned} \tag{10.14}
$$

This 1-D interpolation is then applied separably to the rows and columns within the window to give the interpolated 2-D sample. Here, the filter weights are provided by a lookup table on the fractional component of the pixel. There are two distinct types of interpolation architecture (Boukhtache et al., 2021): homogeneous and heterogeneous.

The simplest is the homogeneous architecture, where the horizontal interpolations are the same for each row, and the vertical interpolations are the same for each column, for example in the case of image scaling or zooming (see, for example, (Nuno-Maganda and Arias-Estrada, 2005; Mahale et al., 2014)). Since there is no

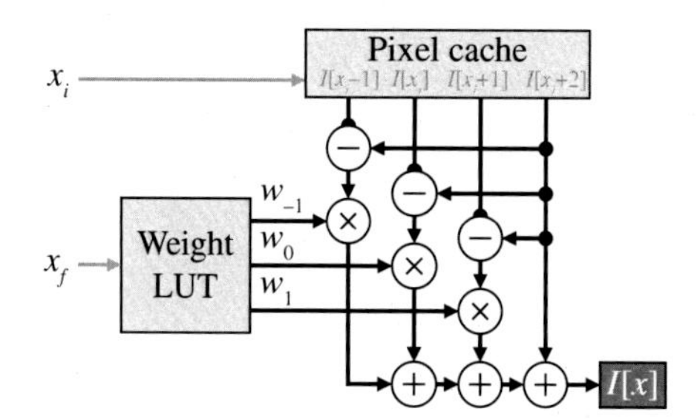

Figure 10.10 Bicubic interpolation in one dimension.

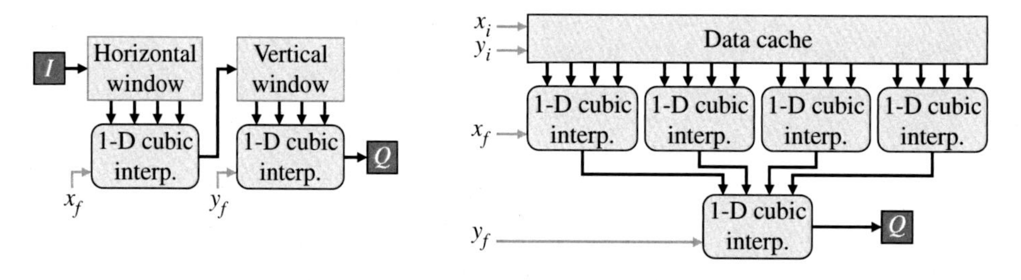

Figure 10.11 Implementation of bicubic interpolation: Left: homogeneous architecture, using separable horizontal and vertical interpolation filters; right: heterogeneous architecture, using a 2-D data cache.

rotation, the regular access pattern enables conventional filter structures to be used, although the coefficients change from one pixel to the next. Since all output pixels from a given column have the same horizontal offset, a single 1-D interpolation unit can be used horizontally with the intermediate values buffered for the separable vertical interpolation, as shown in Figure 10.11.

More complex is the heterogeneous architecture, where each output pixel has a different fractional offset (for example correcting lens distortion (Bellas et al., 2009)). Although separability can still be used, it is no longer possible to reuse partial calculations for adjacent output pixels. Therefore, a total of five 1-D interpolation blocks are required, as demonstrated in Figure 10.11. The coefficient lookup tables for x_f can be shared amongst four of the interpolation blocks.

The larger region of support makes the design of the cache more complex for bicubic interpolation than for bilinear interpolation. In principle, the methods described above for bilinear interpolation may be extended, although memory partitioning will be required to obtain the four pixels on each row within the window. Although bicubic interpolation gives a higher quality output than bilinear interpolation, the added complexity of memory management combined with the extra logic required to perform the filtering for the larger window often makes bilinear interpolation an acceptable compromise between accuracy and computational complexity in many applications.

10.1.2.3 Splines

A cubic spline is the next most commonly used interpolation technique. A cubic spline also defines a piecewise cubic continuous signal that passes through the sample points, with the condition that both the first and second derivatives are continuous at each sample. This has the effect that changing the value of any pixel will affect the segments over quite a wide range. Consequently, interpolating with splines is not a simple convolution with a kernel, because of their wide region of support.

B-splines, on the other hand, do have compact support. They are formed from the successive convolution of a rectangular function, as illustrated in Figure 10.12. B_0 corresponds to nearest-neighbour interpolation, while B_1 gives linear interpolation, as described above.

The cubic (third-order) B-spline interpolation kernel is given by

$$B_3(x) = \begin{cases} \frac{1}{2}|x|^3 - |x|^2 + \frac{2}{3}, & 0 \le |x| < 1, \\ -\frac{1}{6}|x|^3 + |x|^2 - 2|x| + \frac{4}{3}, & 1 \le |x| < 2, \\ 0, & 2 \le |x|. \end{cases} \tag{10.15}$$

The main limitation of higher order B-splines is that they do not go to zero at nonzero integer values of x. This means that Eq. (10.6) cannot be used directly for interpolation, since the continuous surface produced by convolving the image samples with the cubic B-spline kernel will not pass through the data points. Therefore, rather than weight the kernels with the image samples as in Eq. (10.6), it is necessary to derive a set of coefficients, based on the image samples:

$$C(x,y) = \sum_{i,j} c_{i,j} B_3(x-i, y-j). \tag{10.16}$$

These coefficients may be derived by solving the inverse problem, by setting $C[x,y] = I[x,y]$ in Eq. (10.16). The regular structure enables the inverse to be directly implemented using infinite impulse response filters in both the forward and backward directions on each row and column of the image (Unser et al., 1991).

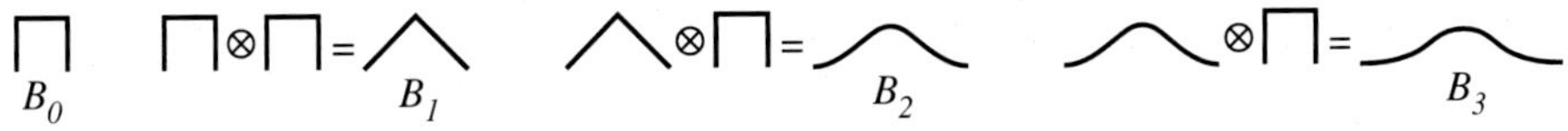

B_0 B_1 B_2 B_3

Figure 10.12 B-splines.

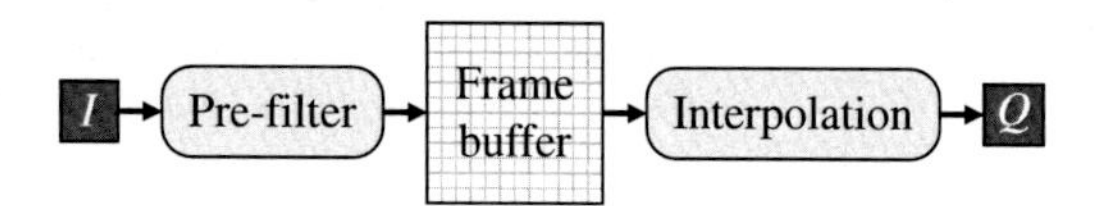

Figure 10.13 Structure of spline interpolation.

Spline interpolation therefore requires two stages as shown in Figure 10.13, pre-filtering the image to obtain the coefficients and then interpolation using the B-spline kernel. When implementing spline interpolation with an arbitrary (non-regular) sampling pattern, the pre-filtering may be applied to the regularly sampled input image, with the result stored in the frame buffer. Then the interpolation stage may proceed as before, using the sampled kernel functions as filter weights to generate the output image.

Direct implementation of the infinite impulse response pre-filtering presents a problem for a streamed processing because each 1-D filter requires two passes, once in the forward direction and once in the reverse direction. However, since the impulse responses of these filters decay relatively quickly, they may be truncated and implemented using a finite impulse response filter (Ferrari and Park, 1997).

Rather than use B-splines, the kernel can be selected so that the coefficients of the pre-filter are successively smaller powers of 2 (Ferrari and Park, 1997) (this has been called a 2-5-2 spline). This enables the filter to be implemented efficiently in hardware, using only shifts for the multiplications, as demonstrated on the left in Figure 10.14. Since each coefficient is shifted successively by one bit, shifting by more than k bits (where k is the number of bits used in the fixed-point representation of $c_{i,j}$) will make the term go to zero. This provides a natural truncation point for the window.

Since each half of the filter is a power series, the sub-filters may be implemented recursively, as shown in the right hand panel of Figure 10.14. The bottom accumulator sums the terms in the first half of the window and needs to be $2k$ bits wide to avoid round-off errors. The upper accumulator sums the terms in the second half of the window. It does not need to subtract out the remainder after a further k delays, because that decays to 0 (this is an infinite impulse response filter).

The kernel for the 2-5-2 spline is

$$K_{2\text{-}5\text{-}2}(x) = \begin{cases} \frac{1}{4}|x|^3 - \frac{7}{12}|x|^2 + \frac{5}{9}, & 0 \le |x| < 1, \\ \frac{1}{36}|x|^3 + \frac{1}{12}|x|^2 - \frac{2}{3}|x| + \frac{7}{9}, & 1 \le |x| < 2, \\ 0, & 2 \le x, \end{cases} \tag{10.17}$$

with the corresponding filter weights given as (Ferrari et al., 1999)

$$\begin{aligned} w_{-1} &= \frac{1}{36}x_f^3 + \frac{1}{6}x_f^2 - \frac{5}{12}x_f + \frac{2}{9}, \\ w_0 &= \frac{1}{4}x_f^3 - \frac{7}{12}x_f^2 + \frac{5}{9}, \\ w_1 &= -\frac{1}{4}x_f^3 + \frac{1}{6}x_f^2 + \frac{5}{12}x_f + \frac{2}{9}, \\ w_2 &= -\frac{1}{36}x_f^3 + \frac{1}{4}x_f^2. \end{aligned} \tag{10.18}$$

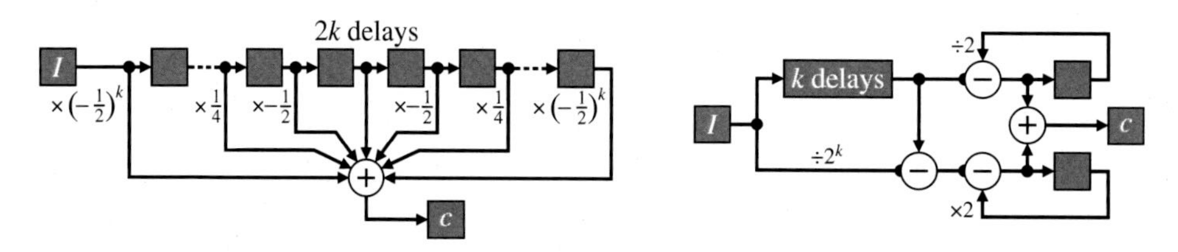

Figure 10.14 Efficient 2-5-2 spline pre-filter. Left: implementation as proposed by Ferrari and Park (1997). Source: Adapted from Ferrari and Park (1997). Right: recursive implementation.

These are used in the same way as for the bicubic interpolation, although the awkwardness of these coefficients makes lookup table implementation better suited than direct calculation.

The filters given here are one-dimensional. For 2-D filtering, the two filters can be used separably on the rows and columns, replacing the delays with row buffers. If the input is coming from a frame buffer, and the memory bandwidth allows, then the k row buffer delay may be replaced by a second access to the frame buffer in a similar manner to the box filter shown in Figure 9.15.

A field-programmable gate array (FPGA) implementation of this cubic spline filter was used to zoom an image by an integer factor by Hudson et al. (1998). The pre-filter shown in the left of Figure 10.14 was used. The interpolation stage was simplified by exploiting the fact that a pure zoom results in a regular access pattern. Since this was fairly early work, the whole algorithm did not fit on a single FPGA. One-dimensional filters were implemented, with the data fed through separately for the horizontal and vertical filters. Dynamic reconfiguration was also used to separate the pre-filtering and interpolation stages.

Several other interpolation kernels are available; a review of the more commonly used kernels for image processing is given by Wolberg (1990). Very few of these have been used for FPGA implementation because of their higher computational complexity than those described in this section.

10.2 Forward Mapping

In contrast to the reverse mapping, the forward mapping determines where each input pixel appears in the output image. Since a point on the image grid does not necessarily map onto the pixel grid in the output image, it is necessary to consider the area occupied in the output image by each input pixel. Therefore, each output pixel value is given by the weighted sum of the input pixels which overlap it, with the weights given by the proportion of the output pixel that is overlapped by the input pixels:

$$Q[u,v] = \sum_{x,y} w_{x,y,u,v} I[x,y], \tag{10.19}$$

where

$$w_{x,y,u,v} = \int Q[u,v] \cap I[x,y] du\, dv, \tag{10.20}$$

as illustrated in Figure 10.15.

For this, it is easiest to map the corners of the pixel, because these define the edges, which in turn define the region occupied by the pixel. On the output, it is necessary to maintain an accumulator image to which fractions of each input pixel are added as pixels are streamed in. There are two difficulties with implementing this approach on an FPGA for a streamed input. First, each input pixel can intersect many output pixels, especially if the magnification of the transform is greater than 1. Therefore, many (depending on the position of the pixel and the magnification) clock cycles are required to adjust all of the output pixels affected by each input pixel. Since each output pixel may be contributed to by several input pixels, a complex caching arrangement is required to overcome the associated memory bandwidth bottleneck. A second problem is that even for relatively simple mappings, determining the intersection as required by Eq. (10.20) is both complex and time consuming.

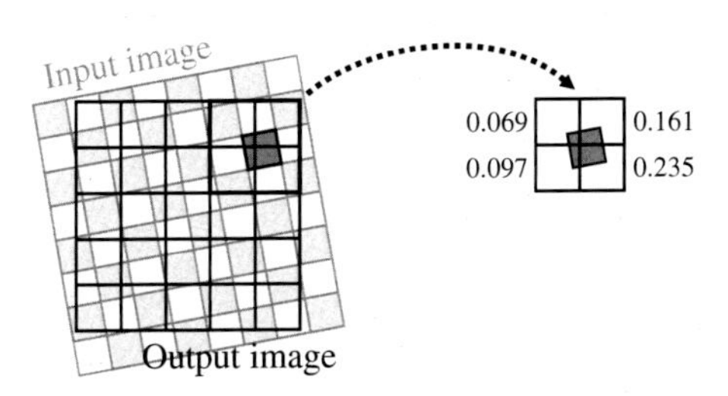

Figure 10.15 Mapping whole pixels from input to output image. The weights of each input pixel to the output pixels are based on the proportion of each output pixel that it overlaps, as illustrated on the right.

10.2.1 Separable Mapping

Both of these problems may be significantly simplified by separating the transformation into two stages or passes (Catmull and Smith, 1980). The key is to operate on each dimension separately:

$$Q[u,v] = Q[u, m_{fv}(u,y)] = T[m_{fu}(x,y), y] = I[x,y]. \tag{10.21}$$

The first pass through the image transforms each row independently to get each pixel on that row into the correct column. This forms the temporary intermediate image, $T[u,y]$. The second pass then operates on each column independently, getting each pixel into the correct row. This process is illustrated with an example in Figure 10.16. The result of the two transformations is that each pixel ends up in the desired output location. Note that the transformation for the second pass is in terms of u and y rather than x and y to reflect the results of the first pass.

Reducing the transformation from two dimensions to one dimension significantly reduces the difficulty in finding the pixel boundaries. In one dimension, each pixel has only two boundaries: with the previous pixel and with the next pixel. The boundary with the previous pixel can be cached from the previous input pixel, so only one boundary needs to be calculated for each new pixel.

Both the input and output from each pass are produced in sequential order enabling stream processing to be used. Although the second pass scans down the columns of the image, stream processing can still be used. However, a frame buffer is required between the two passes to transpose the intermediate image. If the input is streamed from memory, and the bandwidth enables, the image can be processed in place, with the output written back into the same frame buffer.

A single 1-D processor may be used for both passes as shown on the left in Figure 10.17. After processing the rows, the image is read by column (transposing the image) and fed back for column warping. Finally, the image is streamed out by row (transposed back again) while performing row processing on the next frame. The disadvantage of this architecture is that it can only process every second frame.

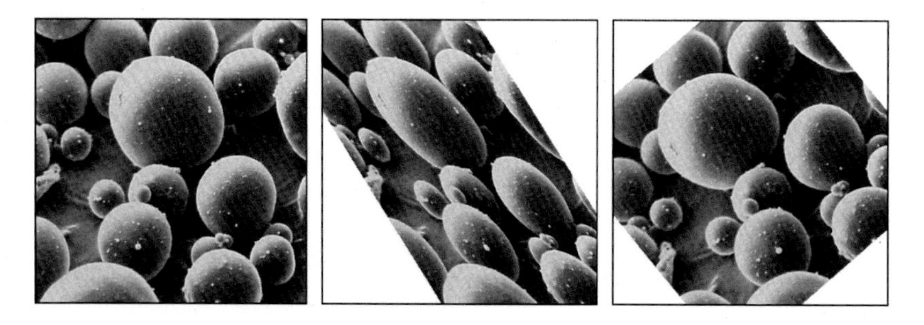

Figure 10.16 Two-pass transformation to rotate an image by 40°. Left: input image; centre: intermediate image after transforming each row; right: transforming each column.

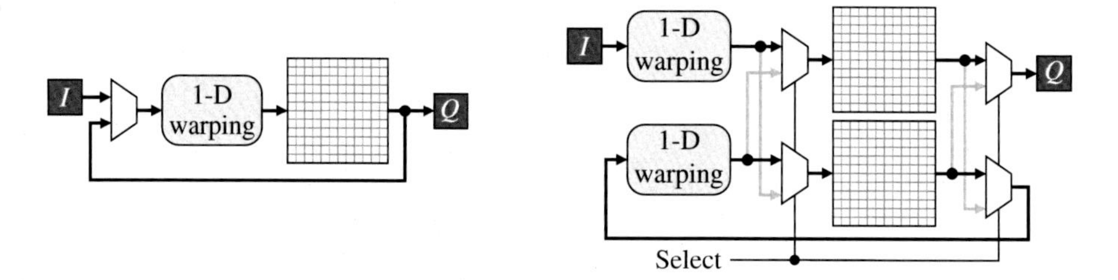

Figure 10.17 Implementation of two-pass warping with forward mapping, using a frame buffer for transposing the image. Left: sequential processing (only processing every second frame); right: using bank-switching to process every frame.

To process every frame, it is necessary to use two 1-D processors, one for the rows and one for the columns, as demonstrated on the right in Figure 10.17. Bank-switching enables the two processors to operate separately, with the banks switched each frame to maintain throughput. The processing latency is a little over two frame periods.

Since each row and column are processed independently, the algorithm may be accelerated by processing multiple rows and columns in parallel. The main limitation on the degree of parallelism is the bandwidth available to the frame buffer.

10.2.1.1 Limitations

The two-pass approach has several limitations. First, consider rotating an image by 90°. All of the pixels on a row in the input will end up in the same column in the output. This maps them to the same point in the intermediate image. The second pass is then unable to map these to separate rows because all the data has been collapsed to a point. This is called the ***bottleneck problem*** and occurs where the local rotation is close to 90°. Even for other angles, there is a reduction in resolution as a result of the compression of data in the first pass, followed by an expansion in the second pass (see Figure 10.16 for an example; the data in the intermediate image occupies a smaller area than either the input or output images). The worst cases may be overcome by rotating the image 90° first and then performing the mapping. This is equivalent to treating the incoming row-wise scanned data as a column scan and performing the column mapping first. While this works for global rotations, many warps may locally have a range of rotation angles in which case some parts of the image should be rotated first and others should not. Wolberg and Boult (1989) overcame this by performing both operations in parallel and selected the output locally based on which had the least compression in the intermediate image. This solution comes at the cost of doubling the hardware required (in particular, the number of frame buffers needed).

The bottleneck problem may be avoided for pure rotations using three passes rather than two (Wolberg, 1990). Each pass is a shear transformation, which merely shifts the pixels along a row or column giving a computationally simple implementation (at the cost of an extra pass).

A second problem occurs if the mappings on each row or column are not monotonically either increasing or decreasing. As a result, the mapping is many-to-one, with data from one part of the image folding over that which has already been calculated. (Such fold-over can occur even if the complete mapping does not fold over; the second pass would map the different folds to different rows.) Catmull and Smith (1980) suggested that such fold-over data from the first pass be held in additional frame buffers, so that it is available for unfolding in the second pass. Without such additional resources, the range of geometric transformations that may be performed using the two-pass approach is reduced.

A third difficulty with the two-pass forward mapping is that the second transformation is in terms of u and y. Since the original mapping is in terms of x and y, it is necessary to invert the first mapping in order to derive the second mapping. Consider, for example, an affine transformation:

$$\begin{aligned} u &= m_{fu}(x, y) = a_{ux}x + a_{uy}y + a_u, \\ v &= m_{fv}(x, y) = a_{vx}x + a_{vy}y + a_v. \end{aligned} \tag{10.22}$$

The first pass is straightforward, giving u in terms of x and y. However, v is also specified in terms of x and y rather than u and y as required for the second pass. Inverting the first mapping gives

$$x = \frac{u - a_{uy}y - a_u}{a_{ux}}, \tag{10.23}$$

and then substituting into the second part of Eq. (10.22) gives the required second mapping:

$$v = m_{fv}(u, y) = \frac{a_{vx}}{a_{ux}}u + \left(a_{vy} - \frac{a_{uy}a_{vx}}{a_{ux}}\right)y + \left(a_v - \frac{a_u a_{vx}}{a_{ux}}\right). \tag{10.24}$$

While inverting the first mapping in this case is straightforward, for many mappings it can be difficult (or even impossible) to derive an analytic mapping for the second pass, and even if it can be derived, it is often computationally expensive to perform. Wolberg and Boult (1989) overcame this by saving the original x in a second frame buffer, enabling $m_{f_y}(x, y)$ from Eq. (10.22) to be used. They also took the process one step further and represented the transformation through a pair of lookup tables rather than directly evaluating the mapping function.

The final issue is that the two-pass separable warp is not identical to the 2-D warp. This is because each 1-D transformation is of an infinitely thin line within the image. However, this line actually consists of a row or column of pixels with a thickness of one pixel. Since each row and column are warped independently, significant differences in the transformations between adjacent rows can lead to jagged edges (Wolberg and Boult, 1989). As long as adjacent rows (and columns) are mapped to within one pixel of one another, this is not a significant problem.

Subject to these limitations, the two-pass separable approach provides an effective approach to implement the forward mapping.

10.2.1.2 One-dimensional Warp

Each pass requires the implementation of a 1-D geometric transformation. While the input data can simply be interpolated to give the output, this can result in aliasing where the magnification is less than 1. To reduce aliasing, it is necessary to filter the input; in general, this filter will be spatially variant, defined by the magnification at each point.

The filtering, interpolation, and resampling can all be performed in a single operation (Fant, 1986). The basic principle is to maintain two fractional valued pointers into the stream of output pixels (see Figure 10.18). *Inseg* indicates how many output pixels that the current input pixel has yet to cover, while *Outseg* indicates how much of the current output pixel remains to be filled. If *Outseg* ≤ *Inseg*, an output pixel can be completed, so an output cycle is triggered. The input value is weighted by *Outseg*, combined with the value accumulated so far in *Accum*, and is output. The accumulator is reset, *Outseg* reset to 1 and *Inseg* decremented by *Outseg* to reflect the remaining input to be used. Otherwise, *Inseg* is smaller, so the current input is insufficient to complete the output pixel. This triggers an input cycle, which accumulates the remainder of the current input pixel (weighted by the fraction of the output pixel it covers), subtracts *Inseg* from *Outseg* to reflect the fraction of the output pixel remaining, and reads a new input pixel value. As each input pixel is loaded, the address is passed to the forward mapping function to get the corresponding pixel boundary in the output. The mapping may be implemented either directly or

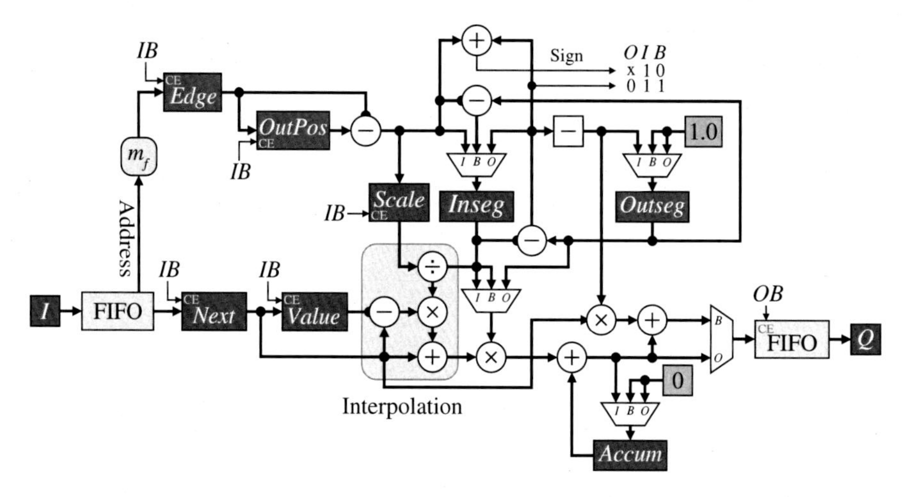

Figure 10.18 One-dimensional warping using resampling interpolation. There are three types of cycle: input (I), output (O), and combined input and output (B). (CE = clock enable.)

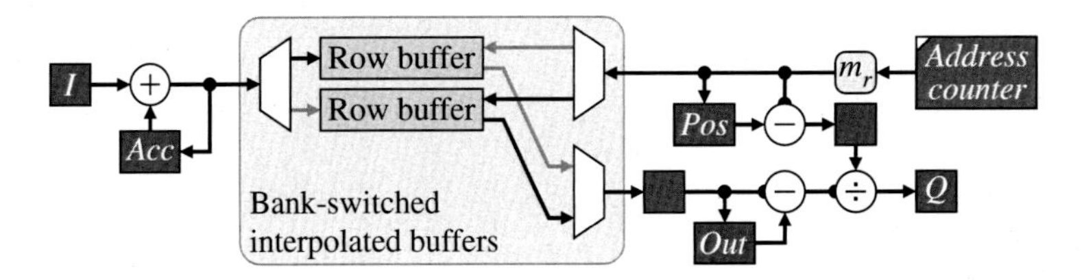

Figure 10.19 Separating the input and output phases using sum tables and interpolating on the output.

via a lookup table (Wolberg and Boult, 1989). The magnification or scale factor for the current pixel is obtained by subtracting the previous pixel boundary, *OutPos*, from the new boundary. This result is used to initialise *Inseg*.

If the input pixel is sufficiently wide to also complete an output pixel, then both (B) an input and output cycle are combined (Wolberg et al., 2000). This will be the case if

$$B = (Inseg < Outseg) \wedge (Edge - Outpos + Inseg > Outseg). \tag{10.25}$$

For the combined cycle, the clock is enabled for both the input and output circuits.

If the magnification is greater than 1, then there will be multiple output pixels for each input pixel, and if the magnification is less than 1, then there will be multiple input pixels for each output pixel. A FIFO buffer is therefore required on both the input and output to smooth the flow of data to allow streaming with one pixel per clock cycle.

The interpolation block is required when the magnification is greater than 1 to prevent a single pixel from simply being replicated in the output, resulting in large flat areas and contouring (Fant, 1986). It performs a linear interpolation between the current input value and the next pixel value in proportion to how much of the input pixel is remaining.

Handling of the image boundaries is not explicitly shown in Figure 10.18. The initial and trailing edge of the input can be detected from the mapping function. If the first pixel is already within the image, a series of empty pixels need to be output. Similarly, at the end of the input row, a series of empty pixels may also be required to complete the output row.

Although the 1-D warping has an average latency of one row, it requires two row times to complete. Therefore, to run at pixel clock rates, two such warping units must be operated in parallel. While many mappings will still take longer than one data row length, the extra time required is small in most cases. Where the magnification is greater than 1, a pixel will be output every clock cycle, with an input cycle when necessary to provide more data. The converse will be true where the magnification is less than 1. Assuming that the input and output are the same size, then extra clock cycles are only required when the magnification is greater than 1 on some parts of the row and less than 1 on the other parts. If the image has horizontal blanking at the end of each row, this would be sufficient to handle most mappings in a single row time.

Since the output pixel value is a sum over a fraction of pixels, Evemy et al. (1990) took a different approach to accelerate the mapping by completely separating the input and output cycles. As the row is input, the image is integrated to create a sum table. This enables any output pixel to be calculated by taking the difference between the two interpolated entries into the sum table corresponding to the edges of the output pixel. After the complete row has been loaded, the output phase begins. The reverse mapping (along the row) is used to determine the position of the edge of the output pixel in the sum table, with the value at the sub-pixel position determined through interpolation (this requires two accesses to the memory; see Figure 5.10). The difference between successive input sums gives the integral from the input corresponding to the output pixel. This is normalised by the width of the output pixel to give the output pixel value. The structure for this is shown in Figure 10.19.

As both input and output take exactly one clock cycle per pixel, regardless of the magnification, this approach can be readily streamed. By bank-switching the summed row buffers, a second row is loaded while the first row is being transformed and output. The latency of this scheme is therefore one row length (plus any pipeline latencies).

10.2.1.3 Two-dimensional Warp

Rather than perform the row and column passes independently, as implied at the start of this section, the two passes can be combined together (Bailey and Bouganis, 2009). In the second pass, rather than process each column separately, they can be processed in parallel, with row buffers used to hold the data. The basic idea is illustrated in Figure 10.20.

The circuit for the first pass row warping can be any of those described above. For the second pass, using sum tables would require an image buffer because the complete column needs to be input before output begins, gaining very little. However, with the interleaved input and output methods, each time a pixel is output from the first pass, it may be passed to the appropriate column in the second pass. Any output pixel from the second pass may be saved into the corresponding location in the output frame buffer. If the magnification is less than 1, then at most one pixel will be written every clock cycle. Note that in general pixels will be written to the frame buffer non-sequentially because the mappings for each column will be different.

If the magnification is greater than 1, then multiple clock cycles may be required to ready the column for the next input pixel before moving onto the next column. This will require that the column warping controls the rate of the row warping, with the FIFO buffer to take up the slack.

To summarise, a forward mapping allows data to be streamed from an input. To save having to maintain an accumulator image, with consequent bandwidth issues, the mapping can be separated into two passes. The first pass operates on each row and warps each pixel into the correct column, whereas the second pass operates on each column, getting each pixel into the correct row.

10.2.2 Hybrid Approach

A hybrid approach proposed by Oh and Kim (2008) is to use a combination of forward and reverse mappings, as shown in Figure 10.21. The corners of each input pixel are forward mapped to define a region in the output image

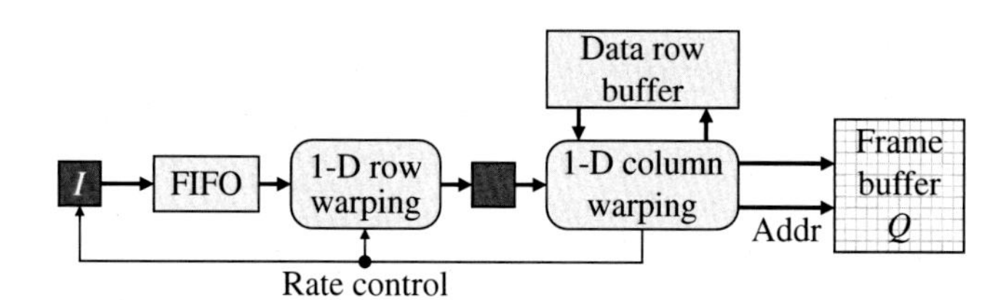

Figure 10.20 Combining the row and column warps by implementing the column warps in parallel.

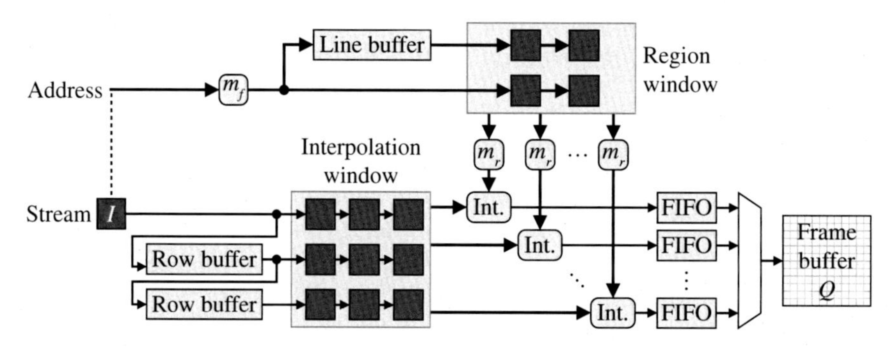

Figure 10.21 Hybrid processing scheme, which uses the forward map to define a region, and reverse mapping to perform interpolation.

(similar to Figure 10.15). This forward mapping only needs to be approximate, so to simplify computation, its size is extended slightly and made rectangular with sides parallel to the output coordinate axes. A single forward map is used by buffering the results of the previous row and pixel in a region window. Then the integer pixel positions within the output region are reverse-mapped back to the input image where they are interpolated to give the corresponding output pixels. Depending on the magnification, multiple reverse mappings may need to be performed in parallel. To facilitate the interpolation, several rows of the input image are cached using row buffers. Only sufficient rows are required to form the interpolation window, because the outputs will come from somewhere within the last input pixel. The interpolation results are stored in output FIFOs for writing to the frame buffer.

10.3 Common Mappings

Some of the common geometric transformations and their optimisation for FPGA implementation will be described in this section.

10.3.1 Affine Transformation

An affine transformation performs a linear mapping between input and output coordinates. It is capable of translation; stretching and scaling (zooming along one or both axes); rotation; and shear transformations. The same form of equations are used for both forward and reverse mapping.

The reverse mapping in matrix form is

$$\begin{bmatrix} x \\ y \end{bmatrix} = \begin{bmatrix} a_{xu} & a_{xv} & a_x \\ a_{yu} & a_{yv} & a_y \end{bmatrix} \begin{bmatrix} u \\ v \\ 1 \end{bmatrix}. \tag{10.26}$$

Implemented directly, this transformation would require four multiplications and four additions. With stream processing, the input to the mapping function is a raster scan. (This will also be the case when using a forward mapping; refer to Figure 10.2 to see that this is so.) This enables incremental calculation to be used to simplify the computation by exploiting the fact that v (or y for the forward mapping) is constant, and $u \leftarrow u + 1$ (or $x \leftarrow x + 1$ for the forward mapping) for successive pixels. Substituting this into Eq. (10.26) gives

$$\begin{bmatrix} x \\ y \end{bmatrix} \leftarrow \begin{bmatrix} a_{xu} & a_{xv} & a_x \\ a_{yu} & a_{yv} & a_y \end{bmatrix} \begin{bmatrix} u+1 \\ v \\ 1 \end{bmatrix} = \begin{bmatrix} x \\ y \end{bmatrix} + \begin{bmatrix} a_{xu} \\ a_{yu} \end{bmatrix}, \tag{10.27}$$

requiring only two additions.

A similar simplification can be used for moving from one row to the next. When implementing these, it is convenient to have a temporary register to hold the position at the start of the row. This may be incremented when moving from one row to the next, and is used to initialise the main register which is updated for stepping along the rows, as shown in Figure 10.22. It is necessary that these registers maintain sufficient guard bits to prevent the accumulation of round-off errors.

10.3.2 Perspective Mapping

Consider a planar object with coordinates (u, v) imaged using a pinhole camera model. The image coordinates are then

$$x = \frac{a_{xu}u + a_{xv}v + a_x}{a_{mu}u + a_{mv}v + a_m}, \qquad y = \frac{a_{yu}u + a_{yv}v + a_y}{a_{mu}u + a_{mv}v + a_m}, \tag{10.28}$$

where the denominator effectively scales the image based on its distance from the camera. The denominator corresponds to the intersection of the object plane with the plane $z = 0$ of the pinhole model. Equation (10.28)

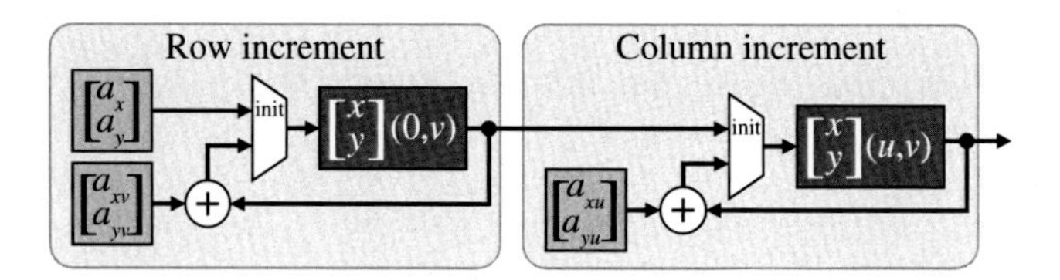

Figure 10.22 Incremental mapping for affine transformation.

may be represented in matrix form by extending Eq. (10.26) using homogeneous coordinates:

$$\begin{bmatrix} mx \\ my \\ m \end{bmatrix} = \begin{bmatrix} a_{xu} & a_{xv} & a_x \\ a_{yu} & a_{yv} & a_y \\ a_{mu} & a_{mv} & a_m \end{bmatrix} \begin{bmatrix} u \\ v \\ 1 \end{bmatrix}, \tag{10.29}$$

where the third row gives the denominator, m. The mx and my terms are then divided by m to give the transformed coordinates.

The forward model may be found by inverting Eq. (10.29). In the forward model, the denominator corresponds to the horizon line (where parallel lines converge) in the image.

The same incremental architecture from Figure 10.22 may be used for the perspective transformation, with the addition of a division at each clock cycle to perform the magnification scaling.

10.3.3 Polynomial Mapping

More complex mappings may be represented using a polynomial mapping, which includes higher order terms. Higher order polynomials have more degrees of freedom but are more complex to calculate and are also more likely to overfit the data. A second-order polynomial forward mapping which includes the second-order terms x^2, xy, and y^2,

$$\begin{aligned} u(x,y) &= a_{uxx}x^2 + a_{uxy}xy + a_{uyy}y^2 + a_{ux}x + a_{uy}y + a_u, \\ v(x,y) &= a_{vxx}x^2 + a_{vxy}xy + a_{vyy}y^2 + a_{vx}x + a_{vy}y + a_v, \end{aligned} \tag{10.30}$$

will be discussed here.

Again incremental update can be used to reduce the processing by reusing calculations from the previous pixel. First, consider the initialisation of each row

$$\begin{aligned} u(0,y) &= a_{uyy}y^2 + a_{uy}y + a_u, \\ v(0,y) &= a_{vyy}y^2 + a_{vy}y + a_v. \end{aligned} \tag{10.31}$$

For the row increment, it is necessary to know the value from the previous row. Looking at the u component:

$$\begin{aligned} u(0,y-1) &= a_{uyy}(y-1)^2 + a_{uy}(y-1) + a_u \\ &= [a_{uyy}y^2 + a_{uy}y + a_u] - a_{uyy}(2y-1) - a_{uy}. \end{aligned} \tag{10.32}$$

Therefore,

$$u(0,y) = u(0,y-1) + a_{uyy}(2y-1) + a_{uy}, \tag{10.33}$$

and similarly

$$v(0,y) = v(0,y-1) + a_{vyy}(2y-1) + a_{vy},$$

with the second terms produced by another accumulator (adding $2a_{uyy}$ and $2a_{vyy}$ each iteration), as shown in Figure 10.23.

For the column increment along the row, it is necessary to know the previous value:

$$\begin{aligned} u(x-1,y) &= a_{uxx}(x-1)^2 + a_{uxy}(x-1)y + a_{uyy}y^2 + a_{ux}(x-1) + a_{uy}y + a_u \\ &= [a_{uxx}x^2 + a_{uxy}xy + a_{uyy}y^2 + a_{ux}x + a_{uy}y + a_u] - a_{uxx}(2x-1) - a_{uxy}y - a_{ux}. \end{aligned} \tag{10.34}$$

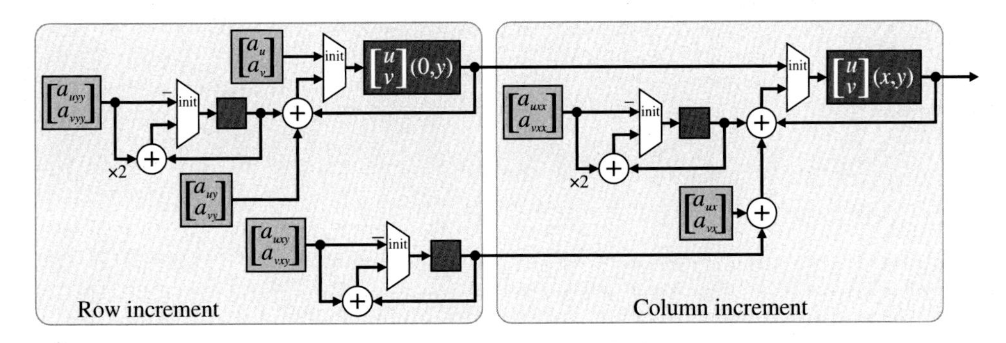

Figure 10.23 Incremental calculation for a second-order polynomial warp.

Therefore,

$$u(x,y) = u(x-1,y) + a_{uxx}(2x-1) + a_{uxy}y + a_{ux},$$
$$v(x,y) = v(x-1,y) + a_{vxx}(2x-1) + a_{vxy}y + a_{vx}, \tag{10.35}$$

where the second terms are produced by a secondary accumulator incremented with each column, and the third terms are provided by a secondary accumulator incremented each row.

A similar approach may be taken for higher order polynomial mappings, with additional layers of accumulators required for the higher order terms.

Unlike the affine and perspective mappings, the inverse of a polynomial mapping is not another polynomial mapping. Therefore, it is important to determine the mapping coefficients for the form that is being used. It is not a simple matter to convert between the forward and reverse mappings for polynomial transformations.

10.3.4 Lens Distortion

Many low-cost lenses suffer from lens distortion, which is defined as any geometric deviation from the image that would be obtained from using a simple pinhole camera model. Such distortion results from the magnification of the lens not being uniform across the field-of-view, an example of which is demonstrated in Figure 10.24. Radial distortion is particularly prevalent with wide angle lenses and is even deliberately introduced by fish-eye lenses to give a wider angle of view.

Lens distortion is usually modelled using a radially dependent magnification from the undistorted radius (as given by the pinhole model), r_u, and the distorted radius in the actual captured image, r_d:

$$r_d = r_u M_d(r_u), \tag{10.36}$$

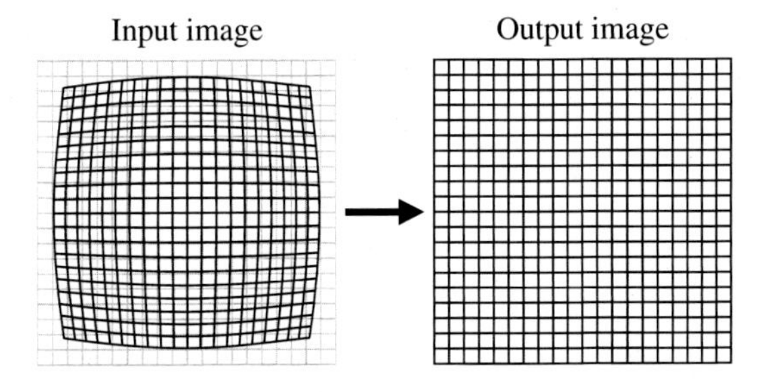

Figure 10.24 Lens distortion correction.

where M_d is the radial magnification caused by the distortion. Note that the centre of distortion (the origin in these equations) is not necessarily in the centre of the image (Willson and Shafer, 1994). Two additional parameters are therefore required to represent the distortion centre, (x_0, y_0).

The model for correcting the distortion is given by the inverse of Eq. (10.36) and is also a radially dependent magnification:

$$r_u = r_d M_u(r_d). \tag{10.37}$$

Three different approaches have been taken to correct radial distortion using FPGAs. The incoming distorted image stream can be transformed on-the-fly using the forward mapping based on Eq. (10.37), with the corrected image being stored in a frame buffer on the output (Oh and Kim, 2008). The transformation can be applied from one frame buffer to another, usually using the reverse map of Eq. (10.36). This allows the image to be divided into tiles to balance between external memory bandwidth and on-chip caching requirements (Bellas et al., 2009). An undistorted output image stream can be produced on-the-fly using the reverse mapping, taking the distorted input from a frame buffer (Eadie et al., 2002; Gribbon et al., 2003).

Several different distortion models (formulations of $M_u(\cdot)$ or $M_d(\cdot)$) have been proposed, particularly for fish-eye lenses (Hughes et al., 2008). One such dedicated fish-eye model was implemented by Bellas et al. (2009). Eadie et al. (2002) used a third-order 2-D polynomial warp. For standard lenses, with mild distortion, the magnification is often modelled using a radial Taylor series (Tsai, 1987):

$$M_u(r_d) = 1 + \kappa_1 r_d^2 + \kappa_2 r_d^4 + \cdots \tag{10.38}$$

Such a model can also include a tangential component and has been implemented by Ngo and Asari (2005) and Oh and Kim (2008). For many lenses, a simple first-order model is sufficient to account for much of the distortion (Li and Lavest, 1996) even when the distortion is quite severe. Such a simple first-order model was implemented by Gribbon et al. (2003).

One disadvantage of the Taylor series model is that it is difficult to invert. For example, Eq. (10.38) is a forward model for distortion correction. The parameters of the corresponding reverse model cannot be derived analytically with more than one parameter and are either approximated (Mallon and Whelan, 2004) or more usually estimated by least squares fitting.

Regardless of the model used, the mapping can either be calculated directly from the model for each pixel, or a small lookup table can be used to hold the mapping. Direct calculation has the advantage of enabling dynamic adjustment of the distortion parameters, whereas a pre-calculated mapping is harder to change dynamically and is best if the lens system does not change.

Most implementations have used bilinear interpolation when resampling, although Bellas et al. (2009) used the more demanding bicubic interpolation.

It will be shown how incremental processing can be used to reduce the computation (Gribbon et al., 2003) when correcting the distortion, using the first-order mapping:

$$r_u = r_d(1 + \kappa r_d^2). \tag{10.39}$$

The processing pipeline is shown in Figure 10.25. To correct a distorted image on-the-fly, producing a streamed output, the reverse mapping is required:

$$r_d = r_u M(\kappa, r_u). \tag{10.40}$$

Since

$$r_u^2 = u^2 + v^2, \tag{10.41}$$

consider moving from one pixel to the next in the output stream:

$$(u + 1)^2 = u^2 + 2u + 1. \tag{10.42}$$

This just requires a single addition (since the 1 can be placed in the least significant bit when u is shifted). A similar calculation can be used for v^2 when moving from one line to the next. At the start of the frame, the v

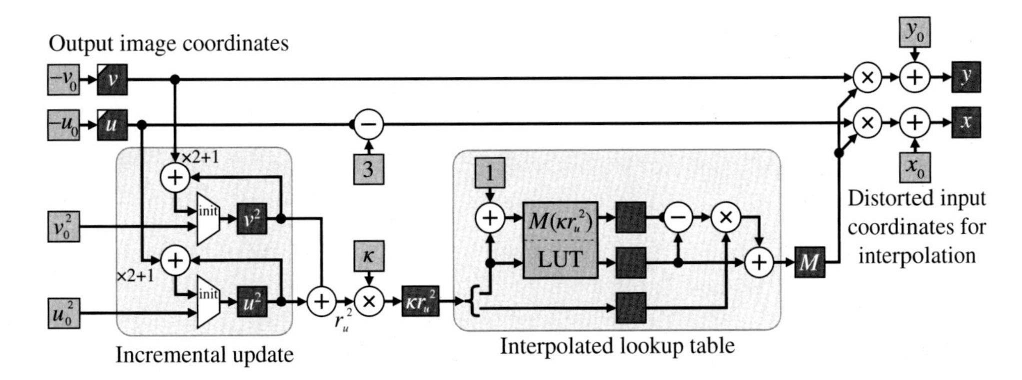

Figure 10.25 Pipeline for correcting lens distortion using incremental calculation and an interpolated lookup table.

and v^2 registers are initialised based on the location of the origin. Similarly, u and u^2 are initialised at the start of each line, with v and v^2 updated for the next line.

Rearranging Eq. (10.40) and substituting in Eq. (10.39) gives

$$M = \frac{r_d}{r_u} = \frac{1}{1+\kappa r_d^2} = \frac{1}{1+\kappa r_u^2 M^2}. \tag{10.43}$$

Therefore,

$$M + \kappa r_u^2 M^3 = 1, \tag{10.44}$$

which can be solved to give $M(\kappa r_u^2)$. The reverse magnification function depends on the product κr_u^2 rather than each of the terms separately. Therefore, a single magnification function can be implemented and used for different values of κ. It is also unnecessary to take the square root of r_u^2 to calculate M.

The solution to Eq. (10.44) is nontrivial to calculate on-the-fly, but is quite smooth (see Figure 10.26), so may be efficiently implemented using either an interpolated lookup table or multipartite tables (described in Chapter 5.2.6). The accuracy of the representation depends on the desired accuracy in the corrected image. For a 1024×1024 image, the maximum radius is approximately 725 pixels in the corner of the image. Therefore, to achieve better than 1 pixel accuracy at the corners of the image, the magnification needs to have at least 10 bits accuracy. To maintain good accuracy, it is desired to represent the magnitude with at least 15 bits accuracy. For κr_u^2 in the range of 0–1, direct lookup table implementation would require 2^{15} entries. An interpolated lookup table only requires 2^8 entries for the same accuracy (Gribbon et al., 2003).

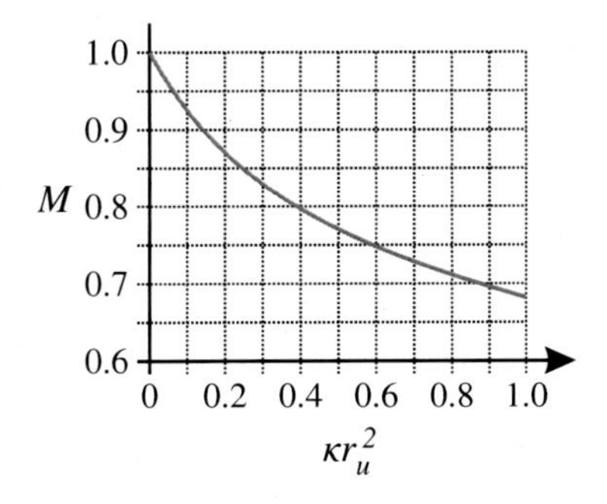

Figure 10.26 Reverse magnification function. Source: Adapted from Gribbon et al. (2003).

The next step is to use the magnification function to scale the radius. One approach for achieving this (Ngo and Asari, 2005) is to use a CORDIC (coordinate rotation digital computer) engine to convert the output Cartesian coordinates to polar coordinates, scale the radius by the magnification function, and then use a second CORDIC engine to convert polar coordinates back to Cartesian coordinates in the input image. However, such transformations are unnecessary because the angle does not change. From similar triangles, scaling the radius can be achieved simply by scaling the u and v components directly:

$$\begin{aligned} x &= uM + x_0, \\ y &= vM + y_0. \end{aligned} \tag{10.45}$$

The −3 in the u line in Figure 10.25 is to account for the pipeline latency; u is incremented three times while determining the magnification factor. (This assumes that there are at least three clock cycles of horizontal blanking, and u continues to be incremented during this period.) An alternative to the subtraction is to add three pipeline registers to the u line before multiplying by the magnification factor.

The last stage is to use the calculated position in the input image to derive the corresponding output pixel value. For this, any of the techniques described in Section 10.1.2 can be used.

10.3.5 Non-parametric Mappings

All of the previous mappings are based on parametric models of the geometric transformation. An alternative is to explicitly store the mapping for each pixel. For a geometric transformation, this map represents the source location (for a reverse mapping) for each pixel in the output image (Wang et al., 2005) and is represented using two 2-D lookup tables (one for x and one for y).

Non-parametric mappings have several advantages over parametric models. First, the mapping can be derived directly from point correspondences (see Section 10.4). In this case, it is not necessary to assume any particular model, enabling systematic imaging errors to be captured, not just the distortions according to the given model (Qui and Ma, 1995). Consequently, it can capture distortions and variations that are difficult or impossible to model with simple parametric models (Bailey et al., 2005). Second, evaluating the map is simply a table lookup, so is generally much faster than evaluating a parametric model. This is the case even if the map was derived from (and used to represent) a parametric model.

However, these advantages come at the cost of the large memory required to represent the map for every pixel. First, the range of numbers within the map can be reduced significantly by representing relative values (offsets) rather than absolute positions. Second, a smooth, slowly varying map can readily be compressed by down-sampling the map and using bilinear interpolation to reconstruct the map for a desired pixel (Di Febbo et al., 2016). The resolution required to represent the down-sampled map obviously depends on the smoothness of the map (which depends on the type and level of distortion) and also the accuracy with which it needs to be recreated.

For stream processing, the required map samples will be aligned with the scan axis, simplifying the access pattern. The relatively small number of weights used can be stored in small RAMs. With a separable mapping, the vertical interpolation should be performed first, since each set of vertical points will be reused for several row calculations. The multiplications associated with bilinear interpolation of the mapping can be eliminated using incremental update (Akeila and Morris, 2008) although this requires maintaining extra guard bits within the accumulators to avoid accumulation of round-off errors.

A similar approach can be used for non-parametric representation of uneven illumination or vignetting. In this case, the map contains the illumination level or correction scale factor directly for each pixel.

10.4 Image Registration

In Sections 10.1–10.3, it was assumed that the geometric transformation was known. Image registration is the process of determining the relationship between the pixels within two images or between parts of images. There is a wide variety of image registration methods (Brown, 1992; Zitova and Flusser, 2003), of which only a selection of the main approaches will be outlined here. There are two broad classes of registration: those based on matching sets of feature points, and those based on matching extended areas within the image. These are quite different in their approach, so they will be described separately. This section concludes with some of the applications of image registration.

10.4.1 Feature-based Methods

The key principle behind feature-based registration is to find and match a set of corresponding points within the images to be registered. It therefore involves the following steps (Zitova and Flusser, 2003):

- detect distinctive, salient feature points within each image,
- match the corresponding feature points between the images,
- estimate the transformation model that maps these correspondences between the images, and
- apply the transformation to register one image with the other.

The last step has been discussed in some detail earlier in this chapter; this section therefore focuses on the other steps.

10.4.1.1 Feature Point Detection

Feature points are points within an image that are locally distinctive, enabling their location to be determined accurately. Ideally, they should also be easy to detect and well distributed throughout the image. The particular features used should also be invariant to the expected transformation between the images to facilitate reliable matching.

If the scene is contrived (as is often the case when performing calibration), then the detection method can be tailored specifically to those expected features. Common calibration patterns include a rectangular array of dots, a checkerboard pattern, or a grid pattern. For an array of dots, adaptive thresholding can reliably segment the dots, with the centre of gravity used to define the feature point location. With a checkerboard pattern, either an edge detector is used to detect the edges, from which the corners are inferred, or a corner detector is used directly. A corner detector just is a particular filter that is tuned to locating corners (Noble, 1988). For grid targets, a line detector can be used, with the location of the intersections identified as the feature point locations.

The regular structure of a contrived target also makes correspondence matching relatively easy because the structure is known in advance. The danger with using a periodic pattern is that the matched locations may be offset by one or more periods between images. This may be overcome by adding a distinctive feature within the image that can act as a key for alignment.

For natural scenes, while edges are relatively easy to detect, there is usually an ambiguity because edges are essentially 1-D features (although there is a high gradient across an edge, it is low along the edge). Corner features are small regions which have high gradients in two perpendicular directions. The Harris corner detector (Harris and Stephens, 1988) detects this by building the structure tensor within a Gaussian weighted window:

$$\mathbf{M} = \sum_{i,j\in\mathbf{W}} w_G[i,j] \begin{bmatrix} I_x^2 & I_xI_y \\ I_xI_y & I_y^2 \end{bmatrix} = \begin{bmatrix} \sum w_G I_x^2 & \sum w_G I_xI_y \\ \sum w_G I_xI_y & \sum w_G I_y^2 \end{bmatrix} = \begin{bmatrix} M_{xx} & M_{xy} \\ M_{xy} & M_{yy} \end{bmatrix}, \tag{10.46}$$

where I_x and I_y are the first derivatives of the image (for example as produced by a Sobel filter). Gaussian weights are commonly used because this effectively makes the window circularly symmetric. The eigenvalues of $\mathbf{M}$, λ_1 and λ_2, are the roots of the characteristic equation:

$$\begin{aligned} P(\lambda) &= (M_{xx} - \lambda)(M_{yy} - \lambda) - M_{xy}^2 \\ &= \lambda^2 - (M_{xx} + M_{yy})\lambda + (M_{xx}M_{yy} - M_{xy}^2) \\ &= \lambda^2 - \text{trace}(\mathbf{M})\lambda + \det(\mathbf{M}) \end{aligned} \tag{10.47}$$

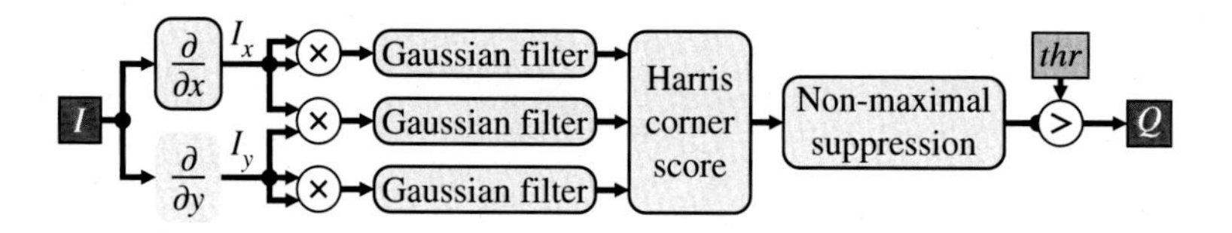

Figure 10.27 Architecture of Harris corner detection.

and correspond to the gradients in two orthogonal directions. Therefore, a point is on a corner when both eigenvalues are large, leading to the corner measure (Harris and Stephens, 1988)

$$C = \det(\mathbf{M}) - \alpha(\text{trace}(\mathbf{M}))^2, \tag{10.48}$$

where α is a constant (typically in the range 0.4–0.6). Variations of this have been implemented in several papers (see, for example, (Lim et al., 2013; Amaricai et al., 2014; Sikka et al., 2021)), with the basic architecture shown in Figure 10.27. Non-maximal suppression selects only the local maximum response as being the best corner candidate.

An alternative corner measure is the minimum eigenvalue (Shi and Tomasi, 1994):

$$C = \min(\lambda_1, \lambda_2) = \frac{1}{2}\text{trace}(\mathbf{M}) - \sqrt{\frac{1}{4}(\text{trace}(\mathbf{M}))^2 - \det(\mathbf{M})}, \tag{10.49}$$

which is a corner when greater than a threshold, $C > \lambda_t$. This calculation requires a square root, so Benedetti and Perona (1998) adapted this to give the corner measure $C = P(\lambda_t)$ (subject to $M_{xx} > \lambda_t$).

Two other corner detectors are SUSAN (smallest univalue segment assimilating nucleus) (Smith and Brady, 1997), which was implemented on an FPGA by Claus et al. (2009), and FAST (features from accelerated segment test) (Rosten and Drummond, 2005), which has been implemented on FPGA by Lim et al. (2013) and Huang et al. (2018).

One limitation of the Harris detector is that it is sensitive to scale. This limitation is overcome using the scale invariant feature transform (SIFT) (Lowe, 1999, 2004). SIFT filters the images with a series of Gaussian filters to obtain a scale-space representation. The difference between successive scales (a difference of Gaussian (DoG) filter) is a form of band-pass filter that responds most strongly to objects or features of that scale. The local maxima and minima both spatially and in scale are considered as candidate feature points and are located to sub-pixel accuracy. Let D be the response of the DoG filter at the local extreme. Feature points which have a weak response ($|D| < 0.03$) are removed. Since the DoG filter also responds strongly to edges, there is a danger that a feature point will be poorly located along the edge. The Hessian of D at the candidate feature point is used to remove potentially unstable edge points (Lowe, 2004), if

$$\frac{(\text{trace}(\mathbf{H}(D)))^2}{\det(\mathbf{H}(D))} = \frac{(D_{xx} + D_{yy})^2}{D_{xx}D_{yy} - D_{xy}^2} > 12.1. \tag{10.50}$$

For SIFT feature detection, most of the computation is spent in filtering the image. Although a Gaussian of a larger scale may be formed by cascading another Gaussian filter over the already processed image, the latency of successive filters makes alignment to calculate D and locate the maxima in scale expensive in hardware. However, the Gaussian filters of different scales may be implemented in parallel, reusing the row buffers (Vourvoulakis et al., 2016). The Gaussian window is separable, so the 2-D filter can be decomposed into a 1-D column filter (sharing the row buffers) followed by a 1-D row filter. After blurring the image sufficiently, it may be safely down-sampled, reducing the volume of processing in the subsequent stages. This arrangement is outlined in Figure 10.28; the gradient filters are used for calculating the feature descriptor (described in Section 10.4.1.2). The same hardware may be used for subsequent octaves, or additional blocks could be used for each octave. After down-sampling, the data volume is reduced by a factor of 4. This allows the hardware to be reduced using multi-phase design or even share the hardware for the remaining octaves (Chang and

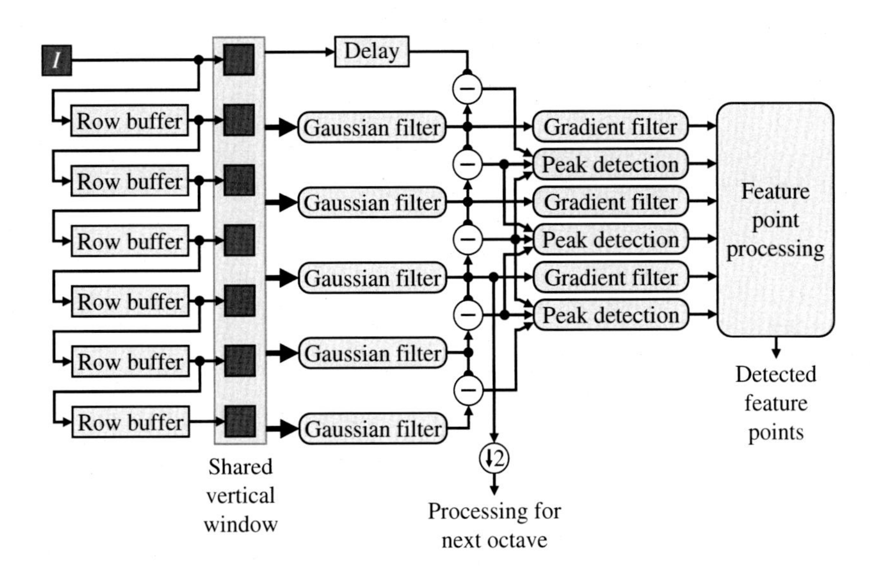

Figure 10.28 SIFT feature point detection for a single octave, with three scales per octave.

Hernandez-Palancar, 2009). Typically fewer than 1% of the pixels within the image are feature points. Therefore, a single instance of the candidate filter of Eq. (10.50) can be shared amongst all scales.

Most hardware implementations follow the software design, with a cascade of Gaussian filters, and compensating delays to align the outputs; the exception is Vourvoulakis et al. (2016) who implement all the filters for an octave in parallel. Most use 7×7 Gaussian filters, although Pettersson and Petersson (2005) used 5×5, and Yao et al. (2009) used 15×15 filters. In detecting the peaks, most implementations only applied non-maximal suppression within an octave, which can lead to multiple detection of features at scales on the boundaries between octaves. This is a trade-off between additional Gaussian filters in each octave and the extra time required to process the additional detected features.

An alternative to SIFT is SURF (speeded up robust features) (Bay et al., 2006, 2008) which uses box filters rather than Gaussian filters. The determinant of the Hessian is used as the feature detector, with the second derivatives at each scale calculated using an integral image. So rather than scaling the images, the filters are scaled, giving constant calculation time regardless of scale. A good example of a SURF implementation is given by Chen et al. (2016).

A detailed discussion of many other feature detectors is beyond the scope of discussion here. A couple worth mentioning are oriented FAST and rotated BRIEF (ORB) (Rublee et al., 2011) and BRISK (Leutenegger et al., 2011), which are both scale-space versions of the FAST corner detector.

10.4.1.2 Feature Descriptors

Once features points have been detected, it is necessary to derive a descriptor for each feature point that uniquely represents the local neighbourhood. An obvious approach is to match the local pixel values within a scaled window between images, but this has been shown to be sensitive to mis-registration and even small local distortions (for example caused by a change in viewpoint). Feature vectors should be as short as possible (to make matching more efficient), while still being distinctive, and ideally should be invariant to the expected transformations, including variations in brightness and contrast. The key idea is that it is more efficient to match a small descriptor than a larger neighbourhood.

The SIFT feature descriptor (Lowe, 2004) is a set of histograms of gradient orientations. First, the orientation of the feature point is determined from the gradient (first derivatives) of the Gaussian filtered image at the detected feature point's scale. A histogram of the gradient orientations within a circular window (weighted by the gradient magnitude) is used to determine the dominant direction. Next, the gradient images are resampled in a 16×16 window, with orientation and scale determined from the detected feature point. From this window, a 4×4 set of local histograms is accumulated of the gradients weighted by edge strength. Each histogram has 8 bins, giving a total of 128 dimensions to the feature descriptor. Basing the descriptor on gradients reduces the sensitivity to absolute light levels, and normalising the final feature to unit length reduces sensitivity to contrast, improving the robustness.

Many hardware implementations simplify this using only a single gradient filter per octave rather than per scale (with a slight reduction in matching robustness), for example by Pettersson and Petersson (2005). Another hardware simplification is to modify the descriptor (including reducing the dimensionality) to make it easier to calculate in hardware, for example by Yao et al. (2009).

SURF (Bay et al., 2008) is similar, although uses Haar wavelets from the integral image to derive the gradients. Rather than the histogram, it effectively measures the average gradient within a 4×4 grid, resulting in a 64-dimensional feature descriptor.

An alternative descriptor that can speed up matching in hardware is a binary descriptor such as BRIEF (binary robust independent elementary features) (Calonder et al., 2010) or BRISK (binary robust invariable scalable keypoints) (Leutenegger et al., 2011). These effectively work by comparing selected pixel values pairwise within a window to create a binary feature vector (1 bit per comparison), giving 128–512 bits total. Feature descriptors can then easily be compared by measuring their Hamming distance (counting the number of different bits between two descriptors).

10.4.1.3 Feature Matching

Once a set of features is found, it is then necessary to match these between images to give a set of point correspondences. Again, there are many matching techniques (Zitova and Flusser, 2003), with a full description beyond the scope of this work.

Feature descriptors are matched by finding the feature in the reference image which has the most similar descriptor. The match is unlikely to be exact because of noise and differences resulting from local distortions. Also, not all feature points detected in one image may be found in the reference image. With a large number of descriptors and a large number of components within each descriptor, a brute force search for the closest match is prohibitive. Therefore, efficient search techniques are required to reduce the search cost.

If the reference is within a database, for example when searching for predefined objects within an image, then two approaches to reduce the search are hashing and tree-based methods. A locality sensitive hash function (Gionis et al., 1999) compresses the dimensionality enabling approximate closest points to be found efficiently. Alternatively, the points within the database may be encoded in a high-dimensional tree-based structure, and using a best bin first heuristic, the tree may be searched efficiently (Beis and Lowe, 1997). Such searches for a particular point are generally sequential. While the search for multiple points may be performed independently, implying that they may be searched in parallel, for large databases this is also impractical because of memory bandwidth bottlenecks. Serial search methods with complex data structures and heuristics are best performed in software, although it may be possible to accelerate key steps using appropriate hardware.

For matching between images, an alternative is to combine the matching and model extraction steps. The first few points are matched using a full search. These are then used to estimate a model of the transformation, which is used to guide the search for correspondences for the remaining points. New matches can then be used to refine the transformation. Again, the search process is largely sequential, although hardware may be used to accelerate the transformation to speed up the search for new matches.

10.4.1.4 Model Extraction

Given a set of points, the transformation model defines the best mapping of the points from one image to the other. For a given model, determining the transformation can be considered as an optimisation problem, where the model parameters are optimised to find the best match.

If the model is sufficiently simple (for example an affine model which is linear in the model parameters), it may be solved directly using least squares minimisation. Consider, for example, the affine mapping of Eq.

(10.22). Each pair of corresponding points provides two constraints. Therefore, the set of equations for N correspondences has $2N$ constraints:

$$\begin{bmatrix} u_1 \\ v_1 \\ \vdots \\ u_N \\ v_N \end{bmatrix} = \begin{bmatrix} x_1 & y_1 & 1 & 0 & 0 & 0 \\ 0 & 0 & 0 & x_1 & y_1 & 1 \\ \vdots & \vdots & \vdots & \vdots & \vdots & \vdots \\ x_N & y_N & 1 & 0 & 0 & 0 \\ 0 & 0 & 0 & x_N & y_N & 1 \end{bmatrix} \begin{bmatrix} a_{ux} \\ a_{uy} \\ a_u \\ a_{vx} \\ a_{vy} \\ a_v \end{bmatrix} \tag{10.51}$$

or

$$\mathbf{u} = \mathbf{Ma}. \tag{10.52}$$

Usually this set of equations is over-determined, and with noise it is impossible to find model parameters which exactly match the constraints. In this case, the transform parameters can be chosen such that the total squared error is minimised by solving

$$\mathbf{M}^\mathsf{T}\mathbf{Ma} = \mathbf{M}^\mathsf{T}\mathbf{u} \tag{10.53}$$

to give

$$\mathbf{a} = (\mathbf{M}^\mathsf{T}\mathbf{M})^{-1}\mathbf{M}^\mathsf{T}\mathbf{u}. \tag{10.54}$$

The regular nature of matrix operations and inversion make them amenable to FPGA implementation (see, for example, (Irturk et al., 2008; Burdeniuk et al., 2010)), although if care is not taken with scaling, the calculation can result in significant quantisation errors if fixed-point arithmetic is used. Least squares optimisation is quite sensitive to outliers (mismatched points in this case), so robust fitting methods can be used, if necessary to identify outliers and remove their contribution from the solution.

If the corresponding points may come from one of several models, then the mixture can cause even robust fitting methods to fail. One approach suggested by Lowe (2004) was to have each pair of corresponding points vote for candidate model parameters using a coarse Hough transform (the principles of the Hough transform are described in Section 13.8). This effectively sorts the pairs of corresponding points based on the particular model, filtering the outliers from each set of points.

With more complex or nonlinear models, a closed-form solution of the parameters is not possible. An initial estimate of the model is iteratively refined as outlined in Figure 10.29. The model is used to transform the reference points, with the error measured against the corresponding target points used to update the model using gradient descent (for example Levenberg–Marquardt error minimisation). Chapter 14 provides a more detailed discussion of model fitting.

10.4.2 Area-based Methods

The alternative to feature-based image registration is to match the images themselves, without explicitly detecting objects within the image. Such area matching may be performed on the image as whole or on small patches extracted from the image.

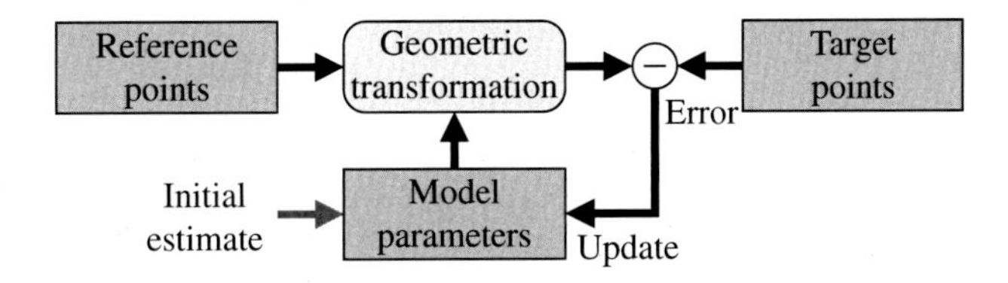

Figure 10.29 Iterative update of registration model.

Area-based matching has a number of limitations compared with feature-based matching (Zitova and Flusser, 2003). Any distortions within the image affect the shape and size of objects within the image and consequently the quality of the match. Similarly, since area matching is generally based on matching pixel values, the matching process can be confounded by any change between the images, for example by varying illumination or using different sensors. When using small image patches, if the patch is textureless without any significant features, it can be matched with just about any other smooth area within the other image, and often the matching is based more on noise or changes in shading rather than a true match. In spite of these limitations, area-based matching is commonly used.

Two metrics commonly used to gauge similarity within a patch **P** between images f and g are the sum of absolute differences:

$$SAD = \sum_{x,y\in\mathbf{P}} |f[x,y] - g[x,y]| \tag{10.55}$$

and the sum of squared differences:

$$SSD = \sum_{x,y\in\mathbf{P}} (f[x,y] - g[x,y])^2. \tag{10.56}$$

By squaring the difference, the SSD effectively gives more weight to large differences. Closely related to SSD is correlation (Jain, 1989). Expanding Eq. (10.56) gives

$$SSD = \sum_{x,y\in\mathbf{P}} f^2[x,y] + \sum_{x,y\in\mathbf{P}} g^2[x,y] - 2\sum_{x,y\in\mathbf{P}} f[x,y]g[x,y], \tag{10.57}$$

where the last term is the correlation

$$COR = \sum_{x,y\in\mathbf{P}} f[x,y]g[x,y]. \tag{10.58}$$

Therefore, maximising the correlation is equivalent to minimising the sum of squared difference.

10.4.2.1 Correlation Methods

When applied to image registration, the spatial offset of one of the images is adjusted to obtain an estimate of the similarity as a function of offset:

$$COR[i,j] = \sum_{x,y\in\mathbf{P}} f[x+i,y+j]g[x,y]. \tag{10.59}$$

This will have a peak at the offset that gives the best match to the nearest pixel.

One problem with simple correlation for image registration is that if the patch has an intensity gradient, this will offset the correlation peak. This is because the first term in Eq. (10.57) is not constant as f shifts. This may be overcome by normalising the correlation by the average pixel value and variance to give the correlation coefficient:

$$CC_N[i,j] = \frac{\sum_{x,y\in\mathbf{P}}(f[x+i,y+j] - \mu_f)(g[x,y] - \mu_g)}{\sqrt{\sum_{x,y\in\mathbf{P}}(f[x+i,y+j] - \mu_f)^2}\sqrt{\sum_{x,y\in\mathbf{P}}(g[x,y] - \mu_g)^2}}. \tag{10.60}$$

When looking for a smaller target within a larger image, this process is often called ***template matching***. The local maxima above a threshold (or minima for difference methods) are the detected locations of the target. If the target is sufficiently small, correlation can be implemented directly using filtering to produce an image of match scores. (See, for example, (Schumacher et al., 2012).) Such a ***matched filter*** consists of a flipped version of the template or target.

When correlating images, calculating the correlation for every offset (to find the maximum) is computationally expensive. The correlation surface is usually not of particular interest and in many cases is relatively smooth, enabling the computation to be reduced by performing a sequential search, optimising to find the peak.

However, there may be multiple local maxima, and search techniques may home in on a sub-optimal local peak, giving the wrong offset.

One common technique to reduce the search complexity is to use a pyramidal approach to the search. An image pyramid is produced by filtering and down-sampling the images to produce a series of progressively lower resolution versions. These enable a wide range of offsets to be searched relatively quickly for three reasons: filtering the images will give a smoother correlation surface, removing many of the sub-optimal local maxima; the lower resolution images are significantly smaller, so measuring the similarity for each offset is faster; and a large step size is used because each pixel is much larger. This enables the global maximum to be found efficiently, even using an exhaustive search at the lowest resolution if necessary. The estimated offset obtained from the low-resolution images can then be refined by finding the maximum at progressively higher resolutions. These latter searches are made more efficient by restricting the search region to the global maximum and can safely perform restricted searches without worry of locking onto the wrong peak.

It is straightforward to perform the correlation for multiple offsets in parallel. One way of achieving this is shown in Figure 10.30, where the 16 correlations within a 4×4 search window are performed in parallel in a single scan through the pair of images. One of the input images is offset relative to the other by a series of shift registers and row buffers, with a multiply-and-accumulate (MAC) at each offset of interest. The multiplication could just as easily be replaced by an absolute difference or squared difference. With a pyramidal search, a block such as this can be reused for each level of the pyramid.

Using Eq. (10.60) will locate the correlation peak to the nearest pixel. To estimate the offset to sub-pixel accuracy, it is necessary to interpolate between adjacent correlation values. Commonly a parabola is fitted between the maximum pixel, $CC[i_0,j_0]$, and those on either side. Assuming the peak is separable, the horizontal sub-pixel offset is then

$$x_0 = i_0 + \frac{CC[i_0+1,j_0] - CC[i_0-1,j_0]}{4CC[i_0,j_0] - 2(CC[i_0+1,j_0] + CC[i_0-1,j_0])}, \tag{10.61}$$

and similarly for the vertical offset (Tian and Huhns, 1986). However, since the autocorrelation of a piecewise constant image has a pyramidal peak, this is often better to fit (Bailey and Lill, 1999; Bailey, 2003). A pyramidal model gives the subpixel offset as

$$x_0 = i_0 + \frac{CC[i_0+1,j_0] - CC[i_0-1,j_0]}{2(CC[i_0,j_0] - \min(CC[i_0+1,j_0], CC[i_0-1,j_0]))}. \tag{10.62}$$

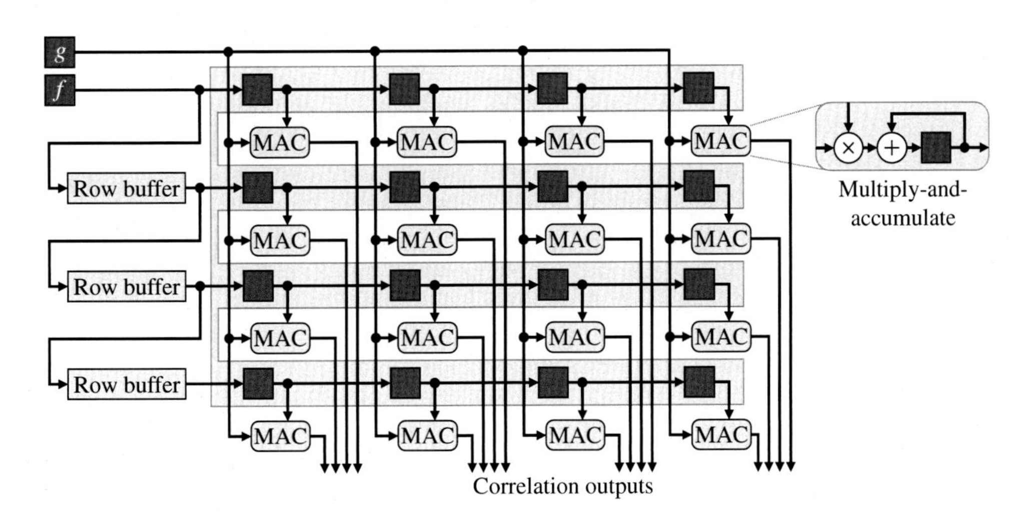

Figure 10.30 Calculating a 4×4 block of cross-correlations in parallel.

One limitation of correlation is that it is generally restricted to a simple translational model. Extending to more complex transformations quickly becomes computationally intractable, because each transformation parameter adds an additional dimension to the search space. One approach to this problem is to correlate small patches and fit the model to the correspondences as outlined in Section 10.4.1.4. Correlating small patches can tolerate small image imperfections, but more complex mappings or those with larger distortions will require some form of gradient search to optimise the match, as described in Sections 10.4.2.3 and 10.4.2.4.

10.4.2.2 Frequency Domain Methods

A correlation in the image domain becomes a product in the frequency domain. Taking the 2-D discrete Fourier transform of Eq. (10.59) gives

$$COR[u,v] = F^*[u,v]G[u,v], \tag{10.63}$$

where u and v are the spatial frequencies in the x and y directions, respectively, F^* is the complex conjugate of F, and F and G are the Fourier transforms of f and g, respectively. The complete correlation surface is found by taking the Fourier transform of each image, calculating the product, and then taking the inverse Fourier transform. (See Section 11.1 for implementing the Fourier transform on an FPGA. Since f and g are both real in general, a single fast Fourier transform (FFT) can be used to calculate both F and G in parallel from Eq. (11.21) and (11.22).) Note that the discrete Fourier transform assumes that the image is periodic, requiring the images to be padded for correct calculation of the correlation.

The normalised correlation of Eq. (10.60) cannot be calculated directly in the frequency domain. However, for rectangular patches, the denominator may be calculated using integral images (Lewis, 1995) enabling the correlation to be efficiently normalised (at the expense of storing the integral images).

Phase correlation reduces the problems associated with lack of normalisation by considering only the phase. If f is the offset version of g,

$$g[x,y] = f[x+x_0, y+y_0], \tag{10.64}$$

then from the shift theorem (Bracewell, 2000), the Fourier transform becomes

$$G[u,v] = F[u,v]e^{j2\pi(ux_0+vy_0)}. \tag{10.65}$$

Taking the phase angle of the ratio of the two spectra gives

$$ux_0 + vy_0 = \frac{1}{2\pi}\angle\frac{G[u,v]}{F[u,v]} = \frac{1}{2\pi}\angle(F^*[u,v]G[u,v]), \tag{10.66}$$

where CORDIC arithmetic can then be used to obtain the magnitude and phase angle of each frequency component.

Three approaches may be used to estimate the offset. The first is to normalise the magnitude to one and take the inverse Fourier transform:

$$\frac{F^*[u,v]G[u,v]}{\|F^*[u,v]G[u,v]\|} \Leftrightarrow \delta[x+x_0, y+y_0], \tag{10.67}$$

where $\delta[x+x_0, y+y_0]$ is a delta function at $(-x_0, -y_0)$. Phase noise, combined with the fact that the offset is not likely to be exactly an integer number of pixels, means that the output is not a delta function. However, a search of the output for the maximum value will usually give the offset to the nearest pixel.

The second approach is to fit a plane to the phase difference in Eq. (10.66) using least squares. This first requires unwrapping the phase, because the arctangent reduces the phase to between $\pm\pi$. From Eq. (10.66), the phase varies linearly with frequency. The simplest approach to phase unwrapping is to consider the phase of adjacent samples. Multiples of 2π are then added to one of the samples until the absolute difference is less than π. In two dimensions, this may be applied both horizontally and vertically, providing additional constraints. A limitation of this simple approach to phase unwrapping is that at some frequencies, the amplitude changes sign (consider, for example, a sinc function). The phase shift between such points should be close to π rather than 0. A second limitation is that when the magnitude is low, the phase angle becomes dominated by noise.

Consequently, only the phases for the low frequencies are reliably unwrapped unless more complex methods are used, and even then the quality of the phase estimate at a given frequency will depend on the amplitude. Fitting a plane to the phase difference may therefore be improved by weighting the phase samples by their amplitudes or masking out those frequencies with low amplitude or those that differ significantly in amplitude between the two images (Stone et al., 1999, 2001).

The third alternative is to consider the phase shear (Stowers et al., 2010). From Eq. (10.66), let $H[u, v] = F^*G$. Then the phase difference between horizontally adjacent frequencies should be

$$\begin{aligned} x_0 &= \frac{1}{2\pi}(\angle H[u+1,v] - \angle H[u,v]) \\ &= \frac{1}{2\pi}\angle(H[u+1,v]H^*[u,v]). \end{aligned} \tag{10.68}$$

An amplitude weighted average of these is given by (Stowers et al., 2010)

$$x_0 = \frac{1}{2\pi}\angle\sum_{u,v} H[u+1,v]H^*[u,v], \tag{10.69}$$

and similarly for the vertical offset. The amplitude weighting will automatically give low weight to those differences where the phase shifts by π, and the problem of phase wrapping is avoided by deferring the arctangent to after the averaging. Alternatively, histograms of the x_0 and y_0 estimated by Eq. (10.68) may be built to enable outliers to be eliminated before averaging.

Using the Fourier transform is limited to pure translation, although the offset may be estimated to sub-pixel accuracy. To extend the registration to include rotation and scaling requires additional processing. Rotating an image will rotate its Fourier transform, and scaling an image will result in an inverse scale within the Fourier transform. Therefore, by converting the Fourier transform from rectangular coordinates to log-polar coordinates (the radial axis is the logarithm of the frequency), rotating the image corresponds to an angular shift, and scaling the image results in a radial shift (Reddy and Chatterji, 1996). The angle and scale factor may be found by correlation in log-polar coordinates.

10.4.2.3 Gradient Methods

Rather than perform an exhaustive search in correlation space, gradient-based methods effectively perform a hill-climbing search using an estimate of the offset based on derivatives of the image gradient. Expanding an offset image, $f(x + x_0, y + y_0)$ about $f(x, y)$ using a 2-D Taylor series gives

$$g(x,y) = f(x,y) + x_0\frac{\partial f}{\partial x} + y_0\frac{\partial f}{\partial y} + \frac{1}{2!}\left(x_0^2\frac{\partial^2 f}{\partial x^2} + 2x_0y_0\frac{\partial^2 f}{\partial x\partial y} + y_0^2\frac{\partial^2 f}{\partial y^2}\right) + \cdots \tag{10.70}$$

For small offsets, the Taylor series can be truncated (the second-order and higher terms are negligible). Therefore, given estimates of the gradients, the offset may be estimated from

$$g(x,y) \approx f(x,y) + x_0\frac{\partial f}{\partial x} + y_0\frac{\partial f}{\partial y} = f + x_0f_x + y_0f_y, \tag{10.71}$$

where the gradients may be obtained from a central difference filter. Each pixel then provides a constraint; since this over-determined system is linear in the offsets x_0 and y_0, it may be solved by least squares (Lucas and Kanade, 1981):

$$\begin{bmatrix} \sum\limits_{i,j} f_x^2 & \sum\limits_{i,j} f_xf_y \\ \sum\limits_{i,j} f_xf_y & \sum\limits_{i,j} f_y^2 \end{bmatrix}\begin{bmatrix} x_0 \\ y_0 \end{bmatrix} = \begin{bmatrix} \sum\limits_{i,j}(g[i,j] - f[i,j])f_x \\ \sum\limits_{i,j}(g[i,j] - f[i,j])f_y \end{bmatrix}. \tag{10.72}$$

The matrix on the left is the Gauss–Newton approximation of the Hessian.

Since the offset is only approximate (it neglects the high-order Taylor series terms), the expansion may be repeated about the approximation. After the first iteration, the offset will be closer, so the truncated Taylor series

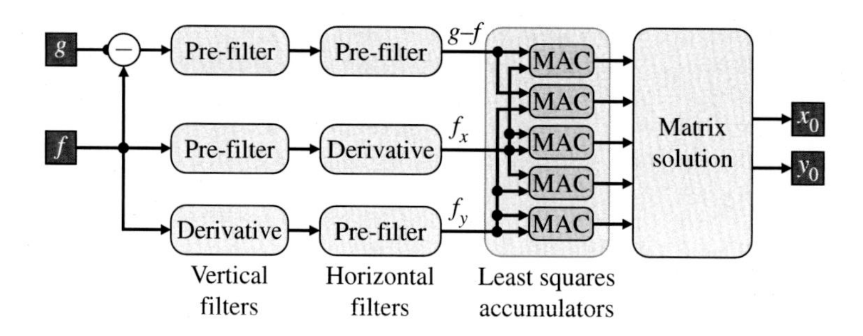

Figure 10.31 Basic approach of gradient-based registration.

will be more accurate, enabling a more accurate estimate of the offset to be obtained. Iterating the formulation following the general pattern of Figure 10.29 will converge to the minimum error (Lucas and Kanade, 1981). If necessary, an initial pyramidal coarse-to-fine search can be used to reduce the chances of becoming locked onto a local minimum (Thevenaz et al., 1998).

The derivatives must be calculated numerically using filters. These may be calculated more accurately using matched derivative filter and pre-filter (Farid and Simoncelli, 2004) as outlined in Section 9.2.3.1. The resulting pipelined implementation would be similar to Figure 10.31.

In principle, this approach can also be readily extended for more complex local warps (Baker and Matthews, 2004).

10.4.2.4 Newton–Raphson Methods

Rather than simply estimating a displacement, consider a local warp between the images where the warped location $(\tilde{x}, \tilde{y})$ is given by

$$\tilde{x} = W_{\tilde{x}}(x, y), \qquad \tilde{y} = W_{\tilde{y}}(x, y), \tag{10.73}$$

where $W_{\tilde{x}}$ and $W_{\tilde{y}}$ are the underlying warp functions. Approximating these warp functions by a first-order Taylor series about the point (x_0, y_0) gives

$$\begin{aligned} \tilde{x} &= x_0 + (x - x_0)W_{\tilde{x}x} + (y - y_0)W_{\tilde{x}y} + W_{\tilde{x}0}, \\ \tilde{y} &= y_0 + (x - x_0)W_{\tilde{y}x} + (y - y_0)W_{\tilde{y}y} + W_{\tilde{y}0}, \end{aligned} \tag{10.74}$$

which corresponds to a local affine warp (see Eq. (10.26)). If the distortion is particularly severe, a second-order Taylor series approximation may be used (Lu and Cary, 2000), which has 12 parameters (see Eq. (10.30)) rather than the 6 of the affine warp, but otherwise the basic procedure is the same.

These local warp parameters may then be optimised using a least squares correlation coefficient (normalising Eq. (10.56) and extending it to match a patch, **P**):

$$SSD_N(\mathbf{W}) = \frac{\sum_{\mathbf{x}\in\mathbf{P}}(f[\tilde{\mathbf{x}}] - g[\mathbf{x}])^2}{\sum_{\mathbf{x}\in\mathbf{P}} g^2[\mathbf{x}]^2}, \tag{10.75}$$

where **W** are the six parameters of the affine warp, with the origin defined as the centre of the patch. The samples of the warped patch $(f[\tilde{\mathbf{x}}])$ are typically found using bicubic interpolation. The minimum will occur when the gradient of SSD_N with respect to each of the warp parameters (the Jacobian) is zero:

$$\nabla SSD_N(\mathbf{W}) = 0. \tag{10.76}$$

Solving this using Newton–Raphson root finding adapts Eq. (5.58) to give the series of iterations:

$$\mathbf{W}_{k+1} = \mathbf{W}_k - (\nabla\nabla SSD_N(\mathbf{W}_k))^{-1}(\nabla SSD_N(\mathbf{W}_k)), \tag{10.77}$$

where $\nabla\nabla SSD_N(\mathbf{W}_k)$ is the Hessian (second partial derivatives) of SSD with respect to the parameters (Bruck et al., 1989). Note that both the Jacobian and the Hessian must be determined numerically, which is quite demanding computationally since the derivatives are with respect to the warp parameters rather than the pixel positions, and they depend on the current warp parameters which change with each iteration.

For the affine warp, the problem can be reformulated using an inverse compositional approach (Baker and Matthews, 2001; Pan et al., 2013). This has two important differences. First, rather than add an update to the parameters (as in Eq. (10.77)), the warp consists of a composition of warps. Second, rather than mapping the reference to the current image, it uses the inverse transform to map the current image to the reference. These differences mean that the same patch is being warped at each step, with the adjustment being combined at each iteration. Therefore, although the Jacobian changes with each iteration, the Hessian stays the same, so that it only needs to be calculated and inverted only once, significantly reducing the computation. The update is then given by

$$\mathbf{W}_{k+1} = \mathbf{W}_k \odot (\mathbf{I} + \Delta\mathbf{W})^{-1}, \tag{10.78}$$

where $\odot$ represents the composition of two warps, $\mathbf{I}$ is the identity, and $\Delta\mathbf{W}$ is the update. For an affine warp represented as a 3×3 matrix, $\odot$ is a matrix multiplication, and $\mathbf{I}$ is the identity matrix. Note that this approach requires that the warp can be readily inverted, and that a warp can be represented as a composition of two warps (Baker and Matthews, 2001). So, although the inverse compositional approach may be applied to the affine mapping, it cannot be used for the higher order mapping.

10.4.2.5 Optimal Filtering

The gradient-based methods effectively create a continuous surface through interpolation, which is then resampled with the desired offset. This is then compared with the target image as in Figure 10.29 to determine the offset that minimises the error. Expressing the interpolation as a convolution in the form of Eq. (10.7) gives

$$(x_0, y_0) = \underset{(x,y)}{\arg\min} \left\| g[i,j] - \sum_{m,n} h_{(x,y)}[m,n]\, f[i-m, j-n] \right\|^2, \tag{10.79}$$

where $h_{(x,y)}[m,n]$ is the interpolation kernel for an offset of (x, y). Finding the offset that minimises this error requires a search because the filter coefficients, and hence the relationship between the two images, depend nonlinearly on the offset.

Predictive interpolation (Bailey and Lill, 1999) turns the problem around to avoid the search. Rather than use an interpolation kernel that makes arbitrary assumptions about the image (such as being band-limited or smooth), the optimal interpolation kernel is derived by predicting the pixel values of the target image from the reference (Gilman and Bailey, 2007). This solves

$$g[i,j] - f[i,j] = \sum_{m,n\neq 0,0} h_{(x_0,y_0)}[m,n](f[i-m, j-n] - f[i,j]), \tag{10.80}$$

where the subtraction of $f[i,j]$ enforces the partition of unity constraint (Unser, 2000) ($\sum h = 1$). Since Eq. (10.80) is linear in the filter coefficients, it may be represented in matrix form as

$$\mathbf{g} = \mathbf{F}\mathbf{h}, \tag{10.81}$$

where $\mathbf{h}$ are the filter coefficients arranged as a vector, and each line in $\mathbf{F}$ and $\mathbf{g}$ corresponds to one pixel in the image. The least squares solution is then

$$\mathbf{h} = (\mathbf{F}^\top\mathbf{F})^{-1}\mathbf{F}^\top\mathbf{g}. \tag{10.82}$$

From the filter coefficients, the offset may be determined by superposition (Gilman and Bailey, 2007):

$$(x_0, y_0) = \sum_{m,n} (m h_{(x_0,y_0)}[m,n], n h_{(x_0,y_0)}[m,n]) \tag{10.83}$$

or

$$\mathbf{x}_0 = \mathbf{w}\mathbf{h} = \mathbf{w}\,(\mathbf{F}^\top\mathbf{F})^{-1}\mathbf{F}^\top\mathbf{g}, \tag{10.84}$$

where **w** is the matrix of constant weights. Predictive interpolation therefore gives a direct (non-iterative) solution to estimating the offset between images to sub-pixel accuracy.

While optimal filtering provides the offset to sub-pixel accuracy, the images must already be registered to the nearest pixel (this can be relaxed slightly with larger windows). Again, a pyramidal coarse-to-fine search or other registration methods may be used to give the initial approximation. Optimal filtering is restricted to translation only and cannot easily be extended to include more general transformations.

10.4.3 Applications

There are many diverse applications of image registration, with variations on some of the techniques described above used to address issues unique to that application. While it is not possible to describe all of the applications in any detail, a brief overview of some of the main applications will be given.

Perhaps the most common is camera calibration, where it is desired to determine the geometric transformation which describes the imaging model of the camera, including its pose. To define the mapping between points within the real world, and pixels within the image, it is usually necessary to have a target object with distinctive features in known physical locations. This simplifies feature detection, and the process becomes one of determining the camera model. Examples of different models and methods of solving for their parameters are given in Brown (1971), Tsai (1987), Heikkila and Silven (1997), Zhang (2000), Devernay and Faugeras (2001), and Kannala and Brandt (2004).

Image block matching is used for video compression, where there can be significant correlation from one frame to the next. The difference can be encoded using significantly fewer bits. Further information on motion compensation for video compression may be found in Section 12.1.3.

Optical flow estimates the motion fields within an image. This requires matching small patches from one frame with the previous frame. Since it is desired to have a coherent flow field rather than treat each patch independently, the motion field is filtered to reduce or remove outliers. FPGA implementations of optical flow have been reported in Diaz et al. (2004), Sosa et al. (2007), Wei et al. (2008), Komorkiewicz et al. (2014), and Seyid et al. (2018).

Camera shake is a common problem with hand-held cameras. Video stabilisation estimates the rigid body motion of the camera and uses this to remove the unwanted motion. Either feature-based (Nou-Shene et al., 2015) or patch-based (Parker et al., 2014) registration can be used. By simplifying the problem to pure translation, Araneda and Figueroa (2014) reduced the hardware by projecting the image horizontally and vertically and used these to give the offsets.

Stereo imaging processes separate images from two spatially separated cameras to extract the disparity or differences in relative position of objects between the two images. Again, either feature points or image patches are matched between images to estimate the disparity and hence determine the range. The matching method must be insensitive to distortion (including photometric effects) resulting from the offset camera positions. More information on stereo imaging can be found in Section 15.4.

Stereo imaging can be extended to structure-from-motion (Ullman and Brenner, 1979), where matching corresponding points between multiple images is used to form a 3-D point point cloud of the object. An FPGA implementation of structure-from-motion has been reported by Komorkiewicz et al. (2015).

Image super-resolution combines several low-resolution images of a scene to create a single image with higher spatial resolution. This typically requires three steps: registration of the input images to sub-pixel accuracy; resampling the ensemble at the higher resolution; and inverse filtering to reduce the blur resulting from the low-resolution imaging process. Again, there are quite a few different techniques used for image super-resolution. Many are iterative and not amenable to real-time implementation, although the algorithms may be accelerated using an FPGA as the computation engine. The optimal filtering techniques have potential, not only for image registration but also for determining efficient resampling (interpolation) filters (Gilman et al., 2009, 2010). Little has been implemented so far on FPGAs. Perhaps one exception is the work of Bowen and Bouganis (2008), although their work assumed that the images were already registered and only considered the image fusion stage.

Registration is required when combining data from different sensor types. This is common in both remote sensing and medical imaging. The correlation or sum of absolute differences matching metrics do not perform well where there are significant differences in contrast between the images. In these instances, it is more effective to find the transformation that maximises mutual information (Viola and Wells, 1997).

Digital image correlation is used in mechanics to measure strain fields (Bruck et al., 1989). The material has a speckle pattern applied, which provides suitable visual texture for measuring subtle displacements. The Newton–Raphson method (Section 10.4.2.4) is then used to estimate the displacements under strain, enabling the strain field to be calculated. An FPGA acceleration of these calculations has been developed by Stokke (2020).

Similar techniques can be applied to X-ray phase-contrast imaging, which measures the phase shift induced by X-rays passing through an object (Zdora, 2018). The X-ray phase shift cannot be measured directly; however, a differential phase shift will distort the wavefront. By introducing a speckle pattern (for example using a sheet of sandpaper), the displacement of the speckle pattern can be measured, and the phase shift inferred.

10.5 Summary

Geometric transformations modify the position of pixels within an image. The transformation is commonly represented in one of two forms: a forward mapping determines where each input pixel maps to in the output image, and a reverse mapping determines where each output pixel comes from in the input image. A forward mapping works best with a streamed input, whereas a reverse mapping works best with a streamed output. The change in geometry means that it is difficult to have both streamed input and output without significant buffering, unless the row offset is relatively small.

Inevitably, the new samples do not coincide with integer pixel locations, so some form of interpolation is required. Bilinear interpolation is relatively easy to achieve, and is resource efficient, although the more complex bicubic interpolation gives a smoother output at the expense of more complex buffering and processing. Spline interpolation is even better, but has a significantly larger region of support requiring more complex preprocessing, and is less suited for stream processing.

With stream processing, many common transformations can use incremental calculation to reuse the results in moving from one pixel to the next in the stream. However, with fixed-point arithmetic, it is important that sufficient guard bits are used to limit the accumulation of rounding errors.

Generally, some form of image registration or calibration is required to determine the mapping for the geometric transformation. Registration may be aligning whole images for processing or small patches within the images for estimating the motion of objects (including the camera).

References

Akeila, H. and Morris, J. (2008). High resolution stereo in real time. In: *2nd International Workshop on Robot Vision*, Auckland, NZ (18–20 February, 2008), *Lecture Notes in Computer Science*, Volume **4931**, 72–84. https://doi.org/10.1007/978-3-540-78157-8_6.

Amaricai, A., Gavriliu, C.E., and Boncalo, O. (2014). An FPGA sliding window-based architecture Harris corner detector. *24th International Conference on Field Programmable Logic and Applications (FPL)*, Munich, Germany (2–4 September 2014), 4 pages. https://doi.org/10.1109/FPL.2014.6927402.

Araneda, L. and Figueroa, M. (2014). Real-time digital video stabilization on an FPGA. *17th Euromicro Conference on Digital System Design (DSD)*, Verona, Italy (27–29 August 2014), 90–97. https://doi.org/10.1109/DSD.2014.26.

Bailey, D.G. (2003). Sub-pixel estimation of local extrema. *Image and Vision Computing New Zealand (IVCNZ'03)*, Palmerston North, NZ (26–28 November, 2003), 414–419.

Bailey, D.G. and Bouganis, C.S. (2009). Implementation of a foveal vision mapping. *International Conference on Field Programmable Technology (FPT'09)*, Sydney, Australia (9–11 December 2009), 22–29. https://doi.org/10.1109/FPT.2009.5377646.

Bailey, D.G. and Lill, T.H. (1999). Image registration methods for resolution improvement. *Image and Vision Computing New Zealand (IVCNZ)*, Christchurch, NZ (30–31 August 1999), 91–96.

Bailey, D.G., Seal, J.R., and Sen Gupta, G. (2005). Nonparametric calibration for catadioptric stereo. *Image and Vision Computing New Zealand (IVCNZ'05)*, Dunedin, NZ (28–29 November 2005), 404–409.

Baker, S. and Matthews, I. (2001). Equivalence and efficiency of image alignment algorithms. *IEEE Computer Society Conference on Computer Vision and Pattern Recognition (CVPR 2001)*, Kauai, HI, USA (8–14 December 2001), Volume **1**, 1090–1097. https://doi.org/10.1109/CVPR.2001.990652.

Baker, S. and Matthews, I. (2004). Lucas-Kanade 20 years on: a unifying framework. *International Journal of Computer Vision* **56** (3): 221–255. https://doi.org/10.1023/B:VISI.0000011205.11775.fd.

Bay, H., Tuytelaars, T., and Gool, L.V. (2006). SURF: speeded up robust features. In: *9th European Conference on Computer Vision*, Graz, Austria (7–13 May 2006), *Lecture Notes in Computer Science*, Volume **3951**, 404–417. https://doi.org/10.1007/11744023_32.

Bay, H., Ess, A., Tuytelaars, T., and Gool, L.V. (2008). Speeded-up robust features (SURF). *Computer Vision and Image Understanding* **110** (3): 346–359. https://doi.org/10.1016/j.cviu.2007.09.014.

Beis, J.S. and Lowe, D.G. (1997). Shape indexing using approximate nearest-neighbour search in high-dimensional spaces. *IEEE Computer Society Conference on Computer Vision and Pattern Recognition*, San Juan, Puerto Rico (17–19 June 1997), 1000–1006. https://doi.org/10.1109/CVPR.1997.609451.

Bellas, N., Chai, S.M., Dwyer, M., and Linzmeier, D. (2009). Real-time fisheye lens distortion correction using automatically generated streaming accelerators. *17th IEEE Symposium on Field Programmable Custom Computing Machines (FCCM)*, Napa, CA, USA (5–7 April 2009), 149–156. https://doi.org/10.1109/FCCM.2009.16.

Benedetti, X. and Perona, P. (1998). Real-time 2-D feature detection on a reconfigurable computer. *IEEE Computer Society Conference on Computer Vision and Pattern Recognition*, Santa Barbara, CA, USA (23–25 June 1998), 586–593. https://doi.org/10.1109/CVPR.1998.698665.

Boukhtache, S., Blaysat, B., Grediac, M., and Berry, F. (2021). FPGA-based architecture for bi-cubic interpolation: the best trade-off between precision and hardware resource consumption. *Journal of Real-Time Image Processing* **18** (3): 901–911. https://doi.org/10.1007/s11554-020-01035-1.

Bowen, O. and Bouganis, C.S. (2008). Real-time image super resolution using an FPGA. *International Conference on Field Programmable Logic and Applications (FPL 2008)*, Heidelberg, Germany (8–10 September 2008), 89–94. https://doi.org/10.1109/FPL.2008.4629913.

Bracewell, R.N. (2000). *The Fourier Transform and its Applications*, 3e. New York: McGraw Hill.

Brown, D.C. (1971). Close range camera calibration. *Photogrammetric Engineering* **37** (8): 855–866.

Brown, L.G. (1992). A survey of image registration techniques. *ACM Computing Surveys* **24** (4): 325–376. https://doi.org/10.1145/146370.146374.

Bruck, H.A., McNeill, S.R., Sutton, M.A., and Peters, W.H. (1989). Digital image correlation using Newton-Raphson method of partial differential correction. *Experimental Mechanics* **29** (3): 261–267. https://doi.org/10.1007/BF02321405.

Burdeniuk, A., To, K.N., Lim, C.C., and Liebelt, M.J. (2010). An event-assisted sequencer to accelerate matrix algorithms. *5th IEEE International Symposium on Electronic Design, Test and Applications (DELTA 2010)*, Ho Chi Minh City, Vietnam (13–15 January 2010), 158–163. https://doi.org/10.1109/DELTA.2010.12.

Calonder, M., Lepetit, V., Strecha, C., and Fua, P. (2010). BRIEF: binary robust independent elementary features. In: *European Conference on Computer Vision (ECCV 2010)*, Heraklion, Crete, Greece (5–11 September 2010), *Lecture Notes in Computer Science*, Volume **6314**, 778–792. https://doi.org/10.1007/978-3-642-15561-1_56.

Catmull, E. and Smith, A.R. (1980). 3-D transformations of images in scanline order. *ACM SIGGRAPH Computer Graphics* **14** (3): 279–285. https://doi.org/10.1145/965105.807505.

Chang, L. and Hernandez-Palancar, J. (2009). A hardware architecture for SIFT candidate keypoints detection. In: *14th Iberoamerican Conference on Pattern Recognition*, Guadalajara, Mexico (15–18 November 2009), *Lecture Notes in Computer Science*, Volume **5856**, 95–102. https://doi.org/10.1007/978-3-642-10268-4_11.

Chen, W., Ding, S., Chai, Z., He, D., Zhang, W., Zhang, G., Peng, Q., and Luo, W. (2016). FPGA-based parallel implementation of SURF algorithm. *IEEE 22nd International Conference on Parallel and Distributed Systems (ICPADS)*, Wuhan, China (13–16 December 2016), 308–315. https://doi.org/10.1109/ICPADS.2016.0049.

Claus, C., Huitl, R., Rausch, J., and Stechele, W. (2009). Optimizing the SUSAN corner detection algorithm for a high speed FPGA implementation. *International Conference on Field Programmable Logic and Applications (FPL)*, Prague, Czech Republic (31 August – 2 September 2009), 138–145. https://doi.org/10.1109/FPL.2009.5272492.

Crow, F.C. (1984). Summed-area tables for texture mapping. *SIGGRAPH'84 International Conference on Computer Graphics and Interactive Techniques*, Minneapolis, MN, USA (23–27 July 1984), 207–212. https://doi.org/10.1145/800031.808600.

Derpanis, K.G., Leung, E.T.H., and Sizintsev, M. (2007). Fast scale-space feature representations by generalized integral images. *IEEE International Conference on Image Processing (ICIP 2007)*, San Antonio, TX, USA (16–19 September 2007), Volume **4**, 521–524. https://doi.org/10.1109/ICIP.2007.4380069.

Devernay, F. and Faugeras, O. (2001). Straight lines have to be straight. *Machine Vision and Applications* **13**: 14–24. https://doi.org/10.1007/PL00013269.

De la Cruz, J.A. (2011). Field-programmable gate array implementation of a scalable integral image architecture based on systolic arrays. Master of Science thesis. Logan, UT, USA: Utah State University.

Diaz, J., Ros, E., Mota, S., Carrillo, R., and Agis, R. (2004). Real time optical flow processing system. In: *14th International Conference on Field Programmable Logic and Application*, Antwerp, Belgium (29 August – 1 September 2004), *Lecture Notes in Computer Science*, Volume **3203**, 617–626. https://doi.org/10.1007/b99787.

Di Febbo, P., Mattoccia, S., and Dal Mutto, C. (2016). Real-time image distortion correction: analysis and evaluation of FPGA-compatible algorithms. *International Conference on ReConFigurable Computing and FPGAs (ReConFig)*, Cancun, Mexico (30 November – 2 December 2016), 6 pages. https://doi.org/10.1109/ReConFig.2016.7857182.

Eadie, D., Shevlin, F.O., and Nisbet, A. (2002). Correction of geometric image distortion using FPGAs. *Opto-Ireland 2002: Optical Metrology, Imaging, and Machine Vision*, Galway, Ireland (5–6 September 2002). SPIE, Volume **4877**, 28–37. https://doi.org/10.1117/12.463765.

Ehsan, S., Clark, A.F., and McDonald-Maier, K.D. (2009). Novel hardware algorithms for row-parallel integral image calculation. *Digital Image Computing: Techniques and Applications (DICTA '09)*, Melbourne, Australia (1–3 December 2009), 61–65. https://doi.org/10.1109/DICTA.2009.20.

Evemy, J.D., Allerton, D.J., and Zaluska, E.J. (1990). Stream processing architecture for real-time implementation of perspective spatial transformations. *IEE Proceedings I: Communications, Speech and Vision* **137** (3): 123–128. https://doi.org/10.1049/ip-i-2.1990.0018.

Fant, K.M. (1986). A nonaliasing, real-time spatial transform technique. *IEEE Computer Graphics and Applications* **6** (1): 71–80. https://doi.org/10.1109/MCG.1986.276613.

Farid, H. and Simoncelli, E.P. (2004). Differentiation of discrete multidimensional signals. *IEEE Transactions on Image Processing* **13** (4): 496–508. https://doi.org/10.1109/TIP.2004.823819.

Ferrari, L.A. and Park, J.H. (1997). An efficient spline basis for multi-dimensional applications: image interpolation. *IEEE International Symposium on Circuits and Systems*, Hong Kong (9–12 June 1997), Volume **1**, 757–760. https://doi.org/10.1109/ISCAS.1997.609003.

Ferrari, L.A., Park, J.H., Healey, A., and Leeman, S. (1999). Interpolation using a fast spline transform (FST). *IEEE Transactions on Circuits and Systems I: Fundamental Theory and Applications* **46** (8): 891–906. https://doi.org/10.1109/81.780371.

Gilman, A. and Bailey, D.G. (2007). Noise characteristics of higher order predictive interpolation for sub-pixel registration. *IEEE Symposium on Signal Processing and Information Technology*, Cairo, Egypt (15–18 December 2007), 269–274. https://doi.org/10.1109/ISSPIT.2007.4458153.

Gilman, A., Bailey, D.G., and Marsland, S.R. (2009). Model-based least squares optimal interpolation. *Image and Vision Computing New Zealand (IVCNZ 2009)*, Wellington, NZ (23–25 November 2009), 124–129. https://doi.org/10.1109/IVCNZ.2009.5378424.

Gilman, A., Bailey, D., and Marsland, S. (2010). Least-squares optimal interpolation for fast image super-resolution. *5th IEEE International Symposium on Electronic Design, Test and Applications (DELTA 2010)*, Ho Chi Minh City, Vietnam (13–15 January 2010), 29–34. https://doi.org/10.1109/DELTA.2010.59.

Gionis, A., Indyk, P., and Motwani, R. (1999). Similarity search in high dimensions via hashing. *25th International Conference on Very Large Data Bases*, Edinburgh, Scotland (7–10 September 1999), 518–529.

Gribbon, K.T. and Bailey, D.G. (2004). A novel approach to real time bilinear interpolation. *2nd IEEE International Workshop on Electronic Design, Test, and Applications (DELTA 2004)*, Perth, Australia (28–30 January 2004), 126–131. https://doi.org/10.1109/DELTA.2004.10055.

Gribbon, K.T., Johnston, C.T., and Bailey, D.G. (2003). A real-time FPGA implementation of a barrel distortion correction algorithm with bilinear interpolation. *Image and Vision Computing New Zealand (IVCNZ'03)*, Palmerston North, NZ (26–28 November 2003), 408–413.

Harris, C. and Stephens, M. (1988). A combined corner and edge detector. *Proceedings of the Alvey Vision Conference*, Manchester, UK (31 August – 2 September 1988), 147–151.

Heikkila, J. and Silven, O. (1997). A four-step camera calibration procedure with implicit correction. *IEEE Computer Society Conference on Computer Vision and Pattern Recognition*, San Juan, Puerto Rico (17–19 June 1997), 1106–1112. https://doi.org/10.1109/CVPR.1997.609468.

Huang, J., Zhou, G., Zhou, X., and Zhang, R. (2018). A new FPGA architecture of FAST and BRIEF algorithm for on-board corner detection and matching. *Sensors* **18** (4): Article ID 1014, 17 pages. https://doi.org/10.3390/s18041014.

Hudson, R.D., Lehn, D.I., and Athanas, P.M. (1998). A run-time reconfigurable engine for image interpolation. *IEEE Symposium on FPGAs for Custom Computing Machines*, Napa Valley, CA, USA (15–17 April 1998), 88–95. https://doi.org/10.1109/FPGA.1998.707886.

Hughes, C., Glavin, M., Jones, E., and Denny, P. (2008). Review of geometric distortion compensation in fish-eye cameras. *Irish Signals and Systems Conference (ISSC 2008)*, Galway, Ireland (18–19 June 2008), 162–167. https://doi.org/10.1049/cp:20080656.

Irturk, A., Benson, B., and Kastner, R. (2008). Automatic generation of decomposition based matrix inversion architectures. *International Conference on Field Programmable Technology*, Taipei, Taiwan (7–10 December 2008), 373–376. https://doi.org/10.1109/FPT.2008.4762421.

Jain, A.K. (1989). *Fundamentals of Image Processing*. Englewood Cliffs, NJ: Prentice Hall.

Kannala, J. and Brandt, S. (2004). A generic camera calibration method for fish-eye lenses. *17th International Conference on Pattern Recognition (ICPR 2004)*, Surrey, UK (23–26 August, 2004), Volume **1**, 10–13. https://doi.org/10.1109/ICPR.2004.1333993.

Keys, R.G. (1981). Cubic convolution interpolation for digital image processing. *IEEE Transactions on Acoustics, Speech and Signal Processing* **29** (6): 1153–1160. https://doi.org/10.1109/TASSP.1981.1163711.

Komorkiewicz, M., Kryjak, T., and Gorgon, M. (2014). Efficient hardware implementation of the Horn-Schunck algorithm for high-resolution real-time dense optical flow sensor. *Sensors* **14** (2): 2860–2891. https://doi.org/10.3390/s140202860.

Komorkiewicz, M., Kryjak, T., Chuchacz-Kowalczyk, K., Skruch, P., and Gorgon, M. (2015). FPGA based system for real-time structure from motion computation. *Conference on Design and Architectures for Signal and Image Processing (DASIP)*, Krakow, Poland (23–25 September 2015), 7 pages. https://doi.org/10.1109/DASIP.2015.7367241.

Leutenegger, S., Chli, M., and Siegwart, R.Y. (2011). BRISK: binary robust invariant scalable keypoints. *IEEE International Conference on Computer Vision (ICCV)*, Barcelona, Spain (6–13 November 2011), 2548–2555. https://doi.org/10.1109/ICCV.2011.6126542.

Lewis, J. (1995). Fast template matching. *Vision Interface*, Quebec City, Canada (15–19 May 1995), 120–123.

Li, M. and Lavest, J.M. (1996). Some aspects of zoom lens camera calibration. *IEEE Transactions on Pattern Analysis and Machine Intelligence* **18** (11): 1105–1110. https://doi.org/10.1109/34.544080.

Lim, Y.K., Kleeman, L., and Drummond, T. (2013). Algorithmic methodologies for FPGA-based vision. *Machine Vision and Applications* **24** (6): 1197–1211. https://doi.org/10.1007/s00138-012-0474-9.

Lowe, D.G. (1999). Object recognition from local scale-invariant features. *7th IEEE International Conference on Computer Vision*, Corfu, Greece (20–27 September 1999), Volume **2**, 1150–1157. https://doi.org/10.1109/ICCV.1999.790410.

Lowe, D.G. (2004). Distinctive image features from scale-invariant keypoints. *International Journal of Computer Vision* **60** (2): 91–110. https://doi.org/10.1023/B:VISI.0000029664.99615.94.

Lu, H. and Cary, P.D. (2000). Deformation measurements by digital image correlation: implementation of a second-order displacement gradient. *Experimental Mechanics* **40** (4): 393–400. https://doi.org/10.1007/BF02326485.

Lucas, B.D. and Kanade, T. (1981). An iterative image registration technique with an application to stereo vision. *7th International Joint Conference on Artificial Intelligence*, Vancouver, Canada (24–28 August 1981), 674–679.

Mahale, G., Mahale, H., Parimi, R.B., Nandy, S.K., and Bhattacharya, S. (2014). Hardware architecture of bi-cubic convolution interpolation for real-time image scaling. *International Conference on Field Programmable Technology*, Shanghai, China (10–12 December 2014), 264–267. https://doi.org/10.1109/FPT.2014.7082790.

Mallon, J. and Whelan, P.F. (2004). Precise radial un-distortion of images. *17th International Conference on Pattern Recognition (ICPR 2004)*, Surrey, UK (23–26 August 2004), Volume **1**, 18–21. https://doi.org/10.1109/ICPR.2004.1333995.

Ngo, H.T. and Asari, V.K. (2005). A pipelined architecture for real-time correction of barrel distortion in wide-angle camera images. *IEEE Transactions on Circuits and Systems for Video Technology* **15** (3): 436–444. https://doi.org/10.1109/TCSVT.2004.842609.

Noble, J.A. (1988). Finding corners. *Image and Vision Computing* **6** (2): 121–128. https://doi.org/10.1016/0262-8856(88)90007-8.

Nou-Shene, T., Pudi, V., Sridharan, K., Thomas, V., and Arthi, J. (2015). Very large-scale integration architecture for video stabilisation and implementation on a field programmable gate array-based autonomous vehicle. *IET Computer Vision* **9** (4): 559–569. https://doi.org/10.1049/iet-cvi.2014.0120.

Nuno-Maganda, M. and Arias-Estrada, M. (2005). Real-time FPGA-based architecture for bicubic interpolation: an application for digital image scaling. *International Conference on Reconfigurable Computing and FPGAs*, Puebla City, Mexico (28–30 September 2005), 8 pages. https://doi.org/10.1109/ReConFig.2005.34.

Oh, S. and Kim, G. (2008). An architecture for on-the-fly correction of radial distortion using FPGA. *Real-Time Image Processing 2008*, San Jose, CA, USA (28–29 January 2008). SPIE, Volume **6811**, Article ID 68110X, 9 pages. https://doi.org/10.1117/12.767125.

Pan, B., Li, K., and Tong, W. (2013). Fast, robust and accurate digital image correlation calculation without redundant computations. *Experimental Mechanics* **53** (7): 1277–1289. https://doi.org/10.1007/s11340-013-9717-6.

Parker, S.C., Hickman, D., and Wu, F. (2014). High-performance electronic image stabilisation for shift and rotation correction. *Airborne Intelligence, Surveillance, Reconnaissance (ISR) Systems and Applications XI*, Baltimore, MD, USA (7–8 May 2014). SPIE, Volume **9076**, Article ID 907602, 18 pages. https://doi.org/10.1117/12.2050267.

Pettersson, N. and Petersson, L. (2005). Online stereo calibration using FPGAs. *IEEE Intelligent Vehicles Symposium*, Las Vegas, NV, USA (6–8 June 2005), 55–60. https://doi.org/10.1109/IVS.2005.1505077.

Qui, M. and Ma, S.D. (1995). The nonparametric approach for camera calibration. *5th International Conference on Computer Vision (ICCV'95)*, Cambridge, MA, USA (20–23 June 1995), 224–229. https://doi.org/10.1109/ICCV.1995.466782.

Reddy, B.S. and Chatterji, B.N. (1996). An FFT-based technique for translation, rotation, and scale-invariant image registration. *IEEE Transactions on Image Processing* **5** (8): 1266–1271. https://doi.org/10.1109/83.506761.

Rosten, E. and Drummond, T. (2005). Fusing points and lines for high performance tracking. *10th IEEE International Conference on Computer Vision (ICCV'05)*, Beijing, China (17–21 October 2005), Volume **2**, 1508–1515. https://doi.org/10.1109/ICCV.2005.104.

Rublee, E., Rabaud, V., Konolige, K., and Bradski, G. (2011). ORB: an efficient alternative to SIFT or SURF. *IEEE International Conference on Computer Vision (ICCV)*, Barcelona, Spain (6–13 November 2011), 2564–2571. https://doi.org/10.1109/ICCV.2011.6126544.

Schumacher, F., Holzer, M., and Greiner, T. (2012). Critical path minimized raster scan hardware architecture for computation of the generalized Hough transform. *19th IEEE International Conference on Electronics, Circuits, and Systems (ICECS 2012)*, Seville, Spain (9–12 December 2012), 681–684. https://doi.org/10.1109/ICECS.2012.6463634.

Seyid, K., Richaud, A., Capoccia, R., and Leblebici, Y. (2018). FPGA based hardware implementation of real-time optical flow calculation. *IEEE Transactions on Circuits and Systems for Video Technology* **28** (1): 206–216. https://doi.org/10.1109/TCSVT.2016.2598703.

Shi, J. and Tomasi, C. (1994). Good features to track. *IEEE Computer Society Conference on Computer Vision and Pattern Recognition (CVPR '94)*, Seattle, WA, USA (21–23 June 1994), 593–600. https://doi.org/10.1109/CVPR.1994.323794.

Sikka, P., Asati, A.R., and Shekhar, C. (2021). Real time FPGA implementation of a high speed and area optimized Harris corner detection algorithm. *Microprocessors and Microsystems* **80**: 103514. https://doi.org/10.1016/j.micpro.2020.103514.

Smith, S.M. and Brady, J.M. (1997). SUSAN—A new approach to low level image processing. *International Journal of Computer Vision* **23** (1): 45–78. https://doi.org/10.1023/A:1007963824710.

Sosa, J.C., Boluda, J.A., Pardo, F., and Gomez-Fabela, R. (2007). Change-driven data flow image processing architecture for optical flow computation. *Journal of Real-Time Image Processing* **2** (4): 259–270. https://doi.org/10.1007/s11554-007-0060-y.

Stokke, K. (2020). An FPGA-based hardware accelerator for the digital image correlation engine. Master of Science in Computer Engineering thesis. Fayetteville, AR, USA: University of Arkansas.

Stone, H., Orchard, M., and Chang, E.C. (1999). Subpixel registration of images. *33rd Asilomar Conference on Signals, Systems, and Computers*, Monterey, CA, USA (24–27 October 1999), Volume **2**, 1446–1452. https://doi.org/10.1109/ACSSC.1999.831945.

Stone, H.S., Orchard, M.T., Chang, E.C., and Martucci, S.A. (2001). A fast direct Fourier-based algorithm for subpixel registration of images. *IEEE Transactions on Geoscience and Remote Sensing* **39** (10): 2235–2243. https://doi.org/10.1109/36.957286.

Stowers, J., Hayes, M., and Bainbridge-Smith, A. (2010). Phase correlation using shear average for image registration. *Image and Vision Computing New Zealand*, Queenstown, NZ (8–9 November 2010), 6 pages. https://doi.org/10.1109/IVCNZ.2010.6148817.

Thevenaz, P., Ruttimann, U.E., and Unser, M. (1998). A pyramid approach to subpixel registration based on intensity. *IEEE Transactions on Image Processing* **7** (1): 27–41. https://doi.org/10.1109/83.650848.

Tian, Q. and Huhns, M.N. (1986). Algorithms for sub-pixel registration. *Computer Vision, Graphics and Image Processing* **35** (2): 220–233. https://doi.org/10.1016/0734-189X(86)90028-9.

Tsai, R.Y. (1987). A versatile camera calibration technique for high-accuracy 3D machine vision metrology using off-the-shelf TV cameras and lenses. *IEEE Journal of Robotics and Automation* **3** (4): 323–344. https://doi.org/10.1109/JRA.1987.1087109.

Ullman, S. and Brenner, S. (1979). The interpretation of structure from motion. *Proceedings of the Royal Society of London. Series B. Biological Sciences* **203** (1153): 405–426. https://doi.org/10.1098/rspb.1979.0006.

Unser, M. (2000). Sampling - 50 years after Shannon. *Proceedings of the IEEE* **88** (4): 569–587. https://doi.org/10.1109/5.843002.

Unser, M., Aldroubi, A., and Eden, M. (1991). Fast B-spline transforms for continuous image representation and interpolation. *IEEE Transactions on Pattern Analysis and Machine Intelligence* **13** (3): 277–285. https://doi.org/10.1109/34.75515.

Viola, P. and Wells, W.M. (1997). Alignment by maximization of mutual information. *International Journal of Computer Vision* **24** (2): 137–154. https://doi.org/10.1023/A:1007958904918.

Vourvoulakis, J., Kalomiros, J., and Lygouras, J. (2016). Fully pipelined FPGA-based architecture for real-time SIFT extraction. *Microprocessors and Microsystems* **40**: 53–73. https://doi.org/10.1016/j.micpro.2015.11.013.

Wang, J., Hall-Holt, O., Konecny, P., and Kaufman, A.E. (2005). Per pixel camera calibration for 3D range scanning. *Videometrics VIII*, San Jose, CA, USA (18–20 January 2005). SPIE, Volume **5665**, 342–352. https://doi.org/10.1117/12.586209.

Wei, Z., Lee, D.J., Nelson, B.E., Archibald, J.K., and Edwards, B.B. (2008). FPGA-based embedded motion estimation sensor. *International Journal of Reconfigurable Computing* **2008**: Article ID 636145, 8 pages. https://doi.org/10.1155/2008/636145.

Williams, L. (1983). Pyramidal parametrics. *ACM SIGGRAPH Computer Graphics* **17** (3): 1–11. https://doi.org/10.1145/964967.801126.

Willson, R.G. and Shafer, S.A. (1994). What is the center of the image? *Journal of the Optical Society of America A* **11** (11): 2946–2955. https://doi.org/10.1364/JOSAA.11.002946.

Wolberg, G. (1990). *Digital Image Warping*. Los Alamitos, CA: IEEE Computer Society Press.

Wolberg, G. and Boult, T.E. (1989). Separable image warping with spatial lookup tables. *ACM SIGGRAPH Computer Graphics* **23** (3): 369–378. https://doi.org/10.1145/74334.74371.

Wolberg, G., Sueyllam, H.M., Ismail, M.A., and Ahmed, K.M. (2000). One-dimensional resampling with inverse and forward mapping functions. *Journal of Graphics Tools* **5** (3): 11–33. https://doi.org/10.1080/10867651.2000.10487525.

Yao, L., Feng, H., Zhu, Y., Jiang, Z., Zhao, D., and Feng, W. (2009). An architecture of optimised SIFT feature detection for an FPGA implementation of an image matcher. *International Conference on Field-Programmable Technology (FPT 2009)*, Sydney, Australia (9–11 December 2009), 30–37. https://doi.org/10.1109/FPT.2009.5377651.

Zdora, M.C. (2018). State of the art of X-ray speckle-based phase-contrast and dark-field imaging. *Journal of Imaging* **4** (5): Article ID 60, 36 pages. https://doi.org/10.3390/jimaging4050060.

Zhang, Z. (2000). A flexible new technique for camera calibration. *IEEE Transactions on Pattern Analysis and Machine Intelligence* **22** (11): 1330–1334. https://doi.org/10.1109/34.888718.

Zitova, B. and Flusser, J. (2003). Image registration methods: a survey. *Image and Vision Computing* **21** (11): 977–1000. https://doi.org/10.1145/146370.146374.

11

Linear Transforms

Linear transformation extends and generalises local linear filtering of Eq. (9.4) by removing the restriction of the window. Each output value is a linear combination of all of the input pixel values:

$$Q[u, v] = \sum_{x,y} w[u, v, x, y]I[x, y]. \tag{11.1}$$

This enables linear transforms to separate different components of an image, for example separating signal from interference or noise.

Direct implementation of Eq. (11.1) is very expensive. For an $N \times N$ input image, each output value requires N^2 multiplications and additions. Therefore, an $N \times N$ output requires N^4 operations. Many useful transforms are separable in that they can be decomposed into separate, independent, one-dimensional transforms on the rows and columns. In this case, Eq. (11.1) simplifies to

$$Q[u, v] = \sum_{y} w[v, y]\left(\sum_{x} w[u, x]I[x, y]\right), \tag{11.2}$$

reducing the number of operations to $2N^3$. Separable transforms can be represented in matrix form as

$$\mathbf{Q} = \mathbf{wIw}^{\top}, \tag{11.3}$$

where $\mathbf{w}$, $\mathbf{I}$, and $\mathbf{Q}$ are the weights and 2-D input and output images represented directly as matrices. All of the transforms considered in this chapter are separable, so the focus will be on the detailed implementation of the 1-D transform. To implement the 2-D transform, the transpose buffer architectures of Figure 10.17 can be adapted.

In many cases, the transform matrix, $\mathbf{w}$, may be further factorised into a cascade of simpler operations. Such factorisations allow the so-called fast transforms, reducing the number of operations to order $N^2 \log N$.

If the rows of $\mathbf{w}$ are orthogonal and normalised so that the vector length is 1, that is

$$\sum_{x} w[u_i, x]w[u_j, x] = \begin{cases} 0, & i \neq j, \\ 1, & i = j, \end{cases} \tag{11.4}$$

then the transformation matrix is unitary, and the transformation can be considered in terms of projecting the input image onto a new set of coordinate axes. Equation (11.1) then becomes

$$Q[u, v] = \sum_{x,y} W_{u,v}[x, y]I[x, y]. \tag{11.5}$$

Design for Embedded Image Processing on FPGAs, Second Edition. Donald G. Bailey.

Companion Website: www.wiley.com/go/bailey/designforembeddedimageprocessc2e

The new axes may be thought of as basis images or component images, $W_{u,v}[x, y]$, with the transformed values, $Q[u, v]$, indicating how much of the corresponding basis is in the original image. Thus,

$$I[x, y] = \sum_{u,v} W_{u,v}[x, y]Q[u, v]. \tag{11.6}$$

In this context, Eq. (11.5) is sometimes called the ***analysis equation*** because it is analysing how much of each basis image is present, and Eq. (11.6) is called the ***synthesis equation*** because it is synthesising the image from the basis images.

11.1 Discrete Fourier Transform

The Fourier transform is one of the most commonly used linear transforms in image and signal processing. It transforms the input image from a function of position (x, y) to a function of spatial frequency (u, v) in the frequency domain. The unitary 2-D discrete Fourier transform is

$$\begin{aligned} w[u, v, x, y] &= \frac{1}{N} e^{-j2\pi(ux+vy)/N} \\ &= \left(\frac{1}{\sqrt{N}} e^{-j2\pi ux/N}\right)\left(\frac{1}{\sqrt{N}} e^{-j2\pi vy/N}\right) \\ &= w[u, x]w[v, y]. \end{aligned} \tag{11.7}$$

The factor of $\frac{1}{\sqrt{N}}$ is often left out of the forward transform, and $\frac{1}{N}$ is applied to the inverse transform. The discrete Fourier transform is clearly separable and is complex-valued in the frequency domain. The periodic nature of the basis functions implies periodicity in both space and frequency. That is,

$$w[-u, x] = w[N - u, x] = w^*[u, x]. \tag{11.8}$$

This is a consequence of both the image and frequency domains being sampled (Bracewell, 2000). Although the origin in the frequency domain is in the corner of the image, it is often shifted to the centre for display purposes by scrolling the image by $N/2$ pixels both horizontally and vertically. Equation (11.8) also implies that the Fourier transform of real images is conjugate symmetric.

Two example images and their Fourier transforms are shown in Figure 11.1. On the top is a sinusoidal intensity pattern. Its Fourier transform has three peaks, one for the DC or average pixel value, and one for each of the positive and negative frequency components. This symmetry results from the Euler identity

$$e^{j2\pi ux/N} = \cos\frac{2\pi ux}{N} + j\sin\frac{2\pi ux}{N}, \tag{11.9}$$

$$\cos\frac{2\pi ux}{N} = \frac{1}{2}(e^{j2\pi ux/N} + e^{-j2\pi ux/N}). \tag{11.10}$$

In general, an image with a periodic spatial pattern will have a set of peaks in the frequency domain, corresponding to the frequency of the pattern along with its harmonics (Bailey, 1993). The position of a peak in the frequency domain indicates the spatial frequency of the corresponding pattern in the image, with the frequency proportional to the distance from the origin, and the direction of the peak from the origin corresponding to the direction of the sinusoidal variation. Rotating an object spatially will rotate its Fourier transform by the same angle. The magnitude of the peaks corresponds to the amplitude of the pattern, and the phase of the frequency components contains information about the position of the pattern in the image (shifting an object will only change the phase in the frequency domain). The shape of the peaks contains information about the shape and size of patterns in the image, and the distribution of the peaks relates to the details of the object. For example, sharp edges within the pattern will result in harmonics with amplitudes that decay in inverse proportion to frequency. In general, the amplitude of natural images in the frequency domain drops approximately in inverse proportion with frequency, primarily as a result of edges and other discontinuities (Millane et al., 2003). In contrast, spatially uncorrelated random noise has a uniform frequency distribution.

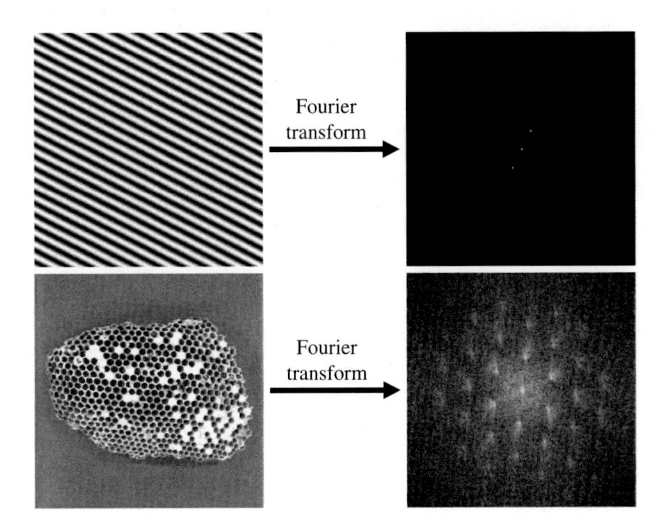

Figure 11.1 Example images and corresponding Fourier transforms (only the magnitude is shown, scaled logarithmically to show the detail more clearly). Top: a sinusoid; bottom: a semi-regular pattern.

Some imaging modalities (for example magnetic resonance imaging and X-ray crystallography) capture their data in the frequency domain, and it must be converted to the spatial domain for visualisation. In other applications, the desired operation may be simpler to perform in the frequency domain, requiring the use of forward and inverse transforms to switch between the spatial and frequency domains.

11.1.1 Fast Fourier Transform (FFT)

A ***fast Fourier transform*** (FFT) is an efficient implementation of the discrete Fourier transform that results from the factorisation of the transformation. For an $N \times N$ transform, the computational complexity is of order $N^2 \log N$. There are several different FFT algorithms (see (Kumar et al., 2019)), only a few of which are described here.

11.1.1.1 Decimation-in-time

Let $I[x]$ be a 1-D signal of length N. First, define

$$W_N = e^{-j2\pi/N}. \tag{11.11}$$

Then, by splitting $I[x]$ into the odd and even samples

$$I_e[x] = I[2x], \qquad I_o[x] = I[2x+1], \tag{11.12}$$

the Fourier transform, $F[u]$, can be expanded as a sum of the Fourier transforms of the odd and even samples:

$$\begin{aligned} F[u] = \sum_{x=0}^{N-1} I[x]W_N^{ux} &= \sum_{x=0}^{\frac{N}{2}-1} I[2x]W_N^{2ux} + \sum_{x=0}^{\frac{N}{2}-1} I[2x+1]W_N^{u(2x+1)} \\ &= \sum_{x=0}^{\frac{N}{2}-1} I_e[x]W_{N/2}^{ux} + W_N^u \sum_{x=0}^{\frac{N}{2}-1} I_o[x]W_{N/2}^{ux} \\ &= F_e[u] + W_N^u F_o[u]. \end{aligned} \tag{11.13}$$

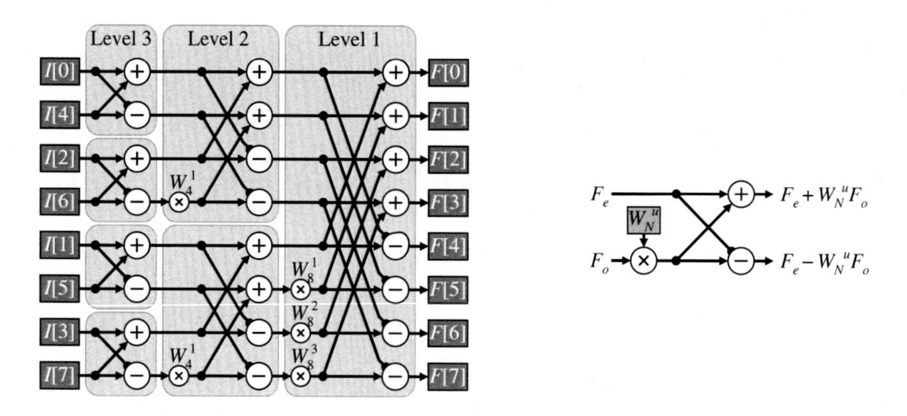

Figure 11.2 Radix-2 decimation-in-time FFT; right: basic butterfly operation.

Since the Fourier transform is periodic, $F_e[u + \frac{N}{2}] = F_e[u]$, and by symmetry, $W_N^{u+\frac{N}{2}} = -W_N^u$. Therefore, the second half of the frequency samples simplifies to

$$F\left[u + \frac{N}{2}\right] = F_e\left[u + \frac{N}{2}\right] + W_N^{u+\frac{N}{2}} F_o\left[u + \frac{N}{2}\right] = F_e[u] - W_N^u F_o[u], \tag{11.14}$$

which reuses the Fourier transforms of the first half but with a change in sign. The W_N^u factors are sometimes called ***twiddle factors***. By recursively applying this same decomposition to F_e and F_o, the $\log_2 N$ levels of the radix-2 decimation-in-time FFT result. This is illustrated for $N = 8$ in Figure 11.2. The computation at each level consists of a set of 'butterfly' computations which can be performed in place in a memory-based system. One consequence of the recursion is that the input samples for the decimation-in-time decomposition are in a bit-reversed order. This will require permuting the input data before performing any calculation in place.

11.1.1.2 Decimation-in-frequency

An alternative decomposition is decimation-in-frequency. Calculating the even and odd frequency components separately gives

$$\begin{aligned} F[2u] &= \sum_{x=0}^{N-1} I[x] W_N^{2ux} = \sum_{x=0}^{\frac{N}{2}-1} I[x] W_N^{2ux} + \sum_{x=0}^{\frac{N}{2}-1} I\left[x + \frac{N}{2}\right] W_N^{2u(x+\frac{N}{2})} \\ &= \sum_{x=0}^{\frac{N}{2}-1} \left(I[x] + I\left[x + \frac{N}{2}\right]\right) W_{N/2}^{ux}. \end{aligned} \tag{11.15}$$

$$\begin{aligned} F[2u+1] &= \sum_{x=0}^{N-1} I[x] W_N^{(2u+1)x} = \sum_{x=0}^{\frac{N}{2}-1} I[x] W_N^{2ux} W_N^x - \sum_{x=0}^{\frac{N}{2}-1} I\left[x + \frac{N}{2}\right] W_N^{2ux} W_N^x \\ &= \sum_{x=0}^{\frac{N}{2}-1} \left(I[x] - I\left[x + \frac{N}{2}\right]\right) W_N^x W_{N/2}^{ux}. \end{aligned} \tag{11.16}$$

These are Fourier transforms of sequences which have half the length of the original sequence. Repeating this recursively gives the decimation-in-frequency algorithm illustrated in Figure 11.3. The input samples are ordered naturally, and the frequency samples are in bit-reversed order as a consequence of recursively separating the outputs into odd and even components. The main differences in the calculation are the order of the groupings, and that the twiddle factor is on the output of the butterfly rather than the input.

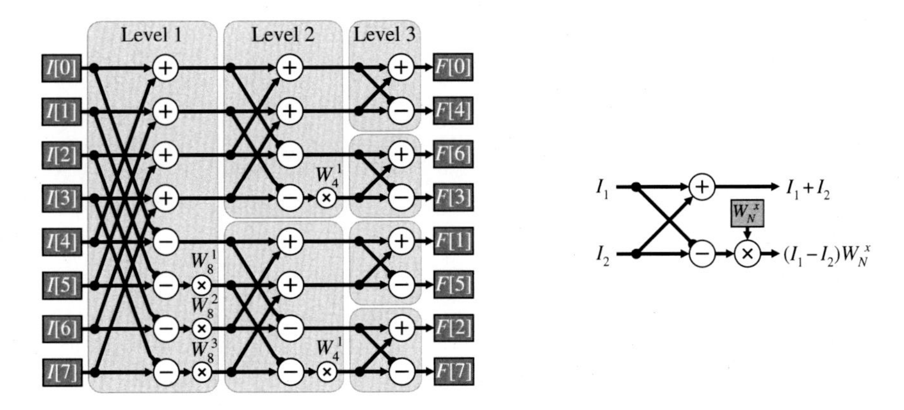

Figure 11.3 Radix-2 decimation-in-frequency FFT; right: basic butterfly operation.

In many applications, the permutation of the frequency samples is of little consequence. If necessary, the samples may be reordered as they are written to memory. If accessing external memory using bursts, the bit-reversed address order can lead to inefficiencies, so it may be best to stage the results using an on-chip RAM buffer and stream the results to external memory in natural order. Alternatively, Lee and Park (2007) proposed using a decimation-in-time final stage to produce the output samples in natural order. This requires a commutation stage to reorder samples in the transition.

11.1.1.3 Complex Multiplication

Note that all of the operations illustrated in Figures 11.2 and 11.3 are on complex numbers. The twiddle factors are

$$W_N^k = e^{-j2\pi k/N} = \cos\frac{2\pi k}{N} - j\sin\frac{2\pi k}{N} = c_k - js_k, \tag{11.17}$$

so each complex multiplication requires four real multiplications and two additions. Let the complex number being multiplied by the twiddle factor be $R + jI$, then through factorisation, the number of multiplications can be reduced to three at the cost of one more addition:

$$\begin{aligned}(R + jI)(c_k - js_k) &= (Rc_k + Is_k) + j(Ic_k - Rs_k)\\ &= (R(c_k + s_k) - (R - I)s_k) + j(I(c_k - s_k) - (R - I)s_k)\end{aligned} \tag{11.18}$$

(since the twiddle factor is a constant) as shown in Figure 11.4.

With multiplication, the number of bits grows, so inevitably there will be round-off or truncation error. The multiplication may be made invertible using lifting (Oraintara et al., 2002), as shown on the right in

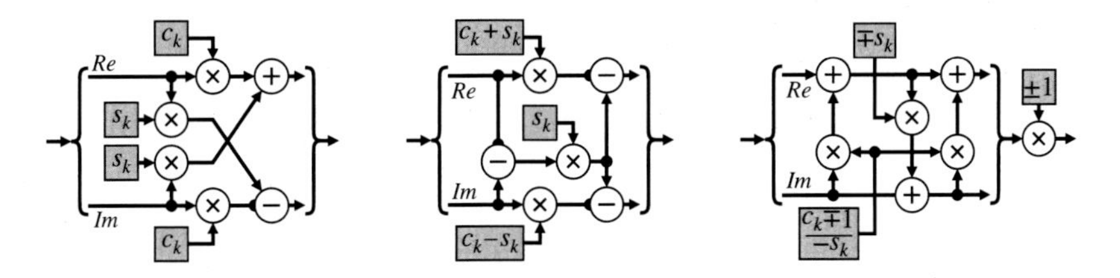

Figure 11.4 Complex multiplication. Left: conventional multiplication; centre: factorised implementation using Eq. (11.18); right: implemented using lifting.

Figure 11.4. Although lifting still has round-off errors, they are reversible if the result of the multiplication is quantised before the associated addition. Running the system in reverse, replacing the additions with subtractions, will give the original result, provided that an identical quantisation is performed. The two alternatives (using the $\pm$) enable the absolute value of the coefficients to be kept less than 1.0 (depending on k), with the final multiplication by ± 1 being a trivial sign change. Chang and Nguyen (2007) showed that even without the reversibility, using lifting results in smaller round-off errors than the other multiplication schemes. The disadvantage of lifting is that the multiplications are performed in series rather than in parallel, increasing the latency.

Since each multiplication is a rotation in the complex plane ($|W_N| = 1$), the complete multiplication can be performed using CORDIC (coordinate rotation digital computer) arithmetic (Despain, 1974; Sansaloni et al., 2003). Efficiencies are gained because the rotation angles are constants, so the decisions at each step of the CORDIC algorithm may be made in advance, and the θ logic (see Section 5.2.2) is eliminated. However, compensated CORDIC is required to prevent the magnitude from changing as the complex number is rotated.

11.1.1.4 Radix-4 Algorithms

While radix-2 algorithms (decomposing the transform to a series of two-point transforms) are the simplest to understand, the number of multiplications may be reduced using a radix-4 algorithm. A four-point Fourier transform is

$$\mathbf{F} = \begin{bmatrix} 1 & 1 & 1 & 1 \\ 1 & -j & -1 & j \\ 1 & -1 & 1 & -1 \\ 1 & j & -1 & -j \end{bmatrix} \mathbf{I}, \tag{11.19}$$

where all the multiplications are trivial. The radix-4 decimation-in-frequency algorithm extends Eqs. (11.15) and (11.16) by splitting the input into four sections, combining them using Eq. (11.19) and taking $\frac{N}{4}$-point transforms of the results to get the frequency samples decimated by four:

$$\begin{aligned}
F[4u] &= \sum_{x=0}^{\frac{N}{4}-1} \left(I[x] + I\left[x + \frac{N}{4}\right] + I\left[x + \frac{N}{2}\right] + I\left[x + \frac{3N}{4}\right] \right) W_{N/4}^{ux}, \\
F[4u+1] &= \sum_{x=0}^{\frac{N}{4}-1} \left(I[x] - jI\left[x + \frac{N}{4}\right] - I\left[x + \frac{N}{2}\right] + jI\left[x + \frac{3N}{4}\right] \right) W_N^{x} W_{N/4}^{ux}, \\
F[4u+2] &= \sum_{x=0}^{\frac{N}{4}-1} \left(I[x] - I\left[x + \frac{N}{4}\right] + I\left[x + \frac{N}{2}\right] - I\left[x + \frac{3N}{4}\right] \right) W_N^{2x} W_{N/4}^{ux}, \\
F[4u+3] &= \sum_{x=0}^{\frac{N}{4}-1} \left(I[x] + jI\left[x + \frac{N}{4}\right] - I\left[x + \frac{N}{2}\right] - jI\left[x + \frac{3N}{4}\right] \right) W_N^{3x} W_{N/4}^{ux}.
\end{aligned} \tag{11.20}$$

The corresponding 'butterfly' is shown in Figure 11.5. Although such decimation can be extended to higher radixes, little is gained from a hardware perspective beyond radix-4.

11.1.1.5 Implementations

There are several ways in which the FFT may be implemented, given the above building blocks. If resources are scarce, a single butterfly unit may be built, with the transformation performed serially in place in memory (see Figure 11.6). The data is first loaded into memory, performing any permutation as necessary. The data is then streamed from memory as needed for each butterfly operation, with the results written back into memory. Memory bandwidth limits the speed to one butterfly every two clock cycles for the radix-2 and one butterfly every four clock cycles for the radix-4 operation. With the serial radix-4 butterfly, the multiplier can be shared

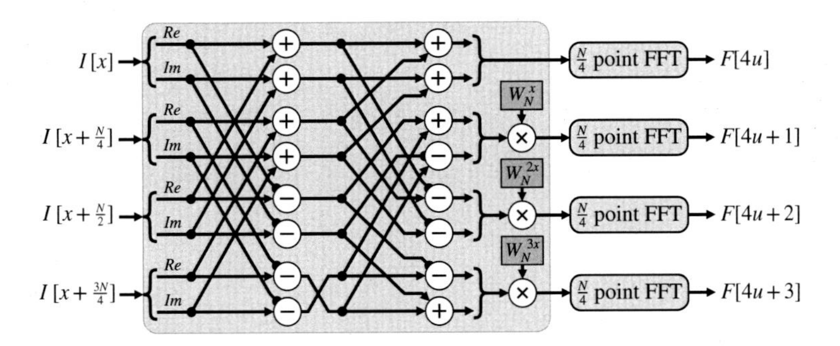

Figure 11.5 Radix-4 decimation-in-frequency with the real and imaginary components shown explicitly.

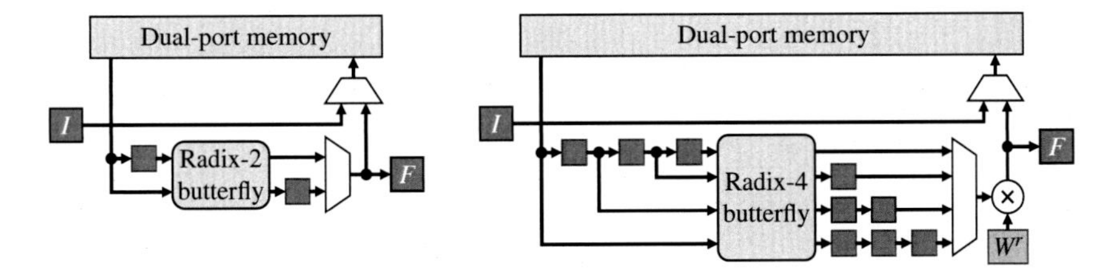

Figure 11.6 Serial FFT operating in-place on memory. Left: radix-2; right: radix-4.

by moving it from within the butterfly to after the multiplexer, as shown in Figure 11.6 (Sansaloni et al., 2003), and the number of adders may also be reduced by multiplexing the input layer of Figure 11.5. The memory addressing required by the scheme in Figure 11.6 can use a standard counter, with the address bits permuted depending on the FFT level.

Uzun et al. (2005) implemented several different butterfly implementations within a generic framework. The speed and resources were compared between the implementations for a number of butterfly units. Not surprisingly, the radix-4 algorithm was the fastest but also consumed the most resources.

The throughput may be improved to one butterfly operation per cycle by partitioning the memory (two blocks for radix-2 and four for radix-4) to enable the data to be fed into the butterfly in parallel. This comes at the cost of more complex memory addressing and a permutation network on both the input and output of the butterfly to route the data to and from the appropriate memories (Kitsakis et al., 2018).

If more resources are available, multiple butterflies may be operated in parallel, with the FFT pipelined. The basic principle is illustrated in Figure 11.7. For a radix-2 implementation, the pipeline consists of $\log_2 N$ butterfly stages. At each stage, the first half of the clock cycles feed the data through into the delay register without processing. The second half combines the incoming data with the delayed data, performing the butterfly. One of the butterfly outputs is fed directly out to the next stage, while the other is fed back into the delay. It is finally output and multiplied by the twiddle factor while the next block is being loaded. Successive stages reduce the delay by a factor of 2 to give the interleaving shown in Figure 11.3. The pipelined FFT takes streamed data

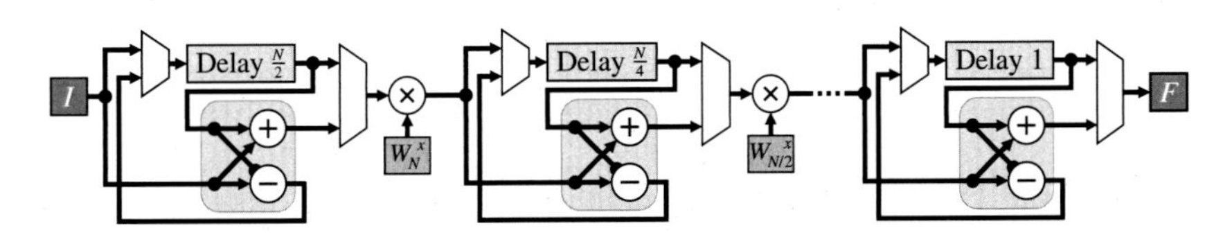

Figure 11.7 Pipelined radix-2 decimation-in-frequency implementation.

as input and produces a streamed output (in bit-reversed order), with a throughput of one sample per clock cycle and a latency of one row.

A selection of related pipelined FFT architectures are reviewed by Garrido (2022). One of the more efficient pipelined implementations is the radix-2^2 design (He and Torkelson, 1996). This has the reduced multipliers of a radix-4 algorithm but effectively forms a radix-4 butterfly as a pair of radix-2 stages. The intermediate twiddle factor is a trivial $-j$. The sign change of the imaginary component can easily be incorporated into the addition/ subtraction of the downstream butterfly. Field-programmable gate array (FPGA) implementations of this scheme are described in Sukhsawas and Benkrid (2004) and Saeed et al. (2009).

Again, if memory bandwidth allows, the early stages of the pipeline may be operated in parallel (Garrido, 2022), significantly reducing the latency.

Although the pipelined approach also extends to two dimensions, the size of the buffers is usually too large for on-chip memory. However, separability allows a 2-D FFT to first take the 1-D Fourier transform of each image row and then a 1-D transform of the columns of the result. This requires storing the intermediate row results into a frame buffer. Care must be taken when accessing the image by column to avoid access latencies (see Section 6.4.2.1). Since each 1-D transform is of independent data, a 2-D transform may be accelerated by performing multiple 1-D transforms in parallel where hardware resources and memory bandwidth allow.

To perform an inverse FFT, the inverse transform matrix is the complex conjugate of the forward transform. This means that the same hardware may be used, with the exception that the complex conjugate of the twiddle factors is used. Alternatively, the inverse FFT may be calculated by taking the complex conjugate of the frequency domain signal and using a standard forward FFT. With either case, the result of each 1-D transform must be divided by N to obtain the original signal.

11.1.1.6 Computation Tricks

With real data, additional efficiencies may be gained by performing two FFTs at once. The Fourier transform of a real signal is conjugate symmetric, that is the real parts of the Fourier transform are even, and the imaginary parts are odd. This allows a single FFT to take the Fourier transform of two sequences (for example two rows) simultaneously. Let two separate rows be I_1 and I_2. A composite signal is formed by combining one signal in the real part and the other in the imaginary:

$$I_c[x] = I_1[x] + jI_2[x]. \tag{11.21}$$

The FFT is taken of the composite signal, and the FFT of the individual components may be extracted as

$$\begin{aligned} F_1[u] &= \frac{1}{2}(F_c[u] + F_c^*[-u]) = \frac{1}{2}(F_c[u] + F_c^*[N-u]), \\ F_2[u] &= \frac{-j}{2}(F_c[u] - F_c^*[-u]) = \frac{-j}{2}(F_c[u] - F_c^*[N-u]). \end{aligned} \tag{11.22}$$

This is effectively another level of butterfly operations, although with a slightly different pattern. Similarly, when performing the inverse FFT of a real image, two conjugate symmetric Fourier transforms may be combined as

$$F_c[u] = F_1[u] + jF_2[u] \tag{11.23}$$

before calculating the inverse.

When calculating the Fourier transform of a 2-D image, symmetry may also be exploited when taking the Fourier transform of the columns. Columns 0 and $\frac{N}{2}$ will only have real data, so may be transformed together with a single FFT. Columns u and $N - u$ will be complex conjugates of one another, so will give related Fourier transforms:

$$F[u, v] = F^*[-u, -v] = F^*[N - u, N - v]. \tag{11.24}$$

Therefore, only one of these needs to be calculated. Exploiting the symmetries of both row and column transforms can reduce the overall computation for the FFT of a real image by a factor of 4.

11.1.1.7 Approximation Error

While an FFT is mathematically invertible, computing using fixed-point arithmetic will only approximate the Fourier transform. Consequently, taking the FFT and then the inverse FFT will generally not give exactly the same result. The three main sources of error are round-off errors in representing the twiddle factors, rounding off the results of the multiplication, and rescaling of the results after each butterfly to avoid overflow.

The transform may be made invertible using lifting to implement the twiddle factor multiplications (Oraintara et al., 2002) (assuming that an extra bit is added after each stage to avoid overflow). However, any operation on the frequency domain data (for example filtering) will mean that round-off errors will no longer cancel when performing the inverse transform.

Chang and Nguyen (2008) analysed the round-off errors for the FFT and showed that each stage of butterflies amplifies the errors. This gave decimation-in-time algorithms an advantage over decimation-in-frequency, because the nontrivial twiddle factors were concentrated closer to the end, reducing the amplification effect. The errors can obviously be reduced using wider bit-width words, but this requires more memory for intermediate storage, giving a trade-off between accuracy and memory requirements. They also analysed the optimum fixed-point bit width for each stage to give the minimum error for a given memory budget.

11.1.1.8 Non-power-of-2 FFTs

The assumption throughout this section has been that the length of the FFT, N, is a power of 2. Other factorisations of N can also result in fast algorithms. In particular, a radix-3 butterfly is relatively simple (Dubois and Venetsanopoulos, 1978; Lofgren and Nilsson, 2011; Winograd, 1978). A radix-5 butterfly is a little more complicated (Lofgren and Nilsson, 2011; Winograd, 1978). Similarly, radix-7 and radix-9 algorithms may be developed (Winograd, 1978; Zohar, 1981). While these require relatively few multiplications, they lack the elegance and computational simplicity of those described above for powers of 2. Therefore, unless there is a particular need for a special size, the complexity of developing the circuitry for such factorisations is generally not worthwhile.

For small radices, one approach is to perform the butterflies directly using a small discrete Fourier transform. These may be implemented efficiently using systolic arrays (Nash, 2014).

An alternative is to express the Fourier transform in terms of a convolution which can be calculated using a power-of-2 FFT (Rabiner et al., 1969). In one dimension:

$$\begin{aligned} F[u] &= \sum_{x=0}^{N-1} I[x]W_N^{ux} = \sum_{x=0}^{N-1} I[x]W_N^{ux+(u^2-u^2+x^2-x^2)/2} \\ &= W_N^{u^2/2} \sum_{x=0}^{N-1} (W_N^{x^2/2} I[x]) W_N^{-(u-x)^2/2}. \end{aligned} \tag{11.25}$$

The summation is now a convolution which may be performed using an FFT of length $\tilde{N} \geq 2N$ of each of the two sequences, taking their product, and calculating the inverse FFT of the result. While this requires calculating three longer FFTs, in many circumstances this is still preferable to calculating the discrete Fourier transform directly (especially if N is prime and cannot be factorised).

11.1.1.9 Border Handling

When considering the discrete Fourier transform of an image, the image is implicitly multiplied by a rectangular window which effectively sets the parts beyond the image borders to zero. A product in the image domain is equivalent to a convolution in the frequency domain, so the true frequency content of the image is convolved by a sinc function, which has relatively high side-lobes. For frequencies that are exactly an integer multiple of the sample spacing in the frequency domain, the frequency samples coincide with the nulls between the side-lobes.

Sampling in the frequency domain implicitly assumes a periodic image. However, the pixel values on opposite edges of most images do not match. These discontinuities mean that the constituent sinusoidal image

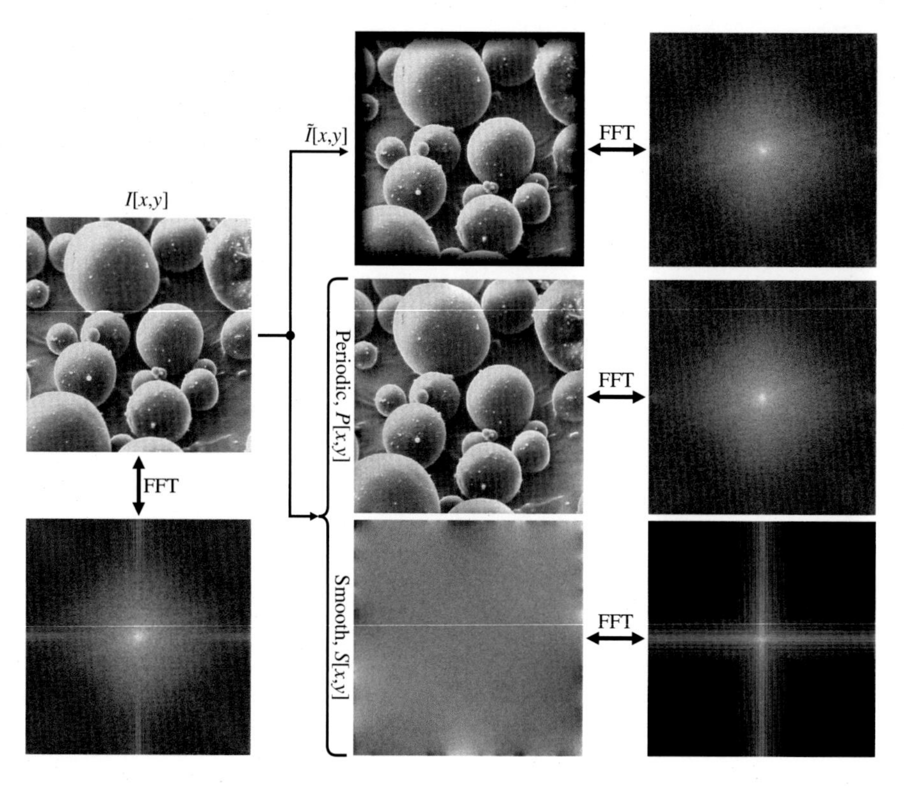

Figure 11.8 Effect of image borders on FFT. Left: an image and its Fourier transform, showing '+' like artefacts from the borders. Top: multiplying by a Tukey apodisation window (with $\alpha = 0.2$) reduces these artefacts. Bottom: periodic plus smooth decomposition; effects of image borders are isolated and separated. Frequency domain images have been logarithmically scaled to make the detail clearer.

components are not exact integer multiples of the frequency sample spacing. Consequently, the frequency samples are not in the nulls, so the information is distributed in the frequency domain based on the side-lobes. This phenomenon is ***spectral leakage***, and often manifests as streaked artefacts in the frequency domain, as demonstrated on the bottom-left in Figure 11.8.

Spectral leakage effects may be reduced by removing the discontinuities at the image edges by multiplying the image by a window or ***apodisation function*** which smoothly tapers to zero at the edges of the image (Harris, 1978):

$$\tilde{I}[x,y] = I[x,y]A[x,y]. \tag{11.26}$$

Ideally, in the frequency domain, the window function should have low side-lobes which decay rapidly with increasing frequency. While there are many apodisation functions that meet this criterion (Harris, 1978), for image processing the image should also be as undistorted as possible, so $A[x,y]$ should be flat over much of the image to minimise loss of information around the edges of the image. This makes the Tukey or tapered cosine window (Bingham et al., 1967)

$$A[x,y] = A[x]A[y], \tag{11.27}$$

with

$$A[x] = \begin{cases} \frac{1}{2}\left(1+\cos\pi\left(\frac{2x}{\alpha N}-1\right)\right), & 0 \le x < \frac{\alpha N}{2}, \\ 1, & \frac{\alpha N}{2} \le x \le \frac{(2-\alpha)N}{2}, \\ \frac{1}{2}\left(1+\cos\pi\left(\frac{2x}{\alpha N}-\frac{2-\alpha}{\alpha}\right)\right), & \frac{(2-\alpha)N}{2} < x < N, \end{cases} \tag{11.28}$$

and similarly for $A[y]$, a good choice for image processing, where α is the proportion of samples in each dimension that are tapered. The reduction in side-lobes directly reduces the spectral leakage, but the wider main lobe blurs the frequency content, resulting in a slightly reduced frequency resolution. The effect of the Tukey window is shown on the top row of Figure 11.8.

For an FPGA implementation, the window function may be stored in a lookup table and multiplied by the incoming image as it is streamed into the FFT. Due to symmetry, only $\frac{N}{2}$ entries are required in the table. For the Tukey window, this is reduced to $\frac{\alpha N}{2}$ entries.

An alternative approach that avoids the image loss is to decompose the image into periodic and smooth components (Moisan, 2011),

$$I[x,y] = P[x,y] + S[x,y], \tag{11.29}$$

where the periodic component $P[x,y]$ has the image content but without the discontinuities at the image borders, and the smooth component $S[x,y]$ contains the border discontinuities but is smooth everywhere else, where smooth is defined by the Laplacian being zero.

Let the border image be defined as the sum of the row and column borders:

$$B_r[x,y] = \begin{cases} I[N-1-x,y] - I[x,y], & x = 0 \text{ or } x = N-1, \\ 0, & \text{elsewhere}, \end{cases} \tag{11.30}$$

$$B_c[x,y] = \begin{cases} I[x,N-1-y] - I[x,y], & y = 0 \text{ or } y = N-1, \\ 0, & \text{elsewhere}. \end{cases} \tag{11.31}$$

Then, the smooth component can be calculated from this by applying an inverse Laplacian filter in the frequency domain:

$$S[u,v] = \frac{B_r[u,v] + B_c[u,v]}{2\cos\frac{2\pi u}{N} + 2\cos\frac{2\pi v}{N} - 4}, \tag{11.32}$$

enabling the Fourier transform of the periodic component to be calculated as

$$P[u,v] = I[u,v] - S[u,v]. \tag{11.33}$$

Mahmood et al. (2015) implemented this on an FPGA using a 2-D FFT for the border image. However, since the border image largely consists of zeros, Bailey et al. (2017) were able to optimise the border processing to use a single 1-D FFT.

11.1.2 Goertzel's Algorithm

In some applications, only one or a few frequencies are of interest. In this case, Goertzel's algorithm (Goertzel, 1958) can require considerably fewer calculations than performing the full Fourier transform. Consider a single frequency component of the Fourier transform

$$\begin{aligned} F_u &= \sum_{x=0}^{N-1} I[x] W_N^{ux} = \sum_{x=0}^{N-1} I[x] (W_N^{-u})^{N-x} \\ &= \left(\sum_{x=0}^{N-2} I[x] (W_N^{-u})^{N-x} + I[N-1] \right) W_N^{-u}. \end{aligned} \tag{11.34}$$

This is effectively a recursive calculation

$$F_u[n] = (F_u[n-1] + I[n]) W_N^{-u}. \tag{11.35}$$

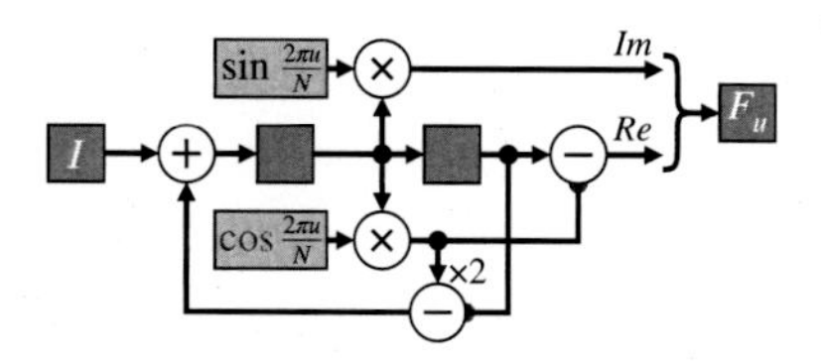

Figure 11.9 Implementation of Goertzel's algorithm.

It still requires one complex multiplication per sample, but all multiplications are now by the same constant. Looking at this recursion in the z-domain:

$$\begin{aligned}\frac{F_u(z)}{I(z)} &= \frac{W_N^{-u}}{1 - W_N^{-u}z^{-1}} = \frac{W_N^{-u}}{1 - W_N^{-u}z^{-1}}\left(\frac{1 - W_N^{u}z^{-1}}{1 - W_N^{u}z^{-1}}\right)\\ &= \frac{W_N^{-u} - z^{-1}}{1 - 2\cos\frac{2\pi u}{N}z^{-1} + z^{-2}}.\end{aligned} \tag{11.36}$$

The denominator coefficients are now real, so this can be implemented as a second-order recursion with real coefficients:

$$S_u[n] = I[n] + 2\cos\frac{2\pi u}{N}S_u[n-1] - S_u[n-2]. \tag{11.37}$$

After the N samples have accumulated, the frequency term may be calculated from Eq. (11.36) as

$$F_u = \left(S_u[N-1]\cos\frac{2\pi u}{N} - S_u[N-2]\right) + jS_u[N-1]\sin\frac{2\pi u}{N}. \tag{11.38}$$

Delaying the output by one cycle enables the same multiplier to be used for the output as for the feedback. An implementation of this is shown in Figure 11.9. Note that the intermediate registers need to be reset to zero before the start of the calculation. The filter processes streamed data and produces the frequency value after N input samples. N is not restricted to be a power of 2; the implementation of the algorithm is independent of the number of samples (apart from the constant multipliers). If multiple frequencies must be calculated, then this circuit can be replicated for each frequency.

11.1.3 Applications

11.1.3.1 Filtering

Linear filters are represented by a convolution in the image domain, using Eq. (9.4). In the frequency domain, this convolution becomes a product:

$$Q[u,v] = K[u,v]I[u,v], \tag{11.39}$$

where $K[u,v]$ is the Fourier transform of the flipped filter kernel. Filtering may therefore be performed in the frequency domain. For large kernel sizes, it is less expensive computationally to take the Fourier transform (using an FFT), multiply by $K[u,v]$, and take the inverse Fourier transform of the result.

The latency of filtering using frequency domain processing can be determined as follows. The row transforms can be performed directly on the input stream, although the column processing cannot begin until the first output of the last row is available, after one frame. Since filtering only requires a point operation in the frequency domain, the forward column FFT can be directly followed by the inverse column FFT. Again, the inverse row transforms must wait until all of the column processing is complete; however, for real images, conjugate symmetry means that only half of the column transforms are actually required, giving a delay of 0.5 frame plus 1 column. The first row of the filtered image can be output after processing the first row, giving a total latency of 1.5 frames plus 1 row and 1 column.

Equation (11.39) implies that a linear filter may be considered in terms of its ***frequency response***, as a weighting of the different frequency components within the image. In this context, a low-pass filter will tend to smooth noise, because it will pass low frequencies with little attenuation, and many images have most of their energy in the low frequency. Uncorrelated noise has uniform frequency content, so attenuating the high frequencies will have only a small effect on the image but a more significant effect on the noise. The loss of high frequencies in the image will result in a loss of fine detail which will, in general, blur the image. Edges and fine detail may be enhanced by boosting the amplitudes of the higher frequencies, but this will also tend to amplify the noise within the image.

One particular application of filtering in the frequency domain is to remove pattern noise from an image. As described earlier, a regular pattern will exhibit a series of peaks within the frequency domain. These peaks may be detected directly in the frequency domain, and masked out, as demonstrated in Figure 11.10. If the pattern is added within the image, then filtering can remove the pattern. Here, the pattern is part of the background, so removing the pattern will also affect the object itself.

The opposite of removing pattern noise is filtering to detect regular patterns within the image. In this case, it is the periodic peaks within the frequency domain that contain the information of interest. One approach to enhancing the regular patterns is to use the image itself as the filter (Bailey, 1993, 1997). Multiplication by the magnitude in the frequency domain:

$$K[u,v] = |I[u,v]| \tag{11.40}$$

is a zero-phase filter, so objects are not shifted. Self-filtering is demonstrated in the bottom of Figure 11.10. The basic self-filter will blur sharp edges, which have frequency content that rolls off inversely proportional to frequency. This may be compensated by weighting the filter with frequency to reduce the attenuation of the harmonics:

$$K[u,v] = \sqrt{u^2+v^2}\,|I[u,v]|. \tag{11.41}$$

In the example on the top in Figure 11.10, the frequency domain was simply masked. This will also remove those frequency components from the objects of interest. An alternative is to design the filter such that it minimises the error after filtering. Consider an image, $F[x,y]$, which has been corrupted by the addition of noise, $N[x,y]$ (uncorrelated with the input image). The ***Wiener filter*** is the optimal filter (in the least squares sense)

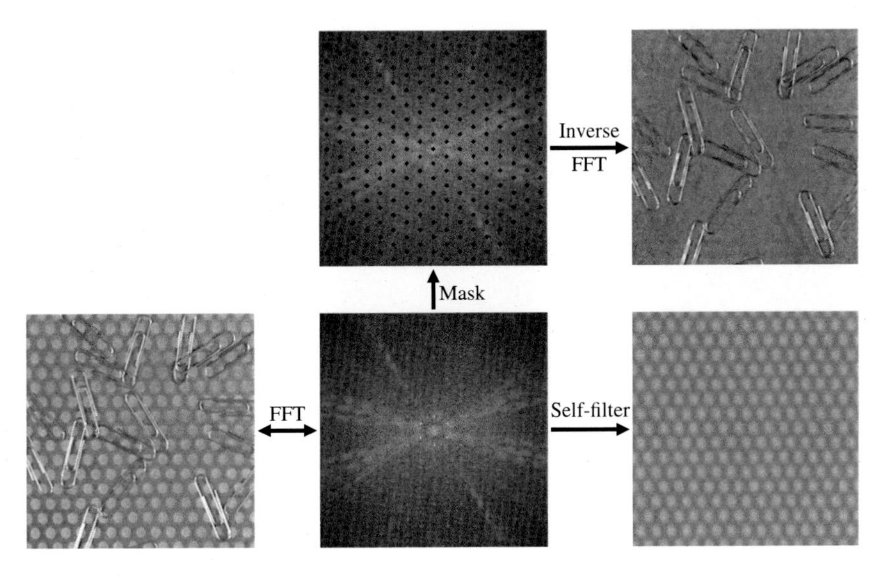

Figure 11.10 Filtering in the frequency domain. Top: masking peaks corresponding to regular patterns to remove the pattern noise. Bottom: self-filtering using Eq. (11.40). The frequency domain is shown logarithmically scaled for clarity.

for recovering the original signal. The analysis is easiest to perform in the frequency domain, where the noisy image is $F[u,v]+N[u,v]$. First, assume that the filter is symmetric so that it does not shift the image; this makes $K[u,v]$ real. The goal is to determine the filter frequency response, $K[u,v]$, that minimises the squared error:

$$\begin{aligned} E^2 &= \sum_{u,v} \|(F+N)K - F\|^2 \\ &= \sum_{u,v} \|F(K-1) - NK\|^2 \\ &= \sum_{u,v} (\|F\|^2(K^2 - 2K + 1) + \|N\|^2K^2) + \sum_{u,v} (FN^* + NF^*)K(K-1). \end{aligned} \tag{11.42}$$

The expected value of the second summation is zero if the image and signal are uncorrelated. The error may be minimised by setting the derivative with respect to the filter to zero

$$\frac{dE^2}{dK} = \sum_{u,v} (2K(\|F\|^2 + \|N\|^2) - 2\|F\|^2) = 0 \tag{11.43}$$

and solving to give the optimal filter as

$$K[u,v] = \frac{\|F[u,v]\|^2}{\|F[u,v]\|^2 + \|N[u,v]\|^2}. \tag{11.44}$$

Given estimates of both the signal and noise spectra, the frequency response of the optimal filter can be derived. If the noise is spatially uncorrelated, it will have a flat spectrum, and N will be a constant. However, for pattern noise, $N[u,v]$ will have peaks corresponding to the noise pattern.

Equation (11.44) may be interpreted as follows: at the frequencies where the signal is much larger than the noise, the frequency response of the filter will approach unity, passing the signal through. However, the frequencies at which the signal is smaller than the noise will be attenuated, effectively reducing the noise power while having little effect on the total signal. The Wiener filter is effective at removing both random noise and pattern noise.

11.1.3.2 Inverse Filtering

Closely related to filtering is inverse filtering. Here, the goal is to remove the effects of filtering from an image. A common application is removing blur, whether from the lens point spread function, or motion blur from a moving object or camera. Let $B[x,y]$ be the blur point spread function, with corresponding Fourier transform, $B[u,v]$. Also, let the desired unblurred image be $Q[x,y]$. In the frequency domain, the blurred input image is given by

$$I[u,v] = Q[u,v]B[u,v]. \tag{11.45}$$

Assuming that the blur function is known, it should be possible to recover the deblurred image as

$$\hat{Q}[u,v] = \frac{I[u,v]}{B[u,v]}. \tag{11.46}$$

Unfortunately, this does not work. The output image from Eq. (11.46) appears to contain only random noise. The reason is that $B[u,v]$ goes to zero at some frequencies with the division by zero undefined. Even if B does not actually go to zero at any of the samples, the scaling of the corresponding frequencies becomes arbitrarily large. For noiseless, infinite precision arithmetic, this would not pose a problem. However, $I[u,v]$ will inevitably have some level of noise (even if just quantisation noise), so applying the inverse filter according to Eq. (11.46) will cause the noise to dominate the output.

One solution to this problem is to manipulate Eq. (11.46) to avoid division by zero:

$$\hat{Q}[u,v] = I[u,v]\frac{1}{B[u,v]} = I\frac{B^*}{\|B\|^2} \approx I\frac{B^*}{\|B\|^2 + k^2}, \tag{11.47}$$

where the first transformation is to make the denominator positive real, and the second adds a positive constant, k^2, in the denominator to prevent division by zero. This effectively limits the gain of the noise. The Wiener filter, which minimises the mean square error, gives the inverse filter as

$$W[u,v] = \frac{B^*[u,v]\,\|Q[u,v]\|^2}{\|B[u,v]\|^2\,\|Q[u,v]\|^2 + \|N[u,v]\|^2} = \frac{B^*[u,v]}{\|B[u,v]\|^2 + \dfrac{\|N[u,v]\|^2}{\|Q[u,v]\|^2}}, \tag{11.48}$$

which sets k^2 optimally according to the signal-to-noise ratio at each frequency.

More complex techniques (usually iterative) are required when the blur function is not known. These are beyond the scope of this book.

11.1.3.3 Interpolation

Images may be interpolated using the frequency domain. If it is assumed that the image is band-limited (and sampled according to the Nyquist sampling criterion), then extending the image in the frequency domain is trivial because the additional higher frequency samples will be zero. Therefore, an image may be interpolated by taking its Fourier transform, padding with zeros, and taking the inverse Fourier transform.

Since padding will make the frequency domain image larger, taking the inverse Fourier transform will take longer. This approach is equivalent to sinc interpolation of the input image. However, in many imaging applications, the input image is not band-limited, therefore sinc interpolation will not necessarily give the best results. Interpolating directly in the image domain (Section 10.1.2) is usually more appropriate.

If multiple, slightly offset, images are available, then a simple approach to image super-resolution is to use the information gleaned from the multiple images to resolve the aliasing, giving an increased resolution (Tsai and Huang, 1984). As a result of aliasing, the value at each frequency sample will result from a mixture of frequencies:

$$F[u,v] = \sum_{k,l} \tilde{F}[u \pm kN, v \pm lN], \tag{11.49}$$

where $\tilde{F}$ are the samples of the continuous frequency distribution of the original unsampled image. Normally, once the frequency information has been aliased, it is impossible to distinguish between the different frequency components that have been added together. However, when an image is offset, the different aliased components will have a different phase shift (since the phase shift is proportional to frequency). Therefore, given multiple offset images, it is possible to untangle the aliasing. Consider enhancing the resolution by a factor of 2. Then, each image may be considered as

$$\begin{aligned} F_i[u,v] \approx{}& \tilde{F}[u,v]e^{j\theta_{i,0,0}} + \tilde{F}[u-N,v]e^{j\theta_{i,1,0}} \\ &+ \tilde{F}[u,v-N]e^{j\theta_{i,0,1}} + \tilde{F}[u-N,v-N]e^{j\theta_{i,1,1}}, \end{aligned} \tag{11.50}$$

where it is assumed that the additional higher order alias terms are negligible. The phase terms are determined directly from the offsets of each of the images (for example using any of the techniques of Section 10.4) and are effectively known (or can be estimated). Given four such images, then in matrix form

$$\begin{bmatrix} F_1[u,v] \\ F_2[u,v] \\ F_3[u,v] \\ F_4[u,v] \end{bmatrix} = \begin{bmatrix} e^{j\theta_{1,0,0}} & e^{j\theta_{1,1,0}} & e^{j\theta_{1,0,1}} & e^{j\theta_{1,1,1}} \\ e^{j\theta_{2,0,0}} & e^{j\theta_{2,1,0}} & e^{j\theta_{2,0,1}} & e^{j\theta_{2,1,1}} \\ e^{j\theta_{3,0,0}} & e^{j\theta_{3,1,0}} & e^{j\theta_{3,0,1}} & e^{j\theta_{3,1,1}} \\ e^{j\theta_{4,0,0}} & e^{j\theta_{4,1,0}} & e^{j\theta_{4,0,1}} & e^{j\theta_{4,1,1}} \end{bmatrix} \begin{bmatrix} \tilde{F}[u,v] \\ \tilde{F}[u-N,v] \\ \tilde{F}[u,v-N] \\ \tilde{F}[u-N,v-N] \end{bmatrix}. \tag{11.51}$$

This equation may be inverted to resolve the aliased higher frequency components. (If more images are available, a least squares inverse may be used to improve the estimate.) By padding the image with these components, and taking the inverse Fourier transform, the resolution may be improved.

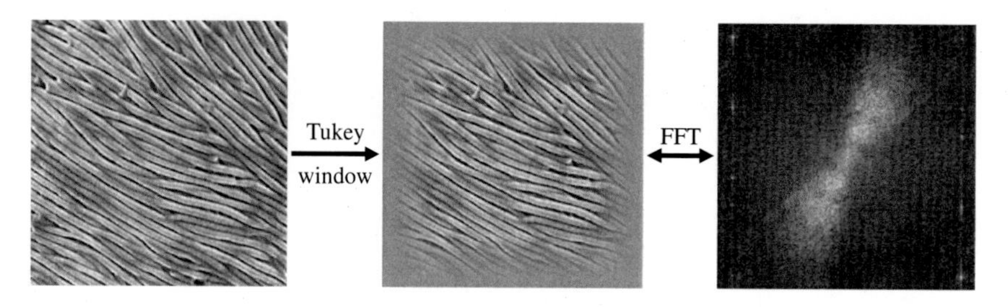

Figure 11.11 Frequency domain features: Left: input image of pollen surface; centre: windowed using a Tukey window with $\alpha = 0.25$; right: Fourier transform showing dominant frequency and angle distribution.

11.1.3.4 Feature Extraction

The frequency domain can provide a rich source of features, particularly in representing image texture. The example in Figure 11.11 shows an electron micrograph of the surface of a pollen grain. From the frequency domain, the dominant frequency and angle variation may be easily measured (Currie, 1995).

11.2 Discrete Cosine Transform (DCT)

Closely related to the Fourier transform is the discrete cosine transform (DCT). While the Fourier transform has both sine and cosine components, the DCT is made purely from cosine terms. This is achieved by enforcing even symmetry and enables the transform to be calculated with real rather than complex numbers.

Mathematically, in one dimension, the DCT is

$$C[u] = \sum_{x=0}^{N-1} \sqrt{\frac{\alpha}{N}} I[x] \cos\left(\frac{\pi u \left(x + \frac{1}{2}\right)}{N}\right), \quad \alpha = \begin{cases} 1, & u = 0, \\ 2, & \text{otherwise}, \end{cases} \tag{11.52}$$

making the transform matrix, $\mathbf{w}$, as

$$w[u,x] = \sqrt{\frac{\alpha}{N}} \cos\left(\frac{\pi u \left(x + \frac{1}{2}\right)}{N}\right). \tag{11.53}$$

The inverse is simply $\mathbf{w}^{\top}$.

Being separable, a 2-D DCT may be calculated by taking the 1-D transform of the rows, saving the results into a transpose buffer, and then use a second processor take the 1-D transform of the columns, as illustrated in Figure 11.12. For small blocks (8×8 or 16×16), the transpose buffer can be readily implemented using a dual-port block RAM. One port is used to write the results of the row transform, while the second port is used to read the values in column order. The block RAM in most FPGAs is also sufficiently large to hold two blocks of data, enabling bank-switching to be incorporated directly within the single memory simply by controlling the most significant address bit.

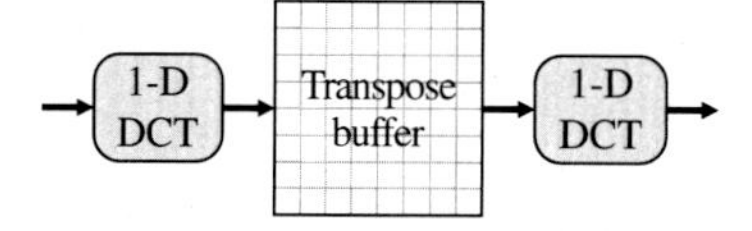

Figure 11.12 2-D discrete cosine transform of a block.

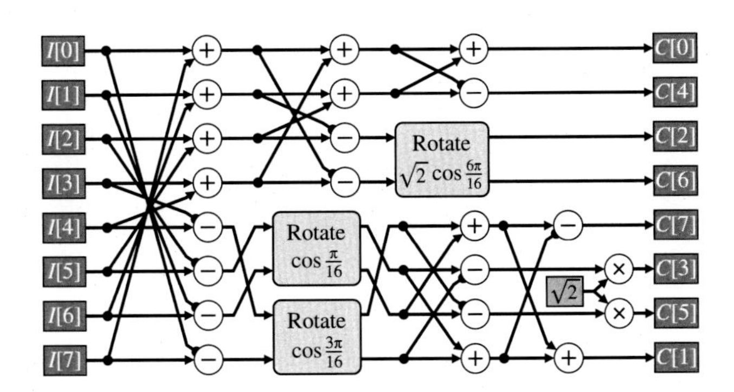

Figure 11.13 Eight-point DCT with 11 multiplications. The rotate block is a scaled rotation requiring three multiplications and additions.

While algorithms for direct computation of the 2-D transform can be developed that require fewer arithmetic operations than the separable transform (Feig and Winograd, 1992), the separable algorithm allows hardware to be reused and results in a simpler implementation for streamed data.

Since it is closely related to the Fourier transform, there are also fast algorithms for calculating the DCT, especially for powers of 2. In fact, the DCT can be calculated using an FFT of length $4N$, with the data in the odd terms (mirroring the data to give even symmetry) and setting the even terms to zero. Removing the unnecessary calculations from the resulting FFT leads to efficient algorithms for the DCT. Algorithms will be discussed here for $N = 8$ since this is the size most commonly used for image and video coding. The most efficient algorithm for an eight-point DCT requires 11 multiplications (Loeffler et al., 1989) and is shown in Figure 11.13.

If a scaled DCT is acceptable, the number of multiplications can be reduced to five (Kovac and Ranganathan, 1995). With a scaled DCT, each frequency coefficient is scaled by an arbitrary factor. When using the DCT for image compression, the coefficients are quantised; scaling the quantisation step size avoids the need for any additional operations. Agostini et al. (2001, 2005) have implemented this on an FPGA for JPEG image compression. Their implementation had a throughput of one pixel per clock cycle but with a latency of 48 clock cycles because they divided the algorithm into six blocks to share hardware. The structure of the DCT is less regular than the FFT, making an elegant pipeline less practical.

Modern FPGAs have plentiful multipliers. This enables a simple and elegant pipelined design as demonstrated in Figure 11.14. It simply multiplies the incoming pixel values by the appropriate DCT matrix coefficients and accumulates the results (Woods et al., 1998). One level of factorisation (the first layer of Figure 11.13) is used to reduce the number of multiplications from eight to four. Each multiply-and-accumulate unit is reset every four clock cycles and calculates a separate output frequency. The factorisation means that the even and odd samples are calculated separately.

Similar circuits may be used for calculating the inverse DCT. Since the inverse uses the transposed coefficient matrix, similar factorisations to the forward transform may also be used.

The application of the DCT to image and video coding will be discussed in more detail in Chapter 12.

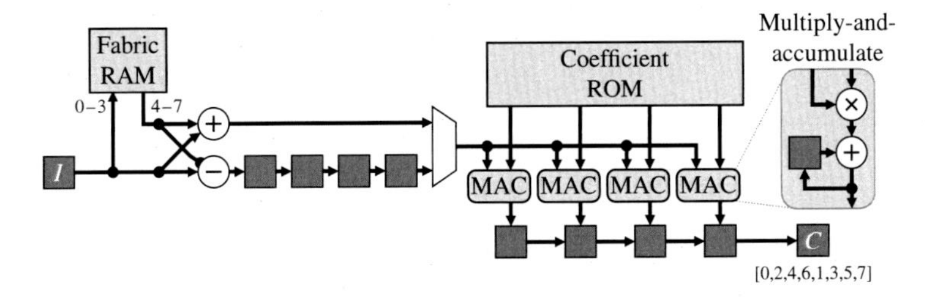

Figure 11.14 Using parallel multipliers to perform an eight-point DCT.

11.3 Wavelet Transform

One of the limitations of frequency domain transformations, such as the Fourier transform and cosine transform, is that the resolution in the frequency domain is inversely proportional to that in the image domain. This is illustrated in one dimension in Figure 11.15. Features which are narrow in space (well localised) have a wide frequency bandwidth, and those with a narrow frequency (for example a sine wave) have a wide spatial extent.

This implies that to obtain good frequency resolution, it is necessary to have a large number of samples in the image. Increasing the frequency resolution requires extending the width of the image. However, in taking the Fourier transform, the direct location of features and objects within the image is lost. Good spatial resolution of an object requires looking within a narrow window within the image, which results in poor frequency resolution.

At low frequencies, the spatial resolution is inevitably poor anyway because of the long period. However, the shorter period of high frequencies makes higher spatial resolution more desirable. This is achieved by dividing the position-frequency space (as shown in the right-hand panel of Figure 11.15) by ***wavelet analysis*** as follows. First, the image is filtered, separating the low- and high-frequency components. As a result of the reduction in bandwidth, these components can be down-sampled by a factor of 2 by discarding every second sample. This gives the position-frequency resolution aspect ratio seen in the figure for the high-frequency components. Repeating this process recursively on the low-frequency components gives the desired position-frequency resolution. The schematic of this process in one dimension is given in Figure 11.16. The wavelet transform effectively looks at the image at a range of scales. For each level of the wavelet transform, the low-pass filter outputs are often called the ***approximation components***, whereas the high-pass filter outputs are the ***detail components***.

With the Fourier transform, every frequency component depends on and is contributed to by every pixel in the image. Conversely, wavelet filters are usually local, finite impulse response filters. Being local, the filters enable spatial location to be maintained, while still separating the frequency components.

The inverse wavelet transform (or ***wavelet synthesis***) reconstructs the data by reversing the process (see Figure 11.16). First, the coefficients are up-sampled by inserting zeros between the samples. These zeros replace

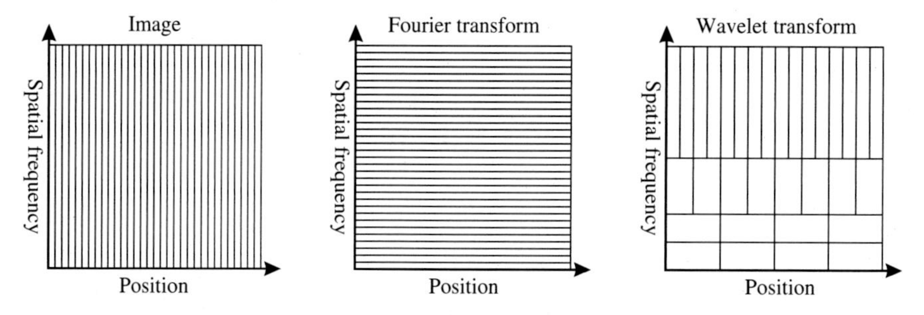

Figure 11.15 Trade-off between spatial and frequency resolution, illustrated in one dimension. Left: image samples; centre: frequency samples from Fourier transform; right: trade-off from the wavelet transform.

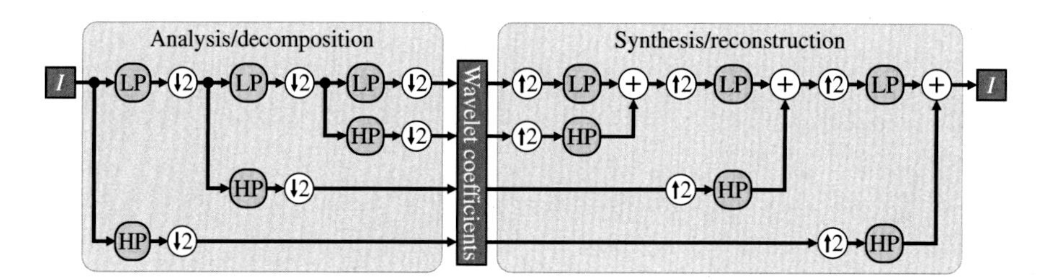

Figure 11.16 Wavelet transform in one dimension, with three decomposition levels shown. LP and HP are low- and high-pass filters, respectively.

the samples discarded in the down-sampling process. The missing samples are then effectively interpolated by the following low- or high-pass filter, depending on which band the signal came from. The components are then added to give the output signal. This is repeated going back through the levels until the final output signal is reconstructed.

Let the low- and high-pass analysis filters in the z-domain, respectively, be

$$a_L(z) = \sum_i a_L[i]z^{-i}, \qquad a_H(z) = \sum_i a_H[i]z^{-i}, \tag{11.54}$$

where $a[i]$ are the filter coefficients, and z^{-1} corresponds to a one sample delay. Similarly, let the synthesis filters be $s_L(z)$ and $s_H(z)$. Since the filters are not ideal low- and high-pass filters, down-sampling will introduce aliasing within the resultant components. However, given appropriately matched filters, the aliasing introduced by each filter is exactly cancelled by the reconstruction filters. These two conditions can be expressed mathematically as (Daubechies and Sweldens, 1998)

$$s_L(z)a_L(z) + s_H(z)a_H(z) = 2, \tag{11.55}$$

$$s_L(z)a_L(-z) + s_H(z)a_H(-z) = 0. \tag{11.56}$$

These imply that the analysis and synthesis filters are related by (Daubechies and Sweldens, 1998)

$$s_L(z) = \mp z^d a_H(-z), \qquad s_H(z) = \pm z^d a_L(-z), \tag{11.57}$$

where d is an integer constant to satisfy the delay requirements of Eq. (11.55).

While it is possible for the analysis and synthesis filters to be the same (apart from delays), and indeed many wavelet families do use the same filters for analysis and synthesis, such filters cannot be symmetrical. With image processing, use of symmetric filters ($a(z) = a(z^{-1})$) is preferred because they prevent movement of features enabling coefficients to be directly related to the position of features within the image. For this reason, the symmetric biorthogonal family of wavelet filters is often used in image processing applications.

Wavelets, unlike the Fourier transform, do not assume anything beyond the end of the data. They are localised, so any edge effects will be localised to the few pixels on the edge of the image. However, since the finite impulse response filter implementing the wavelet transform is a convolution, the output has more samples than the input (it grows at the borders). Since the biorthogonal filters are symmetric, a symmetric extension of the input image results in a symmetric response. Therefore, the data beyond the image borders does not need to be recorded, making the output sequence the same length as the input. The mirroring with duplication technique outlined in Section 9.1.1 can be used to extend the input image.

Wavelets can readily be extended to two dimensions by applying the filter separably to the rows and columns. Each level divides the image into four bands: the LL (low-low) approximation band and HL (high-low), LH (low-high), and HH (high-high) detail bands. The 2-D transform is applied recursively to the LL band, as illustrated in Figure 11.17 for a three-level decomposition. In practice, the high- and low-pass filters are not

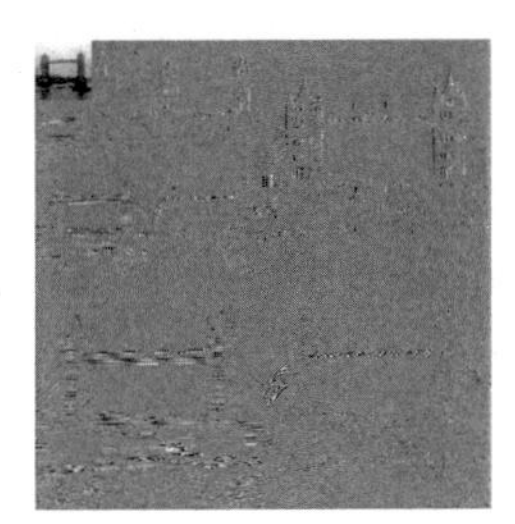

LL_3 HL_3 LH_3 HH_3 HL_2 LH_2 HH_2 HL_1 LH_1 HH_1

Figure 11.17 Two-dimensional wavelet transform of an image. Left: input image; centre: wavelet transform (using the Haar wavelet); right: position of the decomposition components for each level. Source: Photo courtesy of Robyn Bailey.

separated as shown here. Since the filters are implemented in parallel, the outputs are more often interleaved, with the even samples corresponding to the low-pass filter outputs and the odd samples corresponding to the high-pass outputs.

11.3.1 Filter Implementations

11.3.1.1 Direct

The obvious direct approach of implementing the wavelet filters is shown in Figure 11.18. Separate filters are used for analysis and synthesis (they are shown together here for illustration). Down-sampling is achieved by enabling the clock on the filter output registers, L and H, only every second clock cycle. The input is therefore one sample per clock cycle, with the output two samples (one from each of the low- and high-pass filters) every two clock cycles (these can be multiplexed together if required). The corresponding synthesis filter up-samples the wavelet signals by inserting zeros between the samples. Here, a transpose structure is used so that only a single set of registers is required.

Many biorthogonal filters have irrational coefficients. Quantisation of the filter coefficients to enable a fixed-point realisation will generally mean that the filters no longer satisfy Eqs. (11.55) and (11.56); consequently, the reconstructed output will not necessarily match the input exactly. Tay (2000) describes a technique where minor modifications may be made to the filter coefficients to make them rational without violating Eqs. (11.55) and (11.56).

11.3.1.2 Polyphase Filters

The main inefficiency of direct filters is that every second sample calculated by the analysis filter is discarded, and every second sample into the synthesis filter is a zero, so the corresponding operations are redundant. A polyphase representation exploits these redundancies to reduce the computation by a factor of 2. Rearranging Eq. (11.54) to separate the even and odd components,

$$\begin{aligned} a(z) &= \sum_i a[i]z^{-i} = \sum_i a[2i]z^{-2i} + \sum_i a[2i+1]z^{-(2i+1)} \\ &= a_e(z^2) + z^{-1}a_o(z^2), \end{aligned} \tag{11.58}$$

enables the component filters to operate at the lower sample rate, halving the number of operations. The basic idea is shown in Figure 11.19. Note, only a single filter (low or high pass) is shown in Figure 11.19; two such circuits are required, one for each of the low- and high-pass filters.

For the analysis filter, the multiplications only need to be evaluated every second cycle. This allows the hardware to be shared by calculating the even and odd filters in alternate clock cycles as shown in Figure 11.19. The result of the odd filter is held for one cycle and added to that from the even filter.

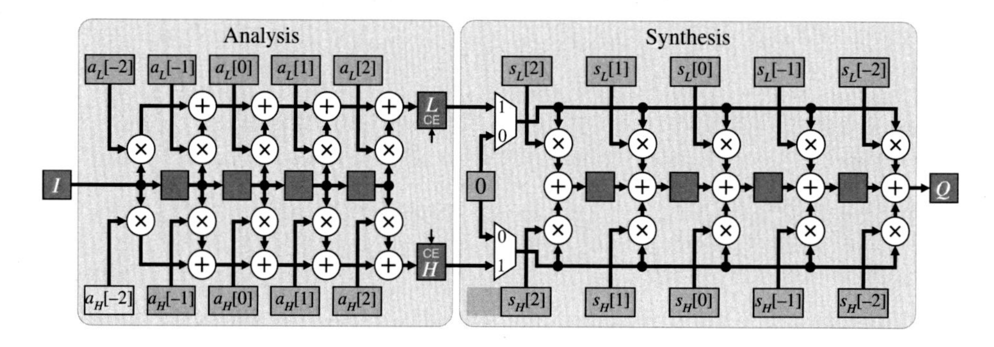

Figure 11.18 Direct implementation of the analysis and synthesis wavelet filters. (CE = clock enable.)

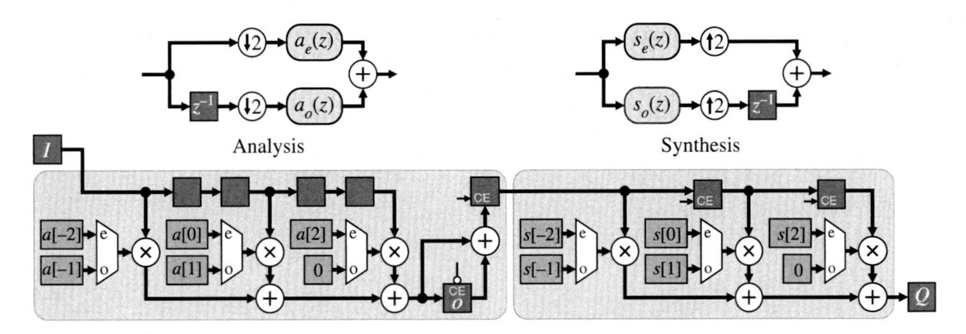

Figure 11.19 Polyphase filter architecture. Top: basic idea of polyphase filters; bottom: hardware implementation. (CE = clock enable.)

For the synthesis filter, the incoming data is only clocked every second cycle. On the output, the results are produced alternately by the even and odd filters. This enables a single filter to be used, by just multiplexing the coefficients.

For a given wavelet, the filter coefficients are constant. Therefore, the multiplexers selecting the filter coefficients can be optimised away, resulting in a very efficient implementation.

11.3.1.3 Lifting

The filtering can be optimised further by exploiting the fact that the low- and high-pass filters are pair-wise complementary. The polyphase filters may be written in matrix form:

$$\begin{bmatrix} L \\ H \end{bmatrix} = \begin{bmatrix} a_{Le}(z^2) & a_{Lo}(z^2) \\ a_{He}(z^2) & a_{Ho}(z^2) \end{bmatrix} \begin{bmatrix} I_e \\ z^{-1}I_o \end{bmatrix}, \tag{11.59}$$

where I_e and I_o are the even- and odd-numbered input samples, respectively. The relationship between the filters allows this filter matrix to be factorised (Daubechies and Sweldens, 1998):

$$\begin{bmatrix} a_{Le}(z) & a_{Lo}(z) \\ a_{He}(z) & a_{Ho}(z) \end{bmatrix} = \begin{bmatrix} k & 0 \\ 0 & k^{-1} \end{bmatrix} \prod_i \begin{bmatrix} a_i(z) & 1 \\ 1 & 0 \end{bmatrix}, \tag{11.60}$$

where k is a constant, and each of the $a_i(z)$ is usually a constant or first-order filter. Note that the factorisation is not unique. Each pair of terms on the right can be converted into a pair of lifting steps:

$$\begin{bmatrix} a_{i+1}(z) & 1 \\ 1 & 0 \end{bmatrix} \begin{bmatrix} a_i(z) & 1 \\ 1 & 0 \end{bmatrix} = \begin{bmatrix} 1 & a_{i+1}(z) \\ 0 & 1 \end{bmatrix} \begin{bmatrix} 1 & 0 \\ a_i(z) & 1 \end{bmatrix}, \tag{11.61}$$

resulting in Figure 11.20. In general, using an appropriate lifting scheme will reduce the number of multiplications by a further factor of 2 over a polyphase implementation.

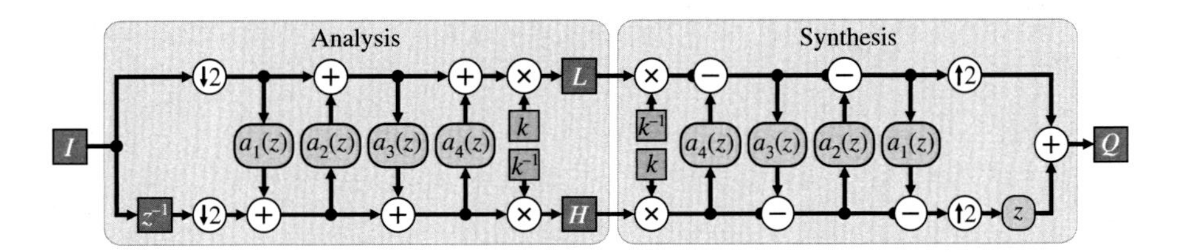

Figure 11.20 Lifting implementation of wavelet filters.

Figure 11.21 5/3 biorthogonal wavelet analysis filters. Left: schematic of lifting as calculated in Eq. (11.63). Right: realisation of the corresponding filters. (CE = clock enable.)

A second advantage of lifting is that even with quantisation of the filter coefficients and rounding of the outputs of the $a_i(z)$ filters, the output is perfectly recoverable. This is because the synthesis filter reverses the steps of the analysis filter, subtracting off the terms that were added in. Any quantisation within the synthesis filter will be exactly the same as that in the analysis filter, recreating the data back along the filter chain.

In this context, the scaling terms (by k and k^{-1} if required) can also be implemented using lifting, to enable rounding errors from the scaling to also be recovered exactly (Daubechies and Sweldens, 1998). If wavelet coding is being used for lossy image compression, the L and H outputs will be quantised. In this case, the scaling terms can be incorporated into the quantisation stage without any additional operations.

Consider the 5/3 biorthogonal wavelet analysis filters, split into polyphase components:

$$\begin{aligned} a_L(z) &= \frac{1}{8}(-z^{-2} + 2z^{-1} + 6 + 2z - z^2) & a_H(z) &= \frac{1}{2}(-z^{-2} + 2z^{-1} - 1) \\ &= \left(\frac{-z^{-2} + 6 - z^2}{8}\right) + z^{-1}\left(\frac{1 + z^2}{4}\right), & &= \left(-\frac{z^{-2} + 1}{2}\right) + z^{-1}. \end{aligned} \tag{11.62}$$

One factorisation of the polyphase matrix into lifting terms gives

$$\begin{bmatrix} \frac{1}{8}(-z^{-1} + 6 - z) & \frac{1}{4}(1 + z) \\ -\frac{1}{2}(z^{-1} + 1) & 1 \end{bmatrix} = \begin{bmatrix} 1 & \frac{1}{4}(1 + z) \\ 0 & 1 \end{bmatrix} \begin{bmatrix} 1 & 0 \\ -\frac{1}{2}(z^{-1} + 1) & 1 \end{bmatrix}, \tag{11.63}$$

with the implementation shown in Figure 11.21. Note that z is one sample advance; it is realised by inserting a delay in all parallel branches.

A more complex example is the 9/7 biorthogonal wavelet, shown in Figure 11.22. (Note that all of the registers are clocked only every second cycle.)

One limitation of using lifting for implementing the wavelet transform is that all of the operations are now in series rather than in parallel as with the direct and polyphase implementations. All of the multiplications are also in series on the critical path. Huang et al. (2004) addressed this problem by replacing one or more multiplications with multiplication by the inverse on the adjacent parallel paths. One implementation which removes all but one of the multiplications and half of the additions from the critical path is shown in the bottom of Figure 11.22. Note that this implementation loses the prefect reconstruction property of the standard lifting scheme.

Rather than rearrange the circuit, the latency can be accepted, with pipelining used to maintain the throughput. All of the filtering is performed after down-sampling, so the filters only process one sample every second clock cycle. Each stage of the filter has the same architecture, allowing the filter to be half the length with the second half using the second clock cycle. Alternatively, the full filter could be implemented with two rows (or columns) transformed in parallel if the input bandwidth allows. Another possibility is to feed the low-pass output back to the input and use the spare cycles for subsequent levels within the decomposition (Liao et al., 2004).

11.3.1.4 Two-dimensional Implementations

The descriptions above were for 1-D filters. When extending to two dimensions, three approaches are commonly used. The simplest approach is to alternate the row and column processing, using a transpose buffer to hold the

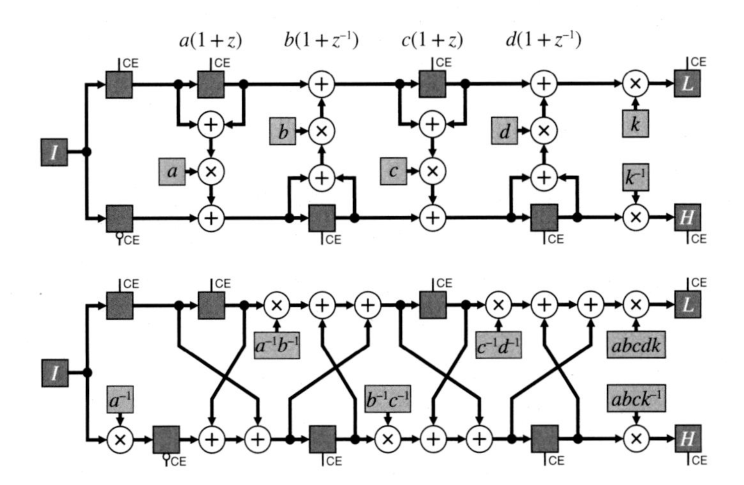

Figure 11.22 9/7 biorthogonal wavelet analysis filter. Top: hardware implementation; bottom: removing multipliers from the critical path. $a = -1.586134342$, $b = -0.052980119$, $c = 0.882911076$, $d = 0.443506852$, and $k = 1.149604398$. (CE = clock enable.)

intermediate results. This requires processing all of the rows first, saving the intermediate results in off-chip RAM. The columns are then processed, returning the results to off-chip memory. This procedure is then repeated for the LL band for each level of decomposition. Since each row and column are processed independently, the direct separable approach is easy to accelerate through multiple parallel processors, subject of course to memory bandwidth limitations.

A second approach is to begin the column processing as the processed row data becomes available. This requires replacing delays within the filter with on-chip row buffers, enabling all of the columns to be processed in parallel (Chrysafis and Ortega, 2000). This fits the stream processing model and completely transforms the image as it is being streamed into the FPGA. If the wavelet transform is the end of the processing, the wavelet coefficients can be streamed to off-chip memory as they are calculated. However, if further processing is required, additional caching is required to hold the results from the lower decomposition levels while the transformation is being performed on the higher levels. The size of the synchronisation memory grows rapidly with the number of decomposition levels (Chrysafis and Ortega, 2000). The line processing approach is also a little harder to parallelise. Similar to the separable approach, multiple rows may be processed in parallel. Processing rows in pairs will also provide the data for column processing two samples at a time. This makes full use of the available clock cycles on the down-sampled data. Alternatively, the spare cycles may be used for processing subsequent decomposition levels (Liao et al., 2004). One aspect to consider is rather than multiplexing individual registers, using dual-port fabric RAM builds in the multiplexers for free and may reduce the resources required.

The on-chip memory requirements of the line processing approach may be reduced by dividing the image into blocks and transforming the image one block at a time. The input blocks need to overlap because data from adjacent blocks is required to correctly calculate the pixels around the block boundaries. The block processing approach is also readily parallelisable, although coordination is required when accessing those pixels which are shared between blocks. However, if the resources allow, it is simpler to use line processing rather than process multiple blocks in parallel.

Angelopoulou et al. (2008) compared these three methods for FPGA implementation. The direct separable approach used the fewest resources because it was the simplest architecture but also took the longest because the data must be accessed multiple times. The best in terms of total time and power was the line buffered approach. Block-based processing has the smallest memory requirement but the most complex logic because of the need to manage the block boundaries.

11.3.2 Applications

As with the Fourier transform, wavelets can be applied to many applications within image processing. While a full discussion is beyond the scope of this book, some of the main applications will be introduced in this section.

One of the most significant applications of the wavelet transform is in image compression. The wavelet transform is good at exploiting correlation within an image, with many of the detail coefficients being zero or close to zero. As an application, image coding is explored in more detail in Chapter 12.

Wavelets can provide the basis for image filtering. One application of note is reducing noise. While noise will affect all of the wavelet coefficients, it will be most noticeable in the detail coefficients which are generally sparse. Setting the coefficients below a threshold to zero will remove the noise from those terms (see Figure 11.23). When the image is reconstructed with an inverse wavelet transform, the noise will be reduced. This is called hard thresholding and has the characteristic that it preserves features such as peak heights (Donoho and Johnstone, 1994). An alternative approach is soft thresholding, where all of the wavelet coefficients are shrunk towards zero. Soft thresholding will tend to give a smoother fit and not result in the same level of lumpiness that normally results from smoothing noise (Donoho, 1995). The threshold for both methods is proportional to the noise standard deviation in the image. These simple approaches assume that all coefficients are independent. However, there is a strong correlation between detail coefficients at different levels of decomposition. More sophisticated denoising techniques take into consideration these correlations (Sendur and Selesnick, 2002), giving better noise reduction.

Multi-scale processing has been mentioned in several contexts throughout this book. Wavelets provide a natural framework for multi-scale object detection (Strickland and Hahn, 1997).

A particular application of multi-scale processing is the fusion of images from different sources. Image fusion combines the data from two or more registered images in a way that provides a higher quality (in some sense of the word) output image. This can vary from simple averaging to reduce noise, through to combining data from disparate imaging modalities (for example positron emission tomography and magnetic resonance images, or multiple spectral bands), and image super-resolution, where the output data has higher resolution than the input. In the context of wavelet processing, each of the registered input images is decomposed through the wavelet transform, the wavelet coefficients are then merged in some way, and then inverse wavelet transformed (Pajares and de la Cruz, 2004).

If the coefficients are simply averaged, then the wavelet transform (being linear) is not necessary unless the images are of different resolution. One method of merging the wavelet coefficients that is surprisingly effective is to simply select the coefficient with the maximum absolute value from the input images. This can readily be extended to filtering within the wavelet domain before selecting or using consistency verification to ensure that groups of adjacent coefficients are consistently selected from the same image (Pajares and de la Cruz, 2004). A classic example of image fusion is focus or depth-of-field extension in microscopy of extended objects. With a thick object, the complete object cannot be brought into focus simultaneously; as the focus is adjusted, different parts of the object come into and out of focus. The in-focus components of each image will have higher local contrast, resulting in a larger magnitude within the detail coefficients. Selecting the components with the maximum absolute values will select the parts of the image that are most in focus within the set, with the fused output image having an extended focus range (Forster et al., 2004; Ambikumar et al., 2016).

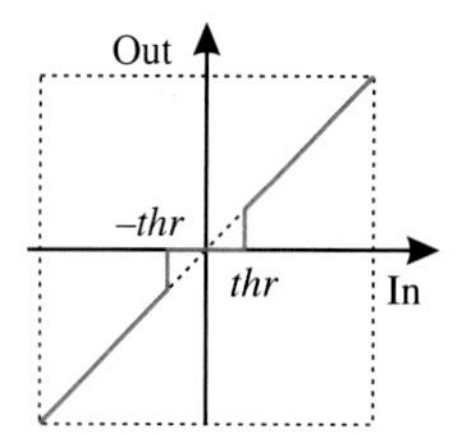

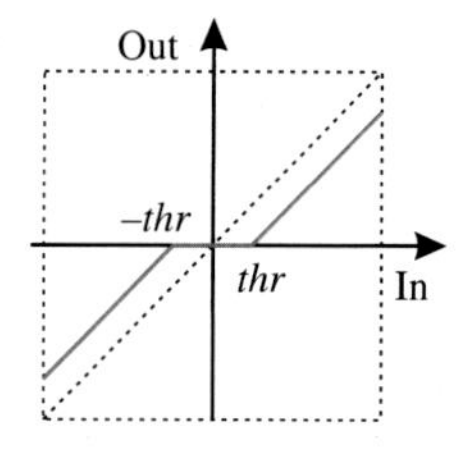

Figure 11.23 Thresholding of wavelet coefficients for denoising. Left: hard thresholding; right: soft thresholding.

11.4 Summary

Linear transformation rearranges the image as a weighted sum of basis functions or images. The goal is to separate different components of an image to improve analysis or to exploit redundancy. A disadvantage of global image transformation is that the results are only available after the whole image has been transformed, which increases latency and has storage implications. However, many applications are simplified by working within the transform domain.

Wavelets provide a natural decomposition for multi-scale processing. This makes them good for feature detection and image fusion.

The Fourier transform is good for filtering images, because it transforms convolution into a multiplication. Filtering is therefore simply a scaling of each of the frequency coefficients.

The discrete cosine and wavelet transforms are good at decorrelating image data, concentrating the energy into a small number of terms, making them good for compression.

References

Agostini, L.V., Silva, I.S., and Bampi, S. (2001). Pipelined fast 2D DCT architecture for JPEG image compression. *14th Symposium on Integrated Circuits and Systems Design*, Pirenopolis, Brazil (10–15 September 2001), 226–231. https://doi.org/10.1109/SBCCI.2001.953032.

Agostini, L.V., Porto, R.C., Bampi, S., and Silva, I.S. (2005). A FPGA based design of a multiplierless and fully pipelined JPEG compressor. *8th Euromicro Conference on Digital System Design*, Porto, Portugal (30 August – 3 September 2005), 210–213. https://doi.org/10.1109/DSD.2005.6.

Ambikumar, A.S., Bailey, D.G., and Sen Gupta, G. (2016). Extending the depth of field in microscopy: a review. *Image and Vision Computing New Zealand (IVCNZ)*, Palmerston North, NZ (21–22 November 2016), 185–190. https://doi.org/10.1109/IVCNZ.2016.7804448.

Angelopoulou, M.E., Cheung, P.Y.K., Masselos, K., and Andreopoulos, Y. (2008). Implementation and comparison of the 5/3 lifting 2D discrete wavelet transform computation schedules on FPGAs. *Journal of Signal Processing Systems* **51** (1): 3–21. https://doi.org/10.1007/s11265-007-0139-5.

Bailey, D.G. (1993). Frequency domain self-filtering for pattern detection. *Proceedings of the 1st New Zealand Conference on Image and Vision Computing*, Auckland, NZ (16–18 August 1993), 237–243.

Bailey, D.G. (1997). Detecting regular patterns using frequency domain self-filtering. *IEEE International Conference on Image Processing*, Santa Barbara, CA, USA (26–29 October 1997), Volume **1**, 440–443. https://doi.org/10.1109/ICIP.1997.647801.

Bailey, D.G., Mahmood, F., and Skoglund, U. (2017). Reducing the cost of removing border artefacts in Fourier transforms. *International Symposium on Highly Efficient Accelerators and Reconfigurable Technologies (HEART 2017)*, Bochum, Germany (7–9 June 2017), Article ID 11, 6 pages. https://doi.org/10.1145/3120895.3120899.

Bingham, C., Godfrey, M., and Tukey, J. (1967). Modern techniques of power spectrum estimation. *IEEE Transactions on Audio and Electroacoustics* **15** (2): 56–66. https://doi.org/10.1109/TAU.1967.1161895.

Bracewell, R.N. (2000). *The Fourier Transform and its Applications*, 3e. New York: McGraw Hill.

Chang, W. and Nguyen, T. (2007). Integer FFT with optimized coefficient sets. *IEEE International Conference on Acoustics, Speech and Signal Processing - ICASSP '07*, Honolulu, HI, USA (15–20 April 2007), Volume **2**, 109–112. https://doi.org/10.1109/ICASSP.2007.366184.

Chang, W. and Nguyen, T.Q. (2008). On the fixed point accuracy analysis of FFT algorithms. *IEEE Transactions on Signal Processing* **56** (10): 4673–4682. https://doi.org/10.1109/TSP.2008.924637.

Chrysafis, C. and Ortega, A. (2000). Line-based, reduced memory, wavelet image compression. *IEEE Transactions on Image Processing* **9** (3): 378–389. https://doi.org/10.1109/83.826776.

Currie, A.J. (1995). Differentiating apple sports by Pollen ultrastructure. Master of Horticultural Science thesis. Palmerston North, NZ: Massey University. https://doi.org/10179/12068.

Daubechies, I. and Sweldens, W. (1998). Factoring wavelet transforms into lifting steps. *Journal of Fourier Analysis and Applications* **4** (3): 247–269. https://doi.org/10.1007/BF02476026.

Despain, A.M. (1974). Fourier transform computers using CORDIC iterations. *IEEE Transactions on Computers* **23** (10): 993–1001. https://doi.org/10.1109/T-C.1974.223800.

Donoho, D.L. (1995). De-noising by soft-thresholding. *IEEE Transactions on Information Theory* **41** (3): 613–627. https://doi.org/10.1109/18.382009.

Donoho, D.L. and Johnstone, I.M. (1994). Threshold selection for wavelet shrinkage of noisy data. *16th Annual International Conference of the IEEE Engineering in Medicine and Biology Society*, Baltimore, MD, USA (3–6 November 1994), Volume **1**, A24–A25. https://doi.org/10.1109/IEMBS.1994.412133.

Dubois, E. and Venetsanopoulos, A. (1978). A new algorithm for the radix-3 FFT. *IEEE Transactions on Acoustics, Speech, and Signal Processing* **26** (3): 222–225. https://doi.org/10.1109/TASSP.1978.1163084.

Feig, E. and Winograd, S. (1992). Fast algorithms for the discrete cosine transform. *IEEE Transactions on Signal Processing* **40** (9): 2174–2193. https://doi.org/10.1109/78.157218.

Forster, B., Van De Ville, D., Berent, J., Sage, D., and Unser, M. (2004). Complex wavelets for extended depth-of-field: a new method for the fusion of multichannel microscopy images. *Microscopy Research and Technique* **65** (1–2): 33–42. https://doi.org/10.1002/jemt.20092.

Garrido, M. (2022). A survey on pipelined FFT hardware architectures. *Journal of Signal Processing Systems* **94**: 1345–1364. https://doi.org/10.1007/s11265-021-01655-1.

Goertzel, G. (1958). An algorithm for the evaluation of finite trigonometric series. *The American Mathematical Monthly* **65** (1): 34–35.

Harris, F.J. (1978). On the use of windows for harmonic analysis with the discrete Fourier transform. *Proceedings of the IEEE* **66** (1): 51–83. https://doi.org/10.1109/PROC.1978.10837.

He, S. and Torkelson, M. (1996). A new approach to pipeline FFT processor. *10th International Parallel Processing Symposium (IPPS'96)*, Honolulu, HI, USA (15–19 April 1996), 766–770. https://doi.org/10.1109/IPPS.1996.508145.

Huang, C.T., Tseng, P.C., and Chen, L.G. (2004). Flipping structure: an efficient VLSI architecture for lifting-based discrete wavelet transform. *IEEE Transactions on Signal Processing* **52** (4): 1080–1089. https://doi.org/10.1109/TSP.2004.823509.

Kitsakis, V., Nakos, K., Reisis, D., and Vlassopoulos, N. (2018). Parallel memory accessing for FFT architectures. *Journal of Signal Processing Systems* **90** (11): 1593–1607. https://doi.org/10.1007/s11265-018-1387-2.

Kovac, M. and Ranganathan, N. (1995). JAGUAR: a fully pipelined VLSI architecture for JPEG image compression standard. *Proceedings of the IEEE* **83** (2): 247–258. https://doi.org/10.1109/5.364464.

Kumar, G.G., Sahoo, S.K., and Meher, P.K. (2019). 50 years of FFT algorithms and applications. *Circuits, Systems, and Signal Processing* **38** (12): 5665–5698. https://doi.org/10.1007/s00034-019-01136-8.

Lee, S. and Park, S. (2007). Modified SDF architecture for mixed DIF/DIT FFT. *IEEE International Symposium on Circuits and Systems*, New Orleans, LA, USA (27–30 May 2007), 2590–2593. https://doi.org/10.1109/ISCAS.2007.377845.

Liao, H., Mandal, M.K., and Cockburn, B.F. (2004). Efficient architectures for 1-D and 2-D lifting-based wavelet transforms. *IEEE Transactions on Signal Processing* **52** (5): 1315–1326. https://doi.org/10.1109/TSP.2004.826175.

Loeffler, C., Ligtenberg, A., and Moschytz, G.S. (1989). Practical fast 1-D DCT algorithms with 11 multiplications. *International Conference on Acoustics, Speech, and Signal Processing (ICASSP-89)*, Glasgow, Scotland (23–26 May 1989), Volume **2**, 988–991. https://doi.org/10.1109/ICASSP.1989.266596.

Lofgren, J. and Nilsson, P. (2011). On hardware implementation of radix 3 and radix 5 FFT kernels for LTE systems. *NORCHIP*, Lund, Sweden (14–15 November 2011), 4 pages. https://doi.org/10.1109/NORCHP.2011.6126703.

Mahmood, F., Toots, M., Ofverstedt, L.G., and Skoglund, U. (2015). 2D discrete Fourier transform with simultaneous edge artifact removal for real-time applications. *International Conference on Field Programmable Technology (FPT)*, Queenstown, NZ (7–9 December 2015), 236–239. https://doi.org/10.1109/FPT.2015.7393157.

Millane, R.P., Alzaidi, S., and Hsiao, W.H. (2003). Scaling and power spectra of natural images. *Image and Vision Computing New Zealand (IVCNZ'03)*, Palmerston North, NZ (26–28 November 2003), 148–153.

Moisan, L. (2011). Periodic plus smooth image decomposition. *Journal of Mathematical Imaging and Vision* **39** (2): 161–179. https://doi.org/10.1007/s10851-010-0227-1.

Nash, J.G. (2014). High-throughput programmable systolic array FFT architecture and FPGA implementations. *International Conference on Computing, Networking and Communications (ICNC)*, Honolulu, HI, USA (3–6 February 2014), 878–884. https://doi.org/10.1109/ICCNC.2014.6785453.

Oraintara, S., Chen, Y.J., and Nguyen, T.Q. (2002). Integer fast Fourier transform. *IEEE Transactions on Signal Processing* **50** (3): 607–618. https://doi.org/10.1109/78.984749.

Pajares, G. and de la Cruz, J.M. (2004). A wavelet-based image fusion tutorial. *Pattern Recognition* **37** (9): 1855–1872. https://doi.org/10.1016/j.patcog.2004.03.010.

Rabiner, L.R., Schafer, R.W., and Rader, C.M. (1969). The chirp z-transform algorithm and its application. *Bell System Technical Journal* **48** (5): 1249–1292. https://doi.org/10.1002/j.1538-7305.1969.tb04268.x.

Saeed, A., Elbably, M., Abdelfadeel, G., and Eladawy, M.I. (2009). Efficient FPGA implementation of FFT / IFFT processor. *International Journal of Circuits, Systems and Signal Processing* **3** (3): 103–110.

Sansaloni, T., Perez-Pascual, A., and Valls, J. (2003). Area-efficient FPGA-based FFT processor. *Electronics Letters* **39** (19): 1369–1370. https://doi.org/10.1049/el:20030892.

Sendur, L. and Selesnick, I.W. (2002). Bivariate shrinkage functions for wavelet-based denoising exploiting interscale dependency. *IEEE Transactions on Signal Processing* **50** (11): 2744–2756. https://doi.org/10.1109/TSP.2002.804091.

Strickland, R.N. and Hahn, H.I. (1997). Wavelet transform methods for object detection and recovery. *IEEE Transactions on Image Processing* **6** (5): 724–735. https://doi.org/10.1109/83.568929.

Sukhsawas, S. and Benkrid, K. (2004). A high-level implementation of a high performance pipeline FFT on Virtex-E FPGAs. *IEEE Computer Society Annual Symposium on VLSI*, Lafayette, LA, USA (19–20 February 2004), 229–232. https://doi.org/10.1109/ISVLSI.2004.1339538.

Tay, D.B.H. (2000). Rationalizing the coefficients of popular biorthogonal wavelet filters. *IEEE Transactions on Circuits and Systems for Video Technology* **10** (6): 998–1005. https://doi.org/10.1109/76.867939.

Tsai, R.Y. and Huang, T.S. (1984). Multiframe image restoration and registration. In: *Advances in Computer Vision and Image Processing*, vol. **1** (ed. T. Huang), 317–339. Greenwich, CT: JAI Press.

Uzun, I.S., Amira, A., and Bouridane, A. (2005). FPGA implementations of fast Fourier transforms for real-time signal and image processing. *IEE Proceedings - Vision, Image and Signal Processing* **152** (3): 283–296. https://doi.org/10.1049/ip-vis:20041114.

Winograd, S. (1978). On computing the discrete Fourier transform. *Mathematics of Computation* **32**: 175–199. https://doi.org/10.1090/S0025-5718-1978-0468306-4.

Woods, R., Trainor, D., and Heron, J.P. (1998). Applying an XC6200 to real-time image processing. *IEEE Design and Test of Computers* **15** (1): 30–38. https://doi.org/10.1109/54.655180.

Zohar, S. (1981). Winograd's discrete Fourier transform algorithm. In: *Two-Dimensional Digital Signal Processing II: Transforms and Median Filters, Topics in Applied Physics*, vol. **43** (ed. T. Huang), 89–160. Berlin, Germany: Springer-Verlag. https://doi.org/10.1007/BFb0057596.

12

Image and Video Coding

The information content within images and video may be compressed to reduce the volume of data for storage, reduce the bandwidth required to transmit data from one point to another (for example over a network), or minimise the power required by the system (both for compression and data transmission). Compression is only possible because images contain significant redundant information. There are at least four types of redundancy within images that can be exploited for compression:

Spatial redundancy results from the high correlation between adjacent pixels. Since adjacent pixels are likely to come from the same object, they are likely to have similar pixel value or colour.

Temporal redundancy comes from the high correlation between the frames of a video sequence. Successive frames tend to be very similar, especially if there is limited movement within the sequence. If any movement can be estimated, then the correlation can be increased further by compensating for the motion.

Spectral redundancy reflects the correlation between the colour components of a standard RGB image, especially when looking at natural scenes. Within hyperspectral images, there is a strong correlation between adjacent spectral bands.

Psychovisual redundancy results from the fact that the human visual system has limited spatial, temporal, and intensity resolution. It is tolerant of errors or noise, especially within textured regions.

A codec (coder–decoder) can be categorised as either lossless or lossy. Lossless codecs reconstruct the exact input pixel values and therefore do not make use of psychovisual redundancy. Lossy codecs only recover an approximation of the original image; the loss is commonly quantified using peak signal-to-noise ratio (PSNR) (Eq. (4.8)), although this does not always correspond to the perceived visual quality.

While there are many different codecs, for still image compression, the most commonly used are PNG (portable network graphics) and the JPEG (Joint Photographic Experts Group) family (including JPEG 2000 and JPEG-LS). For video, the most common are the MPEG (Motion Picture Experts Group) family, many of which also have International Telecommunication Union (ITU) standards (numbered H.26x). Examples are MPEG-2 (H.262), ***advanced video coding*** (H.264 AVC, MPEG-4 part 10), ***high-efficiency video coding*** (H.265 HEVC, MPEG-H), and ***versatile video coding*** (H.266 VVC, MPEG-I). New standards are continually being developed, particularly for video compression. This has in part been driven by the rise in Internet streaming and has led to the release of open-source codecs (Punchihewa and Bailey, 2020). One advantage of using field-programmable gate arrays (FPGAs) for video compression is the ability to update codecs as new techniques and standards emerge. Compression is also computationally intensive, requiring use of hardware to exploit parallelism, especially in real-time applications.

Design for Embedded Image Processing on FPGAs, Second Edition. Donald G. Bailey.

Companion Website: www.wiley.com/go/bailey/designforembeddedimageprocessc2e

12.1 Compression Techniques

Redundancy within the image can be exploited using correlations to predict the pixel values. The resulting residuals, or uncorrelated components, require fewer bits to represent, enabling image data to be compressed. The basic operations commonly used in compressing an image or video sequence are listed in Figure 12.1.

12.1.1 Colour Conversion

Colour images are usually converted from RGB to YCbCr or similar colour space for compression for two reasons. First, most of the variation within images is in intensity. Therefore, the conversion will concentrate most of the signal energy into the luminance component. The weights used for Y component correspond to the relative sensitivities of the human visual system. Second, the lower colour spatial resolution of the human visual system enables the chrominance components to be down-sampled by a factor of 2 without significant loss of subjective quality.

The common sub-sampling formats for YCbCr images are shown in Figure 12.2. For example, in the 4 : 2 : 2 format, there is a single Cb and Cr value for each 2×1 block of Y values. To reproduce the colour signal, the lower resolution Cb and Cr values are combined with multiple luminance values. Within codecs, the 4 : 2 : 0 format is most commonly used. With wavelet-based compression, this down-sampling is managed by discarding the top level of the chrominance components.

Alternative colour conversions are the reversible colour transform (RCT) of Eq. (7.51), used by JPEG 2000 (ISO, 2000), and YCoCg of Eq. (7.54), used by H.264 AVC (Malvar et al., 2008).

12.1.2 Prediction and Transformation

Prediction uses the pixels already available to predict the value of the next pixel (or block of pixels). The residual, or prediction error, has a tighter distribution enabling it to be encoded with fewer bits. It is commonly

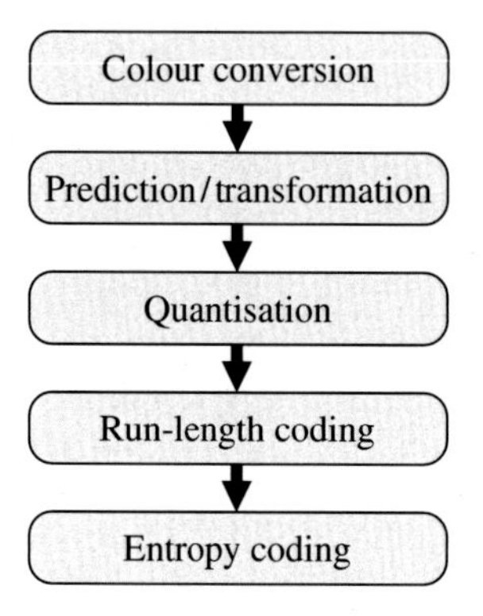

Figure 12.1 Steps within image or video compression.

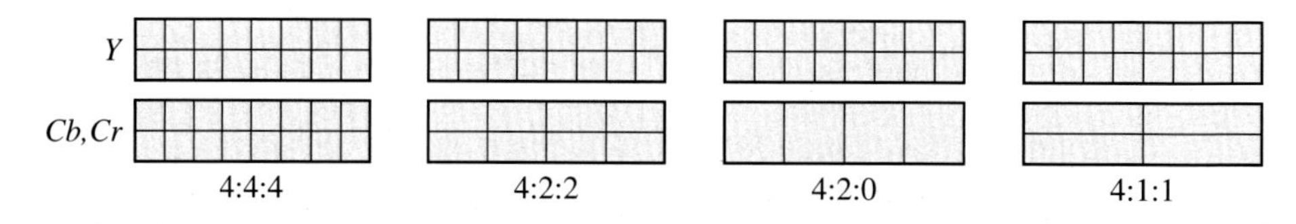

Figure 12.2 Common YCbCr sub-sampling formats, illustrated using an 8×2 block of pixels.

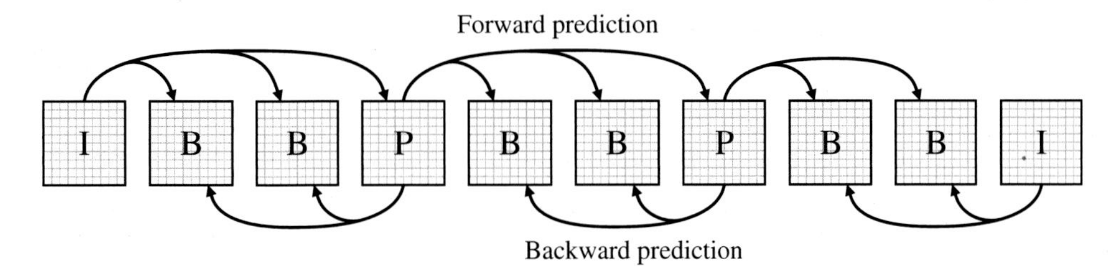

Figure 12.3 Prediction within a group of pictures, showing the relationships among I, P, and B frames.

used with lossless coding methods, where simple prediction of individual pixel values can typically give 2 : 1 compression.

Within video sequences, two distinct types of prediction are commonly used: within a frame (intra-frame) and between frames (inter-frame). Intra-frame prediction usually predicts a block of pixels at a time, with typical block sizes ranging from 4×4 up to 128×128 pixels. Each block is predicted by projecting the neighbouring pixels at a wide range of angles. More recent standards allow blocks to be hierarchically partitioned into smaller blocks using a tree structure (Bross et al., 2021), with each leaf block being predicted using a different mode.

For inter-frame coding, a video sequence is usually divided into groups of pictures (see Figure 12.3). The first frame within the group is coded independently of the other frames using intra-frame coding and is denoted an I-frame. Then, motion-compensated (see Section 12.1.3) forward prediction is used to predict each P-frame from the previous I- or P-frame. Finally, bidirectional (both forward and backward) prediction is then used to predict the intervening frames, denoted B-frames. The B-frames usually require the least data because the backward prediction is effective at predicting the background when objects have moved away. The number of P- and B-frames in each group of pictures depends on the particular codec and application and limits the extent to which a video stream may be accessed randomly.

Since the B-frames require the following P-frame (or I-frame), the frames are not transmitted in their natural order but the order in which they are required to decode the sequence. This requires the encoder to hold several uncompressed B-frames until the following P-frame (or I-frame) is available. The decoder is a little simpler in that it always only needs to hold the two frames used for providing motion-compensated source data.

Closely related to prediction is image transformation. Applying a transform to an image (or block of pixels) concentrates the energy into only a few components by exploiting spatial correlation within the image or block. The reduced variance of many of the components (often with many zero coefficients) enables them to be encoded with fewer bits, giving good compression. Two of the most commonly used transforms are the discrete cosine transform (DCT), used with JPEG and MPEG, and the wavelet transform, used with JPEG 2000.

With lossy codecs, it is necessary for the encoder to decode the frame before using it to make a prediction. This ensures that the prediction is made using the data available at the receiver. Otherwise, any quantisation errors in the prediction will not be taken into account and can propagate from one P-frame to the next. This decoding requires an inverse transform to convert the quantised coefficients back to pixel values. Unfortunately, the encoder has little control over the implementation of the inverse transform in the receiver. With a single image codec, this does not matter, but with video, the differences in round-off error can accumulate from frame to frame, causing the prediction made at the receiver to drift relative to that of the transmitter. Standardisation can help this problem by explicitly defining the behaviour of the decoder. For custom codecs, it requires care with the design of the inverse transform used at both the encoder and the receiver (Reznik et al., 2007).

12.1.3 Motion Estimation and Compensation

Simple inter-frame prediction works well with static objects, but any motion creates a misalignment, reducing the benefits. To achieve good compression, it is necessary to compensate for such motion. This is most commonly achieved by dividing the current frame into small blocks (typically 8×8 or 16×16) and finding the offset between each block and the corresponding data in the previous frame that minimises the prediction error.

The quality of the match is typically measured using the sum of absolute differences (SAD, Eq. (10.55)). Early codecs found the offset to the nearest pixel, although more modern codecs improve performance by estimating fractional offsets. Motion estimation is usually the most computationally intense part of a video codec, so a lot of effort has gone into accelerating this aspect (Huang et al., 2006). To maintain compression efficiency with increasing spatial resolution, it is also necessary to proportionally increase the size of the search window, making this an active research area.

An exhaustive search of every possible offset within the search window will yield the best match but is computationally the most time consuming. Therefore, many shortcuts have been proposed to reduce the search (Huang et al., 2006) by making the assumption that the error surface within the search window is unimodal. These shortcuts usually involve a hierarchical search pattern to avoid needing to calculate large areas within the search window. Although such approaches can reduce the search by an order of magnitude or more (Yu et al., 2004), they can get locked onto a local minimum rather than the global minimum, giving a degradation in compression performance (Huang et al., 2006). This gives trade-offs among accuracy (quality), complexity, and processing time between the various search algorithms.

Two areas of importance for acceleration are the SAD calculation and maximising data reuse during the search (memory bandwidth). The inner loop of all the methods is the calculation of the SAD, so Wong et al. (2002) developed a system that could perform the SAD of a row or column of pixels in parallel. Although more recent FPGAs have considerably more resources, enabling greater parallelism, careful attention needs to be paid to this step to give efficient use of the hardware resources. When performing the search, it is important to efficiently buffer the data to maximise reuse during the search (Tuan et al., 2002; Asano et al., 2010). This is more complicated for fast search methods because the data access patterns are less predictable.

Example implementations of motion estimation are given in Elhamzi et al. (2014), Thang and Dinh (2018), and Alcocer et al. (2019).

12.1.4 Quantisation

Prediction and transformation just rearrange the data, which is completely recoverable via the corresponding inverse transformation. In practice, if the coefficients are not integers, then rounding errors can make the transformation irreversible unless some form of reversible processing scheme is used (for example lifting). The next step in the coding process, quantisation, is lossy in that the reconstructed image will only be an approximation to the original image. However, it is the quantisation of the coefficients that can enable the volume of data used to represent the image to be compressed significantly. With coarser quantisation steps, fewer values are used, and the greater the compression. However, more information will be lost, and this can reduce the quality of the reconstructed image or video.

Quantisation divides the continuous range of input values into a number of discrete bins by specifying a set of increasing threshold levels. A single symbol is then used to represent all the values within each bin. It is this many-to-one mapping that makes quantisation irreversible. When reconstructing the value from the symbol, the centre value for the bin is usually used. The difference between the original value and the reconstructed output is the error or noise introduced by the quantisation process. Since it is usually the transform coefficients that are quantised rather than the input pixel values, the errors do not normally appear as random noise within the image. Typically, quantisation errors show in the image as blurring of fine detail and ringing around sharp edges. In videos, these effects are not stationary, and appear as 'mosquito' noise.

The optimal quantiser adjusts the quantisation thresholds to minimise the mean square error for a given number of quantisation levels or symbols. This makes the bins smaller where there are many similar values, reducing the error, at the expense of increasing the error where there are fewer values. This tends to make the bin occupancy more uniform (Lloyd, 1982), reducing the gains of subsequent entropy coding. The optimal quantiser is data dependent and so requires the levels to be transmitted with the compressed data, reducing compression efficiency. A similar effect may be achieved without the overhead using a predefined nonlinear mapping followed by uniform quantisation (where the threshold levels are equally spaced). For image compression, however, the gains from using optimal quantisation are not sufficiently significant to warrant the additional complexity; it is easier to simply use uniform quantisation with a finer step size and rely on entropy coding to exploit the fact that some bins are relatively empty.

16	11	10	16	24	40	51	61
12	12	14	19	26	58	60	55
14	13	16	24	40	57	69	56
14	17	22	29	51	87	80	62
18	22	37	56	68	109	103	77
24	35	55	64	81	104	113	92
49	64	78	87	103	121	120	101
72	92	95	98	112	100	103	99

17	18	24	47	99	99	99	99
18	21	26	66	99	99	99	99
24	26	56	99	99	99	99	99
47	66	99	99	99	99	99	99
99	99	99	99	99	99	99	99
99	99	99	99	99	99	99	99
99	99	99	99	99	99	99	99
99	99	99	99	99	99	99	99

Figure 12.4 'Standard' quantisation tables for JPEG compression. Left: luminance; right: chrominance.

For image compression, different transformed coefficients often have different visual significance. Psycho-visual redundancy can be exploited using a different quantisation step size for each coefficient (Watson, 1993) and different step sizes for luminance and chrominance components. For example, JPEG has tables specifying the quantisation step sizes for each coefficient (the 'standard' tables (ISO, 1992; Wallace, 1992) are listed in Figure 12.4; although these can be optimised for a particular image for perceived quality or bitrate (Watson, 1993), this is seldom done). If the DCT coefficients are $C[u, v]$, and the quantisation table is $Q[u, v]$, then the quantised coefficients are

$$\hat{C}[u, v] = \text{round}\left(\frac{C[u, v]}{Q[u, v]}\right). \tag{12.1}$$

A single quality factor parameter can be used to scale the quantisation step size to control the compression (and consequently the image quality). The implementation within the open-source IJG (Independent JPEG Group) JPEG codec uses a quality factor, QF, that varies from 1 to 100 to scale the standard quantisation tables as

$$\hat{Q}[u, v] = \begin{cases} \text{round}(\frac{50Q[u, v]}{QF}), & QF < 50, \\ \text{round}\left(Q[u, v]\left(2 - \frac{QF}{50}\right)\right), & 50 \leq QF. \end{cases} \tag{12.2}$$

An alternative approach to quantisation is bit-plane-based coding. The output data is encoded most significant bit first, through to the least significant bit last. For example, the EBCOT algorithm (Taubman, 2000) used by JPEG 2000 divides the wavelet tree into blocks and performs coding in multiple passes at each wavelet layer to ensure that the bits are in order of their significance in terms of contributing to the quality of the final image. This enables the coding for each block to be truncated optimally to minimise the distortion while controlling the bit rate of the final output stream. One FPGA implementation of the EBCOT algorithm is described in Gangadhar and Bhatia (2003).

Vector quantisation works in a similar manner to scalar quantisation in that a set of symbols is used to represent vector quantities. A simple example is palette-coded colour bitmap images, where each pixel is assigned a symbol which represents a colour. One technique commonly used for vector quantisation is K-means clustering (described in more detail in Section 14.3.8.3). Note that in choosing a symbol to represent a colour, it is better to work in a perceptually uniform colour space such as L*a*b* or L*u*v*, where Euclidean distance is more meaningful than in RGB or YCbCr.

12.1.5 Run-length Coding

Line drawings, and other related graphics, have a limited number of distinct pixel values or colours, and these often occur in long runs. Run-length coding represents these by the pixel value (or more generally a symbol) and the count of successive occurrences (Brown and Shepherd, 1995). This makes run-length encoding of such images an effective compression technique. However, for general images, run-length coding is not usually effective on its own because adjacent pixels seldom have identical values.

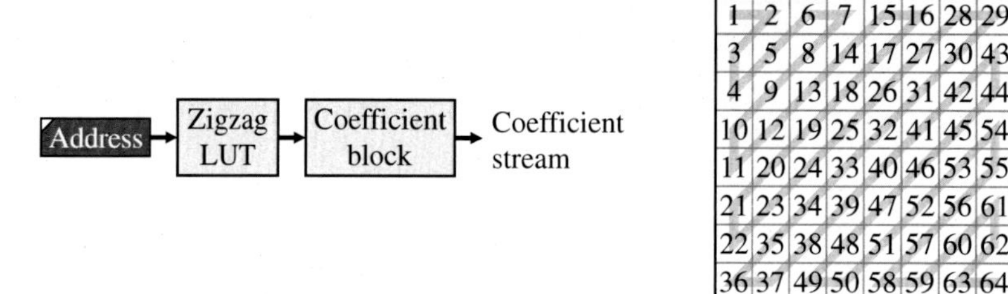

Figure 12.5 Zigzag reordering. Left: implementation, with the lookup table (LUT) translating sequential addresses into zigzag ordered indices; right: the reordering of components within an 8×8 DCT.

Many transform techniques, by concentrating the energy in a few coefficients, have a large proportion of the coefficients close to zero. Having a wider bin for zero (a dead band) gives more zero coefficients and enables run-length coding to be more effective in compressing runs of zeros. A variation in run-length coding is to have a special zero-to-end symbol, which indicates that all of the remaining symbols to the end of a block are zero. This latter approach is effective both with the block-based coding used by JPEG and also with wavelet coding, eliminating whole trees of zero coefficients, especially for the more significant bits when bit-plane coding is used.

With block-based coding (for example JPEG), many of the high-frequency components are zero. These can be placed at the end of the block by reordering the coefficients in zigzag order, enabling many coefficients to be compressed by a single zero-to-end symbol. Zigzag reordering may be implemented efficiently by having a lookup table on the address lines as shown in Figure 12.5. If the coefficients are not in their natural order, for example if the DCT produces a permuted output, then the zigzag lookup table can also take this into account. A similar lookup table-based addressing technique can be used when decoding the images.

12.1.6 Entropy Coding

The final stage within the coding process is entropy coding. This assigns a variable-length code to each symbol in the output stream based on its frequency of occurrence. Symbols which appear frequently are given short codes, while the less common symbols are given longer codes. The entropy of a stream is

$$E = -\sum_{x} p(x) \log_2 p(x), \tag{12.3}$$

where $p(x)$ is the probability of symbol x. The entropy gives a lower bound on the average number of bits required to represent the symbols. For N different symbols, the entropy will be between 0 and $\log_2 N$, with the maximum occurring when all symbols are equally probable. The implementation of entropy coding is complicated by variable-length code words and the need to gather statistics on the frequency distribution of the symbols.

12.1.6.1 Huffman Coding

One of the most common forms of entropy coding is Huffman coding (Huffman, 1952). This uses the optimum integer number of bits for each symbol. The Huffman code is built by sorting the symbols in order of probability and building a tree by successively combining the two symbols with the least probability. The length of each branch is the number of bits needed for the corresponding symbol. Since an integer number of bits is used to represent for each symbol, the average symbol length will be greater than the entropy. In particular, symbols with probability greater than 50% will always require 1 bit, even though the entropy is lower. This may be improved by grouping multiple symbols together and encoding symbol groups. Since the optimum coding is data dependent, an overhead is incurred to represent the coding table (the mapping between a symbol and its

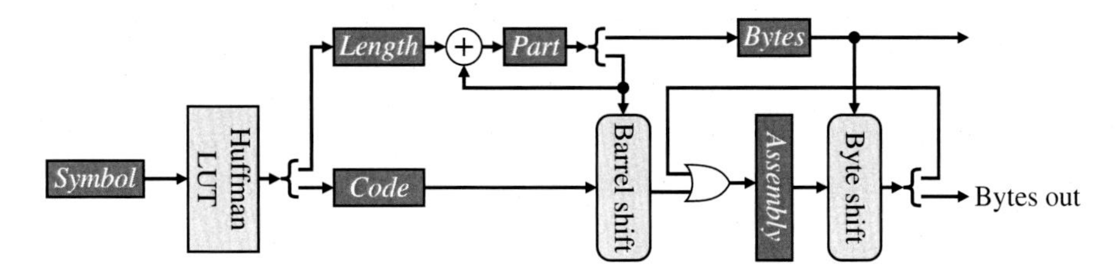

Figure 12.6 Outline of Huffman coding.

code). There are techniques for efficient compression of the coding table (Body and Bailey, 1998), although these are not currently used by any standard.

The optimum Huffman code is based on the statistics of the image, which requires two passes unless a static, predefined code table is used. Building the optimum table requires performing the initial coding steps to generate the symbol stream. The symbols (and any payload) need to be saved in a buffer while a histogram is built to gather the symbol probabilities. The histogram then needs to be sorted in order of probability, and the Huffman tree built. The final step is then to Huffman encode the data from the buffer.

A circuit for a predefined table is outlined in Figure 12.6. It is assumed that the mapping between the symbol and the code (and code length) is stored in the Huffman lookup table. The *Part* register indicates how many bits are used in an incomplete byte. This controls a barrel shift to align the *Code*, with the output being assembled in *Assembly*. *Part* is updated by accumulating the *Length* of the code word. The overflow from *Part* indicates the number of completed *Bytes* in *Assembly*. This is used to shift out the completed bytes. The principle can easily be extended if the symbol has an associated payload, for example with JPEG coding (Agostini et al., 2002).

Conceptually, decoding a bit-stream uses each bit to navigate the Huffman tree to determine the symbol. Clearly, this is impractical for real-time processing, so a table approach can be used to determine the symbol in a single clock cycle. For a maximum code length of l, brute force decoding would require 2^l entries in the table. However, by segmenting the address space (decoding fewer bits at a time), smaller tables can be used although at the expense of more clock cycles (Body et al., 1997). This is only practical if run-length coding is also used, and a value need not be decoded every clock cycle. An alternative approach, described by Sun and Lee (2003), counts the number of leading '1's and uses this to select the table for decoding. This works because longer code words tend to begin with a sequence of '1's.

12.1.6.2 Rice–Golomb Coding

When the symbol probabilities follow a geometric distribution, Golomb (1966) showed that optimal codes take a particular form. In many coding applications, the symbols (for example prediction residuals) can be approximated by a double geometric distribution. These may be converted to a geometric distribution by interleaving the odd and even residuals. Let p be the geometric distribution's decay parameter. Golomb then derived a coding parameter:

$$m = \text{round}\left(\frac{-1}{\log_2 p}\right). \tag{12.4}$$

Let $k = \lceil \log_2 m \rceil$. A symbol, with (non-negative) index S, is then coded as $n = \lfloor \frac{S}{m} \rfloor$ '1's, followed by the binary code for the remainder ($r = S - n \times m$) using k bits (for $r < 2^k - m$) or $r + 2^k - m$ using $k + 1$ bits (for $r \geq 2^k - m$).

Often, p is only approximated, and restricting m to be a power of 2 ($m = 2^k$, giving Rice coding) has a relatively low coding inefficiency, so this is often used for coding and decoding simplicity. The value for k is often chosen adaptively based on the mean index to optimise the compression (Kiely, 2004). For Rice codes, decoding consists of counting the number of leading '1's to give n, with r given as the binary code of the remaining k bits. The symbol is then $S = n \times 2^k + r$.

12.1.6.3 Arithmetic Coding

Arithmetic coding removes the restriction of having an integer number of bits per symbol. It encodes the whole stream as an extended binary number, representing a proportion between 0.0 and 1.0 (Rissanen, 1976; Langdon, 1984; Witten et al., 1987). Each symbol successively divides the remaining space in proportion to its probability. The key to an efficient implementation is to work with finite precision and output the leading bits once they can no longer be affected. Using finite precision (and rounding) introduces a small overhead over the theoretical minimum length. There is also an overhead associated with specifying the symbol probabilities.

The basic structure of an arithmetic coder is shown in Figure 12.7. The *Start* and *Width* registers hold fixed-point representations of the start and width of the current range. At the start of coding, they are initialised to 0.0 and 1.0, respectively. The symbol to be coded is looked up in a table to determine the start and width of the symbol. These are scaled by the current *Width* to give the new *Start* and *Width* remaining. The renormalisation detects the leading bits which can no longer be changed by subsequent symbols and outputs completed bits; this also renormalises the *Start* and *Width* variables The carry propagation records the number of successive '1's (*Ones*) output, because a carry can cause these to switch to '0's (Witten et al., 1987).

12.1.6.4 Adaptive Coding

The overhead associated with determining and specifying the probabilities may be overcome by adaptive coding. This builds the table of probabilities as the data arrives. While adaptive coding can be used with either Huffman or arithmetic coding, it is less convenient with Huffman coding because changing the symbol probabilities requires completely rebuilding the code tree. Since arithmetic coding works directly with the probabilities, this is less of a problem.

There are two approaches to adaptive coding. The first is to assume that each symbol is equally likely (with a count of 1), with the count incremented and probabilities adjusted as the symbols arrive. There is an overhead resulting from overweighting rare symbols at the start. The second approach is to have an empty table, except for an escape symbol, which is used to introduce each new symbol.

Adaptive coding is good for short sequences where the overhead of the table can approach the storage required for the actual sequence itself, especially if there are a large number of possible symbols compared to the length of sequence. It also works well for long sequences because the statistics approach the optimum for the data, although there is an overhead at the start while the table is adapting to the true probabilities. If the statistics are not stationary, then periodic rescaling of the counts (for example dividing all counts by a power of 2) enables the probabilities to track that of the more recent data.

12.1.6.5 Dictionary Methods

One final class of compression methods that will be mentioned briefly is that of methods which build a dynamic dictionary of symbols. These allow sequences of symbols to be combined and represented as a single symbol. The original method in this class is LZ77 (Ziv and Lempel, 1977), which uses as its dictionary a window into the past samples. A sequence of symbols that have already been encountered is coded as an offset plus length.

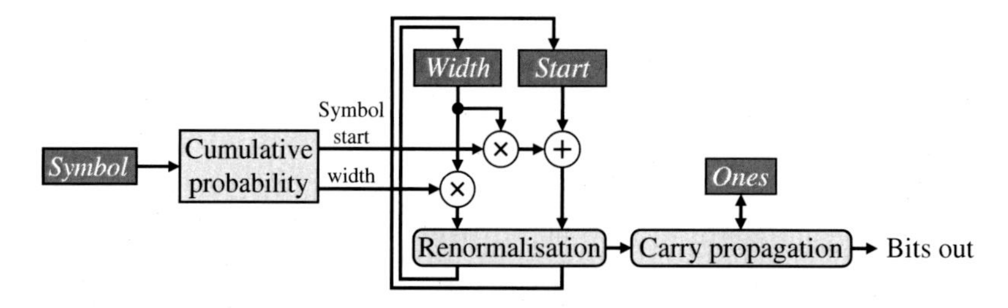

Figure 12.7 Basic structure of arithmetic coding.

LZ77 is combined with Huffman coding to give the 'deflate' algorithm (Deutsch, 1996) used to encode PNG (portable network graphic) image files. The search through past entries can be accelerated through the use of a hash table to locate potential candidate strings (see, for example (Fowers et al., 2015; Kim et al., 2019)). Alternatively, a systolic array may be used to perform the search (Abd Elghany et al., 2007). A related variation is the Lempel-Ziv-Welch (LZW) algorithm (Welch, 1984) which explicitly builds a dictionary on-the-fly by extending the existing sequences each time a new combination is encountered. LZW is used to compress GIF (graphics interchange format) image files.

12.2 DCT-based Codecs

The DCT is close to the optimal linear transform for decorrelating the pixel values (the optimal transform is given by principal components analysis). This enables it to concentrate the energy into a few components, making it an obvious choice for image compression. However, with global transforms, every coefficient depends on every input pixel, making it memory intensive, and also high latency. Therefore, a trade-off is to divide the image into small blocks and transform each block independently. Smaller block sizes exploit the locality within an image but can reduce the compression gains. However, it also makes the algorithm computationally simpler and significantly reduces the memory requirements. Typically, block sizes vary from 4×4 to 32×32 (Sullivan et al., 2012), but the most common is 8×8 used by JPEG.

12.2.1 DCT Block Processing

With block processing, systems operate on a block of pixels at a time. For stream processing, this requires buffering sufficient data to form a block before processing can begin. The buffer must be twice the block height, so that incoming pixels can continue to be buffered while each block is processed (either using dual-port memory or a bank-switched ping-pong buffer). Each block of pixels is then processed asynchronously relative to the incoming data.

DCT processing can exploit the separability of the 2-D DCT, first calculating the DCT of the rows of a block, saving the results into a transpose buffer, and then using a second processor to take the DCT of the columns, as illustrated in Figure 11.12. An alternative approach with streamed data is to calculate the row DCTs on-the-fly as the data is streamed in, storing the intermediate results in an on-chip cache. Once the last row arrives, the data may be read from the cache in column order to perform the column DCTs. With this approach, a transpose buffer is not required, but the cache may need to be a few bits wider to match the required precision of the intermediate result.

One limitation of dividing the image into blocks and encoding each block independently is a loss of coherence at block boundaries. After quantisation, this can lead to blocking artefacts which give the appearance of false edges at block boundaries, especially at high compression rates. These can be reduced to some extent by filtering the image after decoding (for example (Vinh and Kim, 2010)). In fact, many video coding standards incorporate such deblocking filters (for example H.264 AVC (Wiegand et al., 2003), H.265 HEVC (Sullivan et al., 2012), and H.266 VVC (Bross et al., 2021)).

12.2.2 JPEG

Despite its age, the JPEG compression standard (ISO, 1992) remains widely used for image compression. One reason is its relative simplicity. Another is its widespread support among digital cameras and software. JPEG is not actually a single compression method but is a suite of related methods (Wallace, 1992). The most commonly used method, implemented by virtually all systems, is the baseline coding method. It will be described here and is outlined in Figure 12.8.

A colour image is first converted into the YCbCr colour space. The chrominance components are down-sampled by a factor of 2 in each direction, because the spatial sensitivity of the human visual system to chrominance contrast is significantly poorer than to the luminance. Each channel is then processed separately for the remaining steps. Each channel is transformed using block processing, using an 8×8 DCT.

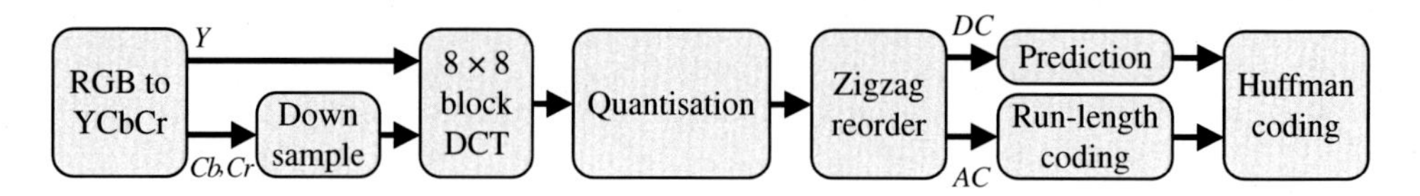

Figure 12.8 Baseline JPEG compression.

The resulting coefficients are then quantised with a separate quantisation step size for each coefficient. The 'standard' tables from Figure 12.4 are commonly scaled, although in principle, any quantisation factors could be chosen. It is usual to have different quantisation factors for the luminance and chrominance components. The coefficients are then reordered using a zigzag reordering. The DC (direct current) coefficient is predicted between horizontally adjacent blocks, using the DC value of the previous block and encoding the difference. Any zeroes in the remaining 63 AC (alternating current) coefficients are then run-length encoded, giving a set of run lengths and associated nonzero coefficients. The run length is combined with the size of the coefficient to form an 8-bit symbol which is Huffman encoded, followed by the actual coefficient stored as payload. A special zero-to-end symbol is used after the last nonzero coefficient. The Huffman tables can be formed by analysing the statistics within the image or simply using the 'standard' typical tables. Finally, the bit-stream is packaged with appropriate headers containing the quantisation tables, Huffman tables, and meta-data using the JPEG file interchange format (Hamilton, 1992).

The compression ratio is controlled by adjusting the quantisation factors, usually by scaling the 'standard' quantisation tables according to Eq. (12.2). The resulting compression rate also depends on the image content.

Several FPGA implementations of JPEG encoders are described in the literature (see, for example (Agostini et al., 2007; De Silva et al., 2012; Scavongelli and Conti, 2016)).

Decoding follows the reverse process (Wallace, 1992), with the Huffman tables and quantisation tables extracted from the file stream. One decoder architecture is reported by Ahmad et al. (2009).

12.2.3 *Video Codecs*

With video codecs, the standards generally define the bit-stream syntax and decoding process rather than the encoding (Wiegand et al., 2003). This allows freedom and flexibility for the encoding process to balance the competing objectives of compression rate, latency, and encoding complexity, depending on the application requirements. Each generation of codecs gives better compression performance than the previous generation, although this is usually at a cost of significantly increased computation. A detailed discussion of the standards is beyond the scope of this work; however, an outline of the techniques will be given here.

The high-level structure of a video coder is illustrated in Figure 12.9. The image is divided into a set of tiles or slices, which can be processed in parallel. An RGB to YCbCr (or YCoCg (Malvar et al., 2008)) colour transform

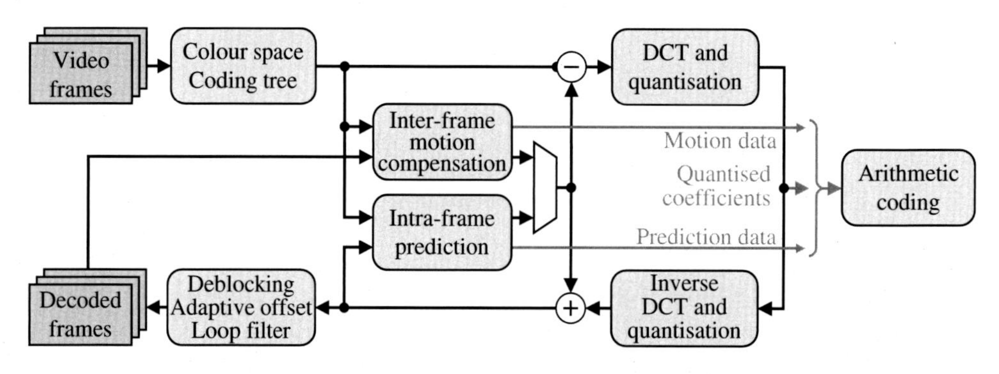

Figure 12.9 Outline of a typical video compression system.

is applied with the chrominance components commonly down-sampled by a factor of 2 (4 : 2 : 0 format). Early standards (up to H.264 AVC) used 16 × 16 macroblocks of luminance samples, although later standards (from H.265 HEVC (Sullivan et al., 2012)) use a coding tree unit of up to 64 × 64 samples. The coding trees can then be partitioned into smaller blocks using a quad-tree structure.

Each slice (group of coding tree units) is predicted using intra-frame (I), unidirectional (P), or bidirectional (B) inter-frame prediction. While early standards applied this on a frame-by-frame basis, more recent standards use a hybrid approach, with intra-frame prediction also considered as an option for P and B coding units. Intra-frame prediction uses the neighbouring decoded pixels to predict a block using directional prediction (H.264 AVC uses 8 directions (Wiegand et al., 2003), H.265 HEVC uses 33 (Sullivan et al., 2012), and H.266 VVC uses 65 (Bross et al., 2021)). H.266 VVC also introduces several other intra-frame prediction modes based on rectangular blocks to give better prediction (Bross et al., 2021).

For inter-frame prediction, blocks are predicted to a fractional pixel offset (typically to the nearest quarter pixel) using separable interpolation filters (Hamzaoglu et al., 2022). Adjacent blocks are used to predict the motion of a block, reducing the data required for sending the motion vector. Rather than predicting the pixels from a single frame, several previously decoded frames can be used to source the prediction block. H.266 VVC extends the prediction to include a more complex affine transform to allow zoom and rotation (Taranto, 2022).

Finally, the residuals after prediction are transformed using an integer DCT to further reduce the spatial redundancy. Transform block sizes vary from 4 × 4 up to 32 × 32 and can even be non-square for H.266 VVC. The resulting coefficients are then quantised using a uniform step size. All of the parameters and coefficients are entropy coded using context-adaptive arithmetic coding, where the statistics of already coded elements provide the context to estimate the symbol probabilities (Schwarz et al., 2021). Having a variable block size complicates the design of efficient hardware because the transform block must be able to handle the requirements for the largest size and yet be fast for the smallest size.

For motion compensation, it is necessary to reconstruct the past frames internally. A deblocking filter reduces the artefacts created using block-based DCT. H.265 HEVC also introduced a sample adaptive offset to apply an adaptive nonlinear mapping to better reconstruct the signal amplitudes (Sullivan et al., 2012). These in-loop filters are described in more detail by Karczewicz et al. (2021).

Much of the published literature on FPGA implementations of video coding and decoding focus on efficient implementation of particular modules or blocks from various standards (see, for example (Saldanha et al., 2020)) rather than on full codecs. A detailed review of the specific techniques is beyond the scope of the discussion here.

A more recent innovation to handle the increasing spatial and temporal resolution is ***low complexity enhancement video coding*** (LCEVC) (Meardi et al., 2020; Battista et al., 2022). The key idea is to use an existing codec to provide a low-resolution base video stream, which is then up-sampled to the required resolution. The residual details are provided by one or two additional coding layers that bring the quality up to the required level.

12.3 Wavelet-based Codecs

The wavelet transform is used primarily to decorrelate the data. Adjacent pixels in many images are similar, and this is exploited by separating the image into approximation and detail components. Most of the energy is compacted into the low-frequency approximation coefficients, giving good energy compaction. The high-frequency detail coefficients are sparse, with a significant number of small or zero coefficients. Truncating these provides an effective means of lossy compression that has a low impact on the perceived image quality. A key advantage of the wavelet transform over the DCT is that each coefficient depends only on local information rather than the whole image. There is no need to split the image into blocks to maintain locality, avoiding the blocking artefacts commonly associated with block processing. Another advantage of using the wavelet transform for compression is that it is easy to reconstruct an output image at a range of different scales simply by choosing which level of wavelet coefficients are used in the reconstruction.

The structure of JPEG 2000 compression is illustrated in Figure 12.10. The input image may be split into multiple rectangular tiles, which are compressed independently (effectively treated as separate images (Skodras et al., 2001)). This reduces the memory requirements when coding and decoding or enables spatial parallelism. Each tile is converted from RGB to YCbCr using either Eq. (7.48) or the RCT of Eq. (7.51). Note that the Cb

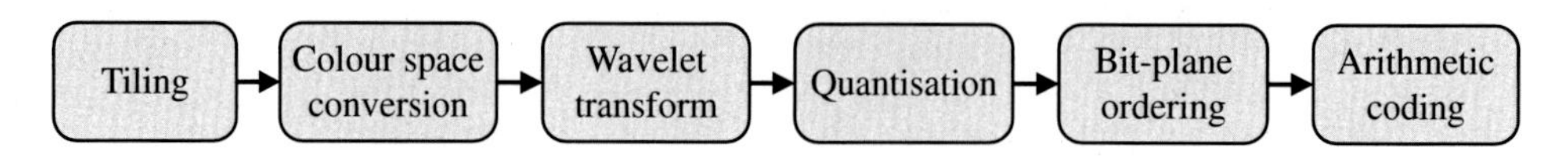

Figure 12.10 Structure of JPEG 2000 compression.

and Cr components are not explicitly down-sampled by a factor of 2; this is accomplished by the first level of the wavelet transform by discarding the detail components.

A 2-D biorthogonal wavelet transform is applied to each component to reduce the redundancy. Either the 5/3 or 9/7 wavelet filters are used. Lossless compression uses 5/3 filters (along with the RCT) because they are integer-to-integer transforms (ISO, 2000). Lossy compression can use either, although better compression is obtained with the 9/7 wavelet (Skodras et al., 2001). Since the biorthogonal wavelets are symmetric, a symmetric extension is used around the borders of the tiles. The wavelet transform is separable, enabling row and column transforms to be performed sequentially. While the row transform may be readily streamed, the column transform requires considerably more storage, and this can become a memory bottleneck, especially when many levels of wavelet transform are used. Three approaches are commonly used: separable processing, line-based processing, and block processing (Angelopoulou et al., 2008), as discussed in Section 11.3.1.4.

The resulting coefficients are then quantised (step size = 1 for lossless compression) using truncation towards 0, so that the zero bin is twice as wide as the others (this dead-band gives more zero coefficients). Then, embedded block coding with optimised truncation (EBCOT) is used to code the coefficients in bit-plane order (Taubman, 2000). A coefficient becomes significant in the bit-plane where its most significant bit is a '1'. For each bit-plane, three passes are used in sequential order: The first pass (significance propagation) encodes insignificant coefficients that have a significant coefficient in their 8-neighbourhood and those that become significant. The second pass (magnitude refinement) encodes the bit of components that became significant in previous bit-planes. The third pass (cleanup) encodes the bits that have not been encoded in the previous two passes. The use of three passes optimises the quality for a given number of bits by ensuring that more important bits are coded first. This enables the compression ratio to be controlled simply by truncating the resulting bit-stream. Regions of interest may be coded with higher quality simply by scaling the region of interest, so that its bits have higher significance and appear earlier in the bit-stream (Skodras et al., 2001). Binary arithmetic encoding (each symbol is either a '1' or '0') is then applied to the bit-stream using the local context to adaptively determine the probability of each symbol. JPEG 2000 uses a simplified version on arithmetic coding scheme that is multiplier free.

Taubman and Marcellin (2002) describe in more detail the algorithms used for coding and decoding. From a hardware implementation perspective, the key issue with the bit-plane and entropy coding is that the steps are largely sequential, with corresponding dependencies. Many researchers have explored various optimisations to accelerate the coding. For example, Gangadhar and Bhatia (2003) reduce the number of memory accesses by running the three passes for each bit-plane in parallel, although this comes at the expense of a small reduction in coding efficiency. Fang et al. (2003) take this one step further and process all of the bit-planes in parallel. Varma et al. (2008) modified the encoding process to reduce dependencies in the context adaption, enabling coding stages to run in parallel without the cost of coding efficiency. Kumar et al. (2012) speed up the arithmetic coder by coding two symbols every clock cycle.

JPEG 2000 typically gives 2 dB improvement in PSNR over JPEG for the same compression ratio (Skodras et al., 2001). However, in spite of this improved performance, the use of JPEG still dominates because of the significant increase in complexity of JPEG 2000.

12.4 Lossless Compression

Many of the image compression standards also provide a lossless compression option. These rely on prediction to reduce the redundancy within the image and avoid the lossy quantisation step. There is an inevitable trade-off between the complexity of the predictor (in terms of number of parameters that must be learned) and either the overhead to represent the trained model, or the compression cost as the model is being adapted

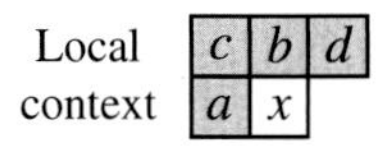

Figure 12.11 Local causal context for the median predictor used by JPEG-LS.

(Weinberger et al., 2000). As a result, the focus for lossless compression has been on low-complexity predictors. A simple adaptive predictor that performs surprisingly well is the median predictor (Martucci, 1990) which takes the median of horizontal, vertical, and planar predictors:

$$x = \text{median}(a, b, a + b - c), \tag{12.5}$$

where the pixel values are those from the causal local neighbourhood, as illustrated in Figure 12.11.

The resulting residuals usually have a Laplacian or two-sided geometrical distribution of amplitudes (O'Neal, 1966). However, the local context (the presence of edges) can result in significant offset or bias of the centre of the distribution (Langdon and Manohar, 1993). For this reason, the JPEG-LS standard adds a context-based correction that is learned by averaging the observed errors. The local context uses the local differences

$$Context = \{d - b, b - c, c - a\} \tag{12.6}$$

to represent the activity around the predicted pixel which affects the statistics of the offsets (Weinberger et al., 2000). Each of these differences is quantised into 9 levels, giving a total of 365 contexts (taking into account symmetry). The average observed offset (based on the context) is added to the prediction from Eq. (12.5) to give the predicted pixel value.

Several implementations and variations are described in the literature. For example, Papadonikolakis et al. (2008) implemented a pipelined version of JPEG-LS. Palaz et al. (2017) implemented a variation of JPEG-LS for embedded raster order compression but without the context correction. They add rate control (making it lossy) but maintain simplicity to enable high throughput. Inatsuki et al. (2015) used a hierarchical 1-D predictor to save having to buffer the previous line, followed by Huffman coding with fixed probability distribution. Although the resulting compression was slightly less than JPEG-LS, it removed the need to buffer a whole line to build the neighbourhood.

Lossless methods can also be adapted for near-lossless compression by quantising the residuals. This can guarantee that the maximum error for each pixel is less than a predefined threshold level.

12.5 Perceptual Coding

Rather than requiring the video to be lossless, perceptual coding focuses on minimising the visual distortion. Different areas within an image can tolerate different levels of distortion based on local context. This threshold of change that is just visible is called ***just noticeable distortion*** (JND). There are two main components to this: luminance masking and chrominance masking. With luminance masking, ΔI_{JND} is given by Weber's law or similar

$$\Delta I_{JND} = kI, \tag{12.7}$$

where I is the local average intensity, and k is a constant. Contrast masking, ΔC_{JND}, indicates where edges and visual texture can hide larger errors. ΔI_{JND} and ΔC_{JND} are combined (usually nonlinearly) to give $JND(x, y)$, which is the visibility threshold as a function of position within an image. Several researchers have proposed models for calculating JND (see, for example (Chou and Li, 1995; Yang et al., 2005; Liu et al., 2010)). Ko et al. (2012) defined a simplified JND measure based on Haar wavelets, which was implemented on an FPGA by Liu et al. (2012).

Since errors less than $JND(x, y)$ cannot be perceived, the simplest approach is to set the quantisation step size equal to, or smaller than, the JND to give visually lossless compression. The main problem with this is that some overhead must be incurred to communicate the step size to the receiver to enable reconstruction.

Two approaches have been adopted to solve this problem. The first is to modify the JND calculation to use only the previously transmitted pixels (Lie and Liu, 2010). This is limited because the calculation window is then one-sided, making it more difficult to accurately detect edges and texture. The second approach is to use a form of two-stage compression (Wang, 2019). In the first stage, some other lossy compression algorithm is used, which provides the data for estimating the JND threshold at the receiver. However, with the lossy compression, there is no guarantee that the errors are smaller than this threshold, so a refinement stream is provided in the second stage to correct or adjust any pixels which exceed the JND threshold. Wang (2019) showed that this worked both with JPEG compression for the first stage, or even simply down-sampling by a factor of 2, and using JPEG-LS.

A further refinement was to use adaptive down-sampling (Wang et al., 2019) to give a low cost (in terms of FPGA resources) and efficient (in terms of compression) perceptually lossless codec.

12.6 Coding Hyperspectral Images

Hyperspectral images contain a lot of redundancy, especially when the spectral bands are closely spaced. This is particularly an issue with satellite-based imaging because of the large volume of data within a hyperspectral cube, and the limited communication bandwidth for transferring the images for ground-based processing. While many possibilities for compressing such images have been explored (Dua et al., 2020), only the standards proposed by the Consultative Committee for Space Data Systems (CCSDS) will be covered here.

The CCSDS 122.1-B-1 standard (CCSDS, 2017) applies a linear spectral transformation to exploit the redundancy between bands. The transformed bands are then coded using standard image compression techniques. Three possible transformations are described:

- Integer wavelet transform along the spectral axis, using 5/3 biorthogonal wavelets.
- Pairwise orthogonal transform, which is close to the optimal given by principal components analysis.
- An arbitrary affine transform, allowing principal components analysis, or indeed any other linear transform of the spectral components to be applied. Note that this also requires the transform matrix to be transmitted with the data, and depending on the transformation, may be lossy.

Fernandez et al. (2019) showed that using principal components analysis (PCA), typically only 10% of the components are required to account for 99.9% of the variance within images. They used an FPGA implementation of PCA for dimensionality reduction of hyperspectral images.

The CCSDS 123.1-B-2 standard (CCSDS, 2019; Hernandez-Cabronero et al., 2021) predicts each sample using the local spatial and spectral samples already transmitted, with the resulting residuals coded using adaptive entropy coding. Similar to the lossless or near-lossless compression methods described earlier, the predictor has low computational cost, making it well suited for hardware implementation. Several researchers have accelerated the computation (or aspects of it) specified by the standard. See, for example Santos et al. (2016), Orlandic et al. (2019), and Bascones et al. (2022).

12.7 Summary

Images can be compressed because much of the data within them is redundant. Adjacent pixels are often part of the same object, giving high spatial correlation (apart from around edges). There is also high spectral correlation, especially for hyperspectral images. Successive video scenes usually contain very similar information, and this can be exploited further by compensating for object (or camera) motion. Finally, not all variations or changes within an image can be perceived by the human visual system, so images compressed for human consumption do not have to be reproduced exactly for visually lossless reproduction.

Transform domain techniques can be used to reduce the correlation between components, mainly spatial and spectral, resulting in many small or zero coefficients. The most commonly used transforms are the DCT and the wavelet transform, although others are also used. An alternative approach is direct prediction of pixel values, in the spectral, spatial, and temporal dimensions. The prediction residuals have a significantly tighter distribution, enabling them to be represented with fewer bits through entropy coding. Where lossy compression is permissible, using a coarse quantisation can significantly reduce the data volume, with only a minor impact on reproduction quality.

Many of the commonly used compression standards have been reviewed in this chapter, showing how the various compression techniques have been exploited. Examples have been provided for how FPGAs have been used to implement systems using the various standards.

References

Abd Elghany, M.A., Salama, A.E., and Khalil, A.H. (2007). Design and implementation of FPGA-based systolic array for LZ data compression. *IEEE International Symposium on Circuits and Systems (ISCAS 2007)*, New Orleans, LA, USA (27–30 May 2007), 3691–3695. https://doi.org/10.1109/ISCAS.2007.378644.

Agostini, L.V., Silva, I.S., and Bampi, S. (2002). Pipelined entropy coders for JPEG compression. *15th Symposium on Integrated Circuits and Systems Design*, Porto Alegre, Brazil (9–14 September 2002), 203–208. https://doi.org/10.1109/SBCCI.2002.1137659.

Agostini, L.V., Silva, I.S., and Bampi, S. (2007). Multiplierless and fully pipelined JPEG compression soft IP targeting FPGAs. *Microprocessors and Microsystems* **31** (8): 487–497. https://doi.org/10.1016/j.micpro.2006.02.002.

Ahmad, J., Raza, K., Ebrahim, M., and Talha, U. (2009). FPGA based implementation of baseline JPEG decoder. *Proceedings of the 7th International Conference on Frontiers of Information Technology*, Abbottabad, Pakistan (16–18 December 2009), 6 pages. https://doi.org/10.1145/1838002.1838035.

Alcocer, E., Gutierrez, R., Lopez-Granado, O., and Malumbres, M.P. (2019). Design and implementation of an efficient hardware integer motion estimator for an HEVC video encoder. *Journal of Real-Time Image Processing* **16** (2): 547–557. https://doi.org/10.1007/s11554-016-0572-4.

Angelopoulou, M.E., Cheung, P.Y.K., Masselos, K., and Andreopoulos, Y. (2008). Implementation and comparison of the 5/3 lifting 2D discrete wavelet transform computation schedules on FPGAs. *Journal of Signal Processing Systems* **51** (1): 3–21. https://doi.org/10.1007/s11265-007-0139-5.

Asano, S., Shun, Z.Z., and Maruyama, T. (2010). An FPGA implementation of full-search variable block size motion estimation. *International Conference on Field Programmable Technology (FPT 2010)*, Beijing, China (8–10 December 2010), 399–402. https://doi.org/10.1109/FPT.2010.5681445.

Bascones, D., Gonzalez, C., and Mozos, D. (2022). A real-time FPGA implementation of the CCSDS 123.0-B-2 standard. *IEEE Transactions on Geoscience and Remote Sensing* **60**: 1–13. https://doi.org/10.1109/TGRS.2022.3160646.

Battista, S., Meardi, G., Ferrara, S., Ciccarelli, L., Maurer, F., Conti, M., and Orcioni, S. (2022). Overview of the low complexity enhancement video coding (LCEVC) standard. *IEEE Transactions on Circuits and Systems for Video Technology* **32** (11): 7983–7995. https://doi.org/10.1109/TCSVT.2022.3182793.

Body, N.B. and Bailey, D.G. (1998). Efficient representation and decoding of static Huffman code tables in a very low bit rate environment. *IEEE International Conference on Image Processing*, Chicago, IL, USA (4–7 October 1998), Volume **3**, 90–94. https://doi.org/10.1109/ICIP.1998.727139.

Body, N.B., Page, W.H., Khan, J.Y., and Hodgson, R.M. (1997). Efficient mapping of image compression algorithms on a modern digital signal processor. *4th Annual New Zealand Engineering and Technology Postgraduate Students Conference*, Hamilton, NZ (3–4 July 1997), 59–64.

Bross, B., Wang, Y.K., Ye, Y., Liu, S., Chen, J., Sullivan, G.J., and Ohm, J.R. (2021). Overview of the versatile video coding (VVC) standard and its applications. *IEEE Transactions on Circuits and Systems for Video Technology* **31** (10): 3736–3764. https://doi.org/10.1109/TCSVT.2021.3101953.

Brown, C.W. and Shepherd, B.J. (1995). *Graphics File Formats: Reference and Guide*. Greenwich, CT: Manning.

CCSDS 122.1-B-1. (2017). *Spectral Preprocessing Transform for Multispectral and Hyperspectral Image Compression*. Consultative Committee on Space Data Systems.

CCSDS 123.0-B-2. (2019). *Low-Complexity Lossless and Near-Lossless Multispectral and Hyperspectral Image Compression*. Consultative Committee on Space Data Systems.

Chou, C.H. and Li, Y.C. (1995). A perceptually tuned subband image coder based on the measure of just-noticeable-distortion profile. *IEEE Transactions on Circuits and Systems for Video Technology* **5** (6): 467–476. https://doi.org/10.1109/76.475889.

Deutsch, P. (1996). *DEFLATE Compressed Data Format Specification version 1.3*. Specification, Aladdin Enterprises.

De Silva, A.M., Bailey, D.G., and Punchihewa, A. (2012). Exploring the implementation of JPEG compression on FPGA. *6th International Conference on Signal Processing and Communication Systems*, Gold Coast, Australia (12–14 December 2012), 9 pages. https://doi.org/10.1109/ICSPCS.2012.6508008.

Dua, Y., Kumar, V., and Singh, R.S. (2020). Comprehensive review of hyperspectral image compression algorithms. *Optical Engineering* **59** (9): Article ID 090902, 39 pages. https://doi.org/10.1117/1.OE.59.9.090902.

Elhamzi, W., Dubois, J., Miteran, J., and Atri, M. (2014). An efficient low-cost FPGA implementation of a configurable motion estimation for H.264 video coding. *Journal of Real-Time Image Processing* **9** (1): 19–30. https://doi.org/10.1007/s11554-012-0274-5.

Fang, H.C., Wang, T.C., Lian, C.J., Chang, T.H., and Chen, L.G. (2003). High speed memory efficient EBCOT architecture for JPEG2000. *IEEE International Symposium on Circuits and Systems (ISCAS)*, Bangkok, Thailand (25–28 May 2003), Volume **2**, 736–739. https://doi.org/10.1109/ISCAS.2003.1206079.

Fernandez, D., Gonzalez, C., Mozos, D., and Lopez, S. (2019). FPGA implementation of the principal component analysis algorithm for dimensionality reduction of hyperspectral images. *Journal of Real-Time Image Processing* **16** (5): 1395–1406. https://doi.org/10.1007/s11554-016-0650-7.

Fowers, J., Kim, J.Y., Burger, D., and Hauck, S. (2015). A scalable high-bandwidth architecture for lossless compression on FPGAs. *IEEE 23rd Annual International Symposium on Field-Programmable Custom Computing Machines*, Vancouver, Canada (2–6 May 2015), 52–59. https://doi.org/10.1109/FCCM.2015.46.

Gangadhar, M. and Bhatia, D. (2003). FPGA based EBCOT architecture for JPEG 2000. *IEEE International Conference on Field-Programmable Technology (FPT)*, Tokyo, Japan (15–17 December 2003), 228–233. https://doi.org/10.1109/FPT.2003.1275752.

Golomb, S. (1966). Run-length encodings. *IEEE Transactions on Information Theory* **12** (3): 399–401. https://doi.org/10.1109/TIT.1966.1053907.

Hamilton, E. (1992). *JPEG File Interchange Format*, Version 1.02. C-Cube Microsystems. http://www.w3.org/Graphics/JPEG/jfif3.pdf.

Hamzaoglu, I., Mahdavi, H., and Taskin, E. (2022). FPGA implementations of VVC fractional interpolation using high-level synthesis. *IEEE International Conference on Consumer Electronics (ICCE)*, Las Vegas, NV, USA (7–9 January 2022), 6 pages. https://doi.org/10.1109/ICCE53296.2022.9730363.

Hernandez-Cabronero, M., Kiely, A.B., Klimesh, M., Blanes, I., Ligo, J., Magli, E., and Serra-Sagrista, J. (2021). The CCSDS 123.0-B-2 "low-complexity lossless and near-lossless multispectral and hyperspectral image compression" standard: a comprehensive review. *IEEE Geoscience and Remote Sensing Magazine* **9** (4): 102–119. https://doi.org/10.1109/MGRS.2020.3048443.

Huang, Y.W., Chen, C.Y., Tsai, C.H., Shen, C.F., and Chen, L.G. (2006). Survey on block matching motion estimation algorithms and architectures with new results. *Journal of VLSI Signal Processing Systems for Signal, Image and Video Technology* **42** (3): 297–320. https://doi.org/10.1007/s11265-006-4190-4.

Huffman, D.A. (1952). A method for the construction of minimum-redundancy codes. *Proceedings of the IRE* **40** (9): 1098–1101. https://doi.org/10.1109/JRPROC.1952.273898.

Inatsuki, T., Matsuura, M., Morinaga, K., Tsutsui, H., and Miyanaga, Y. (2015). An FPGA implementation of low-latency video transmission system using lossless and near-lossless line-based compression. *IEEE International Conference on Digital Signal Processing (DSP)*, Singapore (21–24 July 2015), 1062–1066. https://doi.org/10.1109/ICDSP.2015.7252041.

ISO/IEC 10918-1 (1992). *Digital Compression and Coding of Continuous-Tone Still Images - Requirements and Guidelines.* International Standards Organisation.

ISO/IEC 15444-1:2000 (2000). *JPEG 2000 Image Coding System - Part 1: Core Coding System.* International Standards Organisation.

Karczewicz, M., Hu, N., Taquet, J., Chen, C.Y., Misra, K., Andersson, K., Yin, P., Lu, T., François, E., and Chen, J. (2021). VVC in-loop filters. *IEEE Transactions on Circuits and Systems for Video Technology* **31** (10): 3907–3925. https://doi.org/10.1109/TCSVT.2021.3072297.

Kiely, A. (2004). Selecting the Golomb Parameter in Rice Coding. *Technical report 42-159.* California, USA: Jet Propolusion Laboratory.

Kim, Y., Choi, S., Jeong, J., and Song, Y.H. (2019). Data dependency reduction for high-performance FPGA implementation of DEFLATE compression algorithm. *Journal of Systems Architecture* **98**: 41–52. https://doi.org/10.1016/j.sysarc.2019.06.005.

Ko, L.T., Chen, J.E., Hsin, H.C., Shieh, Y.S., and Sung, T.Y. (2012). Haar-wavelet-based just noticeable distortion model for transparent watermark. *Mathematical Problems in Engineering* **2012**: Article ID 635738, 15 pages. https://doi.org/10.1155/2012/635738.

Kumar, N.R., Xiang, W., and Wang, Y. (2012). Two-symbol FPGA architecture for fast arithmetic encoding in JPEG 2000. *Journal of Signal Processing Systems* **69** (2): 213–224. https://doi.org/10.1007/s11265-011-0655-1.

Langdon, G.G. (1984). An introduction to arithmetic coding. *IBM Journal of Research and Development* **28** (2): 135–149. https://doi.org/10.1147/rd.282.0135.

Langdon, G. and Manohar, M. (1993). Centering of context-dependent components of prediction-error distributions of images. *Applications of Digital Image Processing XVI*, San Diego, CA, USA (14–16 July 1993), SPIE, Volume **2028**. https://doi.org/10.1117/12.158637.

Lie, W.N. and Liu, W.C. (2010). A perceptually lossless image compression scheme based on JND refinement by neural network. *4th Pacific-Rim Symposium on Image and Video Technology*, Singapore (14–17 November 2010), 220–225. https://doi.org/10.1109/PSIVT.2010.44.

Liu, A., Lin, W., Paul, M., Deng, C., and Zhang, F. (2010). Just noticeable difference for images with decomposition model for separating edge and textured regions. *IEEE Transactions on Circuits and Systems for Video Technology* **20** (11): 1648–1652. https://doi.org/10.1109/TCSVT.2010.2087432.

Liu, S., Liu, Z., and Huang, H. (2012). FPGA implementation of a fast pipeline architecture for JND computation. *5th International Congress on Image and Signal Processing (CISP)*, Chongqing, China (16–18 October 2012), 577–581. https://doi.org/10.1109/CISP.2012.6469995.

Lloyd, S.P. (1982). Least squares quantization in PCM. *IEEE Transactions on Information Theory* **28** (2): 129–137. https://doi.org/10.1109/TIT.1982.1056489.

Malvar, H., Sullivan, G., and Srinivasan, S. (2008). Lifting-based reversible color transformations for image compression. *Applications of Digital Image Processing XXXI*, San Diego, CA, USA (11–14 August 2008), SPIE, Volume **7073**, Article ID 707307, 10 pages. https://doi.org/10.1117/12.797091.

Martucci, S.A. (1990). Reversible compression of HDTV images using median adaptive prediction and arithmetic coding. *IEEE International Symposium on Circuits and Systems*, New Orleans, LA, USA (1–3 May 1990), Volume **2**, 1310–1313. https://doi.org/10.1109/ISCAS.1990.112371.

Meardi, G., Ferrara, S., Ciccarelli, L., Cobianchi, G., Poularakis, S., Maurer, F., Battista, S., and Byagowi, A. (2020). MPEG-5 part 2: low complexity enhancement video coding (LCEVC): overview and performance evaluation. *Applications of Digital Image Processing XLIII*, Virtual (24 August – 4 September 2020), SPIE, Volume **11510**, Article ID 115101C, 20 pages. https://doi.org/10.1117/12.2569246.

O'Neal, J.B. (1966). Predictive quantizing systems (differential pulse code modulation) for the transmission of television signals. *The Bell System Technical Journal* **45** (5): 689–721. https://doi.org/10.1002/j.1538-7305.1966.tb01052.x.

Orlandic, M., Fjeldtvedt, J., and Johansen, T.A. (2019). A parallel FPGA implementation of the CCSDS-123 compression algorithm. *Remote Sensing* **11** (6): Article ID 673, 19 pages. https://doi.org/10.3390/rs11060673.

Palaz, O., Ugurdag, H.F., Ozkurt, O., Kertmen, B., and Donmez, F. (2017). RImCom: raster-order image compressor for embedded video applications. *Journal of Signal Processing Systems* **88** (2): 149–165. https://doi.org/10.1007/s11265-016-1211-9.

Papadonikolakis, M.E., Kakarountas, A.P., and Goutis, C.E. (2008). Efficient high-performance implementation of JPEG-LS encoder. *Journal of Real-Time Image Processing* **3** (4): 303–310. https://doi.org/10.1007/s11554-008-0088-7.

Punchihewa, A. and Bailey, D. (2020). A review of emerging video codecs: challenges and opportunities. *35th International Conference on Image and Vision Computing New Zealand*, Wellington, NZ (25–27 November 2020), 181–186. https://doi.org/10.1109/IVCNZ51579.2020.9290536.

Reznik, Y.A., Hinds, A.T., Zhang, C., Yu, L., and Ni, Z. (2007). Efficient fixed-point approximations of the 8x8 inverse discrete cosine transform. *Applications of Digital Image Processing XXX*, San Diego, CA, USA (28–30 August 2007), SPIE, Volume **6696**, 669617. https://doi.org/10.1117/12.740228.

Rissanen, J.J. (1976). Generalized Kraft inequality and arithmetic coding. *IBM Journal of Research and Development* **20** (3): 198–203. https://doi.org/10.1147/rd.203.0198.

Saldanha, M., Correa, M., Correa, G., Palomino, D., Porto, M., Zatt, B., and Agostini, L. (2020). An overview of dedicated hardware designs for state-of-the-art AV1 and H.266/VVC video codecs. *27th IEEE International Conference on Electronics, Circuits and Systems (ICECS)*, Glasgow, Scotland (23–25 November 2020), 4 pages. https://doi.org/10.1109/ICECS49266.2020.9294862.

Santos, L., Berrojo, L., Moreno, J., Lopez, J.F., and Sarmiento, R. (2016). Multispectral and hyperspectral lossless compressor for space applications (HyLoC): a low-complexity FPGA implementation of the CCSDS 123 standard. *IEEE Journal of Selected Topics in Applied Earth Observations and Remote Sensing* **9** (2): 757–770. https://doi.org/10.1109/JSTARS.2015.2497163.

Scavongelli, C. and Conti, M. (2016). FPGA implementation of JPEG encoder architectures for wireless networks. *EURASIP Journal on Embedded Systems* **2017** (1): Article ID 10, 19 pages. https://doi.org/10.1186/s13639-016-0047-5.

Schwarz, H., Coban, M., Karczewicz, M., Chuang, T.D., Bossen, F., Alshin, A., Lainema, J., Helmrich, C.R., and Wiegand, T. (2021). Quantization and entropy coding in the versatile video coding (VVC) standard. *IEEE Transactions on Circuits and Systems for Video Technology* **31** (10): 3891–3906. https://doi.org/10.1109/TCSVT.2021.3072202.

Skodras, A., Christopoulos, C., and Ebrahimi, T. (2001). The JPEG 2000 still image compression standard. *IEEE Signal Processing Magazine* **18** (5): 36–58. https://doi.org/10.1109/79.952804.

Sullivan, G.J., Ohm, J., Han, W., and Wiegand, T. (2012). Overview of the high efficiency video coding (HEVC) standard. *IEEE Transactions on Circuits and Systems for Video Technology* **22** (12): 1649–1668. https://doi.org/10.1109/TCSVT.2012.2221191.

Sun, S.H. and Lee, S.J. (2003). A JPEG chip for image compression and decompression. *The Journal of VLSI Signal Processing* **35** (1): 43–60. https://doi.org/10.1023/A:1023383820503.

Taranto, C. (2022). Simplified affine motion estimation algorithm and architecture for the versatile video coding standard. Master of Science thesis. Turin, Italy: Polytechnic University of Turin.

Taubman, D. (2000). High performance scalable image compression with EBCOT. *IEEE Transactions on Image Processing* **9** (7): 1158–1170. https://doi.org/10.1109/83.847830.

Taubman, D.S. and Marcellin, M.W. (2002). *JPEG2000 Image Compression Fundamentals, Standards and Practice*. New York: Springer. https://doi.org/10.1007/978-1-4615-0799-4.

Thang, N.V. and Dinh, V.N. (2018). High throughput and low cost memory architecture for full search integer motion estimation in HEVC. *International Conference on Advanced Technologies for Communications (ATC)*, Ho Chi Minh City, Vietnam (18–20 October 2018), 174–178. https://doi.org/10.1109/ATC.2018.8587488.

Tuan, J.C., Chang, T.S., and Jen, C.W. (2002). On the data reuse and memory bandwidth analysis for full-search block-matching VLSI architecture. *IEEE Transactions on Circuits and Systems for Video Technology* **12** (1): 61–72. https://doi.org/10.1109/76.981846.

Varma, K., Damecharla, H.B., Bell, A.E., Carletta, J.E., and Back, G.V. (2008). A fast JPEG2000 encoder that preserves coding efficiency: the split arithmetic encoder. *IEEE Transactions on Circuits and Systems I: Regular Papers* **55** (11): 3711–3722. https://doi.org/10.1109/TCSI.2008.927221.

Vinh, T.Q. and Kim, Y.C. (2010). Edge-preserving algorithm for block artifact reduction and its pipelined architecture. *ETRI Journal* **32** (3): 380–389. https://doi.org/10.4218/etrij.10.0109.0290.

Wallace, G.K. (1992). The JPEG still picture compression standard. *IEEE Transactions on Consumer Electronics* **38** (1): 17–34. https://doi.org/10.1109/30.125072.

Wang, Z. (2019). Resource efficient and quality preserving real-time image compression for embedded imaging systems. PhD thesis. Stuttgart, Germany: Stuttgart University.

Wang, Z., Tran, T.H., Muthappa, P.K., and Simon, S. (2019). A JND-based pixel-domain algorithm and hardware architecture for perceptual image coding. *Journal of Imaging* **5** (5): Article ID 50, 29 pages. https://doi.org/10.3390/jimaging5050050.

Watson, A. (1993). DCT quantization matrices visually optimized for individual images. *Human Vision, Visual Processing, and Digital Display IV*, San Diego, CA, USA (1–4 February 1993), SPIE, Volume **1913**, 202–216. https://doi.org/10.1117/12.152694.

Weinberger, M.J., Seroussi, G., and Sapiro, G. (2000). The LOCO-I lossless image compression algorithm: principles and standardization into JPEG-LS. *IEEE Transactions on Image Processing* **9** (8): 1309–1324. https://doi.org/10.1109/83.855427.

Welch, T.A. (1984). A technique for high-performance data compression. *IEEE Computer* **17** (6): 8–19. https://doi.org/10.1109/MC.1984.1659158.

Wiegand, T., Sullivan, G.J., Bjontegaard, G., and Luthra, A. (2003). Overview of the H.264/AVC video coding standard. *IEEE Transactions on Circuits and Systems for Video Technology* **13** (7): 560–576. https://doi.org/10.1109/TCSVT.2003.815165.

Witten, I.H., Neal, R.M., and Cleary, J.G. (1987). Arithmetic coding for data compression. *Communications of the ACM* **30** (6): 520–540. https://doi.org/10.1145/214762.214771.

Wong, S., Vassiliadis, S., and Cotofana, S. (2002). A sum of absolute differences implementation in FPGA hardware. *28th Euromicro Conference*, Dortmund, Germany (4–6 September 2002), 183–188. https://doi.org/10.1109/EURMIC.2002.1046155.

Yang, X.K., Ling, W.S., Lu, Z.K., Ong, E.P., and Yao, S.S. (2005). Just noticeable distortion model and its applications in video coding. *Signal Processing: Image Communication* **20** (7): 662–680. https://doi.org/10.1016/j.image.2005.04.001.

Yu, N., Kim, K., and Salcic, Z. (2004). A new motion estimation algorithm for mobile real-time video and its FPGA implementation. *IEEE Region 10 Conference (TENCON)*, Chiang Mai, Thailand (21–24 November 2004), Volume **1**, 383–386. https://doi.org/10.1109/TENCON.2004.1414437.

Ziv, J. and Lempel, A. (1977). "A universal algorithm for sequential data compression". *IEEE Transactions on Information Theory* **23** (3): 337–343. https://doi.org/10.1109/TIT.1977.1055714.

13

Blob Detection and Labelling

Segmentation divides an image into regions which have a common property. One common approach to segmentation is to enhance the common property through filtering, followed by thresholding to detect the pixels which have the enhanced property. Filtering has been covered in Chapter 9, and various forms of thresholding in Sections 7.1.2, 7.3.3, 8.1.5, 8.2.3, and 9.3.6. However, these operations have processed the pixels individually (using local context in the case of filters). To analyse objects within the image it is necessary to associate groups of related pixels to one another and assign them a common label. This enables object data to be extracted from the corresponding groups of pixels.

The transformation from individual pixels to objects is an intermediate-level operation. Since the input data is still in terms of pixels, it is desirable where possible to use stream-based processing. The output consists of a set of blobs or blob descriptions and may not necessarily be in the form of pixels.

Of the many approaches for blob detection and labelling, only the bounding box, run-length coding, chain coding, and connected component analysis (CCA) will be considered here. Field-programmable gate array (FPGA) implementation of other region-based analysis techniques such as the distance transform, watershed transform, and Hough transform is considered at the end of this chapter.

13.1 Bounding Box

The bounding box of a set of pixels is the smallest rectangular box aligned with the image axes that contains all of the detected points. It is assumed that the input pixel stream has been thresholded or segmented by preprocessing operations. Let C_i be the set of pixels associated with object i. Then, the bounding box of C_i is the tuple $\{x_{min}, x_{max}, y_{min}, y_{max}\}$, where

$$\begin{aligned}
x_{min} &= \min\{x_p \mid (x_p, y_p) \in C_i\},\\
x_{max} &= \max\{x_p \mid (x_p, y_p) \in C_i\},\\
y_{min} &= \min\{y_p \mid (x_p, y_p) \in C_i\},\\
y_{max} &= \max\{y_p \mid (x_p, y_p) \in C_i\}.
\end{aligned} \tag{13.1}$$

Initially assume that the input image is binary; Figure 13.1 shows the basic stream processing implementation. The shaded block is only enabled for detected pixels, so that background pixels are effectively ignored. The *init* register is used to initialise the bounding box when the first pixel is detected. It is preset to 1 during the vertical blanking period before the image is processed, so that it loads the x_{min}, x_{max}, and y_{min} registers when the first object pixel is encountered. Since the image is scanned in raster order, the first pixel detected will provide the y_{min} value. The y value of every detected pixel is clocked into the y_{max} register because the last pixel detected

---

Design for Embedded Image Processing on FPGAs, Second Edition. Donald G. Bailey.

Companion Website: www.wiley.com/go/bailey/designforembeddedimageprocessc2e

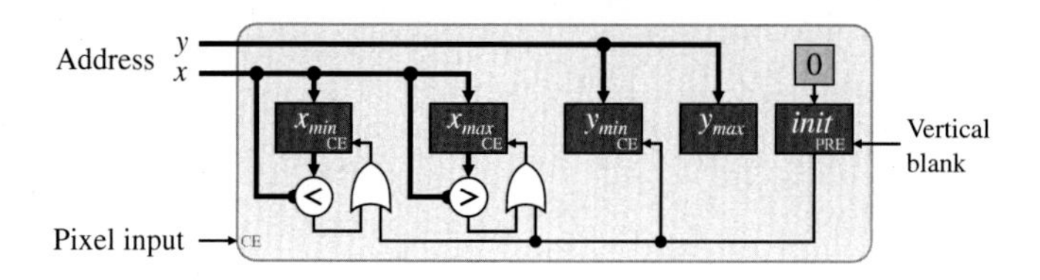

Figure 13.1 Basic bounding box implementation. (CE = clock enable; PRE = preset.)

in the frame will have the largest y. After the first pixel, the x value is compared with x_{min} and x_{max} with the corresponding value updated if the object is extended. At the end of the frame, the four registers contain the bounding box of the object pixels within the image.

The basic implementation can be readily extended to process multiple input pixels per clock cycle. It is easiest if the number of pixels processed simultaneously is a power of 2. The x address then contains only the most significant bits of the address, with the least significant bits provided by two decoders that determine the leftmost (for x_{min}) and rightmost (for x_{max}) object pixels. These decoders can easily be built from the LUTs within the FPGA. A similar approach can be used if multiple rows are processed simultaneously.

If there are multiple input classes, the incoming pixel stream will have a label corresponding to the class. All of the classes can be processed in parallel. However, rather than build separate hardware for each class, the fact that each input pixel can only have one label enables the update logic to be shared (when processing one pixel per clock cycle), with the registers multiplexed by the label of the incoming pixel. An efficient implementation is to use a small fabric RAM block for each register with the address provided by the pixel label. This effectively hides the multiplexing within the address decoding logic. The RAMs for x_{min} and x_{max} need to be dual-port, because the value needs to be read for the comparison. Figure 13.2 assumes a pipelined read, so the label and x value need to be buffered for writing in the following clock cycle. Otherwise, the hardware follows much the same form as the basic implementation in Figure 13.1.

From the bounding box, it is easy to derive the centre position, size, and aspect ratio of the object:

$$centre = \left(\frac{x_{max} + x_{min}}{2}, \frac{y_{max} + y_{min}}{2}\right), \tag{13.2}$$

$$size = (x_{max} - x_{min} + 1, y_{max} - y_{min} + 1), \tag{13.3}$$

$$aspect\ ratio = \frac{y_{max} - y_{min} + 1}{x_{max} - x_{min} + 1}. \tag{13.4}$$

The basic system of Figure 13.1 can be extended to determine orientation of long thin objects if required. Consider an object oriented top-right to bottom-left; the register x_{min} is adjusted on most rows, whereas x_{max} is only adjusted on a few rows at the top. The converse is true for an object oriented top-left to bottom-right. Although they may have the same bounding boxes, these can be discriminated by incrementing an accumulator,

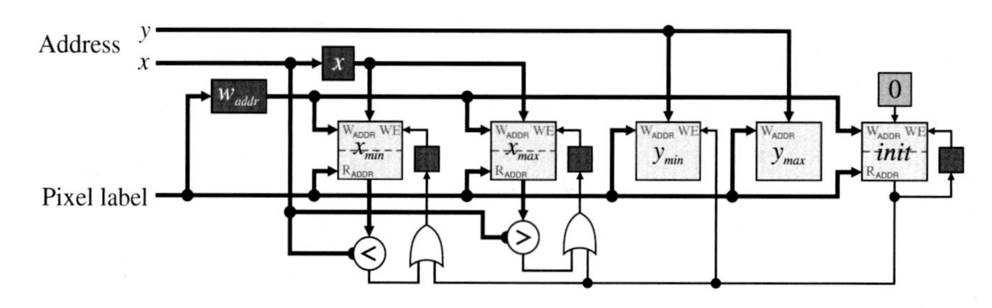

Figure 13.2 Bounding box of multiple labels in parallel. (WE = write enable.)

acc, whenever x_{max} is adjusted and decrementing it whenever x_{min} is adjusted. At the end of the frame, the sign of the accumulator indicates the orientation of the object. The orientation can then be estimated as

$$orientation \approx \text{sign}(acc) \times \tan^{-1}\left(\frac{y_{max} - y_{min}}{x_{max} - x_{min}}\right). \tag{13.5}$$

The main limitation of the bounding box is that it does not ensure or require the pixels to be connected. It is also sensitive to noise. Any noise on the boundary at the extreme points will affect the limits, and consequently the measurements derived in Eqs. (13.2) to (13.5). An isolated noise pixel away from the object will extend the bounding box to include that pixel, giving meaningless results. Therefore, it is essential to apply appropriate filtering to remove noise before using the bounding box. Note that if pre-filtering using a morphological opening, it is sufficient to just perform the erosion, because the effects of the dilation can be taken into account by adjusting Eqs. (13.2) to (13.5).

The biggest advantage of a simple bounding box is the low processing cost. Therefore, if the noise can be successfully filtered, it provides an efficient method of extracting some basic object features.

13.2 Run-length Coding

Many intermediate-level image processing algorithms require more than one pass through the image. In this case, memory bandwidth is often the bottleneck on processing speed. For some image processing operations, this may be overcome by compressing the image and processing runs of pixels as a unit, rather than processing individual pixels. In all but a few pathological cases, there are significantly fewer runs in an object than there are pixels, allowing a significant acceleration of processing.

Several ways of encoding the runs are demonstrated in Figure 13.3. Image compression commonly encodes each run as a (*class*, *length*) pair, although absolute position is only implicit with this representation. An alternative that makes the absolute position explicit is the (*class*, *start*) encoding, with the length implied from the difference between adjacent starts. For binary images, it is only necessary to record (*start*, *end*) of each run of object pixels, and not code the background. It this case, it is also necessary to have an additional symbol (*e*) to indicate the end of a line when it is a background pixel.

Since each row is coded independently, multiple rows may be processed in parallel by building multiple processors.

The application of run-length coding for blob detection and labelling will be considered briefly in Sections 13.3 and 13.4. Run-length coding can also be used for texture characterisation although in general it is not as powerful as other texture analysis methods (Conners and Harlow, 1980). However, the distribution of run-lengths of binary images can be used to discriminate different textures and can provide statistical information on object sizes.

13.3 Chain Coding

One common intermediate-level representation of image regions is to encode each region by its boundary. This reduces each 2-D shape to its 1-D boundary. The reduction in dimensionality can often significantly reduce

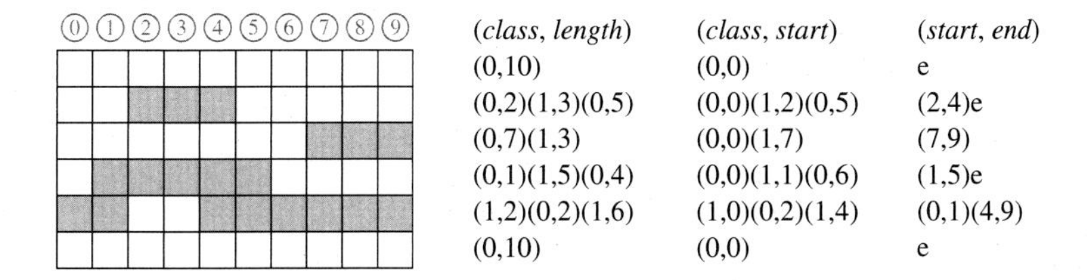

Figure 13.3 Different schemes for run-length encoding.

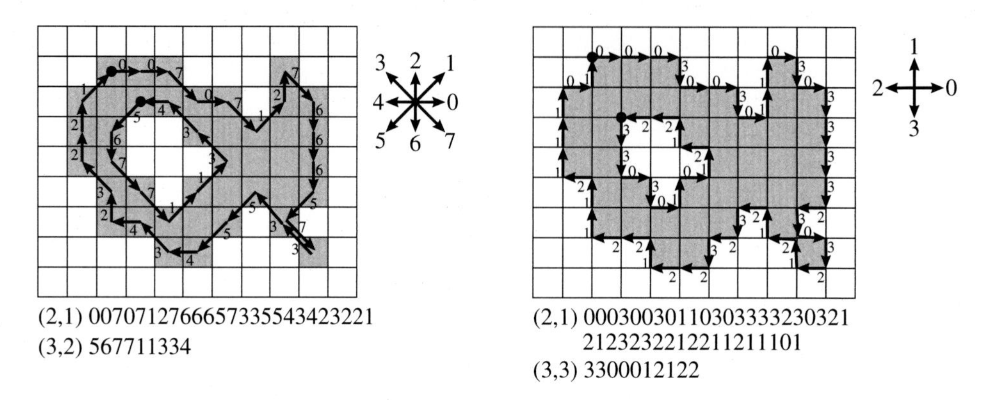

Figure 13.4 Boundary representation codes. Left: chain code; right: crack code. The filled circle represents the start point of the boundary.

the volume of data for representing an object, especially where the regions are large and have relatively smooth boundaries. This can lead to efficient processing algorithms for extracting region features. An image of the region may be reconstructed from the boundary code by drawing the boundary within an empty image, and filling between the boundary pixels using a raster scan, producing a streamed output.

There are several different boundary representations (Wilson, 1997), of which the most common is the Freeman chain code (Freeman, 1961). Each pixel is coded with the direction of the next pixel in the sequence as demonstrated on the left in Figure 13.4, often represented using a linked list. The chain code follows the object pixels on the boundary, keeping the background on the left. Consequently, the outside boundary is traversed clockwise, and holes are coded anti-clockwise. An alternative code is the crack code, which follows the cracks between the object and background pixels, as shown on the right in Figure 13.4.

13.3.1 Sequential Implementation

A software algorithm for chain coding is to perform a raster scan of the image until an unprocessed boundary pixel is encountered. The raster scan is suspended while the object boundary is traced sequentially, adding links to the boundary code. Traced pixels are marked as processed to prevent them from being detected again later when scanning is resumed. Boundary following continues until the complete boundary has been traced. At this point, the raster scan is resumed.

The boundary tracing step requires random access to the image, so it does not fit well within a stream processing framework. The processing time depends on image complexity, with up to six pixels read for each step with chain coding (or two pixels for crack coding). An approach similar to this was implemented on an FPGA by Hedberg et al. (2007). Note that in some applications, the chain codes do not actually have to be saved; features such as area, perimeter, centre of gravity, and the bounding box can be calculated directly as the boundary is scanned.

Scanning may be accelerated using run-length codes as an intermediary (Kim et al., 1988). In the first pass, the image is run-length encoded, and the run adjacency determined. The boundary scan can then be deferred to a second pass, where it is now potentially faster because runs can be processed rather than individual pixels.

13.3.2 Single-pass Stream Processing Algorithms

Even better is to perform the whole process in one pass during the raster scan. The basic principle behind building a chain code in a single raster scan pass is to maintain a set of chain code fragments and grow both ends of each chain fragment as each row is scanned. In this way, all of the chains are built up in parallel as the image is streamed in. The basic principle is illustrated in Figure 13.5. As the image is streamed, each edge

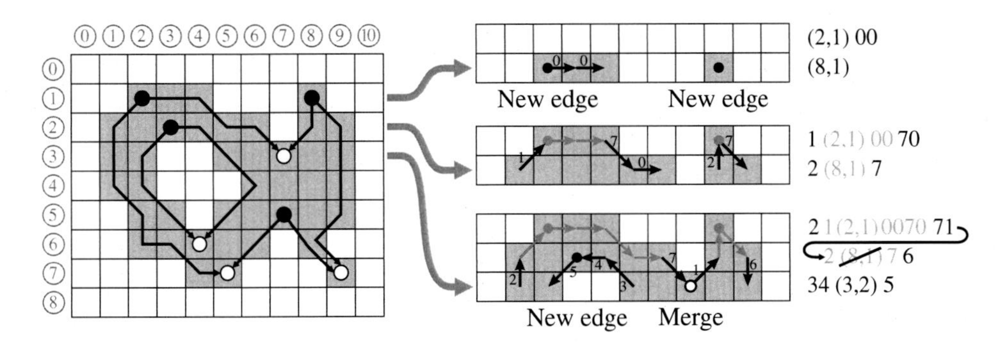

Figure 13.5 Growing a chain code with stream processing. Left: solid circles represent the starting points of new edges, with the lines showing the direction that the fragments are extended from these starting points, until the paths merge at the hollow circles; right: shows how the boundary is built for the first three rows of the object.

is detected, and the chain code links are determined from the local context. Consider the scan of line ① in Figure 13.5. Two separate objects are encountered, and chains begun. For **U**-shaped boundaries, there will be multiple starts for each boundary, occurring at the local tops of the object and background (Cederberg, 1979). As the next line is scanned, the chains are extended on both the left and right ends. On line ③, a new edge is started for the concavity that will eventually become the hole. The two fragments from the original two chains also merge at the bottom of the **U**, and the corresponding chains linked.

Such processing complicates the memory management. When only one chain is encoded at a time and is built in one direction only (by tracking around the boundary of the object), the chain links can simply be stored sequentially in an array. Linking from one chain link to the next is implicit in the sequential addressing. However, building multiple fragments simultaneously, with both the start and the end being extended, requires some form of linked list. The storage space required for the linking is often significantly larger than the storage required for the chain code itself, and the multiple memory accesses required for adding links (see Section 5.4.4) require careful memory system design.

Several software approaches using this basic technique are described in the literature. Cederberg (1979) scanned a 3×3 window through the image to generate the chain links and only coded the left and right fragments as illustrated by the arrows on the left in Figure 13.5. He did not consider linking the fragments. Chakravarty (1981) also scanned a 3×3 window through the image and added segments built to a list. The algorithm involved a search to match corresponding segments as they are constructed. Shih and Wong (1992) used a similar approach to extend mid-crack codes, using a lookup table of the pixels within a 3×3 window to derive the segments to be added. Mandler and Oberlander (1990) used a 2×2 window to detect the corners in a crack code, which were then linked together in a single scan through the image. Their technique did not require thresholding the image first and derived the boundaries of each connected region of the same pixel value. This meant that each boundary is coded twice: once for the pixel value on each side of the crack. Zingaretti et al. (1998) performed a similar encoding of greyscale images. Their technique used a form of run-length coding, looking up a 3×2 block of run-lengths to create temporary chain fragments, and linking these together as the image was scanned. Unfortunately, memory management complexity makes these algorithms difficult to implement efficiently on an FPGA.

Based on stream processing on an FPGA, Bailey (2010) proposed run-length coding the cracks between pixels, with a simple linking mechanism from one row to the next. This hybrid crack run-length code is demonstrated on the left in Figure 13.6. Each run of pixels consists of the length of the run plus one bit indicating whether the run links up to the previous row or down to the next row. A 2×2 window is scanned through the image, with the start of a run on the previous row generating a $\uparrow$, and the end of a run on the current row generating a $\downarrow$. Row buffers are used to maintain link context from one row to the next, so that the chain links are only written to memory once during coding.

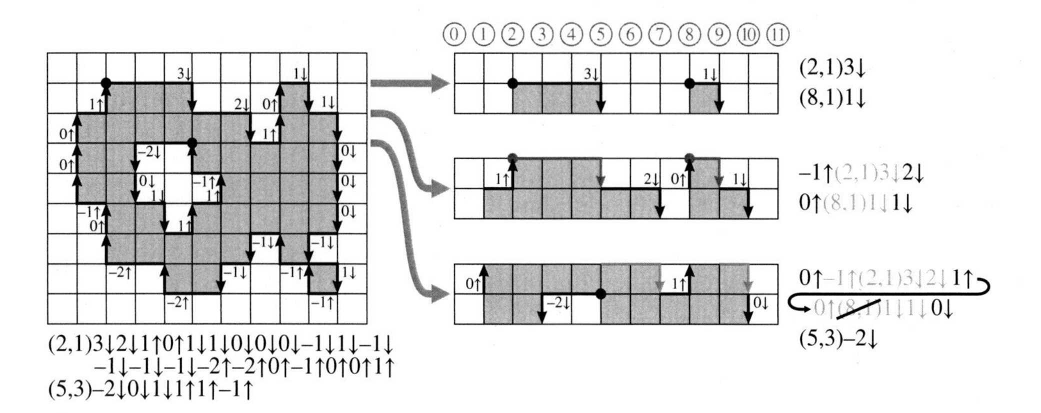

Figure 13.6 Raster-based crack run-length coding. Left: the coding of the regions; right: building the linked lists representing the boundary, showing the first three lines of cracks.

If necessary, the crack run-length code can be converted to a conventional crack code or Freeman chain code for subsequent processing. However, it is also possible to perform much of the processing directly on the crack run-length code.

13.3.3 Feature Extraction

Since the boundary contains information about the shape of a region, many shape features may be derived directly from the chain code (or other boundary code). Note, however, that most of these algorithms are largely sequential, making their FPGA implementation not particularly efficient.

The most obvious measurement from the chain code is the perimeter of the object. Unfortunately, simply counting the number of steps within the chain or crack code will overestimate the length because of quantisation effects. The error in the estimate also depends on the angle of the boundary within the image. To overcome this problem, it is necessary to smooth the boundary by measuring the distance between points k steps on either side, and assigning the average as the step size at each link (Proffitt and Rosen, 1979):

$$Perimeter = \frac{1}{k}\sum_i \sqrt{(y_{i+k} - y_i)^2 + (x_{i+k} - x_i)^2}. \tag{13.6}$$

For long straight boundaries, increasing k can reduce the error in the estimate of the perimeter to an arbitrarily low level. However, this will result in cutting corners, so making k too large will under-estimate the perimeter. For objects with rough boundaries, the perimeter will depend on the scale of the measurement, enabling a fractal dimension to be calculated as a feature.

Corners can be defined as regions of extreme curvature. Again, the quantisation effects on the boundary mean that the local curvature is quite noisy and can be smoothed by considering points k links apart. A simple corner measure is the difference in angle between the points k links on either side of the current point (see Figure 13.7):

$$\Delta\theta_i = \tan^{-1}\frac{y_{i+k} - y_i}{x_{i+k} - x_i} - \tan^{-1}\frac{y_i - y_{i-k}}{x_i - x_{i-k}}. \tag{13.7}$$

Since k is a small integer, and the steps are integers, the angles may be determined by table lookup. (For example, with $k = 3$, $-3 \leq y_{i+3} - y_i, x_{i+3} - x_i \leq 3$, requiring only 4 bits from each of Δx and Δy, or 8 input bits total for the lookup table to implement the arctangent.) Corners can be defined as peaks in $\Delta\theta_i$ that are over a predefined angle. A more sophisticated corner measure that takes into account the length over which the angle changes is given by Freeman and Davis (1977).

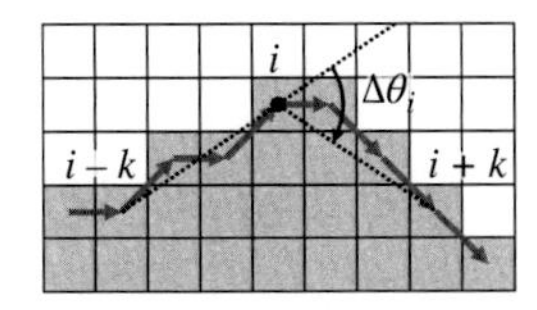

Figure 13.7 Measuring curvature for detecting corners.

Measuring the area of a region consists of counting the number of pixels within the boundary. This may be obtained directly from the boundary using contour integration (Freeman, 1961). It is important to note that the chain code is on one side of the boundary, so it will underestimate the area of regions and overestimate the area of holes unless the part pixels outside the chain are taken into account. The crack code does not have the same problem. Expressing the area in terms of the crack codes,

$$Area = \oint_{chain} 1\, dx\, dy = \sum_j \left(\sum_i^j \Delta x_i \right) \Delta y_j. \tag{13.8}$$

This is amenable to direct calculation from the crack run-length codes. The area is generally much less sensitive than the perimeter to digitisation errors and noise.

Blobs will have a positive area, while holes will have a negative area. This enables external and internal boundaries to be distinguished, enabling the number of holes, and the Euler number to be calculated.

The centre of gravity may be determined from the contour integral for the first moment:

$$\begin{aligned} x_{COG} &= \frac{S_x}{Area}, \quad S_x = \oint_{chain} x\, dx\, dy = \frac{1}{2} \sum_i x_i^2 \Delta y_i, \\ y_{COG} &= \frac{S_y}{Area}, \quad S_y = \oint_{chain} y\, dx\, dy = \frac{1}{2} \sum_i x_i (y_{i-1} + y_i) \Delta y_i. \end{aligned} \tag{13.9}$$

Higher order moments may be calculated in a similar way.

The convex hull may also be determined from the boundary code. Consider each link consisting of an arbitrary offset, $(\Delta x_i, \Delta y_i)$. Point i on the boundary is concave if

$$\Delta x_i \Delta y_{i-1} \leq \Delta x_{i-1} \Delta y_i. \tag{13.10}$$

A concave point may then be eliminated by combining it with the previous point:

$$(\Delta x_{i-1}, \Delta y_{i-1}) \Leftarrow (\Delta x_{i-1}, \Delta y_{i-1}) + (\Delta x_i, \Delta y_i). \tag{13.11}$$

If this is performed for every pair of adjacent points until no further points can be eliminated, then the remaining points are the vertices of the minimum convex polygon containing the region. (Note that holes will reduce to nothing since they are concave. To obtain the convex hull of a hole, the left- and right-hand-sides of Eq. (13.10) need to be swapped.)

From the convex hull, the depth of the bays may be determined, enabling a concavity analysis approach to separating touching convex objects (Bailey, 1992). The principle is that a convex object should have no significant concavities; so, a concavity indicates multiple touching objects. From the deepest concavity, an opposite concavity is found, and the object separated along the line connecting them. This is repeated until no significant concavities remain.

From the convex hull, the minimum area enclosing rectangle may be determined (Freeman and Shapira, 1975). The minimum area enclosing rectangle has one side coincident with the minimum convex polygon, so each edge can be tested in turn to find the extent of the region both parallel and perpendicular to that edge. The extents that give the smallest area correspond to the minimum area enclosing rectangle.

The bounding box (aligned with the pixel axes) can be found from a simple scan around the boundary, finding the extreme points in each direction.

Chain codes may also be used for boundary or shape matching (Freeman, 1974); however, a full description is beyond the scope of this work.

One set of shape features that may be extracted from boundary codes are Fourier descriptors. For a closed boundary, the shape is periodic, enabling analysis by Fourier series (Zhang and Lu, 2002). While any function of the boundary points could be used, the simplest is the radius from the centre of gravity as a function of angle. The main limitation of this is that the shape must be fairly simple for the radius function to be single-valued. An arbitrary shape can be represented directly by its x and y coordinates, as a complex number: $x(t) + jy(t)$, as a function of length along the perimeter, t. Alternatively, the angle of the tangent as a function of length can be used, normalised to make it periodic (Zahn and Roskies, 1972). Whichever features are used, the shape may be represented by the low-order Fourier series coefficients. The method is also relatively insensitive to quantisation noise and with normalisation can be used for rotation and scale invariant matching. The Fourier descriptors also allow a smoothed boundary to be reconstructed from a few coefficients.

13.4 Connected Component Labelling (CCL)

While chain coding concentrates on the boundaries, an alternative approach labels the pixels within the connected components. Each set of connected pixels is assigned a unique label, enabling the pixels associated with a region to be isolated and features of that region to be extracted. There are several distinct approaches to connected component labelling (CCL) (He et al., 2017).

13.4.1 Random Access Algorithms

The first extends the ideas of chain coding by determining the boundary and then filling in the object pixels inside the boundary with a unique label (Chang et al., 2004). Similar to chain coding, the image is scanned in a raster fashion until an unlabelled object pixel is encountered. A new label is assigned to this object, and the boundary is scanned. Rather than extract chain codes of the boundary, the boundary pixels are assigned the new label. When the raster scan is resumed, these boundary labels are propagated internally to label the pixels associated with the object. Such an approach was implemented on an FPGA by Hedberg et al. (2007). The boundary scanning, being random access, requires the complete image to be available in a frame buffer. Little auxiliary memory structures are required, as the image can be labelled in place directly within the frame buffer. Features that can be extracted directly from chain codes may also be extracted during the boundary scan phase.

An alternative to scanning around the boundary of an object is to perform a flood-fill operation when an unlabelled pixel is encountered during the raster scan (AbuBaker et al., 2007). A flood-fill suspends the raster scan and defines the unlabelled pixel as a seed pixel for the new region. It then uses a form of region growing to fill all pixels connected to the seed pixel with the allocated label. A row can be filled easily, with unlabelled connected pixels in the rows immediately above and below added to a stack. Then, when no more unlabelled pixels are found adjacent to the current row, the seed is replaced by the top of stack, and flooding continues. By carefully choosing which pixels are pushed onto the stack, an efficient algorithm can be obtained (AbuBaker et al., 2007).

13.4.2 Multiple Pass Algorithms

To avoid the need for random access processing, the alternative is to propagate labels from previously labelled connected pixels. The basic idea is illustrated in Figure 13.8. When an object pixel is encountered during the initial raster scan, its neighbours to the left and above are examined. If all are background pixels, a new label is assigned. Otherwise, the minimum nonzero (non-background) label is propagated to the current pixel. This is accomplished by having the input multiplexer select the *Label* counter during the first pass. If *Label* is the minimum nonzero value, then the new label is used, and *Label* is incremented. For **U**-shaped objects, separate

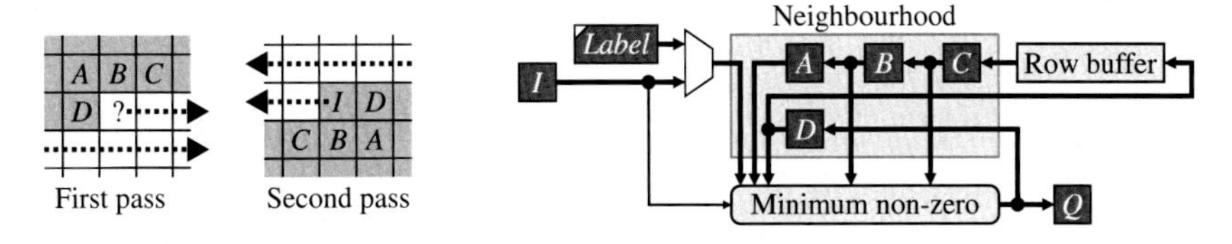

Figure 13.8 Multi-pass connected component labelling. Left: neighbourhood context, with the dotted lines showing the scan direction; right: label propagation circuit.

labels will be assigned for each branch of the **U**. At the bottom of the **U**, the minimum of the two labels is selected and propagated. However, the pixels that have already been labelled with the non-minimum label must be relabelled. The simplest method of accomplishing this is to perform a second, reverse, raster scan through the image, this time from the bottom-right corner back up through the image. In the second pass (and subsequent passes), the input multiplexer is switched to the input pixel (the current label), effectively including it within the neighbourhood. The second pass propagates the minimum nonzero label back up the branch, replacing the larger initial label. For **W**-shaped objects and spirals, additional forward and reverse passes through the image may be required to propagate the minimum label to all pixels within the connected component. Therefore, the process is iterated until there are no further changes within the image.

This iterative, multi-pass labelling approach was implemented on an FPGA by Crookes and Benkrid (1999) and Benkrid et al. (2003). The circuitry is very simple and quite small, although a frame buffer is required for intermediate storage (with the processing performed in place). However, the number of passes through the image depends on the complexity of the connected components and is indeterminate at the start of processing.

13.4.3 Two-pass Algorithms

Label propagation may be reduced to two passes by explicitly keeping track of the pairs of equivalent labels whenever two parts of a component merge. The classic software algorithm (Rosenfeld and Pfaltz, 1966) does just this and is illustrated with a simple example in Figure 13.9. The first pass is similar to the first pass of the multi-pass algorithm as described above. At the bottom of **U**-shaped regions, where sub-components merge, there will be two different labels within the neighbourhood (in Figure 13.10, *B* will be background, and the label of *C* will be different to that of *A* or *D*). These labels are equivalent in that they are associated with the same connected component. The smaller label is retained, and all instances of the larger label must be converted to the smaller label. Rather than use multiple additional passes to propagate these changes through the image, whenever two sub-components merge, their labels are recorded as equivalent. The issue of equivalence resolution has received much attention in the literature and is the main area of difference between many related algorithms (see, for example, (Wu et al., 2005; He et al., 2007)).

A wide variety of techniques has been used to reduce the memory requirements of the data structures used to represent the sets of equivalent labels and to speed the search for the representative label of a set. One common

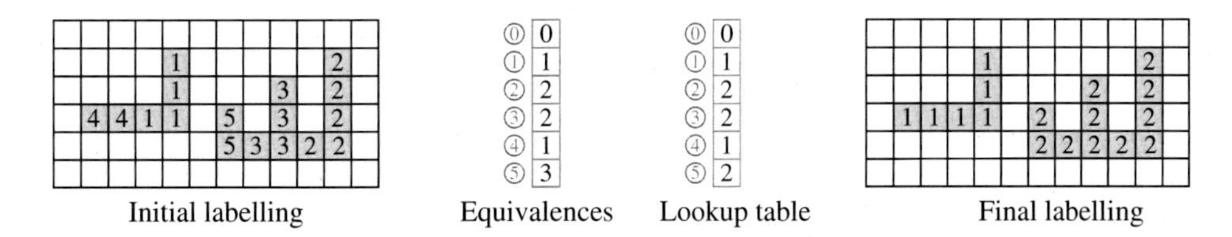

Figure 13.9 Connected component labelling process. From the left: the initial labelling after the first pass; the equivalence or merger table; the lookup table after resolving equivalences; the final labelling after the second pass.

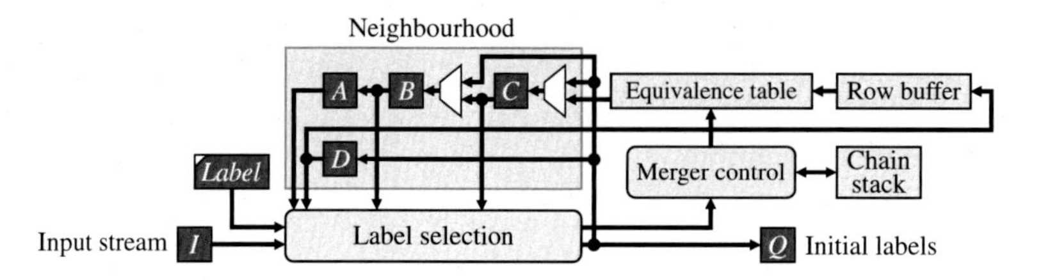

Figure 13.10 First pass logic required to efficiently implement merging.

representation is to use an equivalence table to represent a tree structure between equivalent labels. The links in the tree always point from a larger label to a smaller label, with the smallest label in the tree (the root) being the representative label. When a new label is created, its entry in the equivalence table is set to point to itself, creating a new tree. Whenever a merger occurs, the two trees are merged, usually using some form of union-find algorithm (Klaiber et al., 2019a). If using a simple equivalence table, and a required link already has an entry (as a result of a prior merger), multiple clock cycles may be required to update the equivalence table to keep it consistent, limiting its use within a streamed implementation. One approach is to partially resolve the mergers with each row (Bailey and Johnston, 2007), as outlined in Figure 13.10. The basic idea is to prevent old labels from propagating further. To accomplish this, it is necessary to look up the output of the row buffer in the equivalence table to replace old labels with the current minimum (representative) label for a sub-component. Klaiber et al. (2019a) show that this is possible with a single lookup, enabling streamed operation. The merger control block performs the following steps to maintain the equivalence table:

- Initialises the entry in the table when a new label is allocated, so that the label is returned unchanged.
- In the event of a merger, updates the table to translate the larger label to return the smaller label. Such an update requires the equivalence table to be implemented using dual-port memory. (The multiplexers on the input of *B* and *C* also ensure that the old label within the neighbourhood is replaced by the minimum label.)
- Records links where the smaller label is in position *C* into a chain stack (Bailey and Johnston, 2007). Chains of mergers (such as the $5 \rightarrow 3 \rightarrow 2$ in Figure 13.9) can then be unlinked (also called path compression (Klaiber et al., 2019a)) by propagating links in the equivalence table in reverse scan order during the horizontal blanking period. With pipelining, unchaining requires one clock cycle per stack entry (Bailey and Johnston, 2007).

At the end of the first pass, it is necessary to reconcile chains of mergers that have taken place on different rows, by following links to the minimum label for the component. A single pass through the equivalence table from the smallest label through to the largest is sufficient to flatten the tree structure, so that each label maps to the final label. If desired, the final labels may also be renumbered consecutively during this pass.

The second pass through the image simply uses the flattened equivalence table as a lookup table to translate the initial labels to the final label used for each connected component.

An early FPGA implementation of the standard two-pass algorithm (Rachakonda et al., 1995) used a bank-switched memory architecture to enable every frame to be processed. As the initial labelling pass was performed on one image, the relabelling pass was performed on the previous frame. An implementation of the standard two-pass algorithm using Handel-C is described by Jablonski and Gorgon (2004), where they show how the software algorithm is ported and refined to exploit parallelism.

In software, the second pass can be accelerated using run-length coding (He et al., 2008), because in general there will be fewer runs than pixels. In hardware though, little is gained unless the image is already run-length encoded by a previous operation. Otherwise, the pixels still have to be read one by one, and all of the processing for each pixel takes place in a single clock cycle (although several rows can be run-length encoded in parallel (Trein et al., 2007)). If a streamed output image is required, the only advantage of run-length encoding is to reduce the size of the intermediate frame buffer. However, if the output is an image in a frame buffer rather than streamed, then run-length coding can provide additional acceleration (Appiah et al., 2008). During the first pass, in addition to storing the temporary labels in a frame buffer, the image is run-length coded. Then, during the second relabelling pass, whole runs can be relabelled at a time, skipping over background pixels.

Ito and Nakano (2010) relaxed the definition of connected components by limiting the maximum depth of **U** concavities to k rows (giving so-called k-convex connected components). This removes the need for an intermediate frame buffer, reducing it to k row buffers to hold the preliminary labels. The relabelling pass begins k rows after the initial preliminary labelling, significantly reducing the latency, and enabling the whole process to operate on streamed images. For many applications, the k-convexity restriction is quite realistic (for an appropriate k).

13.4.4 Parallel Algorithms

The algorithms considered so far are largely sequential. Although connectivity is defined locally, labelling is not local as distant pixels could potentially be connected, requiring the same label. This has led to several different parallel architectures to speed up the label propagation (Alnuweiri and Prasanna, 1992). Some of these approaches use massive parallelism, making their FPGA implementation very resource intensive.

One technique that may be exploited by an FPGA implementation is to divide the image into non-overlapping blocks. A separate processor is then used to label each of the blocks in parallel. In the merger resolution step, it is also necessary to consider links between adjacent blocks. This requires local communication between the processors for adjacent blocks. The second pass, relabelling each block with the final label, can also be performed on each block in parallel.

Another useful acceleration approach is to process blocks of pixels at a time. All of the object pixels within a 2×2 block are connected and will therefore have the same label. This enables 2×2 blocks of pixels to be processed as a single unit (Grana et al., 2010; Chang et al., 2015). Label propagation follows similar principles to that described above, although a little more complex because of the increased connectivity complexity between blocks. Most of the advances in software have been to optimise the order in which the neighbours are searched to reduce the number of tests required to make a decision for the current block (Bolelli et al., 2020). However, many of these optimisations are less applicable for a hardware implementation. Block processing was implemented on an FPGA by Kowalczyk et al. (2021), although this is in the context of CCA rather than labelling.

13.4.5 Hysteresis Thresholding

A task closely related to CCL is that of hysteresis thresholding (for example as used by the Canny edge detection filter). Two threshold levels are used:

- All pixels above the upper threshold are considered detected.
- Pixels above a lower threshold are detected only if they are connected to a pixel above the upper threshold (hence the hysteresis).

Many FPGA implementations of the Canny filter simplify this to either using only a single threshold or only checking if a neighbouring pixel (within a 3×3 window) is above the upper threshold. Jang et al. (2019) adapted the two-pass CCL approach to implement true hysteresis thresholding for the Canny filter.

13.5 Connected Component Analysis (CCA)

The global nature of CCL requires a minimum of two passes through the image to produce a labelled image. Frequently, CCL is followed by feature extraction from each region, with the labelling operation used to distinguish between separate regions. In such cases, it is not actually necessary to produce the labelled image as long as the correct feature vector for each connected component can be determined.

The basic principle behind single-pass CCA approaches is to extract the data required to calculate the features during the first pass (Bailey, 1991). This requires a data table to accumulate the feature vector for each connected component. When two sub-components merge, the data associated with the sub-components is also merged. Therefore, at the end of the image scan, feature data has been accumulated for each connected component.

The data that needs to be maintained depends on the features that are being extracted from each connected component. The features that can be accumulating on-the-fly are limited by the ability to associatively combine the data for each region when a merger occurs. Obviously, some features are unable to be calculated incrementally; for these, it is necessary to derive the labelled image prior to feature extraction. Some of the key features that can be determined in a single pass are:

- The number of components; this can be determined with a single global counter. Each time a new label is assigned, the counter is incremented, and when two sub-components merge, the counter is decremented.
- The area of each region requires a separate counter for each label. When two regions are merged, the corresponding counts are combined.
- The centre of gravity of the region requires maintaining sums of the x and y coordinates. The corresponding sums are simply added on merging to give the new sum for the combined region. At the end of the image, the centre of gravity may be obtained by normalising the sums by the area:

$$(x_{cog}, y_{cog}) = \left(\frac{\sum x}{\sum 1}, \frac{\sum y}{\sum 1} \right). \tag{13.12}$$

- Similarly, higher order moments may be obtained by accumulating appropriate powers of the x and y coordinates. Features that can be derived from the moments include the orientation and size of the best fit ellipse.
- The bounding box of each component records the extreme coordinates of the component pixels. On merging, the combined bounds may be obtained using the minimum or maximum as necessary of the sub-components.
- The average colour or pixel value may be calculated by adding the pixel values associated with each region into an accumulator and normalising by the area at the end of the scan. Similarly, higher order statistics, such as variance (or covariance for colour images), skew, and kurtosis, may be determined by summing powers of each pixel value and deriving the feature value from the accumulated data at the end of the image.
- Other, more complex, features such as perimeter may require additional operations (such as edge detection) prior to accumulating data.

Single-pass stream processing algorithms are well suited for FPGA implementation, making CCA an active research area in recent years (Bailey, 2020).

13.5.1 Basic Single-pass Algorithm

The architecture of Figure 13.10 can be extended to give Figure 13.11 (Bailey and Johnston, 2007; Bailey et al., 2008). The data table can be implemented with single-port memory by reading the data associated with a label into a register and directly updating the register for each pixel processed. On a merger, the merged entry can be read from the data table and combined with the data in the register. Then, when a background pixel is encountered during the raster scan, the data may be written back to the data table (Bailey and Johnston, 2007). The object data is then available from the data table at the end of the frame.

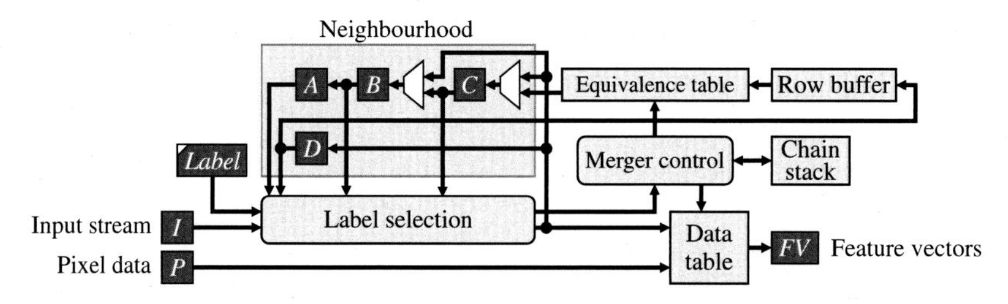

Figure 13.11 Single-pass connected component analysis. The input pixel data are the x, y, pixel value, and so on as needed for feature calculation.

The main limitation with this approach is that the size of the data and merger tables depends on the number of initial labels within the image. In the worst case, this is proportional to the area of the image, where the maximum number of labels is one quarter of the number of pixels.

13.5.2 Reducing Memory Requirements

Since the initial labels are only saved in the row buffer for one row period, the number of labels in active use at any one time is proportional to the width of the image not the area. Therefore, the memory used by the data and equivalence tables is being used inefficiently. The memory requirements of the system can be optimised by recycling labels that are no longer used (Lumia et al., 1983; Khanna et al., 2002).

The approach taken by Ma et al. (2008) was to completely relabel each row. This requires the addition of a translation table to convert the labels used in the previous row to those used in the current row. Data is transferred from the old label to the new label with each row (requiring separate data tables for the current and previous rows). Any data not transferred therefore belongs to a completed object and can be output at the end of the row.

Trein et al. (2007, 2008) run-length encoded up to 32 pixels in parallel, enabling a slower clock speed to be used to sequentially process the run-length encoded data. When sub-components merged, a pointer from the old label was linked to the label that was kept; however, chaining was not considered. To keep the data table small, a garbage collector detected when a region had been completely scanned, outputted the resulting region data, and queued the released label for reuse.

Klaiber et al. (2016) adapted the basic algorithm and took an alternative approach to recycling labels, replacing the label counter of Figure 13.11 with a recycling queue. The resulting algorithm was formally proved correct (Klaiber et al., 2019a). Some of the complexities of the algorithm were simplified by performing a double lookup in the equivalence table (Klaiber et al., 2019a).

13.5.3 Eliminating End-of-row Overheads

Most of the early algorithms required processing at the end of each row to ensure that all of the merged labels were linked to the current representative label to allow the current label to be obtained with a single lookup (path compression). This requires a blanking period at the end of each row, and in the worst case requires up to 20% overhead per row or up to 50% overhead on a single row (Bailey and Johnston, 2007). Several methods have been proposed to eliminate this end-of-row overhead.

The most obvious solution is to immediately replace all old labels with the current label (Jeong et al., 2016). While this is not practical for CCL, CCA only has one row of labels (in the row buffer) to update. The row buffer and equivalence table in Figure 13.11 are replaced with a shift register where all locations of the old label are updated in parallel with the new label whenever a merger occurs. This makes such an approach quite resource intensive and not particularly practical on an FPGA.

Since the path compression overhead is functionally equivalent to a reverse scan of each image row, this can be incorporated into the regular processing using a zigzag scan pattern (Bailey and Klaiber, 2019). The previous line is then updated in parallel with the processing of the next line. This requires zigzag reordering of the input stream and replaces the row buffer with a zigzag row buffer.

Rather than use a standard lookup table to represent the equivalences between labels, Tang et al. (2018) proposed run-length encoding and used linked lists to manage the linkages. Labelling of each run is deterministic (in that labels are allocated sequentially, independent of connectivity; see Figure 13.12), which means that the row buffer does not need to save the labels between rows but can recreate them from the binary input image using counters. The linked lists were implemented with three tables, containing pointers to the *next* label in the list, the *head* of the list which links from a label in the previous row to the head of the list in the current row, and the *tail* of the list which is set for *head* labels to identify the end of the list to facilitate merging. Merging occurs when there is an overlap between runs in the previous and current rows. Data is accumulated in the label corresponding to the *head* of the list and passed onto the *next* entry when the head label drops out of the scanning window. If there is no *next* entry, then the connected component has been completely scanned, and the data can be output.

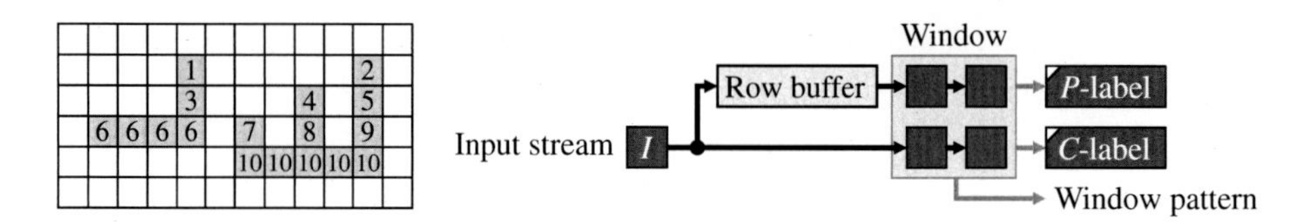

Figure 13.12 Deterministic labelling: Left: labels for the example in Figure 13.9; right: label assignment circuit; the two label counters indicate the most recent labels in the previous (*P*-label) and current (*C*-label) rows of the window, respectively.

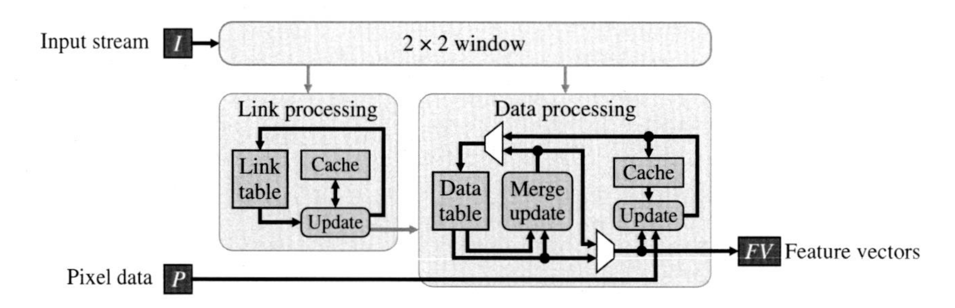

Figure 13.13 Simplified architecture of CCA based on union-retire.

Bailey and Klaiber (2021) made the observation that most existing algorithms focus on the labelling rather than the connectivity. For CCA, the labels only represent an intermediate data structure, and it is the feature vector that is accumulated as a result of the connectivity that is of primary importance. This led to the union-retire algorithm (Bailey and Klaiber, 2021), which they then implemented on an FPGA (Bailey and Klaiber, 2022), as illustrated in Figure 13.13. Like Tang et al. (2018), it was based on deterministic labelling and used a linked list to represent the connectivity; however, it only required two links, rather than three, reducing the memory requirements further. Rather than following a linear structure, the links represent a directed acyclic graph corresponding to the connectivity between the runs of pixels. A union operation adds a link between sub-graphs when the runs of pixels overlap in the window between the previous and current rows. The link is always directed from an earlier label to a later label. When a label leaves the previous row of the window, it is retired. The corresponding node is removed from the graph, and if it has two links, a link is added between those nodes to ensure that connectivity is retained. Caching is used to reduce the memory bandwidth, enabling the link and data tables to be implemented using dual-port RAM.

13.5.4 Parallel Algorithms

As with CCL, CCA is difficult to parallelise. Kumar et al. (2010) divided the image into horizontal slices which were processed in parallel. A separate coalescing unit combined the feature vectors of regions between slices at the end of the frame. Horizontal slices require the data to be streamed from a frame buffer, rather than processing a single streamed input image.

Klaiber et al. (2012) overcame this by dividing the image into vertical slices but was otherwise similar in its operation. This was then extended (Klaiber et al., 2013, 2019b) to coalesce the data between the vertical slices in parallel with the processing within each slice, significantly reducing both memory and latency.

To process 4K video streams, Kowalczyk et al. (2021) used block-based processing, processing a 2×2 block of pixels as a single pixel. Neighbourhood processing is more complex, because of the larger number of configurations for propagation of labels into the current super-pixel, and it is also possible to have two mergers (three different labels) within a neighbourhood. Otherwise, their solution followed a similar approach to the basic algorithm, including using a label stack to unlink chains of mergers at the end of each line.

13.5.5 Further Considerations and Optimisations

If the regions being labelled or extracted are convex, then further simplifications are possible. Since convex shapes have no **U**-shaped boundaries, merger logic is no longer required. Simple label propagation is sufficient to completely label the image in a single pass. With pixel-based processing, mergers can only occur with **J**-shaped regions (such as the left-hand object in Figure 13.9) along the top-left edge of convex objects. This can be overcome by deferring the assignment of a new label until the end of a run of pixels, managed by run-length coding the pixels stored in the row buffer.

In all of the discussion in this section, it has been assumed that the input image is binary. If, instead, the input was a labelled image, for example from multi-level thresholding or colour segmentation, then the processing outlined above needs to be extended. Since each incoming pixel can only have one label, the hardware of a single processor may be augmented to handle multiple input labels. The main change is that the labelling process should only propagate labels within regions with the same input label; any different input label is considered as background. This requires that the input label must also be cached in the row buffer. Equality comparisons with the input label are required to create the neighbourhood mask based on the incoming pixel. Since both fore- and background are being labelled, potentially twice as many labels are required. Consequently, the merger table and data tables need to be twice as large. There may also be bandwidth implications for some of the more advanced methods. Otherwise, the processing is the same as the single-label case.

13.6 Distance Transform

Another form of labelling object pixels is provided by the distance transform. This labels each object pixel with its distance to the nearest background pixel. Different distance metrics result in different distance transforms. A common distance metric is the Minkowski distance, which for parameter p in two dimensions gives

$$d_p = \|\Delta x, \Delta y\|_{L_p} = (|\Delta x|^p + |\Delta y|^p)^{\frac{1}{p}}. \tag{13.13}$$

The Euclidean distance ($p = 2$) corresponds with how distances are measured in the real world:

$$d_2 = \|\Delta x, \Delta y\|_{L_2} = \sqrt{(\Delta x)^2 + (\Delta y)^2}. \tag{13.14}$$

Although the Euclidean distance is isotropic, it can be difficult to determine which pixel is actually the closest background pixel without testing them all. This is complicated by the fact that images are sampled with a rectangular sampling grid.

Two other metrics that are easier to calculate are the Manhattan or city-block distance ($p = 1$):

$$d_1 = \|\Delta x, \Delta y\|_{L_1} = |\Delta x| + |\Delta y|, \tag{13.15}$$

which counts pixels steps, considering neighbouring pixels to be 4-connected, and the Chebyshev or chessboard distance ($p = \infty$):

$$d_\infty = \|\Delta x, \Delta y\|_{L_\infty} = \max(|\Delta x|, |\Delta y|), \tag{13.16}$$

which considers pixels to be 8-connected. It is clear from the comparison in Figure 13.14 that the non-Euclidean metrics are anisotropic. This has led to several other metrics that give better approximation to the Euclidean distance, while retaining the ease of calculation of the Manhattan and Chebyshev distances. Several different methods have been developed to efficiently calculate the distance transform of a binary image.

13.6.1 Morphological Approaches

One approach is to successively erode the object using a series of morphological filters. If each pass removes one layer of pixels, then the distance of a pixel from the boundary is the number of passes required for an object

Figure 13.14 Comparison of common distance metrics (contour lines have been added for illustration). Left: Euclidean, d_2; centre: Manhattan or city-block, d_1; right: Chebyshev or chessboard, d_∞.

pixel to become part of the background. The Chebyshev distance is obtained by eroding with a 3×3 square structuring element, whereas the Manhattan distance is obtained by eroding with a five-element '+'-shaped structuring element. A closer approximation to Euclidean distance may be obtained by alternating the square and cross elements with successive iterations (Russ, 2002), which results in an octagonal-shaped pattern.

The main limitation of such algorithms is that many iterations are required to completely label an image. On an FPGA, the iterations may be implemented as a pipeline up to a certain distance (determined by the number of processors which are constructed). Beyond that distance, it is necessary to save the result as an intermediate image and feed the image back through.

To improve the approximation to a Euclidean distance, a larger structuring element needs to be used. This may be accomplished by having a series of different sized circular structuring elements applied in parallel (Waltz and Garnaoui, 1994). Such a stack of structuring elements is equivalent to using greyscale morphology with a conical structuring element. While using a conical structuring element directly is expensive, by decomposing it into a sequence of smaller structuring elements, it can be made computationally more feasible (Huang and Mitchell, 1994). However, for arbitrary sized objects, it is still necessary to iterate using a finite sized structuring element.

13.6.2 Chamfer Distance

The iterative approach of morphological filters may be simplified by making the observation that successive layers will be adjacent. This allows the distance to be calculated recursively through local propagations using stream processing (Borgefors, 1986; Butt and Maragos, 1998). The result is the chamfer distance transform.

Two raster scanned passes through the image are required. The first pass is a normal raster scan and propagates the distances from the top and left boundaries into the object:

$$Q_1[x,y] = \begin{cases} 0, & I[x,y] = 0, \\ \min\begin{pmatrix} Q_1[x,y-1]+a, Q_1[x-1,y]+a, \\ Q_1[x-1,y-1]+b, Q_1[x+1,y-1]+b \end{pmatrix}, & \text{otherwise.} \end{cases} \tag{13.17}$$

The second pass is a reverse raster scan, propagating the distance from the bottom and right boundaries:

$$Q_2[x,y] = \min\begin{pmatrix} Q_1[x,y], Q_2[x,y+1]+a, Q_2[x+1,y]+a, \\ Q_2[x+1,y+1]+b, Q_2[x-1,y+1]+b \end{pmatrix}. \tag{13.18}$$

An implementation of the chamfer distance transform is given in Figure 13.15. A frame buffer is required to hold the distances between passes. Note that the streamed output is in a reverse raster scan format. Hezel et al. (2002) implemented such an algorithm on an FPGA. Higher throughput can be obtained using separate hardware for each pass, with a bank-switched frame buffer in between.

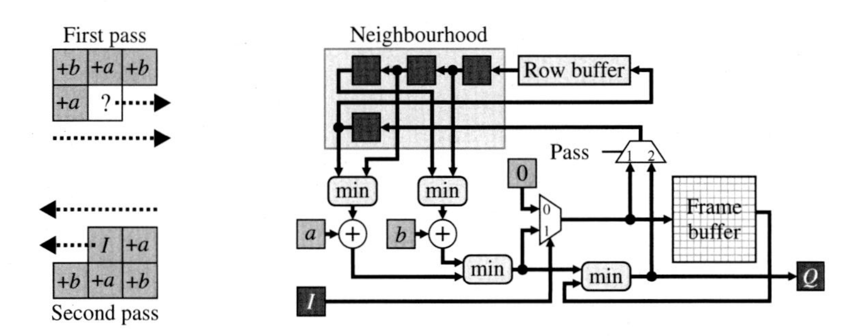

Figure 13.15 3×3 chamfer distance transform. Left: propagation increments for each pass; right: hardware implementation reusing the circuitry between the passes.

Different values of a and b define different distance metrics. The Chebyshev distance results when $a = b = 1$. Using only 4-connected points ($a = 1, b = \infty$) gives the Manhattan distance. The best approximation to the Euclidean distance using small integers is for $a = 3$ and $b = 4$ and dividing the result by 3. Since the division by 3 is not easy in hardware, the result can be left undivided. Alternatively, using $a = 2$ and $a = 3$, the result can be divided by 2 (a right shift). The chamfer distance is given by

$$\begin{aligned} d_{c_{a,b}} &= b \min(\Delta x, \Delta y) + a|\Delta x - \Delta y| \\ &= (2a+b)d_\infty - (a+b)d_1. \end{aligned} \tag{13.19}$$

A closer approximation to the Euclidean distance may be obtained using a larger window size. A 5×5 window, with weights as shown in Figure 13.16, with $a = 5$, $b = 7$, and $c = 11$, gives the best approximation with small integer weights (Borgefors, 1986; Butt and Maragos, 1998).

The chamfer distance transform may be accelerated by processing multiple lines in parallel (Trieu and Maruyama, 2006). This requires sufficient memory bandwidth to read data for several lines simultaneously. (Horizontal data packing is more usual, but the same effect can be achieved by staggering the horizontal accesses and using short first-in first-out (FIFO) buffers.) The processor for each successive line must be offset, as shown in Figure 13.17, so that it is working with available data from the previous line. Depending on the frame buffer access pattern, short FIFO buffers may be required to delay the data for successive rows and to assemble the processed data for saving back to memory. Only the single row that overlaps between the strips needs to be cached in a row buffer. The same approach can also be applied to the reverse scan through the image.

The chamfer distance algorithm may be adapted to measure Euclidean distance by propagating vectors $(\Delta x, \Delta y)$ instead of scalar distances. However, to correctly propagate the distance to all points requires three passes through the image (Danielsson, 1980; Ragnemalm, 1993; Bailey, 2004). The intermediate frame buffer also needs to be wider because it needs to store the vector. A further limitation is that because of the discrete sampling of images, local propagation of vectors can leave some points with a small error (less than one pixel) relative to the Euclidean distance (Cuisenaire and Macq, 1999).

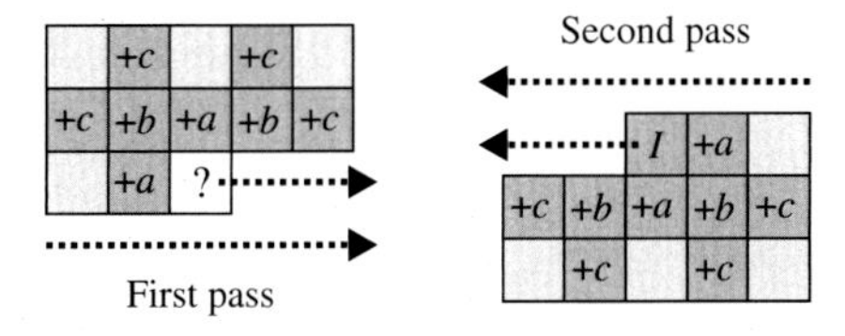

Figure 13.16 Masks for a 5×5 chamfer distance transform. Weights are not required for the blank spaces because these are multiples of smaller increments.

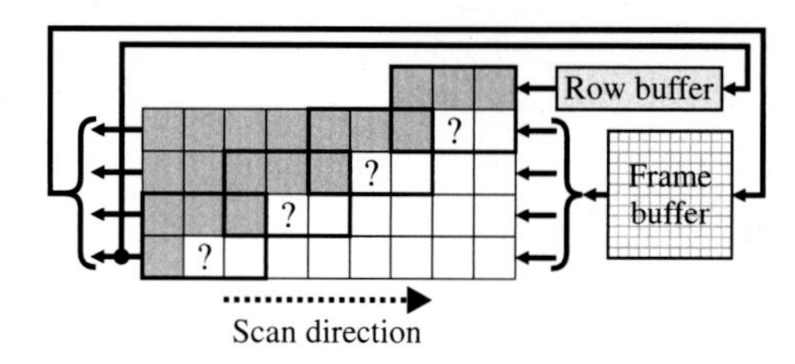

Figure 13.17 Processing multiple lines in parallel with the chamfer distance.

13.6.3 Euclidean Distance

In many applications, the chamfer distance is sufficiently close to the Euclidean distance to be acceptable. If, however, the true Euclidean distance transform is required, it may be obtained using a separable algorithm (Hirata, 1996; Maurer et al., 2003; Bailey, 2004). This relies on the fact that squaring Eq. (13.14) gives

$$d_2^2 = (\Delta x)^2 + (\Delta y)^2, \tag{13.20}$$

which is separable. This allows the 1-D distance-squared transform to be obtained of each row first, and then of each column of the result. The processing can also be performed using integer arithmetic since Δx and Δy are integers. If the actual distance is required (rather than distance squared), then a square root operation may be pipelined on the output of the column processing. Since each row (and then column) is processed independently, the algorithm is readily parallelisable and is also easily extendable to three or higher dimensional images.

Row processing requires two passes, first in the forward direction and then in the reverse direction. The forward pass determines the distance from the left edge of the object, while the reverse pass updates with the distance from the right edge. The processing for each of the two passes is

$$Q_1[x,y] = \begin{cases} 0, & I[x,y] = 0, \\ Q_1[x-1,y]+1, & \text{otherwise}, \end{cases} \tag{13.21}$$

$$Q_2[x,y] = \min(Q_1[x,y], Q_1[x+1,y]+1). \tag{13.22}$$

At the start of each pass, Q_1 and Q_2 need to be initialised based on the boundary condition. If it is assumed that outside the image is background, they need to be initialised to 0, otherwise they need to be initialised to the maximum value. The circuit for implementing the scanning is given in Figure 13.18; the zigzag row buffers read the output in reverse order. The addition of 1 in the feedback loop needs to be performed with saturation, especially if the outside of the image is considered to be object. Since the column processing stage requires the distances squared, these can be calculated directly using incremental processing to avoid the multiplication (see Section 5.3.1) as shown in the bottom of Figure 13.18.

Column processing is a little more complex. Finding the minimum distance at a point requires evaluating Eq. (13.20) for each row and determining the minimum:

$$D^2[x,y] = \min_{y_i} \left(Q_2^2[x,y_i] + (y-y_i)^2 \right). \tag{13.23}$$

Rather than test every possible row within the column to find the minimum, observe that the contributions from any two rows (say y_1 and y_2) are ordered based on the perpendicular bisector between the corresponding background points. Let $D_1^2 = Q_2^2[x,y_1]$ and $D_2^2 = Q_2^2[x,y_2]$. Then, as shown in Figure 13.19, the perpendicular bisector divides the column into two at

$$y' = \frac{1}{2}\left(\frac{D_2^2 - D_1^2}{y_2 - y_1} + y_2 + y_1 \right). \tag{13.24}$$

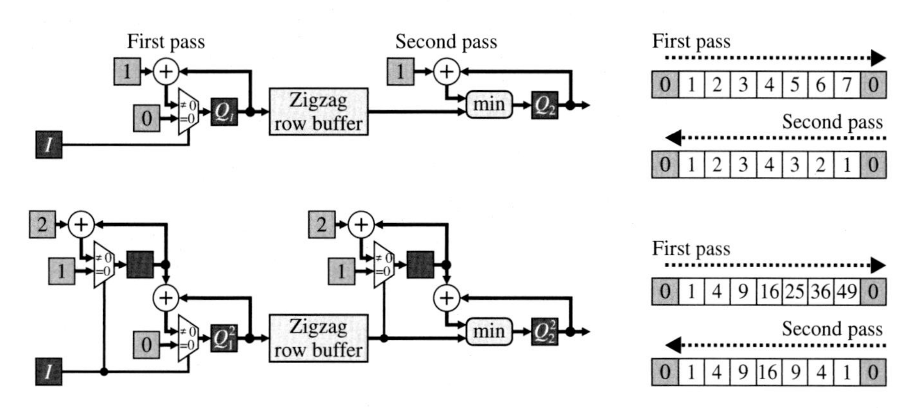

Figure 13.18 Row transform for the separable Euclidean distance transform. Top: performing the distance transform on each row; bottom: the distance-squared transform using incremental update; right: the result of processing a simple example.

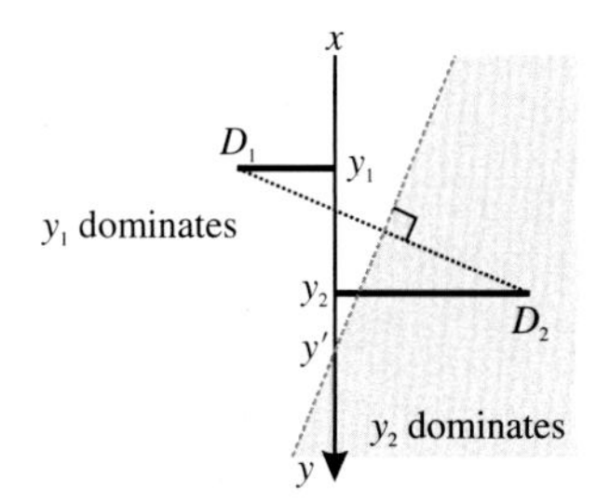

Figure 13.19 Regions of influence of two rows on the minimum distance within a column.

Therefore, a single traverse down the column determines which rows are dominated by the background point associated with each row (Bailey, 2004). The logic for this is shown on the left in Figure 13.20. The row number is used to read the row distance squared from the frame buffer. These are combined with the *Previous* values to give the boundary of influence using Eq. (13.24). Note that integer division is sufficient to determine which two rows the boundary falls between. If the boundary is greater than that on the top of the stack (cached in the y'_{TOS} register), then the *Previous* values are pushed onto the stack, y is incremented, and the next value read from memory. If less than or equal, then the *Previous* values have no influence, so the top of the stack is popped off into the *Previous* and y'_{TOS} registers, and the comparison repeated.

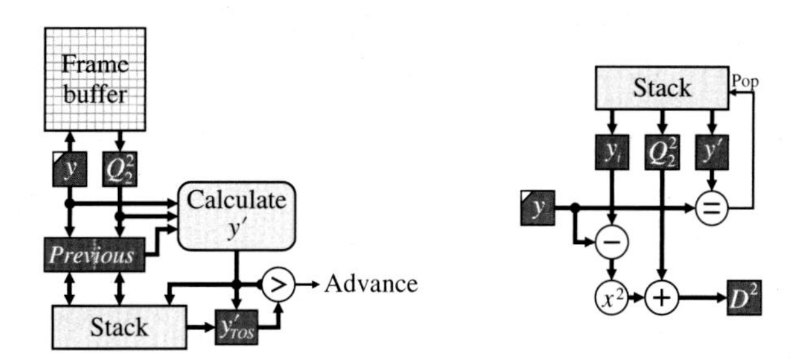

Figure 13.20 Column transform for the separable Euclidean distance transform. Left: first pass, determining the regions of influence; right: second pass, expanding the regions of influence to give the distance.

Although y is not necessarily incremented with every clock cycle, the number of clock cycles required for the first pass through the column is at most twice the number of rows. Therefore, to process one pixel per clock cycle, two such units must operate in parallel. The sporadic nature of frame buffer reads can be smoothed by reading the column from the frame buffer into a FIFO buffer and accessing the distance values from the FIFO buffer as necessary.

At the end of the first pass down the column, the stack contains an ordered list of all of the regions of influence. It is then simply a case of treating the stack as a queue in a second pass (right panel of Figure 13.20), with each section used to directly calculate the distance using Eq. (13.23). When the end of the region of influence is reached, the next section of the column is simply pulled from the queue, until the bottom of the column is reached. The result is a streamed output image, with a vertical scan pattern rather than the conventional raster scan.

As described here, the row processing has a latency of two row lengths as a result of the combination of the forward and reverse passes. The latency of processing a column, is two column times (the worst case time it takes for the first pass down the column). This may be reduced by beginning the second pass down the column before the first pass has been completed.

The separability of the transform implies that all of the rows must be processed before beginning any of the columns. However, this is not necessarily the case. As with separable warping (Section 10.2.1), the two stages can be combined. Rather than process each column separately, they can be processed in parallel, with the stacks built in parallel as the data is streamed from the row processing. This is more complex with the distance transform because the volume of data required to hold the stacks would almost certainly require use of external memory. If the outside of the image can be considered background, the worst case latency would be just over half a frame, with conventional raster scanned input and output streams. The improved latency comes at the cost of more complex memory management. However, if outside the image is considered object, then it is necessary to process the complete frame in the worst case before any pixels can be output (for example if the only background pixel is in the bottom right corner in the image).

Since each row and each column are processed independently, an approach to accelerating the algorithm is by operating several processors in parallel. The main limitation to processing speed is governed by memory bandwidth for the input, intermediate storage, and output streams.

13.6.4 Applications

The distance transform has several applications in image processing (Fabbri et al., 2008). One use is to derive features or shape descriptors of objects. Statistics of the distances within an object, such as mean and standard deviation, are effective size and shape measures. Ridges within the distance transform correspond to the medial axis or skeleton of the object (see, for example, (Arcelli and Di Baja, 1985)). The distance to the medial axis is therefore half the width or thickness of the object at that point. Extending this further, local maxima correspond to the centres of the largest inscribed 'circles'.

If the distance transform of objects has only a single local maximum, then it can be used to provide a count of the number of objects within the image even if adjacent objects are touching. This concept may be extended to the separation of touching objects through the watershed transform (see Section 13.7).

Fast binary morphological filtering can be accomplished by thresholding the distance transformed image. Rather than perform several iterations with a small structuring element, the distance transform effectively performs these all in one step, with the thresholding selecting the particular iteration.

The distance transform can be used for robust template matching (Hezel et al., 2002) or shape matching (Arias-Estrada and Rodriguez-Palacios, 2002). The basic principle here is to compare the distance transformed images rather than simple binary images. This makes the matching less sensitive to small differences in shape, resulting, for example, from minor errors in segmentation.

13.6.5 Geodesic Distance Transform

A variation in the distance transform is the geodesic distance. The ***geodesic distance*** between two points within a region is defined as the shortest path between those two points that lies completely within the region. Any of the previously defined distance metrics may be used, but rather than measure the distance from the boundary, the distance from an arbitrary set of points is used. The previously described methods need to be modified to measure the geodesic distance.

Morphological methods use a series of constrained dilations beginning with the starting or seed pixels. The input is dilated from the seed pixels, with each iteration ANDed with the shape of the region to constrain the path to lie within the object. The distance from a seed point to an arbitrary point is the number of dilations required to include that point within the growing region. Using larger conical structuring elements to perform multiple iterations in a single step does not work with this approach because a larger step may violate the constraint of remaining within the region.

The chamfer distance transform can also be used to calculate the geodesic distance. In general, two passes will not be sufficient for the distances to propagate around concavities in the region. The back and forth passes must be iterated until there is no further change.

Unfortunately, the constrained distance propagation cannot be easily achieved using stream processing. An approach that requires random access to the image can be built on a variation of Dijkstra's shortest path algorithm (Dijkstra, 1959). The seed pixels are marked within the image and are inserted into a FIFO queue. Then, pixels are extracted from the queue, and their unmarked neighbours determined. These neighbours are marked with the incremented distance, and their locations added to the end of the queue. The queue therefore implements a breadth-first search which ensures that all pixels of one distance are processed and labelled before any of a greater distance. A Euclidean distance transform may be approximated by selecting four- or eight-neighbour propagation depending on the level. A potentially large FIFO queue is required to correctly process an arbitrary image. Pixels are accessed in a random order, and searching for the unmarked neighbours also requires multiple accesses to memory. An FPGA implementation of this algorithm was described by Trieu and Maruyama (2008).

Rather than use a FIFO queue, it is also possible to use chain codes as an intermediate step (Vincent, 1991). Propagating chain codes reduces the number of memory accesses because the location of unmarked pixels can come directly from the chain code, and only a single access is required to verify that the pixel is unmarked and apply the distance label.

One application of the geodesic distance is for navigation and path planning for robotics (Sudha and Mohan, 2008). The geodesic distance transform enables both obstacle avoidance and determining the shortest or lowest-cost path from source to destination.

13.7 Watershed Transform

The watershed transform considers the pixel values of an image as elevations in a topographical map and segments the image based on the topographical watersheds. For example, after applying an edge detection operator, the edges become ridges in the image which become the watersheds separating the regions. The watershed will tend to divide the image along the edges between regions, as illustrated in Figure 13.21. One limitation is that any noise within the image will create false edges, with consequent over-segmentation. This can be partially addressed by smoothing the image before segmentation and ignoring the edges below a threshold height.

Watershed segmentation can also be used to separate touching objects, as demonstrated in Figure 13.22. The basic principle is that convex objects (and some non-convex objects) will have a single peak in their distance transform. Therefore, if multiple objects are touching, then the connected component of the combination will have multiple peaks within its distance transform. Inverting the distance will convert the peaks to basins, and the watershed transform will associate each pixel within a region with its corresponding basin. The one failure in Figure 13.22 is for the non-convex object which has two peaks. Again, it is necessary to use a small threshold because with discrete images the distance transform can have small peaks along a diagonal ridge resulting in over-segmentation.

The watershed transform can also be used where detecting peaks is of interest. One example of this is detecting the peaks following a Hough transform (described in Section 13.8). Inverting the image will convert peaks to basins, allowing them to be individually labelled.

Watershed-based segmentation can be considered a form of region growing. There are two main approaches to determining the watersheds: flow and immersion algorithms. Flow algorithms consider a droplet of water

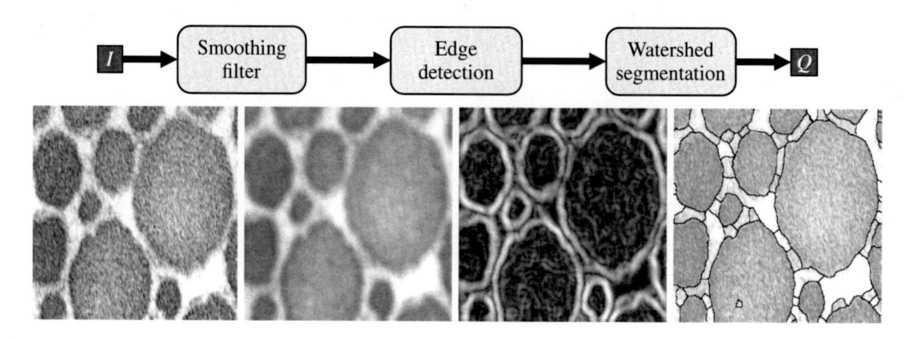

Figure 13.21 Watershed segmentation. Images from the left: original input image; Gaussian filtered with $\sigma = 2.3$; edge detection using the range within a 3×3 window; watershed segmentation merging adjacent regions with a threshold of 5.

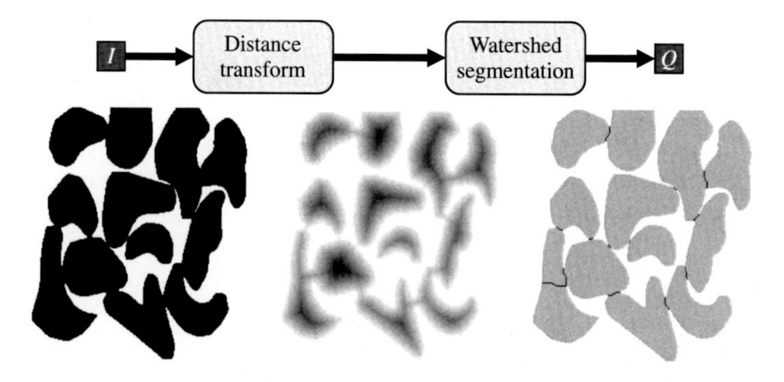

Figure 13.22 Watershed segmentation for separating touching objects. Left: original image; centre: applying the 3,4 ($a = 3$, $b = 4$) chamfer distance transform; right: original with watershed boundaries overlaid.

falling on each pixel within the image. The droplet will flow downhill until it reaches a basin where it will stop. The image is segmented based on grouping together all of the pixels that flow into the same basin. The immersion approach reverses this process, and starts with the basins, gradually raising the water level, creating lakes of pixels associated with each basin. When two separate lakes meet, a barrier is built between them; this barrier is the watershed (Vincent and Soille, 1991).

13.7.1 Flow Algorithms

Flow algorithms for the watershed transform are closely related to CCL (Bailey, 1991). The key principle is to connect each pixel to its minimum neighbour less than itself. The structure of such an implementation is illustrated in Figure 13.23. The input stream of pixel values is augmented with an initial label of zero (unlabelled). The 3×3 neighbourhood is examined to determine the direction of the minimum neighbour. If there are multiple pixels with the minimum value, one can be chosen arbitrarily (or the gradient within the neighbourhood could be used to resolve the ambiguity (Bailey, 1991)). If no pixels are less than the central pixel, it is a new local minimum (basin) and is assigned a label if it does not already have one. Otherwise, a pixel is connected to its neighbourhood minimum. If neither are labelled, a new label is assigned to both pixels. If one is labelled, the label is propagated to the other pixel. If both are labelled, then a merger occurs; this is recorded in a merger table, in the same way as with CCL.

To enable the two read accesses (to update the labels from the row buffer), the merger table can be duplicated (or run at twice the clock frequency). A further possibility is a lazy caching arrangement, where a label is translated only if necessary (if it is the centre pixel or the neighbourhood minimum).

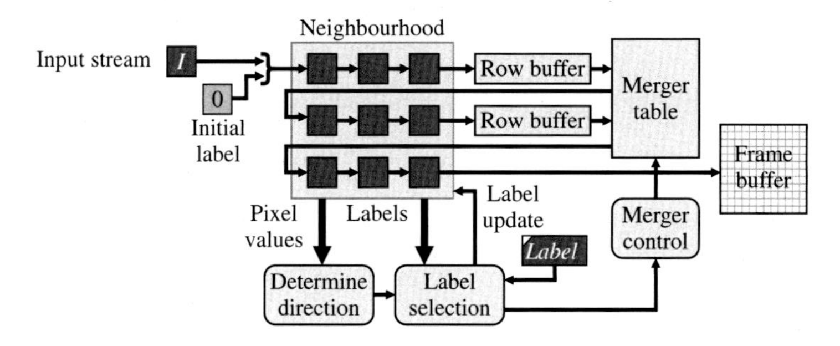

Figure 13.23 Stream-based implementation of watershed segmentation.

If a labelled image is required, then two passes through the image are necessary. The first pass performs an initial labelling and determines the connectivity. Initial labels are saved in a frame buffer, as shown in Figure 13.23, and the merger table is used to provide a final consistently labelled image in the second pass. Alternatively, if the data for each region can be extracted in the first pass (as with CCA in Section 13.5), then the frame buffer is not required but can be replaced with a data table (Bailey, 1991).

One limitation of the above algorithm is that it does not evenly divide plateaus which lie on a watershed (the whole plateau gets assigned a single label). Generally it is better to divide such plateaus down the 'middle' with half being assigned to each watershed. This can be accomplished by distance transforming any plateaus, assigning pixels to the minimum that is the closest. This is a geodesic distance transform, which requires two or more passes through the image. This approach was taken by Trieu and Maruyama (2006) and Trieu and Maruyama (2007), where multiple back- and forward scans are taken to calculate the geodesic distance transform of the plateaus where necessary. Because of the large volume of data that needs to be maintained, all of the data structures were kept in external memory. To accelerate their algorithm, they had multiple memory banks and processed multiple rows in parallel using the scheme introduced in Figure 13.17. This approach also has its own limitations, in particular, the number of back- and forward scans through the image.

Trieu and Maruyama (2008) adapted the distance transform described in Section 13.6. The new scheme requires four processing phases. The first phase is a raster scan to identify the plateau boundaries with smaller pixel values. The locations of these boundary pixels are placed in a FIFO queue. In the second phase, the queue is used to propagate the boundary pixels, distance transforming the plateaus. The third phase is another raster scan, which locates local minima and unlabelled plateaus, which must correspond to the basins. When a basin is found, the raster scan is suspended, and the FIFO queue is used again to propagate the labels for each region (this is the fourth phase). Once the region is completed, all pixels which drain into that basin have been segmented, so the raster scan of phase three is resumed.

Yeong et al. (2009) also implemented a flow-based algorithm on FPGA using 4-connectivity. To get parallel access to the pixels within a neighbourhood, they divided the image into multiple banks. When a plateau is encountered, all plateau pixels are added to a queue and then processed to split the plateau.

13.7.2 Immersion Algorithms

The immersion approach to determining the watershed begins with the basins and increases the water level sequentially throughout the whole image, adding pixels to adjacent connected regions. This process considers the pixels within the image in strict pixel value order by first sorting the pixels into increasing order.

Traditional sorting algorithms do not scale well with the size of the image, but since the pixel values are discrete, fast methods that are linear with the number of pixels are possible. A bin-sort can sort all of the pixels in a single pass through the image. This requires having 2^N bins, where N is the number of bits per pixel, and as the image is scanned, the coordinates of each pixel are added to the corresponding bin. To determine how many bin entries to allocate for each pixel value, the histogram of the image is obtained as the image is streamed into a frame buffer. The cumulative histogram then gives the address in the bin memory of the end of each bin. If the

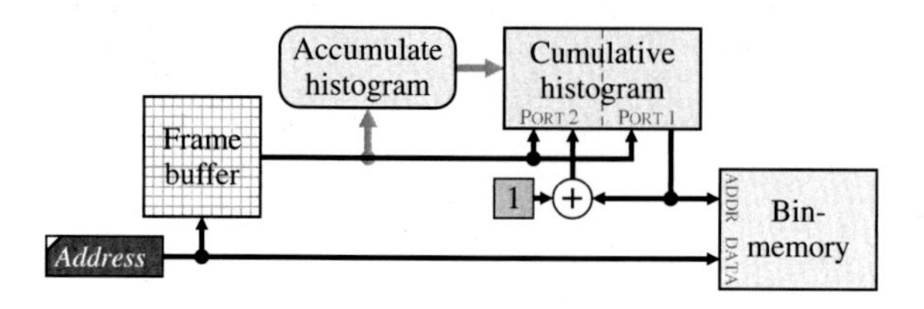

Figure 13.24 Bin-sort, using the cumulative histogram for bin addressing.

cumulative histogram is offset by one pixel value, so the count for pixel 0 is address 1, and so on, then it will act as address translation for the incoming pixel values. As shown in Figure 13.24, the incoming pixel value is looked up in the cumulative histogram and used to store the pixel address in the bin memory. The bin memory address from the cumulative histogram is incremented, so that it will point to the next location. At the end of the image, the cumulative translation memory again contains the cumulative histogram, and all of the pixels are sorted.

While the histogram-based bin-sort is memory efficient, it requires two passes through the image. An alternative approach is to use an array of linked lists, one for each bin (De Smet, 2010). With a linked list, it is not necessary to know the bin size in advance. A circuit for the construction of the linked lists is shown in Figure 13.25. Prior to processing the input stream, the entries in the *Tail* array need to be set to 0. When the first pixel value of a particular value comes in, the corresponding *Tail* entry will be 0, so the pixel address is stored in the corresponding *Head* entry. Otherwise, the address is stored in the linked list at the entry pointed to by the *Tail*. In both cases, *Tail* is updated with the new address. If the *Tail* is initialised to 0, a little bit of additional logic is required to distinguish the case when a true 0 address is written into the *Tail* register. This may be done as illustrated in Figure 13.25, or alternatively an additional bit could be added to the width of *Tail* to indicate this.

Since only one *Head* or *Tail* register is accessed at a time, they can be stored in either fabric RAM or block RAM. *Head* only needs to be single port, while *Tail* will need to be dual-port. The image addresses do not need to be stored explicitly because the linked list is structured such that the bin address corresponds to the image address. The pixel values also do not need to be saved within the bins because this is implicit with the linked list structure as each different pixel value has a separate list. However, since the neighbours need to be examined, this will require storing the image within a frame buffer. Construction of the linked list is easily parallelisable if memory bandwidth allows. Since bin addressing corresponds to image addressing, the image may be split vertically into a number of blocks, which are processed in parallel. At the end, it is then necessary to add the links between the blocks for each pixel value.

When performing the immersion, pixels are accessed in the order of pixel value from the sorted list. Pixels are added to the existing boundaries if an adjacent pixel has a lower level, with the label propagated from the adjacent labelled pixel. If two differently labelled neighbours exist, the pixel is a watershed pixel. (A threshold may be implemented by merging adjacent regions if the basin depth is less than the threshold below the current immersion level.) If no smaller labelled pixels are in the neighbourhood, the pixels are added to a queue to perform the geodesic distance transform of the plateau. Any unlabelled pixels remaining after this process are new basins, so are assigned a new label.

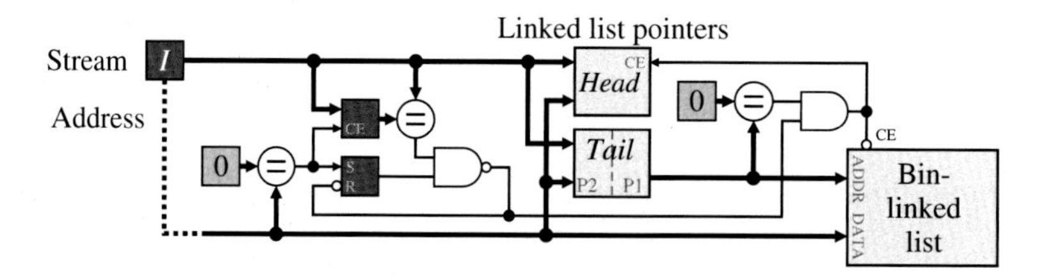

Figure 13.25 Bin-sort using an array of linked lists. (CE = clock enable; S = set; R = reset.)

An FPGA implementation of this process is described by Rambabu et al. (2002) and Rambabu and Chakrabarti (2007). Kuo et al. (2001) accelerated the standard immersion algorithm by labelling a 3×3 (or larger) block of pixels at a time, which significantly reduced the total number of memory accesses required. Roy et al. (2016) implemented a hill-climbing version of the immersion algorithm, which was parallelised using an array of cells to do the processing. Unfortunately, the paper provides limited architectural details.

13.8 Hough Transform

The Hough transform is a technique for detecting objects or features from detected edges within an image (Illingworth and Kittler, 1988; Leavers, 1993). While originally proposed for lines (Hough, 1962), it can readily be extended to any shape that can be parameterised. The basic idea behind the Hough transform is represented in Figure 13.26. First, the image is filtered to detect edges within the image. Each edge pixel then votes for all of the sets of parameters with which it is compatible; that is, an object with those parameters would have an edge pixel at that point. This voting process is effectively accumulating a multi-dimensional histogram within the parameter space. Sets of parameters which have a large number of votes have significant support for the corresponding objects within the image. Therefore, detecting peaks within the parameter space determines potential objects within the image. Candidate objects can then be reconstructed from their parameters and verified.

The idea of voting means that the complete object does not need to be present within the image. There only needs to be enough of the object to create a significant peak. The Hough transform therefore copes well with occlusion. Noise and other false edges with the image will vote for random sets of parameters which, overall, will receive relatively little support, making the Hough transform relatively insensitive to noise. In fact, the voting process can be considered as a form of maximum-likelihood object or feature detection (Stephens, 1991).

13.8.1 Line Hough Transform

Lines are arguably the simplest image feature, being described by only two parameters. The standard equation for a line is

$$y = mx + c, \tag{13.25}$$

with parameters m and c. A detected edge point, (x, y), will therefore vote for points along the line

$$c = y - mx \tag{13.26}$$

in $\{m, c\}$ parameter space. Such a parameterisation is suitable for lines which are approximately horizontal, but vertical lines have large (potentially infinite) values of m and c. Such non-uniformity makes both vote recording and peak detection expensive and impractical, although this may be overcome by having two voting planes, one for $|m| \leq 1$ and the other for $|m| > 1$ (using $c = x - \frac{1}{m}y$).

A more usual parameterisation of lines (Duda and Hart, 1972) is

$$x \cos \theta + y \sin \theta = \rho, \tag{13.27}$$

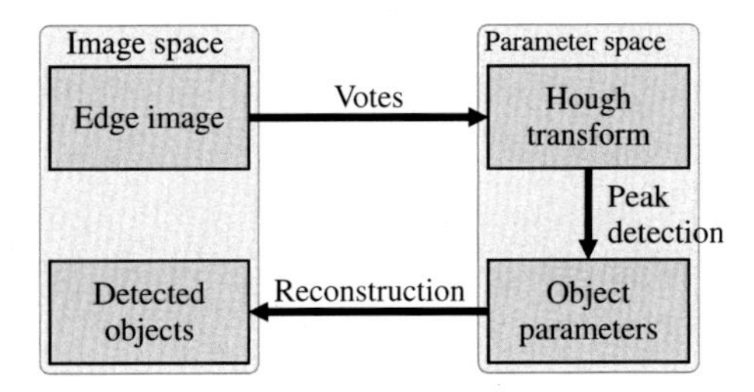

Figure 13.26 The principle behind using the Hough transform to detect objects.

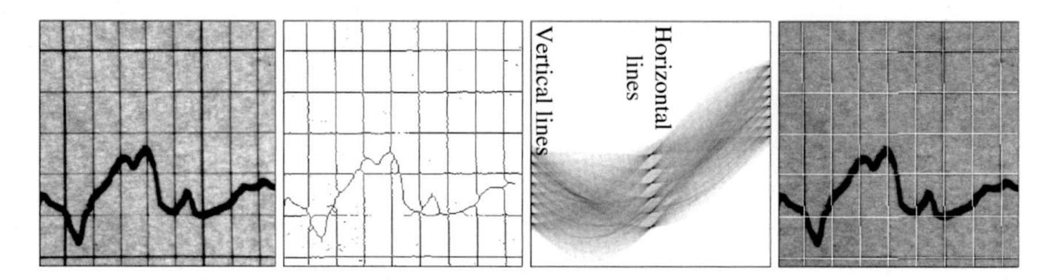

Figure 13.27 Using the Hough transform to detect the gridlines of a graph. From the left: input image; after thresholding and thinning the lines; Hough transform; detected gridlines overlaid on the original image.

where θ is the angle of the line, and ρ represents the closest distance between the line and the image origin. A point (x, y) votes for points along sinusoids in $\{\theta, \rho\}$ parameter space. This process is illustrated with an example in Figure 13.27.

Another useful line parameterisation uses parallel coordinates (Inselberg et al., 1997; Dubska et al., 2011). The principles are similar, so will not be discussed further here.

13.8.1.1 Parameter Calculation

The main problem with the Hough transform is its computational expense. This comes from two sources. The first is the calculation of the sine and cosine to determine the sets of parameters where each point votes. Normally θ is incremented from 0 to π (or $-\frac{\pi}{2}$ to $\frac{\pi}{2}$), and the corresponding ρ calculated from Eq. (13.27). Since the values of θ are fixed by the bin spacing, the values of $\sin\theta$ and $\cos\theta$ can be obtained from a lookup table. These then need to be multiplied by the detected pixel location as shown in Figure 13.28. The size of the lookup tables can be reduced by a factor of 4 (for θ, in the range from 0 to $\frac{\pi}{4}$) with a little extra logic by exploiting trigonometric identities (Cucchiara et al., 1998). Although the proportion of detected pixels is relatively low for normal images, these pixels tend to be clustered rather than evenly distributed. The FIFO buffer on the incoming detected pixel locations allows the accumulation process to run independently of the pixel detection process.

Mayasandra et al. (2005) used a bit-serial distributed arithmetic implementation that enabled the addition to be combined with the lookup table, with a shift-and-add accumulator to perform the multiplication. This significantly reduces the hardware at the cost of taking several clock cycles to calculate each point. (This was addressed by having multiple processors to calculate all angles in parallel.) Lee and Evagelos (2008) used the architecture of Figure 13.28 using the logarithmic number system to avoid multiplications. They then converted from the logarithmic to standard binary for performing the final addition. Alternatively, a CORDIC (coordinate rotation digital computer) rotator of Eq. (5.22) could be used to perform the whole computation of Eq. (13.27) (Karabernou et al., 2005).

The parameter space could also be modified to simplify the calculation of the accumulator points. This is one of the advantages of the linear parameterisation of Eq. (13.26) or parallel coordinates, as incremental update can be used to calculate the intercepts for successive slopes (Bailey, 2011). Tagzout et al. (2001) used a modified version of Eq. (13.27) and used the small angle approximation of sine and cosine to enable incremental calculations

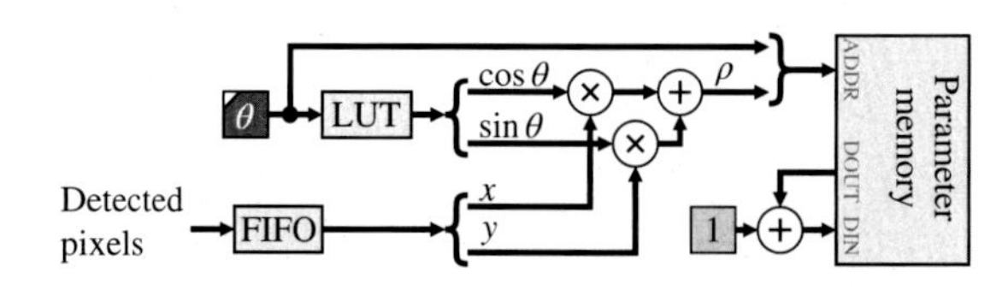

Figure 13.28 Vote accumulator for line Hough transform.

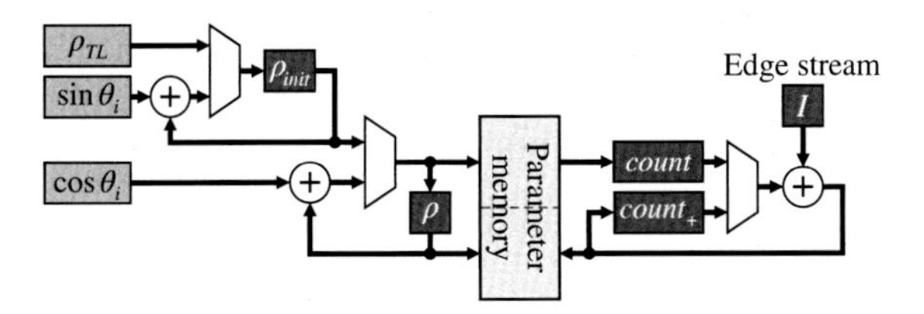

Figure 13.29 Vote accumulation using incremental calculation of ρ for a streamed input. A separate memory and processor is used for each θ. The bin *count* is cached in $count_+$ because successive accesses along a row may increment the same bin. Source: Adapted from Bailey (2017).

to calculate the parameters for a series of angles. However, they still needed to reinitialise the calculation about every 10° to prevent accumulation of errors. Lu et al. (2013) used a similar incremental calculation.

13.8.1.2 Vote Accumulation

A finely quantised parameter space will usually need to be implemented off-chip, which creates a bandwidth bottleneck. Each detected pixel votes for many bins within the parameter space, with each vote requiring two memory accesses (a read and a write).

Chen et al. (2012) overcame this by observing that ρ and θ for closely spaced pixels are similar. So, the image is processed a block (up to 4×4 or 4×8 pixels) at a time, with the votes for a block consolidated on chip before adding to accumulator memory.

With a more coarsely quantised parameter space, the voting array may fit within distributed on-chip block RAMs. This would allow multiple bins to be updated in parallel, at the expense of multiple processors to calculate the line parameters. Lu et al. (2013) pipelined the incremental calculations over θ to separate memories. Chern and Lu (2005) and Bailey (2017) used a separate block RAM for each value of θ and calculated ρ using incremental calculation based on the stream input (see Figure 13.29):

$$\rho_{x+1} = (x + 1)\cos\theta + y\sin\theta = \rho_x + \cos\theta. \tag{13.28}$$

Memory bandwidth may also be reduced by not accumulating votes for every angle. Edge pixels are usually detected with an orientation. Knowing the edge orientation restricts the accumulation to a single angle (or small range of angles), giving the gradient Hough transform (O'Gorman and Clowes, 1976). This reduces not only the number of calculations but also the clutter within the parameter space, making peak detection more reliable. Several researchers used this approach to accelerate their FPGA implementations (Cucchiara et al., 1998; Karabernou et al., 2005; Zhou et al., 2014a; Bailey, 2017).

The adaptive Hough transform reduces the memory requirements by performing the Hough transform in two or more passes (Illingworth and Kittler, 1987). The first pass uses a relatively coarse quantisation to find the significant peaks but gives the parameters with low accuracy. A second pass uses a higher resolution quantisation within those specific detected areas to refine the parameters. The adaptive approach significantly reduces both the total computational burden and the memory requirements, enabling it to be implemented with on-chip memory. Obviously the edge data needs to be buffered between passes.

13.8.1.3 Finding Peaks and Object Parameters

A single scan through the parameter space with a window is suitable for finding the peaks. A peak detection filter can be implemented with a small (3×3 or 5×5) window. If necessary, adjacent counts can be combined to give the centre of gravity of the peak with increased resolution. During the peak scan pass, it is also useful to clear the accumulator memory to prepare it for the next image. This is easy to do with dual-port on-chip memory but is a little harder with off-chip memory.

While the location of a peak gives the parameters of the detected line, often this corresponds with only a line segment within the image. It is therefore necessary to search along the length of each line detected to find the start and end points. These extended operations are not discussed in many FPGA implementations. One exception is that of Nagata and Maruyama (2004). To give good processing throughput, the stages of edge detection, parameter accumulation, peak detection, and line endpoint detection were pipelined.

In one application, Bailey et al. (2018) used the shape of peaks of distorted lines in an image to correct for lens distortion in the image.

13.8.1.4 Reconstruction

If the application only requires the parameters of the line, it is unnecessary to reconstruct the final lines in an output image. While stream processing can easily be used for voting, it is difficult to use it to show the detected lines. Lu et al. (2013) only display the longest line (corresponding to the highest peak in the parameter space). Both Karabernou et al. (2005) and Kim et al. (2008) reconstruct the lines sequentially into a frame buffer for subsequent display.

Bailey (2017) marked the detected peaks with a flag and used back-projection to display the detected lines while accumulating the data for the next frame. The key principle is that if a bin could potentially be incremented, then the current pixel lies on a line with those parameters. Therefore, if that bin has the flag set, it is on a detected line, so the current pixel is displayed. This approach requires no additional time and can potentially display a large number of detected lines.

13.8.2 Circle Hough Transform

Detecting circles (or circular arcs) within an image requires a parameterisation of the circle (Yuen et al., 1990). The commonly used parameterisation is in terms of the circle centre and radius $\{x_c, y_c, r\}$:

$$(x - x_c)^2 + (y - y_c)^2 = r^2. \tag{13.29}$$

The three parameters require a 3-D parameter space, with each detected point in the image voting for parameters on the surface of a cone. The increased dimensionality exacerbates the computational and bandwidth problems of the Hough transform.

Since two of the parameters relate directly to the position of the circle, the 3-D Hough transform may be recast as a 3-D convolution (Hollitt, 2009). Accumulating votes is equivalent to convolving the 2-D image of detected points with a 3-D conical surface. The convolution may be implemented in the frequency domain by taking the 2-D Fourier transform of the input image, multiplying it by the precalculated 3-D Fourier transform of the voting cone, and then calculating the inverse Fourier transform.

As with the line Hough transform, using the edge orientation helps to reduce the computational dimensionality. The direction of steepest gradient will point either towards or away from the centre of the circle. Rather than accumulate on the surface of a cone, this reduces to accumulating points along a ray in 3-D space. The reduction in the number of spurious points accumulated also reduces the clutter within the parameter space. However, small errors in estimating the edge orientation make the peak become less focused, particularly at larger radii.

Rather than accumulate the rays in 3-D space, the parameter space can be reduced to 2-D by projecting along the radius axis (Bailey, 1992). Determining the three parameters then requires two steps (Illingworth and Kittler, 1987; Davies, 1988), as illustrated with an example in Figure 13.30. First, a 2-D Hough transform accumulates the projected rays. These will reinforce at the circle centre, enabling x_c and y_c to be determined from the location of the peak. Then, the radius may be determined from a 1-D Hough transform (a histogram) of the distances of each point from the detected centres.

The two-pass approach requires buffering the list of detected points between passes. However, this is typically only a small fraction of the whole image, so may be stored efficiently (using chain codes is one possibility). Projecting the lines results in a fuzzy blob around the circle centres. When the peaks are found, these need to be smoothed (for example by calculating the centre of gravity) to obtain a better estimate of the peak location.

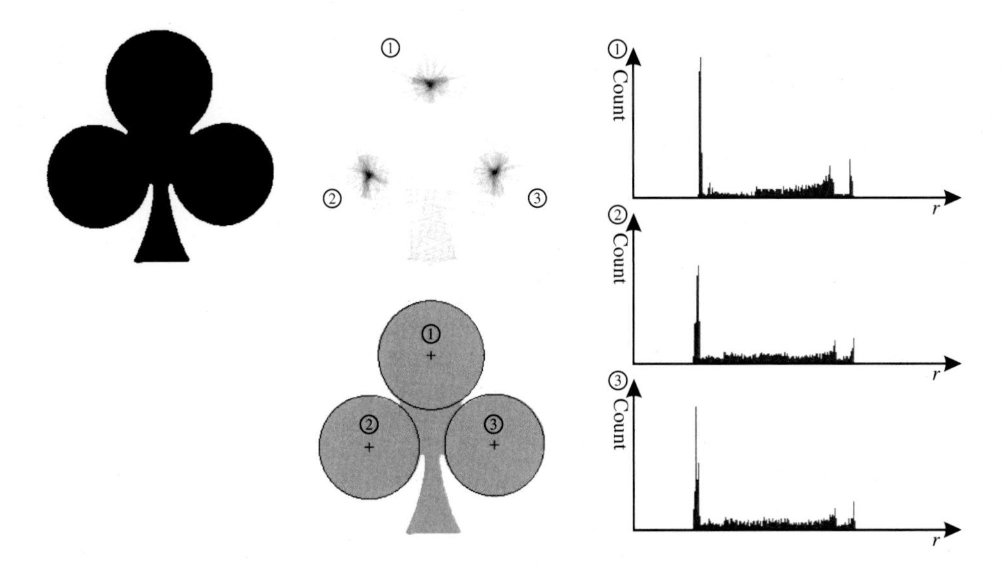

Figure 13.30 Two-step circle Hough transform. Left: original image; centre top: two-dimensional transform to detect circle centres; right: radius histograms about the detected centres; centre bottom: detected circles overlaid on the original image.

In the second pass, several radius histograms may be accumulated in parallel, with the peak and the few radii on either side averaged to obtain the circle radius. Zhou et al. (2014b) and Orlando et al. (2018) broke this down further into three one-dimensional Hough transforms. First, pairs of points along rows and columns are used to calculate candidates for x_c and y_c, respectively. The principle is that if these points are on the same circle, the centre has the corresponding coordinate half-way between them. Each combination of x_c and y_c is then verified using a radius Hough transform.

The coarse-to-fine approach can also be applied to the circular Hough transform. Kumar et al. (2018) took a novel approach to reduce the memory requirements, with three sets of accumulators for each radius plane. First, a reduced resolution 2-D accumulator is used to obtain a coarse estimate of the circle centre (in that plane). In parallel, two 1-D x and y accumulators are also used. The low-resolution estimate is refined by finding the corresponding peaks within the 1-D arrays. This approach is effective as long as there are not too many circles of the same size aligned to the image rows and columns.

Obviously, if the radius is known in advance, then this reduces to a 2-D problem. Alim et al. (2012) used a 3-D parameter space (but with a limited range of radii) and CORDIC arithmetic to vote along the circular pattern on each radius plane.

13.8.3 Generalised Hough Transform

The techniques described above can be generalised to arbitrary shaped objects, initially in 2-D (Merlin and Farber, 1975) and later in 4-D (Ballard, 1981) (parameterised by position, scale, and orientation). The object being detected is represented by a template, with the generalised Hough transform being a form of edge-based template matching. Indeed, Schumacher et al. (2012) determined the position (not scale or orientation) of an arbitrary shape directly through template matching (a linear filter).

Maruyama (2004) proposed a two-stage process for implementing a generalised Hough transform. The first stage uses reduced resolution (by a factor of 16×16) to locate and identify a shape approximately. The second stage uses full resolution but focused on the particular area and shape. This enabled over 100 shapes per image to be detected at frame rates between 13 and 26 frames per second for VGA resolution images (640×480).

Geninatti et al. (2009) used a variation in the generalised Hough transform to estimate the relative rotation between successive frames within a video sequence. To reduce the processing, only the DC components of MPEG compressed images were used, making the effective resolution 44 × 36 pixels.

A full 4-D parameter space was implemented by Kiefer et al. (2016). By dividing the processing over several processing cells, the system could easily be configured in terms of parallelisation and parameter resolution depending on the available on-chip memory. Coarse-to-fine refinement was proposed to overcome the problem of the size of the parameter space, making the full transform tractable. In the refinement stage, the approximate locations of targets of interest are known, enabling cropped images to be processed sequentially. Sarcher et al. (2018) generalised this to enable the same hardware to be used for both the general Hough transform and the line Hough transform.

13.9 Summary

The processing required for blob detection and labelling falls within the intermediate processing level of the image processing pyramid. The input is in the form of image data, with the output in terms of segmented regions or even data extracted from those regions. While there are many possible algorithms for all of these operations, the primary focus in this chapter has been on stream processing. However, the non-local nature of labelling means that stream processing is not always possible.

The bounding box focuses purely on extracting size and shape information from each object based on simple pixel labelling. Consequently, non-connected pixels are considered to belong to the same object if they have the same label. Full CCL is required to take into consideration the connectivity. Producing a labelled image requires at least two passes, although if all of the required region data can be extracted in the first pass, then the second pass is unnecessary.

Chain coding is an alternative region labelling approach, which defines regions by their boundaries, rather than explicitly associating pixels with regions. Chain coding can be accomplished in a single streamed pass, although this requires some quite complex memory management. Subsequent data extraction then works with the 1-D chains rather than the 2-D image.

Related intermediate-level operations explored in this chapter were the distance transform, the watershed transform, and the Hough transform. All can be used as part of segmentation or to extract data from an image. These require either multiple streamed passes through the image or random access processing (with the Hough transform, the random access is within the parameter space).

While other intermediate- and high-level image processing operations can also be implemented in hardware on an FPGA, these operations were selected as representative operations where significant benefits can be gained by exploiting parallelism. This does not mean that other image processing operations cannot be implemented in hardware. However, with many higher level operations, the processing often becomes data dependent. If implemented in hardware, separate hardware must be developed to handle each of the cases. As a result, the hardware is often sitting idle, waiting for the specific cases required to trigger its operation. There comes a point when the acceleration of the algorithm through hardware implementation becomes too expensive and complex relative to a software implementation.

References

AbuBaker, A., Qahwaji, R., Ipson, S., and Saleh, M. (2007). One scan connected component labeling technique. *IEEE International Conference on Signal Processing and Communications (ICSPC 2007)*, Dubai, United Arab Emirates (24–27 November 2007), 1283–1286. https://doi.org/10.1109/ICSPC.2007.4728561.

Alim, F.F., Messaoudi, K., Seddiki, S., and Kerdjidj, O. (2012). Modified circular Hough transform using FPGA. *24th International Conference on Microelectronics (ICM)*, Algiers, Algeria (16–20 December 2012), 4 pages. https://doi.org/10.1109/ICM.2012.6471412.

Alnuweiri, H.M. and Prasanna, V.K. (1992). Parallel architectures and algorithms for image component labeling. *IEEE Transactions on Pattern Analysis and Machine Intelligence* **14** (10): 1014–1034. https://doi.org/10.1109/34.159904.

Appiah, K., Hunter, A., Dickenson, P., and Owens, J. (2008). A run-length based connected component algorithm for FPGA implementation. *International Conference on Field Programmable Technology*, Taipei, Taiwan (8–10 December 2008), 177–184. https://doi.org/10.1109/FPT.2008.4762381.

Arcelli, C. and Di Baja, G.S. (1985). A width independent fast thinning algorithm. *IEEE Transactions on Pattern Analysis and Machine Intelligence* **7** (4): 463–474. https://doi.org/10.1109/TPAMI.1985.4767685.

Arias-Estrada, M. and Rodriguez-Palacios, E. (2002). An FPGA co-processor for real-time visual tracking. In: *International Conference on Field Programmable Logic and Applications*, Montpellier, France (2–4 September 2002), *Lecture Notes in Computer Science*, Volume **2438**, 710–719. https://doi.org/10.1007/3-540-46117-5_73.

Bailey, D.G. (1991). Raster based region growing. *6th New Zealand Image Processing Workshop*, Lower Hutt, NZ (29–30 August 1991), 21–26.

Bailey, D.G. (1992). Segmentation of touching objects. *7th New Zealand Image Processing Workshop*, Christchurch, NZ (26–28 August 1992), 195–200.

Bailey, D.G. (2004). An efficient Euclidean distance transform. In: *International Workshop on Combinatorial Image Analysis*, Auckland, NZ (1–3 December 2004), *Lecture Notes in Computer Science*, Volume **3322**, 394–408. https://doi.org/10.1007/b103936.

Bailey, D.G. (2010). Chain coding streamed images through crack run-length encoding. *Image and Vision Computing New Zealand (IVCNZ 2010)*, Queenstown, NZ (8–9 November 2010), 6 pages. https://doi.org/10.1109/IVCNZ.2010.6148812.

Bailey, D. (2011). Considerations for hardware Hough transforms. *Image and Vision Computing New Zealand (IVCNZ)*, Auckland, NZ (29 November – 1 December 2011), 120–125.

Bailey, D. (2017). Streamed Hough transform and line reconstruction on FPGA. *Image and Vision Computing New Zealand*, Christchurch, NZ (4–6 December 2017), 6 pages. https://doi.org/10.1109/IVCNZ.2017.8402473.

Bailey, D.G. (2020). History and evolution of single pass connected component analysis. *35th International Conference on Image and Vision Computing New Zealand*, Wellington, NZ (25–27 November 2020), 317–322. https://doi.org/10.1109/IVCNZ51579.2020.9290585.

Bailey, D.G. and Johnston, C.T. (2007). Single pass connected components analysis. *Image and Vision Computing New Zealand (IVCNZ)*, Hamilton, NZ (5–7 December 2007), 282–287.

Bailey, D.G. and Klaiber, M.J. (2019). Zig-zag based single pass connected components analysis. *Journal of Imaging* **5** (4): Article ID 45, 26 pages. https://doi.org/10.3390/jimaging5040045.

Bailey, D.G. and Klaiber, M.J. (2021). Union-retire: a new paradigm for single-pass connected component analysis. In: *International Symposium on Geometry and Vision (ISGV 2021)*, Auckland, NZ (28–29 January, 2021), *Communications in Computer and Information Science*, Volume **1386**, 273–287. https://doi.org/10.1007/978-3-030-72073-5_21.

Bailey, D.G. and Klaiber, M.J. (2022). Union-retire for connected components analysis on FPGA. *Journal of Imaging* **8** (4): Article ID 89, 21 pages. https://doi.org/10.3390/jimaging8040089.

Bailey, D.G., Johnston, C.T., and Ma, N. (2008). Connected components analysis of streamed images. *International Conference on Field Programmable Logic and Applications (FPL 2008)*, Heidelberg, Germany (8–10 September 2008), 679–682. https://doi.org/10.1109/FPL.2008.4630038.

Bailey, D.G., Chang, Y., and Le Moan, S. (2018). Lens distortion self-calibration using the Hough transform. *International Conference on Field Programmable Technology*, Naha, Okinawa, Japan (12–14 December 2018), 388–391. https://doi.org/10.1109/FPT.2018.00080.

Ballard, D.H. (1981). Generalizing the Hough transform to detect arbitrary shapes. *Pattern Recognition* **13** (2): 111–122. https://doi.org/10.1016/0031-3203(81)90009-1.

Benkrid, K., Sukhsawas, S., Crookes, D., and Benkrid, A. (2003). An FPGA-based image connected component labeller. In: *International Conference on Field Programmable Logic and Applications (FPL 2003)*, Lisbon, Portugal (1–3 September 2003), *Lecture Notes in Computer Science*, Volume **2778**, 1012–1015. https://doi.org/10.1007/b12007.

Bolelli, F., Allegretti, S., Baraldi, L., and Grana, C. (2020). Spaghetti labeling: directed acyclic graphs for block-based connected components labeling. *IEEE Transactions on Image Processing* **29**: 1999–2012. https://doi.org/10.1109/TIP.2019.2946979.

Borgefors, G. (1986). Distance transformations in digital images. *Computer Vision, Graphics, and Image Processing* **34** (3): 344–371. https://doi.org/10.1016/S0734-189X(86)80047-0.

Butt, M.A. and Maragos, P. (1998). Optimal design of chamfer distance transforms. *IEEE Transactions on Image Processing* **7** (10): 1477–1484. https://doi.org/10.1109/83.718487.

Cederberg, R.T.L. (1979). Chain link coding and segmentation for raster scan devices. *Computer Graphics and Image Processing* **10** (3): 224–234. https://doi.org/10.1016/0146-664X(79)90002-9.

Chakravarty, I. (1981). A single-pass, chain generating algorithm for region boundaries. *Computer Graphics and Image Processing* **15** (2): 182–193. https://doi.org/10.1016/0146-664X(81)90078-2.

Chang, F., Chen, C.J., and Lu, C.J. (2004). A linear-time component-labeling algorithm using contour tracing technique. *Computer Vision and Image Understanding* **93** (2): 206–220. https://doi.org/10.1016/j.cviu.2003.09.002.

Chang, W.Y., Chiu, C.C., and Yang, J.H. (2015). Block-based connected-component labeling algorithm using binary decision trees. *Sensors* **15** (9): 23763–23787. https://doi.org/10.3390/s150923763.

Chen, Z.H., Su, A.W.Y., and Sun, M.T. (2012). Resource-efficient FPGA architecture and implementation of Hough transform. *IEEE Transactions on Very Large Scale Integration (VLSI) Systems* **20** (8): 1419–1428. https://doi.org/10.1109/TVLSI.2011.2160002.

Chern, M.Y. and Lu, Y.H. (2005). Design and integration of parallel Hough-transform chips for high-speed line detection. *11th International Conference on Parallel and Distributed Systems (ICPADS'05)*, Fukuoka, Japan (22–22 July 2005), Volume **2**, 42–46. https://doi.org/10.1109/ICPADS.2005.126.

Conners, R.W. and Harlow, C.A. (1980). A theoretical comparison of texture algorithms. *IEEE Transactions on Pattern Analysis and Machine Intelligence* **2** (3): 204–223. https://doi.org/10.1109/TPAMI.1980.4767008.

Crookes, D. and Benkrid, K. (1999). An FPGA implementation of image component labelling. *Reconfigurable Technology: FPGAs for Computing and Applications*, Boston, MA, USA (20–21 September 1999). SPIE, Volume **3844**, 17–23. https://doi.org/10.1117/12.359538.

Cucchiara, R., Neri, G., and Piccardi, M. (1998). A real-time hardware implementation of the Hough transform. *Journal of Systems Architecture* **45** (1): 31–45. https://doi.org/10.1016/S1383-7621(97)00071-4.

Cuisenaire, O. and Macq, B. (1999). Fast Euclidean distance transformation by propagation using multiple neighbourhoods. *Computer Vision and Image Understanding* **76** (2): 163–172. https://doi.org/10.1006/cviu.1999.0783.

Danielsson, P.E. (1980). Euclidean distance mapping. *Computer Graphics and Image Processing* **14** (3): 227–248. https://doi.org/10.1016/0146-664X(80)90054-4.

Davies, E.R. (1988). A modified Hough scheme for general circle location. *Pattern Recognition Letters* **7** (1): 37–43. https://doi.org/10.1016/0167-8655(88)90042-6.

De Smet, P. (2010). Optimized high speed pixel sorting and its application in watershed based image segmentation. *Pattern Recognition* **43** (7): 2359–2366. https://doi.org/10.1016/j.patcog.2010.01.014.

Dijkstra, E.W. (1959). A note on two problems in connexion with graphs. *Numerische Mathematik* **1** (1): 269–271. https://doi.org/10.1007/BF01386390.

Dubska, M., Herout, A., and Havel, J. (2011). PClines - line detection using parallel coordinates. *IEEE Conference on Computer Vision and Pattern Recognition (CVPR)*, Colorado Springs, CO, USA (20–25 June 2011), 1489–1494. https://doi.org/10.1109/CVPR.2011.5995501.

Duda, R.O. and Hart, P.E. (1972). Use of the Hough transformation to detect lines and curves in pictures. *Communications of the ACM* **15** (1): 11–15. https://doi.org/10.1145/361237.361242.

Fabbri, R., Costa, L.D.F., Torelli, J.C., and Bruno, O.M. (2008). 2D Euclidean distance transform algorithms: a comparative survey. *ACM Computing Surveys* **40** (1): Article ID 2, 44 pages. https://doi.org/10.1145/1322432.1322434.

Freeman, H. (1961). On the encoding of arbitrary geometric configurations. *IRE Transactions on Electronic Computers* **10** (2): 260–268. https://doi.org/10.1109/TEC.1961.5219197.

Freeman, H. (1974). Computer processing of line-drawing images. *ACM Computing Surveys* **6** (1): 57–97. https://doi.org/10.1145/356625.356627.

Freeman, H. and Davis, L.S. (1977). A corner finding algorithm for chain coded curves. *IEEE Transactions on Computers* **26** (3): 297–303. https://doi.org/10.1109/TC.1977.1674825.

Freeman, H. and Shapira, R. (1975). Determining the minimum area encasing rectangle for an arbitrary enclosed curve. *Communications of the ACM* **18** (7): 409–413. https://doi.org/10.1145/360881.360919.

Geninatti, S.R., Benitez, J.I.B., Calvio, M.H., Mata, N.G., and Luna, J.G. (2009). FPGA implementation of the generalized Hough transform. *International Conference on Reconfigurable Computing and FPGAs (ReConFig '09)*, Cancun, Mexico (9–11 December 2009), 172–177. https://doi.org/10.1109/ReConFig.2009.78.

Grana, C., Borghesani, D., and Cucchiara, R. (2010). Optimized block-based connected components labeling with decision trees. *IEEE Transactions on Image Processing* **19** (6): 1596–1609. https://doi.org/10.1109/TIP.2010.2044963.

He, L., Chao, Y., and Suzuki, K. (2007). A linear-time two-scan labelling algorithm. *IEEE International Conference on Image Processing (ICIP 2007)*, San Antonio, TX, USA (16–19 September 2007), Volume **5**, 241–244. https://doi.org/10.1109/ICIP.2007.4379810.

He, L., Chao, Y., and Suzuki, K. (2008). A run-based two-scan labeling algorithm. *IEEE Transactions on Image Processing* **17** (5): 749–756. https://doi.org/10.1109/TIP.2008.919369.

He, L., Ren, X., Gao, Q., Zhao, X., Yao, B., and Chao, Y. (2017). The connected-component labeling problem: a review of state-of-the-art algorithms. *Pattern Recognition* **70**: 25–43. https://doi.org/10.1016/j.patcog.2017.04.018.

Hedberg, H., Kristensen, F., and Owall, V. (2007). Implementation of a labeling algorithm based on contour tracing with feature extraction. *IEEE International Symposium on Circuits and Systems (ISCAS 2007)*, New Orleans, LA, USA (27–30 May 2007), 1101–1104. https://doi.org/10.1109/ISCAS.2007.378202.

Hezel, S., Kugel, A., Manner, R., and Gavrila, D.M. (2002). FPGA-based template matching using distance transforms. *Symposium on Field-Programmable Custom Computing Machines*, Napa, CA, USA (22–24 April 2002), 89–97. https://doi.org/10.1109/FPGA.2002.1106664.

Hirata, T. (1996). A unified linear-time algorithm for computing distance maps. *Information Processing Letters* **58** (3): 129–133. https://doi.org/10.1016/0020-0190(96)00049-X.

Hollitt, C. (2009). Reduction of computational complexity of Hough transforms using a convolution approach. *24th International Conference Image and Vision Computing New Zealand (IVCNZ '09)*, Wellington, NZ (23–25 November 2009), 373–378. https://doi.org/10.1109/IVCNZ.2009.5378379.

Hough, P.V.C. (1962). Method and means for recognizing complex patterns. United States of America Patent 3069654.

Huang, C.T. and Mitchell, O.R. (1994). A Euclidean distance transform using grayscale morphology decomposition. *IEEE Transactions on Pattern Analysis and Machine Intelligence* **16** (4): 443–448. https://doi.org/10.1109/34.277600.

Illingworth, J. and Kittler, J. (1987). The adaptive Hough transform. *IEEE Transactions on Pattern Analysis and Machine Intelligence* **9** (5): 690–698. https://doi.org/10.1109/TPAMI.1987.4767964.

Illingworth, J. and Kittler, J. (1988). A survey of the Hough transform. *Computer Vision, Graphics, and Image Processing* **44** (1): 87–116. https://doi.org/10.1016/S0734-189X(88)80033-1.

Inselberg, A., Chatterjee, A., and Dimsdale, B. (1997). System using parallel coordinates for automated line detection in noisy images. United States of America Patent 5631982.

Ito, Y. and Nakano, K. (2010). Low-latency connected component labeling using an FPGA. *International Journal of Foundations of Computer Science* **21** (3): 405–425. https://doi.org/10.1142/S0129054110007337.

Jablonski, M. and Gorgon, M. (2004). Handel-C implementation of classical component labelling algorithm. *Euromicro Symposium on Digital System Design (DSD 2004)*, Rennes, France (31 August – 3 September 2004), 387–393. https://doi .org/10.1109/DSD.2004.1333301.

Jang, Y., Mun, J., Nam, Y., and Kim, J. (2019). Novel hysteresis thresholding FPGA architecture for accurate Canny edge map. *34th International Technical Conference on Circuits/Systems, Computers and Communications (ITC-CSCC)*, Jeju, Republic of Korea (23–26 June 2019), 3 pages. https://doi.org/10.1109/ITC-CSCC.2019.8793293.

Jeong, J.W., Lee, G.B., Lee, M.J., and Kim, J.G. (2016). A single-pass connected component labeler without label merging period. *Journal of Signal Processing Systems* **84** (2): 211–223. https://doi.org/10.1007/s11265-015-1048-7.

Karabernou, S.M., Kessal, L., and Terranti, F. (2005). Real-time FPGA implementation of Hough transform using gradient and CORDIC algorithm. *Image and Vision Computing* **23** (11): 1009–1017. https://doi.org/10.1016/j.imavis.2005.07.004.

Khanna, V., Gupta, P., and Hwang, C. (2002). Finding connected components in digital images by aggressive reuse of labels. *Image and Vision Computing* **20** (8): 557–568. https://doi.org/10.1016/S0262-8856(02)00044-6.

Kiefer, G., Vahl, M., Sarcher, J., and Schaeferling, M. (2016). A configurable architecture for the generalized Hough transform applied to the analysis of huge aerial images and to traffic sign detection. *International Conference on ReConFigurable Computing and FPGAs (ReConFig)*, Cancun, Mexico (30 November – 2 December 2016), 7 pages. https://doi.org/10 .1109/ReConFig.2016.7857143.

Kim, S.D., Lee, J.H., and Kim, J.K. (1988). A new chain-coding algorithm for binary images using run-length codes. *Computer Vision, Graphics, and Image Processing* **41** (1): 114–128. https://doi.org/10.1016/0734-189X(88)90121-1.

Kim, D., Jin, S.H., Thuy, N.T., Kim, K.H., and Jeon, J.W. (2008). A real-time finite line detection system based on FPGA. *6th IEEE International Conference on Industrial Informatics*, Daejeon, Republic of Korea (13–16 July 2008), 655–660. https://doi.org/10.1109/INDIN.2008.4618183.

Klaiber, M., Rockstroh, L., Wang, Z., Baroud, Y., and Simon, S. (2012). A memory efficient parallel single pass architecture for connected component labeling of streamed images. *International Conference on Field Programmable Technology (ICFPT)*, Seoul, Republic of Korea (10–12 December 2012), 159–165. https://doi.org/10.1109/FPT.2012.6412129.

Klaiber, M.J., Bailey, D.G., Ahmed, S., Baroud, Y., and Simon, S. (2013). A high-throughput FPGA architecture for parallel connected components analysis based on label reuse. *International Conference on Field Programmable Technology*, Kyoto, Japan (9–11 December 2013), 302–305. https://doi.org/10.1109/FPT.2013.6718372.

Klaiber, M.J., Bailey, D.G., Baroud, Y.O., and Simon, S. (2016). A resource-efficient hardware architecture for connected component analysis. *IEEE Transactions on Circuits and Systems for Video Technology* **26** (7): 1334–1349. https://doi.org/ 10.1109/TCSVT.2015.2450371.

Klaiber, M., Bailey, D.G., and Simon, S. (2019a). Comparative study and proof of single-pass connected components algorithms. *Journal of Mathematical Imaging and Vision* **61** (8): 1112–1134. https://doi.org/10.1007/s10851-019-00891-2.

Klaiber, M.J., Bailey, D.G., and Simon, S. (2019b). A single-cycle parallel multi-slice connected components analysis hardware architecture. *Journal of Real-Time Image Processing* **16** (4): 1165–1175. https://doi.org/10.1007/s11554-016-0610-2.

Kowalczyk, M., Ciarach, P., Przewlocka-Rus, D., Szolc, H., and Kryjak, T. (2021). Real-time FPGA implementation of parallel connected component labelling for a 4K video stream. *Journal of Signal Processing Systems* **93** (5): 481–498. https://doi.org/10.1007/s11265-021-01636-4.

Kumar, V.S., Irick, K., Maashri, A.A., and Vijaykrishnan, N. (2010). A scalable bandwidth aware architecture for connected component labeling. *IEEE Computer Society Annual Symposium on VLSI (ISVLSI)*, Lixouri, Kefalonia, Greece (5–7 July 2010), 116–121. https://doi.org/10.1109/ISVLSI.2010.89.

Kumar, V., Asati, A., and Gupta, A. (2018). Memory-efficient architecture of circle Hough transform and its FPGA implementation for iris localisation. *IET Image Processing* **12** (10): 1753–1761. https://doi.org/10.1049/iet-ipr.2017.1167.

Kuo, C.J., Odeh, S.F., and Huang, M.C. (2001). Image segmentation with improved watershed algorithm and its FPGA implementation. *IEEE International Symposium on Circuits and Systems (ISCAS 2001)*, Sydney, Australia (6–9 May 2001), Volume 2, 753–756. https://doi.org/10.1109/ISCAS.2001.921180.

Leavers, V.F. (1993). Which Hough transform? *Computer Vision, Graphics, and Image Processing: Image Understanding* **58** (2): 250–264. https://doi.org/10.1006/ciun.1993.1041.

Lee, P. and Evagelos, A. (2008). An implementation of a multiplierless Hough transform on an FPGA platform using hybrid-log arithmetic. *Real-Time Image Processing 2008*, San Jose, CA, USA (28–29 January 2008). SPIE, Volume **6811**, Article ID 68110G, 10 pages. https://doi.org/10.1117/12.766459.

Lu, X., Song, L., Shen, S., He, K., Yu, S., and Ling, N. (2013). Parallel Hough transform-based straight line detection and its FPGA implementation in embedded vision. *Sensors* **13** (7): 9223–9247. https://doi.org/10.3390/s130709223.

Lumia, R., Shapiro, L., and Zuniga, O. (1983). A new connected components algorithm for virtual memory computers. *Computer Vision, Graphics and Image Processing* **22** (2): 287–300. https://doi.org/10.1016/0734-189X(83)90071-3.

Ma, N., Bailey, D., and Johnston, C. (2008). Optimised single pass connected components analysis. *International Conference on Field Programmable Technology*, Taipei, Taiwan (8–10 December 2008), 185–192. https://doi.org/10.1109/FPT.2008.4762382.

Mandler, E. and Oberlander, M.F. (1990). One-pass encoding of connected components in multivalued images. *10th International Conference on Pattern Recognition*, Atlantic City, NJ, USA (16–21 June 1990), Volume **2**, 64–69. https://doi.org/10.1109/ICPR.1990.119331.

Maruyama, T. (2004). Real-time computation of the generalized Hough transform. In: *14th International Conference on Field Programmable Logic and Application*, Antwerp, Belgium (29 August – 1 September 2004), *Lecture Notes in Computer Science*, Volume **3203**, 980–985. https://doi.org/10.1007/b99787.

Maurer, C.R., Qi, R., and Raghavan, V. (2003). A linear time algorithm for computing exact Euclidean distance transforms of binary images in arbitrary dimensions. *IEEE Transactions on Pattern Analysis and Machine Intelligence* **25** (2): 265–270. https://doi.org/10.1109/TPAMI.2003.1177156.

Mayasandra, K., Salehi, S., Wang, W., and Ladak, H.M. (2005). A distributed arithmetic hardware architecture for real-time Hough-transform-based segmentation. *Canadian Journal of Electrical and Computer Engineering* **30** (4): 201–205. https://doi.org/10.1109/CJECE.2005.1541752.

Merlin, P.M. and Farber, D.J. (1975). A parallel mechanism for detecting curves in pictures. *IEEE Transactions on Computers* **C-24** (1): 96–98. https://doi.org/10.1109/T-C.1975.224087.

Nagata, N. and Maruyama, T. (2004). Real-time detection of line segments using the line Hough transform. *IEEE International Conference on Field-Programmable Technology*, Brisbane, Australia (6–8 December 2004), 89–96. https://doi.org/10.1109/FPT.2004.1393255.

O'Gorman, F. and Clowes, M.B. (1976). Finding picture edges through collinearity of feature points. *IEEE Transactions on Computers* **25** (4): 449–456. https://doi.org/10.1109/TC.1976.1674627.

Orlando, C., Andrea, P., Christophel, M., Xavier, D., and Granado, B. (2018). FPGA-based real time embedded Hough transform architecture for circles detection. *Conference on Design and Architectures for Signal and Image Processing (DASIP)*, Porto, Portugal (10–12 October 2018), 31–36. https://doi.org/10.1109/DASIP.2018.8597174.

Proffitt, D. and Rosen, D. (1979). Metrication errors and coding efficiency of chain encoding schemes for the representation of lines and edges. *Computer Graphics and Image Processing* **10** (4): 318–332. https://doi.org/10.1016/S0146-664X(79)80041-6.

Rachakonda, R.V., Athanas, P.M., and Abbott, A.L. (1995). High-speed region detection and labeling using an FPGA-based custom computing platform. In: *Field Programmable Logic and Applications*, Oxford, UK (29 August – 1 September 1995), *Lecture Notes in Computer Science*, Volume **975**, 86–93. https://doi.org/10.1007/3-540-60294-1_101.

Ragnemalm, I. (1993). The Euclidean distance transformation in arbitrary dimensions. *Pattern Recognition Letters* **14** (11): 883–888. https://doi.org/10.1016/0167-8655(93)90152-4.

Rambabu, C. and Chakrabarti, I. (2007). An efficient immersion-based watershed transform method and its prototype architecture. *Journal of Systems Architecture* **53** (4): 210–226. https://doi.org/10.1016/j.sysarc.2005.12.005.

Rambabu, C., Chakrabarti, I., and Mahanta, A. (2002). An efficient architecture for an improved watershed algorithm and its FPGA implementation. *IEEE International Conference on Field-Programmable Technology (FPT)*, Hong Kong (16–18 December 2002), 370–373. https://doi.org/10.1109/FPT.2002.1188713.

Rosenfeld, A. and Pfaltz, J. (1966). Sequential operations in digital picture processing. *Journal of the Association for Computing Machinery* **13** (4): 471–494. https://doi.org/10.1145/321356.321357.

Roy, P., Biswas, P.K., and Das, B.K. (2016). A modified hill climbing based watershed algorithm and its real time FPGA implementation. *29th International Conference on VLSI Design and 2016 15th International Conference on Embedded Systems (VLSID)*, Kolkata, India (4–8 January 2016), 451–456. https://doi.org/10.1109/VLSID.2016.103.

Russ, J.C. (2002). *The Image Processing Handbook*, 4e. Boca Raton, FL: CRC Press.

Sarcher, J., Scheglmann, C., Zoellner, A., Dolereit, T., Schaeferling, M., Vahl, M., and Kiefer, G. (2018). A configurable framework for Hough-transform-based embedded object recognition systems. *IEEE 29th International Conference on Application-specific Systems, Architectures and Processors (ASAP)*, Milan, Italy (10–12 July 2018), 8 pages. https://doi.org/10.1109/ASAP.2018.8445086.

Schumacher, F., Holzer, M., and Greiner, T. (2012). Critical path minimized raster scan hardware architecture for computation of the generalized Hough transform. *19th IEEE International Conference on Electronics, Circuits, and Systems (ICECS 2012)*, Seville, Spain (9–12 December 2012), 681–684. https://doi.org/10.1109/ICECS.2012.6463634.

Shih, F.Y. and Wong, W.T. (1992). A new single-pass algorithm for extracting the mid-crack codes of multiple regions. *Journal of Visual Communication and Image Representation* **3** (3): 217–224. https://doi.org/10.1016/1047-3203(92)90018-O.

Stephens, R.S. (1991). Probabilistic approach to the Hough transform. *Image and Vision Computing* **9** (1): 66–71. https://doi.org/10.1016/0262-8856(91)90051-P.

Sudha, N. and Mohan, A.R. (2008). Design of a hardware accelerator for path planning on the Euclidean distance transform. *Journal of Systems Architecture* **54** (1–2): 253–264. https://doi.org/10.1016/j.sysarc.2007.06.003.

Tagzout, S., Achour, K., and Djekoune, O. (2001). Hough transform algorithm for FPGA implementation. *Signal Processing* **81** (6): 1295–1301. https://doi.org/10.1016/S0165-1684(00)00248-6.

Tang, J.W., Shaikh-Husin, N., Sheikh, U.U., and Marsono, M.N. (2018). A linked list run-length-based single-pass connected component analysis for real-time embedded hardware. *Journal of Real-Time Image Processing* **15** (1): 197–215. https://doi.org/10.1007/s11554-016-0590-2.

Trein, J., Schwarzbacher, A.T., Hoppe, B., Noffz, K.H., and Trenschel, T. (2007). Development of a FPGA based real-time blob analysis circuit. *Irish Signals and Systems Conference*, Derry, Northern Ireland (13–14 September 2007), 121–126.

Trein, J., Schwarzbacher, A.T., and Hoppe, B. (2008). FPGA implementation of a single pass real-time blob analysis using run length encoding. *MPC-Workshop*, Ravensburg-Weingarten, Germany (1 February 2008), 71–77.

Trieu, D.B.K. and Maruyama, T. (2006). Implementation of a parallel and pipelined watershed algorithm on FPGA. *International Conference on Field Programmable Logic and Applications (FPL'06)*, Madrid, Spain (28–30 August 2006), 561–566. https://doi.org/10.1109/FPL.2006.311267.

Trieu, D.B.K. and Maruyama, T. (2007). A pipeline implementation of a watershed algorithm on FPGA. *International Conference on Field Programmable Logic and Applications (FPL 2007)*, Amsterdam, The Netherlands (27–29 August 2007), 714–717. https://doi.org/10.1109/FPL.2007.4380752.

Trieu, D.B.K. and Maruyama, T. (2008). An implementation of a watershed algorithm based on connected components on FPGA. *International Conference on Field Programmable Technology*, Taipei, Taiwan (7–10 December 2008), 253–256. https://doi.org/10.1109/FPT.2008.4762391.

Vincent, L. (1991). Exact Euclidean distance function by chain propagations. *IEEE Computer Society Conference on Computer Vision and Pattern Recognition*, Maui, HI, USA (3–6 June 1991), 520–525. https://doi.org/10.1109/CVPR.1991.139746.

Vincent, L. and Soille, P. (1991). Watersheds in digital spaces: an efficient algorithm based on immersion simulations. *IEEE Transactions on Pattern Analysis and Machine Intelligence* **13** (6): 583–598. https://doi.org/10.1109/34.87344.

Waltz, F.M. and Garnaoui, H.H. (1994). Fast computation of the grassfire transform using SKIPSM. *Machine Vision Applications, Architectures and System Integration III*, Boston, MA, USA (31 October – 2 November 1994). SPIE, Volume **2347**, 408–416. https://doi.org/10.1117/12.188751.

Wilson, G.R. (1997). Properties of contour codes. *IEE Proceedings Vision, Image and Signal Processing* **144** (3): 145–149. https://doi.org/10.1049/ip-vis:19971159.

Wu, K., Otoo, E., and Shoshani, A. (2005). Optimizing connected component labelling algorithms. *Medical Imaging 2005: Image Processing*, San Diego, CA, USA (15–17 February 2005). SPIE, Volume **5747**, 1965–1976. https://doi.org/10.1117/12.596105.

Yeong, L.S., Ngau, C.W.H., Ang, L.M., and Seng, K.P. (2009). Efficient processing of a rainfall simulation watershed on an FPGA-based architecture with fast access to neighbourhood pixels. *EURASIP Journal on Embedded Systems* **2009** (1): Article ID 318654, 19 pages. https://doi.org/10.1155/2009/318654.

Yuen, H.K., Princen, J., Illingworth, J., and Kittler, J. (1990). Comparative study of Hough transform methods for circle finding. *Image and Vision Computing* **8** (1): 71–77. https://doi.org/10.1016/0262-8856(90)90059-E.

Zahn, C.T. and Roskies, R.Z. (1972). Fourier descriptors for plane closed curves. *IEEE Transactions on Computers* **21** (3): 269–281. https://doi.org/10.1109/TC.1972.5008949.

Zhang, D. and Lu, G. (2002). A comparative study of Fourier descriptors for shape representation and retrieval. *5th Asian Conference on Computer Vision*, Melbourne, Australia (23–25 January 2002), 646–651.

Zhou, X., Ito, Y., and Nakano, K. (2014a). An efficient implementation of the gradient-based Hough transform using DSP slices and block RAMs on the FPGA. *IEEE International Parallel and Distributed Processing Symposium Workshops*, Phoenix, AZ, USA (19–23 May 2014), 762–770. https://doi.org/10.1109/IPDPSW.2014.88.

Zhou, X., Ito, Y., and Nakano, K. (2014b). An efficient implementation of the one-dimensional Hough transform algorithm for circle detection on the FPGA. *2nd International Symposium on Computing and Networking*, Shizuoka, Japan (10–12 December 2014), 447–452. https://doi.org/10.1109/CANDAR.2014.32.

Zingaretti, P., Gasparroni, M., and Vecci, L. (1998). Fast chain coding of region boundaries. *IEEE Transactions on Pattern Analysis and Machine Intelligence* **20** (4): 407–415. https://doi.org/10.1109/34.677272.

14

Machine Learning

Chapters 7–13 have focused primarily on operations from the low and intermediate levels of the image processing pyramid. Images have been processed to identify regions, and from those regions various discriminating features have been extracted. The final stages in any application are usually classification and recognition. These generally involve models that associate a class or object with the underlying features that have been extracted. The outputs of classification models are usually discrete, with a class label assigned for each possibility that the model has been trained for. Other modelling tasks require a continuous numeric output, for example the quantification of a parameter associated with an object, in which case the modelling is called regression. Machine learning models are often used for these tasks.

The generic architecture of using a machine learning model is illustrated in Figure 14.1. There are two phases to the use of such a model. The first phase is ***training***, where the model parameters for the task are determined. Training is based on datasets that consist of many example images or feature vectors. Training can be categorised as supervised or unsupervised learning. With supervised learning, training uses ***labelled data***, where the corresponding classification or regression outputs for each training example are known. Unsupervised learning looks for patterns within the training dataset and uses these to represent or categorise the features. After training comes the inference or prediction phase, where the model is applied to new data not in the training dataset. In the final implementation, only model inference needs to be considered, unless the model is trained online.

Some of the basic principles associated with supervised training are discussed in Section 14.1, before describing some of the different machine learning models in more depth.

14.1 Training

Supervised learning is an optimisation process, where the model parameters are optimised to give the best accuracy on the labelled training data, in terms of either the value output from the regression or the classification accuracy. Such an optimisation requires some measure of performance that can be optimised. The ***loss function*** measures the loss in performance or accuracy as a result of the error in an individual training sample or image. Adjusting the model parameters to improve the performance for one image will often make the performance worse for other images. Therefore, averaging the loss over the M samples within the training dataset gives the ***cost function*** which measures the overall performance of the model:

$$cost = \frac{1}{M}\sum_{i=1}^{M} loss_i. \tag{14.1}$$

Averaging gives a measure that is independent of the size of the dataset, enabling comparison, although the scaling by M is not important from an optimisation perspective (it is just a constant factor). Training therefore closes the loop in Figure 14.2 to optimise the model parameters to minimise the cost.

Design for Embedded Image Processing on FPGAs, Second Edition. Donald G. Bailey.

Companion Website: www.wiley.com/go/bailey/designforembeddedimageprocessc2e

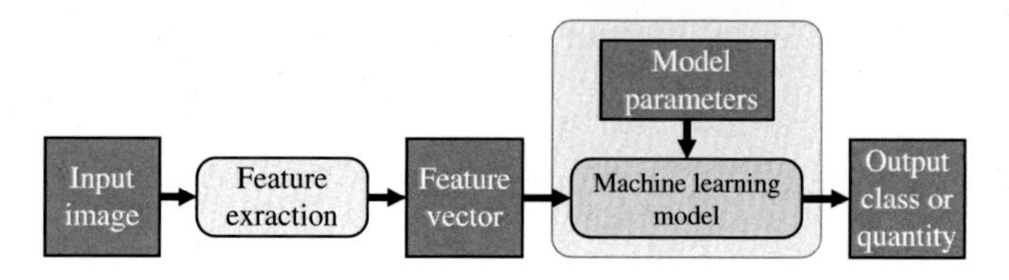

Figure 14.1 Using a machine learning model for classification or regression.

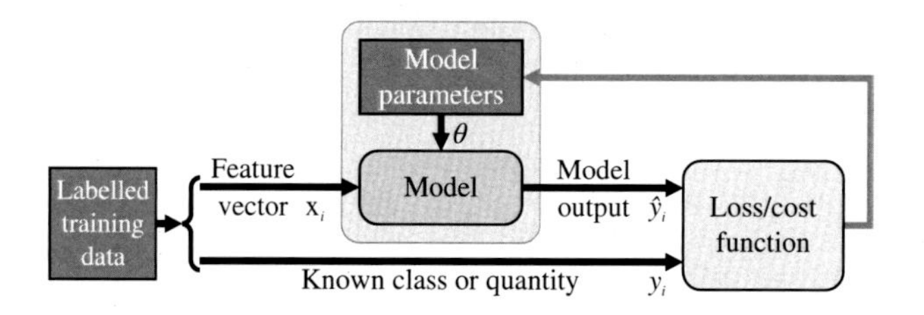

Figure 14.2 Training a machine learning model.

14.1.1 Loss and Cost Functions

While there are many possible loss functions, only the more commonly used ones are outlined here.

First consider regression tasks, where the model gives a continuous output, $y(\mathbf{x}, \boldsymbol{\theta})$, as a function of the input feature vector, $\mathbf{x}$, and model parameters, $\boldsymbol{\theta}$. Let the model output for the ith training sample be $\hat{y}_i = y(\mathbf{x}_i, \boldsymbol{\theta})$, and the known or desired output be y_i. The most common loss function is the squared error:

$$loss(\hat{y}_i, y_i) = (\hat{y}_i - y_i)^2, \tag{14.2}$$

with the corresponding cost function being the mean square error (MSE). Minimising this cost gives the least squares optimisation, which is best suited when the losses have a Gaussian distribution. The main limitation of MSE is its sensitivity to samples where the loss is very large (statistical outliers) (Besl et al., 1989). A single outlier can have undue influence over the final model, introducing bias and causing its performance to deteriorate for new, unseen samples.

Less weight is given to the outliers using an absolute error loss function:

$$loss(\hat{y}_i, y_i) = |\hat{y}_i - y_i|. \tag{14.3}$$

The mean absolute error (MAE) cost function is best suited when the losses have a double exponential distribution. Cort and Kenji (2005) argue that MAE provides a more natural estimate of the average error than MSE (or root mean square error (RMSE)). One limitation of MAE is that it is not differentiable at zero loss, making it harder to use gradient descent algorithms for optimisation. This may be overcome using the Huber loss (Huber, 1964), which is a hybrid of squared error (for small errors) and absolute error (for large errors):

$$loss_\delta(\hat{y}_i, y_i) = \begin{cases} \frac{1}{2}(\hat{y}_i - y_i)^2, & |\hat{y}_i - y_i| \le \delta, \\ \delta\left(|\hat{y}_i - y_i| - \frac{1}{2}\delta\right), & \text{otherwise,} \end{cases} \tag{14.4}$$

where δ defines the boundary between the two components. For a contaminated Gaussian distribution with spread σ, setting $\sigma \le \delta \le 2\sigma$ (Huber, 1964) gives a good estimate of the mean without being strongly biased by outliers.

The size of MAE relates to the scale of the data. This scale may be removed by scaling the absolute error by the actual value giving the relative error as the loss function (de Myttenaere et al., 2016):

$$loss(\hat{y}_i, y_i) = \frac{|\hat{y}_i - y_i|}{|y_i|}. \tag{14.5}$$

The corresponding cost function is often scaled by 100% to give the mean absolute percentage error (MAPE). Using the relative error in this way has two limitations (Tofallis, 2015). First, the scaling effectively weights each absolute error by $\frac{1}{y_i}$. This tends to underestimate the amplitude of the resulting output because $|y_i| < |\hat{y}_i|$ has more weight in percentage terms than $|\hat{y}_i| < |y_i|$. Second, and more serious, is that estimates which have a small true value ($y_i \approx 0$) will be given undue weight. These can be improved by weighting the values within the data by y_i to give MAE over mean cost function (Kolassa and Schutz, 2007):

$$MAE/M = \frac{\frac{1}{M}\sum_{i=1}^{M}|y_i|\frac{|\hat{y}_i - y_i|}{|y_i|}}{\frac{1}{M}\sum_{i=1}^{M}|y_i|} = \frac{\frac{1}{M}\sum_{i=1}^{M}|\hat{y}_i - y_i|}{\frac{1}{M}\sum_{i=1}^{M}|y_i|} = \frac{MAE}{\overline{|y|}}, \tag{14.6}$$

which is a scale-free relative cost measure.

For classification, where the output is discrete, forming a loss function becomes a little more complicated. The usual approach is to model the mapping from the feature vector to a probability distribution for each class, C, that is

$$y_C(\mathbf{x}) = p(C|\mathbf{x}), \tag{14.7}$$

with the final output given as the class that is the most likely

$$C = \arg\max_o\{y_{C_o}(\mathbf{x})\} = \arg\max_o\{p(C_o|\mathbf{x})\}. \tag{14.8}$$

For example, binary classification (pass, $y = 1$; fail, $y = 0$) commonly uses logistic regression to model the probability distribution:

$$y_{pass}(\mathbf{x}) = \frac{1}{1 + e^{-f(\mathbf{x})}}, \tag{14.9}$$

where $f(\mathbf{x})$ is often a linear combination of the features within the feature vector. For this, a suitable loss function is the cross-entropy

$$loss(\hat{y}_i, y_i) = -y_i \ln \hat{y}_i - (1 - y_i)\ln(1 - \hat{y}_i), \tag{14.10}$$

which decreases as the modelled distribution becomes closer to the distribution represented by the samples. This can readily be generalised to multi-class classification.

An alternative loss function for classification (with target outputs $y = \pm 1$) is the hinge loss:

$$loss(\hat{y}_i, y_i) = \max(0, 1 - \hat{y}_i y_i). \tag{14.11}$$

This penalises wrong and also right predictions where the margin is insufficient. In this way, it effectively penalises samples based on their distance from the decision boundary or, equivalently, shifts the decision boundaries to minimise the cost.

14.1.2 Model Optimisation

The model parameters are often optimised numerically using some form of iterative gradient descent algorithm. Note that such algorithms will converge to a local minimum of the error surface, depending on the starting values of the model parameters, $\boldsymbol{\theta}_0$. Iterative training is often computationally intensive, and when applied to large training sets, memory bandwidth constraints can limit how fast data can be processed.

Let the cost function be $J(\boldsymbol{\theta}; \langle \mathbf{x}_i, y_i \rangle)$, where $\langle \mathbf{x}_i, y_i \rangle$ represents the labelled training dataset. ***Gradient descent*** uses the gradient information (the vector of partial derivatives of the cost function with respect to each model parameter, $\nabla J(\boldsymbol{\theta})$) to adjust the parameters in a direction which reduces the cost. The gradient may be calculated algebraically, from the partial derivatives of the cost function, or numerically, by taking a small change in each parameter, and calculating the change in the cost function. The kth iteration is then

$$\boldsymbol{\theta}_k = \boldsymbol{\theta}_{k-1} - \eta \nabla J(\boldsymbol{\theta}_{k-1}), \tag{14.12}$$

where η is the step size or learning rate. Selection of the learning rate is critical. If it is too large, successive iterations repeatedly overshoot the minimum, and the algorithm fails to converge by jumping around randomly. If too small, convergence is very slow, and it is more likely to converge into a poor local minimum close to $\boldsymbol{\theta}_0$.

The gradient calculations can be quite expensive, especially when there are many parameters, and the training set is large, since the cost function represents the average loss from the whole dataset (batch gradient descent). Therefore, one approach is ***stochastic gradient descent***, which updates the parameters according to Eq. (14.12) for each sample. Each step will improve the loss for that data sample but may be in the wrong direction for the overall cost. Consequently, the parameter updates will be all over the place, although on average will be in the right direction (hence the name stochastic). A compromise is often used by updating the parameters based on an estimate of the cost function derived from the average loss of a mini-batch of training samples. This gives more stable convergence by reducing the variance of the parameter updates.

There are many variations to the basic gradient descent described here, including automatically updating the learning rate, η, adding momentum, and other optimisations to improve the convergence (Ruder, 2016), and reduce the chance of getting stuck in a sub-optimal local minimum.

Some research has explored the use of field-programmable gate arrays (FPGAs) for accelerating the computation for gradient descent. Much has focused on mini-batch training to overcome the problems with data bandwidth for large training sets. The parallelism of hardware can provide scalability, giving good acceleration, although the limited on-chip memory gives a trade-off between the number of parallel blocks and batch size (Rasoori and Akella, 2018). De Sa et al. (2017) showed that gradient descent could be implemented on an FPGA with limited precision arithmetic.

Where the cost function has many local minima, there is no guarantee that the gradient descent will find the global optimum. In such cases, better performance can often be obtained by methods based on a population of candidate solutions, which can explore parameter space more thoroughly. Examples of such methods are evolutionary algorithms and particle swarm optimisation. The parallelism of FPGAs can readily be used to accelerate population-based methods such as particle swarm optimisation (see, for example, (Munoz et al., 2010; Da Costa et al., 2019)).

14.1.3 Considerations

Generally, cost functions cannot distinguish between systematic and random errors. Systematic errors can result from the measurement process, or through inappropriate processing, or using an unsuitable model or loss function. Systematic errors in measurement can often be mitigated through appropriate calibration. In fitting a model to measured data, outliers can also result in systematic errors. Systematic errors introduce bias and reduce the accuracy of the results. Random errors can result from both measurement noise and quantisation noise. They contribute to the variance and reduce the precision of the results. The distinction between these is illustrated in Figure 14.3.

There are several factors that can lead to a poor machine learning model. These can give a low quality estimate as a result of regression or a faulty or inaccurate segmentation.

First, the selection of features is critical to the success of any machine learning model. For regression, the model is trying to identify the underlying dependency between the measured features and the output. If unsuitable features are used, then it may not be possible to accurately model the underlying relationship. Similarly for classification, the feature vector must have sufficient discrimination power, otherwise the probability distributions for two or more classes can have significant overlap. Both of these cases result in an unacceptably high cost of the final optimisation and may be remedied by measuring additional features. Having too many features can also be a problem (Khalid et al., 2014), especially if there are strong correlations. Features which are irrelevant, redundant, or even misleading not only slow the convergence but can also lead to a poorer model. Khalid et al. (2014) reviewed some of the techniques that have been used for feature selection and dimensionality reduction.

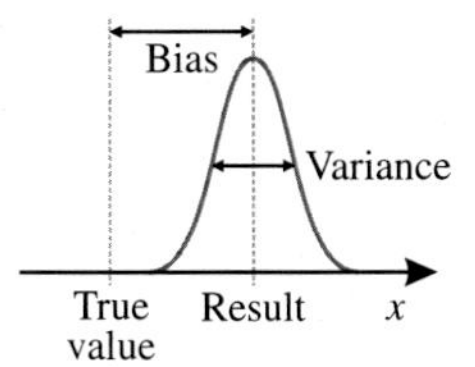

Figure 14.3 The effects of systematic and random errors on a resultant measurement. Systematic errors introduce bias, and random errors contribute to the variance.

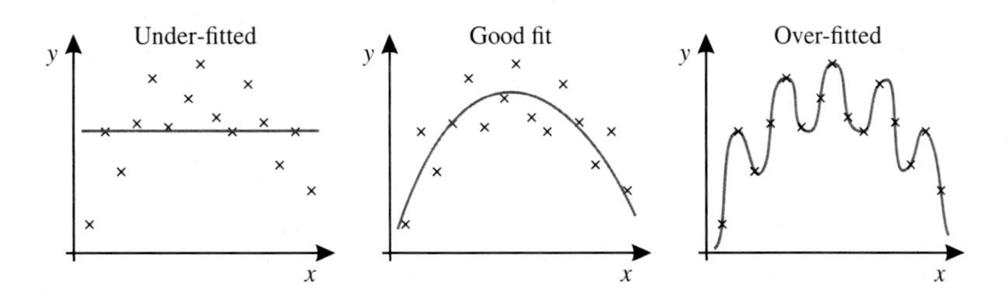

Figure 14.4 Comparison of under- and over-fitting. Left: under-fitting has a high fit cost; centre: a good fit has a moderate cost that reflects the expected level of noise in the data; right: over-fitting has very low or zero cost but generalises poorly by giving a significantly larger error on new data.

The samples in the training dataset must be representative, so that the set of feature vectors obtained do not bias the model. They must provide adequate coverage of feature-space where measurements or classifications must be inferred. For classification, this means that each class must be adequately represented by the training samples. In unbalanced classification tasks, where one class is significantly more likely than another, having fewer examples of the rarer class effectively gives more weight to the more popular class. Modelling may be balanced by weighted optimisation, giving more weight to samples from the rarer class. Care must also be taken to identify any outliers, as these can unduly influence the resulting model parameters. If necessary, outliers could be eliminated before training.

Third, the model must have sufficient power and be appropriate for the modelling task. Examples of such under-fitting are trying to model a curve with a straight line, using a neural network with insufficient nodes in the hidden layer to adequately model the data, or using linear discriminant analysis where the classes are unable to be separated using hyperplanes. Under-fitting can usually be identified by a high cost of the final optimisation (see Figure 14.4). This may be remedied using a more complex model, or for classification, adding additional discriminating features and associated model parameters.

The opposite of under-fitting is over-fitting (see Figure 14.4). This is where the model has too many parameters for the underlying task and as a result starts to fit the noise present in the training data. Consequently, the model gives excellent performance for the set of images used for training but poor results on previously unseen images. Similar issues arise from over-training. An over-trained classifier recognises the training examples but does not generalise well to other examples of the same class. For these reasons, it is important that the data available for training be separated into two independent sets. The first is used for actually training the classifier. After training is completed, the second set is used to test or provide an unbiased evaluation of the classifier performance and identify any training issues. It is important that both the training and test sets have sufficient samples to be representative (Xu and Goodacre, 2018). Significantly poorer performance on the test set than on the training set is an indication of over-training.

Many models have variations in the model that are controlled by hyperparameters. They can be structural, for example the number of nodes in the hidden layer of a neural network or even selecting between different classification methods, or procedural, for example the particular cost function used, or the learning rate. The best values for the hyperparameters also need to be determined as part of the training process. In such cases, it is

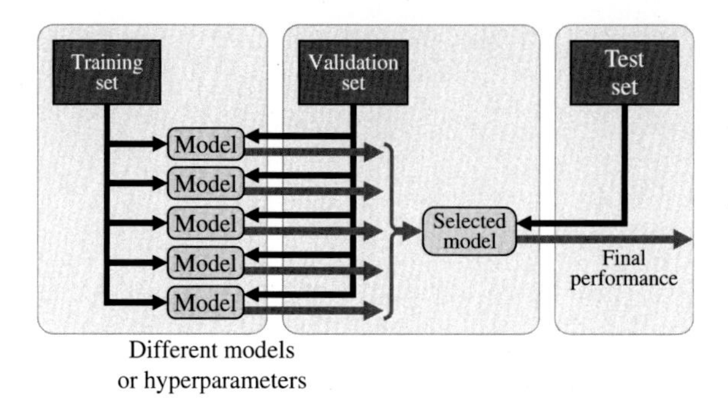

Figure 14.5 Three phases of evaluating a model: training the model parameters of one or more models; validation to determine when training is complete and to select a particular model; testing the performance of the final model.

usual to divide the available data into three sets: training, validation, and test, which are then applied as shown in Figure 14.5. During training, each of the different models (or each model with different values for the hyperparameters) is trained on the training dataset. The validation dataset is used to select the particular model or set of hyperparameters that give the best performance. It can also be used to stop training early, for example deteriorating performance on the validation set can indicate that the model is starting to be over-trained (Prechelt, 1998). These uses of the validation dataset make it part of the training, so it is not independent of the training process. Finally, the test set provides the final independent unbiased evaluation of the model.

The training process makes use of noisy training data, so inevitably there will be noise associated with the final model. The loss function describes how well the model fits the data, but to avoid over-fitting, it may be desirable to impose additional constraints. These can be added to the cost function as an explicit ***regularisation*** term:

$$cost = \frac{1}{M}\sum_{i=1}^{M} loss(y(\mathbf{x_i}, \theta), \mathbf{x_i}) + \lambda R(y), \tag{14.13}$$

where $R(y)$ is the regularisation constraint on the model, and λ controls how much weight is given to the regularisation relative to the data. Common regularisation constraints include:

- **Smoothness**: Requiring a fit to be smooth can reduce over-fitting by preferring a simpler model over a more complex model. It can also help reduce the number of local minima, enabling convergence on a better solution.
- **Minimum norm**: Prefers solutions with smaller norms. Again it can help to reduce the number of local minima, leading to a better solution.
- **Prior knowledge**: Incorporates additional known information into the cost function that can serve as additional constraints. For example, Bayesian modelling can incorporate a prior distribution, with the resulting output combining the training data with the prior.
- **Sparsity**: Prefers solutions where many of the model parameters are zero. This can lead to simpler, and more interpretable, models and is commonly used with dictionary-based learning models.

When using FPGAs for training, the training can often be accelerated using multiple parallel processes. The training data, because of its volume, is usually stored off-chip, with the memory bandwidth limiting the overall training speed. To maximise the acceleration, it is necessary to make best use of the available bandwidth by maximising reuse of the loaded data. As outlined earlier, loading the data into on-chip caches can enable multiple accesses. However, if these accesses are from multiple parallel processes, the on-chip access conflicts may counteract the expected performance gains. This requires that either these processes be coordinated to operate on the same data simultaneously, or that the data is replicated into separate parallel caches, local to each processor (Liu et al., 2008).

14.2 Regression

With regression, the goal is to model one or more continuous outputs as a function of the inputs. This is commonly used for geometric transformation, where the output coordinates are modelled as a function of the input coordinates for a forward mapping (or vice versa for a reverse mapping). It is also used to estimate a value from an image (for example crop yield) on the basis of the measured features.

14.2.1 Linear Regression

The simplest empirical model is a weighted sum of the feature values. Let the feature vector have N components, and for convenience the feature vector is augmented with a constant term ($x_0 = 1$). The model parameters, $\boldsymbol{\theta}$, are the weights, θ_i, given to each of the feature values. The weighted sum is therefore a dot-product between the weight vector and feature vector:

$$y = \sum_{i=0}^{N} \theta_i x_i = \boldsymbol{\theta}^{\top}\mathbf{x} = \boldsymbol{\theta} \cdot \mathbf{x}. \tag{14.14}$$

When using a squared-error loss function, the cost function is quadratic in $\boldsymbol{\theta}$ with a single global minimum. Such least squares problems can be solved analytically to give the model parameters.

If necessary, a higher order linear model may be formulated which includes products and powers (or indeed any function) of the individual feature values. If each of these terms is considered as a separate feature value, with its corresponding weight, the resulting regression is still linear in the model weights (for example the polynomial mapping in Section 10.3.3).

Implementation of a linear regression model consists of multiplications and additions, as in Eq. (14.14). There is little point in calculating the output to a significantly higher accuracy than the minimum of the cost function. Therefore, with appropriate scaling of the feature values and model weights, most models can be implemented to the desired precision using fixed-point arithmetic using the digital signal processing (DSP) blocks on an FPGA.

14.2.2 Nonlinear Regression

Any other model is nonlinear and, even with a squared-error loss function, cannot in general be solved analytically. In many cases there is a mathematical model where the model parameters are determined by fitting to available data. One common example of nonlinear regression is determining the coordinate transformation associated with camera calibration where there is a combination of lens and perspective distortions.

14.2.3 Neural Networks

A neural network is made of an interconnected set of neurons, inspired by biological neural networks. Feed-forward networks (see Figure 14.6) consist of an input layer, containing the inputs to the network, one or more hidden layers of neurons, and an output layer of neurons which produce the network output. Each neuron consists of a weighted sum, s, of its inputs (plus a constant offset, w_0) followed by a nonlinear activation function, $a(s)$, to produce the activation:

$$y = a(s) = a(\mathbf{w}^{\top}\mathbf{x}). \tag{14.15}$$

Traditionally, hyperbolic tangent or sigmoidal activation functions have been used (Kalman and Kwasny, 1992), either

$$a(s) = \tanh s = \frac{e^s - e^{-s}}{e^s + e^{-s}} \tag{14.16}$$

or

$$a(s) = \frac{1}{1+e^{-s}} = \frac{1}{2} + \frac{1}{2}\tanh\frac{s}{2}, \tag{14.17}$$

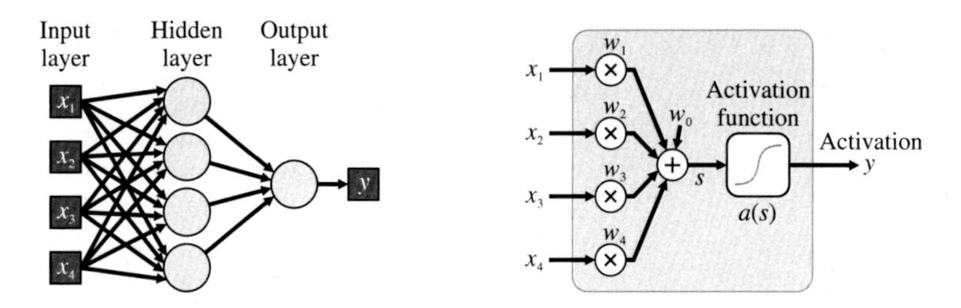

Figure 14.6 Neural network architecture. Left: structure of a feed-forward network, consisting of a set of interconnected neurons. Right: each neuron consists of a weighted sum of its inputs, followed by a nonlinear activation function.

to compress large values into the range $-1 < y < 1$ or $0 < y < 1$, respectively. However, Ramachandran et al. (2017) showed that almost any nonlinear activation function works, although some are better than others. With deep networks (many layers), a rectified linear unit (ReLU) activation function is popular:

$$a(s) = \max(0, s). \tag{14.18}$$

Neural networks can model any arbitrary mapping (given sufficient depth, width of hidden layers, and appropriate training) (Hornik et al., 1989). The challenge with using neural networks is to select appropriate model hyperparameters (number of layers and number of neurons in each hidden layer) to find a good local minimum without over-fitting to the training data.

Neural networks are trained using gradient descent, with the partial derivatives of the cost function with respect to each weight evaluated using the chain rule. Back-propagation is a technique where the effects of changes in the output of a neuron are propagated to the inputs, enabling efficient use of the chain rule. The use of sigmoidal activation functions helps in this regard because they are easily differentiated; for the hyperbolic tangent:

$$\frac{da(s)}{ds} = 1 - \tanh^2 s = (1 + y)(1 - y), \tag{14.19}$$

and for the sigmoid:

$$\frac{da(s)}{ds} = y(1 - y). \tag{14.20}$$

The ReLU is even more trivial,

$$\frac{da(s)}{ds} = \begin{cases} 0, & s < 0, \\ 1, & s > 0, \end{cases} \tag{14.21}$$

apart from not being differentiable at $s = 0$.

Many researchers have implemented basic neural networks on FPGAs (Misra and Saha, 2010). Computationally, each neuron consists of a dot-product, the same as for linear regression (Section 14.2.1), followed by the nonlinear activation function. Much of the early work used floating-point arithmetic, but error analysis shows that, even for large networks, fixed-point can be used (Choi et al., 1993). The activation outputs of each neuron (with a sigmoid activation function) are $|y| \leq 1.0$, and these become the inputs of the next layer. For consistency, the input layer can normalise the features to within this range as well. Quantisation errors introduce noise into the result in three ways: quantisation of the weights, rounding the results of the multiplication, and approximation errors of the sigmoid function. Of these, weight quantisation has the least effect, and multiplication rounding has the greatest (Choi et al., 1993). To limit the buildup of error, it is important to maintain the precision from each multiplication and perform any rounding on the final sum, s.

Approaches for realising the sigmoid activation function include using CORDIC (coordinate rotation digital computer) arithmetic (Chen et al., 2006), lookup tables (Baptista and Dias, 2012), piece-wise polynomial approximation (Baptista and Dias, 2012), and rational polynomial approximation (Hajduk, 2018).

The smoothness of the activation function makes it ideal for implementation using an interpolated lookup table, although this becomes expensive with large numbers of neurons.

For large networks, there may be insufficient resources to implement the whole network in parallel. Sequential implementations can work at two levels. One is to use a processing element that implements a single neuron, and use it sequentially for each neuron in the model, using block RAM to hold the network weights and intermediate values. The main limitation of this approach is the bandwidth in getting the data and weights in parallel to the inputs of the processing elements. This may be overcome by serialising the data and weights and calculating each of the multiplications of the dot-product sequentially. If necessary, the reduced operating speed can be compensated by implementing several processing elements in parallel. Overall, there is a trade-off among parallelism, speed, and power requirements. A given application will dictate the required speed and accuracy, and hence the degree of parallelism required.

Networks are usually trained using 32-bit floating-point. Deployment on FPGAs can work successfully using fixed-point, although this often comes with a slight degradation in accuracy.

14.3 Classification

The goal of classification is to recognise or assign meaning to an object within an image, or even the whole image, based on the feature vector measured from that object or image. The actual classification may be as simple as 'pass' or 'fail', assigning the object a quality grade, or in image recognition tasks, determining what the object is from a predefined list of classes. The feature vector defines a point in feature-space, with feature vectors of objects from the same class ideally being similar, forming clusters. Classifier training can be thought of as identifying the cluster for each class, or more importantly, determining the boundaries between the clusters. Training therefore determines the model parameters that minimise the classification errors, as defined by the loss function. During the inference phase, the final classification model takes as its input the feature vector and outputs the associated class. Saidi et al. (2021) surveyed the use of FPGAs both in training classification models and using the models for inference.

14.3.1 Decision Trees

A decision tree determines the class through a series of (usually binary) decisions. Each decision divides feature-space into two partitions by thresholding on a single feature value. Decision trees provide good interpretability, especially for shallow trees, because the tree provides the reasoning behind the mapping from a feature vector to a class.

Where there are only a few features, feature-space can be manually partitioned; however, this becomes more difficult when there is a large number of features or the clusters are not well separated. Usually, however, the tree is built algorithmically (Suthaharan, 2016), starting at the root of the tree and successively splitting the nodes. Splitting stops when all of the leaf nodes represent a single class. To keep the tree as simple as possible, we want to choose the split which gains the most information or has the most consistent child nodes and best separates the classes. Two of the measures commonly used for information gain are the entropy and the Gini impurity.

Let the proportion of training samples from class C present in the region of feature-space represented by node k be $p_k(C)$. For J classes, the entropy of node k is

$$H_k = -\sum_{j=1}^{J} p_k(C_j)\log_2 p_k(C_j). \tag{14.22}$$

The information gained by splitting node k into k_1 and k_2 is

$$IG_k = H_k - (p(k_1)H_{k_1} + p(k_2)H_{k_2}), \tag{14.23}$$

where $p(k_1)$ and $p(k_2)$ are the proportions of the training samples from node k partitioned into the two child nodes. The split that maximises the information gain at each node is chosen as the decision criterion.

Since H_k is constant for a particular split, maximising the information gain in Eq. (14.23) is equivalent to choosing the split that minimises the second term:

$$decision = \arg\min_{k_1 \wr k_2}(p(k_1)H_{k_1} + p(k_2)H_{k_2}), \tag{14.24}$$

where $k_1 \wr k_2$ represents the threshold that splits node k into two partitions.

The Gini impurity is used in a similar way. The Gini impurity of node k is

$$G_k = \sum_{j=1}^{J} 1 - p_k^2(C_j) \tag{14.25}$$

and measures how consistent the classes are within the node. Maximising the information gain chooses the split that minimises the weighted Gini impurity:

$$decision = \arg\min_{k_1 \wr k_2}(p(k_1)G_{k_1} + p(k_2)G_{k_2}). \tag{14.26}$$

The Gini impurity has the advantage that it does not require calculating logarithms.

Decision trees work best when the classes are well separated. Otherwise the tree can become quite deep as a result of making each leaf node represent a single class. This over-fits the training data and leads to poor generalisation when applied to new data.

14.3.2 Random Forests

Random forests improve the classification performance of decision trees, by training many independent trees and using a majority vote of the results (Ho, 1995). Two approaches can be used for training. One is to train each tree using a random subset of the training data. The second is to use the whole dataset for training each tree but only consider a random subset of the features at each split. This latter approach reduces the correlation between trees by forcing different features to be used.

FPGAs can be used both for accelerating training and inference. Cheng and Bouganis (2013) used a merge sort algorithm to sort the training instances within a node in terms of the feature being used for the split. The output from the sorter was used directly to calculate the weighted Gini impurity, giving the split in a single pass through the data for the node. For inference, Lin et al. (2017) compared memory-, comparator-, and synthesis-centric implementations. The memory-centric approach represented each tree within memory and sequentially stepped through based on the decision at each node. The comparator-centric approach built a custom comparator for each decision and a multiplexer tree to select the corresponding comparison result at each level. The synthesis-centric approach reduced each tree to stateless Boolean function, which was synthesised directly. The memory-centric approach had the most flexibility, in that the forest could be changed without reprogramming the FPGA, but it also used the most resources.

14.3.3 Bayesian Classification

A Bayesian classifier uses probability models of the distributions of each of the features to estimate the most likely class associated with a feature vector. Training builds both the prior probability of seeing each class, $p(C_j)$, and the probability model for each feature vector, given the class, $p(\mathbf{x} \mid C_j)$. For classification, the opposite is wanted: given a feature vector, $\mathbf{x}$, determine the most likely class. This can be obtained from Bayes' theorem as

$$p(C_j \mid \mathbf{x}) = \frac{p(C_j)\,p(\mathbf{x} \mid C_j)}{p(\mathbf{x})}. \tag{14.27}$$

Since the denominator does not depend on the class, it is a scale factor that will scale the results for each class equally. Therefore, the ***discriminant function*** for each class becomes

$$d_j(\mathbf{x}) = p(C_j)\,p(\mathbf{x} \mid C_j). \tag{14.28}$$

Note that adding a constant or scaling by a positive constant will not affect the class order, nor will taking a monotonically increasing function of the discriminant function.

The Bayesian classifier therefore selects the class which has the largest value from the discriminant function:

$$C = \arg\max_j\{d_j(\mathbf{x})\} = \arg\max_j\{p(C_j)p(\mathbf{x} \mid C_j)\}. \tag{14.29}$$

It is also possible to obtain an estimate of the confidence in the classification by comparing the relative probabilities of the maximum and second maximum class; if the second highest score is too similar to the highest, then there is potentially ambiguity in the classification.

Several techniques can be used for modelling the probability distributions $p(\mathbf{x} \mid C_j)$. If the feature-space has only a few dimensions, then building a multi-dimensional histogram can capture the joint probability distribution. For example, this is used in Section 8.2.3.1 for detecting skin colour. Unfortunately, for more than about three features the memory requirements become significant unless a coarse bin-size is used, and this works against accurate classification, especially at the boundaries between classes.

It is more usual to assume that the features are conditionally independent,

$$p(\mathbf{x} \mid C_j) = \prod_{i=1}^{N} p(x_i \mid C_j). \tag{14.30}$$

Such an assumption is not usually valid; hence, this is called a ***naive Bayes classifier***:

$$C = \arg\max_j \left\{ p(C_j)\prod_{i=1}^{N} p(x_i \mid C_j) \right\}. \tag{14.31}$$

Using a logarithmic number system simplifies the computation. Taking the logarithm of the discriminant function in Eq. (14.31) gives

$$d_j(\mathbf{x}) = \log_2 p(C_j) + \sum_{i=1}^{N} \log_2 p(x_i \mid C_j). \tag{14.32}$$

In spite of violating the independence assumption, experiments have shown that the naive Bayes classifier still works surprisingly well (John and Langley, 1995; Rish, 2001).

Assuming independence allows a separate one-dimensional histogram to be constructed for each feature, for each class, to represent $p(x_i \mid C_j)$, rather than a multi-dimensional joint histogram. Rather than quantising the feature value (through histogram binning), a continuous probability density function can be modelled using kernel density estimation (John and Langley, 1995), which accumulates the convolution of each sample with a Gaussian kernel.

To reduce the number of parameters required to represent the distribution, a Gaussian mixture model may be used (Reynolds, 2009). Either the full joint distribution could be modelled (allowing the use of the full Bayes classifier of Eq. (14.29)) or the distribution of each feature could be modelled (with the inference made using a naive Bayesian classifier).

Meng et al. (2011) used binary features based on colour histograms of the target being tracked. Separate red, green and blue histograms were accumulated for the region of interest, and a 768-bit binary signature was created by thresholding each bin at the average occupancy. The statistics were then accumulated using an array of counters for each class during training. At the completion of training, a lookup table was used to calculate the logarithms of the probabilities which were stored in an array. For inference, the colour histogram was thresholded to give the binary feature vector. For each class, the corresponding log-probabilities were accessed from the array, and accumulated, with the final selection being the class with the maximum total. Xue et al. (2020) extended this to use continuous-valued features (which were binned to enable the probability distribution to be calculated). To calculate the log probability, a bin count was normalised by shifting so that the most significant bit (MSB) was 1 (giving the integer part of the logarithm), with the normalised count indexing a lookup table to get the fractional component.

14.3.4 Quadratic Discriminant Analysis

If each cluster is relatively compact, then the probability distribution for each cluster may be modelled as a multi-dimensional Gaussian distribution. For an N-dimensional feature vector, each class is parameterised by a mean vector, $\boldsymbol{\mu}_j$, and covariance matrix, $\boldsymbol{\Sigma}_j$:

$$p(\mathbf{x} \mid C_j) = \frac{e^{-\frac{1}{2}(\mathbf{x}-\boldsymbol{\mu}_j)^{\mathsf{T}}\boldsymbol{\Sigma}_j^{-1}(\mathbf{x}-\boldsymbol{\mu}_j)}}{\sqrt{(2\pi)^N |\boldsymbol{\Sigma}_j|}}, \tag{14.33}$$

where $|\boldsymbol{\Sigma}_j|$ is the determinant of $\boldsymbol{\Sigma}_j$. Taking the logarithm gives the Bayesian discriminant function:

$$d_j(\mathbf{x}) = -\frac{1}{2}(\mathbf{x}-\boldsymbol{\mu}_j)^{\mathsf{T}}\boldsymbol{\Sigma}_j^{-1}(\mathbf{x}-\boldsymbol{\mu}_j) - \frac{1}{2}\ln|\boldsymbol{\Sigma}_j| + \ln p(C_j). \tag{14.34}$$

Note that the first term represents the Mahalanobis distance, which measures the distance of a vector from a distribution in terms of the number of standard deviations. Therefore,

$$C = \arg\max_j\{d_j(\mathbf{x})\} = \arg\min_j\{d^2_{Mahalanobis} + \ln|\boldsymbol{\Sigma}_j| - 2\ln p(C_j)\}, \tag{14.35}$$

which effectively selects the nearest class (with a scaling offset to account for the prior).

The boundary separating two classes, i and j, will occur when discriminant functions for the two classes give the same value:

$$d_i(\mathbf{x}) = d_j(\mathbf{x}). \tag{14.36}$$

Since Eq. (14.34) is a quadratic form in $\mathbf{x}$, solving Eq. (14.36) for $\mathbf{x}$ will give a generalised quadratic decision surface in feature-space. Therefore, this technique is known as quadratic discriminant analysis.

14.3.5 Linear Discriminant Analysis

If it is further assumed that all clusters (classes) have the same covariance matrix, then when considering Eq. (14.36) the quadratic term cancels. The discrimination boundary becomes linear in $\mathbf{x}$ (a hyperplane in feature-space):

$$(\boldsymbol{\Sigma}^{-1}(\boldsymbol{\mu}_i - \boldsymbol{\mu}_j))^{\mathsf{T}}\mathbf{x} = (\boldsymbol{\Sigma}^{-1}(\boldsymbol{\mu}_i - \boldsymbol{\mu}_j))^{\mathsf{T}}\left(\frac{\boldsymbol{\mu}_i + \boldsymbol{\mu}_j}{2}\right) + \frac{1}{2}\ln\frac{p(C_j)}{p(C_i)}. \tag{14.37}$$

The left-hand side represents the projection onto the axis normal to the separating hyperplane, $\mathbf{w}$, that gives best separation between classes:

$$\mathbf{w} = \boldsymbol{\Sigma}^{-1}(\boldsymbol{\mu}_i - \boldsymbol{\mu}_j); \tag{14.38}$$

the right-hand side is a constant threshold that represents the projection of the point midway between the centres of the distributions, offset by the relative likelihoods of the two classes:

$$\mathbf{w}^{\mathsf{T}}\mathbf{x} = c \quad = \mathbf{w}^{\mathsf{T}}\left(\frac{\boldsymbol{\mu}_i + \boldsymbol{\mu}_j}{2}\right) + \frac{1}{2}\ln\frac{p(C_j)}{p(C_i)}. \tag{14.39}$$

This lends itself to straightforward implementation. Each unknown feature vector is projected (a dot-product) onto the discrimination vector, $\mathbf{w}$, and thresholding determines which of the two classes is assigned.

There are two approaches to handling more than two classes. The first is to perform a pairwise comparison between each pair of classes. This requires $\frac{1}{2}N(N-1)$ discrimination vectors, although only $N-1$ need to be used for any particular classification (each successive comparison eliminates one of the classes). The alternative is to compare one class against the rest, requiring N discrimination vectors and N comparisons. If required, confidence in final classification can be gauged from the distance to the classification threshold.

Tahir et al. (2004) used linear discriminant analysis on an FPGA to detect cancer cells. Programmed in Handel-C, the application measured texture features from the grey-level co-occurrence matrix (see Section 8.2.5) which were then classified into cancerous or non-cancerous. Having only two classes significantly simplified the classifier design. Piyasena et al. (2021) performed on-line linear discriminant analysis, where the mean and covariance matrix were updated after each sample. To simplify the computation, they avoided the need for matrix inversion to calculate $\mathbf{\Sigma}^{-1}$ by constraining the covariance matrix to being diagonal (effectively implementing a naive classifier). The system was quite small, with only four features and three classes.

14.3.6 Support Vector Machines

Support vector machines (SVMs) extend the idea behind linear discriminant analysis and attempt to find the hyperplane that gives the largest separation or margin between classes. Where it differs, however, is that it does not assume any particular probability distribution for the values within a cluster but instead works directly with the samples themselves.

First, assume that the classes are linearly separable, with the discriminant function based on a hyperplane:

$$d(\mathbf{x}) = \mathbf{w}^{\top}\mathbf{x} - c, \tag{14.40}$$

with output

$$y = \text{sign}(d(\mathbf{x})). \tag{14.41}$$

The closest feature vectors from the separating hyperplane define the location of the plane and hence are called the ***support vectors*** (see Figure 14.7). The training feature vectors therefore satisfy

$$y_i d(\mathbf{x}_i) \geq 1, \tag{14.42}$$

where equality occurs for the support vectors. The distance of the support vectors from the discriminating hyperplane is $\frac{1}{\|\mathbf{w}\|_2}$. To maximise the margin, it is necessary to minimise $\|\mathbf{w}\|_2$. The optimisation is therefore

$$\theta = \min_{\mathbf{w},c} \|\mathbf{w}\|_2^2 \quad \text{subject to} \quad y_i(\mathbf{w}^{\top}\mathbf{x}_i - c) \geq 1, \tag{14.43}$$

which can be solved using quadratic programming.

However, in the case that the classes are not linearly separable, we want to define the hyperplane that does the best, given the training data. Feature vectors that satisfy Eq. (14.42) are classified correctly and incur no loss. The hinge loss function from Eq. (14.11) does this by giving no penalty to samples that are classified correctly, but gives increasing loss for those that do not have sufficient margin, or are classified incorrectly. Equation (14.43) becomes

$$\theta = \min_{\mathbf{w},c} \left\{ \lambda \|\mathbf{w}\|_2^2 + \frac{1}{M} \sum_{i=1}^{M} \max(0, 1 - y_i(\mathbf{w}^{\top}\mathbf{x}_i - c)) \right\}, \tag{14.44}$$

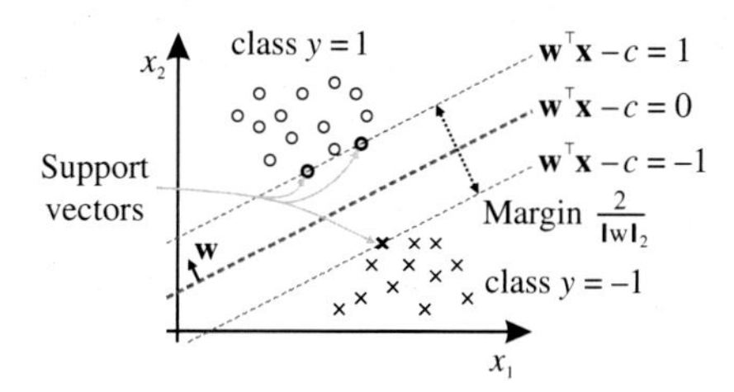

Figure 14.7 Support vector machine classification using the hyperplane that maximises the margin between classes.

where λ governs the balance between maximising the margin and ensuring that all points are classified correctly. This can be optimised using a gradient descent algorithm. The solution to this has the normal vector **w** as a weighted sum of the support vectors:

$$\mathbf{w} = \sum_{i=1}^{M} \alpha_i y_i \mathbf{x}_i, \tag{14.45}$$

where $\alpha_i = 0$ for feature vectors that have sufficient margin.

Nonlinear SVMs use a mapping function $\varphi(\mathbf{x})$ to map the feature vector to a higher dimensional space where the classes are linearly separable. However, rather than directly finding the mapping, a kernel function is used to measure distances between feature vectors in the higher dimensional space that enables the output to be calculated directly in feature-space. To achieve this, the kernel must satisfy

$$K(\mathbf{x}_i, \mathbf{x}_j) = \varphi(\mathbf{x}_i) \cdot \varphi(\mathbf{x}_j), \tag{14.46}$$

and this enables the mapping to be used implicitly. Commonly, the radial basis function is used as the kernel function:

$$k(\mathbf{x}_i, \mathbf{x}_j) = e^{-\gamma \|\mathbf{x}_i - \mathbf{x}_j\|^2}, \tag{14.47}$$

although other kernels are also used (Afifi et al., 2020). In particular, a closely related kernel

$$k(\mathbf{x}_i, \mathbf{x}_j) = 2^{-\gamma \|\mathbf{x}_i - \mathbf{x}_j\|_1} \tag{14.48}$$

(with γ being a power of 2) is hardware friendly and can be computed without multiplication using a CORDIC-like algorithm (Anguita et al., 2006). The output of a nonlinear SVM is given by

$$y(\mathbf{x}) = \text{sign}\left(\sum_{i=1}^{M} \alpha_i y_i K(\mathbf{x}_i, \mathbf{x}) - c\right), \tag{14.49}$$

where α_i are parameters determined as a result of training. Note that $\alpha_i = 0$ for feature vectors that are not support vectors, so the key role of training is to identify (and weight) the support vectors.

The SVM was designed for binary decisions. Where there are more than two classes, the easiest approach is to perform pairwise comparisons and select the class that has the most 'wins' (Hsu and Lin, 2002).

FPGAs have been used both for accelerating training and deployment of SVMs within embedded systems. Papadonikolakis and Bouganis (2008a) investigated training a linear SVM using quadratic programming. Their design had several processing elements for calculating dot-products, which were then cached for iterative training. They then extended their design to include nonlinear SVMs, replacing the dot-product with evaluation of the kernel function (Papadonikolakis and Bouganis, 2008b).

The many designs implementing SVM classifiers for inference are reviewed by Afifi et al. (2020). Two common architectures are pipelined parallel processing and systolic array processing. Some make use of partial reconfigurability to dynamically change SVM parameters. However, it is still a challenge dealing with high feature-space dimensionality, with many support vectors.

14.3.7 Neural Networks

Neural networks can be used for classification as well as regression. For classification, the network has one output for each class. The network is trained with a 1 output for the desired class, and 0 (or −1 depending on the activation function) for the other classes. The classification output of the network is the class which has the highest score (winner takes all). Another alternative is to apply a soft-max to the outputs to turn them into probabilities for each class. Let c_k be the output of the network for class k; the probabilities are then calculated as

$$p_k = \frac{e^{c_k}}{\sum_k e^{c_k}}. \tag{14.50}$$

Otherwise, neural networks follow the development described in Section 14.2.3. It is possible to also gauge some confidence in the result from the second highest score; if it, too, is high, then there is potential ambiguity in the result.

14.3.8 Clustering

Some applications have little or no labelled training data (because such annotation is time consuming and expensive). In this case, an unsupervised classification method is required. These identify clusters of similar feature vectors with the assumption that each cluster represents a separate object class. Clustering methods therefore partition feature-space based on the density of feature vectors.

There are two limitations of this approach. First, it requires a large number of training data samples, especially if feature-space has high dimensionality. Clustering and density estimation suffer from the curse of dimensionality. As the number of dimensions increase, samples become sparse, and the relative distances within each class and between classes start becoming similar. This makes it more difficult to group similar feature vectors and in particular find the boundaries between clusters. Second, even if the clustering is accurate, there is the problem of assigning meaning to each cluster.

It is important to normalise the data before performing any form of clustering, so that distances in the different dimensions have similar meaning. One common approach is to scale each feature to have unity variance:

$$\hat{x}_i = \frac{x_i}{\sigma_{x_i}}. \tag{14.51}$$

A computationally simpler method is to scale by the maximum absolute value of the feature:

$$\hat{x}_i = \frac{x_i}{\max|x_i|}, \tag{14.52}$$

although this can be sensitive to outliers.

There are three broad classes of clustering algorithms: connectivity-, density-, and centroid based. Common to all of these methods is measuring the distances within feature-space. Since making distance measurements is often the most time-consuming part of many clustering algorithms, Wang et al. (2021) developed an accelerator for this purpose. Most of the actual clustering algorithm is written is software, with custom instructions used to access the hardware accelerator.

14.3.8.1 Connectivity-based Clustering

This approach groups samples based on the distances between samples within the training dataset. While there are many possible distance functions that could be used, the simplest is the Euclidean distance. Samples within a threshold distance from one another are considered to belong to the same cluster. Therefore, between any two samples within a cluster there will be a chain of samples, with each pair within the chain closer than the threshold distance. Note that the square root associated with the Euclidean distance can be avoided by thresholding the distance squared.

The search for neighbouring samples can be accelerated by storing the feature vectors in a kd-tree. This is a binary tree which successively partitions each dimension, providing a mechanism for locating feature vectors within a region of feature-space without having to search through the whole dataset.

Connectivity-based methods are quite sensitive to outliers, which can often appear as isolated clusters of one or two samples. One danger of outliers between clusters is that they can create a chain which can cause clusters to merge.

Niamat et al. (1998) used a systolic array to implement connectivity-based clustering on FPGAs. The small devices of the time meant they required 8 FPGAs, although the design would easily fit on a single modern FPGA.

14.3.8.2 Density-based Clustering

Density-based methods are similar to connectivity-based methods, with the added requirement that the points within a cluster must satisfy a minimum density requirement. The most widely known density-based method is DBSCAN (density-based spatial clustering of applications with noise) (Ester et al., 1996). The density requirement is met by having a minimum of k samples closer than a distance threshold, ϵ. The basic algorithm is to add neighbouring points to a cluster if they satisfy the inclusion criterion.

One limitation of this approach is that it assumes that all clusters have the same density (which can be a problem with unbalanced classes), and that there are gaps between the clusters with density lower than the threshold. These factors can make the clustering quite dependent on the k and ϵ parameters.

Shi et al. (2014) parallelised this algorithm on an FPGA and worked on building multiple clusters in parallel. They had a collision mechanism to detect when two clusters joined. To avoid memory bandwidth issues, the incoming points were streamed to the processing elements arranged in a pipelined structure. Scicluna and Bouganis (2015) instead used parallelism to build the neighbours of one cluster at a time and also simplified the design to use the computationally less-expensive Manhattan distance rather than Euclidean distance. Rather than store the points themselves in a queue, the address of the data was used to make more efficient use of the memory (independent of the dimensionality and precision of the feature data). The addresses effectively indexed the points to enable more efficient access in retrieving the neighbours.

14.3.8.3 Centroid-based Clustering

The basic principle behind centroid-based clustering is to assign points to a cluster based on the distance to the cluster centre. The most common method within this category is K-means clustering, which defines the number of clusters, K, in advance. K-means is an iterative algorithm that first randomly selects K cluster centres. All data points are assigned to their closest cluster, and the cluster centres updated to be the average feature vector of the assigned points. This is then iterated until there is no further change. Having to specify K in advance is one limitation of the algorithm, although this makes it well suited for vector quantisation.

The algorithm can be sensitive to the selection of the initial centres (both in terms of the clusters produced and also the time it takes to converge). Therefore, many modifications to the basic algorithm have been proposed to make it more efficient and robust. However, while the basic algorithm is easily implemented in hardware, these modifications can make the hardware design significantly more complex (Maruyama, 2006).

Leeser et al. (2001) applied K-means to cluster hyperspectral data. To reduce the computation, they used the Manhattan distance and also used reduced precision on the incoming hyperspectral data. Maruyama (2006) applied K-means clustering to VGA resolution RGB images at up to 30 frames per second. The key to achieving this speed was to use a kd-tree to filter out many of the points that do not change clusters during the recursive update process. However, the kd-tree leaves stored the cluster centres associated with the node, so needed to be rebuilt every iteration as the centres were updated. Winterstein et al. (2013) also used a kd-tree, but instead stored the samples, so that the tree only needed to be built once. The associated filtering during later iterations gave significant speedup compared to not filtering. Rather than use a kd-tree for filtering, Lin et al. (2012) used the triangle inequality to filter out points that do not change clusters. This did not have the memory overheads associated with the kd-tree, but still gave significant speed improvement, especially in the later stages where the cluster centres only moved a small distance. Badawi and Bilal (2019) took a different approach and used on-line clustering of colour images. For this, the iterative update was eliminated, and a simple recursive filter was used to update the cluster centre, C_i, of the closest cluster:

$$\{C_i = \alpha C_i + (1 - \alpha)P, \tag{14.53}$$

where P is the incoming colour pixel value, and $\alpha \approx 0.999$. This avoids the time-consuming division operation and enables the cluster centres to track changes in the distribution. In fact, setting $\alpha = 1 - 2^{-10}$ saves even having to do a multiplication. It enables real-time stream processing with low latency, although it has the disadvantage that the cluster centres are not constant within an image.

14.4 Deep Learning

One of the limitations of classical machine learning models is their dependence on manually selected or hand-crafted features. This requires not only the manual selection of the feature vector but also the design of the image processing necessary to accurately extract the features from the images. In contrast, deep neural network models overcome this limitation by automatically learning and extracting the features as well as the classification. In this context, 'deep' refers to a neural network with more than one hidden layer (Sze et al., 2017). The series of layers learns a hierarchy of features, starting with edges and corners in the early layers, and combining these into increasingly complex abstract and shape features in successive layers. This hierarchical feature learning has led to impressive accuracy over many different applications. This makes deep learning models a popular choice in many image processing tasks, not just to classification and segmentation.

This section is not intended as a detailed discussion of deep learning, but an overview of the basic techniques as applied to computer vision, and in particular some of the techniques used to accelerate their deployment on FPGAs. For a more detailed analysis on deep learning, how it works, and how to apply it, there are several good books (see, for example, Goodfellow et al. (2016) and Zhang et al. (2021)).

The utilisation of deep learning in various applications is faced with several challenges, one of the most prominent being its computational cost. The training process involves adjusting the vast number of weights within the network, which can be demanding on computing resources. Additionally, deploying the trained network to embedded devices, which is often necessary in real-world applications, can also be challenging due to its computational needs. This makes FPGAs a good choice where low-power, high-speed inference is required. However, the large size of modern deep learning networks makes mapping them onto FPGAs challenging; therefore, the main focus of this section will be on some of the techniques that have been used to address this. For many image processing tasks, the computational cost of deep learning is overkill, and traditional image processing techniques (or a hybrid between traditional image processing and deep learning) may be more appropriate.

Another limitation of deep learning is the requirement for a substantial amount of labelled or annotated training data. Annotation can be a time-consuming and laborious process.

A third challenge is the 'black box' nature of deep learning models, which can make it difficult to understand how the model arrived at its final output. This can lead to a situation where people use deep learning models without fully understanding the problem they are trying to solve or without performing common sense preprocessing to simplify the model. Many researchers simply apply deep learning models to their datasets and tweak the hyperparameters until they achieve state-of-the-art performance. Unfortunately, such an approach provides little insight into the problem being solved.

The building blocks of deep neural networks are introduced in Section 14.4.1 before outlining some of the common network architectures in Section 14.4.2. These provide the context for understanding some of the problems associated with realising the networks on an FPGA.

14.4.1 Building Blocks

14.4.1.1 Fully Connected Layer

A fully connected layer has each neuron connected to all of the inputs for the layer, through appropriate weights, followed by a nonlinear activation function. As such, it is much the same as the neural network described in Section 14.2.3. Generally, all of the weights for each neuron are independent. Within deep networks, they are typically only used in the last one to three layers where the number of neurons is smaller; otherwise, the large number of parameters becomes unwieldy.

14.4.1.2 Convolution Layer

Simply extending traditional, fully connected, neural networks to have many layers does not work well for images. The large number of input pixels would make even a single layer network prohibitive because of the large number of independent weights required. Ideally, when detecting an object, we want to have some degree of translation invariance, which does not come with a fully connected image layer. Both problems are solved by convolutional layers.

A linear convolutional filter (see Section 9.2) is applied to the input, but with the difference that the filter weights are trained as a neural network (Pugmire et al., 1995). This shares a small number of weights over all positions within the image using a convolution. The nonlinear activation function enables the modelling power of neural networks to be brought to bear. Thus, each neuron is simply a filter, with only a few trainable parameters, and this is applied across the whole image, producing a filtered image as the output activation. The activation is usually referred to as a ***feature map***.

In practice, it is a little more complex than this, because at each layer there are several feature maps or channels. The summation is therefore 3-D, with the inputs coming from all of the input feature maps, and there are multiple feature maps output. With M input feature maps, $I[x, y]$, and N output feature maps, the convolution outputs become

$$S[n, x, y] = w_0[n] + \sum_{m=1}^{M} \sum_{i,j \in \mathbf{W}} w[m, i, j]I[m, x + i, y + j] \quad \text{for } 1 \leq n \leq N. \tag{14.54}$$

Often $N = M$, except after pooling where it is common to have $N = 2M$.

Almost all convolutions are square, with 3×3 being the most common. Szegedy et al. (2016) suggested that for larger filter sizes, the parameter count can be reduced by a series of smaller filters. For example, a 5×5 filter could be replaced by a series of two 3×3 filters, or a $k \times k$ filter could be replaced by a $k \times 1$ filter followed by a $1 \times k$ filter. Note that with such filter cascades, keeping the nonlinear activation function between the filters gives improved representational capability than simply chaining the filters (Szegedy et al., 2016).

A 1×1 filter has no spatial extent but does combine the input feature maps. They can be used to reduce the number of activations (Iandola et al., 2017) that are input to the following layer by reorganising the information across the feature maps ($N < M$). This can significantly reduce the number of parameters required in the following layers.

14.4.1.3 Activation Function

The output feature map, or activation, is given by applying a nonlinear activation function, $a(\cdot)$, to the filter outputs:

$$Q[n, x, y] = a(S[n, x, y]). \tag{14.55}$$

For simplicity of computation, the most common nonlinear activation function in deep learning is the ReLU from Eq. (14.18).

For back-propagation, the constant gradient allows the gradient to propagate more easily than with a conventional saturating sigmoidal activation function (Glorot et al., 2011). Another advantage is that the sparsity generated by the true zero output enables faster training because whole neurons can be ignored when they do not contribute to the output (Glorot et al., 2011). However, the sharp transition at the origin is not differentiable, so sometimes other smoother variations are used. The ReLU has the big advantage that it is trivial to implement in hardware.

14.4.1.4 Batch Normalisation

Training deep networks can be difficult and slow because the statistical distribution of the signals at the internal nodes changes during training, a phenomenon known as internal covariate shift. To address this, a solution is to normalise the distribution of each input within the hidden layers after each batch of training (Ioffe and Szegedy, 2015).

Where traditional sigmoidal activation functions are used, batch normalisation is applied to the neuron inputs. However, with the ReLU and related unbalanced activation functions, it is usually applied on the filter outputs before the activation function (Ioffe and Szegedy, 2015). In the latter case, it scales the filter outputs, S, to have zero mean and unit variance which is then further scaled and offset:

$$\hat{S} = \frac{S - \mu_S}{\sqrt{\sigma_S^2 + \epsilon}}\gamma + \beta, \tag{14.56}$$

where μ_S and σ_S^2 are the mean and variance from the training mini-batch, ϵ is a small constant to avoid numerical problems, and γ and β are additional trained parameters. Once the network has been trained, μ_S and σ_S^2 are derived from the whole training dataset to give a scale and offset.

However, Santurkar et al. (2018) showed that the benefits of batch normalisation result not so much from reducing the internal covariate shift as from smoothing the optimisation landscape. Thus, it is effectively a form of regularisation that enables higher learning rates with improved stability and convergence.

Note that once the network has been trained, the scale and offset associated with batch normalisation can be incorporated into the neuron weights. Therefore, for inference, batch normalisation has no additional cost (Sze et al., 2020).

14.4.1.5 Pooling

Pooling is a technique used to reduce the resolution of each feature map in a deep network. Down-sampling is used to effectively construct a hierarchy of complex features and to reduce the amount of data before the final fully connected layers. By reducing the resolution, larger area features can be constructed more efficiently with smaller filters, and an information bottleneck can be created to force the network to extract important features. This is used, for example, in auto-encoders to reduce the image content into a few latent variables. For classification tasks, it is necessary to summarise the whole image into one of a few classes. Pooling uses a small window, typically 2×2, and selects either the maximum value (max-pooling) or the average value (average pooling) from the window to produce the output.

The opposite of pooling is unpooling, which increases the resolution of each feature map. Unpooling can up-sample a feature map through two methods: zero insertion and interpolation. Zero insertion results in the next convolution layer effectively interpolating the feature map. Note that the computation in the following layer can be optimised by skipping the redundant multiplication by zeroes.

14.4.1.6 Soft-Max

Networks that provide a classification output typically have a soft-max layer on their output. This converts the numerical output for each class into a probability using Eq. (14.50).

14.4.2 Architectures and Applications

There is a wide variety of network architectures used for deep learning, and new variations are continually being developed and tested. This section briefly outlines the structure of some of the more common architectures and some of their applications. However, the design of deep learning networks for various applications is beyond the scope of this brief summary.

14.4.2.1 Convolutional Neural Networks

Most deep networks are made up primarily of convolution layers. Each convolution block, see Figure 14.8, is composed of a convolution layer followed by a nonlinear activation function. Virtually all modern networks use batch normalisation during training since it has proved effective; however, during inference, the operations can be combined with the convolution filters. The pooling layer at the end is optional, in that not every convolution block has pooling.

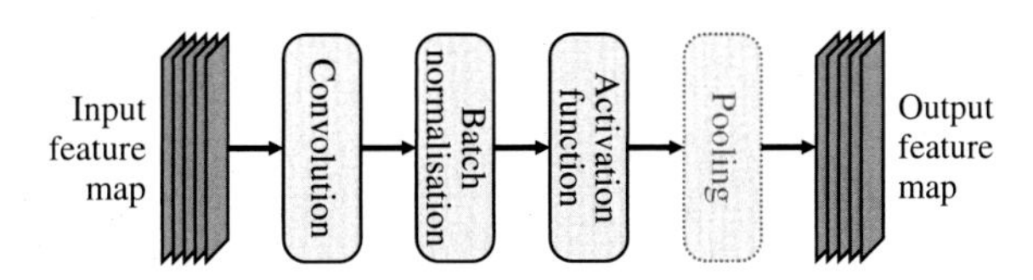

Figure 14.8 A convolution layer within a convolutional neural network. The pooling layer is optional.

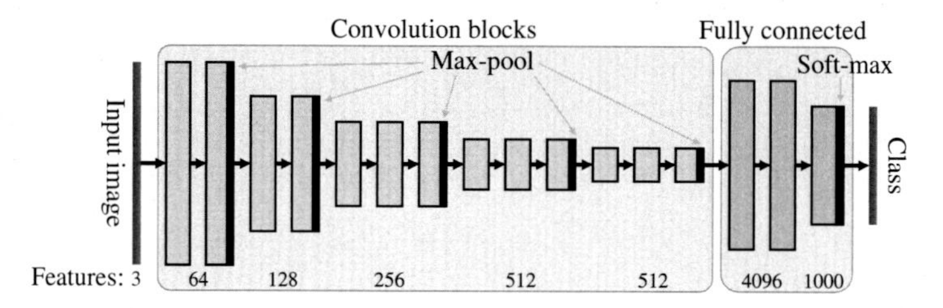

Figure 14.9 A typical convolutional neural network for image classification. Shown here is a representation of VGG-16; all of the convolution filters are 3 × 3, and the network has 138 million parameters. Source: Adapted from Simonyan and Zisserman (2015).

A convolutional neural network consists of several (between 5 and 1000 (Sze et al., 2017)) layers of convolution blocks. These are often followed by two or three fully connected layers. A typical example, shown in Figure 14.9, is that of VGG-16 for image classification (Simonyan and Zisserman, 2015). It is common after pooling to increase the number of features, as this maintains the representation expressiveness. In general, deeper networks give better classification performance, but this comes at the expense of an increased number of parameters.

In addition to classification, convolutional networks are also used for object detection, which not only identifies objects but locates them within the image. The output is numerical, providing the coordinates of the bounding box around a detected object. Such networks often use non-maximal suppression to select the best candidate box for each object. This is extended further by semantic segmentation, which assigns a semantic label to each pixel in the input image.

Convolutional networks are also used for image restoration, which removes degradation such as noise or blur from an image. Such networks learn the mapping from the degraded input image to the clean output image. For example, image denoising (Zhang et al., 2018; Remez et al., 2017) trains using a noisy input image with the target being the clean output. Such networks may not have pooling layers, with all the feature maps being the same size as the input image. As with any noise smoothing operation, there is a balance between smoothing noise and loss of fine details.

The trend in more recent convolutional neural networks is to reduce the number of fully connected layers at the end and compensate by increasing the depth of the network with more convolution blocks.

14.4.2.2 Residual Networks

In deep networks, the gradients that are used for training can become very small as they are back-propagated through the many layers of the network. The problem of vanishing gradients is especially pronounced in very deep networks, where the gradients can become vanishingly small and not provide enough information to update the network weights effectively. This makes it difficult for the network to learn, especially in the earlier layers.

Residual networks address this problem by adding residual connections which bypass one or more layers. The residual connections allow the gradients to propagate more easily back through the network, making it possible for the network to learn, even for very deep networks. He et al. (2016) demonstrated successful training even with over 1000 convolution layers!

A residual block is composed of two or three convolution blocks with a bypass connection that is added in just before the last activation function of the block, as illustrated in Figure 14.10 (He et al., 2016). If the mapping being learned by the residual block is $H(\mathbf{x})$, then adding in $\mathbf{x}$ forces the convolution layers within the block to learn the residual mapping, $H(\mathbf{x}) - \mathbf{x}$. This makes it possible for the network to learn more complex functions, as the residuals can be learned in a more gradual manner, layer by layer.

With deep networks, to avoid the explosion in the number of parameters, He et al. (2016) also proposed the bottleneck residual block shown on the right in Figure 14.10. This uses two 1 × 1 convolutions, one at the start to reduce the number of feature maps by a factor of 4, and the other at the end to restore the number of feature

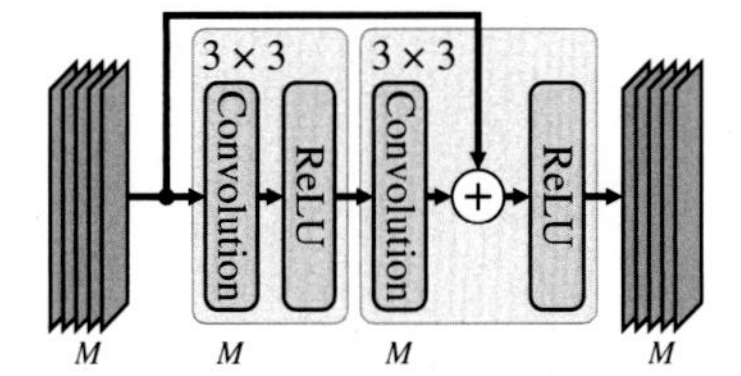

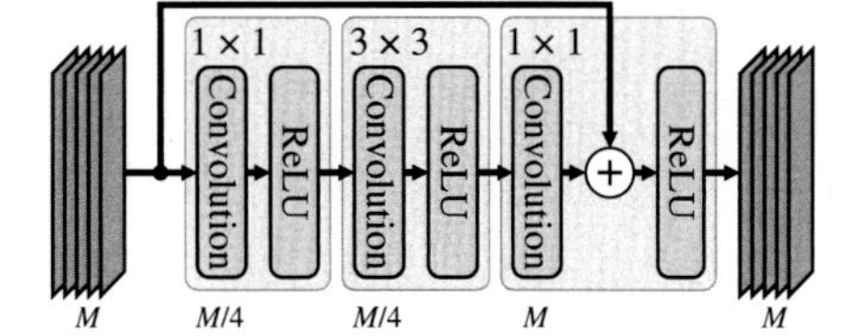

Figure 14.10 Residual blocks within a convolutional neural network. Left: simple residual block, where the number of feature maps is the same throughout (M); right: residual block with bottleneck. The first 1×1 convolution reduces the number of feature maps to $\frac{M}{4}$; the second 1×1 convolution restores the number of feature maps back to M.

maps again. This reduces the number of parameters in the middle 3×3 layer by a factor of 16. ResNet is an example of a network that uses these residual blocks (He et al., 2016).

Huang et al. (2017) extended this idea further with many bypass connections. Within a 'dense' block, rather than add the bypass connection in, the bypass connections are concatenated in parallel, effectively increasing the number of feature maps available for the next layer. This reuse of earlier feature maps allows for more efficient inference, with significantly fewer parameters, because the number of feature maps added by each layer can be relatively small.

14.4.2.3 Auto-encoders

Auto-encoders are a type of network architecture that is used for unsupervised learning. The principle behind auto-encoders is to learn a compact, low-dimensional representation of the input images, typically referred to as the encoding. The encoding is then used to reconstruct the original input, which serves as the output of the network. The auto-encoder is trained by minimising the reconstruction error between the original input and the output of the network.

As shown in Figure 14.11, an auto-encoder consists of two main components: an encoder and a decoder. The encoder takes the input image and compresses it into a lower dimensional representation, h, while the decoder takes the encoding and expands it back into the original input shape. The encoding and decoding are performed by convolution layers, and the objective is to learn the encoding that allows for the best reconstruction of the original input. The reduced size of h forces the auto-encoder to learn the key characteristics, or latent variables, of the input images seen during training, and forces a dimensionality reduction (Hinton and Salakhutdinov, 2006). As such, an auto-encoder performs a form of nonlinear PCA (principal components analysis); indeed if the activation functions were all linear and a squared-error loss function is used, the encoder would learn the PCA subspace (Kunin et al., 2019).

It is important that the size of h is not too small; otherwise, it has insufficient capacity to be able to represent and reconstruct the image. Conversely, if h has too much capacity, the auto-encoder can easily reproduce the

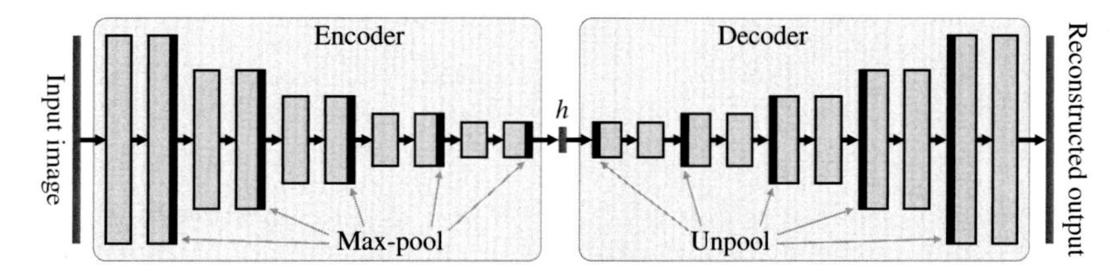

Figure 14.11 Illustration of an auto-encoder network. The encoder uses a series of convolution layers to give a compressed representation, h, of the input image. The decoder reconstructs the image from the compressed representation.

input without extracting useful information. Optimally sized, the auto-encoder learns the key features or latent representation from the set of input images. This is especially so if h is made to be sparse by imposing a suitable regularising term to the loss function.

Some of the applications of auto-encoders include:

- **Noise removal**: Rather than train the auto-encoder to reproduce the input, if the input has noise added, and the output is clean, then the auto-encoder can learn to remove noise from the image (Cho, 2013). Again, this is helped by imposing sparsity on h.
- **Semantic segmentation**: Semantically related features are close to one another within the code space. Therefore, rather than reconstruct an image, it is possible to reconstruct a semantic segmentation of the input image (Noh et al., 2015).
- **Image compression**: h is usually significantly smaller than the input image.
- **Anomaly detection**: By training the network to identify normal patterns in the input images, any deviations from normal indicate the presence of an anomaly. This is indicated by the reconstructed output differing significantly from the input.
- **Depth estimation**: Wofk et al. (2019) estimated the depth or range from a single input image.
- **Super-resolution**: Zeng et al. (2017) used an auto-encoder for single-image super-resolution.

14.4.2.4 Generative Adversarial Networks

Generative adversarial networks (GANs) are a type of network architecture that is used for generative modelling. The principle behind GANs is to train two separate neural networks, a generator and a discriminator, in a competition with each other, as shown in Figure 14.12. The generator uses the input to produce synthetic images, while the discriminator determines whether the samples are real or fake (Creswell et al., 2018). The score from the discriminator is fed back to the generator during training, so that the generator learns to produce more realistic images.

Although the generator network can have virtually any structure, functionally it is similar to the decoder part of an auto-encoder, creating an image from a latent representation. The input is mapped from the latent variables to produce the synthetic output image that is meant to resemble the true data distribution. The discriminator network takes an input, a sample either from the generator or from a real image, and outputs a probability that the sample is real. During training, random inputs are provided as input to the generator. Note that the generator has no access to real images; it is trained by the score it receives from the discriminator. So, rather than learning the latent representation through an encoder, the latent representation and reconstruction is learned through trying to outwit the discriminator. The discriminator does have access to real images and produces as its output a probability indicating how likely the input is real. The generator and discriminator are trained simultaneously in an adversarial manner, where the generator tries to produce samples that the discriminator cannot distinguish from real samples, and the discriminator tries to correctly identify whether the samples are real or synthetic. As the quality of the generator improves, the discriminator must learn the key features of real images that distinguish them from fakes.

One of the key advantages of GANs is that they can generate high-quality synthetic images, even for complex distributions. By training the generator to produce synthetic images that are similar to the true images, GANs

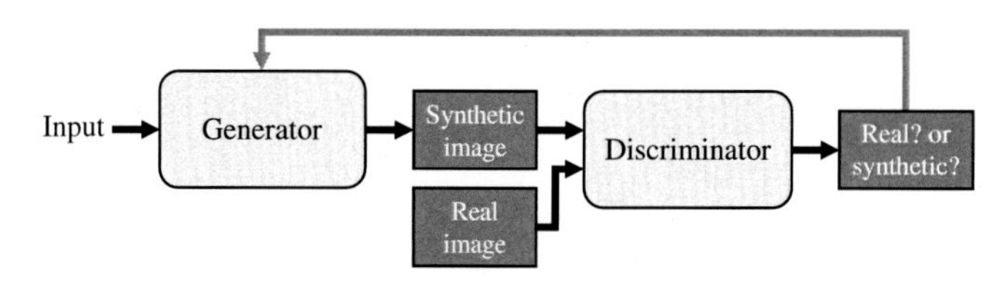

Figure 14.12 Basic structure of a generative adversarial network. The generator learns to generate better synthetic images, while the discriminator learns to distinguish synthetic images from real images.

can be used to generate new content or data, or to augment existing data sets. After training, either the generator or the discriminator can be used. While image synthesis is a core capability of GANs, they can be adapted to perform any image-to-image transformation.

For example, for single-image super-resolution (Ledig et al., 2017), the input is a low-resolution input image, and the discriminator trains the network to generate realistic details as part of the up-sampling process. Zhu et al. (2017) extended this idea further and introduced the idea of cycle consistency. A forward and reverse transform are developed in parallel between pairs of images, and the cycle loss is used to enforce consistency between the original image and the reverse transformed image. This can be used, for example, to transfer an artistic style to an image or to convert a greyscale image into colour.

14.4.2.5 Recurrent Networks

A recurrent network is a type of architecture designed for processing time-sequential data. The networks described so far are feed-forward networks, where the signals propagate unidirectionally from the input to the output. In a recurrent neural network, the inputs for a neuron come not only from the previous layer but also from the neuron activation at the previous time step. This mechanism provides state information that is passed from one time step to the next, where the state information summarises the past history. This hidden state allows a recurrent network to capture long-term dependencies between the inputs, with the final output being a prediction or decision based on the entire sequence of inputs. Such networks are well suited for processing sequences of data such as video.

One common recurrent building block is the long short-term memory (LSTM), as illustrated in Figure 14.13. At the heart of the LSTM is a cell, which remembers a state vector, $\mathbf{c}_t$; the feedback connection here gives the cell memory. The input comes from the system inputs and the previous LSTM activations. These are combined by a set of input neurons, one for each component of the state vector. An additional three sets of neurons provide gate signals for each component of the state vector. The input gate controls the flow of information into the cell, the output gate controls the visibility of the cell memory on the output, and the forget gate controls the memory, determining how much of the cell contents is remembered from one cycle to the next. The hidden state of the LSTM network contains the information that has been accumulated over the previous time steps, and it is used as input to the next time step. This allows the LSTM network to remember information over a longer period of time and to capture relationships between input samples.

A network with LSTMs is usually fully connected. Therefore, for video processing, a convolutional network first reduces the number of features, before feeding into a few LSTM layers at the end (Zhang et al., 2017). LSTMs model the temporal relationships between frames and capture the long-term dependencies in video data.

LSTMs can be used for video classification, where the network takes a sequence of frames as input and outputs a prediction of the video class. The hidden state in the LSTM can capture the temporal evolution of the video and the relationships between the frames, which are used to make the final prediction. LSTMs can

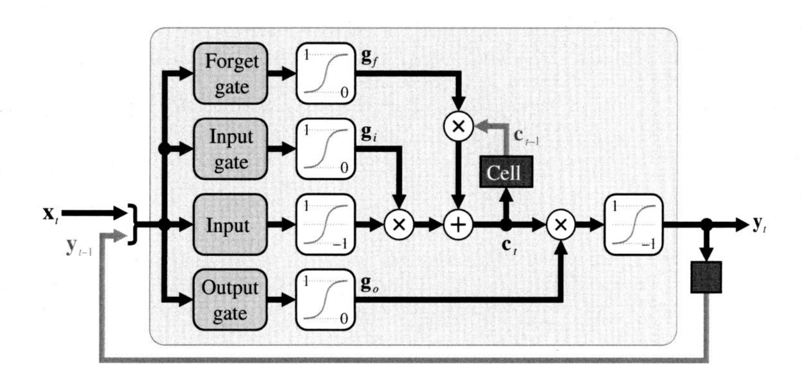

Figure 14.13 A basic LSTM (long short-term memory) block.

also be used for video segmentation, where the goal is to segment the video into meaningful regions, such as object regions or background regions. In this task, the LSTM can be used to model the temporal dependencies between the frames and the relationships between the regions, which are used to make the final segmentation decisions. For example, Zhang et al. (2017) implemented LSTM-based video content recognition on an FPGA.

14.4.2.6 Transformers

Transformers are a relatively recent addition to deep learning architectures. Transformers grew out of recurrent networks primarily developed for natural language processing (Vaswani et al., 2017), where the temporal context was captured within the states of an LSTM. However, remembering the right contextual information and accessing it when required is indirect, and some form of attention mechanism is required to help this. Transformers grew out of this work, using self-attention to directly access the required context (Vaswani et al., 2017).

Vision transformers are inspired by this architecture (Dosovitskiy et al., 2021). The basic idea behind vision transformers is to treat an image as a sequence of 'tokens' and then use the transformer architecture to process these tokens. The sequence of tokens can be thought of as a grid of cells or regions that cover the image. Each token is represented as a vector, or 'embedding', and the transformer then performs self-attention operations to identify the relative importance of different parts of the image and provide context for the current token in making a prediction. Finally, the token embeddings are passed through one or more fully connected layers to produce a final prediction for the image as a whole. In simple terms, vision transformers work by breaking an image into small parts, understanding the relationships between these parts, and using this understanding to make a prediction about the whole image.

Self-attention is a mechanism used in the transformer architecture to allow the network to focus on different parts of the input as it processes them. In the case of vision transformers, self-attention allows the network to focus on different parts of the image as it processes each token. Self-attention works by computing a set of attention scores for each token, indicating the importance of each token with respect to the others. These scores are computed using a series of dot-products between each token and a set of 'query', 'key', and 'value' vectors. The query vector represents the current focus of the network, what the network is looking for, and the key vectors represent the features of each token in the input sequence. The dot-product between the query and each key vector produces a scalar value representing the compatibility between the query and key, and these are used to determine which options are most relevant to the current task. The value vectors are used to weight the contribution of each token to the output of the self-attention mechanism. This information is then passed through a feed-forward neural network to produce the final prediction. In essence, self-attention allows the network to dynamically focus on different parts of the input as it processes them, allowing it to identify the most important information for making a prediction.

Attention therefore decides which other parts of the sequence are important or provide context for the current element. Transformers use self-attention to identify the relationship or relevance between elements within a sequence. For vision tasks, self-attention maps capture the interactions between pixels in both space and time and effectively map features from the current frame or region to all of the relevant features captured in the transformer model. The self-attention map therefore models global contextual information for interpreting the current frame or patch.

The tokens in a vision transformer can be derived from using a convolutional network or directly by applying a learned linear transformation to the pixels in each region (Dosovitskiy et al., 2021). The self-attention relationships in the transformer are determined by analysing large-scale datasets, which is done during the pre-training process. A significant advantage of this approach over traditional convolutional neural networks is that the attention pre-training process can be unsupervised, meaning that the network can learn without the need for labelled data. When processing a new image, the vision transformer maps it onto the global context learned during pre-training, drawing information from all relevant training data. This information is used to make predictions, such as deriving an image label, recognising an action, or transforming the image (for example super-resolution) (Khan et al., 2022).

While convolutional networks have been the mainstay of deep learning algorithms, transformers are becoming an alternative, achieving comparable results in a wide variety of image processing applications

(Han et al., 2023; Khan et al., 2022). The topic of vision transformers is quite new and is currently undergoing strong evolution.

14.4.2.7 Graph Neural Networks

Graph neural networks (GNNs) are a class of deep learning models that operate on graph-structured data. In the context of vision, GNNs can be used to process graph-structured representations of image data, where each node in the graph represents a region or a feature in the image and edges represent relationships between nodes. The goal of GNNs is to learn a compact representation of the graph structure that can be used to make predictions, such as image classification or segmentation. This is done by passing messages between nodes in the graph, where each node aggregates information from its neighbours to form its final representation.

GNNs can capture complex relationships between regions or features in an image that are not easily represented in traditional grid-based image representations used by convolutional neural networks. This allows GNNs to effectively model global context, which is crucial for many vision tasks. These models have been applied to several different vision tasks, including object recognition, semantic segmentation, pose estimation, and object tracking (Jiao et al., 2022).

14.4.3 Training

A detailed discussion of the training process for deep networks is beyond the scope of this work. In principle, training deep neural networks is similar to training any other neural network, although the deeper architectures can make back-propagation less effective, particularly in the earlier layers. Due to the large number of parameters that need to be optimised, training can take a significant amount of time. To address this issue, several techniques have been developed including:

- Using residual connections, which improve the flow of gradients during back-propagation, allowing for faster convergence.
- Network pruning, which removes redundant or unimportant neurons, thereby reducing the number of parameters that need to be trained.
- Dropout, which helps reduce over-fitting by randomly dropping out neurons during training and forcing all the parallel features to learn different feature sets.
- Regularisation, which helps smooth the error surface, making it easier to escape sub-optimal local minima. Regularisation also reduces over-fitting by applying constraints that simplify the model.
- Transfer learning, which uses the pre-trained weights of a parent network trained on another task to speed up training on the current task. This approach takes advantage of the early layers of the parent network, which are often slow to converge but are generic to many image processing tasks.

Training also requires large amounts of data. For many tasks, the data must be annotated, which is both time consuming and potentially error prone. Data augmentation techniques are often used to reuse existing annotated data. They work by transforming the input images (and if necessary the target outputs) in a variety of ways to provide additional data. Transformations can be geometric (rotation, zoom, and reflection), photometric (adjusting brightness and contrast), and adding noise. One side effect of data augmentation is that it tends to make the trained network more robust to variations in the input and reduce over-fitting by increasing the diversity of the data. GANs have also been used to generate realistic artificial data for training. However, generating high-quality data using GANs can be challenging and requires careful tuning of the network architecture and training process.

There are many platforms for developing, experimenting with, and training deep learning networks (see, for example, (Hatcher and Yu, 2018)). Some of the better known platforms are Caffe, Keras, Tensorflow, and Pytorch. The back-propagation used to train most deep learning algorithms usually requires floating-point to give the dynamic range required for calculating the gradient updates. As a result, FPGAs are not usually used for training deep learning models. The considerable computation resources required for training are often provided by graphics processing units (GPUs), either on a local computer or using cloud-based servers.

14.4.4 Implementation Issues

The focus within the research community has almost exclusively been on increasing the accuracy. This has been spurred on by a successive series of challenges. This focus has led to deeper and more complex networks, with an increasing number of parameters. Such models are difficult to implement on embedded devices such as FPGAs (Iandola and Keutzer, 2017). Other metrics of importance within real-time embedded systems are latency, throughput, power, and cost (Sze et al., 2020), with an inevitable trade-off among these various criteria. There are several good surveys that review FPGA implementation issues; see, for example, Guo et al. (2019), Deng et al. (2020), Mittal (2020), and Sze et al. (2020).

The basic operation of most networks is the multiply-and-accumulate used for both convolution and dot-products in fully connected layers. The DSP blocks of FPGAs are well suited to this task; however, the problem is the sheer number of them required for implementing deep learning-based inference. Resource constraints require that these be shared and reused. Most models are represented using floating-point, so the algorithms need to be converted to use fixed-point for FPGA implementation. However, some FPGAs are beginning to appear that provide hardware floating-point operations (Sinha, 2017). Architectural changes that support deep learning are also starting to appear in FPGAs. For example, Intel in their Stratix 10 NX architecture have introduced a tensor block aimed specifically at dot-products, with the block able to calculate three 10-element dot-products every clock cycle. Although the multipliers are 8-bit integer multiplies, they can be configured as block floating-point (with a shared exponent) with the outputs converted to floating-point within the block. To support the high throughput, one set of coefficients is pre-loaded into the tensor block, and applied to a stream of input values (Langhammer et al., 2021), making the tensor block well suited for matrix-vector multiplications.

For almost all deep models, there is insufficient memory on the FPGA to hold all of the parameters and the intermediate feature maps. Resource constraints require both data and parameters to be streamed from external memory. Again, FPGA architectural changes are starting to appear that support this, with the introduction of in-package high-bandwidth memory in the Xilinx Ultrascale+ HBM (Wissolik et al., 2019). As emphasised throughout this book, it is desirable to perform as much computation on the FPGA as possible between streaming the data in and streaming intermediate results back out.

Inevitably, there is a trade-off between computational capability and memory bandwidth. This is often represented by the roofline model (Siracusa et al., 2022), illustrated in Figure 14.14. The horizontal axis gives the computation intensity in operations per byte loaded (multiplications per byte for deep models) and represents the degree of parallelism used by the design. The vertical axis is computational performance in operations per second. The roofline is bounded on the left by peak memory bandwidth; actual memory bandwidth will be lower than this because overheads mean that data is unable to be transferred every cycle (due to refresh cycles, bus turnaround, memory controller inefficiencies, and so on). The roofline is bounded on the top by the availability of multipliers on the FPGA. Again, the actual compute bound will be lower than this peak because it may not be possible to use every multiplier every clock cycle (because of vector sizes, overheads in performing other operations, and so on). The goal is to match the compute resources with the memory bandwidth by operating close to the point of intersection. Use of the model can help with exploring the design-space in mapping the design to the FPGA by identifying the limitations or weaknesses of a particular design.

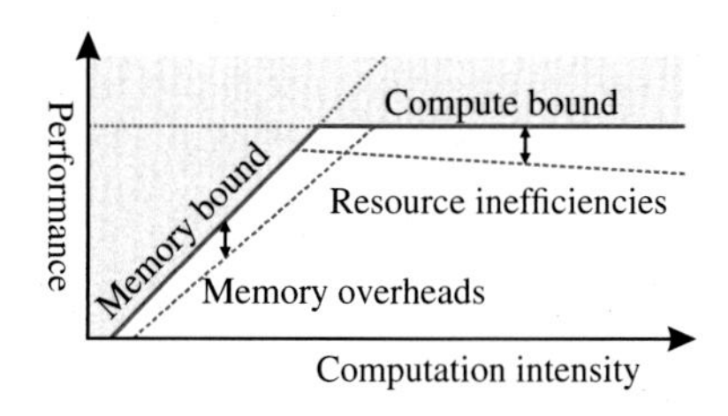

Figure 14.14 Roofline model indicating memory and compute limits to performance. Both axes use a logarithmic scale.

14.4.4.1 Architecture Optimisation

Virtually all deep learning architectures are over-parameterised and contain significant redundancy (Cheng et al., 2015; Srinivas and Babu, 2015). Therefore, the biggest optimisation effort can be made by modifying the architecture to use fewer parameters (Iandola and Keutzer, 2017). This can significantly reduce both the computation and external memory bandwidth required. Some of the techniques that have been used here are smaller filters, fewer layers, fewer feature maps, and model compression.

In convolution layers, rather than use large filters, they can be replaced by more layers of smaller filters (Szegedy et al., 2016; He et al., 2016). The size also depends on the number of feature maps being filtered. This can be reduced using 1×1 convolutions with fewer output features, followed by a convolution, and then an output 1×1 filter to increase the number of feature maps again (He et al., 2016; Iandola et al., 2017). This is shown on the right in Figure 14.10.

A deep learning model can use fewer feature maps in each layer (Huang et al., 2017). Alternatively, the computation and parameters may be reduced by not using every feature map in each convolution layer (Iandola and Keutzer, 2017). For example, by having half of the convolutions in a layer use one half of the feature maps, and the other half using the remaining maps.

The fully connected layers at the end often require most of the parameters. By reducing the number of fully connected layers and replacing them with convolutions, the number of parameters can be significantly reduced (Iandola and Keutzer, 2017), with limited effect on accuracy.

Finally, the model can be compressed by pruning unimportant weights or neurons from the network (Srinivas and Babu, 2015). Using appropriate regularisation during training, a sparse network may be designed, with up to 90% of the weights pruned (Liu et al., 2015). It can also be compressed by decomposing the low rank matrices used for the convolution and fully connected layers using singular value decomposition or low-rank factorisation (Kim et al., 2016; Qiu et al., 2016).

While these architectural simplifications can usually reduce both the computation and storage required, they may also reduce the network accuracy. The applicability of the trade-offs must be evaluated based on the initial requirements analysis.

14.4.4.2 Reducing the Computation

For deep learning, most of the computation consists of multiply-and-accumulate operations. These come from the convolutions in the convolutional layers and the dot-products in the fully connected layers. The computations can be transformed to reduce the number of multiplications required, for example using an fast Fourier transform (FFT) or the Winograd transform.

A convolution in the spatial domain becomes a product after Fourier transform. Therefore, for larger sized convolution kernels, the computation may be more efficiently performed in the frequency domain. For a single convolution, this has little benefit, but each feature map is used as input for several convolutions, so only needs to be converted into the frequency domain once (Mathieu et al., 2014). The kernels can be pre-transformed, and the sums of each of the feature maps can be performed in the frequency domain. Unfortunately, it is necessary to perform an inverse transform before the nonlinear activation function. The main limitation of using FFTs is the large storage size required.

Winograd (1980) devised a minimal complexity filtering algorithm that rearranges the convolution arithmetic to significantly reduce the number of multiplications required for each output sample. Similar to the Fourier transform it transforms a block in the input and the filter kernel, multiplies the results, and then performs an inverse transform to get the output (see Figure 14.15). An advantage over the FFT approach is that the Winograd transform works well for small filters. For calculating a 2×2 output patch, filtering with a 3×3 kernel, all of the transforms are additions and subtractions (and trivial multiplications by $\frac{1}{2}$). In this case, the number of multiplications is reduced from 36 down to 16, a reduction in complexity of 2.25 (Lavin and Gray, 2016). For larger patch sizes, the savings are even greater (Liang et al., 2020), although the transformation complexity also increases. Liang et al. (2020) compared the FPGA implementations of small FFTs (8×8) and the Winograd transform and found that both gave significant throughput improvements. Lu et al. (2017) pipelined a Winograd processing unit that produced a 4×4 patch each clock cycle.

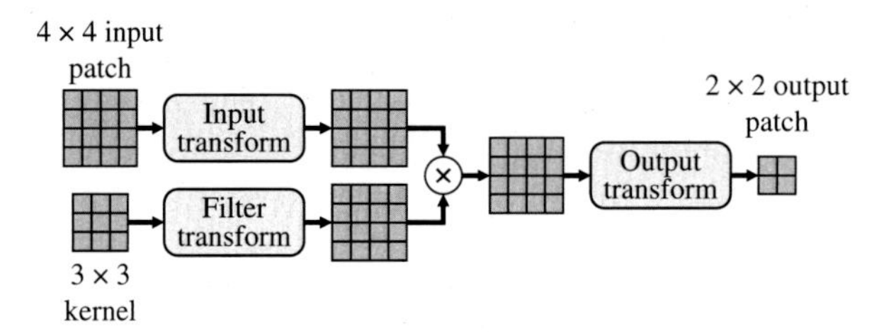

Figure 14.15 Winograd transform for producing a 2 × 2 convolution patch. The input, filter, and output transformations all consist of additions and subtractions.

Another obvious approach to reducing the complexity is exploiting sparsity (Deng et al., 2020). This comes from two sources: through pruning the network (Srinivas and Babu, 2015; Liu et al., 2015; Zhu et al., 2020) and through the ReLU activation function producing a zero output (Kurtz et al., 2020). In both cases, one of the multiplicands is zero, making the multiplication unnecessary. When the sparsity comes through pruning, it is known in advance, and the network can be compressed and the computations performed efficiently (Liu et al., 2015; Zhu et al., 2020). However, activation sparsity occurs during inference and is different from one image to the next. ReLU activation typically produces 50% of the values within feature maps as zero, although this is slightly lower in the earlier layers and higher in the later layers (Kurtz et al., 2020). Activation sparsity can also be increased through regularisation during training (Kurtz et al., 2020). While significant savings can be gained from exploiting sparsity (in either form), it disrupts the regular flow of data through the processing units and requires careful design to maximise the utilisation of the multiply-and-accumulate blocks.

14.4.4.3 Quantisation

Although deep networks are usually trained using floating-point, when deployed for inference, the precision can be significantly reduced without a significant change in the accuracy of the results. Quantisation can be applied in two places: the parameters or network weights and the activations or feature map. The result is savings in a number of places (Colangelo et al., 2018):

- memory required to store the weights and feature maps,
- memory bandwidth in reading and writing these from external memory,
- hardware resources required for the computation,
- power, given the fewer resources, less routing, and lower memory bandwidth used.

Gholami et al. (2022) and Deng et al. (2020) surveyed the many quantisation techniques used when implementing deep networks on resource-constrained devices. Commonly used data types are 16-bit floating-point, 8-bit integer, and 4-bit integer (with block exponent). This section gives a brief overview of some of the key techniques.

On central processing unit (CPU) and GPU platforms, there is little incentive to reduce the precision below 8 bits, because that is the minimum native data granularity. Therefore, most research in the past has focused on 8-bit fixed-point implementations. Quantising the weights will generally reduce the accuracy of the network commensurately with the reduction in precision. Therefore, it is often necessary to fine-tune the network weights after quantisation by retraining.

For compression, some researchers are exploring the use of the posit number system (see Section 5.1.7). Posits have the advantage of better precision for numbers close to 1.0 than floating-point and wider dynamic range. This means that significantly narrower posits can be used to achieve the same accuracy than floating-point numbers. For example, Langroudi et al. (2018) saved the network weights as 8-bit posits and converted them back to 32-bit floating-point as they were loaded from memory. For computation, Carmichael et al. (2019) showed that posits in the 5-bit to 8-bit range for both the weights and activations could be used for inference.

The better precision meant that 8-bit posits out-performed 8-bit floating-point, and the increased dynamic range gave a considerable advantage over the 8-bit fixed-point representation.

To solve the problem of the limited dynamic range of fixed-point arithmetic, one compromise is block floating-point. All of the numbers within a block share the same exponent, so the computations within the block can be performed using integer arithmetic, with the numbers converted at block boundaries. For example, Song et al. (2018) showed that this could be applied to convolution blocks without needing to retrain the network for the reduced precision. Block floating-point is used in the majority of neural networks with quantisation.

When using fixed-point arithmetic, it is not necessary to use the same precision throughout the network. Qiu et al. (2016) demonstrated that the accuracy could be improved by having a different number of fraction bits in different layers. Fully connected layers are generally less sensitive to representation with fewer bits than the convolutional layers (Qiu et al., 2016).

Reducing activations below 8 bits generally results in deteriorating accuracy. This can be compensated by making the network wider (more feature maps) (Colangelo et al., 2018). Note that with narrow multipliers, a single DSP block is able to perform several multiplications in parallel.

Going to the extreme are binary and ternary neural networks, where the network weights are either $\{-1, +1\}$ or $\{-1, 0, +1\}$, respectively. The multiplications can be eliminated completely, reducing the computation to simply adding or subtracting the input activations. A ternary network can be considered as a sparse binary network. Deng and Zhang (2022) used regularisation to increase the sparsity of ternary weighted networks.

If the activations are also made binary (or ternary), the multiply-and-accumulate becomes a logic gate and counter, resulting in an extremely low-resource, low-power system. A lot of research has gone into this space (see, for example, (Qin et al., 2020)). Simple quantisation of an existing network generally gives unacceptable accuracy, so a lot of research effort has been applied to efficiently retraining the network after quantisation, or even training the quantised network (Gholami et al., 2022).

14.4.4.4 Performing the Computation

So far, techniques for reducing the computation have been reviewed. This section looks at how to map the computations onto the processing units. The convolution layers and fully connected layers will be considered separately, because the convolutions typically require over 90% of the computations even if they use fewer than 10% of the network parameters (Guo et al., 2019).

First, consider the convolution layers of the network. The computation for this part consists of five nested loops as represented in Figure 14.16. Each convolution itself, from Eq. (14.54), consists of a 2D loop through the filter kernel, with another loop for each of the input channels. This is repeated for each feature map output from a layer and then for each layer within the network. Even exploiting temporal parallelism, and processing streamed data with deep pipelining, there are insufficient resources on even the largest FPGAs to be able to implement all but trivial networks as a single pipeline. Therefore, it is necessary to partially unroll some of the loops and tile the input image and feature maps. The goal, of course, is to perform as many operations on the data while it is on the FPGA before having to swap the partial results back out to memory. Another consideration is the on-chip storage available for queuing data for processing, and between processors. This is a complex

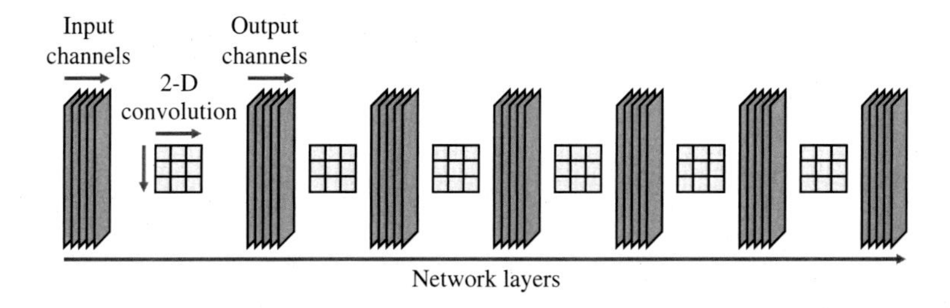

Figure 14.16 The nested loops within a convolutional neural network.

design-space exploration issue, especially since the different layers within the convolutional network may have different sizes of feature maps and number of channels. The potential areas of data reuse are (Shahshahani et al., 2018):

- **Kernel weights**: The filter weights of a particular kernel contribute to multiple output samples. Reuse is governed by the tile size used for processing.
- **Input channels**: Each input channel is filtered by several filters, each contributing to different output feature maps. Reuse is governed by the width of the output feature map slice that is computed in a block.
- **Output channels**: The output for each feature map is made from multiple filtered input channels. If the full input feature map is not processed at once, it is necessary to hold the output data for a tile temporarily until all of the input slices have been processed. Reuse is governed by the width of the input feature map slice that is computed.
- **Between layers**: Consideration can also be given to processing several layers while the data is on-chip (Erdem et al., 2019).

The choices will also depend on the target FPGA and the particular resources it has available. A key principle is to design the processing engines so that they may be used flexibly; then much of the design-space exploration is applied to the scheduling. Care must be taken because different layers require different computing patterns (Vestias et al., 2020). The roofline model can be used to guide the schedule and optimise the design (Zhang et al., 2015; Ma et al., 2018).

Fully connected layers pose different design constraints. There are only three nested loops for the output layer computation as illustrated in Figure 14.17, and the network layers dimension is small (especially if there is only a single fully connected layer!). Typically over 90% of the network parameters are used by the fully connected layers, and each parameter is only used once. Therefore, fully connected layers are often memory bound, especially with reading the weights. Really the only solution to this is to compress the weights, both through quantisation and sparsity. It is also worth considering Huffman coding the weights because they do not have a uniform distribution. Potential areas of data reuse in fully connected layers include:

- **Input activations**: Each input will contribute to each output. If the weights are dynamically loaded, then it is possible to exploit activation sparsity and not load the weights for zero activations. Reuse is governed by the width of the output that is computed in a block.
- **Output activations**: Each output is contributed to by each input. Partially accumulated outputs may need to be buffered while the remainder of the layer is accumulated. Reuse is governed by the width of the input that is computed in a block.

Although the access pattern is quite different from convolutional layers, consideration should be given to reusing the hardware for both, which adds another layer to the design-space exploration.

In addition to the convolution and fully connected layers, there are also the activation functions, pooling, soft-max, and other computation layers to consider. These generally do not have the same computation intensity and can readily be pipelined on the output of the other layers.

Several papers have shown how even large networks can be implemented efficiently on a small FPGA. See, for example, Zhang et al. (2015) and Vestias et al. (2020).

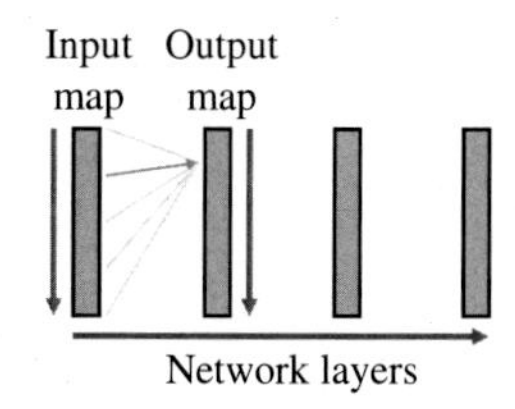

Figure 14.17 The nested loops within the fully connected section of the network.

14.4.4.5 Tools

There are a number of tools now becoming available that automate some of the FPGA mapping process. These usually include managing quantisation and design-space exploration for mapping the design onto processing units and scheduling the resulting network execution.

Both Intel and Xilinx provide frameworks and platforms for optimising deep learning designs for their respective FPGAs. These include OpenVino (Demidovskij et al., 2020) and Vitis AI (Kathail, 2020).

Tensor virtual machine (TVM) (Chen et al., 2018) is a compiler that takes a network design from a machine learning library and optimises it for execution on diverse computation platforms (CPU, GPU, mobile devices, and FPGAs). It performs high-level operation fusion, mapping to primitives available on the target platform, and schedules operations to hide memory latency.

CNN2Gate (Ghaffari and Savaria, 2020) takes a network design from a machine learning library and converts it to an OpenCL design for synthesis on the FPGA. It performs design-space exploration to match the design to the resources available on the FPGA. Design-space exploration is via reinforcement learning, with the compiler in the loop.

TensorIR (Feng et al., 2023) is another compiler that optimises deep networks for a range of platforms. It explicitly makes tensors and tensor operations the main class of operations, rather than the lower level nest of loops described earlier. Again, it automatically optimises the design for the primitives available on the target platform. While not specifically targeting FPGA, it could be used as a front end for FPGA-based implementation.

This is an active research area, and with new hardware platforms being developed targeting deep learning (including architectural changes within FPGAs), this is likely to be a dynamic landscape for some time to come.

14.5 Summary

A wide selection of machine learning methods, and their implementation on FPGAs, has been described in this chapter. Traditional machine learning approaches used for regression and classification work with features derived from images using classical image processing techniques.

One limitation of classical techniques is the need to select appropriate features, and design the image processing algorithms in detail. The advantage is that this can lead to very small, compact, and efficient implementations. Deep learning overcomes this by 'learning from examples' and trains a deep neural network to both develop a hierarchy of features and produce the desired output. While deep learning can give excellent accuracy, the cost, however, is the significant computational burden. Several of the key deep learning architectures have been reviewed, and the challenges posed for efficient implementation outlined.

FPGAs are increasingly being used to accelerate machine learning and deep learning inference due to their ability to provide high performance, low power consumption, and flexibility. They are well suited for this application because they allow for the efficient implementation of the matrix operations required by the algorithms, and their reconfigurable nature means that they can be optimised for specific use cases. The demand from deep learning is leading to architectural changes in FPGAs to support the new workload, for example tensor processing blocks, and access to high-bandwidth memory.

The machine learning landscape is evolving rapidly, particularly with deep learning where new models and algorithms are continually being developed. The reconfigurability of FPGAs allows them to be reprogrammed to support these new models or algorithms, making it possible to maintain the same hardware over time while still benefiting from advances in deep learning.

References

Afifi, S., GholamHosseini, H., and Sinha, R. (2020). FPGA implementations of SVM classifiers: a review. *SN Computer Science* **1** (3): Article ID 133, 17 pages. https://doi.org/10.1007/s42979-020-00128-9.

Anguita, D., Pischiutta, S., Ridella, S., and Sterpi, D. (2006). Feed-forward support vector machine without multipliers. *IEEE Transactions on Neural Networks* **17** (5): 1328–1331. https://doi.org/10.1109/TNN.2006.877537.

Badawi, A. and Bilal, M. (2019). High-level synthesis of online K-means clustering hardware for a real-time image processing pipeline. *Journal of Imaging* **5** (3): Article ID 38, 17 pages. https://doi.org/10.3390/jimaging5030038.

Baptista, D. and Dias, M. (2012). On the implementation of different hyperbolic tangent solutions in FPGA. *10th Portugese Conference on Automatic Control (CONTROLO'2012)*, Funchal, Madeira Island, Portugal (16–18 July 2012), 6 pages.

Besl, P.J., Birch, J.B., and Watson, L.T. (1989). Robust window operators. *Machine Vision and Applications* **2** (4): 179–191. https://doi.org/10.1007/BF01215874.

Carmichael, Z., Langroudi, H.F., Khazanov, C., Lillie, J., Gustafson, J.L., and Kudithipudi, D. (2019). Performance-efficiency trade-off of low-precision numerical formats in deep neural networks. *Conference for Next Generation Arithmetic 2019*, Singapore, Singapore (13–14 March 2019), Article ID 3, 9 pages. https://doi.org/10.1145/3316279.3316282.

Chen, X., Wang, G., Zhou, W., Chang, S., and Sun, S. (2006). Efficient sigmoid function for neural networks based FPGA design. In: *International Conference on Intelligent Computing*, Kunming, China (16–19 August 2006), *Lecture Notes in Computer Science*, Volume **4113**, 672–677. https://doi.org/10.1007/11816157_80.

Chen, T., Moreau, T., Jiang, Z., Zheng, L., Yan, E., Cowan, M., Shen, H., Wang, L., Hu, Y., Ceze, L., Guestrin, C., and Krishnamurthy, A. (2018). TVM: an automated end-to-end optimizing compiler for deep learning. *13th USENIX conference on Operating Systems Design and Implementation*, Carlsbad, CA, USA (8–10 October 2018), 579–594. https://doi.org/10.48550/arXiv.1802.04799.

Cheng, C. and Bouganis, C.S. (2013). Accelerating random forest training process using FPGA. *23rd International Conference on Field programmable Logic and Applications*, Porto, Portugal (2–4 September 2013), 7 pages. https://doi.org/10.1109/FPL.2013.6645500.

Cheng, Y., Yu, F.X., Feris, R.S., Kumar, S., Choudhary, A., and Chang, S.F. (2015). An exploration of parameter redundancy in deep networks with circulant projections. *IEEE International Conference on Computer Vision (ICCV)*, Santiago, Chile (7–13 December 2015), 2857–2865. https://doi.org/10.1109/ICCV.2015.327.

Cho, K. (2013). Simple sparsification improves sparse denoising autoencoders in denoising highly corrupted images. In: *30th International Conference on Machine Learning*, Atlanta, GA, USA (16–21 June 2013), *Proceeding of Machine Learning Research*, Volume **28**, 432–440.

Choi, H., Burleson, W.P., and Phatak, D.S. (1993). Fixed-point roundoff error analysis of large feedforward neural networks. *International Conference on Neural Networks*, Nagoya, Japan (25–29 October 1993), Volume **2**, 1947–1950. https://doi.org/10.1109/IJCNN.1993.717037.

Colangelo, P., Nasiri, N., Nurvitadhi, E., Mishra, A., Margala, M., and Nealis, K. (2018). Exploration of low numeric precision deep learning inference using Intel FPGAs. *IEEE 26th Annual International Symposium on Field-Programmable Custom Computing Machines (FCCM)*, Boulder, CO, USA (29 April–1 May 2018), 73–80. https://doi.org/10.1109/FCCM.2018.00020.

Cort, J.W. and Kenji, M. (2005). Advantages of the mean absolute error (MAE) over the root mean square error (RMSE) in assessing average model performance. *Climate Research* **30** (1): 79–82. https://doi.org/10.3354/cr030079.

Creswell, A., White, T., Dumoulin, V., Arulkumaran, K., Sengupta, B., and Bharath, A.A. (2018). Generative adversarial networks: an overview. *IEEE Signal Processing Magazine* **35** (1): 53–65. https://doi.org/10.1109/MSP.2017.2765202.

Da Costa, A.L.X., Silva, C.A.D., Torquato, M.F., and Fernandes, M.A.C. (2019). Parallel implementation of particle swarm optimization on FPGA. *IEEE Transactions on Circuits and Systems II: Express Briefs* **66** (11): 1875–1879. https://doi.org/10.1109/TCSII.2019.2895343.

de Myttenaere, A., Golden, B., Le Grand, B., and Rossi, F. (2016). Mean absolute percentage error for regression models. *Neurocomputing* **192**: 38–48. https://doi.org/10.1016/j.neucom.2015.12.114.

Demidovskij, A., Tugaryov, A., Suvorov, A., Tarkan, Y., Fatekhov, M., Salnikov, I., Kashchikhin, A., Golubenko, V., Dedyukhina, G., Alborova, A., Palmer, R., Fedorov, M., and Gorbachev, Y. (2020). OpenVINO deep learning workbench: a platform for model optimization, analysis and deployment. *IEEE 32nd International Conference on Tools with Artificial Intelligence (ICTAI)*, Baltimore, MD, USA (9–11 November 2020), 661–668. https://doi.org/10.1109/ICTAI50040.2020.00106.

Deng, X. and Zhang, Z. (2022). Sparsity-control ternary weight networks. *Neural Networks* **145**: 221–232. https://doi.org/10.1016/j.neunet.2021.10.018.

Deng, L., Li, G., Han, S., Shi, L., and Xie, Y. (2020). Model compression and hardware acceleration for neural networks: a comprehensive survey. *Proceedings of the IEEE* **108** (4): 485–532. https://doi.org/10.1109/JPROC.2020.2976475.

De Sa, C., Feldman, M., Re, C., and Olukotun, K. (2017). Understanding and optimizing asynchronous low-precision stochastic gradient descent. *4th Annual International Symposium on Computer Architecture*, Toronto, Canada (24–28 June 2017), 561–574. https://doi.org/10.1145/3079856.3080248.

Dosovitskiy, A., Beyer, L., Kolesnikov, A., Weissenborn, D., Zhai, X., Unterthiner, T., Dehghani, M., Minderer, M., Heigold, G., Gelly, S., Uszkoreit, J., and Houlsby, N. (2021). An image is worth 16x16 words: transformers for image recognition at scale. *9th International Conference on Learning Representations*, Virtual (3–7 May 2021), 22 pages. https://doi.org/10.48550/arXiv.2010.11929.

Erdem, A., Babic, D., and Silvano, C. (2019). A tile-based fused-layer approach to accelerate DCNNs on low-density FPGAs. *26th IEEE International Conference on Electronics, Circuits and Systems (ICECS)*, Genoa, Italy (27–29 November 2019), 37–40. https://doi.org/10.1109/ICECS46596.2019.8964870.

Ester, M., Kriegel, H.P., Sander, J., and Xu, X. (1996). A density-based algorithm for discovering clusters in large spatial databases with noise. *2nd International Conference on Knowledge Discovery and Data Mining*, Portland, OR, USA (2–4 August 1996), 226–231.

Feng, S., Hou, B., Jin, H., Lin, W., Shao, J., Lai, R., Ye, Z., Zheng, L., Yu, C.H., Yu, Y., and Chen, T. (2023). TensorIR: an abstraction for automatic tensorized program optimization. *28th ACM International Conference on Architectural Support for Programming Languages and Operating Systems*, Volume **2**, Vancouver, Canada (25–29 March 2023), 804–817. https://doi.org/10.1145/3575693.3576933.

Ghaffari, A. and Savaria, Y. (2020). CNN2Gate: an implementation of convolutional neural networks inference on FPGAs with automated design space exploration. *Electronics* **9** (12): Article ID 2200, 23 pages. https://doi.org/10.3390/electronics9122200.

Gholami, A., Kim, S., Dong, Z., Yao, Z., Mahoney, M.W., and Keutzer, K. (2022). A survey of quantization methods for efficient neural network inference. In: *Low-Power Computer Vision: Improve the Efficiency of Artificial Intelligence*, Chapter 13 (ed. G. Thiruvathukal, Y. Lu, J. Kim, Y. Chen, and B. Chen), 291–326. New York: Chapman and Hall/CRC. https://doi.org/10.1201/9781003162810.

Glorot, X., Bordes, A., and Bengio, Y. (2011). Deep sparse rectifier neural networks. *14th International Conference on Artificial Intelligence and Statistics*, Fort Lauderdale, FL, USA (11–13 April 2011), 315–323.

Goodfellow, I., Bengio, Y., and Courville, A. (2016). *Deep Learning*. Cambridge, MA: MIT Press.

Guo, K., Zeng, S., Yu, J., Wang, Y., and Yang, H. (2019). A survey of FPGA-based neural network inference accelerators. *ACM Transactions on Reconfigurable Technologies and Systems* **12** (1): Article ID 2, 26 pages. https://doi.org/10.1145/3289185.

Hajduk, Z. (2018). Reconfigurable FPGA implementation of neural networks. *Neurocomputing* **308**: 227–234. https://doi.org/10.1016/j.neucom.2018.04.077.

Han, K., Wang, Y., Chen, H., Chen, X., Guo, J., Liu, Z., Tang, Y., Xiao, A., Xu, C., Xu, Y., Yang, Z., Zhang, Y., and Tao, D. (2023). A survey on vision transformer. *IEEE Transactions on Pattern Analysis and Machine Intelligence* **45** (1): 87–110. https://doi.org/10.1109/TPAMI.2022.3152247.

Hatcher, W.G. and Yu, W. (2018). A survey of deep learning: platforms, applications and emerging research trends. *IEEE Access* **6**: 24411–24432. https://doi.org/10.1109/ACCESS.2018.2830661.

He, K., Zhang, X., Ren, S., and Sun, J. (2016). Deep residual learning for image recognition. *IEEE Conference on Computer Vision and Pattern Recognition (CVPR)*, Las Vegas, NV, USA (27–30 June 2016), 770–778. https://doi.org/10.1109/CVPR.2016.90.

Hinton, G.E. and Salakhutdinov, R.R. (2006). Reducing the dimensionality of data with neural networks. *Science* **313** (5786): 504–507. https://doi.org/10.1126/science.1127647.

Ho, T.K. (1995). Random decision forests. *3rd International Conference on Document Analysis and Recognition*, Montreal, Quebec, Canada (14–16 August 1995), Volume **1**, 278–282. https://doi.org/10.1109/ICDAR.1995.598994.

Hornik, K., Stinchcombe, M., and White, H. (1989). Multilayer feedforward networks are universal approximators. *Neural Networks* **2** (5): 359–366. https://doi.org/10.1016/0893-6080(89)90020-8.

Hsu, C.W. and Lin, C.J. (2002). A comparison of methods for multiclass support vector machines. *IEEE Transactions on Neural Networks* **13** (2): 415–425. https://doi.org/10.1109/72.991427.

Huang, G., Liu, Z., Maaten, L.V.D., and Weinberger, K.Q. (2017). Densely connected convolutional networks. *IEEE Conference on Computer Vision and Pattern Recognition (CVPR)*, Honolulu, HI, USA (21–26 July 2017), 2261–2269. https://doi.org/10.1109/CVPR.2017.243.

Huber, P.J. (1964). Robust estimation of a location parameter. *The Annals of Mathematical Statistics* **35** (1): 73–101. https://doi.org/10.1214/aoms/1177703732.

Iandola, F. and Keutzer, K. (2017). Small neural nets are beautiful: enabling embedded systems with small deep-neural-network architectures. *12th IEEE/ACM/IFIP International Conference on Hardware/Software Codesign and System Synthesis Companion*, Seoul, Republic of Korea, 10 pages. https://doi.org/10.1145/3125502.3125606.

Iandola, F.N., Han, S., Moskewicz, M.W., Ashraf, K., Dally, W.J., and Keutzer, K. (2017). SqueezeNet: AlexNet-level accuracy with 50x fewer parameters and <0.5MB model size. *5th International Conference on Learning Representations (ICLR 2017)*, Toulon, France (24–26 April 2017). https://doi.org/10.48550/arXiv.1602.07360.

Ioffe, S. and Szegedy, C. (2015). Batch normalization: accelerating deep network training by reducing internal covariate shift. In: *32nd International Conference on Machine Learning*, Lille, France (6–11 July 2015), *Proceeding of Machine Learning Research*, Volume **37**, 448–456.

Jiao, L., Chen, J., Liu, F., Yang, S., You, C., Liu, X., Li, L., and Hou, B. (2022). Graph representation learning meets computer vision: a survey. *IEEE Transactions on Artificial Intelligence* **4** (1): 1–22. https://doi.org/10.1109/TAI.2022.3194869.

John, G.H. and Langley, P. (1995). Estimating continuous distributions in Bayesian classifiers. *11th Conference on Uncertainty in Artificial Intelligence*, Montreal, Quebec, Canada (18–20 August 1995), 338–345.

Kalman, B.L. and Kwasny, S.C. (1992). Why tanh: choosing a sigmoidal function. *International Joint Conference on Neural Networks (IJCNN)*, Baltimore, MD, USA (7–11 June 1992), Volume **4**, 578–581. https://doi.org/10.1109/IJCNN.1992.227257.

Kathail, V. (2020). Xilinx Vitis unified software platform. *ACM/SIGDA International Symposium on Field-Programmable Gate Arrays*, Seaside, CA, USA, 173–174. https://doi.org/10.1145/3373087.3375887.

Khalid, S., Khalil, T., and Nasreen, S. (2014). A survey of feature selection and feature extraction techniques in machine learning. *Science and Information Conference*, London, UK (27–29 August 2014), 372–378. https://doi.org/10.1109/SAI.2014.6918213.

Khan, S., Naseer, M., Hayat, M., Zamir, S.W., Khan, F.S., and Shah, M. (2022). Transformers in vision: a survey. *ACM Computing Surveys* **54** (10s): Article ID 200, 41 pages. https://doi.org/10.1145/3505244.

Kim, Y.D., Park, E., Yoo, S., Choi, T., Yang, L., and Shin, D. (2016). Compression of deep convolutional neural networks for fast and low power mobile applications. *4th International Conference on Learning Representations (ICLR 2016)*, San Juan, Puerto Rico (2–4 May 2016), 16 pages. https://doi.org/10.48550/arXiv.1511.06530.

Kolassa, S. and Schutz, W. (2007). Advantages of the MAD/MEAN ratio over the MAPE. *Foresight: The International Journal of Applied Forecasting* (6): 40–43.

Kunin, D., Bloom, J., Goeva, A., and Seed, C. (2019). Loss landscapes of regularized linear autoencoders. In: *36th International Conference on Machine Learning*, Long Beach, CA, USA (10–15 June 2019), *Proceeding of Machine Learning Research*, Volume **97**, 3560–3569.

Kurtz, M., Kopinsky, J., Gelashvili, R., Matveev, A., Carr, J., Goin, M., Leiserson, W., Moore, S., Shavit, N., and Alistarh, D. (2020). Inducing and exploiting activation sparsity for fast inference on deep neural networks. In: *37th International Conference on Machine Learning*, Virtual (13–18 July, 2020), *Proceedings of Machine Learning Research*, Volume **119**, 5533–5543.

Langhammer, M., Nurvitadhi, E., Pasca, B., and Gribok, S. (2021). Stratix 10 NX architecture and applications. *ACM/SIGDA International Symposium on Field-Programmable Gate Arrays (FPGA '21)*, Virtual (28 February – 2 March 2021), 57–67. https://doi.org/10.1145/3431920.3439293.

Langroudi, S.H.F., Pandit, T., and Kudithipudi, D. (2018). Deep learning inference on embedded devices: fixed-point vs posit. *1st Workshop on Energy Efficient Machine Learning and Cognitive Computing for Embedded Applications (EMC2)*, Williamsburg, VA, USA (25–25 March 2018), 19–23. https://doi.org/10.1109/EMC2.2018.00012.

Lavin, A. and Gray, S. (2016). Fast algorithms for convolutional neural networks. *IEEE Conference on Computer Vision and Pattern Recognition (CVPR)*, Las Vegas, NV, USA (27–30 June 2016), 4013–4021. https://doi.org/10.1109/CVPR.2016.435.

Ledig, C., Theis, L., Huszar, F., Caballero, J., Cunningham, A., Acosta, A., Aitken, A., Tejani, A., Totz, J., Wang, Z., and Shi, W. (2017). Photo-realistic single image super-resolution using a generative adversarial network. *IEEE Conference on Computer Vision and Pattern Recognition (CVPR)*, Honolulu, HI, USA (21–26 July 2017), 105–114. https://doi.org/10.1109/CVPR.2017.19.

Leeser, M.E., Belanovic, P., Estlick, M., Gokhale, M., Szymanski, J.J., and Theiler, J.P. (2001). Applying reconfigurable hardware to the analysis of multispectral and hyperspectral imagery. *Imaging Spectrometry VII*, San Diego, CA, USA (29 July – 3 August 2001). SPIE, Volume 4480, 8 pages. https://doi.org/10.1117/12.453329.

Liang, Y., Lu, L., Xiao, Q., and Yan, S. (2020). Evaluating fast algorithms for convolutional neural networks on FPGAs. *IEEE Transactions on Computer-Aided Design of Integrated Circuits and Systems* **39** (4): 857–870. https://doi.org/10.1109/TCAD.2019.2897701.

Lin, Z., Lo, C., and Chow, P. (2012). K-means implementation on FPGA for high-dimensional data using triangle inequality. *22nd International Conference on Field Programmable Logic and Applications (FPL)*, Oslo, Norway (29–31 August 2012), 437–442. https://doi.org/10.1109/FPL.2012.6339141.

Lin, X., Blanton, R.S., and Thomas, D.E. (2017). Random forest architectures on FPGA for multiple applications. *Great Lakes Symposium on VLSI 2017*, Banff, Alberta, Canada (10–12 May 2017), 415–418. https://doi.org/10.1145/3060403.3060416.

Liu, Q., Constantinides, G.A., Masselos, K., and Cheung, P.Y.K. (2008). Combining data reuse exploitation with data-level parallelization for FPGA targeted hardware compilation: a geometric programming framework. *International Conference on Field Programmable Logic and Applications (FPL 2008)*, Heidelberg, Germany (8–10 September 2008), 179–184. https://doi.org/10.1109/FPL.2008.4629928.

Liu, B., Wang, M., Foroosh, H., Tappen, M., and Penksy, M. (2015). Sparse convolutional neural networks. *IEEE Conference on Computer Vision and Pattern Recognition (CVPR)*, Boston, MA, USA (7–12 June 2015), 806–814. https://doi.org/10.1109/CVPR.2015.7298681.

Lu, L., Liang, Y., Xiao, Q., and Yan, S. (2017). Evaluating fast algorithms for convolutional neural networks on FPGAs. *IEEE 25th Annual International Symposium on Field-Programmable Custom Computing Machines (FCCM)*, Napa, CA, USA (30 April – 2 May 2017), 101–108. https://doi.org/10.1109/FCCM.2017.64.

Ma, Y., Cao, Y., Vrudhula, S., and Seo, J. (2018). Optimizing the convolution operation to accelerate deep neural networks on FPGA. *IEEE Transactions on Very Large Scale Integration (VLSI) Systems* **26** (7): 1354–1367. https://doi.org/10.1109/TVLSI.2018.2815603.

Maruyama, T. (2006). Real-time K-means clustering for color images on reconfigurable hardware. *18th International Conference on Pattern Recognition*, Hong Kong (20–24 August 2006), Volume **2**, 816–819. https://doi.org/10.1109/ICPR.2006.961.

Mathieu, M., Henaff, M., and LeCun, Y. (2014). Fast training of convolutional networks through FFTS. *2nd International Conference on Learning Representations, ICLR 2014*, Banff, Canada (14–16 April 2014), 9 pages. https://doi.org/10.48550/arXiv.1312.5851.

Meng, H., Appiah, K., Hunter, A., and Dickinson, P. (2011). FPGA implementation of naive Bayes classifier for visual object recognition. *7th IEEE Workshop on Embedded Computer Vision*, Colorado Springs, CO, USA (20–25 June 2011), 123–128. https://doi.org/10.1109/CVPRW.2011.5981831.

Misra, J. and Saha, I. (2010). Artificial neural networks in hardware: a survey of two decades of progress. *Neurocomputing* **74** (1): 239–255. https://doi.org/10.1016/j.neucom.2010.03.021.

Mittal, S. (2020). A survey of FPGA-based accelerators for convolutional neural networks. *Neural Computing and Applications* **32** (4): 1109–1139. https://doi.org/10.1007/s00521-018-3761-1.

Munoz, D.M., Llanos, C.H., Coelho, L.d.S., and Ayala-Rincon, M. (2010). Comparison between two FPGA implementations of the particle swarm optimization algorithm for high-performance embedded applications. *IEEE 5th International Conference on Bio-Inspired Computing: Theories and Applications*, Changsha, China (23–26 September 2010), 1637–1645. https://doi.org/10.1109/BICTA.2010.5645256.

Niamat, M.Y., Bitter, D., and Jamali, M.M. (1998). FPGA implementation of hierarchical clustering algorithms. *IEEE International Symposium on Circuits and Systems (ISCAS '98)*, Monterey, CA, USA (31 May–3 June 1998), Volume **5**, 70–73. https://doi.org/10.1109/ISCAS.1998.694410.

Noh, H., Hong, S., and Han, B. (2015). Learning deconvolution network for semantic segmentation. *IEEE International Conference on Computer Vision (ICCV)*, Santiago, Chile (7–13 December 2015), 1520–1528. https://doi.org/10.1109/ICCV.2015.178.

Papadonikolakis, M. and Bouganis, C.S. (2008a). Efficient FPGA mapping of Gilbert's algorithm for SVM training on large-scale classification problems. *International Conference on Field Programmable Logic and Applications*, Heidelberg, Germany (8–10 September 2008), 385–390. https://doi.org/10.1109/FPL.2008.4629968.

Papadonikolakis, M. and Bouganis, C.S. (2008b). A scalable FPGA architecture for non-linear SVM training. *International Conference on Field Programmable Technology*, Taipei, Taiwan (8–10 December 2008), 337–340. https://doi.org/10.1109/FPT.2008.4762412.

Piyasena, D., Lam, S.K., and Wu, M. (2021). Accelerating continual learning on edge FPGA. *31st International Conference on Field-Programmable Logic and Applications (FPL)*, Dresden, Germany (30 August –3 September 2021), 294–300. https://doi.org/10.1109/FPL53798.2021.00059.

Prechelt, L. (1998). Early stopping - but when? In: *Neural Networks: Tricks of the Trade*, Chapter 2 (ed. G.B. Orr and K.R. Muller), 55–69. Berlin, Germany: Springer-Verlag. https://doi.org/10.1007/3-540-49430-8_3.

Pugmire, R.H., Hodgson, R.M., and Chaplin, R.I. (1995). The properties and training of a neural network based universal window filter (UWF). *5th International Conference on Image Processing and its Applications*, Edinburgh, Scotland (4–6 July 1995), 642–646. https://doi.org/10.1049/cp:19950738.

Qin, H., Gong, R., Liu, X., Bai, X., Song, J., and Sebe, N. (2020). Binary neural networks: a survey. *Pattern Recognition* **105**: 107281. https://doi.org/10.1016/j.patcog.2020.107281.

Qiu, J., Wang, J., Yao, S., Guo, K., Li, B., Zhou, E., Yu, J., Tang, T., Xu, N., Song, S., Wang, Y., and Yang, H. (2016). Going deeper with embedded FPGA platform for convolutional neural network. *ACM/SIGDA International Symposium on Field-Programmable Gate Arrays*, Monterey, CA, USA, 26–35. https://doi.org/10.1145/2847263.2847265.

Ramachandran, P., Zoph, B., and Le, Q.V. (2017). Searching for activation functions. *arXiv preprint*, (1710.05941), 13 pages. https://doi.org/10.48550/arxiv.1710.05941.

Rasoori, S. and Akella, V. (2018). Scalable hardware accelerator for mini-batch gradient descent. *Great Lakes Symposium on VLSI*, Chicago, IL, USA (23–25 May 2018), 159–164. https://doi.org/10.1145/3194554.3194559.

Remez, T., Litany, O., Giryes, R., and Bronstein, A.M. (2017). Deep convolutional denoising of low-light images. *arXiv preprint*, (1701.01687), 11 pages. https://doi.org/10.48550/arXiv.1701.01687.

Reynolds, D. (2009). Gaussian mixture models. In: *Encyclopedia of Biometrics* (ed. S.Z. Li and A. Jain), 659–663. Boston, MA: Springer. https://doi.org/10.1007/978-0-387-73003-5_196.

Rish, I. (2001). An empirical study of the naive Bayes classifier. *IJCAI 2001 Workshop on Empirical Methods in Artificial Iintelligence* (4 August 2001), Volume **3**, 41–46.

Ruder, S. (2016). An overview of gradient descent optimization algorithms. *ArXiv preprint*, (1609.04747), 14 pages. https://doi.org/10.48550/arXiv.1609.04747.

Saidi, A., Othman, S.B., Dhouibi, M., and Saoud, S.B. (2021). FPGA-based implementation of classification techniques: a survey. *Integration* **81**: 280–299. https://doi.org/10.1016/j.vlsi.2021.08.004.

Santurkar, S., Tsipras, D., Ilyas, A., and Madry, A. (2018). How does batch normalization help optimization? *32nd Conference on Neural Information Processing Systems (NeurIPS 2018)*, Montreal, Quebec, Canada (3–8 December 2018), 2483–2493.

Scicluna, N. and Bouganis, C.S. (2015). A multidimensional FPGA-based parallel DBSCAN architecture. *ACM Transactions on Reconfigurable Technologies and Systems* **9** (1): Article ID 2, 15 pages. https://doi.org/10.1145/2724722.

Shahshahani, M., Goswami, P., and Bhatia, D. (2018). Memory optimization techniques for FPGA based CNN implementations. *IEEE 13th Dallas Circuits and Systems Conference (DCAS)*, Dallas, TX, USA (12–12 November 2018), 6 pages. https://doi.org/10.1109/DCAS.2018.8620112.

Shi, Q., Yue, S., and Wang, Q. (2014). FPGA based accelerator for parallel DBSCAN algorithm. *Computer Modelling and New Technologies* **18** (2): 135–142.

Simonyan, K. and Zisserman, A. (2015). Very deep convolutional networks for large-scale image recognition. *3rd International Conference on Learning Representations (ICLR 2015)*, San Diego, CA, USA (7–9 May 2015), 14 pages. https://doi.org/10.48550/arXiv.1409.1556.

Sinha, U. (2017). Enabling impactful DSP designs on FPGAs with hardened floating-point implementation. White paper, Intel Corporation, USA.

Siracusa, M., Sozzo, E.D., Rabozzi, M., Tucci, L.D., Williams, S., Sciuto, D., and Santambrogio, M.D. (2022). A comprehensive methodology to optimize FPGA designs via the roofline model. *IEEE Transactions on Computers* **71** (8): 1903–1915. https://doi.org/10.1109/TC.2021.3111761.

Song, Z., Liu, Z., and Wang, D. (2018). Computation error analysis of block floating point arithmetic oriented convolution neural network accelerator design. *AAAI Conference on Artificial Intelligence*, New Orleans, LA, USA (2–7 February 2018), Volume **32**, 816–823. https://doi.org/10.1609/aaai.v32i1.11334.

Srinivas, S. and Babu, R.V. (2015). Data-free parameter pruning for deep neural networks. *British Machine Vision Conference (BMVC 2015)*, Swansea, UK (7–10 September 2015), Article ID 31, 12 pages. https://doi.org/10.5244/C.29.31.

Suthaharan, S. (2016). Decision tree learning. In: *Machine Learning Models and Algorithms for Big Data Classification, Integrated Series in Information Systems*, Chapter 10, vol. **36**, 237–269. Boston, MA: Springer. https://doi.org/10.1007/978-1-4899-7641-3_10.

Sze, V., Chen, Y.H., Yang, T.J., and Emer, J.S. (2017). Efficient processing of deep neural networks: a tutorial and survey. *Proceedings of the IEEE* **105** (12): 2295–2329. https://doi.org/10.1109/JPROC.2017.2761740.

Sze, V., Chen, Y.H., Yang, T.J., and Emer, J.S. (2020). *Efficient Processing of Deep Neural Networks, Synthesis Lectures on Computer Architecture*, vol. **50**. San Rafael, CA: Morgan and Claypool. https://doi.org/10.2200/S01004ED1V01Y202004CAC050.

Szegedy, C., Vanhoucke, V., Ioffe, S., Shlens, J., and Wojna, Z. (2016). Rethinking the Inception architecture for computer vision. *IEEE Conference on Computer Vision and Pattern Recognition (CVPR)*, Las Vegas, NV, USA (27–30 June 2016), 2818–2826. https://doi.org/10.1109/CVPR.2016.308.

Tahir, M.A., Bouridane, A., and Kurugollu, F. (2004). An FPGA based coprocessor for the classification of tissue patterns in prostatic cancer. In: *Field Programmable Logic and Applications (FPL 2004)*, Leuven, Belgium (30 August – 1 September 2004), *Lecture Notes in Computer Science*, vol. **3203**, 771–780. https://doi.org/10.1007/978-3-540-30117-2_78.

Tofallis, C. (2015). A better measure of relative prediction accuracy for model selection and model estimation. *Journal of the Operational Research Society* **66** (8): 1352–1362. https://doi.org/10.1057/jors.2014.103.

Vaswani, A., Shazeer, N., Parmar, N., Uszkoreit, J., Jones, L., Gomez, A.N., Kaiser, L., and Polosukhin, I. (2017). Attention is all you need. In: *31st Conference on Neural Information Processing Systems*, Long Beach, CA, USA (4–9 December 2017), *Advances in Neural Information Processing Systems*, Volume **30**, 11 pages.

Vestias, M.P., Duarte, R.P., de Sousa, J.T., and Neto, H.C. (2020). A fast and scalable architecture to run convolutional neural networks in low density FPGAs. *Microprocessors and Microsystems* **77**: Article ID 103136, 15 pages. https://doi.org/10.1016/j.micpro.2020.103136.

Wang, C., Gong, L., Jia, F., and Zhou, X. (2021). An FPGA based accelerator for clustering algorithms with custom instructions. *IEEE Transactions on Computers* **70** (5): 725–732. https://doi.org/10.1109/TC.2020.2995761.

Winograd, S. (1980). *Arithmetic Complexity of Computations, CBMS-NSF Regional Conference Series in Applied Mathematics*, vol. **33**. Philadelphia, PA: Society for Industrial and Applied Mathematics. https://doi.org/10.1137/1.9781611970364.

Winterstein, F., Bayliss, S., and Constantinides, G.A. (2013). FPGA-based K-means clustering using tree-based data structures. *23rd International Conference on Field programmable Logic and Applications*, Porto, Portugal (2–4 September 2013), 6 pages. https://doi.org/10.1109/FPL.2013.6645501.

Wissolik, M., Zacher, D., Torza, A., and Day, B. (2019). Virtex UltraScale+ HBM FPGA: a revolutionary increase in memory performance. White paper. Xilinx Corporation, USA.

Wofk, D., Ma, F., Yang, T.J., Karaman, S., and Sze, V. (2019). FastDepth: fast monocular depth estimation on embedded systems. *International Conference on Robotics and Automation (ICRA)*, Montreal, Quebec, Canada (20–24 May 2019), 6101–6108. https://doi.org/10.1109/ICRA.2019.8794182.

Xu, Y. and Goodacre, R. (2018). On splitting training and validation set: a comparative study of cross-validation, bootstrap and systematic sampling for estimating the generalization performance of supervised learning. *Journal of Analysis and Testing* **2** (3): 249–262. https://doi.org/10.1007/s41664-018-0068-2.

Xue, Z., Wei, J., and Guo, W. (2020). A real-time naive Bayes classifier accelerator on FPGA. *IEEE Access* **8**: 40755–40766. https://doi.org/10.1109/ACCESS.2020.2976879.

Zeng, K., Yu, J., Wang, R., Li, C., and Tao, D. (2017). Coupled deep autoencoder for single image super-resolution. *IEEE Transactions on Cybernetics* **47** (1): 27–37. https://doi.org/10.1109/TCYB.2015.2501373.

Zhang, C., Li, P., Sun, G., Guan, Y., Xiao, B., and Cong, J. (2015). Optimizing FPGA-based accelerator design for deep convolutional neural networks. *ACM/SIGDA International Symposium on Field-Programmable Gate Arrays*, Monterey, CA, USA, 161–170. https://doi.org/10.1145/2684746.2689060.

Zhang, X., Liu, X., Ramachandran, A., Zhuge, C., Tang, S., Ouyang, P., Cheng, Z., Rupnow, K., and Chen, D. (2017). High-performance video content recognition with long-term recurrent convolutional network for FPGA. *27th International Conference on Field Programmable Logic and Applications (FPL)*, Ghent, Belgium (4–8 September 2017), 4 pages. https://doi.org/10.23919/FPL.2017.8056833.

Zhang, K., Zuo, W., and Zhang, L. (2018). FFDNet: toward a fast and flexible solution for CNN-based image denoising. *IEEE Transactions on Image Processing* **27** (9): 4608–4622. https://doi.org/10.1109/TIP.2018.2839891.

Zhang, A., Lipton, Z.C., Li, M., and Smola, A.J. (2021). Dive into deep learning, volume 2106.11342. arXiv e-prints. https://doi.org/10.48550/arXiv.2106.11342.

Zhu, J.Y., Park, T., Isola, P., and Efros, A.A. (2017). Unpaired image-to-image translation using cycle-consistent adversarial networks. *IEEE International Conference on Computer Vision (ICCV)*, Venice, Italy (22–29 October 2017), 2242–2251. https://doi.org/10.1109/ICCV.2017.244.

Zhu, C., Huang, K., Yang, S., Zhu, Z., Zhang, H., and Shen, H. (2020). An efficient hardware accelerator for structured sparse convolutional neural networks on FPGAs. *IEEE Transactions on Very Large Scale Integration (VLSI) Systems* **28** (9): 1953–1965. https://doi.org/10.1109/TVLSI.2020.3002779.

15

Example Applications

In the earlier chapters, much of the focus in describing field-programmable gate array (FPGA) implementation was on individual image processing operations, rather than complete applications. This final chapter shows how the individual operations tie together within an application. Of particular interest are some of the optimisations used to reduce hardware or processing time.

15.1 Coloured Region Tracking

The desire in this application was to use coloured paddles to create a gesture-based user input (Johnston et al., 2005b). A secondary goal was to minimise resource utilisation to enable the remainder of the FPGA to be used for the application being controlled (Johnston et al., 2005a).

The basic structure of the algorithm is shown in Figure 15.1. The distinctive colours of the paddles are used to segment them from the background. A simple bounding box is then used to determine the paddle locations. The remainder of this section details the algorithm and the optimisations made to reduce the size of the implementation.

This application was originally implemented on an RC100 board from Celoxica Ltd. The board used a Xilinx XC2S200 Spartan II FPGA, quite small, with very limited resources by modern standards. The live video was captured using a video decoder chip which digitises the composite video signal from the camera and provides a stream of 16-bit colour (RGB565) pixels to the FPGA. The decoder provides a 27-MHz pixel clock, with one pixel streamed every two clock cycles (two-phase clocking).

Two counters were implemented to keep track of the input, one to count the pixels on the row, x, and the other to count the row number, y. Both were reset at the end of the respective blanking intervals (on the start of active video). The control signals for driving the rest of the application were derived from these counters.

The first stage is to identify the colour of the pixels in the incoming pixel stream. As described in Section 7.3.3, the RGB colour space is not the most suitable for colour segmentation. Since the output is not intended for human viewing, the simplified YCoCg colour space of Eq. (7.54) was used, which can be implemented with only four additions as shown in Figure 7.32. The fact that G has one more bit than R and B does not matter as all three values are treated at binary fractions. The resultant YCoCg components each have 7 bits. However, since the RGB to YCoCg transformation is linear, scaling the intensity will scale all three components equally. Therefore, to normalise the intensity, Co and Cg can be divided by Y:

$$Co_n = \frac{Co}{Y}, \qquad Cg_n = \frac{Cg}{Y}. \tag{15.1}$$

Removing the intensity dependence in this way does have the side effect of reducing colour specificity, since all colours with a similar hue are mapped to the same range. Normalisation therefore introduces a trade-off between insensitivity to intensity and colour selectivity.

Design for Embedded Image Processing on FPGAs, Second Edition. Donald G. Bailey.

Companion Website: www.wiley.com/go/bailey/designforembeddedimageprocessc2e

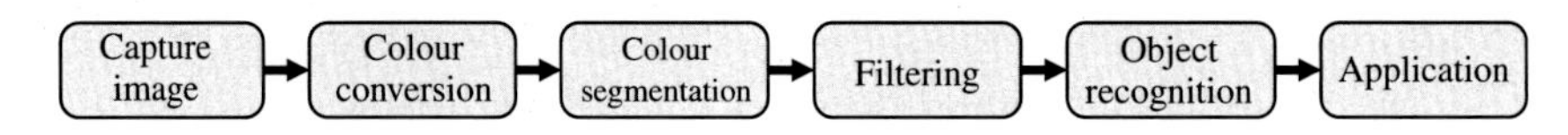

Figure 15.1 Steps within a colour tracking algorithm.

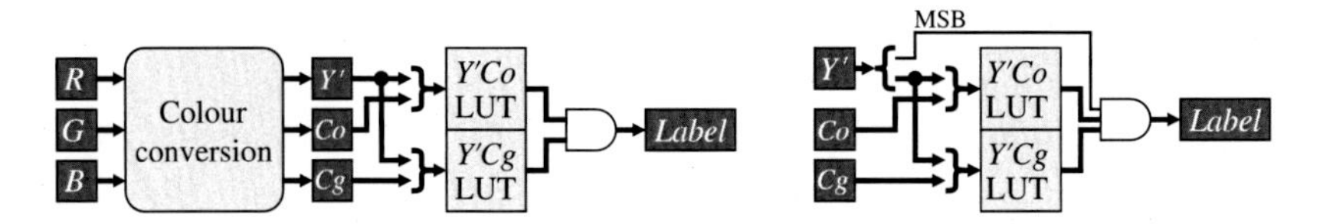

Figure 15.2 Colour segmentation and labelling. Left: basic idea of joint lookup tables; right: gaining an extra bit of precision for Y'.

Unfortunately with the definition in Eq. (7.54), Co can exceed Y for colours dominant in red or blue, with Co_n exceeding the range −1 to 1. Therefore, an alternative definition was used for the luminance:

$$Y' = \max(R, G, B). \tag{15.2}$$

This normalises Co_n and Cg_n to the range $-\frac{1}{2}$ to $\frac{1}{2}$.

The next step in the process is colour segmentation. This detects pixels within a rectangular box within $Y'Co_nCg_n$ colour space. Rather than build a set of comparators for each colour detected, a variation on the lookup table approach of Figure 7.37 is used. The divisions of Eq. 15.1 are relatively expensive (although with the two-phase input clock, a single divider could be shared between the two channels as shown in Figure 4.20). The divisions are avoided using the lookup table to perform the division. This is accomplished by concatenating the Y' with Co and Cg, respectively, as shown on the left in Figure 15.2 and setting the table contents based on Eq. (15.1). Of course, during tuning when the colour thresholds are set, the division is still required to initialise the lookup tables (LUTs).

To minimise the required resources, it was decided to fit both lookup tables in a single dual-port block RAM (Johnston et al., 2005a). On the target FPGA, each block RAM is 4 Kbits in size. To detect four colours in parallel, the memory needs to be 4-bits wide. Therefore, a 4-Kbit RAM allows 512 entries for each table, or 9 address bits to be divided between the Y' and $CoCg$ components. Since colour resolution is more important than precise intensity normalisation, 6 bits were used for each of Co and Cg, and 3 bits were allocated to Y'. Since the paddles are always lighter coloured, the minimum threshold for Y' is always above the 50% level. This allows an effective 4 bits to be used from Y', with the most significant bit used directly to control the AND gate, and the next 3 bits passed to the lookup table (as illustrated on the right in Figure 15.2). While this gives only an approximate result, the 4 bits are sufficient to give some degree of intensity normalisation.

Each coloured object was recognised by finding the bounding box of the corresponding colour label. However, since the bounding box is sensitive to noise, a 3 × 3 erosion filter was used to remove isolated noise pixels. These commonly occur on the boundaries between colours or around specular highlights where the pixel values saturate. The erosion filter is separable, decomposing into a 3 × 1 horizontal filter followed by a 1 × 3 vertical filter. The row buffers are indexed directly by the pixel counter, x, rather than needing an additional counter. Although erosion shrinks the size of each region, it is not necessary to dilate the image again; the erosion does not affect the centre of the bounding box, and if necessary this can be taken into account when interpreting the bounding box dimensions.

The bounding box implementation of Figure 13.2 is used to determine the bounding box of all the labels in parallel. Fabric RAM is used to avoid the explicit need for a multiplexer. With a two-phase clock, only a single-port fabric RAM is required, reducing the resource requirements. Each 4-LUT in the FPGA gives a 16-bit deep RAM. With four bits used to represent the label, one bit for each colour, it is unnecessary to explicitly convert this to binary. The raw 4-bit label is used to directly address the RAM. As described in Section 13.1, each bounding box is augmented with a counter to determine the orientation of long thin objects.

The timing for the coloured region tracking pipeline is shown in Figure 15.3. Signals from the video decoder are valid on the rising edge of the 27-MHz codec produced clock. The clock reference (CREF) signal produced

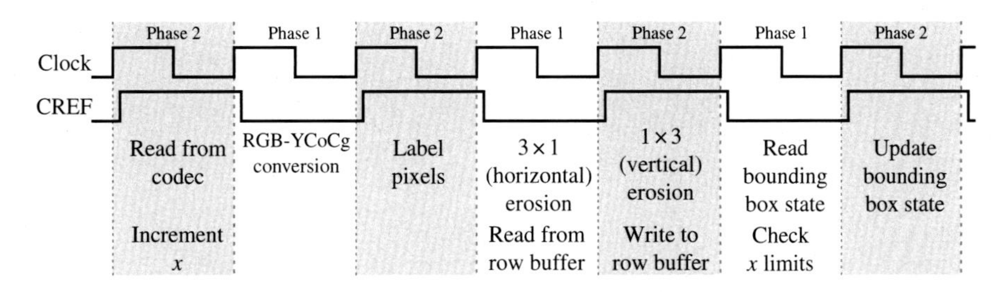

Figure 15.3 Pixel timing for coloured region tracking.

by the decoder is high when pixel data is available (every second cycle). The x register is also incremented in this phase, so that it remains constant between phase 1 when the row buffer and bounding box are read, and when they are written in phase 2.

The pipeline timing in Figure 15.3 shows the pixel processing. Additional processing is performed at the end of each line and frame. After each row, the pixel counter, x, is reset, and the row counter, y, incremented. The state variable used for determining the orientation is also reset. At the end of each field, the data for each bounding box is transmitted to the main clock domain over a channel (all of the image processing is performed directly in the video decoder's clock domain). After transmission, the bounding box data is reset for the next field.

The tracking algorithm locates up to four programmed colours in each field at either 50 or 60 fields per second (depending on whether a PAL (phase alternating line) or NTSC (National Television Standards Committee) camera is connected). All of the processing is performed on-the-fly as the data is streamed in. The only pixel buffering is two block RAMs used as row buffers for the vertical erosion filter. The complete tracking system used less than 10% of the logic resources of the FPGA (and 3 of the 14 block RAMs), leaving the remainder of the system to be used for the application.

15.2 Foveal Sensor

Low-cost high-resolution sensors (up to 25 megapixels and larger) are now becoming commonplace. This presents both a bonus and a problem for image processing. The increased resolution means that more accurate and higher quality results can be produced. However, there is also a significantly increased computational cost to achieve these results, especially when real-time constraints have to be met. Therefore, there is often a trade-off between quality of results and processing requirements.

Real-time processing of single high-resolution images is feasible on an FPGA if stream processing can be used. However, any algorithm that requires multiple frames will require significant off-chip memory and large associated memory bandwidth. Simply reducing the data volume by reducing the resolution can result in a critical loss of information. In many applications (for example tracking and pattern recognition), high resolution is only required in a small region of the image, although a wide field-of-view is important to maintain context. One solution to this dilemma is to use a foveated window. Inspired by the human visual system, this maintains high resolution in the fovea, with resolution decreasing in the periphery. It has been shown that such a variable spatial resolution can reduce the volume of data in tracking applications by a factor of 22 (Martinez and Altamirano, 2006) to 64 (Bailey and Bouganis, 2009b) without severely affecting the tracking performance. Several researchers have investigated foveal mappings using FPGAs.

Traditionally, a log-polar mapping has been used for foveal vision in software to mimic the change of resolution within the human visual system (Wilson and Hodgson, 1992; Traver and Pla, 2003). The log-polar map is both rotation and scale invariant once the fovea is positioned on a common key point. Arribas and Macia (1999) implemented a log-polar mapping using an FPGA that could achieve real-time performance. They represented the mapping as a full lookup table from each input pixel to the corresponding output pixel. The image was mapped from a frame buffer rather than directly as it was streamed from the camera.

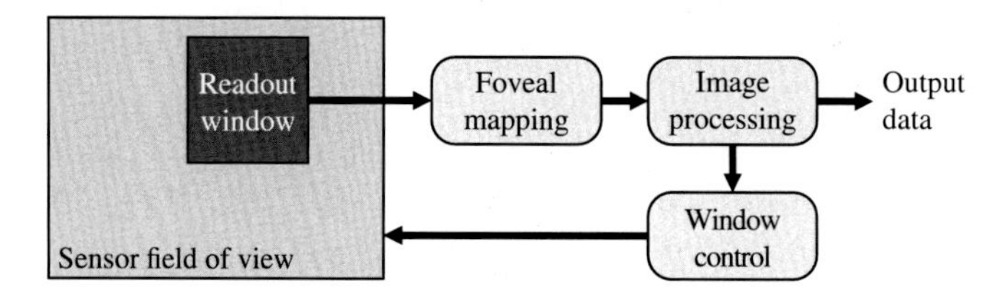

Figure 15.4 Active vision architecture using a foveal mapping.

The complexities of processing a log-polar image have led many researchers to work with Cartesian approximations. Camacho et al. (1998) used a pyramidal-based fovea with each successive level decreasing the resolution by a factor of 2 towards the periphery, giving a form of stepped log-Cartesian topology. Ovod et al. (2005) implemented multiple fovea, allowing multiple objects to be tracked simultaneously, although their system only allowed two levels of resolution: fovea and periphery. Martinez and Altamirano (2006) used high-resolution rectangular sampling within the fovea and approximated a log-polar mapping in the periphery by mapping multiple square rings to a rectangular output image using simple sub-sampling. Bailey and Bouganis (2008) introduced a family of Cartesian foveal mappings, with continuously variable resolution. To minimise latency, the foveal image was mapped on-the-fly as the pixels were streamed from the camera.

An important part of foveal vision systems is the use of active vision techniques (either mechanical or electronic systems) to ensure that the region of the sensor with the high resolution corresponds to the region of the scene where it can be most effective. Active vision requires minimising the latency between image capture and repositioning the fovea, because delays may degrade the performance of the application and limit the stability of this control loop. The lower resolution of the foveal images can significantly increase their processing rate, enabling real-time camera control.

A CMOS (complementary metal oxide semiconductor) sensor with a wide-angle lens enables a wide field-of-view to be captured. CMOS sensors also allow an arbitrary rectangular window of the sensor to be selectively read out; this window may be positioned anywhere within the active area of the camera. Repositioning the readout window from one frame to the next provides the digital equivalent of pan and tilt, without the physical latency and motion blur associated with physically moving the camera. An active foveal vision system based on this mechanism (Bailey and Bouganis, 2008, 2009a,c) is illustrated in Figure 15.4. The pixels within the readout window undergo a foveal mapping to give a reduced resolution image. This image is then processed to extract the required data and to determine the best position to relocate the window to within the CMOS sensor for the next frame.

15.2.1 Foveal Mapping

A separable foveal mapping (Bailey and Bouganis, 2009a) will be used because it is easier to implement, and it has been demonstrated through simulations (Bailey and Bouganis, 2009b) that its performance in tracking applications is similar to more complex radial mappings. An example reverse mapping between a 512×512 window (coordinates (x, y)) and a 64×64 foveal image (coordinates (u, v)) is given by

$$\begin{aligned} x &= u + \frac{7}{32}u^2 \operatorname{sign}(u), \\ y &= v + \frac{7}{32}v^2 \operatorname{sign}(v). \end{aligned} \tag{15.3}$$

Note that due to the symmetry of the mapping, signed coordinates are used, with the origin in the centre of the images (Figure 15.5).

In this application, it is desirable to use a forward mapping for several reasons. First, it saves the need for a frame buffer to hold the much larger input image. Second, the forward mapping enables the image to be transformed on-the-fly as it is streamed from the camera, significantly reducing the latency. Third, the reduction

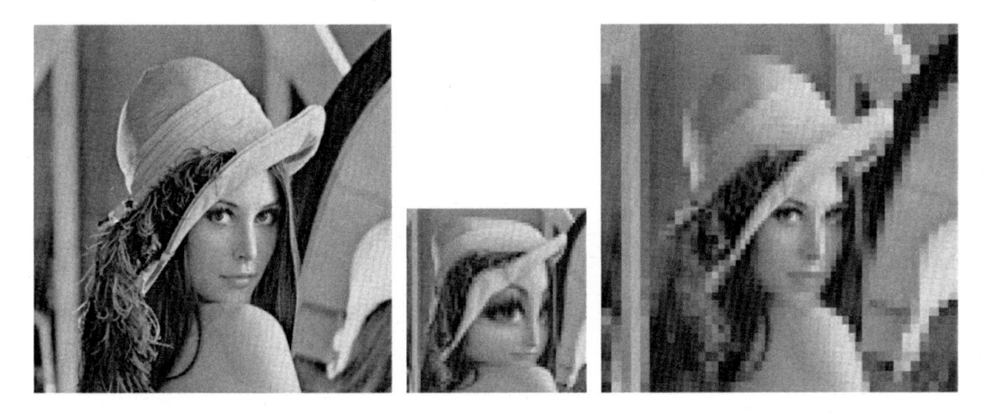

Figure 15.5 The foveal mapping of Eq. (15.3). Left: original 512 × 512 image; centre: after the foveal map to 64 × 64; right: reconstructed to the original size (using nearest-neighbour interpolation), showing the reduction in resolution in the periphery. Source: Reproduced with permission from Bailey and Bouganis (2008) © 2008 IEEE.

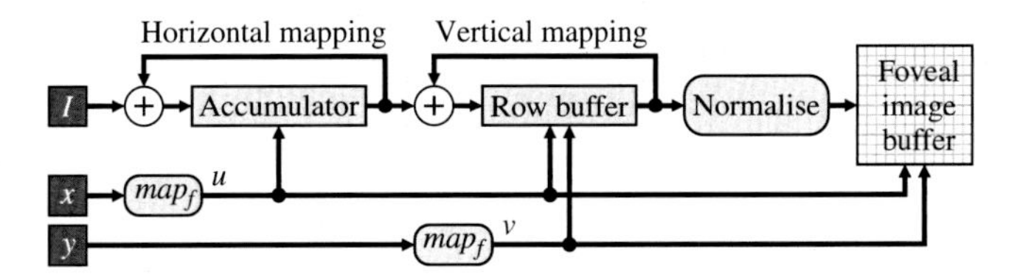

Figure 15.6 Architecture of the separable foveal mapping. Source: Reproduced with permission from Bailey and Bouganis (2009a) © 2009 IEEE.

in resolution in the periphery requires a spatially variant anti-aliasing filter. This can easily be incorporated into the forward mapping as described in Section 10.2.

The basic architecture of the forward mapping is shown in Figure 15.6. In the first stage, incoming pixels are mapped into their correct column and accumulated until the output pixel is completed horizontally. The horizontal pixel is then passed to the vertical mapping, which adds it to the appropriate column accumulator (stored in the row buffer). Once each row is completed vertically, the completed output pixels are normalised (the accumulated total is divided by the area) and saved to the foveal image buffer. Since each output pixel is contributed to by many input pixels, it is necessary for each pixel accumulator, Acc, to consist of two components: one to accumulate the pixel value, $Acc.V$, and one to accumulate the area (or number of pixels), $Acc.A$.

For a continuously variable resolution function, some input pixels will need to be split between adjacent output pixels. Although Figure 15.6 indicates a forward mapping, it is actually more natural to implement this using a reverse mapping function (Bailey and Bouganis, 2009a). The index of the current output pixel being accumulated is looked up in a table to obtain the position of its edge in the input image (this is a reverse mapping). The input pixels can then be accumulated until this edge is reached, with the last fraction of a pixel added before the accumulated pixel is passed to the next stage. Another advantage of using the reverse mapping function is that the output image has fewer pixels, so the table is significantly smaller. The circuit for implementing the foveal mapping is shown in Figure 15.7.

A token passing mechanism is used to indicate that valid data is available. T_{in} indicates a valid incoming pixel.

Consider first the horizontal mapping. The incoming pixel is augmented with 1 (the area of the pixel). If the pixel is not split, it is simply added to the accumulator. A split pixel is detected if the input pixel position

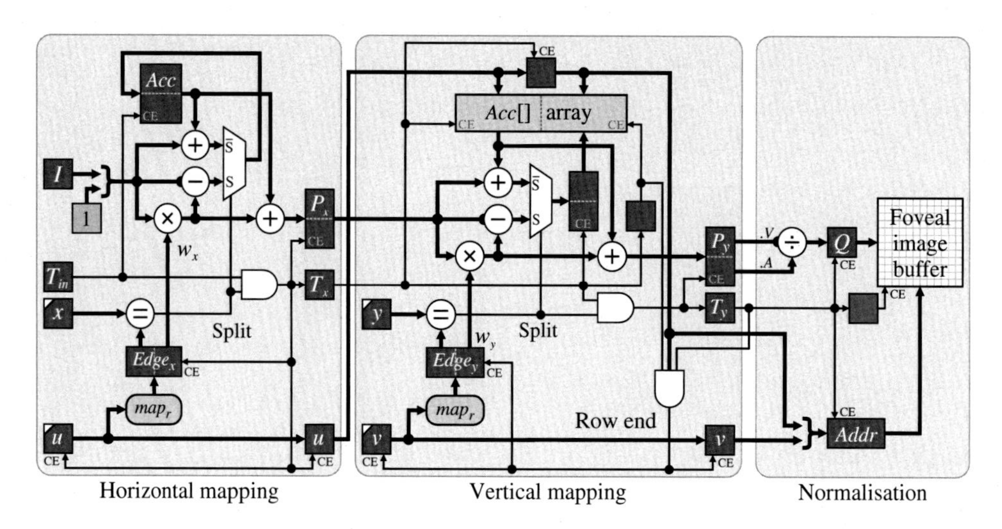

Figure 15.7 Detailed foveal mapping circuit. (CE = clock enable.) Source: Adapted from Bailey and Bouganis (2009a).

matches the integer part of the edge of the output pixel. In that case the fractional part of the edge, w_x, is used to weight the incoming pixel which is combined with the accumulator and passed to the output register P_x. The remainder of the pixel is used to update the accumulator:

$$Acc = \begin{cases} Acc + I, & \text{if not split,} \\ (1 - w_x)I, & \text{if split,} \end{cases} \tag{15.4}$$

and

$$p_x = Acc + w_x I. \tag{15.5}$$

Since w_x is typically only a few bits, it can be implemented efficiently in the FPGA fabric using adders rather than dedicated multipliers. The incoming pixel area is always 1, so a multiplier for that is not required. Whenever a split occurs, T_x is asserted to indicate that a pixel is completed, and the current value of u is registered with it.

The vertical mapping is very similar to the horizontal mapping logic. The difference is that the pixels are accumulated in columns, although the data is still streamed horizontally. Therefore, all columns are mapped in parallel, with the accumulators maintained in a memory-based array. The incoming u selects the accumulator to add the new data to, with the accumulated data written back to the array in the following clock cycle:

$$Acc[u] = \begin{cases} Acc[u] + P_x, & \text{if not split,} \\ (1 - w_y)P_x, & \text{if split,} \end{cases} \tag{15.6}$$

and

$$P_y = Acc[u] + w_y P_x, \tag{15.7}$$

with the corresponding output valid token, T_y. Since the mapping is separable, on a split row, all of the pixels will be split. The v counter is only incremented at the end of split rows, with an AND gate over the bits of the u address used to detect the end of row.

The final processing stage is to normalise the accumulated pixel value by the area to obtain the average input value within the area spanned by each output pixel.

$$Q[u, v] = \frac{P_y.V}{P_y.A}. \tag{15.8}$$

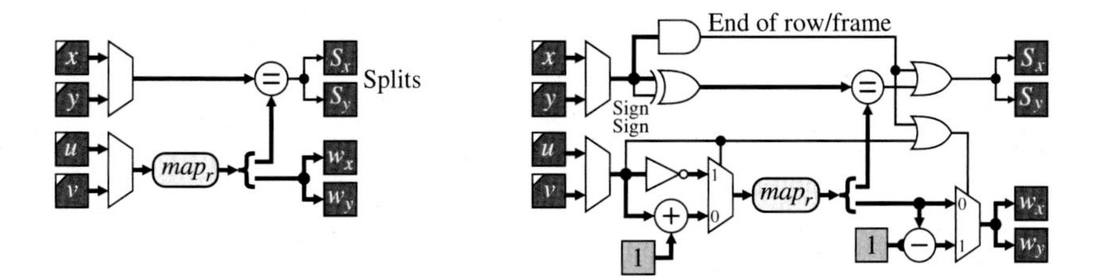

Figure 15.8 Sharing the mapping between the horizontal and vertical mapping sections. Left: simple sharing; right: reducing the map LUT size by mirroring.

The division may be implemented efficiently using eight iterations (for 8-bit output) of a non-restoring division algorithm (Bailey, 2006). If necessary, the division can be pipelined to achieve the desired clock frequency.

If desired, the circuit can be optimised further (Bailey and Bouganis, 2009a). Since the vertical mapping only uses the map lookup table at the end of each row, a single table can readily be shared between the horizontal and vertical mapping sections, as demonstrated on the left in Figure 15.8. The mapping is also symmetrical; therefore, only one half needs to be stored in the table. This requires additional logic to mirror the input and foveal coordinates. The fraction also needs to be subtracted from 1 for the first half of the table to reflect this mirroring. With mirroring, the end of the row is no longer detected by the mapping, so this must be detected explicitly. The mirroring logic is shown on the right in Figure 15.8.

These optimisations reduce the memory required by the lookup table by a factor of 4. An additional benefit is that if the foveal mapping itself is changed dynamically, it only needs to be changed in one place.

15.2.2 Using the Sensor

The primary applications of a foveal sensor are in pattern recognition and tracking. In this section, the use of the sensor for object tracking will be described briefly. When tracking an object, it is desired to centre the fovea on the target as it moves through the scene. The following stages are required within any tracking system (Bailey and Bouganis, 2009b):

1. The high-resolution fovea is positioned on the expected position of the target within the image. Precise positioning is not essential because the lower resolution periphery enables a small image to be used while still maintaining a wider field-of-view.
2. The target is detected within the foveal image. Any standard object detection technique, such as filtering, thresholding, or colour detection, can be used. Processing is relatively quick because of the small image size.
3. The true location of the target is estimated, based on the fact that the image is distorted. Note that standard correlation and registration-based techniques will give a biased result because of the distortion introduced in the mapping. However, for a given mapping, this bias may be estimated and compensated for. If using the centre of gravity, it is necessary to take into account the size and position of each pixel and weight accordingly (Bailey and Bouganis, 2009b).
4. The detected target is fed into a tracking system (for example a Kalman filter (Lee and Salcic, 1997; Liu et al., 2007)), and the next location of the target is predicted. This is then used to reposition the window for the next image captured.

Using a foveal sensor has a number of advantages over a conventional sensor in this application. It allows a lower resolution image to be processed, which allows more complex processing with reduced latency. For tracking, the latency requirements make it important that the input image is converted to a foveal image on-the-fly. Multiple targets can be tracked by alternating the fovea between the targets with successive images captured (this reduces the effective frame rate for each target).

15.3 Real-time Produce Grading

This application will describe a machine vision system for the grading of asparagus by weight. The goal was to estimate the weight of each spear and perform simple quality grading at up to 15 pieces per second. While the original algorithms (Bailey et al., 2004; Bailey, 2012) were written in C and executed sufficiently quickly on a desktop machine for real-time operation, this application would be an ideal candidate for FPGA implementation within a smart camera.

While there is significantly more to a machine vision application than just the image processing, the mechanical handling, lighting, and image capture arrangement are beyond the scope of the discussion here. The focus here will be solely on implementing the image processing algorithm on an FPGA; the other aspects of the project are described in more detail by Bailey (2012). The basic principle behind estimating the weight is to treat each asparagus spear as a generalised cylinder and estimate its volume by measuring its projected diameter in two perpendicular views. The weight is then estimated as

$$Weight \propto \sum_{x} D_1^2[x] + D_2^2[x], \tag{15.9}$$

where x is the pixel position along the length of the spear, and D_1 and D_2 are the diameters as a function of length in the two views. The constant of proportionality is determined through calibration. The long, thin nature of asparagus enables multiple views to be captured in a single image using a series of mirrors to divide the field-of-view.

15.3.1 Software Algorithm

The structure of the software algorithm is shown in Figure 15.9 along with key images at various stages. Images are captured asynchronously, triggered by each cup (which carries the asparagus) passing over a sensor. Each cup has a sequential identifier, with cup 0 causing an indicator LED to light up (which is used in software to reset the cup counter). The first processing steps were to check for the presence of this indicator and for the presence of a spear on the cup. The spear was detected by calculating the mean and standard deviation within the one-pixel wide window indicated in image ① in Figure 15.9. If no spear was present, no further processing was performed for that image.

If a spear was detected, the exposure is checked by building a histogram of the input image and determining the pixel value associated with the 99.6th percentile, essentially using the algorithm described in Section 8.1.4. This was used to automatically adjust the exposure by a small increment.

The contrast is enhanced using a lookup table (see image ②). Specular highlights were filtered out by a 9×1 horizontal greyscale morphological closing. To segment the spear from the background, an adaptive threshold was used. Within each view, the background intensity was estimated by averaging the pixels in each column. This background was subtracted from each pixel in the column, and the contrast expanded using

$$\hat{p}[x,y] = \begin{cases} 0, & p_i[x,y] \le \mu_i[x], \\ \dfrac{p_i[x,y] - \mu_i[x]}{255 - \mu_i[x]}, & p_i[x,y] > \mu_i[x], \end{cases} \tag{15.10}$$

where p_i is a pixel value in the ith view, and μ_i is the corresponding column mean. A binary image of the spear was obtained by thresholding this image with a constant threshold level. The binary image was then cleaned using a series of binary morphological filters: a vertical 1×9 opening was followed by a horizontal 5×1 opening and a 5×1 closing. The pixels in each column were counted for two perpendicular views to enable the volume to be estimated using Eq. (15.9).

Finally, the quality was graded using four simple tests. First, the ovality of the spear was estimated by measuring the diameter at the base of the spear in each of the three views. The ovality ratio was the minimum diameter divided by the maximum diameter, after compensating for the change in magnification in the centre view. Second, the angle of the base was checked in each of the three views for the presence of a field cut, with

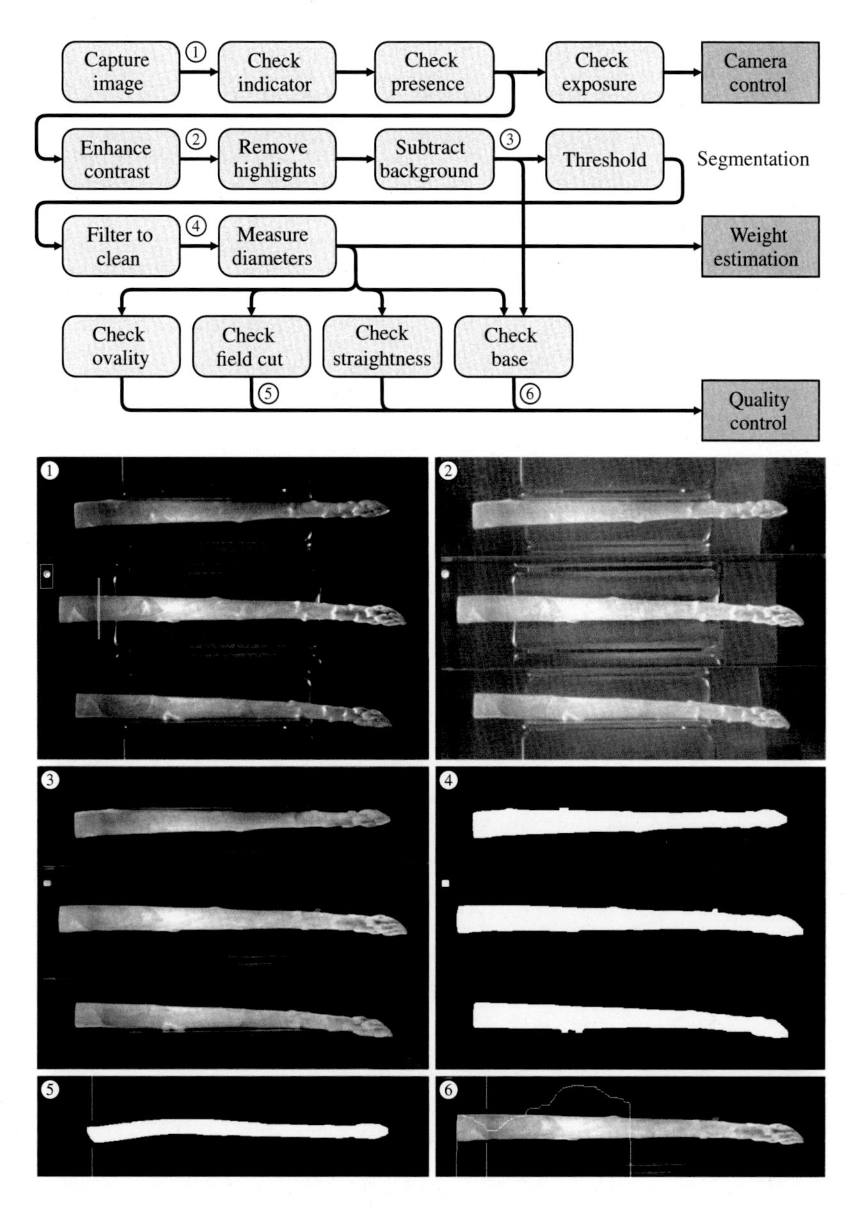

Figure 15.9 Top: software machine vision algorithm for grading asparagus. bottom: images at various stages through the grading algorithm: ① captured image with indicator region and asparagus detection window overlaid; ② after contrast enhancement; ③ after background subtraction; ④ after thresholding and cleaning; ⑤ side image showing field cut at base; ⑥ detecting any white stalk at the base of the spear.

the pixel position required to remove the cut determined. A diagonal field cut is clearly seen in image ⑤ of Figure 15.9. Third, the straightness of the spear is evaluated by measuring the variance of the spear centreline in the y-coordinate. Fourth, the presence of a white base is detected in the central view. The greyscale image is filtered horizontally with an 11×1 minimum filter before reducing the intensity profile to one dimension by calculating the median of each column. This profile is then checked for a dip, indicating the presence of a white base, as illustrated in image ⑥ of Figure 15.9.

15.3.2 Hardware Implementation

Several design optimisations and simplifications may be made to exploit parallelism when mapping the algorithm to hardware. Only the image processing steps for the weight estimation and the filtering required for checking the base of the spear will be implemented in hardware. The output from measuring the diameter is a set of one-dimensional data, consisting of the edge positions of the spear in each view. The processing for the base check is a 1-D intensity profile from the central view. The quality control checks only need to be performed once per frame and are more control oriented, so can be implemented in software using the preprocessed 1-D data.

To minimise the memory requirements, as much as possible of the processing will be performed as the data is streamed in from the camera. However, since the processing consists of a mix of row- and column-oriented processing, a frame buffer will be necessary to hold each view. The detailed implementation of the algorithm is shown in Figure 15.10 and is described in the following paragraphs.

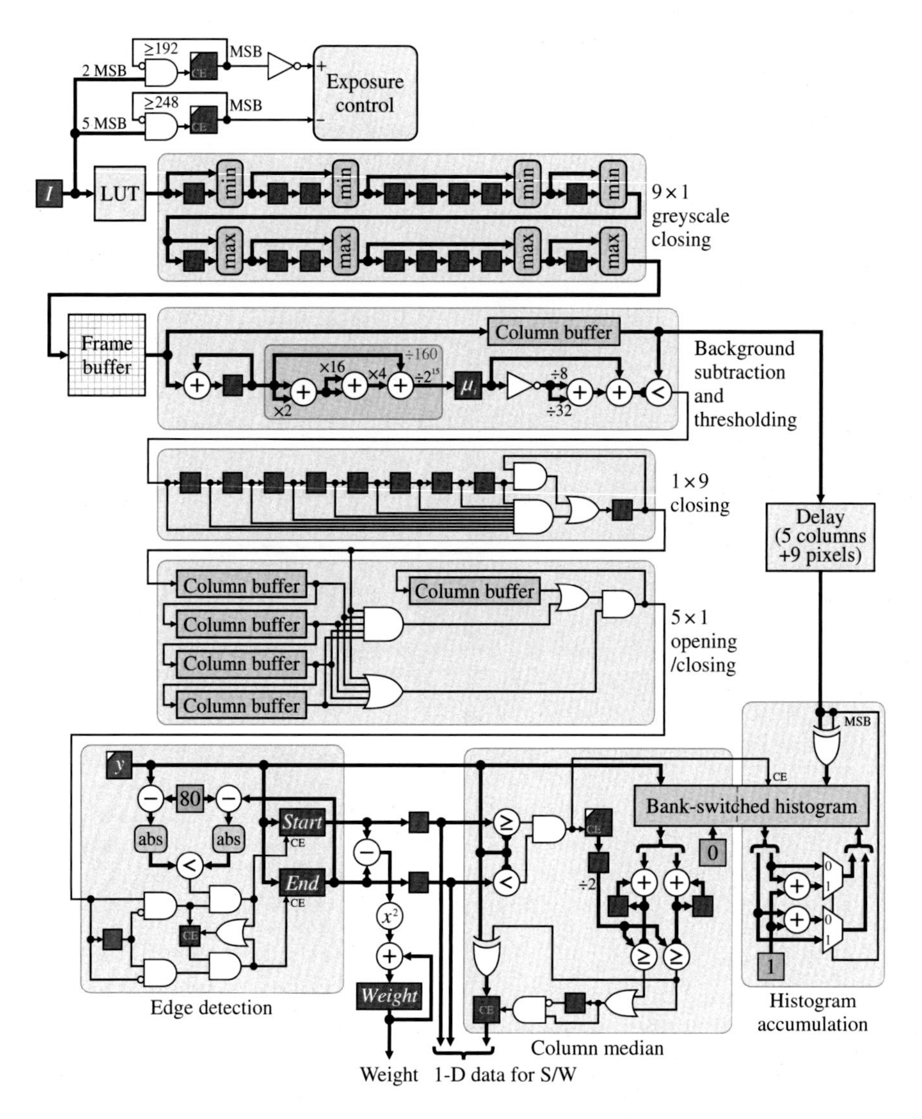

Figure 15.10 Schematic of the FPGA implementation. (CE = clock enable; MSB = most significant bit.)

Of the initial processing checks, the check for the indicator LED is no longer necessary. Since the images are processed directly as they are streamed from the camera, this signal can be fed directly to the FPGA rather than first capturing it via the image. The check for the presence of an asparagus spear will need to be deferred because it relies on the middle view. Processing will need to begin assuming that a spear is present and discard the results if it is later determined that no spear is present. For exposure control, rather than building a histogram to determine whether the 99.6th percentile is within a specified range, the processing can be simplified by checking whether the incoming pixels are within the range and accumulating them accordingly. In software, the range was set to between 192 and 250. By adjusting the threshold levels to 192 (binary '11000000') and 248 (binary '11111000'), comparison of the pixel value with the threshold levels reduces to ANDing the upper two and upper five bits, respectively, of the pixel value. These signals are used to enable the clocks of two counters to count the pixels above the respective thresholds. The 99.6th percentile is somewhat arbitrary; it is equivalent to two lines of pixels in the 640×480 image. Adjusting this to 1024 (a power of 2) simplifies the checking logic; if the most significant bit (MSB) is set, this disables the counter, effectively making it saturate. At the end of the frame, if a spear is present, a simple check of the MSB of the counters can be used to increment or decrement the exposure.

The contrast enhancement used a fixed lookup table in software. The incoming stream may be processed in exactly the same way, with a lookup table implemented using a block RAM. The 9×1 greyscale closing may be pipelined from the output of the lookup table. The erosion and dilation are implemented using the filter structure from Figure 9.48. Subsequent processing requires a column of data to remove the average. A bank of 160 row buffers could be used, but to fit this on-chip would require a large FPGA. Therefore, the image is streamed to an off-chip frame buffer. To enable continuous processing, the frame buffer would need to be bank-switched. Alternatively, the RAM could be clocked at twice the pixel clock (which is only 25 MHz for VGA (video graphics array) resolution), or two pixels could be packed into each memory location, although this would require a row buffer on the input or column buffer on the output.

After each image panel has been loaded into the frame buffer (160 columns), the column processing may begin. This streams the data out of the frame buffer by column for each third of the image. The background subtraction and thresholding for the adaptive threshold can be considerably simplified. To calculate the column average, the pixel data is accumulated, and at the end of the column, the total is divided by 160 and clocked into the mean register, μ_i. This division by a constant may be implemented efficiently with three adders (multiplying by $205/2^{15}$ and eliminating a common sub-expression; $205 = 12 \times 17 + 1$) as shown in Figure 15.10. Since only a binary image is required, the normalisation division of Eq. (15.10) is avoided by rearranging as

$$\frac{p_i[x,y] - \mu_i[x]}{255 - \mu_i[x]} > Thr,$$
$$p_i[x,y] > \mu_i[x] + (255 - \mu_i[x])Thr. \tag{15.11}$$

The subtraction from 255 is simply a one's complement, and multiplication by the constant threshold ($Thr = \frac{40}{256}$) may be implemented by a single addition ($\frac{40}{256} = \frac{1}{8} + \frac{1}{32}$). The pixels for each column (delayed by a column buffer) are compared with this adaptive threshold to give the binary image.

The presence of a spear may be checked by testing μ_i in the appropriate column of the central panel. If it is above a predefined threshold, then a spear is present (this is not shown in Figure 15.10).

The binary morphological filters for cleaning the image may also be significantly simplified because they are one dimensional. The 1×9 vertical closing is normally implemented as an erosion followed by a dilation. The 9-input AND gate performs the erosion and sets the output register when nine consecutive '1's are detected. The output is held at '1' until the '0' propagates through to the end of the chain; this is effectively performing the dilation, using the same shift registers.

Rather than use another frame buffer to switch back to row processing, the 5×1 horizontal cleaning filters are implemented in parallel using column buffers to cache the previous columns along the row. Rather than perform the opening and closing as two separate operations, both operations are performed in parallel: a sequence of five consecutive '1's will switch the output to a '1', and a sequence of five '0's will switch the output to a '0'. While this filter is not identical to that used in software, it has the same intent of removing sequences of fewer than five '1's or '0's.

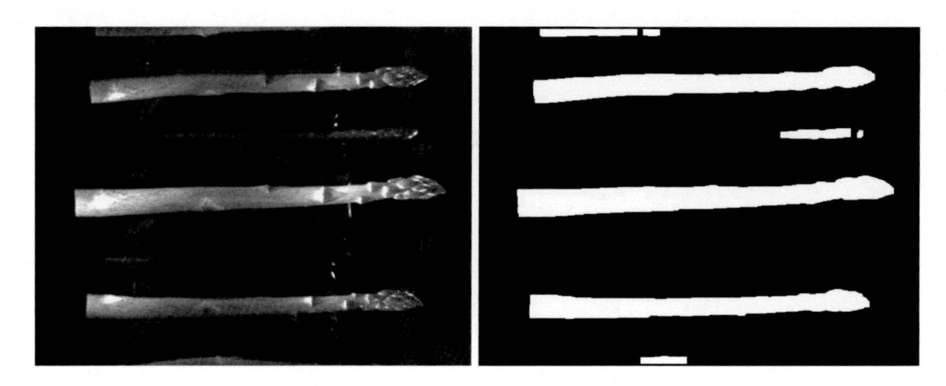

Figure 15.11 Spears from adjacent cups are sometimes visible and detected.

The next step is to determine the row numbers corresponding to the top and bottom edges of the asparagus spear in each column. A binary transition detector is used to clock the row counter, y, into the *start* and *end* registers. This is complicated slightly by the fact that occasionally spears in adjacent cups are also detected near the edge of the image, as illustrated in Figure 15.11. The region detected near the centre of the panel corresponds to the correct region to be measured. A second detected region replaces the first only if it is nearer to the centre of the panel, that is if

$$|y - 80| < |end - 80|. \tag{15.12}$$

At the end of the column, the edge values are transferred to a pair of output registers. The squared difference is also added to the weight accumulator only for the side images, evaluating Eq. (15.9) for the current spear.

If desired, additional accumulators could be used to calculate the column variance for each view for evaluating the spear straightness. This would require two accumulators and a counter, and a squarer and a division:

$$straightness = \frac{\sum (start + end)^2}{\sum 1} - \left(\frac{\sum(start + end)}{\sum 1}\right)^2. \tag{15.13}$$

Since the data is accumulated one item per column, if necessary, the squarer and division can use a sequential algorithm over several clock cycles (and be shared between the parts of the calculation).

The final processing step is to calculate the median of the asparagus pixel values in each column of the central view. The greyscale values are delayed to compensate for the latency of the filters and the edge detection block. The most efficient way of calculating the median of a variable number of pixels is through accumulating a histogram and finding the 50th percentile. Only pixels within the spear are accumulated by comparing the row counter, y, against spear edges to enable the clock of the histogram accumulator. A counter also counts the pixels which are accumulated. When the column is complete, the histogram is summed to find the 50th percentile. This requires two histogram banks to be used; while the median is being found from one bank, the other bank is accumulating the histogram of the next column.

The problem here is that only 160 clock cycles are available for each column, requiring two bins to be summed every clock cycle. In the implementation in Figure 15.10, two histogram bins are stored in each memory location. Address n holds the counts for histogram bins n and $255 - n$. When accumulating the histogram, if the MSB of the pixel value is set, the least significant bits are inverted to get the address. The MSB also controls the multiplexers which determine which bin is incremented. When determining the median, the two bins are accumulated separately, using the y counter as the address until half the total count is reached. This effectively scans from both ends of the histogram in parallel until the 50th percentile is located. If the upper bin is reached first, y is inverted to obtain the correct median value. While scanning for the median, the write port is used to reset the histogram count to zero to ready it for the next column.

Not shown in Figure 15.10 is much of the control logic. Various registers need to be reset at the start of the row (or column or image), and others are only latched at the end of a column. In controlling these, allowance must also be made for the latencies of the preceding stages within the pipeline. It is also necessary to account for the horizontal and vertical blanking periods when controlling the various modules and determining the latencies. Also not shown are how the outputs are sent to the serial processor. This depends on whether the processor is implemented on the FPGA or is external. However, because of the relatively low data rate (three numbers per column), this is straightforward.

While quite a complex algorithm, it can still readily be implemented on an FPGA by successively designing each image processing component and piecing these together. A hardware implementation has allowed a number of optimisations that were not available in software, including the simplification of several of the filters. This example demonstrates that even quite complex machine vision algorithms can be targeted to an FPGA implementation.

15.4 Stereo Imaging

Stereo imaging is one method of measuring 3-D depth or range information. One of the motivating applications is its use by autonomous vehicles for the real-time extraction of range data for obstacle detection and collision avoidance. It is also commonly used in mobile robotics for determining the 3-D environment in front of, and around, the robot, and this is the focus of this application.

Stereo vision works by capturing images of a scene from two spatially separated cameras and uses the disparity or difference in relative position of an object within the two images to calculate the depth. By finding the point in each image corresponding to an object, the rays are back-projected through the camera centres to determine the 3-D location of the point (at least relative to the cameras). This is illustrated in Figure 15.12 for canonical parallel-axis stereo vision, where the principal axes of the cameras are parallel to each other and perpendicular to the baseline. Let two cameras, both with effective focal distance, f, be positioned at O_L and O_R, separated by the baseline, B. The image of object point, P, is found in both images, with disparity d. Assuming a pinhole camera model, the depth or range of the object point from the baseline can therefore be found through similar triangles to be

$$D = \frac{Bf}{d}. \tag{15.14}$$

Note that if the principal axes of the camera do not conform to the geometry of Figure 15.12, then each of the images can be transformed to match that geometry; this process is called stereo rectification.

Of importance is the depth resolution, which is the smallest distance that can be distinguished between two objects in the scene. If the minimum disparity that can be reliably detected between the images is p, then the range resolution may be derived from Eq. (15.14) as

$$\Delta D = \frac{Bf}{d} - \frac{Bf}{d+p} = \frac{Bfp}{d(d+p)} \approx \frac{Dp}{d} = \frac{D^2p}{Bf}, \tag{15.15}$$

since $p \ll d$ at the working range. This implies that the range resolution deteriorates significantly with range, which is usually workable, since there is more time to take evasive action if an obstacle is in the distance. The

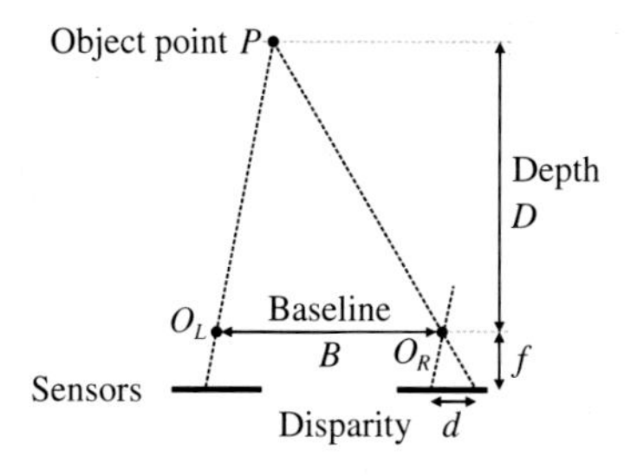

Figure 15.12 Principle behind stereo imaging. The image of object point, P, is back-projected enabling the range to be calculated.

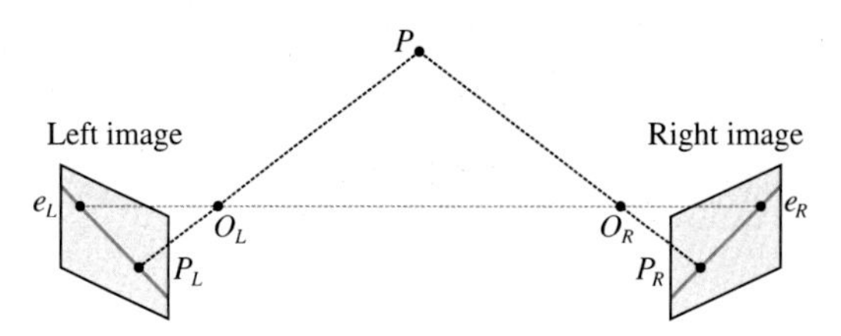

Figure 15.13 Epipolar geometry constrains matching points to lie on a line.

range resolution improves with an increased baseline separation between the cameras. However, an increased baseline separation reduces the overlap within the field-of-view of the two cameras, making it harder to match corresponding points. The change in appearance between the two views of an object can also pose a challenge for matching.

The target application in this section is a vision system for a mobile indoor robot with the following specifications. A 90° horizontal field-of-view is required, with a working range between 500 and 5000 mm. At a range of 1000 mm, the depth resolution needs to be 10 mm. The sensors within the cameras have a resolution of 800 × 600 pixels with 5-μm pixels.

To achieve a 90° field-of-view with the specified sensors requires using 2-mm focal length lenses. To maximise the focus over the working range, each lens should be focussed at about 900-mm distance. A lens with $f/2$ aperture has more than adequate depth of field to cover the working range (using Eq. (1.4), the circle of confusion is significantly smaller than one pixel through the whole working range). From Eq. (15.15), and assuming that the minimum disparity is one pixel, the baseline needs to be 250 mm to achieve the specified depth resolution. With this arrangement, the disparity will vary from 20 pixels at 5000 mm (with a depth resolution of 250 mm) up to 200 pixels at 500 mm (with a depth resolution of 2.5 mm).

Two approaches can be used to find the correspondences between images. The first is to detect distinctive feature points in each image (see Section 10.4.1) and match these features between the images. The resulting sparse set of matches, when back-projected, give a 3-D point cloud. This approach is commonly used during camera calibration (Pettersson and Petersson, 2005), where a distinctive target pattern is used to make feature points easy to find and match.

The second approach is to match small patches of pixel values between the pair of images (see Section 10.4.2). By matching a patch centred at each pixel, a range measurement may be made for each pixel, giving a dense range image. The search for matches in a pair of stereo images may be simplified by exploiting epipolar geometry, as demonstrated in Figure 15.13. When two cameras image an object point, the camera and object locations define a plane in 3-D space, and all points on that plane will be imaged to a single line in each image (the epipolar lines). This restricts the search for the corresponding point to a 1-D search along these epipolar lines. The epipolar constraint can then be represented through the fundamental matrix, **F**, as (Hartley and Zisserman, 2000)

$$\mathbf{P}_L^\mathsf{T}\mathbf{F}\mathbf{P}_R = 0, \tag{15.16}$$

where $\mathbf{P}_L$ and $\mathbf{P}_R$ are the image points in the left and right images, respectively, represented using homogeneous coordinates. All epipolar lines pass through the epipoles, e_L and e_R, which are the projections of the other camera centre onto the respective image planes. Note that any image distortion will make the epipolar lines curved.

15.4.1 Rectification

To further simplify the search, the images can be rectified to align the epipolar lines with the image scan-lines. Even if the cameras are set up according to the canonical geometry, there will inevitably be minor physical misalignment between the cameras. Any lens distortion will cause the cameras to deviate from the pinhole model, both complicating the search and also reducing the accuracy of the resultant range measurements.

Camera calibration determines the relationship between the cameras and the geometric transformation for each camera required to:

- correct lens distortion to ensure that the geometry follows the pinhole model;
- rotate each camera viewpoint, so that it is perpendicular to the baseline and makes the principal axes parallel; and
- rotate each camera about its principal axis, so that the corrected scan-lines correspond to epipolar lines.

The geometric transformations are then used to rectify the images before stereo matching.

For stereo imaging, the cameras can be aligned reasonably accurately, especially if using a grid target while setting up the cameras. For 0.5° of misalignment, the errors are on the order of 7 scan-lines ($800 \times \tan 0.5°$). (While this calculation is for camera roll, yaw and pitch will give similar errors.) A reasonable quality lens will also have a worst case distortion of about five to six scan-lines. Therefore, using the reverse mapping architecture of Figure 10.3 with a buffer of 16 lines should be adequate to correct the distortion as the images are streamed from the camera. Using a power of 2 significantly simplifies address calculations for both reading and writing. Lines are buffered storing odd and even rows and columns in separate memory blocks (M_{ee}, M_{eo}, M_{oe}, and M_{oo} in Figure 15.14) to enable four pixels to be read in parallel for bilinear interpolation.

The distortion map is usually quite smooth, so rather than model it analytically, the actual distortion map is represented non-parametrically, and down-sampled by a factor of 16, like Akeila and Morris (2008). This reduces the size of the mapping from 800×600 down to 50×39 samples. Also, rather than save the actual mapping, the relative mapping is used:

$$m_{rx}[u, v] = u - m_{rx}(u, v), \qquad m_{ry}[u, v] = u - m_{ry}(u, v) \tag{15.17}$$

since it requires fewer bits to represent. The mapping is split into two tables, as shown in Figure 15.14. First, a 1-D interpolated lookup table is used to obtain m_{rx} and m_{ry} for the first column ($u = 0$). In parallel, a 2-D interpolated lookup table obtains the differential offsets. This stores

$$dx = m_{rx}[u, v + 16] - m_{rx}[u, v], \tag{15.18}$$

and similarly for dy, which is interpolated vertically to give the increment for the current row. The offset is then computed incrementally for each output by adding dx and dy into the respective accumulators. The 2-D table is

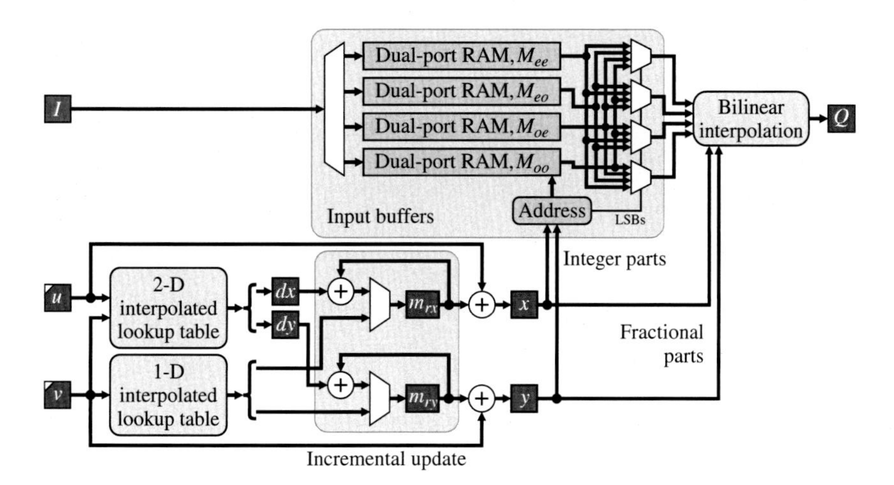

Figure 15.14 Rectification of the input image stream. Interpolated lookup tables give the mapping offset. Direct interpolation is used vertically, and incremental update is used horizontally. The integer parts access the pixel values within the 2×2 window, while the fractional parts perform bilinear interpolation. (LSB = least significant bit.)

looked up every 16 pixels along the row to give the incremental updates for each tile. Finally, u and v are added back in to give the source address for the output pixel. The integer parts are used to access the pixels in 2×2 window, with the fractional parts used to perform the interpolation.

15.4.2 Calculating the Depth

From Eq. (15.14), the depth is inversely proportional to the disparity. Therefore, rather than actually calculate the depth, the disparity is often output directly (from which the depth can be derived if necessary).

Several different stereo matching algorithms can be used; a selection of those based on an FPGA implementation will be reviewed here. Most algorithms can be divided into four (not necessarily distinct) steps (Scharstein and Szeliski, 2002): computing the matching cost, aggregating the cost within a region of support, optimising to compute the disparity, and refining the disparity.

Several different cost matching functions can be used. On an FPGA, the sum of absolute differences (SAD) is common because it is easy to calculate, but other related matching functions are the sum of squared differences and normalised cross-correlation. Aggregation is often performed directly by accumulating the differences within a window. Using a larger window generally gives better match accuracy but also runs into problems on the best way to reuse data once it is loaded onto the FPGA. For a $W \times W$ match window and D disparity levels, each pixel is involved in W^2D difference calculations, of which W^2 calculate the same difference. Therefore, the issue becomes one of design-space exploration, trading off memory for storing differences or partial sums of differences, with logic required for recalculation (Shan et al., 2012; Colodro-Conde et al., 2014). Colodro-Conde et al. (2014) showed that reusing column sums significantly reduces both logic and memory requirements. One variation is to preprocess the image by applying a Sobel filter to detect edges before matching. Ttofis et al. (2013) thresholded the edge image, making it binary, so that the absolute difference is reduced to an exclusive-OR. SAD was then used with an 11×11 window, with 128 disparity levels.

Masrani and MacLean (2006) used phase correlation for disparity estimation. They assumed that the disparity did not change significantly between successive frames and used a multi-resolution search centred on the disparity detected in the previous frame to update the disparity. This significantly reduced the search space but also limited the maximum velocity towards or away from the camera per frame.

Another common cost function for FPGAs is the census transform (Colodro-Conde et al., 2014), which assigns each pixel a binary code based on the sign of the differences of neighbouring pixels relative to the centre pixel. The similarity between two pixels is then calculated based on the Hamming distance between their binary codes. This takes the difference between two codes (using an exclusive-OR) and counts the differences. Related to the census transform is the rank transform (Zabih and Woodfill, 1994), which assigns a value to a pixel based on its rank within its local neighbourhood. The difference between rank values is then measured using SAD. One advantage of the census and rank transforms is their relative insensitivity to light levels and noise because they do not use the actual pixel values but instead work on an ordering of the pixel values. Longfield and Chang (2009) described a system using the census transform, parameterised by image size, window size, and number of disparities. All of the disparities are calculated in parallel, choosing as output the match with the smallest Hamming distance (winner takes all). Devy et al. (2011) used a 7×7 window for the census transform, with 64 disparities, using a comparator tree to find the best match. The disparity maps produced by the census transform often have distinctive point-noise errors, which can be removed by post-processing with a median filter (Colodro-Conde et al., 2014).

Michalik et al. (2017) took a hybrid approach and used both the absolute difference of the Sobel filtered image (using 4 bits) and a sparse census transform (eight pixels within a 5×5 window). Differences were accumulated using a 17×17 window. To reuse hardware, a higher clock frequency was used internally; however, details are light on how they applied hardware parallelism. Wang et al. (2013) also combined the absolute difference and census transform matching.

The local matches effectively form a 3-D volume of the matching costs over all pixels and disparities. The goal then is to select the true matches from this cost space. Neighbouring pixels or regions in the image are likely to have similar disparities, so averaging the matching costs of neighbouring regions can help to reduce noise and errors in the disparity map. Commonly, a simple box filter is used, although more complex techniques use a weighted average (for example a bilateral filter) or variable-sized cross-shaped windows and other adaptive support-weighted averaging. For example, Wang et al. (2013) used cross-based aggregation to smooth the results of a combined absolute difference and census transform.

Disparity optimisation methods can be divided into local and global methods. Locally, each pixel selects the disparity with the lowest cost (winner takes all) for the reference pixel (Scharstein and Szeliski, 2002). A limitation of this is that several reference pixels may select the same target pixel. One extension is to enforce the consistency of mapping between the left and right images rather than a one-way map. A left/right consistency check matches both from the left image to the right image and vice versa, discarding a match if it is not the optimum in both directions. For example, Devy et al. (2011) used a left/right consistency check to validate matches.

Global optimisation optimises the costs globally (often using smoothness constraints) to give the best match for all pixels. However, global methods are quite computationally expensive and may not scale well for larger images. Other optimisation techniques, such as graph-cut-based or belief propagation-based methods, can also be used.

Dynamic programming is one approach of optimising each row of the disparity map. Usually the absolute difference is used directly with dynamic programming, although a sum using a small window could be used. Dynamic programming considers all possible paths through the disparity map for an image row, accumulating the cost from one side of the image to the other. Then the minimum total cost path is selected and backtracked to give the optimum disparity map for the image row. The absolute difference is used as the cost of a match, with an additional penalty for occlusions (where a point is visible in only one image). Calderon et al. (2008) accelerated the dynamic stereo calculations by reading and writing from 16 memory banks in parallel for optimising over 16 disparity levels. Morris et al. (2009) used dynamic programming to produce a symmetric stereo match, with the final disparity map viewed from mid-way between the two cameras. Taking a symmetric approach simplifies the state transitions between pixels which are visible on only the left image, both images, and the right image, so that only 3 bits need to be stored per disparity for the backtracking (Morris et al., 2009). The main limitation of row-based dynamic programming is that each scan-line is processed independently, and is not always consistent from one row to the next, giving distinctive horizontal streaks. These can be reduced by post-processing using a vertical median filter (Shi and Gimel'farb, 2000).

Semi-global matching estimates a globally smooth map by penalising changes in disparity. Rather than applying the smoothness globally, this is approximated by averaging over lines in several directions (typically 16) (Hirschmuller, 2008). On an FPGA though, four directions are commonly used to make the processing easier (Wang et al., 2013; Cambuim et al., 2017). Cambuim et al. (2017) used a systolic array for the path calculations and also found that the matches were improved by preprocessing the input images with a Sobel filter. Wang et al. (2013) found that the semi-global matching gave significant improvements in the match quality but required careful buffer design to reuse data, so that it would not be the main processing bottleneck.

Another approach to global matching is to use a cross-tree representation (Zha et al., 2016). However, to reduce memory, Zha et al. (2016) split the image into 128×128 tiles for processing (although this limits the disparity range). Jin and Maruyama (2012) used a tree-based dynamic programming method, which enforces 2-D consistency of the disparity map.

Finally, disparity refinement cleans up the resulting disparity map, filtering noise or filling in missing points and occlusions, or using sub-pixel techniques to give a finer disparity map (Scharstein and Szeliski, 2002). In recent years, deep learning methods have increasingly been applied to the stereo correspondence problem (Zhou et al., 2020), with success rivalling the traditional approaches. The main limitation, as was seen in Section 14.4, is that deep learning approaches are quite resource hungry, currently making them less suited for FPGA implementation, although this may change as research in this area continues.

15.4.3 Stereo Matching Design

For the mobile robotics application in this section, the census transform approach will be used, because it is well suited to hardware implementation. It also allows a single window to be applied to each input stream, as demonstrated in Figure 15.15, rather than one window calculation for each disparity. There is a trade-off

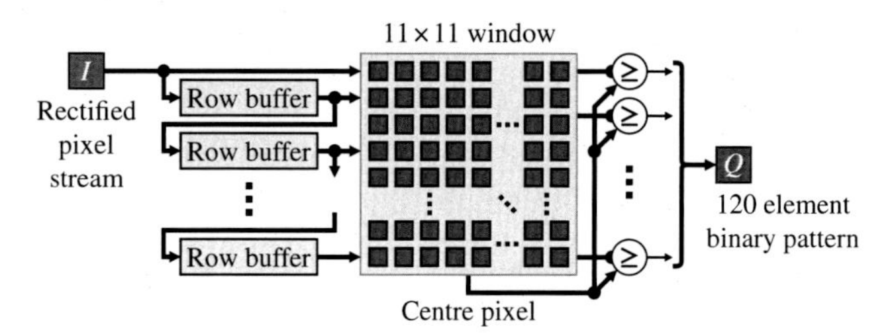

Figure 15.15 Census transform compares each window pixel with the window centre.

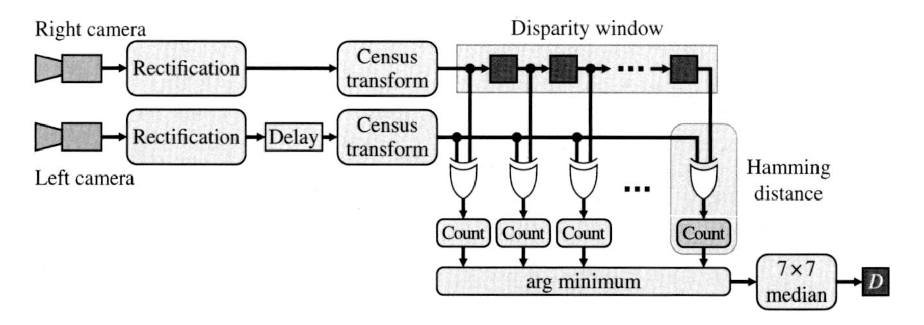

Figure 15.16 Schematic of the full stereo imaging system.

between accuracy (using a large window and more resources) and retention of fine depth features (requires a small window). A compromise was chosen, making the census window 11 × 11 pixels, producing a 120-bit binary pattern.

The resultant stereo imaging system is shown in Figure 15.16. The two cameras, separated by 250 mm, are synchronised so that they capture the scene at the same time. The pixel stream from each camera is rectified to correct for lens distortion and any small camera misalignments, so that the resultant scan-lines correspond to epipolar lines. The census transform is applied to the rectified image streams from each camera. The stream from the left camera is delayed by the width of the disparity window to allow the disparity window to fill up.

The Hamming distance is calculated in parallel between the left census pattern and each of the patterns from the right camera in the disparity window. Each difference is calculated using a 120-bit wide exclusive-OR gate and counting the resulting number of '1's. The counters are implemented using a pipelined adder-tree structure to minimise the propagation delay. Although there could be up to 120 '1's, requiring 7 bits to be used, in practice, any counts over 15 represent poor-quality matches. Therefore, 4-bit saturating counters are sufficient for these. The next stage is to determine which disparity has the lowest cost for the window in the left image. This is accomplished by the arg minimum function, which is also implemented using a pipelined tree structure to minimise the propagation delay. When there is more than one disparity with the same minimum, the largest disparity is chosen (corresponding to a point closer to the camera). Limiting the cost to 4 bits also significantly reduces the logic required to determine the minimum. The output is 8 bits representing the disparity, ranging from 0 (at infinite distance) up to 200 (corresponding to a range of 500 mm).

Finally, the output disparity map is postprocessed using a 7 × 7 median filter. Rather than use a full median, a separable median is used to give a similar result, at significantly lower cost. For the separable median, the 7 × 1 column filter is followed by the 1 × 7 row filter. The output frame is 786 × 586 pixels, because of pixels lost at the image borders due to the census transform, and the median filter.

Overall, 63 block RAMs are required: 18 for rectifying each image (16 for input buffers and 2 for the interpolated lookup tables), 10 for row buffers for each census transform, 1 for the delay line on the left camera signal, and 6 for the vertical median filter. Most of the logic is dedicated to calculating the Hamming distances.

Although each calculation is relatively simple, 201 matches are required, one for each disparity level. This is helped by limiting the counters to 4 bits wide.

While not state of the art, the design meets the requirement of measuring 200 disparity levels between 500 and 5000 mm. Although the design actually measures out to infinite distance, beyond 5000 mm the depth resolution is very poor. It is capable of real-time processing of 800×600 input images, producing a 786×586 disparity map at 60 frames per second.

15.5 Face Detection

Face detection is important in a wide range of applications including image focusing, surveillance, and as a precursor to further processing that makes use of the face, such as face recognition, biometric authentication, emotion recognition, and monitoring attention and engagement. This section focuses on the face detection step of the processing. The goal of face detection is to determine the location and extent of all faces within an image.

Yang et al. (2002) identified four categories of methods for face detection:

- Knowledge-based methods use rules that capture relationships between facial features to localise faces. These follow a top-down approach starting with high-level search to find potential face regions, and then verify that those regions contain features such as symmetry, and eyes, nose, and mouth in appropriate locations.
- Feature invariant approaches use structural features that are independent of pose and lighting. These take the opposite approach and work from the bottom up, detecting features such as edges, skin colour, and texture, and piece these together to find faces. Skin colour has been a common feature (Chen and Grecos, 2005); although faces vary in colour, this is mainly in intensity. While skin colour alone is not sufficient to detect faces, it can eliminate many non-faces and be used to provide candidate regions for other local features to be applied for detecting eyes, eyebrows, nose, and mouth. Feature approaches tend to be quite sensitive to noise, occlusion, and changes in lighting.
- Template matching methods correlate a series of predetermined face templates with the image to detect and locate faces. The templates are usually applied to a gradient image (for example from a Sobel filter). These start with whole face templates to identify the basic face outline and are followed by secondary templates associated with eyes, nose, and mouth. One limitation of template matching approaches is their sensitivity to scale and pose.
- Appearance-based methods use models that have been trained on faces to find face-like regions within an image. These are based on machine learning approaches to identify key features and their relationships associated with faces.

Many face detection methods fit into more than one category. Most recent algorithms have come from the appearance-based category, since these have demonstrated robustness. Most approaches are based on a window which is scanned through the image, and the detector determines whether or not the window contains a face. To detect faces at a range of scales, a multi-resolution approach is required, where the image is processed at a range of resolutions.

One appearance-based method that has become classic for face detection is the Viola–Jones algorithm (Viola and Jones, 2001, 2004). It is based on Haar-like features to identify edges and contrasts within an image that correspond to facial features. These features are based on sums of pixels within rectangular areas, which can be calculated quickly using an integral image (see Figure 10.5). Haar-like features consist of differences between adjacent rectangles. These features are then passed to a cascade of weak classifiers, where each classifier successively eliminates regions that are not faces. Several FPGA implementations have been developed for face detection using this approach, as demonstrated by studies such as (Cho et al., 2009; Irgens et al., 2017; Gajjar et al., 2018). Cheng et al. (2012) used a hybrid approach between traditional segmentation and Viola–Jones classification. To accelerate processing, it uses a heavily down-sampled image to segment foreground from background and detect skin-coloured pixels in the foreground. Then in the original image, a Viola–Jones type classifier is used to process only around the detected skin areas.

Neural networks are another appearance-based approach that has more recently been applied to face detection. For example, Irick et al. (2007) used a 20×20 neural network to determine if the window contained a face. By having a step of several pixels between each window location, a small amount of hardware was able to be reused over multiple windows as the data was streamed in. With convolutional neural networks coming of age within the past decade, these have been applied to face detection, also with considerable success. These enable a relatively simple network (in terms of number of parameters) to be used for detecting faces, see, for example, Li et al. (2015), Xu et al. (2019), and Dalal et al. (2022).

15.5.1 Design

Lim et al. (2013) argues that rather than adapt a software algorithm for implementation on an FPGA, the solution should take the opposite approach and design an algorithm that explicitly exploits the capabilities of FPGAs. In the case of face detection, consider the Viola–Jones algorithm. The integral image may be constructed on an FPGA efficiently, but it must be stored off-chip because of the volume of data it holds. Although calculating the sum of pixel values of any rectangular region can be performed in constant time, on an FPGA these require access to external memory and must be performed sequentially. The whole classification algorithm requires a sequential cascade of classifiers, after measuring the corresponding Haar-like features. While this sequential access is well suited to a software implementation on a serial processor, it is poorly adapted to acceleration by parallel hardware, such as on an FPGA.

Instead, Lim (2012) asked: 'What is the best way to exploit the capabilities of an FPGA for this task?' Since FPGAs are readily able to work in parallel (given sufficient resources), the goal in designing an algorithm on an FPGA should focus on how to make the best use of parallelism. For face detection, Lim (2012) chose a hybrid between a feature-invariant and appearance-based approach, selecting four simple features that are empirically useful for detecting faces: skin colour, horizontal edges (characterising features around the eyes and mouth), vertical edges (characterising the nose), and brightness (characterising areas such as eye sockets). These four features are able to be measured independently, so they can be calculated in parallel, enabling a window to be classified as containing a face or not at video rates.

A more detailed overview of the proposed algorithm (as adapted from (Lim, 2012; Lim et al., 2013)) is shown in Figure 15.17. A 19 × 19 window is used for face detection. The binary skin feature map is formed by thresholding a simple R/G ratio (Brand and Mason, 2000). Rather than use a division, the test can be rearranged to use multiplications:

$$skin = \begin{cases} 1, & k_1 G \leq R \leq k_2 G, \\ 0, & \text{otherwise}, \end{cases} \tag{15.19}$$

where k_1 and k_2 are constant scale factors defining the colour space boundaries (Lim, 2012). The skin pixels within the window are counted, and if greater than a threshold, *Thr*, the window potentially contains a face.

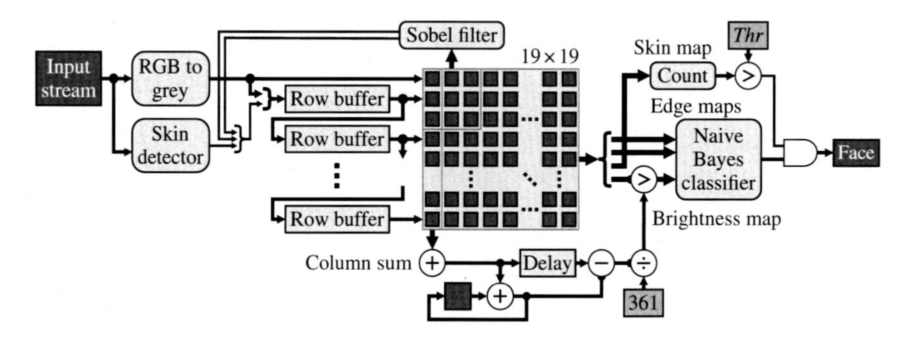

Figure 15.17 Face detection on streamed data, using a 19 × 19 window. Four feature maps are created: skin map, horizontal edge map, vertical edge map, and brightness map. A naive Bayes classifier determines if the maps correspond to a face.

In parallel, the colour image can be converted to greyscale, with the simple addition:

$$Y = \frac{1}{4}(R + 2G + B). \tag{15.20}$$

The binary horizontal and vertical edge maps may be formed by thresholding the outputs of a Sobel edge filter using an appropriate threshold level. To simplify processing, the window border pixels were ignored, giving 17×17 edge maps.

The binary brightness map is formed by comparing each greyscale pixel in the window with the window average. Lim (2012) found that truncating Y to 4 bits gave sufficient accuracy for the classification. Although the window average could be formed via an integral image (as described in (Lim et al., 2013)), this requires extra width for the row buffers, significantly increasing the memory requirements. A less memory-intensive approach is to directly calculate the sum of the first column entering the window, and using a 1-D integral image horizontally, as shown on the bottom in Figure 15.17. The division by 361 may be approximated by multiplication by $45/2^{14}$, which can be implemented with two additions (5×9).

Finally, the brightness map and edge maps are used to classify the window using a naive Bayesian classifier. The naive classifier treats each bit of the feature maps as a separate feature, effectively giving a 939-input classifier. The logarithmic discrimination function of Eq. (14.32) was used to avoid multiplications. A series of small lookup tables contained the log probabilities for several bits (Lim, 2012); these were looked up in parallel with the results added to determine the discrimination score for the window. Finally, a 5×5 non-maximal suppression filter was used to eliminate multiple overlapping detections.

Since it is necessary to detect faces at a range of scales (from distant to close up), several instances of Figure 15.17 were built in parallel, each operating on down-scaled versions of the input image. An image pyramid was constructed by down-sampling using nearest-neighbour interpolation. Down-sampling of the input pixel stream is easily accomplished using a 'valid' flag to indicate which pixels are part of the lower resolution image. When overlapping faces have been detected at different scales, the smallest scale (with the largest window) is retained, and the other discarded.

The algorithm was trained using the MIT-CBCL (Massachusetts Institute of Technology Center for Biological and Computational Learning) face dataset, which consists of 2429 faces and 4548 non-faces. To reduce the number of false positives, a bootstrapping approach was applied (Sung and Poggio, 1998), where false detections were added to the non-face training set.

Overall, the detection rate was a little low (approximately 80% (Lim et al., 2013)); however, the detector was able to run on VGA resolution images at over 60 frames per second, demonstrating real-time performance on a low-end FPGA.

15.6 Summary

This chapter has illustrated the design of a number of embedded image processing systems implemented using FPGAs. They have varied from the relatively simple to the complex. There has not been sufficient space available to go through the complete design process, including the false starts and mistakes. What is presented here is an overview of some of the aspects of the designs. It is hoped that these have demonstrated several of the techniques covered in this book.

Overall, the author trusts that this book has provided both a solid foundation and will continue to provide a useful reference in your design. It is now over to you to build on this foundation, in the development of your own embedded image processing applications based on FPGAs.

References

Akeila, H. and Morris, J. (2008). High resolution stereo in real time. In: *2nd International Workshop on Robot Vision*, Auckland, NZ (18–20 February 2008), *Lecture Notes in Computer Science*, Volume **4931**, 72–84. https://doi.org/10.1007/978-3-540-78157-8_6.

Arribas, P.C. and Macia, F.M.H. (1999). FPGA implementation of a log-polar algorithm for real time applications. *Conference on Design of Circuits and Integrated Systems*, Mallorca, Spain (16–19 November 1999), 63–68.

Bailey, D.G. (2006). Space efficient division on FPGAs. *Electronics New Zealand Conference (ENZCon'06)*, Christchurch, NZ (13–14 November 2006), 206–211.

Bailey, D.G. (2012). Automatic produce grading system. In: *Machine Vision Handbook*, Chapter 37 (ed. B. Batchelor), 1289–1316. London: Springer-Verlag. https://doi.org/10.1007/978-1-84996-169-1_37.

Bailey, D.G. and Bouganis, C.S. (2008). Reconfigurable foveated active vision system. *International Conference on Sensing Technology*, Tainan, Taiwan (30 November – 3 December 2008), 162–169. https://doi.org/10.1109/ICSENST.2008.4757093.

Bailey, D.G. and Bouganis, C.S. (2009a). Implementation of a foveal vision mapping. *International Conference on Field Programmable Technology (FPT'09)*, Sydney, Australia (9–11 December 2009), 22–29. https://doi.org/10.1109/FPT.2009.5377646.

Bailey, D.G. and Bouganis, C.S. (2009b). Tracking performance of a foveated vision system. *International Conference on Autonomous Robots and Agents (ICARA 2009)*, Wellington, NZ (10–12 February 2009), 414–419. https://doi.org/10.1109/ICARA.2000.4804029.

Bailey, D.G. and Bouganis, C.S. (2009c). Vision sensor with an active digital fovea. In: *Recent Advances in Sensing Technology, Lecture Notes in Electrical Engineering*, vol. **49** (ed. S. Mukhopadhyay, G. Sen Gupta, and R. Huang), 91–111. Berlin, Germany: Springer-Verlag. https://doi.org/10.1007/978-3-642-00578-7_6.

Bailey, D.G., Mercer, K.A., Plaw, C., Ball, R., and Barraclough, H. (2004). High speed weight estimation by image analysis. *New Zealand National Conference on Non Destructive Testing*, Palmerston North, NZ (27–29 July 2004), 89–96.

Brand, J.D. and Mason, J.S.D. (2000). A comparative assessment of three approaches to pixel-level human skin-detection. *15th International Conference on Pattern Recognition*, Barcelona, Spain (3–7 September 2000), vol. **1**, 1056–1059. https://doi.org/10.1109/ICPR.2000.905653.

Calderon, H., Ortiz, J., and Fontaine, J.G. (2008). Disparity map hardware accelerator. *International Conference on Reconfigurable Computing and FPGAs (ReConFig '08)*, Cancun, Mexico (3–5 December 2008), 295–300. https://doi.org/10.1109/ReConFig.2008.29.

Camacho, P., Arrebola, F., and Sadoval, F. (1998). Multiresolution sensors with adaptive structure. *24th Annual Conference of the IEEE Industrial Electronics Society (IECON '98)*, Aachen, Germany (31 August – 4 September 1998), Volume **2**, 1230–1235. https://doi.org/10.1109/IECON.1998.724279.

Cambuim, L.F.S., Barbosa, J.P.F., and Barros, E.N.S. (2017). Hardware module for low-resource and real-time stereo vision engine using semi-global matching approach. *30th Symposium on Integrated Circuits and Systems Design (SBCCI)*, Fortaleza, Brazil (28 August – 1 September 2017), 53–58.

Chen, L. and Grecos, C. (2005). Fast skin color detector for face extraction. *Real-Time Imaging IX*, San Jose, CA, USA (18–20 January 2005), SPIE, Volume **5671**, 93–101. https://doi.org/10.1117/12.585341.

Cheng, X., Lakemond, R., Fookes, C., and Sridharan, S. (2012). Efficient real-time face detection for high resolution surveillance applications. *6th International Conference on Signal Processing and Communication Systems*, Gold Coast, Australia (12–14 December 2012), 6 pages. https://doi.org/10.1109/ICSPCS.2012.6508005.

Cho, J., Mirzaei, S., Oberg, J., and Kastner, R. (2009). FPGA-based face detection system using Haar classifiers. *ACM/SIGDA International Symposium on Field Programmable Gate Arrays*, Monterey, CA, USA, 103–112. https://doi.org/10.1145/1508128.1508144.

Colodro-Conde, C., Toledo-Moreo, F.J., Toledo-Moreo, R., Martinez-Alvarez, J.J., Garrigos Guerrero, J., and Ferrandez-Vicente, J.M. (2014). Evaluation of stereo correspondence algorithms and their implementation on FPGA. *Journal of Systems Architecture*, **60** (1): 22–31. https://doi.org/10.1016/j.sysarc.2013.11.006.

Dalal, A., Choudhary, M., Aakash, V., and Balamurugan, S. (2022). Design of hardware accelerator for facial recognition system using convolutional neural networks based on FPGA. *International Conference on Microelectronic Devices, Circuits and Systems (ICMDCS 2022)*, Vellore, India (11–13 August 2022), *Communications in Computer and Information Science*, Volume **1743**, 15–26. https://doi.org/10.1007/978-3-031-23973-1_2.

Devy, M., Boizard, J.L., Galeano, D.B., Lindado, H.C., Manzano, M.I., Irki, Z., Naoulou, A., Lacroix, P., Fillatreau, P., Fourniols, J.Y., and Parra, C. (2011). Stereovision algorithm to be executed at 100Hz on a FPGA-based architecture. In: *Advances in Theory and Applications of Stereo Vision*, Chapter 17 (ed. A. Bhatti), 327–352. Vukovar, Croatia: InTech. https://doi.org/10.5772/14037.

Gajjar, A., Yang, X., Wu, L., Koc, H., Unwala, I., Zhang, Y., and Feng, Y. (2018). An FPGA synthesis of face detection algorithm using Haar classifier. *2nd International Conference on Algorithms, Computing and Systems*, Beijing, China (27–29 July 2018), 133–137. https://doi.org/10.1145/3242840.3242851.

Hartley, R. and Zisserman, A. (2000). *Multiple View Geometry in Computer Vision*. Cambridge: Cambridge University Press.

Hirschmuller, H. (2008). Stereo processing by semiglobal matching and mutual information. *IEEE Transactions on Pattern Analysis and Machine Intelligence* **30** (2): 328–341. https://doi.org/10.1109/TPAMI.2007.1166.

Irgens, P., Bader, C., Le, T., Saxena, D., and Ababei, C. (2017). An efficient and cost effective FPGA based implementation of the Viola-Jones face detection algorithm. *HardwareX* **1**: 68–75. https://doi.org/10.1016/j.ohx.2017.03.002.

Irick, K., DeBole, M., Narayanan, V., Sharma, R., Moon, H., and Mummareddy, S. (2007). A unified streaming architecture for real time face detection and gender classification. *International Conference on Field Programmable Logic and Applications (FPL 2007)*, Amsterdam, The Netherlands (27–29 August 2007), 267–272. https://doi.org/10.1109/FPL.2007.4380658.

Jin, M. and Maruyama, T. (2012). A real-time stereo vision system using a tree-structured dynamic programming on FPGA. *ACM/SIGDA International Symposium on Field Programmable Gate Arrays*, Monterey, CA, USA (22–24 February 2012), 21–24. https://doi.org/10.1145/2145694.2145698.

Johnston, C.T., Bailey, D.G., and Gribbon, K.T. (2005a). Optimisation of a colour segmentation and tracking algorithm for real-time FPGA implementation. *Image and Vision Computing New Zealand (IVCNZ'05)*, Dunedin, NZ (28–29 November 2005), 422–427.

Johnston, C.T., Gribbon, K.T., and Bailey, D.G. (2005b). FPGA based remote object tracking for real-time control. *International Conference on Sensing Technology*, Palmerston North, NZ (21–23 November 2005), 66–71.

Lee, C.R. and Salcic, Z. (1997). High-performance FPGA-based implementation of Kalman filter. *Microprocessors and Microsystems* **21** (4): 257–265. https://doi.org/10.1016/S0141-9331(97)00040-9.

Li, H., Lin, Z., Shen, X., Brandt, J., and Hua, G. (2015). A convolutional neural network cascade for face detection. *IEEE Conference on Computer Vision and Pattern Recognition (CVPR)*, Boston, MA, USA (7–12 June 2015), 5325–5334. https://doi.org/10.1109/CVPR.2015.7299170.

Lim, Y.K. (2012). Algorithmic strategies for FPGA-based vision. Master of Engineering Science thesis. Melbourne, Australia: Monash University. https://doi.org/10.4225/03/583d038fa90c6.

Lim, Y.K., Kleeman, L., and Drummond, T. (2013). Algorithmic methodologies for FPGA-based vision. *Machine Vision and Applications* **24** (6): 1197–1211. https://doi.org/10.1007/s00138-012-0474-9.

Liu, Y., Bouganis, C.S., and Cheung, P.Y.K. (2007). Efficient mapping of a Kalman filter into an FPGA using Taylor expansion. *International Conference on Field Programmable Logic and Applications (FPL 2007)*, Amsterdam, The Netherlands (27–29 August 2007), 345–350. https://doi.org/10.1109/FPL.2007.4380670.

Longfield, S. and Chang, M.L. (2009). A parameterized stereo vision core for FPGAs. *17th IEEE Symposium on Field Programmable Custom Computing Machines (FCCM)*, Napa, CA, USA (5–7 April 2009), 263–265. https://doi.org/10.1109/FCCM.2009.32.

Martinez, J. and Altamirano, L. (2006). FPGA-based pipeline architecture to transform Cartesian images into foveal images by using a new foveation approach. *IEEE International Conference on Reconfigurable Computing and FPGA's*, San Luis Potosi, Mexico (20–22 September 2006), 227–236. https://doi.org/10.1109/RECONF.2006.307774.

Masrani, D.K. and MacLean, W.J. (2006). A real-time large disparity range stereo-system using FPGAs. *7th Asian Conference on Computer Vision (ACCV 2006)*, Hyderabad, India (13–16 January 2006), *Lecture Notes in Computer Science*, Volume **3852**, 42–51. https://doi.org/10.1007/11612704_5.

Michalik, S., Michalik, S., Naghmouchi, J., and Berekovic, M. (2017). Real-time smart stereo camera based on FPGA-SoC. *IEEE-RAS 17th International Conference on Humanoid Robotics (Humanoids)*, Birmingham, UK (15–17 November 2017), 311–317. https://doi.org/10.1109/HUMANOIDS.2017.8246891.

Morris, J., Jawed, K., Gimel'farb, G., and Khan, T. (2009). Breaking the 'Ton': achieving 1% depth accuracy from stereo in real time. *24th International Conference Image and Vision Computing New Zealand (IVCNZ '09)*, Wellington, NZ (23–25 November 2009), 142–147. https://doi.org/10.1109/IVCNZ.2009.5378423.

Ovod, V.I., Baxter, C.R., Massie, M.A., and McCarley, P.L. (2005). Advanced image processing package for FPGA-based re-programmable miniature electronics. *Infrared Technology and Applications XXXI*, Orlando, FL, USA (28 March – 1 April 2005), SPIE, Volume **5783**, 304–315. https://doi.org/10.1117/12.603019.

Pettersson, N. and Petersson, L. (2005). Online stereo calibration using FPGAs. *IEEE Intelligent Vehicles Symposium*, Las Vegas, NV, USA (6–8 June 2005), 55–60. https://doi.org/10.1109/IVS.2005.1505077.

Scharstein, D. and Szeliski, R. (2002). A taxonomy and evaluation of dense two-frame stereo correspondence algorithms. *International Journal of Computer Vision* **47** (1): 7–42. https://doi.org/10.1023/A:1014573219977.

Shan, Y., Wang, Z., Wang, W., Hao, Y., Wang, Y., Tsoi, K., Luk, W., and Yang, H. (2012). FPGA based memory efficient high resolution stereo vision system for video tolling. *International Conference on Field Programmable Technology (FPT)*, Seoul, Republic of Korea (10–12 December 2012), 29–32. https://doi.org/10.1109/FPT.2012.6412106.

Shi, Q. and Gimel'farb, G. (2000). Postprocessing for dynamic programming stereo. *Image and Vision Computing New Zealand (IVCNZ'00)*, Hamilton, NZ (27–29 November 2000), 210–215.

Sung, K.K. and Poggio, T. (1998). Example-based learning for view-based human face detection. *IEEE Transactions on Pattern Analysis and Machine Intelligence* **20** (1): 39–51. https://doi.org/10.1109/34.655648.

Traver, V.J. and Pla, F. (2003). The log-polar image representation in pattern recognition tasks. *1st Iberian Conference on Pattern Recognition and Image Analysis*, Mallorca, Spain (4–6 June 2003), 1032–1040. https://doi.org/10.1007/b12122.

Ttofis, C., Hadjitheophanous, S., Georghiades, A.S., and Theocharides, T. (2013). Edge-directed hardware architecture for real-time disparity map computation. *IEEE Transactions on Computers* **62** (4): 690–704. https://doi.org/10.1109/TC.2012.32.

Viola, P. and Jones, M. (2001). Rapid object detection using a boosted cascade of simple features. *IEEE Computer Society Conference on Computer Vision and Pattern Recognition (CVPR 2001)*, Kauai, HI, USA (8–14 December 2001), Volume **1**, 511–518. https://doi.org/10.1109/CVPR.2001.990517.

Viola, P. and Jones, M.J. (2004). Robust real time face detection. *International Journal of Computer Vision* **57** (2): 137–154. https://doi.org/10.1023/B:VISI.0000013087.49260.fb.

Wang, W., Yan, J., Xu, N., Wang, Y., and Hsu, F.H. (2013). Real-time high-quality stereo vision system in FPGA. *International Conference on Field Programmable Technology*, Kyoto, Japan (9–11 December 2013), 358–361. https://doi.org/10.1109/FPT.2013.6718387.

Wilson, J.C. and Hodgson, R.M. (1992). A pattern recognition system based on models of aspects of the human visual system. *International Conference on Image Processing and its Applications*, Maastricht, The Netherlands (7–9 April 1992), 258–261.

Xu, H., Wu, Z., Ding, J., Li, B., Lin, L., Zhu, J., and Hao, Z. (2019). FPGA based real-time multi-face detection system with convolution neural network. *8th International Symposium on Next Generation Electronics (ISNE)*, Zhengzhou, China (9–10 October 2019), 3 pages. https://doi.org/10.1109/ISNE.2019.8896551.

Yang, M.H., Kriegman, D.J., and Ahuja, N. (2002). Detecting faces in images: a survey. *IEEE Transactions on Pattern Analysis and Machine Intelligence* **24** (1): 34–58. https://doi.org/10.1109/34.982883.

Zabih, R. and Woodfill, J. (1994). Non-parametric local transforms for computing visual correspondence. In: *3rd European Conference on Computer Vision (ECCV'94)*, Stockholm, Sweden, *Lecture Notes in Computer Science*, Volume **801**, 151–158. https://doi.org/10.1007/BFb0028345.

Zha, D., Jin, X., and Xiang, T. (2016). A real-time global stereo-matching on FPGA. *Microprocessors and Microsystems* **47**: 419–428. https://doi.org/10.1016/j.micpro.2016.08.005.

Zhou, K., Meng, X., and Cheng, B. (2020). Review of stereo matching algorithms based on deep learning. *Computational Intelligence and Neuroscience* **2020**: Article ID 8562323, 12 pages. https://doi.org/10.1155/2020/8562323.

Index

Page numbers appearing in bold refer to definitions or significant sections within the document where the topic is covered in some depth (for example section headings).

Design for Embedded Image Processing on FPGAs, Second Edition. Donald G. Bailey.

Companion Website: www.wiley.com/go/bailey/designforembeddedimageprocessc2e